Handbook of Research on Military, Aeronautical, and Maritime Logistics and Operations

Alberto Ochoa-Zezzatti
Juarez City University, Mexico

Jöns Sánchez
Consejo Nacional De Ciencie Y Tecnologia (CONACYT), Mexico

Gastón Cedillo-Campos
Transportation and Logistics Systems National Laboratory; Mexican Institute of Transportation, Mexico

Margain de Lourdes
Polytechnic University of Aguascalientes, Mexico

A volume in the Advances in Logistics, Operations, and Management Science (ALOMS) Book Series

BUSINESS SCIENCE Reference
An Imprint of IGI Global

Published in the United States of America by
Business Science Reference (an imprint of IGI Global)
701 E. Chocolate Avenue
Hershey PA, USA 17033
Tel: 717-533-8845
Fax: 717-533-8661
E-mail: cust@igi-global.com
Web site: http://www.igi-global.com

Library of Congress Cataloging-in-Publication Data

Names: Ochoa Ortiz Zezzatti, Carlos Alberto, 1974- editor.
Title: Handbook of research on military, aeronautical, and maritime logistics and operations / Alberto Ochoa-Zezzatti, Jons Sanchez, Miguel Gaston Cedillo-Campos, and Margain de Lourdes, editors.
Description: Hershey, PA : Business Science Reference, [2016] | Includes bibliographical references and index.
Identifiers: LCCN 2015042005| ISBN 9781466697799 (hardcover) | ISBN 9781466697805 (ebook)
Subjects: LCSH: Logistics. | Business logistics. | Operations research.
Classification: LCC U168 .H365 2016 | DDC 355.4/11--dc23 LC record available at http://lccn.loc.gov/2015042005

This book is published in the IGI Global book series Advances in Logistics, Operations, and Management Science (ALOMS) (ISSN: 2327-350X; eISSN: 2327-3518)

British Cataloguing in Publication Data
A Cataloguing in Publication record for this book is available from the British Library.

Advances in Logistics, Operations, and Management Science (ALOMS) Book Series

John Wang
Montclair State University, USA

ISSN: 2327-350X
EISSN: 2327-3518

Mission

Operations research and management science continue to influence business processes, administration, and management information systems, particularly in covering the application methods for decision-making processes. New case studies and applications on management science, operations management, social sciences, and other behavioral sciences have been incorporated into business and organizations real-world objectives.

The **Advances in Logistics, Operations, and Management Science** (ALOMS) Book Series provides a collection of reference publications on the current trends, applications, theories, and practices in the management science field. Providing relevant and current research, this series and its individual publications would be useful for academics, researchers, scholars, and practitioners interested in improving decision making models and business functions.

Coverage

- Finance
- Risk Management
- Marketing engineering
- Computing and information technologies
- Services management
- Organizational Behavior
- Information management
- Decision analysis and decision support
- Production management
- Political Science

IGI Global is currently accepting manuscripts for publication within this series. To submit a proposal for a volume in this series, please contact our Acquisition Editors at Acquisitions@igi-global.com or visit: http://www.igi-global.com/publish/.

The Advances in Logistics, Operations, and Management Science (ALOMS) Book Series (ISSN 2327-350X) is published by IGI Global, 701 E. Chocolate Avenue, Hershey, PA 17033-1240, USA, www.igi-global.com. This series is composed of titles available for purchase individually; each title is edited to be contextually exclusive from any other title within the series. For pricing and ordering information please visit http://www.igi-global.com/book-series/advances-logistics-operations-management-science/37170. Postmaster: Send all address changes to above address.

Titles in this Series

For a list of additional titles in this series, please visit: www.igi-global.com

Managerial Strategies and Practice in the Asian Business Sector
U Zeyar Myo Aung (Tourism Promotion and International Relations Department, Myanmar) and Patricia Ordoñez de Pablos (The University of Oviedo, Spain)
Business Science Reference • copyright 2016 • 395pp • H/C (ISBN: 9781466697584) • US $195.00 (our price)

Designing and Implementing Global Supply Chain Management
Sudhanshu Joshi (School of Management, Doon University, India) and Rohit Joshi (Indian Institute of Management IIM, Shillong, India)
Business Science Reference • copyright 2016 • 342pp • H/C (ISBN: 9781466697201) • US $195.00 (our price)

Innovative Management Education Pedagogies for Preparing Next-Generation Leaders
Shalini Rahul Tiwari (Institute of Management Technology, India) and Lubna Nafees (Institute of Management Technology, India)
Business Science Reference • copyright 2016 • 313pp • H/C (ISBN: 9781466696914) • US $195.00 (our price)

Handbook of Research on Global Supply Chain Management
Bryan Christiansen (PryMarke, LLC, USA)
Business Science Reference • copyright 2016 • 689pp • H/C (ISBN: 9781466696396) • US $310.00 (our price)

Analyzing Risk through Probabilistic Modeling in Operations Research
Dariusz Jacek Jakóbczak (Technical University of Koszalin, Poland)
Business Science Reference • copyright 2016 • 441pp • H/C (ISBN: 9781466694583) • US $225.00 (our price)

Organizational Change Management Strategies in Modern Business
Aslı Goksoy (American University in Bulgaria, Bulgaria)
Business Science Reference • copyright 2016 • 460pp • H/C (ISBN: 9781466695337) • US $215.00 (our price)

Maximizing Management Performance and Quality with Service Analytics
Yixin Diao (IBM T. J. Watson Research Center, USA) and Daniela Rosu (IBM T. J. Watson Research Center, USA)
Business Science Reference • copyright 2015 • 465pp • H/C (ISBN: 9781466684966) • US $255.00 (our price)

Optimization of Supply Chain Management in Contemporary Organizations
Ehap Sabri (JDA Software, USA & University of Texas at Dallas, USA)
Business Science Reference • copyright 2015 • 337pp • H/C (ISBN: 9781466682283) • US $205.00 (our price)

IGI GLOBAL
DISSEMINATOR OF KNOWLEDGE
www.igi-global.com

701 E. Chocolate Ave., Hershey, PA 17033
Order online at www.igi-global.com or call 717-533-8845 x100
To place a standing order for titles released in this series, contact: cust@igi-global.com
Mon-Fri 8:00 am - 5:00 pm (est) or fax 24 hours a day 717-533-8661

List of Contributors

Table of Contents

Detailed Table of Contents

bioinformatics. Up to our knowledge, there are only two trajectory-based metaheuristic algorithms for VSP documented in the literature. The main contribution of this chapter is that we extend the available heuristics to solve VSP by proposing a genetic algorithm (GA). It is of particular interest to study the impact of four different crossover operators in the algorithm performance. The experimental results showed that the order-based crossover is the best. Moreover, the best GA variant was compared with the best algorithm for VSP: GVNS. The results of this comparison showed that GVNS outperforms our best GA variant by approximately 1.54 times in solution quality.

Contemporary prosperity depends on effective and secure supply chain networks that support economic competitiveness. Disruptions in global supply chains would have critical consequences on economies. The lack of technical studies and quantitative data concerning security that affects supply chain operations in Latin America, motivated to develop an exploratory study. Considering the complexity of the question studied, this paper details a set of case studies that explore, from a qualitative research approach, to what extent fulfilling security international standards now necessary to access mature markets such as the U.S and Canada allows export companies located in emerging countries as Mexico to face effectively the different types and levels of local risk. These results should help both academics and practitioners to more readily understand, first, the key logistics components now taken into account when improving security in export-oriented supply chains is required, and second, decision-makers' perspectives regarding supply chain security standards (SCSS) available on the market. A discussion of results is exposed and finally, discussion and future research are presented.

As globalization increases, supply chains are becoming increasingly complex and vulnerable, making the management of their security an important part of corporate management. The need to adopt control and prevention measures that allow for guaranteeing security in international commercial operations is essential to prevent becoming a target of these threats. Some of these threats are terrorism, piracy and theft. Within the food industry, food terrorism has gained relevance due to the extreme consequences it may have on both public health and the market. Food terrorism refers to the intentional contamination of food in any link of the supply chain, which covers everything from production to consumption. This chapter deals with the problems of food terrorism and makes a proposal that allows to establish a management system to prevent intentional food contamination in the supply chain through managing risks and improving security in the supply chain on this aspect.

improved runways, terminal expansions, and entirely new airports. However, this effort is sometimes limited due to their geographic location. In this work, two main objectives are pursued: first, to highlight the importance of the industry by exposing the current situation and future trends all over the world focusing in the Mexican industry; and second, to introduce a simulation model which can be used as a decision making tool for the upcoming demand. The analysis of the scenarios illustrates how to develop strategies to cope with the different airspace user's needs.

In this chapter, a scientific tool designed to facilitate fair comparisons of heuristics is introduced. Making a fair comparison of the performance of different algorithms is a general problem for the heuristic community. Most of the works on experimental analysis of heuristic algorithms have been focused on tabular comparisons of experimental results over standard sets of benchmark instances. However, from a statistical point of view, and according to the experimental design theory, a minimum requirement to compare heuristic algorithms is the use of non-parametric tests. Non-parametric tests can be used for comparing algorithms whose results represent average values, in spite of the inexistence of relationships between them, and explicit conditions of normality, among others. The proposed tool, referred to as VisTHAA, incorporates four non-parametric statistical tests to facilitate the comparative analysis of heuristics. As a case study, VisTHAA is applied to analyze the published results for the best state-of-the-art algorithms that solve the one-dimensional Bin Packing Problem.

The issue of route choice is a key factor for the freight transport performance. Congestion at roads encourages hauliers to change routes to minimize the delays and keep lead times in a reliable range. In the context of transport planning, the route choice problem can be assessed by modelling the travel times needed to reach a destination through the different routes in a road network. Is in this point where the volume-delay functions become relevant. A Volume-Delay Function (VDF) is a mathematical representation of the increase of the travel time as more and more vehicles utilize the routes, causing congestion on the road networks. The related literature and practitioners report on the use of some known functional forms, as the BPR function, the Conical volume-delay function or the Akcelik's function, which are widely utilized in flow's assignation modelling in transport planning. A successful application of VDFs requires a proper fitting of the function's parameters. In a classical focus these parameters can be deduced from speed-flow surveys carried out at the routes or links of interest; these surveys generally require time money and personnel. As an alternative to this classical focus, particularly when facing scarcity of resources, this work carries out a mathematical analysis of the VDF functional forms, as well with an interpretation of their parameters in relation to road's operation. The results of these analyses clarifies the meaning of the functional forms for the VDFs and their parameters, and suggest other ways to assess those parameters which may be more practical for the purpose of modelling the choice of route in freight transport. Some considerations to put into practice this in Mexico are discussed at the end of this work.

The results obtained in this research confirm that currently manufacturing for SMEs by the nature of its operational activities, the handling of Supply Chain Management (SCM), is significant in the performance of this type of organizations. Because of that, the answers of the business man were analyzed with the support of polls through a field work, finding that for them, is essential that the SCM considering its complexity and impact in the supply of material resources, which is significantly influence for the manufacturer SMEs to have a bigger performance. This research, of empiric origin,, and with support of the statistic software EQS where the structural equations have been used, permit conclude that the SCM is essential for the manufacturer SMEs performance. For the investigation the sample of 288 enterprises of the state Aguascalientes has been used, obtained between August and December 0f 2013, which contain from 11 to 250 workers.

There exist two general approaches to solve multiple objective problems. The first approach belongs to the classical mathematical methods: The weighted sum method, goal programming, or utility functions methods pertain to this approach. The output of mathematical methods is a single optimal solution. In the second approach are the heuristic methods, like the multiple objective evolutionary algorithms that offer the decision maker a set of optimal solutions usually called non- dominated or, Pareto-optimal solutions. This set is usually very large and the decision maker faces the problem of reducing the size of this set to a manageable number of solutions to analyze. In this paper the second approach is used to reduce the Pareto front using two weights generator for the non-numerical ranking preferences method and their performance is compared.

This chapter proposes a multi-dimensional patterns recognition model for Configuration Management's Risk Identification in Aerospace Safety Critical Systems. This work has been designed for Aerospace software systems where companies require full compliance with the Aerospace Standard DO-178b. The solution focuses on Risk Identification for the Configuration Management Process Area. An Anomaly Detection Solution has been designed through the modeling of statistics and artificial intelligence algorithms, following CRISP-DM model standard for data mining solutions. A dimensional architecture was designed to model the problem through three dependent and interconnected dimensions. The first dimension, Behavioral Biometrics, which this model has extended to Human Behavioral Patterns. The second dimension is Infrastructure, which represents all physical specialized equipment, environments, networking, and its configurations. The third dimension is space-time, which in this model represents a time dimension against all geographical information project related (code, files, among others).

In this chapter, a hybrid meta-heuristic algorithm is proposed, which is based on simulated annealing in order to solve the Vehicle Routing Problem (VRP) with stochastic demands. The proposed algorithm has several annealing phases, which are named Simulated Multi-annealing (SMA). This algorithm is combined with the Clarke and Wright saving algorithm (CW). This last algorithm was placed inside the Metropolis Cycle of Simulated Multi-annealing algorithm. The hybrid algorithm is applied to obtain optimal routes for a cleaning distribution company. The initial solution of VRP is generated by the CW Algorithm, and it was improved by the multi-annealing phases. The solutions into the metropolis cycle were generated by a solution perturbation approach, each solution was compared with a previous one in order to obtain an optimal one.

In the literature, mostly, under the heading of maritime logistics, classical sea transportation issues have been studied. In actual fact, with broader scope, maritime logistics is a different field under the logistics theme and sea transportation is just one of the functions that must be carried effectively and in an integrated manner with other logistics functions. This chapter focuses on the maritime logistics functions which were selected by the authors in accordance with the gap in the literature, namely transportation and traffic management, inventory management, demand forecasting, material handling, and packaging and labelling. In addition to the maritime logistics functions; port logistics, green maritime logistics and decision problems in maritime logistics topics are handled deeply with the aim of completing the skeleton of the maritime logistics concept by which an understanding of the clear definition of maritime logistics can be achieved. The practitioners can gain more integrated and holistic viewpoint about maritime logistics with the help of this chapter.

In this chapter we discussed the application of Towers of Hanoi in logistics management applications, for this purpose. Firstly, we discussed how pile problems applications impact in logistics, later we discussed the Hanoi Towers application in Logistics Management, its mathematical model and common solutions applying different paradigms (iterative, recursive, and heuristics), we present an in depth analysis in genetic algorithms, later we present the analysis of a genetic algorithm applied to solve basic Hanoi Tower game (three pegs three discs) implemented in C language, and later we discuss its generalization from three to four discs. Finally, we discussed preliminary results and present our conclusions. Main contribution of this chapter is demonstrate game theory, specifically Towers of Hanoi, can be applied to solve logistics problems, and an approach for the generalization of basic Hanoi Tower form, three discs to four discs, by means of genetic algorithms implemented in C language.

Travel in Chihuahua is a very complicated situation to realize because the distances between land routes is very far because the poor quality of road. The importance of this research is to understand from a Multivariable optimization associated with the path of a group of airplanes associated with an aero taxi company and determine the optimal flight route involve speed, storage and travel resources for determining the cost benefit have partnered with a travel plan, which has as principal basis the orography airstrip restriction, although this problem has been studied on several occasions by the literature failed to establish by supporting ubiquitous computing for interacting with the various values associated with the achievement of the group of airplanes -Cessna 208 Caravan- and their cost-benefit of each issue of the company and comparing their individual trips for the rest of group. There are several factors that can influence in the achievement of a group of Cessna 208 Caravan group for our research we propose to use Bat Algorithm, which has proven to be efficient for the convergence of several issues (artificial bats) when they have such restrictions and obstacles should use this energy to avoid in our case, a gain resource as food which in our case is represented as the use of travel passengers and goods optimally for the duration of a long trip with the uncertainty of not knowing when you have a resupply or reduce this quantity.

The aim of this chapter is about the inclusion of real world scenarios, viewed as a Generalized Vehicle Routing Problem (GVRP) model problem, and treated by bio inspired algorithms in order to find optimum routing of product delivery. GVRP is the generalization of the classical Vehicle Routing Problem (VRP) that is well known NP-hard as generalized combinatorial optimization problem with a number of real world applications and a variety of different versions. Due to its complexity, large instances of VRP are hard to solve using exact methods. Thus a solution by a soft computing technique is desired. From a methodological standpoint, the chapter includes four bio inspired algorithms, ant colony optimization and firefly. From an application standpoint, several factors of the generalized vehicle routing are considered from a real world scenario.

Mixed flow impellers are extensively used in turbomachines either to convert mechanical energy to fluid energy or to convert fluid energy to mechanical energy. According to the geometry of flow passage, turbo machines can be classified as radial, axial and mixed flow. Mixed flow turbomachines are widely used for engineering applications like cooling water duties, water intake impellers for maritime applications, flood water draining, irrigation and other application fields. The design of mixed flow impellers of high specific speed is a direct extension of the well-established methods of the designing of radial flow impellers but the introduction of near diagonal flow layout at a still larger specific speed stimulated the incorporation of axial impeller design techniques in mixed flow impeller technology. Here, an attempt has been made to design a mixed flow turbo machine blade from the basic principle of turbomachinery and fluid mechanics. On the basis of stress analysis, the blade positioning in the meridional annulus was selected and validated using artificial neural network.

In the supply chain and the planning and control of warehouse processes, the order picking is an aspect critical. Combining customer orders into picking orders to minimize the picking time is known such order batching. Extensive evolutionary algorithms haven been proposed to build better batches for the order picking. The authors think that any algorithm should preserve batches that appear frequently in all members of the population in order to keep track and inherit these characteristics exhibited by the parents to the next generation. However, the traditional evolutionary operators used in current research sometimes lose the characteristics mentioned. In order to describe the characteristics exhibited by the parents as a distribution of the solution space, the authors build a probability model. An acceptable performance using the model proposed is shown against different evolutionary algorithms known in the literature in a series of extensive numerical experiments.

Foreword

Business competitiveness and Logistics is a complex phenomenon that has become critical for companies to reach their development locally and internationally. On the one hand, macro factors and market structure influence in business competitiveness, but also in a regional or sector context. The internal aspects and the use of various business and Logistics tools contribute to the ability to create value in an organization.

This edited collection of papers entitled *Handbook of Research on Military, Aeronautical, and Maritime Logistics and Operations* presents a set of applications which purpose is to solve a wide variety of modern and real world problems on logistics management and optimization by means of Hybrid Artificial Intelligence Systems. This book consists of twenty seven chapters, grouped in four main areas: Theoretical foundations of Logistics, Implementations of Logistics, Applications of Innovative Logistics, and Optimization in Logistics, where are discussed several applications using Hybrid Artificial Intelligence Systems that result of integrating different Artificial Intelligence Techniques on different Logistics fields. Chapters describe important relatively new or emerging areas of work in which the authors are personally involved; most of them hot topics of discussion in Scientific and Academic Community.

THEORETICAL FOUNDATIONS OF LOGISTICS

This section consists of six chapters, Chapter 1, "The Premises of Logistics: The Organization of Warships in France in the 17th and 18th Centuries," which aboard the theoretical fundamentals of an adequate decision support systems under uncertainty, Chapter 2, "Optimization of the Vertex Separation Problem with Genetic Algorithms" is presented a mathematical resolution to a business company. Chapter 3, "Supply Chains under Security Threat: The First National Exploratory Study in México" determines actions to improvement an innovative model of supply chain. Chapter 4, "Intentional Food Contamination in the Food Supply Chain: Proposal of a Management System for its Prevention" explains different models to prevent accidents with contaminated food associated with their transportation. Chapter 5, "Clio-Combinatorics: A Novel Framework to Analyze Military Logistics Choices Using Operations Research Techniques" detailed different strategies to models uncertainty scenarios. Chapter 6 "Optimization of Utility Functions in an Admissible Space of Higher Dimension: Optimization of Utility Functions in an Admissible Space of Higher Dimension" explains different models to organize transportation with extra dimensions to reduce spaces.

IMPLEMENTATIONS OF LOGISTICS

The second section, the most industrial on the book, consists of six chapters: Chapter 7, "Analyzing Airport Capacity by Simulation: A Mexican Case Study" explains a model related with their routes of delay using Mathematical models and specialized software. Chapter 8, "VisTHAA: A Statistical Tool for Comparison of Heuristics", proposes an innovative Visualization Tool to analyze heurists related with Intelligent Logistics, Chapter 9, "Modelling the Route Choice: The Role of Volume-Delay Functions in Transport Planning" describe different models associated with the reactive Logistics under uncertainty situations. Chapter 10, "The Influence and Management of the Supply Chain Performance of Manufacturing SMEs in Aguascalientes" explains a real problem about the logistics to organize times in a novel SME. Chapter 11, "Comparison of Two Random Weight Generators for Multi-Objective Optimization" details novel specification to reduce times to distribution in a Mathematical Model. Chapter 12, "A Hybrid Metaheuristic Algorithm for the Quay Crane Scheduling Problem" explains a model hybrid to solve problems with time and space. Chapter 13, "Temperature Modeling of a Greenhouse Environment" which analyzes a model to control the different exogenous aspects in a Greenhouse to grow food in semi-desert areas.

APPLICATIONS OF INNOVATIVE LOGISTICS

The Applications of Innovative Logistics consists of seven chapters: Chapter 14, "Multi-Objective Simulated Annealing Algorithms for General Problems" which explains different models to solve general problems in Logistics. Chapter 15, "Emergency Department Logistic Optimization Using Design of Experiments: From Triage to High Quality of Service" detailed a novel model of design of experiments to reduce costs in Logistics of service. Chapter 16, "Implementing a Container Ship Stowage Problem for Humanitarian Aid in Palestine Based on Cultural Algorithms" explains a real problem with the support to an isolated society and the best options to organize this service on the time take as important factor the locations and ubiquities in the Container Ship and problems of each point to send the correct goods. Chapter 17, "A Hybrid Intelligent Risk Identification Model for Configuration Management in Aerospace Systems" describes a model to improve times and costs associated with management in an Aerospatiale project. Chapter 18, "Hybrid Multi-Annealing Simulated Annealing Applied to Vehicle Routing Problem: A Case of Study" explains a specific topic related with problems on the exogenous costs in a Vehicle routing problem in a Logistics System. Chapter 19, "Maritime Logistics" detailed a model to improve Logistics in ships. Chapter 20, "A Mobile Application for Helping Urban Public Transport and Its Logistics," this chapter proposes new ideas related with the delivery of products and detailed a novel technique to analyze the restriction of Time Windows.

OPTIMIZATION IN LOGISTICS

This section consists of seven chapters: Chapter 21, "A Review of the Main Options of Tools for Optimizing Operations (in Companies, Manufacturing, and Supply Chains)" explains the way to obtain best models to routes in different kind of companies. Chapter 22, "The Application of Hanoi Towers Game in Logistics Management" describes the way to collect thematic objects and improve these collections using a parallel hybrid algorithm. Chapter 23, "Determining the relationship between Time of Construction and Cranes Organization by Applying the Bees' Algorithm for Reducing the Time of Construction" explains different strategies to model and improve the reduction of time when a building is constructed. Chapter 24, "Use of GVRP as a Model of Two Specific Real World Problems and Its Bioinspired Solution" are explained different models to improve the solution in a real word problem to transportation perishing food in a largest city. Chapter 25, "Strategic Designing and Optimization of Mixed Flow Impeller Blades for Maritime Applications" is presented a Model to reduce the costs in Maritime Transportation. Chapter 26, "An Estimation of Distribution Algorithm-Based Approach for the Order Batching Problem: An Experimental Study" is showed a model based on Bioinspired Algorithms to improve a specific stochastic model demand., and Chapter 27, "Determining Maximum Load of Passengers and Goods to an Aero Taxi in Southwestern Chihuahua" is a novel technique used on Aeronautical Models to improve the feature of a system based on aerial transportation to adequate the necessities of Transportation sector on Mexico.

All these chapters provide an interesting prospective on where this important field is going at the beginning of the second decade of the twenty-first century.

Oscar Castillo
Mexican Technological Institute, Mexico

Ocotlán Díaz-Parra
Hidalgo University, Mexico

Jorge Ruíz-Vanoye
Hidalgo University, Mexico

Oscar Castillo *holds the Doctor in Science degree (Doctor Habilitatus) in Computer Science from the Polish Academy of Sciences (with the Dissertation "Soft Computing and Fractal Theory for Intelligent Manufacturing"). He is a Professor of Computer Science in the Graduate Division, Tijuana Institute of Technology, Tijuana, Mexico. In addition, he is serving as Research Director of Computer Science and head of the research group on Hybrid Fuzzy Intelligent Systems. Currently, he is Vice-President of HAFSA (Hispanic American Fuzzy Systems Association) and President of IFSA (International Fuzzy Systems Association). Prof. Castillo is also Chair of the Mexican Chapter of the Computational Intelligence Society (IEEE). He also belongs to the Technical Committee on Fuzzy Systems of IEEE and to the Task Force on "Extensions to Type-1 Fuzzy Systems". He is also a member of NAFIPS, IFSA and IEEE. He belongs to the Mexican Research System (SNI Level 3). His research interests are in Type-2 Fuzzy Logic, Fuzzy Control, Neuro-Fuzzy and Genetic-Fuzzy hybrid approaches. He has published over 100 journal papers, 7 authored books, 16 edited books, and 500 papers in conference proceedings. He has been Guest Editor of several successful Special Issues in the past, like in the following journals: Applied Soft Computing, Intelligent Systems, Information Sciences, Non-Linear Studies, Fuzzy Sets and Systems, JAMRIS and Engineering Letters. He is currently Associate Editor of the Information Sciences Journal and the IEEE Transactions on Fuzzy Systems.*

Ocotlán Díaz-Parra, *Ph.D. of Engineering & Applied Sciences, 2006-2008 from Centro de Investigación en Ingeniería y Ciencias Aplicadas (CIICAp). She is expert in Combinatorial Optimization, Genetic Algorithms, Vehicle Routing, Artificial Intelligence, Heuristic Algorithms and Algorithms. She belongs to the Mexican Research System (SNI Level 1).*

Jorge Ruíz-Vanoye, *Ph.D. on Computational Sciences, 2004-2008 from Cenidet. He is Editor to IJCOPI and expert in Lineal Programming, Algorithmic, Evolving algorithm, Software engineering, Computational Intelligence, Computer Science Security, Computational Educational and Parallel & Distributed Computing. He belongs to the Mexican Research System (SNI Level 1).*

Preface

Handbook of Research on Military, Aeronautical, and Maritime Logistics and Operations is an novel, innovative and adequate source of information that compiles interdisciplinary perspectives about diverse issues related with Logistics on different ways about Intelligent Optimization, Industrial Applications on real world, Social applications and Technology applications each one with a different perspective about the correct solution of this kind of methodologies. This book is a collective effort to introduce new ideas and paradigms from a variety of perspectives using innovative techniques related with Bioinspired Algorithms and methodologies associated with Artificial Intelligence. An innovative Handbook specialized on optimization considers different aspects to realize this "Intelligent Optimization" tries to improve with innovative techniques and methodologies different daily aspects of our lives, in each one of them is possible understand the necessity of improve distances, time, costs, spaces and a plethora of features associated with the modern life (labor associated with delivery of goods, materials or products). We received manuscripts from renowned researchers from all around the world associated with Theoretical foundations of Logistics to understand many paradigms on different Optimization implementation kind. In addition we receive many manuscripts with expertise on improving optimization related with Logistics of products and services, Optimization of different elements in the time and location, Social Applications to enjoy our life of a better way, and finally, Technologies Applications of diverse ways to increase our Life Quality. The book starts with a section entitled Theoretical foundations of Logistics, featuring six chapters on the theoretical ideas related with the correct implementation of diverse range of Logistics applications on real world. The first chapter of this section is "The premises of logistics: The organization of warships in France in the 17th and 18th Centuries," which aboard the theoretical fundamentals of an adequate decision support systems under uncertainty, The Chapter: "Optimization of the Vertex Separation Problem with Genetic Algorithms" is presented a mathematical resolution to a business company. The Chapter: "Supply Chains under Security Threat: The First National Exploratory Study in México" determines actions to improvement an innovative model of supply chain. The Chapter entitled: "Intentional Food Contamination in the Food Supply Chain: Proposal of a Management System for its Prevention" explains different models to prevent accidents with contaminated food associated with their transportation. The Chapter: "Clio-combinatorics: A novel framework to analyze military logistics choices using operations research techniques" detailed different strategies to models uncertainty scenarios. Finally the last chapter of this section is entitled: "Optimization of Utility Functions in an Admissible Space of Higher Dimension: Optimization of Utility Functions in an Admissible Space of Higher Dimension" explains different models to organize transportation with extra dimensions to reduce spaces.

The next section is named Implementations of Logistics, featuring six chapters related with different comparatives of Logistics in the search to improve resources in diverse aspects of our transportations or improve process to this. The Chapter entitled: "Analyzing airport capacity by simulation: A Mexican case study" explains a model related with their routes of delay using Mathematical models and specialized software. The Chapter: "VisTHAA: A Statistical Tool for Comparison of Heuristics", proposes an innovative Visualization Tool to analyze heurists related with Intelligent Logistics, The Chapter entitled: "Modelling the route choice: The role of volume-delay functions in transport planning" describe different models associated with the reactive Logistics under uncertainty situations. The Chapter: "The influence and management of the supply chain performance of manufacturing SMEs in Aguascalientes" explains a real problem about the logistics to organize times in a novel SME. The Chapter: "Comparison of two random weight generators for Multi-objective Optimization: Comparison of two random weight generators for Multi-objective Optimization" details novel specification to reduce times to distribution in a Mathematical Model. The Chapter entitled: "A Hybrid metaheuristic algorithm for the quay crane scheduling problem" explains a model hybrid to solve problems with time and space. Finally the last chapter in this section is: "Temperature Modeling of a Greenhouse Environment" which analyzes a model to control the different exogenous aspects in a Greenhouse to grow food in semi-desert areas.

The third section is named Applications of Innovative Logistics, featuring six chapters related with different comparatives of Logistics Models in the search to improve resources in diverse aspects of companies and to improve our lives. The first chapter of this section is: "Multi-Objective Simulated Annealing Algorithms for General Problems" which explains different models to solve general problems in Logistics. The Chapter entitled: "Emergency Department Logistic Optimization using Design of Experiments: From Triage to High Quality of Service" detailed a novel model of design of experiments to reduce costs in Logistics of service. The Chapter: "Implementing a Container Ship Stowage Problem for humanitarian aid in Palestine based on cultural algorithms" explains a real problem with the support to an isolated society and the best options to organize this service on the time take as important factor the locations and ubiquities in the Container Ship and problems of each point to send the correct goods. The Chapter: "A Hybrid Intelligent Risk Identification Model for Configuration Management in Aerospace Systems" describes a model to improve times and costs associated with management in an Aerospatiale project. The Chapter: "Hybrid Multi-annealing simulated annealing applied to vehicle routing problem a case of study" explains a specific topic related with problems on the exogenous costs in a Vehicle routing problem in a Logistics System. The Chapter: "Maritime Logistics" detailed a model to improve Logistics in ships. Finally in this section is presented "A Mobile Application for Helping Urban Public Transport and Its Logistics," this chapter proposes new ideas related with the delivery of products and detailed a novel technique to analyze the restriction of Time Windows.

In the section entitled Optimization in Logistics grouped six different papers related with solutions derived of specific aspects try to improve daily activities on Optimization with real applications to social topics. In the chapter "A review of the main options of tools for optimizing operations (in companies, manufacturing and supply chains)" explains the way to obtain best models to routes in different kind of companies. "The application of Hanoi towers game in logistics management" describes the way to collect thematic objects and improve these collections using a parallel hybrid algorithm. The chapter entitled ""Determining the relationship between time of construction and cranes organization by applying the bees' algorithm for reducing the time of construction" explains different strategies to model and improve the reduction of time when a building is constructed. In "Use of GVRP as a model of two specific real world problems and its bioinspired solution" are explained different models to improve the solution in a real word problem to transportation perishing food in a largest city. Meanwhile in: "Strategic Designing and Optimization of Mixed Flow Impeller Blades for Maritime Applications" is presented a Model to reduce the costs in Maritime Transportation. In the Chapter: "An estimation of distribution algorithm-based approach for the order batching problem. An experimental study" is showed a model based on Bioinspired Algorithms to improve a specific stochastic model demand. Finally the chapter "Determining maximum load of passengers and goods to an aero taxi in southwestern Chihuahua"" is a novel technique used on Aeronautical Models to improve the feature of a system based on aerial transportation to adequate the necessities of Transportation sector on Mexico.

It is important to state that the chapters were selected following a rigorous analysis done by the book editors, and each chapter was double or triple-blind peer-reviewed by at least two experts in the area. This would not have been possible without the valuable help of the Editorial Advisory Board; our sincere appreciation to: Dr. Alberto Hernández, Dr. Alfredo Bueno (Mexican Institute of Transportation), Dr. Julio Ponce, Dr. Francisco Ornelas, Dr. Christian Correa, and several Ph.D. students from UJAT, UAA, University of Nuevo Leon and CIATEC (National Council of Science and Technology); finally, to Dr. Oscar Castillo (Mexican Technological Institute); Ocotlán Parra and Jorge Ruíz-Vanoye (Hidalgo University) by their dedication to write the Foreword. A special thanks to Hermes Cedillo (YoLogistico.COM) who provided us useful support related to technological updates about collective intelligence in logistics and supply chain. We also thank many other researchers from around the world as Chloë Malépart, UQAM Québec; Simonné Suarent, Technical Univesity of Mauritius; Taardemi Suaromi, Technical University of Oulu, Finland; Czongor Sziladzi, Pécs University Hungary; Antonio Padméterakiris, Nicosia University Cyprus; Aliya Tatkedhek, Kalmykya University, Russia Federation, that helped with the peer-reviewing process. We also wish to sincerely thank Janine Haughton for her clerical assistance during the critical final stages in the preparation of this book.

A book that aims to improve our lives would not be complete without giving itself to a well-intentioned cause. Therefore, from the very beginning of the project, we decided to humbly donate all of our revenue generated by this book to World Wild Foundation. The content of the chapters included in this book is the sole responsibility of the authors. The views, opinions or positions expressed by the chapter authors are solely those of the authors, and do not necessarily reflect the views, opinions or positions of the editors. All trademarks, trade names, service marks, and logos referenced in the chapters of this book belong to their respective companies.

Alberto Ochoa-Zezzatti
Juarez City University, Mexico

Jöns Sánchez
Consejo Nacional De Ciencie Y Tecnologia (CONACYT), Mexico

Miguel Gastón Cedillo-Campos
Mexican Institute of Transportation, Mexico

Margain de Lourdes
Polytechnic University of Aguascalientes, Mexico

Section 1
Theoretical Foundations of Logistics

Featuring six chapters on the theoretical ideas related with the correct implementation of diverse range of logistics applications on real world.

Chapter 1

The Premises of Logistics:
The Organisation of Warships in France in the 17th and 18th Centuries

Jacques Colin
Aix Marseille University, France & ESC Rennes School of Management Aix-en-Provence, France

ABSTRACT

The objective of this paper is to show how the French Royal Navy, faced with multiple challenges of different kinds, has built in the 17th and 18th centuries a particularly complex military-industrial and organisational tool, which foreshadows the most up-to-date industrial and logistical organisations. By reinterpreting this pre-industrial episode, one could not only set out the major principles that constitute the foundations of the logistical and SCM backgrounds (anticipation, reactivity, standardisation, normalisation, productivity, modularity, flexibility, interoperability, fluidity, continuity), but also some logistical archetypes (strategic control of space and strategic control of time, transport infrastructures and accessibility, global sourcing and suppliers' networks, nomenclatures and production ranges, warehouses and stocks availability).

INTRODUCTION

From the 17th Century, the main European Nations, including France under the influence of Richelieu, have created permanent war fleets. These consisted of dedicated ships and replaced the use of merchant ships, equipped urgently and with no preparation for war. Nonetheless, during peacetime, and for obvious economic reasons we will discuss later, these war fleets were not ready for action: even though the warships were well built, they remained laid-up in ports in an "*organized abandonment state*" (Meyer & Acerra, 1994, p.7). When necessary, the Navy would requisition merchant and fishing "seafarers" (to have the right to work, "seafarers" had to serve in the Royal Navy) and order to equip a war fleet by mobilizing all the necessary resources taken from the Royal Navy arsenals. Rodger (1995, p. 365) notes "*the main navies were permanent forces since they counted with a well established administration, appropriate infrastructures, and a ship fleet*". He also underlines that "*war has always depended a lot on*

DOI: 10.4018/978-1-4666-9779-9.ch001

logistics and finance, but the 18th Century demobilized navies were especially subject to administrative systems able to put in action the huge and expensive navy force as fast and as efficiently as possible".

Unlike the army, which can use the supplies provided by the lands it goes through, a navy squadron moves in an empty space: "since *the Navy could not be supplied every day, it had to thoroughly anticipate stocks of supplies and material, organizing storage and packing for transport*" (Corvisier, 1995, P. 134). The conclusion of this paper wishes to show how the French Royal Navy, facing multiple and different challenges, was very early able to establish a "military-industrial" and organisational tool, particularly complex, that prefigures the state-of-the-art industrial and logistic current organisations.

THE NAVY FACING POLITICAL, ADMINISTRATIVE, AND STRATEGIC CHALLENGES

In case of war, especially against England, the main challenge the Navy had to deal with was a fast increasing power of a very technical military tool, able to face the "structural" numerical superiority of the English Royal Navy, which was nonetheless slower to reach its full strength. In peacetime, the fleet would only count with a reduced number of ships, less than ten, with 4.000 seafarers on the eve of the American Revolutionary war in 1778 (Meyer & Acerra, 1994). In 1790, it "theoretically" reached 81 vessels and 69 frigates with 78.000 seafarers (Acerra & Meyer, 1998), and almost 8.000 cannons according to the *Tableau de la Marine Royale* (1789) and regulated by almost 1.700 officers (Acerra & Meyer, 1998). By comparison, the Great Army of Napoleon in 1812 counted with less than 1.400 cannons, for more than 400.000 men (Rey, 2012). Most vessels from this fleet in 1790 were indeed already built. Some were being built and others in "kit" in arsenals[1].

As such, the equipment of the fleet depended completely on the "naval munitions" (structural members, masts, sails, lines, cannons, munitions, powder, pitch, tar, food supplies, etc.) that were bought and stocked in advance in special warehouses to face any possibility. The Navy counted on a long-term strategic vision to enable the acquisition of all the necessary raw materials, and would it be required, to transform and package them. Simultaneously, the class system enabled to identify and quickly call the "seafarers" necessary to serve on the vessels (*Ordonnance de Louis XIV*, 1689, Chapter Eight).

France counted with many resources such as wood to build ships, but forestry was strictly regulated. The Ordonnance established by Colbert regarding Waters and Forests in 1669 was giving the Navy a pre-emptive right on any tree located at less than ten leagues away from the sea or two leagues away from a navigable river (this perimeter would then be broadened to the whole Kingdom in 1743). Any landowner had to ask an authorization to fell a tree to the Bureau of the Navy, which had 6 months to give its answer (Beyeler, 1997). Some trees, with ideal shapes to make specific structural members (see Figure 1) were hot-ironed branded with an anchor and a fleur-de-lys above it to the exclusive use of the Navy (Ballu, 2000). Such measure implied of course that the location of these trees was precisely known and recorded so they could be felled would it be necessary.

Other resources were mainly lacking and had to be imported from very far away, such as hemp, tar and timber for masts. Masts must meet "*at their highest degree three main qualities: flexibility, elasticity and lightness*" (Forfait, 1788, p. 182), and are mainly located in northern Europe (Russia, Norway, Ukraine, Livonia, etc. but also in Acadia and Canada, even though such timber with a lower quality could be found in Auvergne or in the Pyrenees (Vial du Clairbois, 1787). Importing these goods was a

Figure 1. Nomenclature of parts related to a type of ship
Source: Encyclopédie méthodique marine (1786), Pancoucke Librairie, Paris, Volume 2, p. 41.

42 DEV DEV

MATIÈRES.		N°. 1.			N°. 2.		
Bois de chêne : pour la construction.	*Prix des matières, marchandises ou munitions.*	*Nombre des pièces.*	*Quantités, poids & mesures.*	*Sommes.*	*Nombre des pièces.*	*Quantités, poids & mesures.*	*Sommes.*
				liv. s. d.			liv. s. d.
Montant de l'autre part..	....	.....	....	37374..7.5	.....	....	38597..8.10
Première espèce........	à 57 s. le pi. cub.	64	1933p.c.	5509..1.0	65	2192p.c.	6247..4..0
Deuxième *idem*........	à 52........	299	5991	15576.12.0	381	7498¼	19495..9..0
Troisième *idem*........	à 47........	414	5427	12755..9.0	349	4708¼	11065.11..0
Quatrième *idem*........	à 42........	618	5806	12192.12.0	559	5385¼	11310..1..6
Cinquième *idem*.......	à 36........	16	56¼	102..3.0	17	60	108..0..0
28 pièces.............	à 40........	28	194½	389..0.0	16	192½	259..0..0
22 *idem.* de démolition..	à 30........	22	201½	302..5.0	12	106	159..0..0
Bordages de 4 à 5 pouces.	à 3 liv.......	831	7072½	21218..5.0	912	7354½	22063.10..0
Bouts de bordages.....	à 55 s......	40	48	132..0.0	20	24	66..0..0
Croûtes de chêne......	à 20........	60	40	40..0.0	63	44¼	44.15..0
Cabrions, au pied courant.	à 50........	569	35½	88.15.0	843	59½	148.15..0
Barrotins..............	à *idem*......	230	10	25..0.0	240	11	27.10..0
Tringles de chêne, au pi. cu.	à 3 liv......	1300	56	168..0.0	1200	45	135..0..0
Pour la Menuiserie.							
Bordage de chêne, au pi. }	à 3 liv.......	214	51		155	35½	
Planches de chêne, à *id.* }		161	31½		219	27¼	
			82½	248..5.0		62¼	188..5..0
Barrotins, au pied.... }	à 50 s........	641	3:½		680	52	
Cabrions de chêne, à *id.* }		284	21		327	25½	
			53½	133.15.0		57½	143.15..0
Lattes de chêne........	à 55 s.......	676	7	19..5.0		9	24.15..0
Bois de sap : pour la construction.							
Bordage de sap........	à 40 s.......	435½	2216½	4433..0.0	325	2075½	4151..0..0
Planches des Pyrennées..	à *idem*.......	4	5	10..0.0	132	95½	191..0..0
Idem...............	à 30........	225	162½	243.15.0	331	274¼	411..7..6
Lisses de sap, au pi. cour^t..	à 40........	440	74	148..0.0	446	75½	151..0..0
Planches de demi-Prusse..	à 36........	40	166	298.16.0			
Idem. de Cooperwick...	à 34........	525	287	487.18.0	650	332¼	14.16..6
Idem. de Bergue.......	à 28........	150	124½	173.19.0	186	149	208.12..0
Cabrions de sap au pied courant..............	à 45........	1101	68¾	154.13.9	2055	183	411.15..0
Bordages de sap p^r la scul^re.	à 40........	4	56	112..0.0	3	42	84..0..0
Pour la Menuiserie.							
Bordage de sap........	à 40........	1[illegible]4	34¾	69.10.0	190	38½	77..0..0
Planches de demi-prusse.	à 36........	4	16	29.14.0			
Id. du nord de 19 à 26 p^ds de long^r sur 18 p^o de larg^r.	à 42........	7	20	42..0.0			
Cabrions de sap, au pied courant..........	à 45........	161	10	22.10.0	145	9½	21..7..6
Planches de Cooperwick.	à 34........	776	741	1259.14.0	930	888	1509.12..0
Idem. de Bergue........	à 28........	489	244½	342..6.0	599	299½	419..6..0
Idem. sciées en tringles..	à 28........	88	44	61.12.0	71	35[illegible]	49..4..0
TOTAL..................		.....	......	114164..2.2	.....	......	117783.19.10

real issue for the French Navy during wartime. Indeed, the English Royal Navy, thanks to its numerical superiority, could easily block French ports and intercept northern supply routes that would transit through the Sound, between Denmark and Sweden.

To overcome its inferiority, the French Navy adopted two very innovative strategies. It chose to mainly supply itself in peacetime, when the budget of the Navy was making it possible. As such, they could count with all the necessary stock to rapidly equip the vessels. It all implied a very strict organisation, qualified by Zysberg (1995, p. 455) as an "*administrative monarchy*" which origin was the 1689 Ordonnance. This Ordonnance describes in 23 books (chapters) the structure and the administration of the Navy: nine books deal with men and hierarchy; 5 with ports, arsenals, and hospitals; six with building and equipping vessels; one with finances, one with prizes and the last one with felling and stocking marine timber. Later, the 1786 Ordonnance de Castries will complete this bureaucratic measure and Lacour-Gayet (1905, p. 564) writes "*it covers all the navy administrative issues; port tools, shipbuilding, military men, the Secretary sees it all*". It must not be forgotten that these Ordonnances do not deal at all with naval tactics and even less with naval war strategies, but only with a "naval system".

The second strategy was to open the Southern route, which, through the Volga basin, the Black and Mediterranean Seas, a zone the British Royal Navy did not control, would enable to supply in Toulon timber and naval munitions from Northern Europe. Louis XIV also had the Canal du Midi built to enable convoys to reach without any danger the arsenal of Rochefort and Brest along the French costs. Later on, Napoleon even had the strange canal from Nantes to Brest dug, parallel to the Atlantic Ocean, to supply Brest without being bothered by the blockade of this major French Navy port. It also seems that the reason why Louis XV annexed Corsica to France was to prevent the English from settling there, right in front of Toulon, second French war port and arsenal; and from foresting the *laricio* pine trees, which could make good masts.

THE NAVY FACING FINANCIAL AND ORGANISATIONAL CHALLENGES

To face a crisis, the Secretary of State of the Navy had to assure the fast increasing power of the fleet, which implied a budgetary priority granted to the Navy so it could count with the necessary funds to operate the "bureaucratic machine" underlying this war effort. Building a battleship, with its masts and cannons cost at the end of the 18th Century 500.000 pounds for a frigate, 1.350.000 pounds for first-rate vessel and 850.000 pounds for a 74-gun vessel, which was the most built ship of the line at this time (Loir, 1893). Of course, had to be added to this cost all costs related to the equipment (sails, lines, munitions, replacement parts, anchors, etc.) and food supplies for the ships, as well as expenses related to the wages of the crew, that is 775 men for a 74-gun ship (Boudriot, 1988) and up to 1.300 men for a 118-gun vessel (Toudouze *et al.*, 1959), only 62 meter long and 16 meter wide (Loir, 1893).

In the 17th and 18th centuries, most of the nation budgets were dedicated to military expenses. In the 1687 France, this budget was divided between the Navy (20%), which counted with the Flotte du Levant in the Mediterranean and the Flotte du Ponant (Atlantic Fleet); and the War department (80%), which counted with the Army and the fortifications (Corvisier, 1995). This was not the case in England, which always gave priority to its Royal Navy because of its being an island. Nevertheless, the importance of the allowances to the Navy depended on the circumstances and always on the influence the Secretary of Navy had on the King. This is why between 1778 and 1782, during the American Revolutionary War, the budget of the Navy, with 693 millions of pounds, largely exceeded the War's budget with 532 millions of pounds (Meyer et Acerra, 1994). The financial priority given to the naval tool enabled France to quickly reach a powerful fleet that competed equally with the British Royal Navy (Jenkins, 1977). As such, France was able to supply without any problems the American Insurgents all the material and military help necessary to win the decisive battle of Yorktown in 1781. Indeed, the main British Army commanded by Cornwallis, besieged in Yorktown by the Franco-Americans, had to surrender after the French fleet, during the Battle of the Chesapeake, intercepted and turned away the British fleet that had come to help Cornwallis.

How to explain this victorious war in a context of Great-Britain naval superiority? Let's remember that this was the only naval battle France won against England. This victory was particularly decisive for world history since it not only precipitated the fall of the French monarchy, overwhelmed by a colossal war debt of a conflict that cost 1.500 millions of pounds (Meyer & Acerra, 1994), but also led to the Independence of the United States of America in 1783. Even though France was theoretically inferior, it was able to equip a battle fleet faster than the English, thanks to the anticipation of its needs and to the industrial power of its arsenals largely doted with State skilled workers and with all kinds of naval munitions. The English did not count with such powerful royal arsenals; they subcontracted about half the construction of their ships to private shipyards (Knight, 1998). They trusted their "historical" naval superiority as well as the control they had on the main shipping routes, and would get their supplies on a daily basis according to their needs, without trying to stock up in advance.

This little planned and "go with the flow" naval policy meant very long construction and equipment times. Moreover, the British system of "impressment", meaning taking men by force to form crews, turned out to be less effective than the French class system, which made the French Secretary of Navy La Luzerne say that this system "*enabled us to get our fleets out before the English ones*" (Acerra & Meyer, 1988). Acerra and Meyer (1988, p. 31) believe that this quick mobilisation of the French fleet "*gives it, for more or less 18 months, an undeniable leadership over the English fleets and also explains the French success at the beginning of the wars*". This is why we consider that the French Navy, to fight its obvious strategic inferiority against an England controlling the maritime spaces, knew very early how to innovate by anticipating its logistic needs and by increasing its timing, based on the performance of its naval and military-industrial complex. As such, it developed an obvious strategic superiority over the British Royal Navy, based on an improved management of times related to all the necessary processes to conceive, build, organise and raise a naval system born from the "Reason".

SCIENTIFIC, TECHNOLOGIC, INDUSTRIAL, AND LOGISTIC CHALLENGES

Sailboats at the end of the 18th Century are probably the climax of a technical system coming directly from the Age of Enlightenment, which largely took advantage of the state-of-the-art scientific and technologic skills of the time. Bernoulli's principle (1738) sets the basis of fluid mechanics and more precisely of hydrodynamics. Thanks to it, the engineers-constructors of the Navy established by the Ordonnance of the 26th of March 1765 (*Encyclopédie méthodique marine*, 1783, volume 1) could calculate the optimal theoretical profile of the vessel's hull to reduce ship resistance in the water. Before, ships use to be empirically built according to the experience of the master-carpenters, which gave them unpredictable performances. Some ships could turn out to be "bons marcheurs" (good sailing ships) others not at all. Duhamel du Monceau (1752) dedicated the ninth chapter of his treatise on naval architecture to a "Method to calculate the water resistance on the vessel's bow". Euler, very dedicated to the infinitesimal calculus, published in Saint Petersburg a *Théorie complète de la construction et de la manœuvre des vaisseaux* (*Complete theory of the construction and steering of ships*) (1773) and his *Théorème sur la courbure des surfaces* (*Theorem on the curvature of curves on a surface*) (1760) enabled him to calculate the forces on sails, to optimize its shape. This calculation was recently used for the wings of the Airbus A380!

At the same time, the Navy generated many technological innovations. The problem of longitude that is determining the East/West position of a ship, could only be solved thanks to the improvements of micromechanics that enabled to build perfect "time keepers" chronometers. The French watchmaker Leroy tested in 1767 a marine chronometer during a trip entirely financed by a benefactor, the Marquis de Courtanvaux, who had a frigate specially built for the occasion (Pingre, 1768). To improve the speed of their ships, the English added a copper sheet to the hull to prevent seaweeds and molluscs from attaching. The French experienced it for the first time in 1778 on the Iphigénie (*Encyclopédie méthodique marine*, 1786, volume 2), before extending its use on battleships. Learned societies and schools were founded to stimulate and transmit to the marine the scientific applications and emerging technics. The Naval Academy was founded in Brest in 1752 to become the Royal Naval Academy in 1767 in Paris. In 1741, a Naval School was created in Paris. It became in 1765 the School of Student Engineer Constructors (now ENSTA - Superior National School of Advanced Techniques).

Unger (1985, p. 32) reveals "*that during the 18th Century conceiving and shipbuilding separated*". In the case of France, this becomes particularly true by the conceiving work of the engineers constructors (among which Sané and Borda are the most famous) who realize the ship blueprints and supervise the construction by the Navy carpenters, well before the industrial era and Taylorism. The Grande Ordonnance of 1689 standardises the vessels in five rates, from the longest (163 feet for a first-rate vessel) to the shortest (110 feet for a fifth-rate vessel). In the mid-18th Century, Choisel had a new standardisation adopted, with only three rates (Jenkins, 1977). At the end of the American Revolutionary War, de Castries defined very precisely the characteristics of the three rates (first-rate three-deckers of 110 to 120 guns, second-rate two-deckers of 80 guns and especially third-rate of 74 guns) and the two types of frigates of the French fleet (Jenkins, 1977). All these ships were built from the same models, with some slights changes. Acerra (1993) reveals that this rate normalisation turned out to be quite theoretical and suffered many infidelities. Only the smaller ships (corvettes and avisos) did not follow models and were often used as test-beds for such or such innovation (Acerra & Meyer, 1988) that could potentially be generalised to the all fleet, as for the copper sheet.

Of course, one of the tactical advantages expected from this standardisation is to count with a homogeneous fleet, with ships with identical performances and able to move synchronously, in a battle or a convoy. Another advantage is, even though this standardisation might not always be strictly respected, to be able to define normalised parts from each model. This normalisation for parts enables to establish very accurate nomenclature of all parts related to a type de ship (see table 1). As such, parts could be mass-produced (cost reduction) and in advance (immediately available). Figure 1 also shows how extremely rigorous and thorough the work of the engineers constructors was. They foresee consuming 831 planks of four to five inches, that is 7.072,25 cubic feet of timber (1/28.000 thousandth accuracy!) for a cost of

21.218 pounds and five sols[2]... Other parts can be used for different types of ships (masts, sails, structural members) and the equipment is also normalised to be assigned to different types of ships, such as cannons that only count with three models (36 pounds, 24 pounds, and 18 pounds, referring to the weight of the cannonball). This very modular design of both the ships and the parts, at least partially interchangeable, makes maintenance easier (let's not forget that a ship is often damaged during a storm or a battle) and leads to an interoperability of the naval means. It also enables some flexibility in the fleet management by making it possible to build or equip such or such type of ships according to the context.

The fact of defining models of ships also enables to stabilise the construction processes of the ships as well as their different "components" and to establish actual, very rational and productive "production ranges". Figure 2's presentation of the different tasks related to shipbuilding is extraordinarily accurate. It gives the cost and the quantity of work necessary for each profession. As an example, we see that to produce a type 1 frigate, 2.903,5 workdays of a driller are necessary, which represents 3.390 pounds and 11 sols! The final cost of this frigate is of 166.032 pounds 15 sols and four deniers[3]: Could it be more accurate and better organized? Defining all the phases of shipbuilding and the availability of all the material and human resources in an arsenal could also enable, according to the means used, a great reactivity: in 1780, in the middle of the American Revolutionary war, the 100 gun vessel le Majestueux was built in seven months instead of twelve normally. On the other hand, in peacetime, the arsenal could use several years to build a ship in order to retain the skilled workforce.

It clearly appears that the arsenal, where three authorities coexist-military, administrative and corporative (Acerra, 1993) represents the central piece of this enormous machinery that becomes the backbone of the war navy's operation: "*It's a closed area with a port belonging to the government, where it has its vessels and all that is right to build, maintain, equip, lay-up, refit*" (*Encyclopédie méthodique marine*, 1783, volume 1, p. 65). The arsenal is an industrial and logistic centre; it's a closed area, with a restricted access to protect the buildings and what is inside them; very rationally organised to ensure the greatest fluidity of circulation of the "naval munitions" from the workshops and/or warehouses to thoroughly packaged and precisely kept in books stocks. The arsenals also represent the greatest workforce concentrations of the time: in the middle of the American Revolutionary war, Brest counted with more than 10.000 workers only in the arsenal, divided in about thirty very specific professions (Acerra, 1993); and the arsenals of Brest, Toulon and Rochefort counted at the end of the 18th Century from 8,000 to 16,000 workers (Toudouze *et al.*, 1959).

The arsenal is also the key element of a very dense network of suppliers, whether locals, nationals and continentals, even worldwide as seen previously. Even though the timber for mast came from very far away, transiting often through Le Havre or Toulon, most of the structural timber came from French woods, strictly looked after by the Navy. The Navy bought this timber from tenderers and munitions suppliers according to prices fixed by the Navy (Acerra, 1993). Even though some subcontractors could be located next to the arsenal, they would receive their material from external factories. For example, the cannons would come, among others, from foundries in Ruelle or from Indret (a 36 pound cannon would weight 3.5 tons); the anchors from Cosne-sur-Loire (a anchor could weight up to five tons); the salted meat from France but also from Ireland and Denmark for the finest qualities; the wine from the South-western part of France or from the Loire Valley; the eau de vie (spirits) from Cognac, cheese from Holland (Boudriot, 1988), etc. This would imply many transportation issues and waterborne transport was massively used since it was the only one that could carry such heavy a load as a 5 ton-anchor.

The arsenal is also an industrial complex where very diverse industrial buildings are found, such as line makers (the one in Rochefort, still in place, is 300 meter long), sail makers, carpenters, sculptors, bakers to make cookies (seamen's staple food), essential coopers to stock provisions (water, wine, eau de vie, cookies, salted meat, etc.). It also counts with spectacular industrial infrastructures such as slipways

Figure 2. The different tasks related to ship building
Source: Encyclopédie méthodique marine (1786), Pancoucke Librairie, Paris, Volume2, p. 42.

DEV DEV 41

DEVIS *ou état de la dépenſe en journées d'ouvriers, façon d'ouvrages, matières & munitions de toutes eſpèces employées à la conſtruction des frégates* N°. 1 & N°. 2, *de* 32 *canons, exécutées en* 1777 & 1778, *y compris leur mâture.*

SAVOIR:

JOURNÉES D'OUVRIERS. *Eſpèces d'Ouvriers : pour la Conſtruction.*	*Taxes.*	N°. 1. *Journées.*	N°. 1. *Sommes.*	N°. 2. *Journées.*	N°. 2. *Sommes.*
	ſ.		liv. ſ. d.		liv. ſ. d.
Contre-maîtres charpentiers	32 à 45	288	615...8..0	323½	617...9..0
Charpentiers	14 à 30	13124	17264..12..6	14003	18239..14..0
Idem. du petit entretien	18 à 26	249	340...2..0	255	348..10..0
Idem. ſculpteurs	20 à 30	230	273...3..0	200	264...3..0
Idem. mâteurs	14 à 30	1171	1615...0..0	1098	1410...0..0
Apprentifs charpentiers	8 à 12	1320½	702...0..0	1369	732..10..0
Contre-maîtres perceurs	30 à 42	147½	278...8..0	172½	295...3..0
Perceurs	20 à 26	2903½	3390..11..0	3304½	4119..13..6
Journaliers	18 à 20	470	485..12..0	433	426...8..0
Menuiſiers	12 à 30	606½	753..14..6	657¼	845...4..0
Serruriers	20 à 24	26	30..18..0	24	28..18..0
Calfats	14 à 28	2087	2713...2..0	2011	2614...6..0
Chaloupiers des bois	20 à 22	1431½	1520..18..0	1327	1496...8..0
Broyeurs de couleurs	à 20	110½	110..10..0	114	114...0..0
Forçats pour ſupplément de ration	à 2 ſ. 3 d. la ration.	6150	691..17..6	6170	694...2..6
Sous-comites	à 20	133	133...0..0	128	128...0..0
Pour la miſe à l'eau.					
Apprentifs canonniers, y compris un cap.	12 à 20	180	112...0..0	182	113...4..0
Gardiens volans, y compris un patron	20 à 30	186	195...0..0	184	193...0..0
Forçats, pour ſupplément de ration	à 2 ſ. 3 d.	280	31..10..0	272	30..12..0
Pour garniture de poulies.					
Journées de poulieurs	14 à 20	2159½	1722..10..0	2156¼	1719...5..0
TOTAL		33253½	32979..16..6	34384	34430..10..0

Façon d'ouvrages : pour la conſtruction.	*Prix des façons.*	*Quantités, poids & meſures.*	*Sommes.*	*Quantités, poids & meſures.*	*Sommes.*
			liv. ſ. d.		liv. ſ. d.
Montant tiré de ci-deſſus			32979.16.6		34430.10.0
Sciage des bois, au pied quarré	à 7 den. ½ le pi.	21726pds.	678.18.7	23794	743.11.3
Fonte pour la cloche, façon compriſe	à 34 ſ. 3 d. la l. 95 }	146 liv.	396.0..6	87 } 132	368..1.0
Id. pour la ferrure du gouvernail	à *idem* 29 }			24 }	
Id. pour celle de l'étambot	à *idem* 22 }			21 }	
Plomb coulé p[r] tuyaux & dalots	à 1 ſ. 6 d. la l.	1174	88..1.0	1418	106..7.0
Sculpture eſtimée à la ſomme de			900..0.0		800..0.0
Pour la Menuiſerie.					
Sciage de planch. de Cooperwick	à 2 ſ. 6 d. la pi.	403	50..7.6	490	61..5.0
Id. de planches de Bergues	à 3 ſ.	73	10.19.0	79	11.17.0
Id. en tringles	à 1 ſ.	88	4..8.0	71	3.11.0
Feuilles de planches de Cooperwick travaillées	à 3 d. ½ le pi.	807...8877pds.	129..9.0	1077.11847pds.	172.15.4
Planches d'*idem. idem*	à *idem*	275...3025	44..2.3	330..3630	52.18.9
Feuilles de *idem.* de Bergues *id.*	à *idem*	65...586½	8.11.2	141..1260	18..7.6
Planches de *idem. idem*	à *idem*	310...2790	40.13.9	414..3720	54..5.0
Main-d'œuvre			2043..0.0		1774..0.0
Total des journées d'ouvriers & façon d'ouvrage			37374..7.5		38597.8.10

Marine. Tome II. F

(the ones in Rochefort are still intact and a reproduction of the 1779 Hermione, which ferried Lafayette to America was recently built there), and dry docks for maintenance, but also industrial equipment such as mating machines.

Finally, the arsenal is a key logistic site where essential and multiple naval munitions to build, maintain and equip a battle fleet are stocked in warehouses or special fields. Structural members are stocked outside in logs, or in warehouses if the parts are already shaped. Figure 4 in appendix presents structural elements that might be stocked in warehouses, but that could also be "kept on foot", for free, and in the best storage conditions, as shown in Figure 3, remaining in their original wood. Part "d" shaped as a square in Figure 4 can be found as part "M" in Figure 3; and part "T" shaped as a Y in Figure 4 can be found as part "V" in Figure 3.

The amount of timber necessary to build a ship was enormous. A 74 gun-vessel required 2.000 big oaks of at least 150 years old, which represents 60 hectares of wood; the largest vessels required 4.000 of them. (Ballu, 2000). Vial du Clairbois (1787) considers that building a 74 gun-vessel required 77.790 cubic feet (2.680 m^3) of oak and 7.835 cubic feet (270 m^3) of fir. Since the average ship lasted about fifteen years, it was necessary to build at least ten ships per year and maintain on a regular basis the ships in service to count on a fleet of 80 vessels and 70 frigates (the French fleet on the eve of the Revolution). Let us not forget that France, always threatened by an English blockade, anticipated its need and would create very important stocks to be able to count with superior forces than the Royal Navy's, always slower to equip.

Timber for masts was very particularly stocked: it was soaked in slightly slated water pools to keep almost indefinitely its mechanic proprieties of flexibility and resistance. To deal with any possible issue, the Navy did not hesitate to count with enormous safety stock: in 1785, it counted with 1.900 large

Figure 3.
Source: Encyclopédie méthodique marine, planche n° 103, Pancoucke Librairie, Paris, 1783 to 1787.

Figure 4.
Source: From Duhamel du Monceau (1752), p. 133.

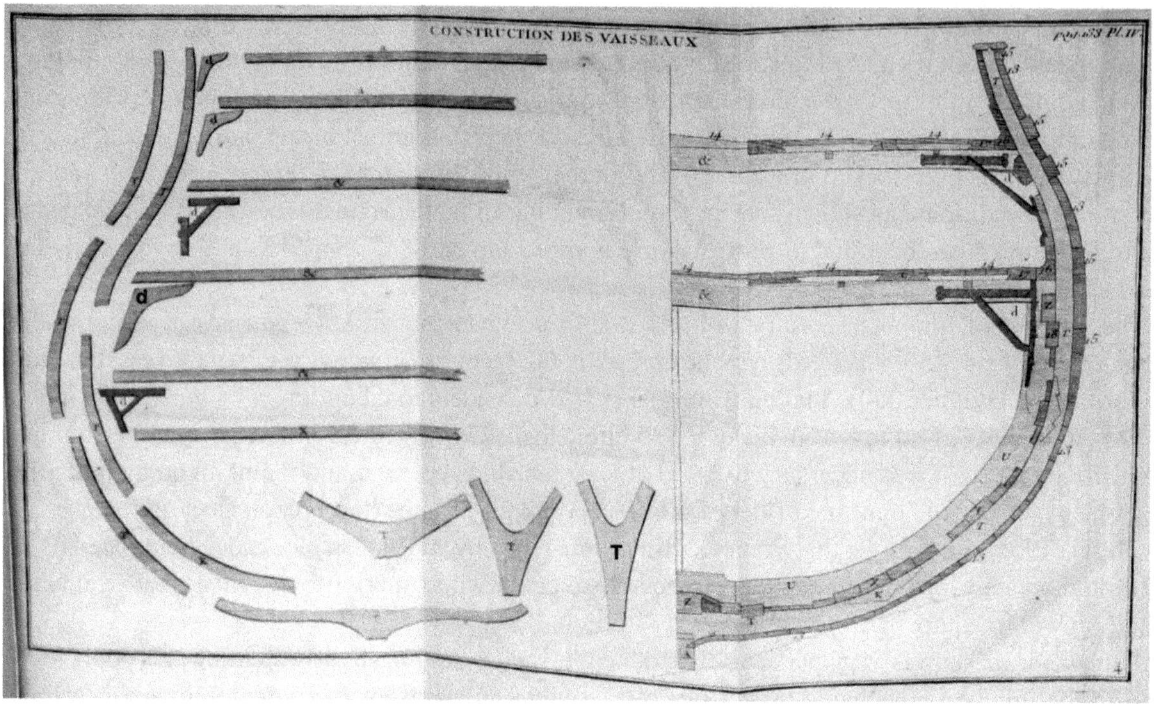

masts and with 622 smaller spars (Acerra, 1988) to re-equip almost the whole fleet available! Flammable (tar, pitch, etc.) or dangerous (powder) materials were stocked in dedicated warehouses (as a powder keg), far away from timber fields to avoid any damage of a possible fire. In 1793, the English, besieged in Toulon by the French, did not hesitate to burn all the arsenal's supplies to weaken the French Navy in the long run.

Cannons, anchors and cannonballs were stocked outdoors, whereas sails, lines, and light weapons were stocked in dedicated warehouses, sheltered from bad weather. In peacetime, all these naval munitions were transported and handled normally to assure the building and/or the maintenance of the vessels. On the other hand, when a fleet required being equipped, the arsenal had to deal with an intense activity to rapidly count with ships ready to go. When squadrons were back, ships had to be laid-up, all parts and vulnerable naval munitions (sails, lines, masts, barrels, light weapons, powder, etc.) as well as the heavy ones (cannons, cannonballs, anchors) had to be disassembled or taken out from the ships. Afterwards, they were re-packed and stocked again in the arsenal for future equipment. The now bare ships could hibernate without damage in ports and wait for another campaign.

CONCLUSION: THE DECLINING SAILING FLEETS AND SHADOWS ON LOGISTICS

The French Navy knew its climax under the reign of Louis XVI and we can find in this administrative and industrial organisation the main current principles of logistics; nonetheless these were not named as such at the time and were thus shadowed for a long time. In spite of the technological leadership of the French "naval system", England, thanks to a much more important population of seamen, conquered back its naval supremacy over France during the Revolution and the Empire. But France almost outflanked the British by proceeding, quite early, in the mid-19th Century, to all kinds of technical innovations that could have made the very powerful English war Navy obsolete and destabilised it. Indeed, the English Navy only counted with sail boats, with an optimal technical but rigid system, of which all the needs were predictable.

Indeed, building a fleet of iron steam battleships, armed with smoothbores and rifled cannons, shooting explosive shells, would enable to outperform a fleet made of wooden sailing ships. Nonetheless, Great Britain did not fall behind and rapidly adopted these new principles of shipbuilding. An actual innovation race started: each ship was already obsolete the moment it was launched and the Navies became real and very heterogeneous "sample fleets". It was now impossible to anticipate the always-renewed material needs and the arsenals had to adopt a reactive policy on a daily basis. The consequence was obvious and brutal for the great principles of logistics: they were forgotten for a century.

REFERENCES

Acerra, M. (1993). *Rochefort et la construction navale française, 1661-1815*. Paris: Librairie de l'Inde Éditeur.

Acerra, M., & Meyer, J. (1988). *Marines et révolution*. Rennes: Éditions Ouest France.

Ballu, J.-M. (2000). *Bois de marine: les bateaux naissent en forêt*. Paris: Marine Nationale et Éditions du Gerfaut.

Beyeler, C. (Ed.). (1997). Du bois dont on fait les vaisseaux… de lárbre en sa futaie à la figure de proue sculptée 1650 – 1850. Vincennes: Service historique de la Marine.

Boudriot, J. (1988). *Le vaisseau de 74 canons*. Paris: Éditions Jean Boudriot.

Corvisier, A. (1995). Quelques réflexions sur les relations entre armée et marine sous l Ancien Regime. In État, marine et société. Presses de l'Université de Paris-Sorbonne.

Duhamel du Monceau, H.-L. (1752). *Élements d l'architecture navale*. Paris: Charles-Antoine Jombert.

Forfait, P. (1788). *Traité élémentaire de la mâture des vaisseaux*. Paris: Clousier.

Jenkins, H. (1977). *Histoire de la marine française*. Paris: Albin Michel.

Knight, R. (1998). Construction et entretien de la flotte anglaise durant les guerres anglo-françaises. In Les marines de guerre européennes: XVII-XVIII siècles. Presses de l'Université de Paris-Sorbonne.

Lacour-Gayet, G. (1905). *La marine militaire de la France sous le règne de Louis XVI*. Paris: Librairie Honoré Champion.

Loir, M. (1893). *La marine française*. Paris: Librairie Hachette.

Meyer, J., & Acerra, M. (1994). *Histoire de la marine française*. Rennes, France: Éditions Ouest.

Ordonnance de Louis XIV touchant la Marine. (1681). Août.

Pingre, A. (1768). *Journal du voyage de M. le Marquis de Courtanvaux*. Paris: Imprimerie Royale.

Rey, M.-P. (2012). *L'effroyable tragédie: une nouvelle histoire de la campagne de Russie*. Paris: Flammarion.

Rodger, N. (1995). La mobilisation navale au XVII siècle. In État, marine et société. Presses de l'Université de Paris-Sorbonne.

Toudouze, G.-G., De la Roncière, C., Tramond, J., Rondeleux, M., Dollfus, C., & Dubard, P. (1959). Histoire de la marine. Éditions de l'illustration.

Unger, R. (1988). Conception et construction des vaisseaux de guerre européennes au XVII et XVIII siècles. In Les marines de guerre européennes: XVII-XVIII siècles. Presses de l'Université de Paris-Sorbonne.

Zysberg, A. (1998). La soumission du rivage aux volontés de l'État Royal. In Les marines de guerre européennes: XVII-XVIII siècles. Presses de l'Université de Paris-Sorbonne.

ENDNOTES

1 The strength of the navy of a country depended completely on the amount of "seafarers" able to be part of the fleet crews. England has always been able to count with a stock of seamen twice as important as the French one.

2 The "sol" was a French currency until the 18th Century.

3 The denier was also a French currency

Chapter 2
Optimization of the Vertex Separation Problem with Genetic Algorithms

Héctor J. Fraire Huacuja
Tecnológico Nacional de México, Instituto Tecnológico de Ciudad Madero, Mexico

Norberto Castillo-García
Tecnológico Nacional de México, Instituto Tecnológico de Ciudad Madero, Mexico

ABSTRACT

The Vertex Separation Problem (VSP) is an NP-hard combinatorial optimization problem in the context of graph theory. The importance of studying VSP lies in its close relation with other problems. Thus, VSP has important practical applications in the contexts of very large scale integration design, computer language compiler design, natural language processing, order processing of manufactured products and bioinformatics. Up to our knowledge, there are only two trajectory-based metaheuristic algorithms for VSP documented in the literature. The main contribution of this chapter is that we extend the available heuristics to solve VSP by proposing a genetic algorithm (GA). It is of particular interest to study the impact of four different crossover operators in the algorithm performance. The experimental results showed that the order-based crossover is the best. Moreover, the best GA variant was compared with the best algorithm for VSP: GVNS. The results of this comparison showed that GVNS outperforms our best GA variant by approximately 1.54 times in solution quality.

INTRODUCTION

The vertex separation problem (VSP) is an NP-hard combinatorial optimization problem (Lengauer, 1981) that asks for a linear ordering of the vertices of an input graph that minimizes the maximum number of vertex separators at each position of the ordering (Díaz, Petit, & Serna, 2002). VSP has important practical applications since it is strongly related to other important problems in a variety of contexts such as VLSI design (Leiserson, 1980; Linhares & Yanasse, 2002), computer language compiler design (Bodlaender, Gustedt, & Telle, 1998), natural language processing (Kornai & Tuza, 1992), order

DOI: 10.4018/978-1-4666-9779-9.ch002

processing of manufactured products (Lopes & de Carvalho, 2010) and bioinformatics (Luque & Alba, 2005). Perhaps, one of the most important applications of VSP is the optimization of the layout area required to assemble a set of logical gates on an electronic circuit. Specifically, this application requires to assign a set of circuit nodes (gates) in an optimal sequence such that the number of tracks needed to cover the interconnection of the gates is minimized (De Oliveira & Lorena, 2002). In order to illustrate this application, consider the example depicted in Figure 1.

In this example there are nine gates and seven nets. The nets represent the interconnection of the gates. The gate numbers are presented at the top of each figure. In Figure 1a is presented a configuration (ordering) of the gates such that the number of tracks required is seven. In other words, each net occupies one track in the circuit. Conversely, Figure 1b presents another configuration in which only five tracks are needed to connect the circuit completely. In this configuration two nets can share the same track. More precisely, the nets $\{4,6,9\}$ and $\{1,2,5,8\}$ (highlighted in red) occupy the second track and the nets $\{6,9\}$ and $\{5,7,8\}$ (highlighted in blue) share the fifth track. Notice that this example exhibits the importance of looking for a linear ordering of the gates such that the maximum number of tracks is minimized, and hence, the layout area is optimized.

The main contribution of this chapter is the extension of the available heuristics to solve VSP by proposing a genetic algorithm. This is the first population-based metaheuristic algorithm to solve stochastically VSP. It is also proposed a crossover operator based on the path relinking methodology. Three different crossover operators from the literature are compared with our crossover proposed. In order to assess the performance of our algorithm proposed, three computational experiments were conducted. The objective of the first two experiments is to evaluate the impact of each crossover operator. In both experiments, our crossover operator was ranked number two while the well-known order-based crossover was the best one. In addition, our best genetic algorithm variant, the one with the order-based crossover, was compared to the best algorithm for VSP available in the literature, GVNS. The results clearly

Figure 1. a) One possible ordering of the gates that requires seven tracks to allocate all the nets. b) Another possible ordering of the gates, this ordering only requires five tracks (optimal ordering). The gate numbers are at the top of the figures. This example was taken from (De Oliveira & Lorena, 2002)

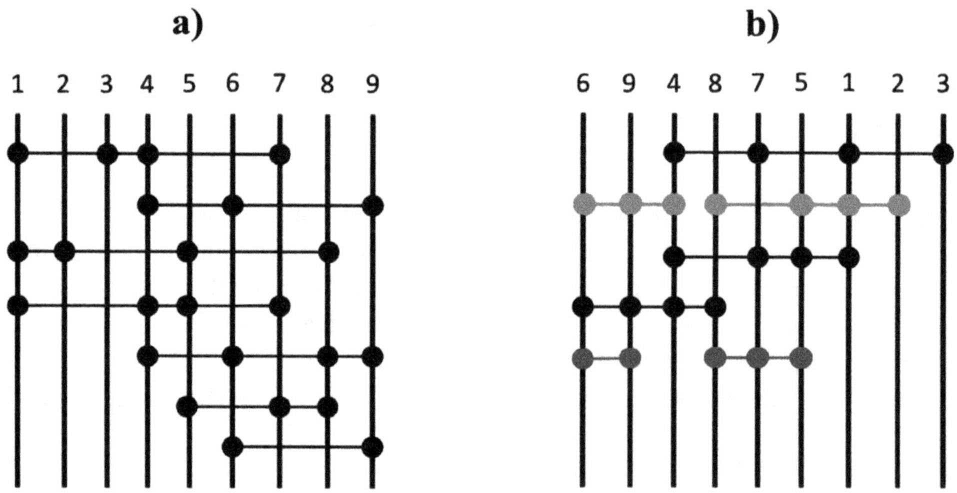

show that GVNS outperforms our best GA variant. Specifically, GVNS was able to find 50 best-known solutions out of 62 instances evaluated, which gives it an effectiveness of about 80.64%. Our best GA variant found 5 best-known solutions and its effectiveness is about 8.06%. These results might suggest that trajectory algorithms perform much better than population-based algorithms in the context of VSP. However, a more robust experimentation with a large number of both types of algorithms must be conducted in order to either prove or disprove the aforementioned hypothesis.

The remaining of this chapter is organized as follows. Section 2 formally defines the vertex separation problem. Section 3 elucidates how VSP is related to military logistics. Section 4 details our genetic algorithm proposed and all its fundamental parts. Section 5 presents the experiments conducted to assess the impact of the crossover operators in the genetic algorithm and to evaluate the performance of our best GA variant with respect to the best algorithm for VSP. Finally, Section 6 discusses the major conclusions of this research.

FORMAL DEFINITION OF VSP

Let $G=\left(V,E\right)$ be a connected simple graph with discrete and finite sets of vertices and edges V and E, respectively. The graph has .. vertices and $m=\left|E\right|$ edges. A linear ordering of the vertices is an arrangement of the vertices in a straight line. Formally, a linear ordering is a bijective function $\varphi:V\rightarrow\left\{1,2,\ldots,n\right\}$ that assigns one integer number between 1 and n to every element of the set of vertices. The integer number assigned to the vertex $u\in V$ is known as *label* and is denoted by $\varphi\left(u\right)$. Notice that the label of some vertex indicates the position of the vertex in the linear ordering. Thus, a linear ordering is actually a permutation of the vertices. An input graph with n vertices has $n!$ feasible solutions for the problem.

Let p be an integer number between 1 and n that indicates some position in the solution φ (permutation). The set of left vertices $L\left(p,\varphi,G\right)=\left\{u\in V\middle|\varphi\left(u\right)\leq p\right\}$ contains those vertices whose label is less than or equal to the current position p. Symmetrically, $R\left(p,\varphi,G\right)=\left\{u\in V\middle|\varphi\left(u\right)>p\right\}$ is the set of right vertices and contains the vertices whose label is greater than p. The set of vertex separators at position p contains the vertices in the set of left vertices with at least one adjacent vertex in the set of right vertices, that is:

$$\delta\left(p,\varphi,G\right)=\left\{v\in L\left(p,\varphi,G\right)\middle|\exists w\in R\left(p,\varphi,G\right):\left(u,v\right)\in E\right\}.$$

The number of vertices in the set $\delta\left(p,\varphi,G\right)$ is the *cut value* at position p. Since there are n positions for each permutation of vertices, there are also n cut values. The largest cut value represents the objective value of the solution φ. Formally, the objective value is given by:

$$vs\left(\varphi,G\right)=\max_{p=1,\ldots,n}\left\{\left|\delta\left(p,\varphi,G\right)\right|\right\}.$$

The Vertex Separation Problem is to find the permutation of the vertices whose objective value is minimal (Díaz et al., 2002).

VSP IN MILITARY LOGISTICS

According to Lopes and de Carvalho (2010), the Vertex Separation Problem is directly related to the Minimization of Open Stacks Problem (MOSP). MOSP consists in determining the best sequence of products to be manufactured in a production line in which only one product can be fabricated at a time. The goal of MOSP is to minimize the maximum number of simultaneous incomplete orders for rapidly fulfilling the customers' orders.

In the context of military logistics, the relation between VSP and MOSP could be used to optimize the storage space in mass production of war equipment. This optimization leads to a rapid fulfillment of the troop requirements (orders). In order to illustrate how this optimization can be performed, let us consider the following example. Suppose that five troops require a set of ten products (war equipment). The requirements are detailed in Table 1. The rows of this table specify the equipment required by the troops. The troops are identified by letters and the equipment by numbers.

The example presented in Table 1 can be represented by the graph $G = (V,E)$ in which the vertices represent the troops. In this graph, there is an edge joining two vertices if and only if their associated troops require the same equipment. For example, vertices A and B (representing troops A and B, respectively) must be joined by an edge since both require the equipment number 7. The complete graph representing this example is presented in Figure 2.

Table 1. Comparison of all the variants of our GA with an improved initial population over the set of 62 instances in HB

Equipment	1	2	3	4	5	6	7	8	9	10
Troop A			✓	✓			✓			
Troop B	✓						✓		✓	
Troop C		✓						✓		✓
Troop D			✓		✓					✓
Troop E	✓				✓	✓				

Figure 2. Graph G with five vertices and five edges corresponding to the example presented in Table 1

$G = (V,E)$:

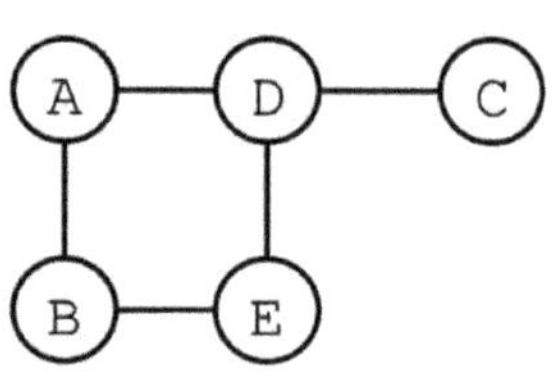

$V = \{A, B, C, D, E\}$

$E = \{(A,B), (A,D), (B,E), (C,D), (D,E)\}$

Figure 3. Optimal linear ordering of the vertices for the graph of the example

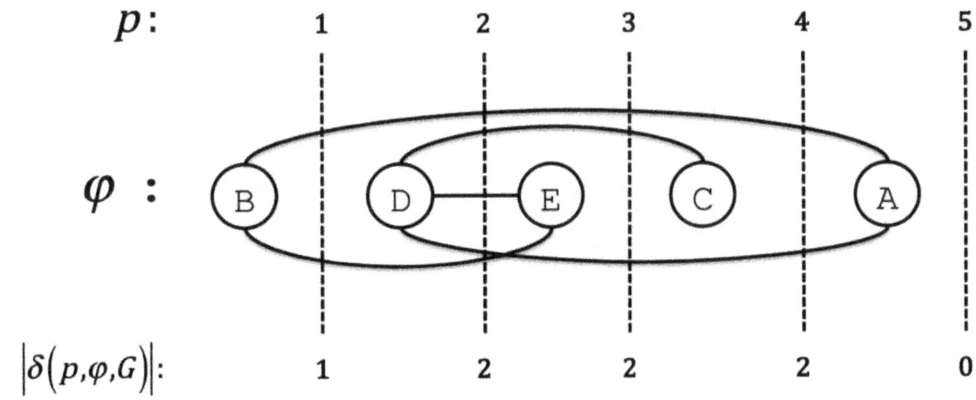

Notice that finding the best sequence for manufacturing the equipment required by the troops is tricky. This is so because the equipment does not appear explicitly in the graph. Thus, the best sequence to process the troop requirements must be found. The sequence for manufacturing the equipment can be obtained from the order of the troops in consequence.

In order to determine the best sequence, this example was modeled by the integer linear programming formulation for VSP designed by Fraire et al. (2015) and solved by the well-known optimization engine CPLEX (CPLEX, 2012). The best sequence to process the troop requirements is $\varphi^* = \left(B, D, E, C, A\right)$. This means that the equipment required by troop B should be manufactured first, then the equipment for troop D, and so on. Figure 3 presents the optimal linear ordering of the vertices φ^*. In this figure, the positions of the permutation are presented at the top and the cut values at the bottom. The vertex separation number (objective value of the optimal solution) is given by the maximum cut value of the optimal solution, i.e., $vs\left(G\right) = vs(\varphi^*, G) = \max\left\{1, 2, 2, 2, 0\right\} = 2$.

The maximum number of open stacks at each time instant is the vertex separation number plus one. This fact was confirmed by modeling the example through the integer linear programming formulation for MOSP designed by Lopes & de Carvalho (2010) and optimally solving the resulting program with CPLEX. Thus, the maximum number of simultaneously open stacks is $k = 3$ for this example. As mentioned before, the sequence for manufacturing the equipment can be obtained from the sequence for the troops. A possible sequence is presented in Table 2. This table has the same structure than that of Table 1. The highlighted sections of each row denote the lifetime of the orders. For example, the order of troop E has a lifetime of three iterations while the lifetime of the order of troop B corresponds to the entire manufacture process (ten iterations). The manufacture of one product is considered one iteration.

The solution presented in Table 2 is optimal since it needs at most $k = 3$ open stacks at each time instant. This solution requires the manufacture of equipment 9 first, which opens a new stack for troop B. Manufacturing equipment 5 opens two new stacks: one stack for troop D and the other for troop E. At this time there are three stacks simultaneously open (B, D and E). The manufacture of equipment 1 does not open a new stack since this equipment is exclusively required by troops B and E, which have their corresponding stacks active. When equipment 6 is manufactured, the order of troop E is completed. This means that the stack belonging to troop E is not longer active, and hence, only the stacks for troops B and D remain open. It also implies that the order of troop E is ready to be sent.

Table 2. Optimal sequence for manufacturing the equipment required by the troops

Equipment:	9	5	1	6	10	3	2	8	4	7
Troop A						✓			✓	✓
Troop B	✓		✓							✓
Troop C					✓		✓	✓		
Troop D		✓			✓	✓				
Troop E		✓	✓	✓						

Manufacturing equipment 10 opens a stack for troop C since this equipment is the first of its order. At this time, there are three open stacks (B, C and D). The manufacture of equipment 3 closes the stack for troop D but opens a new stack for troop A. Thus, at this time there are three stacks active simultaneously (A, B and C). Besides, the order of troop D is completed and ready to be delivered. Manufacturing equipment 2 does not increase or decrease the current number of open stacks. Nevertheless, when equipment 8 is manufactured, the stack belonging to troop C is closed and its order is finished. At this time there are only two open stacks (A and B). Equipment 4 is only required by troop A, therefore, its manufacture does not affect the current number of open stacks. Finally, the entire process finishes when equipment 7 is manufactured.

The previous example exhibits the importance of finding good solutions for the Vertex Separation Problem in the context of military logistics. Unfortunately, not all the instances are as small as the example presented, and hence, stochastic methods should be developed for dealing with large instances.

GENETIC ALGORITHM

Genetic Algorithms (GAs) are population-based metaheuristic algorithms frequently used to solve NP-hard combinatorial optimization problems. GAs were originally invented by John Holland and his students and colleagues at the University of Michigan in the sixties (Mitchell, 1999). This section presents the outline of our GA proposed as well as all its parts.

Genetic Algorithm Proposed

According to Mitchell (1999), a genetic algorithm is a method for evolving one population of *chromosomes* into a new population by applying genetics-inspired operators. The chromosomes encode the solutions for the problem by using a set of *genes*. Each gene is an instance of a particular *allele*, which is the value of its corresponding gene. The value of the gene exclusively depends on the solution representation used. In our case, VSP represents the solution by means of a permutation of vertices. Therefore, the chromosome consists of $n = |V|$ genes, one for each position of the permutation, and the alleles are the vertices of the graph, that is, integer numbers. Thus, each chromosome represents an *individual* and hence the *population* is actually a pool of solutions.

Each individual of the population is related to a *fitness* value, which quantifies how good an individual is with respect to another individuals. A natural fitness value for a combinatorial optimization problem is the objective value. In GAs, the individuals with better fitness values have a larger probability of being selected for reproduction. A typical GA uses three main genetic operators, namely, *selection*, *crossover* and *mutation*. The selection operator chooses those individuals from the population that will be allowed to reproduce. The crossover operator recombines the genetic code of two parents to generate new individuals. The new individuals generated are known as *offspring*. The mutation operator alters the value of some genes of the offspring in order to induce variability in the population. Algorithm 1 shows the high-level pseudocode of our genetic algorithm proposed.

Algorithm 1. *Pseudocode of our genetic algorithm proposed*

1. Generate the initial population P.
2. **while** stopping condition not met
3. $Evaluate(P)$.
4. $Q \leftarrow SelectParents(P)$.
5. $R \leftarrow Crossover(Q)$.
6. $P \leftarrow SelectTheBest(P \cup R)$.
7. **End while**

The algorithm starts by generating the initial population. This can be performed either at random or by executing a heuristic method (line 1). Then, all the individuals from the population must be evaluated in order to know their fitness values (line 3). Next, the algorithm selects a subset $Q \subseteq P$ of individuals from the population for reproduction (line 4). The individuals selected (parents) are recombined to generate the set of offspring R by using a crossover operator (line 5). In this research four different crossover operators are studied. Finally, the best individuals from both the population and the offspring are selected to be in the population of the next generation (line 6). The whole process is repeated until the stopping condition is met (lines 2—7). The only stopping condition of our GA is the time limit. Notice that our GA does not have any mutation operator. This is because our study is exclusively focused on the impact of the crossover operator. Therefore, it is desirable to reduce the number of independent variables of the experiment as much as possible since they might affect the results.

Chromosome

One key point when designing GAs is the representation of the individuals (solutions). Every individual is represented by a chromosome, which consists of a collection of genes. The value of each gene is known as allele.

Depending on the problem, the type of chromosome might vary. For example, the chromosome for problems involving decisions variables might be a string of binary values. Nevertheless, the most suitable chromosome for linear ordering problems such as VSP is that consisting of an array of integer numbers. Thus, the chromosome for VSP is a one-dimensional array of integers of size n. In other words, the chromosome for VSP is a permutation of vertices of the graph. Figure 4 shows an example of the chromosome and its fundamental parts for a graph with six vertices.

Figure 4. Example of a chromosome for a graph with $n = 6$ vertices. The chromosome is an array of integers of size n. Each position of the chromosome (shown in square brackets) is known as gene and the value of the gene is known as allele

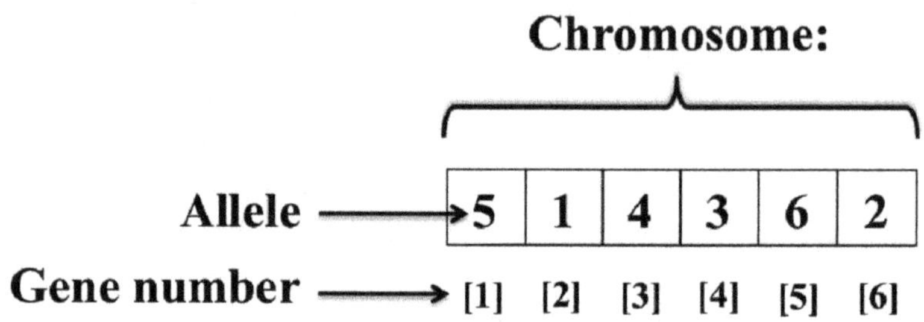

The chromosome shown in Figure 4 consists of six genes, one for each vertex of the input graph. The number of each gene corresponds to the position of the permutation and the allele of the gene is the vertex placed at the aforementioned position. Thus, the first gene has the allele 5, the second gene has the allele 1 and so on. This means that the vertex 5 occupies the first position of the permutation, the vertex 1 occupies the second position and so on.

Selection Operator

The selection is the genetic operator used to choose the individuals from the current population to be recombined in order to generate new individuals. The chosen individuals from the population are called the parents while the new individuals produced by the recombination of the parents are called the offspring.

In this research, the selection method used to choose the parents is the well-known roulette-wheel selection. This operator proportionally quantifies the probability of each individual from the population to be selected according to its own fitness (objective value). Thus, the best individuals have the largest probabilities of being selected whereas the worst individuals have the smallest probabilities. The roulette-wheel method gives all the individuals the opportunity to become a parent (even the worst individual) and at the same time it does not guarantee that the best individual will be selected for reproduction. In order to illustrate how the roulette-wheel selection works, consider the example shown in Figure 5.

In this example, the population consists of ten different individuals numbered from 1 to 10. Each individual is related to a fitness value, which is the objective value. For example, the first individual has a fitness value of 75, the second individual has a fitness of 93 and so on. The roulette-wheel method computes the total sum of fitness values and the corresponding percentage of each individual with respect to the total sum. In the example, the total sum of fitness values is 833. The percentage of the first individual is 9%, the percentage of the second one is 11.16% and so on.

However, the percentages recently calculated cannot be used to select the parents since the individuals with the largest fitness values will have the largest probabilities of being chosen. For example, the eighth individual has a fitness value of 112 units and a percentage of 13.45%, which is the largest one. This means that this individual will have the largest chance of being selected for reproduction. However, this is not suitable since VSP is a minimization problem, which implies that the individuals with low fitness values must be considered the best ones. In order to correctly quantify the probabilities of each individual, it is necessary to invert the current probabilities. This is achieved by interchanging the

Figure 5. Example of the roulette-wheel selection operator

Individual	Fitness	Percentage	Percentage inverted	Range	
				Lower	Upper
1	75	9.00	11.16	0.00	11.16
2	93	11.16	9.00	11.16	20.16
3	87	10.44	9.48	20.16	29.64
4	**62**	7.44	13.45	**29.64**	**43.09**
5	83	9.96	9.96	43.09	53.05
6	66	7.92	11.16	53.05	64.21
7	79	9.48	10.44	64.21	74.65
8	112	13.45	7.44	74.65	82.09
9	83	9.96	9.96	82.09	92.05
10	93	11.16	7.92	92.05	100.00
TOTAL	**833**	**100.00**	-	-	-

Roulette-wheel:

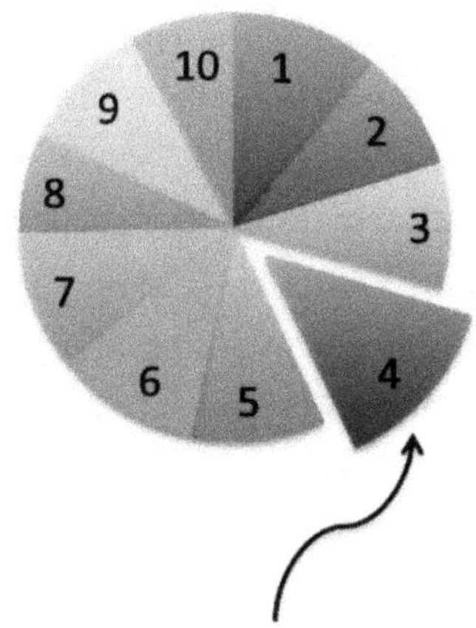

Random number $r = 30.64$

probability of the best individual with that of the worst, the probability of the second best with that of the second worst and so on. For example, the fourth individual (the best) and the eighth individual (the worst) must interchange their corresponding percentages.

Once inverted the probabilities, it is necessary to proportionally divide the interval $[0,100]$ among the probabilities of the individuals. Thus, the portion of the first individual ranges from 0.00 to 11.16, the portion of the second individual ranges from 11.16 to $11.16 + 9.00 = 20.16$, the portion of the third individual ranges from 20.16 to $20.16 + 9.48 = 29.64$ and so on. In order to select one individual from the population, a random number between 0 and 100 must be generated. Then, the roulette-wheel method selects the individual whose range surrounds the random number. For example, if the random number generated is $r = 30.64$, then the fourth individual must be selected since $29.64 < r \leq 43.09$. There is no limit on the number of individuals to be selected. Nevertheless, one random number must be generated to select each individual.

Crossover Operators

A crossover operator is a method to recombine the genetic code of two parents for generating the offspring. In this section four different crossover operators are described. The first operator is based on the path relinking strategy. The other three operators are directly taken from the literature. All the crossover operators studied here are specially designed for chromosomes of integer numbers and generate one single child.

Crossover Based on Path Relinking

Given two different solutions φ_1 and φ_2, the main idea of the path relinking (PR) strategy is to generate a pool of solutions by transforming the solution φ_1 into the solution φ_2. Solutions φ_1 and $ \ var phi _ {2} $ are known as the *initiating* solution and the *guiding* solution, respectively. The solutions generated in the transformation process are the *intermediate* solutions (Durte, Martí, Resende, & Silva, 2011).

In the context of our proposed crossover PR, the solutions φ_1 and φ_2 represent the parents while the intermediate solutions represent the offspring. The initiating parent φ_1 is the one with the best objective value while the guiding parent φ_2 is the other. In case of both parents have the same objective value, the initiating and guiding parents are selected randomly. In order for PR to generate the offspring, a subset of alleles A is constructed by selecting the alleles whose gene numbers in the chromosomes of both parents are different. In mathematical terms, $A = \left\{ v \in V \middle| \varphi_1\left(v\right) \neq \varphi_2\left(v\right) \right\}$.

Thus, A is considered the set of candidate alleles to be examined. Then, PR performs a set of interchange moves between pairs of alleles. Specifically, PR selects at each step one allele u from A and searches for its gene number in the guiding parent, i.e., $\varphi_2\left(u\right)$. Then, PR interchanges the positions of the allele u and the allele w whose gene number in the initiating parent is $\varphi_1\left(w\right) = \varphi_2\left(u\right)$. Finally, the allele u is removed from A, the initiating solution is updated with the current intermediate solution and the next candidate allele is selected. This process is repeated until all the candidate alleles in A have been selected, i.e., $A = \varnothing$.

Each interchange move between a pair of alleles generates a new child. In general, the number of k children that can be generated is at most $n - \left|A\right|$. However, this number can be modified for reducing the computational effort of PR. In particular, PR evaluates a number of $k = \left\lceil \left(n - \left|A\right|\right) \times \alpha \right\rceil$ children, where $\alpha \in \left[0,1\right]$ is a real number set to 0.1. In other words, PR evaluates the 10% of the total offspring. In case of both parents have the same genes, i.e., $A = \varnothing$, PR induces a mutation in one of the parents by inversing a subset of consecutive genes. Specifically, PR inverts at least 50% of the genes of the parent selected. The size of the inversion and the subset of consecutive genes are determined at random. Finally, the crossover PR keeps the child with the best objective value. Figure 6 shows how the crossover PR works for a graph with six vertices.

Figure 6. Example of the crossover based on path relinking (PR)

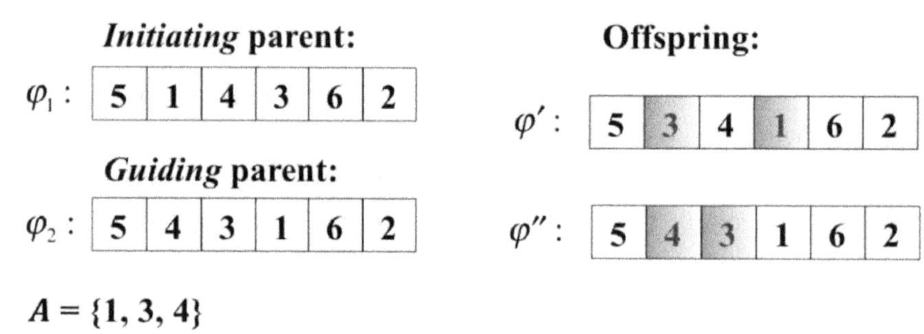

PR starts by computing the set $A = \{1,3,4\}$. This set does not contain the alleles 2, 5 and 6 since they appear in the same gene of both parents. More precisely, the allele 2 is in the sixth gene of both parents, the allele 5 is in the first gene of both parents and the allele 6 is in the fifth gene of both parents. Afterwards, PR selects the first allele u from A, i.e., $u = 1$. Then, PR searches for the allele w such that $\varphi_1(w) = \varphi_2(u)$. Thus, $w = 3$ since $\varphi_1(3) = \varphi_2(1) = 4$. PR then interchanges the positions of alleles 3 and 1 in the initiating parent, generating the first child φ'. Once the first child is generated, the set A must be updated, i.e., $A = A \setminus \{u\} = \{3,4\}$ and the initiating solution must be updated, i.e., $\varphi_1 \leftarrow \varphi'$. The next allele to be evaluated is $u = 3$ and the allele w is 4 since $\varphi_1(4) = \varphi_2(3) = 3$. PR interchanges the positions of alleles 3 and 4 in the initiating parent, which generates the second child φ''. The set A and the initiating parent are updated, that is, $A = A \setminus \{u\} = \{4\}$ and $\varphi_1 \leftarrow \varphi''$. Notice that the second child is identical to the guiding parent. Thus, when PR evaluates the last allele from A, $u = 4$, the third child has exactly the same genes as the guiding parent and the second child since $u = w = 4$.

Order–Based Crossover

The order-based crossover (OB) is a specific crossover operator for chromosomes of integer numbers that consists of two main phases. In the first phase the entire genetic code of the first parent is copied to the offspring. In the second phase OB selects k genes randomly (with $0 < k < n$) and orders the alleles in these genes with respect to the order in the second parent (Schiavinotto & Stützle, 2004). In the context of this research, the number of genes selected randomly has been set to $k = \lceil n/2 \rceil$. Figure 7 shows an example of the crossover OB.

The first parent is $\varphi_1 = (5,1,4,3,6,2)$ and the second parent is $\varphi_2 = (5,4,3,1,6,2)$. First, OB copies the genetic code of $\$ \mathrm{var}\, phi\_\{1\}\$$ to the offspring, i.e., $\varphi' = (5,1,4,3,6,2)$. Next, OB selects $k = \lceil 6/2 \rceil = 3$ genes randomly. In our example the genes selected are 2, 4 and 6. Then, the alleles of these genes are re-ordered in φ' according to the order of these alleles in φ_2. More precisely, the alleles of the genes 2, 4 and 6 in φ' are 1, 3 and 2, respectively. These alleles appear in a different order in φ_2, that is, allele 3 appears before alleles 1 and 2 (at the third gene), allele 1 appears before allele 2 (at the fourth gene) and allele 2 appears at the end of the chromosome. Thus, the new order of the alleles is $(3,1,2)$. These alleles must be placed at the positions of φ' selected randomly in the first phase, that is,

Figure 7. Example of the order-based crossover (OB)

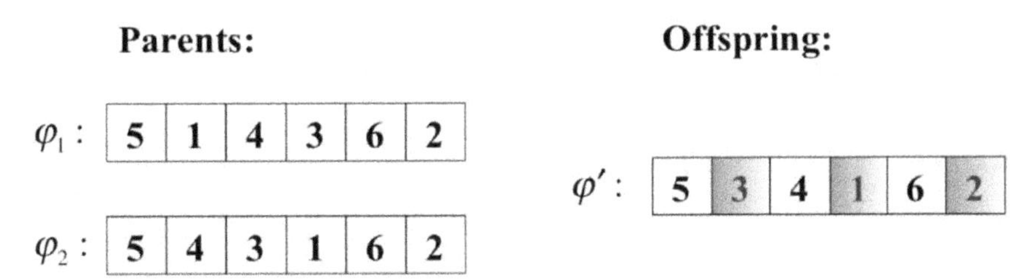

the allele 3 at the second gene, the allele 1 at the fourth gene and the allele 2 at the sixth gene. Finally, the offspring resulting from this process is $\varphi' = (5,3,4,1,6,2)$.

DPX Crossover

The DPX crossover is another genetic operator for recombining the genes of two parents and it works as follows. The offspring inherits the alleles whose gene numbers are the same in both parents. The alleles are put into the same gene numbers as in the parents. Then, the remaining alleles are randomly assigned to the remaining genes (Schiavinotto & Stützle, 2004). Figure 8 presents an example of this genetic operator.

First, the DPX crossover looks for the alleles whose gene number are the same in both parents. In the example those alleles are 5, 6 and 2 since they are in the first, fifth and sixth genes of both parents. These alleles are inherited to the offspring in exactly the same genes. The remaining alleles, 1, 3 and 4, are randomly assigned to the available genes in the offspring. In the example, the allele 3 was assigned to the second gene, the allele 1 to the third gene and the allele 4 to the fourth gene.

Rank Crossover

The Rank crossover recombines two parents to generate one child. Specifically, the rank crossover computes the average ranking of the alleles over the two parents. Then, the alleles are sorted by their average ranking and are assigned to the offspring. In case of ties, they are broken at random (Schiavinotto & Stützle, 2004). An example of this genetic operator is shown in Figure 9.

The rank crossover computes the average ranking of each allele over the two parents. Thus, the average ranking for allele 1 is 3, for allele 2 is 6, for allele 3 is 3.5, for allele 4 is 2.5, for allele 5 is 1 and finally for allele 6 is 5. Then, the alleles are sorted by their average ranking in ascending order. The resulting positions correspond to the gene numbers in the offspring. More precisely, the sorted average rankings are: 1, 2.5, 3, 3.5, 5 and 6. This means that the allele whose average ranking is 1 (allele 5) must occupy the first position, the allele with ranking 2.5 (allele 4) must occupy the second position and so on. Finally, the resulting offspring is $\varphi' = (5,4,1,3,6,2)$.

COMPUTATIONAL EXPERIMENTS AND RESULTS

In this section three computational experiments are presented. The first two experiments are designed to measure the impact of the crossover operator. In particular, the first experiment evaluates the perfor-

Figure 8. Example of the DPX crossover (DPX)

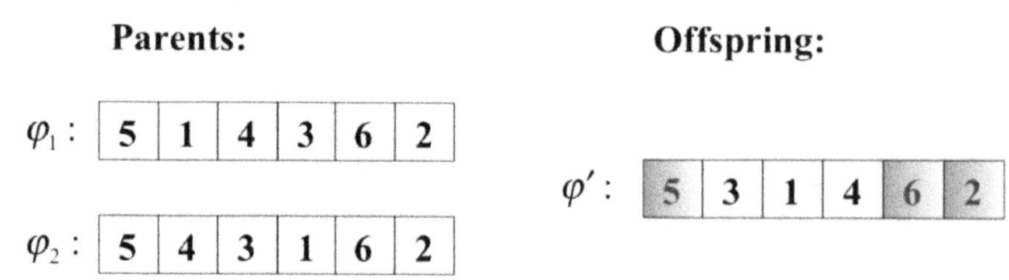

Figure 9. Example of the Rank crossover (Rank)

Parents:

φ_1 : | 5 | 1 | 4 | 3 | 6 | 2 |

φ_2 : | 5 | 4 | 3 | 1 | 6 | 2 |

Offspring:

φ' : | 5 | 4 | 1 | 3 | 6 | 2 |

Computations:

Allele	Gene number		Average Ranking
	First Parent	Second Parent	
1	2	4	3
2	6	6	6
3	4	3	3.5
4	3	2	2.5
5	1	1	1
6	5	5	5

mance of our GA with a random initial population. In the second experiment four elite individuals are inserted into the initial population. The results of both experiments are compared with each other and the key findings are discussed. The last experiment was conducted to compare the results of our best GA variant with the current best algorithm for VSP.

Experimental Conditions

All the experiments were conducted on a standard computer with clock-rate of 2.27 GHz and 4 GB of RAM. All the variants of the algorithm were implemented in Java JRE 1.7.0_07. The set of instances selected to test the performance of the variants of the GA is the Harwell-Boeing (HB) collection, which is considered the most difficult dataset for the Vertex Separation Problem. This dataset contains 62 graphs whose numbers of vertices and edges range from 24 to 960 and from 34 to 3721, respectively (Duarte et al., 2012). Currently, the optimal values for the instances in HB remain unknown.

In the first two experiments a time limit of 300 seconds for each instance was used. In the final comparison a time limit of 1000 seconds for instance was set to each algorithm. Thus, the experiments approximately required a total of 75 hours of uninterrupted computations to be completed. The only genetic operator to be changed was the crossover operator. The remaining operators and parameters maintained unaltered along the experiments. Therefore, the variable of interest for our study was isolated. The values and operators used in the experiments are presented in Table 3.

Experiment 1: Random Initial Population

In this experiment a group of one hundred random individuals was used to populate the first generation of our GA. The results of this experiment are summarized in Table 4. The table reports the average objective value obtained by each variant (O.V.). The table does not report the execution time since all the variants have exactly the same time limit, 300 seconds.

From the experimental results it can be observed that the variant GAOB (with the order-based crossover operator) achieved the best solution quality. The quality of the solutions found by GAOB outperforms those of GAPR, GARANK and GADPX by 29.5%, 66.1% and 72.5%, respectively. GAPR

Table 3. Values assigned to the parameters and genetic operators used in the experiments

Parameter / Operator	Value
Chromosome size	n
Population size	100
Number of descendants	50
Selection	Roulette-wheel
Crossover	PR / OB / DPX / Rank
Mutation	Inversion (only used in the crossover PR)

Table 4. Comparison of all the variants of our GA with a random initial population over the set of 62 instances in HB

GA Variant	O.V.
GAPR	149.52
GAOB	115.42
GADPX	199.10
GARANK	191.73

was the second best variant in solution quality since it outperformed those of GARANK and GADPX by 28.2% and 33.2%, respectively. The worst variants in solution quality were GARANK and GADPX, which were ranked number three and four, respectively.

In order to graphically illustrate the performance of all the variants, let us consider the plot showed in Figure 10. The x -axis presents all the instances in the dataset HB. The y -axis represents the average objective value obtained by each variant. From the plot, it can be observed that the variants GAOB and GAPR (ranked number one and two) are not too close to each other. This means that the objective values found by GAOB and GAPR were different in most cases. Conversely, the variants GADPX and GARANK obtained very similar objective values for the HB instances, which makes them appear really close to each other in the plot. It can also be observed that the variant GAOB frequently appear below the other variants in the plot. This is so because the objective values obtained by GAOB were the smallest for almost all the instances evaluated.

Experiment 2: Improved Initial Population

In this experiment the initial population consisted of four elite individuals and ninety six random individuals. The set of elite individuals were obtained by the constructive algorithms HA and HN proposed by Castillo-García et al. (2015). In particular, the variants HA_1, HA_3, HN_1 and HN_3 were used in this experiment. The experimental results are presented in Table 5. The structure of this table is the same as that of Table 2.

Figure 10. Performance of the four variants of our GA with a random initial population over the set of 62 HB instances

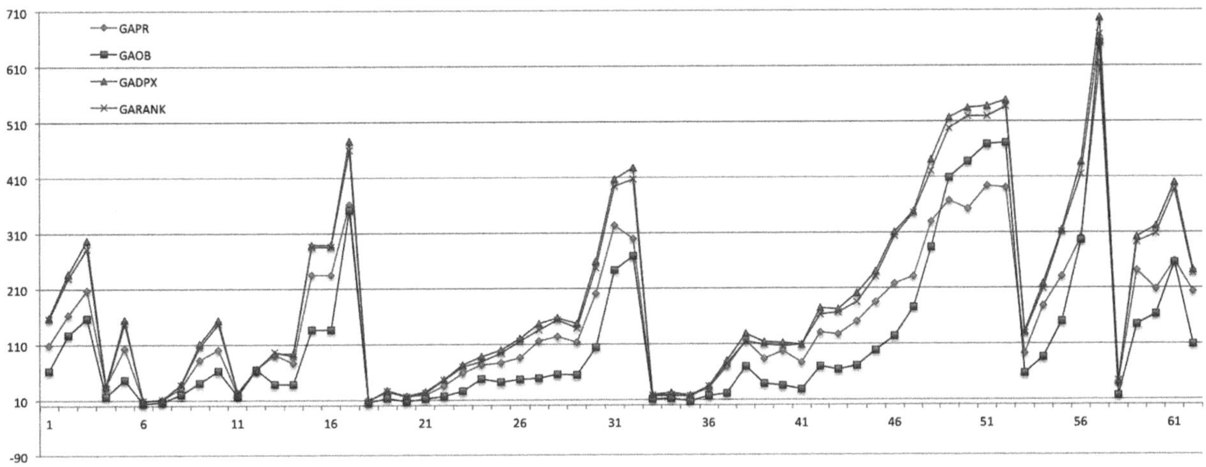

Table 5. Comparison of all the variants of our GA with an improved initial population over the set of 62 instances in HB

Variant of the GA	O.V.
GAPR	40.52
GAOB	38.50
GADPX	43.73
GARANK	42.37

From the results of this experiment it can be observed that all the variants achieved the same position in the ranking. More precisely, in this experiment GAOB, GAPR, GARANK and GADPX were also ranked number one, two, three and four, respectively. However, it can also be observed that there is a huge difference in the solution quality of all the variants with respect to the first experiment. Specifically, GAOB improved its own quality by 66.6%, GAPR by 72.9%, GARANK by 77.9% and GADPX by 78%. These new results of the variants are also closer to each other than the previous results from the first experiment. This fact can be observed clearly in Figure 11, where a plot similar to that from the previous experiment is presented. Notice that the objective values are so similar that it is really difficult to distinguish each variant in the plot. Moreover, it is also difficult to determine which is the best variant by only observing the plot.

Final Experiment: Comparison with GVNS

The goal of the last experiment is to compare the best variant of GA (GAOB) with the best algorithm for VSP available in the literature: GVNS (Sánchez-Oro, Pantrigo, & Duarte, 2014). For this experiment, the original source code of GVNS was used. Thus, the comparison is as fair as possible since the experiment was performed under the same hardware and software conditions. In order for this experiment to be fairly carried out, the time limit for GAOB was set to 1000 CPU seconds for instance. This is so

Figure 11. Performance of the four variants of our GA with an improved initial population over the set of 62 HB instances

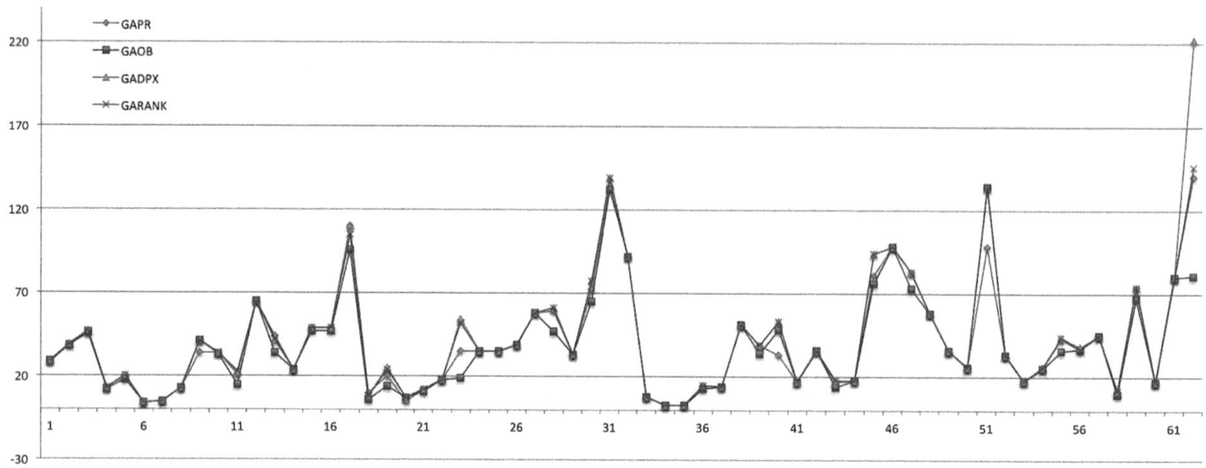

because the time limit for GVNS is set to 1000 CPU seconds and cannot be modified. Both algorithms solved the most difficult set of instances for VSP, the 62 instances from the dataset HB. The results of this experiment are summarized in Table 6. This table reports two statistics: the average objective value obtained by each algorithm (O. V.) and the number of instances in which one algorithm defeats the other in solution quality (# Wins).

From Table 6 it can be observed that GVNS obtains both the best average solution quality and the largest number of best solutions found. Specifically, the effectiveness of GVNS is about 80.64% whereas the effectiveness of GAOB is about 8.06%. In solution quality, GVNS is approximately 1.54 times better than GAOB. On the positive side, GAOB was able to find five new best-known solutions for the standard benchmark of VSP.

CONCLUSION

In this chapter the available heuristics to solve the Vertex Separation Problem (VSP) has been extended to include a genetic algorithm (GA). A comparative study on the impact of four different crossover operators was presented. Three of these operators were taken from the literature, namely, the order-based crossover (OB), the DPX crossover (DPX) and the rank-based crossover (RANK) (Schiavinotto & Stützle, 2004). The other crossover is proposed in this chapter and is based on the Path Relinking (PR) strategy. The four crossover operators were coupled with the GA, generating four variants: GAOB, GADPX, GARANK and GAPR, respectively.

Table 6. Final comparison between our best variant GAOB and the best algorithm for VSP found in the literature GVNS

	O. V.	# Wins
GAOB	38.50	5
GVNS	24.92	50

In this research, three computational experiments were conducted. The first two experiments were conducted to assess the impact of the crossover operator in the performance of the genetic algorithm. This is considered a preliminary experimentation and its goal is to determine which is the best variant for the GA proposed. Particularly, in the first experiment the initial population was generated completely at random. In the second experiment four elite individuals generated by four efficient constructive heuristics designed by Castillo-García et al. (2015) were inserted into the initial population. In the last experiment we compare the best variant of GA with the best algorithm for VSP in the state of the art: GVNS.

The results of the preliminary experimentation indicate that the order-based crossover operator was the best option. In particular, GAOB outperformed the second best variant of GA (GAPR) by 29.5% in the first experiment and by 5.2% in the second experiment. The reduction in the improvement is directly related to the elite solutions. The elite solutions naturally lead to low objective values since their genes are eventually inherited to the members of the population.

The results of the final experimentation clearly show that GVNS is superior to GAOB in solution quality. In fact, GVNS found 50 best-known solutions out of 62 instances evaluated. This gives GVNS an effectiveness of about 80.64%. GAOB found 5 best-known solutions, which gives it an effectiveness of about 8.06%. Although there is no sufficient evidence, these results suggest that the performance of trajectory algorithms is better than that of population-based algorithms when solving the Vertex Separation Problem.

As a future work, an efficient local search procedure will be used to convert the genetic algorithm into a memetic algorithm. This partially allows us to prove (or disprove) the hypothesis previously stated about the trajectory algorithms versus population-based algorithms in the context of VSP.

ACKNOWLEDGMENT

This research was partially supported by the National Council of Science and Technology of Mexico (CONACyT). The second author would like to thank CONACyT for his Ph.D. scholarship (229698). The authors of this research thank Jesús Sánchez Oro and his colleagues for kindly sharing the executable source code of GVNS. We are deeply grateful to IBM Academic Initiative for allowing us to use the optimization engine CPLEX. The authors also wish to thank the two anonymous reviewers for their time and valuable suggestions to improve the quality of this chapter.

REFERENCES

Bodlaender, H. L., Gustedt, J., & Telle, J. A. (1998). Linear-time register allocation for a fixed number of registers. In *Proceedings of the ninth annual ACM-SIAM Symposium on Discrete Algorithms.*

Castillo-García, N., Fraire, H. J., Flores, J. A., Pazos, R. A., González, J. J., & Carpio, J. M. (2015). Comparative study on constructive heuristics for the vertex separation problem. In P. Melín, O. Castillo, & J. Kacprzyk (Eds.), *Design of Intelligent Systems Based on Fuzzy Logic, Neural Networks and Nature-Inspired Optimization* (pp. 465–474). Springer. doi:10.1007/978-3-319-17747-2_35

De Oliveira, A., & Lorena, L. A. N. (2002). A constructive genetic algorithm for gate matrix layout problems. *IEEE Transactions on Computer-Aided Design of Integrated Circuits and Systems*, *21*(8), 969–974. doi:10.1109/TCAD.2002.800454

Díaz, J., Petit, J., & Serna, M. (2002). A survey of graph layout problems. *ACM Computing Surveys*, *34*(3), 313–356. doi:10.1145/568522.568523

Duarte, A., Escudero, L. F., Martí, R., Mladenovic, N., Pantrigo, J. J., & Sánchez-Oro, J. (2012). Variable neighborhood search for the vertex separation problem. *Computers & Operations Research*, *39*(12), 3247–3255. doi:10.1016/j.cor.2012.04.017

Duarte, A., Martí, R., Resende, M. G., & Silva, R. M. (2011). Grasp with path relinking heuristics for the antibandwidth problem. *Networks*, *58*(3), 171–189. doi:10.1002/net.20418

Fraire, H. J., Castillo-García, N., Pazos, R. A., Flores, J. A., González, J. J., & Carpio, J. M. (2015). Two new exact methods for the vertex separation problem. *International Journal of Combinatorial Optimization Problems and Informatics*, *6*(1), 31–41.

Kornai, A., & Tuza, Z. (1992). Narrowness, path-width, and their application in natural language processing. *Discrete Applied Mathematics*, *36*(1), 87–92. doi:10.1016/0166-218X(92)90208-R

Leiserson, C. E. (1980). Area-efficient graph layouts. In *the 21st Annual Symposium on foundations of Computer Science*.

Lengauer, T. (1981). Black-white pebbles and graph separation. *Acta Informatica*, *16*(4), 465–475. doi:10.1007/BF00264496

Linhares, A., & Yanasse, H. H. (2002). Connections between cutting-pattern sequencing, vlsi design, and reflexive machines. *Computers & Operations Research*, *29*(12), 1759–1772. doi:10.1016/S0305-0548(01)00054-5

Lopes, I. C., & de Carvalho, J. M. V. (2010). Minimization of open orders using interval graphs. *IAENG International Journal of Applied Mathematics*, *40*(4), 297–306.

Luque, G., & Alba, E. (2005). Metaheuristics for the dna fragment assembly problem. *International Journal of Computational Intelligence Research*, *1*(2), 98–108. doi:10.5019/j.ijcir.2005.28

Mitchell, M. (1999). *An introduction to genetic algorithms*. The MIT Press.

Optimization Studio, C. P. L. E. X. v12.5. (2012). *IBM Academic Initiative*. Retrieved from https://www.ibm.com/developerworks/downloads/ws/ilogcplex/

Sánchez-Oro, J., José Pantrigo, J., & Duarte, A. (2014). Combining intensification and diversification strategies in vns. An application to the vertex separation problem. *Computers & Operations Research*, *52*, 209–219.

Schiavinotto, T., & Stützle, T. (2004). The linear ordering problem: Instances, search space analysis and algorithms. *Journal of Mathematical Modelling and Algorithms*, *3*(4), 367–402. doi:10.1023/B:JMMA.0000049426.06305.d8

KEY TERMS AND DEFINITIONS

Combinatorial Optimization Problem: An optimization problem consists in finding the best solution from a set of feasible solutions. An optimization problem whose variables are discrete (not continuous) is known as a combinatorial optimization problem. The goal of a combinatorial optimization problem is to find a mathematical object, e.g., a permutation, from a finite (or countable infinite) set.

Crossover Operator: The crossover operator is perhaps the main method of the genetic algorithms. The crossover operator combines the characteristics of two solutions (parents) in order to generate a new solution (offspring). This operator is inspired in the way in which the genetic code of one individual is inherited to its descendants in nature.

Genetic Algorithm: A genetic algorithm is a population-based metaheuristic algorithm that uses genetics-inspired operators to sample the solution space. This means that this algorithm applies some kind of genetic operators to a population of individuals (solutions) in order to evolve (improve) them throughout the generations (iterations).

Graph: A graph is a mathematical abstraction of a collection of objects called vertices where some pairs of objects are connected by links called edges. In other words, a graph represents binary relations among the objects.

Linear Ordering: A linear ordering consists in arranging the vertices of the graph in a horizontal line. Mathematically, a linear ordering is the one-to-one mapping function $\varphi : V \to \{1,\ldots,n\}$, where V is the set of vertices and $n = |V|$ is the number of vertices in the graph. In the linear ordering $\varphi = (v_1, v_2, \ldots, v_n)$, the vertex v_1 is placed at the first position, the vertex v_2 at the second position and so on.

Metaheuristic Algorithm: A metaheuristic algorithm is general framework to tackle NP-hard combinatorial optimization problems. Basically, a metaheuristic algorithm samples the solution space by visiting those regions with larger probabilities of having the best solution.

Solution Space: The solution space is the set of all possible solutions for the combinatorial optimization problem. The solution space of the Vertex Separation Problem contains all the linear orderings of the vertices. Thus, the cardinality of the solution space of VSP is $n!$. Notice that the number of feasible solutions for the problem grows exponentially with respect to the size of the instance (number of vertices).

Chapter 3
Supply Chains under Security Threat:
The First National Exploratory Study in Mexico

Miguel Gastón Cedillo-Campos
Transportation Systems and Logistics National Laboratory, Mexican Institute of Transportation, Mexico

A. Bueno-Solano
Mexican Institute of Transportation (IMT), Mexico

R. G. González-Ramírez
Pontificia Universidad Católica de Valparaíso, Chile

E. Jiménez-Sánchez
Mexican Institute of Transportation (IMT), Mexico

G. Pérez-Salas
Natural Resources and Infrastructure Division, ECLAC - United Nations, Chile.

ABSTRACT

Contemporary prosperity depends on effective and secure supply chain networks that support economic competitiveness. Disruptions in global supply chains would have critical consequences on economies. The lack of technical studies and quantitative data concerning security that affects supply chain operations in Latin America, motivated to develop an exploratory study. Considering the complexity of the question studied, this paper details a set of case studies that explore, from a qualitative research approach, to what extent fulfilling security international standards now necessary to access mature markets such as the U.S and Canada allows export companies located in emerging countries as Mexico to face effectively the different types and levels of local risk. These results should help both academics and practitioners to more readily understand, first, the key logistics components now taken into account when improving security in export-oriented supply chains is required, and second, decision-makers' perspectives regarding supply chain security standards (SCSS) available on the market. A discussion of results is exposed and finally, discussion and future research are presented.

DOI: 10.4018/978-1-4666-9779-9.ch003

1. INTRODUCTION

During the last decade, Mexico and other Latin American emerging countries, logistics systems have experienced a dynamic change. Several years ago, in many countries there was a low perception of risk in supply chains - especially insecurity - as an important element of competitiveness (Giunipero & Eltantawy 2003). However, currently in emerging markets as Mexico, the Private Sector's Centre for Economic Studies (CEESP) affirms that the cost of insecurity exceeds 15 per cent of gross domestic product (GDP). In this sense, according to the Mexican Association of Insurance Institutions (AMIS), approximately 10,000 transportation vehicles are robbed annually. Furthermore, AMIS notes that from 2006 to 2010, transportation vehicle thefts increased 108 percent (Cruz 2011). Consequently, since theft and other felonies on Mexican highways, the cost of transportation services increased 40 per cent (Cervantes 2011), forcing companies to change operation systems driving toward multimodal approaches (Jiménez & De la Torre 2011).

Actually, security deficits at national and regional levels (in North America and Latin America) and the current lack of a risk management approach when designing supply chains lead to a loss of regional competitiveness because of operational costs for cargo owners and transportation service providers. At the same time, the reputation of freight companies is damaged since stolen products are often sold on black markets without control or traceability. Thus, as security costs increase, direct and indirect costs of freight transportation increase due to preventive measures that have to be taken, resulting in investments and additional costs (Hintsa et al. 2010). Moreover, there is a strong impact on the performance of various actors caused by delays and noncompliance with clients, generating greater variability in delivery schedules that also generate greater financial costs for want of higher levels of safety stock.

In this context, since we observe a broad variety of definitions on how supply chain security (SCS) should be analysed, our research adopts the definition proposed by Closs and McGarrell (2004, pg. 8):

Supply chain security management is the application of policies, procedures, and technology to protect supply chain assets (product, facilities, equipment, information, and personnel) from theft, damage, or terrorism, and to prevent the introduction of unauthorized contraband, people, or weapons of mass destruction into the supply chain.

In this sense, the purpose of this paper is to identify to what extent fulfilling security international standards now necessary to access mature markets such as the U.S and Canada allows export companies located in emerging countries as Mexico to face effectively the different types and levels of local risk. These results should help both academics and practitioners to more readily understand, first, the key logistics components now taken into account when improving security in export-oriented supply chains is required, and second, decision-makers' perspectives regarding supply chain security standards (SCSS) available in the market (see Table 1 in Appendix).

While there is much published research on security supply chain and risk management, research on how international security practices approved to promote safe global trade also respond to the local security needs of companies located in emerging markets has largely been ignored. By studying the case of export-oriented Mexican companies facing important security challenges at a local level, we gained greater insight into factors as: i) Guidelines about current security policies and practices; ii) Level of added value provided by current security initiatives (public and private); and iii) Future security issues impacting supply chain design.

The findings of our research should assist decision makers recognize and understand how local risk factors affect global trade operations. In that sense, effective decisions and actions at a local level can be taken before they become problems to regional or global operations. The results may improve or strengthen security practices and policies in the supply chain domain.

This paper is organized as follows: Section 2 provides the background concerning supply chain security issues, and a brief description of the current supply chain security standards. In Section 3, the research design and methodology are presented. In Section 4, the research findings will be discussed. Finally, Section 5 presents provides conclusions and suggestions for future research work.

2. BACKGROUND

International trade continues to be a key element of Mexico's economic growth. In the last few decades, corporations located in Mexico invested in designing and operating more efficient (lean) supply chains, which derived in their current logistics vulnerability. Since organizations designed supply chains from a standardized perspective, they are now facing greater variability or even rupture in primary operations because supply chain engineers and managers did not consider the consequences of security threats in international trade operations (Christopher 2003; Jackson 2011; Waters 2007).

In 2010, Barbero conducted a study funded by the Inter-American Development Bank (IDB) to evaluate freight logistics performance in Latin America. For Mexico, the author points out the importance of road infrastructures as the primary mode of transportation for international trade. However, modernization of both maintenance programs and investment in road infrastructures is required. Furthermore, road transportation is highly atomized and inefficient, especially at borders crossing points to the United States. With respect to port systems, inefficient control inspections by customs agents and a lack of information systems were observed. There were also corruption issues and a need of public policies to enhance sector competitiveness. The findings highlighted also problems related to theft on roads.

Improving SCS is not only a goal of decision-makers operating in Mexico; at present it is possible to identify worldwide a wide range of SCSS proposed by public and private entities. Under a globalized environment, success of global supply chains (GSC) depends on guaranteeing a timely and efficient international flow of goods. However, during supply chain design, most of the key variables are approached with a partial vision; decision-makers often implement decisions to reduce uncertainty based only on information available. Thus, the variety of SCSS not only fails to present a unified conceptual focus on SCS, it also generates confusion and economic distortions because of the diversity of perspective analysis and actions proposed. The importance and complexity of this issue are pushing governments to shift from safety to broad-risk analyses (The White House 2012). This is why a systems perspective is required (Cedillo-Campos 2010; Cedillo-Campos & Donnadieu 2012).

Several authors (Gaonkar & Viswanadham 2009; Pickett 2003; Sheffi et al. 2003) argue that to reduce analysis complexity, risk management should not focus on the nature of catastrophic or disruptive events. Since decision-makers cannot anticipate every threat—much less determine the likelihood that a threat will occur—efforts should focus on evaluating the impact generated by the disruptions and catastrophes in supply chain operations. The authors conclude that a variety of events such as natural disasters and terrorist attacks tend to produce similar effects in a supply chain. Stecke and Kumar (2006) agree that there is impact similarity, but only when events share characteristics that originate the event.

Before the 9/11 attacks in the USA, academic research and best practices centred on security measures to reduce disruptions of supply chains because of terrorist attacks (Rice et al., 2003; Lee & Wolfe 2003; Sheffi et al. 2003; Rice 2007; Rosoff & Von Winterfeldt 2007). However, all these security analysis have not only demonstrated the ineffective designs used by organizations. Thus, modern analyses integrate a systemic perspective to address threats early and foster resilient supply chains that absorb and recover from unanticipated disruptions (Bueno & Cedillo-Campos, 2014).

Consequently, several international organizations point out the global importance of this topic. The National Board of Trade, a governmental agency and the central administrative body in Sweden dealing with foreign trade and trade policy, in its publication *Supply Chain Security Initiatives: A Trade Facilitation Perspective* (Kommerskollegium, 2008) describes a comprehensive study of development of security strategies for global supply chains. Its results suggest that commercial activities with the great trade blocks will not be facilitated until governments and corporations invest in development of regulations and other initiatives that improve security to global supply chains.

2.1 Supply Chain Risk Management

As global supply chains perform operations in a greater number of international marketplaces, these have become more complex. Consequently, supply chain risk management increased its importance almost at the same rate as globalization economic paradigm involved more regions. Thus, in 2003, Jüttner et al. outlined a research agenda for supply chain risk management research, for which they present a literature review and also an empirical study based on semi-structured interviews to capture experiences and perspective of practitioners regarding supply chain risk perceptions and management strategies. The authors identified four critical research issues: (1) assessing the risk sources for the supply chain; (2) identifying the risk concept of the supply chain by defining the most relevant risk consequences; (3) tracking the risk drivers in the supply chain strategy; and (4) mitigating risks in the supply chain.

The next year, Finch (2004) exposes a literature review based on case studies for supply chain risk management with the aim of determining the increasing exposure to risk for large companies by having small and medium-size enterprises (SMEs) as partners in business critical positions in the supply chain. The authors considered 2,000 articles, papers, surveys and case studies and the methodology was based on an information system framework defined by Bandyopahyay, (1999). The results showed that exposure to risk appeared to be increased by inter-organisational networking, but also that having SMEs as partners in the supply chain further increased the risk exposure and that SMEs increase their own exposure to risk by becoming partners in a supply chain.

In that sense, Blackhurst et al. (2005) presents an agenda of critical research issues for managing supply-chain disruptions based on a multi-industry and a multi-methodology empirical study. The authors present several critical issues related to the analysis and mitigation of the detrimental impact of supply chain disruptions in a global environment based on three elements: disruption discovery, disruption recovery and supply-chain redesign.

Thus, Peck, (2005) proposes an integrated framework of the drivers of supply chain vulnerability and reports on findings of a cross-sector empirical study of the sources and drivers of supply chain vulnerability based on an in-depth exploratory case study of commercial supply chain engaged in the manufacture and assembly of high performance military aircraft. Data collection involved semi-structured interviews with 47 managers, representing five tiers of the network involved in the production of four aircraft types. The interviewers were selected using snowball sampling (Jankowicz, 1995).

Autry and Bobbit, (2008) perform an exploratory study addressing security awareness within firms. The authors conceptualize the SCSO construct based on a review of the relevant literature and assess this conceptualization through an examination of qualitative questionnaire data provided by supply chain managers. Also, the authors present a general framework that includes antecedents, outcomes and moderating conditions associated with or influencing SCSO. Finally, they discuss future directions for research and practice. On the other hand, Williams et al., (2008) propose a categorization of supply chain security into three organizational approaches: (1) intra-organizational, (2) inter-organizational, and (3) a combination of intra-organizational and inter-organizational. The study develops a focused research agenda for future, primary, and empirical research on SCS.

Wagner and Bode, (2008) present an empirical study to investigate the relationship between supply chain risks and supply chain performance, motivated by the fact that firms may implement supply chain risk management activities only if supply chain risk interferes with supply chain performance. The authors consider five classes of supply chain risk sources: (1) demand side; (2) supply side; (3) regulatory, legal and bureaucratic; (4) infrastructure; and (5) catastrophic. Based on a survey in which they received 760 questionnaires from top-level executives in logistics they concluded that there are negative associations between supply and demand side risks and supply chain performance. Actually, Manuj and Mentzer (2008) analyse risk management and risk management strategies in global supply chains, recognizing the fact that global supply chains are more risky than domestic supply chains due to numerous links interconnecting a wide network of firms.

Rao and Goldsby (2009) perform a literature review of supply chain risks management and develop a typology of risk sources, consisting of environmental, industry, organizational, problem-specific and decision-maker related factors. They consider as a key implication to focus risk management efforts on the links that show the highest vulnerability. The authors demonstrate that any narrow-minded risk management strategy attempting to deal with specific risk in the supply chain may turn out to be suboptimal if it fails to recognize and quantify the overall risks impacting supply chain activity, and they propose as a method to quantify the inherent risks in the supply chain to look at another dimension along with the sources of risk and the impacts of risk.

In fact, a supply chain is only as secure as its weakest link. Cedillo-Campos and Bueno-Solano (2014) argue that impacts of a disruption derived from a security breach and its following propagation throughout the supply chain cause other critical phenomena as reverse bullwhip effect. Hence, SCS is increasingly a key topic for any organization, and more for all for exporters.

2.2 Impacts for Global Trade

Currently, the global supply chain system that supports international trade is essential to countries' economies and is a critical global asset (The White House 2012). In that sense, Cipoletta et al. (2010) suggests that to drive competitiveness, trade, and economic and social growth of a nation, an integral strategy for public policies of infrastructure, transportation, and logistics systems with a national and regional perspective are needed. This approach, called triad, should not only be present and clearly implemented, it should also be shared between public and private sectors from a SCS viewpoint.

Current global supply chains are vulnerable, facing a variety of risks, threats, and catastrophes. Risks to a supply chain are unforeseen events that can interrupt the smooth flow of materials (Waters 2007;

Figure 1. The disruption profile (Sheffi, 2005)

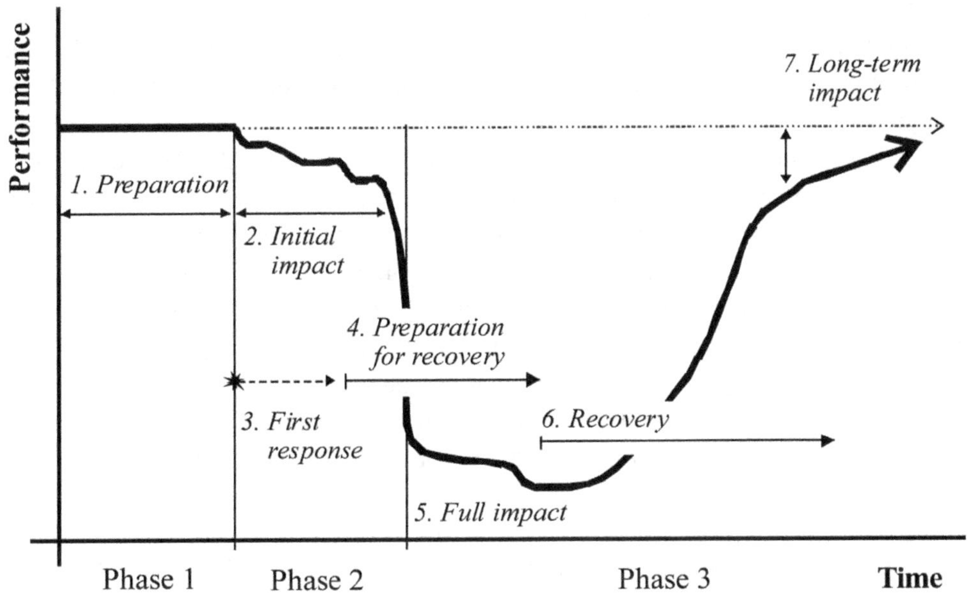

Wilson, 2007). It could be categorized as: 1) risks, threats, and catastrophes caused by natural phenomena, 2) risks, threats, and catastrophes derived from human actions generated by mistakes or carelessness, and 3) criminal activity and terrorism.

According to The American Chamber of Commerce of Mexico (2011) and focusing on criminal activity, *The Impact of Security in Mexico in the Private Sector* demonstrates that 27 per cent of businesses reconsidered investment in Mexico because of insecurity. Although 90 per cent of respondents expressed their companies were willing to keep operating in Mexico, they recognized investment in security reached between 2 and 10 per cent of operation costs.

As external sources of risk, natural disasters, terrorist attacks, and criminal assaults are not new (Pfohl et al. 2010). However, supply chain managers recognize them because of the current globalized situation where the export-oriented supply chains are most susceptible to such disruptions. Wishing to understand better the impact a disruptive event has on supply chain performance over time, Sheffi and Rice (2005) propose analysing disruptions on seven phases (see figure 1): i) Preparation; ii) Initial impact; iii) First response; iv) Preparation for recovery; v) Full impact; vi) Recovery; vii) Long-term impact.

Thus, Knemeyer et al. (2009) and Sodhi and Tang (2012) argue that most global companies lack a formal process to estimate the probability of catastrophic events, which restricts them to evaluate a limited number of catastrophes. They highlight that there are methods available to estimate the probability of occurrence concerning catastrophic events, but these methods estimate only local risk for each catastrophe. Consequently, it is necessary to implement reliable databases and quantitative models that allow decision-makers to evaluate risk of combined disruptive events for each link in the supply chain. Since there often exist high inter-organizational connections, a disruptive event not only affects a corporation in its facilities or operations; the risk propagation effect (RPE) also upsets suppliers, manufacturers, customers, logistics services, and others.

In this sense, Jones et al. (2011) proposes a modelling tool for analysing flows of U.S. imports and exports of containerized freight, and the potential changes in those flows under a variety of conditions (e.g., port disruptions, extensive security-related delays, etc.). The authors conducted several experiments and the results obtained highlighted how changes or disruptions in one part of the system can affect flows in geographically disparate parts of the system.

Contrary to being an inherently problematic, Marsden and Docherty (2013), present an analysis on how disruptive events can be an opportunity through which to construct and accelerate transport policy changes which are commonly slow and incremental over long periods of times.

More focused on an economic analysis of port security policies, Yeo et al. (2013) presents an analysis of dynamic effects on seaports adopting security policies in order to derive recommendations for effective implementation of security measures. And in a more recent paper, Yang et al. (2014) introduces a novel fuzzy evidential reasoning approach to facilitate the quantitative analysis of port facility security assessment (PFSA). The authors identified the key security performance indicators in port security practice and effectively organized them in a hierarchy for risk inference and applied the methodology to a real case study.

Jankowicz (2000) actually proved that the most pertinent research methods and techniques depend on the problems and fixed objectives. As such, the study regarding security in supply chain identified that there are as many different variables involved, as there are logistic operation contexts. As a consequence, the diverse situations brought together with the lack of statistical information made necessary the need an exploratory research technique that investigates in depth a few similar situations to extract general conclusions.

In this sense, to identify to what extend fulfilling security international standards now necessary to access mature markets such as the U.S and Canada allows export companies located in emerging countries as Mexico to face effectively the different types and levels of local risk, two research questions were established (RQ):

RQ1: What are the logistic components taken into account when trying to improve the security of export-oriented supply chains?

RQ2: What are the decision makers' perspective regarding the security standards availables on the market?

3. METHODOLOGY

Focusing only on the security of each node of the manufacturing supply chains is unlikely to guarantee enduring measures of overall security for global trade flows. To achieve the objective stated above, this study was designed based on an exploratory, qualitative process from a systems viewpoint (Cedillo-Campos & Sanchez 2013). For this paper, an exploratory research is understood as an early research achieved to elucidate and outline the nature of security issues for export-oriented supply chains running operations in emerging markets as Mexico. Thus, an exploratory research is assumed as a series of case studies to provide background evidence.

Since there are only scarce statistical information on how international security practices approved to promote safe global trade also respond to the local security needs of companies located in emerging markets, investigating this matter based on a series of case studies was helpful.

According to the theoretical toolbox proposed by Shook et al. (2009), the research model was built on systems theory. Bertalanffy (1968) explains that from a systems theory approach, organizations are a group of interconnected entities that interact to deliver products. This study considers that interactions among every entity affect the entire system. Consequently, interdependence between firms determines the nature of these effects.

In accord with Eisenhardt (1989) and Yin (2003), the decision to use a case study was driven by two factors. First, case-based research in operations management and supply chain research is well established as Ellram (1996), Meredith (1998), and Voss et al. (2002) demonstrate. Second, there is a lack of extant SCS research in emerging markets linking feedback between manufacturing needs and national-regional operational contexts from a global standpoint. As mentioned by Yin (2003), Ghauri and Gronhaug (2005) and Maxwell (1998), a qualitative research approach is more appropriate for delving deeper into the aspects of this new focus of analysis. Meredith (1998) argues an exploratory case approach is an appropriate alternative where "*the variables are still unknown and the phenomenon not at all understood*". Yin (2003) discusses that "*a case study is an empirical enquiry that (1) investigates a contemporary phenomenon within its real life context, especially when (2) the boundaries between phenomenon and context are not clearly evident*". Statistical gaps in SCS made a quantitative analysis of the phenomenon impossible. While there are some theoretical and practical papers on SCS, there are not enough technical and operational bases to establish clear quantitative models of analysis. This is an evolving phenomenon; adopting a rigid epistemological position for analysis is inappropriate.

We originally relied on the methodological propositions of Thiétart (1999). The research began with an abductive or hybrid process suitable for decoding complex systems. As Jick (1989) proposes, a multi-method approach was developed using both qualitative and quantitative data. Integration of the area of study observed and research into theoretical knowledge took place simultaneously throughout the investigation. Thus, the problem statement was directed toward exploration, description, and understanding rather than predicting behaviours of variables. A quantitative approach is highly valued when the variables of the phenomenon are set and a broad base of reliable information is available. Data collection was based on a small sample but reliable cases and experiences rather than database analysis.

Based on the proposed methodology of Kovács and Spends (2005), the process began with a set of theoretical information that served to reveal the reality observed. According to the approach proposed by Dubois and Gadde (2002), while data collection was carried out, we developed the theoretical foundations of the project. Thus, an exploratory analysis was justified in the absence of extant research.

3.1 Case Study

The use of case studies as methodology when performing exploratory research is useful to detect, elucidate, and explore phenomena within their real-life setting (Yin, 2003). Hence, researchers gain a better understanding of why something happened as it did, and what else might be important for further analysis (Host & Runeson, 2008; Runeson & Höst, 2009; Verner et al., 2009). For our research, this methodology provided us with a systematic way of looking at events, collecting data, analysing information, and reporting results. It allowed us to investigate about two research questions. Because of the importance of the operational logistics context of the matter here analysed, it was critical to take into account quantitative variables as well as qualitative ones (as perception of risk).

Because of the exploratory nature of our research, qualitative approach was an important basis. Thus, our research focused on the natural settings of the SCS in Mexico. In fact, detailed information was generated since "richer" answers to questions were obtained (Leedy & Ormrod, 2005). Thus, the case study methodology we use is organized into several phases, and most of the actions in the first three phases took place before the case(s) were really chosen. The methodology included six phases: i) pre-planning; ii) planning; iii) design; iv) data collection; v) data analysis; and vi) reporting.

A detailed validation and data analysis were performed, enabling development of conclusions, detecting areas of opportunity to improve professional practices, and guiding research activities and technological innovations. When contradictions were identified, follow-up telephone calls were made to verify the discrepancy or obtain the missing data. Although this study's topic is an issue of broad national impact, the number of archived cases was modest due to an image of trustworthiness respondents wished to preserve. The importance of voicing concerns as a means to participate in designing solutions was made known to participants.

3.2 Industrial Sector

Accounting for 81.9% of total Mexican exports, manufacturing sector was selected as part of our study (INEGI, 2012; Lawrence and Leon 2010). In Mexico, manufacturing supply chains are leaders in logistics innovations transmitted to other sectors. Thus, focusing on this type of supply chain over others resulted effective. Although current supply chains in manufacturing tend to exhibit global aspects—including complex interactions and flows among multiple companies and facilities geographically distributed among many regions and countries—most are structurally similar (Chopra & Meindl 2013). Consequently, this study identifies significant elements when designing secure supply chains operating in emerging economies that could be generalized to other sectors.

3.3 Limitations and Bias

This exploratory study reports the perceptions of practitioners working in international manufacturing companies operating in Mexico. In our knowledge, there were not studies on how practitioners perceive security risk in supply chains operating in Mexico, and in Latin America to provide high officials and academics information to develop solutions for this relatively new regional challenge.

Supply chain security is now a global interconnected topic. Consequently, since this exploratory study only investigated the perceptions of industry practitioners in Mexico during a specific time, the study's findings may be limited because the sample was not representative of industry practitioners across an entire region, NAFTA (North American Trade Agreement) or Latin America for instance. This study only takes into account Mexican case studies, bordered by restrictive information policies since security is a very sensitive topic for companies located in the country.

Practitioners' perceptions may differ depending on the industry sector and at the same time, the senior management perceptions influence them. Therefore, attention should be exercised in interpreting the results. The exploratory nature of the study, which only focuses on the manufacturing sector, may skew the results. Thus, future studies should use larger, more diverse samples from various geographic regions and more industry sectors to improve the results.

4. RESEARCH FINDINGS

4.1 Case Studies' Participants Profile

While participants were related to export operations in the manufacturing sector, from 70 validated case studies, 42 per cent of practitioners providing information to shape case studies were decision-makers working directly in manufacturing. 16 per cent were professionals providing transportation and logistics services, 17 per cent provided services to foreign trade, 17 per cent offered local distribution services and storage, 1 per cent offered services related to information technology, and 7 per cent provided advisory services and training. More than 60 per cent of the decision-makers take part in activities directly related to logistics. Regarding the companies' size involved in the case studies, 72.9 per cent were large companies (more than 250 employees according to the classification proposed by the Mexican Secretariat of Economy).

In terms of geographic coverage of the functions' participants, their position provides them a comprehensive view of companies' supply chain. Regarding importing and exporting operations oriented to the United States, we found several interesting issues. While about 58.8 per cent of respondents import from the U.S., 97.1 per cent export to this country. This validates strong commercial relationships between Mexico and the U.S., but it primarily identifies manufacturing and logistics as levers for greater regional integration.

A lower trade rate with Canada was observed. Only about 35 per cent of companies export to this country, which is worrying when we consider that 50 per cent of companies have irrelevant or no exports to other countries besides the U.S. and Canada. At the same time, it was possible to confirm a high dependence of the Mexican foreign trade on a single port of entry. The most important point of entry for export transactions is Laredo, Texas, through which runs nearly 50 per cent of goods exported by companies taking part of the case studies. As for maritime exports, Manzanillo is used most frequently with 11.4 per cent of participation.

4.2 Importance of Security Standards

The Swedish National Board of Trade suggests that trade with major trading blocs will not be facilitated if governments and companies from other countries do not invest in development of regulations and other initiatives for improving security export supply chains (Kommerskollegium 2008). In this regard, companies that are part of the analysed case studies implemented only one security standard. They argued the high cost of simultaneously maintaining different standards. On the other hand, 38.6 per cent of companies have not implemented any security standard (see figure 2). Consequently, they are exposing their export operations to risk, as well as the operations of other companies part of their supply chain. In that case, the decision-makers are assuming a reactive attitude toward security breaches. They tend to ignore the risk, which will force them to react swiftly, not always in a well-planned way before a disruptive event.

It is highly likely that the low rate of hiring of special logistic services (i.e., three-party logistics) is part of the reason why only some of the companies enjoy access to specialized knowledge that allow them to appreciate the value of implementing reliable security initiatives. In fact, only 52.9 per cent of the analysed cases studies employ specialized logistics operators for their import and/or export operations. Despite this lack of access to security expertise, pressures from international trading partners are

Figure 2. Supply Chain Security Standards implemented by respondents

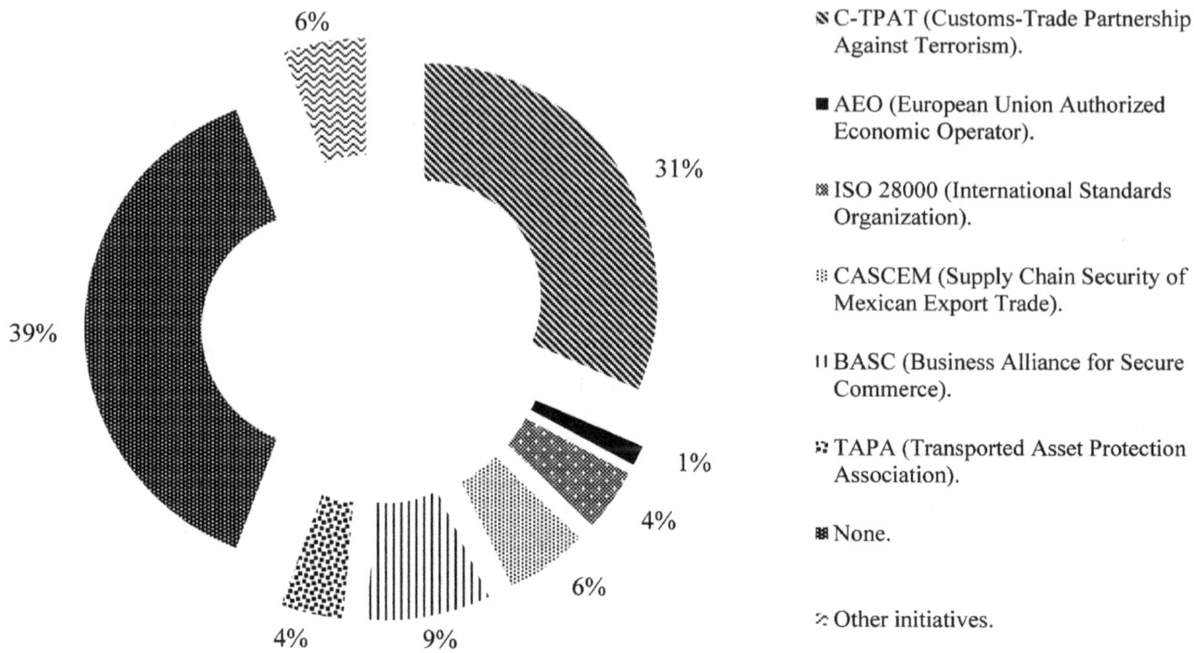

becoming an important driver for companies to implement security. In fact, more than 70 per cent of the cases studies implemented some security because of requirements and/or conditions imposed by international partners.

Findings suggest that in the near future, companies that already implement securities plan not only to maintain current initiatives but also to seek specialized knowledge. Based on the analysed case studies, some companies do not perceive the benefits of ensuring SCS as a competitive element in the export process. This is a serious deficiency, especially if we assume that a supply chain is only as secure as the security of its most vulnerable link. Individual strategies without proper communication to others in the supply chain—coupled with low efficiency in investment, misunderstanding of the purpose of the measures, and lack of trust among members of the supply chain—lead to opportunistic, local savings by taking advantage of the security investments of other actors but that actually reduce overall security.

In accordance with Bueno and Cedillo-Campos (2014), our findings suggest that by ensuring a link in the supply chain, risk migrates to weaker links (Risk Propagation Effect). Thus, higher hierarchical levels are beginning to recognize security as a critical matter as well as its impact on all the supply chain members (see figure 3).

An important point to note is a concern expressed by decision-makers regarding the variety of commercially available supply chain security standards (SCSS). Respondents stressed the importance of "*modular*" SCSS that allow greater flexibility during implementation. In that sense, a "*modular SCSS*" is understood as a standard, which is composed of homogenous units or sections for easy assembly or flexible arrangement with other SCS standards. That is to say, to decrease complexity to maintain diverse SCSS, while raising possibilities to improve SCS. The assumption is that the use of generic modules interchangeable in a number of different SCSS could contribute to more efficient mitigation in response to specific risk a company should face to.

Figure 3. Attention of senior management levels regarding SCS

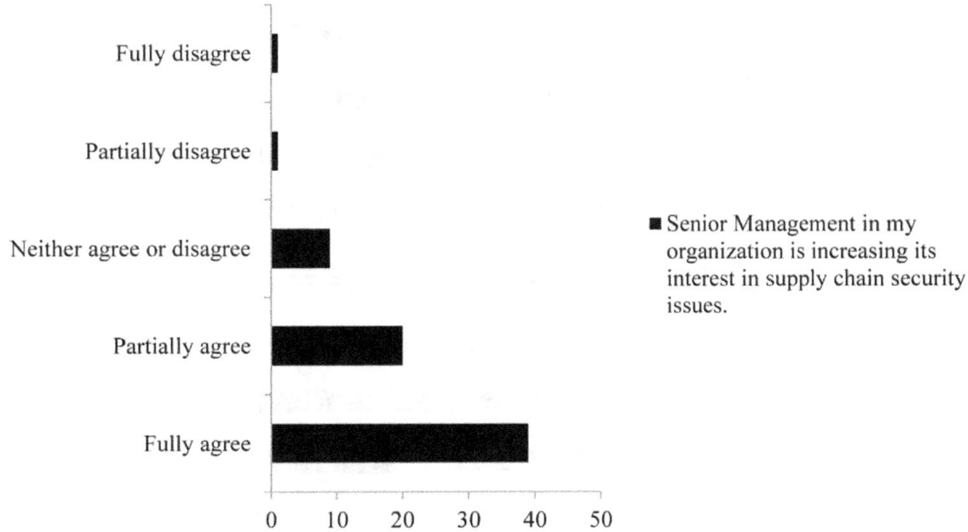

At the same time, decision-makers agreed most standards include gaps that should be covered by complementary risk analysis. They perceive the latter approach is more effective considering the operating range of contexts in which supply chains run. Most agreed there is a need for greater training and awareness-rising concerning the importance of SCS.

Companies developing schemes to improve security, gradually recognize the importance of all supply chain members sharing the same concern for achieving significant results. Involvement of each organization in a supply chain lowers vulnerability with respect to increasing globalization risks. Since local criminal organizations have developed global abilities, solutions must also have at the same time, a global reach but connection to the local reality.

4.3 Risk Perception

Regarding risk perception of decision-makers, case studies suggest that a local critical issue is the infiltration of drug trafficking as a primary source of vulnerability or disruption. For most of decision-makers, this issue concerns contamination of shipments by drug trafficking organizations. A second level of perceived risk includes theft of goods, at a third level, the use of supply chains for smuggling goods, and at a fourth level, the use of supply chains for human trafficking (see figure 4).

Decision-makers in Mexico did not perceive terrorism as a risk that undermines export logistics operations; they considered it a factor only as part of their check-list when exporting to North America and Europe. This indicates that from a SCS perspective, Mexican companies face important differences regarding the type of critical risk to mitigate.

4.4 Benefits of Complying with SCSS

To mitigate risk, a significant number of case studies implement SCSS. When questioned about the overall benefits they perceive, responses were generally positive. Most of them noticed implementa-

Figure 4. Risk perceptions in Mexico

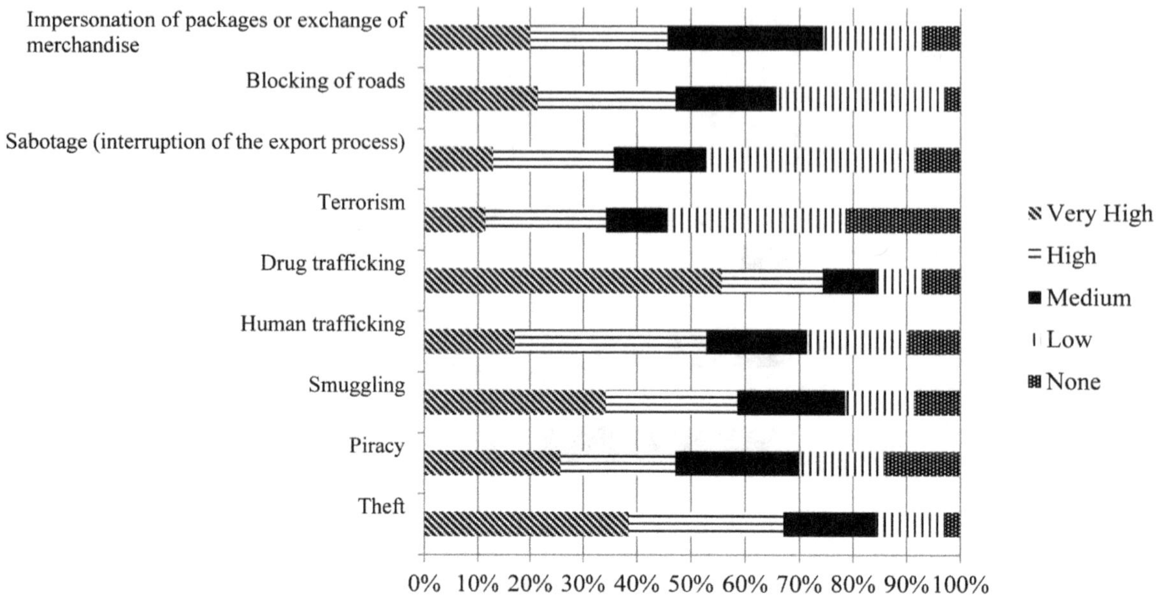

tion of SCSS not only offers SCS benefits, it also enhances effectiveness of supply chain operations. Among the most important benefits achieved, employers reported the following in order of importance: 1) co-benefits in logistics (e.g., better visibility throughout supply chains and improved reactivity in case of disruptions); 2) reduction in the number and/or value of incidents (e.g., theft, contamination of shipments); 3) standardization of operations allowing better interaction with business partners; 4) reduction in insurance premiums by providing evidence of compliance with one or more standards; and 5) operational benefits provided by U.S. and Mexican government agencies.

Decision-makers stated that although one of the advantages of SCSS is an operational benefit provided by a relatively agile access to the U.S. market, there is a huge gap between expected and real benefits (see figure 5).

4.5 Balance of Security

Regarding the balance of security incidents in supply chains over the past years, decision-makers not only detected an increase in criminal incidences within the companies. Perceived crime in the supply chain is also increasingly becoming a multinational problem. These results suggest that the most widely accepted solution for development of mitigation strategies is a national approach, but with effective regional coordination (i.e., NAFTA, Latin America.

Although decision-makers confirm the benefits of having implemented a security standard, and the growing need to maintain and update it, they perceive this type of initiative includes inherent risks and creates dilemmas. Decision-makers also consider SCSS tools that reveal information about vulnerabilities and reaction plans to criminals and terrorists. They assume that given the sensitivity of the information, employees, suppliers, and customers could communicate this information to criminal organizations and/or terrorists.

Figure 5. Perceived benefits of complying with SCSS

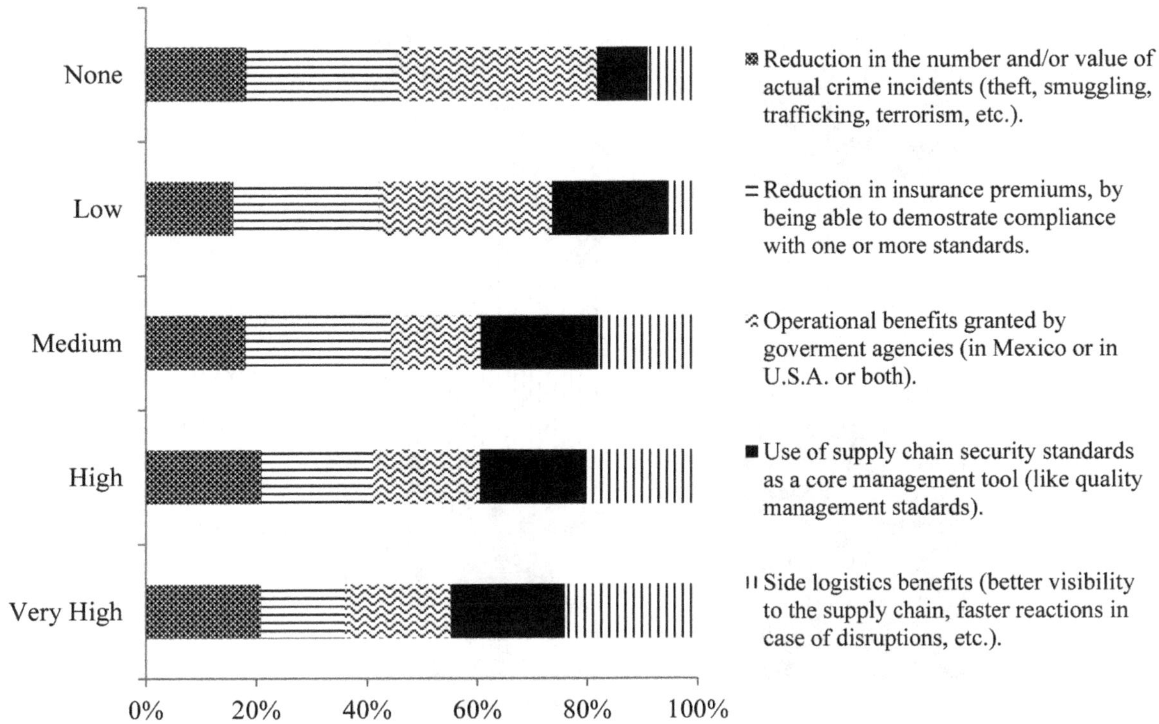

Decision-makers stressed that once initiated a SCSS should not be just another exercise. Using it only as a trophy certification without proper monitoring not only generates excess confidence, it accumulates sensitive information without real benefits to the company.

4.6 Opportunities

Cases studies pointed out an opportunity to promote a multimodal approach to SCSS (see figure 6).

From a broader perspective, another opportunity mentioned was the development of information services providing statistics classified by incidents and geographic locations of incident risk types—assuming constant updating, data reliability, confidentiality, and controlled access—to make better supply chain decisions. Similarly, a tool that decision-makers considered necessary was a database to check if their business partners were reliable companies, avoiding dealings with counterfeit companies. Another contribution to security improvement was developing a standardized approach to incident indicators in North America.

Decision-makers perceived their companies could improve export logistic operations with the creation of secure parking lots for freight at strategic points along major NAFTA transport corridors, as well as having a trust seal recognized by NAFTA that enables maintaining of operations during and after periods of maximum alerts at borders.

Regarding a question on how decision-makers perceive the future, participants expressed that within five years, they believed very likely that SMEs engaged in providing logistics services would request a trust seal or security certification. They considered it very likely that in the next future, a trust seal or

Figure 6. Multimodal approach about SCSS

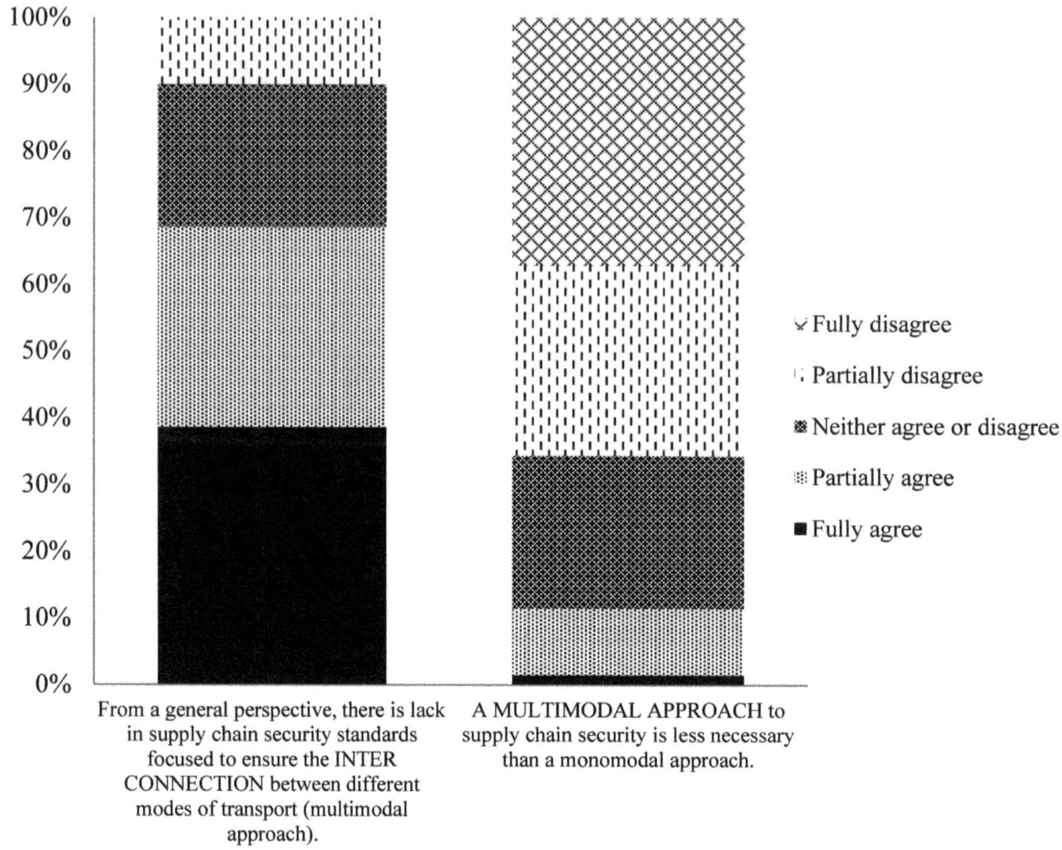

security certification would be required for non-manufacturing companies and enterprises not engaged in logistics (e.g., cleaning services, office supplies, etc.). Thus, they strengthen the idea that having a secure supply chain is essential, and that every participant in the value stream should be sensitive to the challenges of ensuring security along the entire global supply chain.

In fact, in the NAFTA region, a lack of coordination concerning SCS is evident. On 23 January 2012, The White House issued a document titled: "*National strategy for global supply chain security*", which does not identify the importance of Mexico and Canada for establishment of a secure regional trade bloc. Mexico does not have a national-regional logistics plan integrating security guidelines that foster binding national and regional efforts. Sensitivity about this issue is highly diverse in the country.

In that sense, based on the analysed case studies, four approaches used by supply chain decision-makers when facing security issues were detected. At the first level, from a "*denial*" approach, companies reject security risks. These companies develop operating activities under normal conditions, ignoring the degradation of the operational context and jeopardising not only their operations but also those of their partners. At the second level, from a "*reactive*" approach, companies recognize the problem but prefer to take only responsive measures to events (e.g., insurance, fines). This response to damage is ineffective since disruptions later propagate to supply chain partners. At the third level, from a "*proactive*" approach, companies perceive some sensitivity to security risks; consequently, they develop and implement preventive measures and maintain certifications, improve best practices, training processes, and recruit personnel qualified in intelligence activities. At the last level, from a "*smart*" approach, companies demonstrate advanced perceptions of security risks, establishing decisive and coordinated actions by integrating networks of executives responsible for security to establish security committees. Such is the case in the Mexican Silicon Valley in Guadalajara, which members perform better SCS practices than other companies in the remainder of the country. In security committees, the impulse for unification of risk criteria is highlighted. Development of joint databases for better decision-making, coordination of actions among companies and authorities, and voluntary dissemination of best practices among members is forefront as a security mitigation program. Active collaboration is paramount in the design of technological supply chain solutions.

While security is part of the global effort against terrorist threats, aspects related to local crime must not be ignored. To be effective, rather than only seeking a global reach, standards should address effectiveness at the regional or local level. Simultaneously, in most of the case studies, decision-makers expressed a preference for implementing SCSS proposed by private associations rather than government ones, with the exception of C-TPAT, which is perceived as a key element for export operations to the U.S. They perceived that SCSS proposed by private associations meets their needs more closely.

5. CONCLUSION

The purpose of this study was to perform an exploratory study to identify the key components influencing security in manufacturing supply chains for export operations in Mexico and at the same time, to analyse decision-makers' perspectives regarding supply chain security standards (SCSS) available in the market. As a result, it was identified that a dialogue among various organizations proposing standards is necessary to unify criteria concerning what constitutes good security strategies for export-oriented supply chains. The importance of building consensus on risks to all the participants in a global supply chain is now critical. For this reason, it is important to consider remarks expressed by case studies participants who believe that while for the U.S. it is important to approach security with emphasis on terrorism, initiatives implemented in Mexico should also respond (and prioritize) prevention of other criminal acts present in emerging markets such as those in Latin American.

A highly demanded aspect is mutual recognition among security initiatives, but this aspect appears complex in light of the variety of certifications offered. As a result, decision-makers suggest modular schemes allowing equivalences on key points among the standards. Most respondents preferred a standardized approach that enables them to be clear about the certification services on security they need to improve their supply chain export operations. The benefits they want to access when seeking security include: i) access to international markets; ii) attracting new customers; and iii) benefits for brand positioning. To clarify the value propositions of each security initiative, decision-makers considered it is important to develop a differentiated offer based on a value-added approach. Security specialization could be more advantageous than generalist or global-based approaches that currently create confusion.

Since enjoying a truly secure supply chain implies each actor to have a secure system, in the future, security must integrate new players who are currently ignored (e.g., SME logistics providers, cleaning service companies, financial services). In terms of policy recommendations, the case studies allow us to detect the need to create national observatories of freight transport and logistics to provide systematized, periodic, standardized and harmonized field data on comprehensive security and logistics costs. Another important key component to improve supply chain security is the creation of national public-private committees to work on the coordinated implementation of specific mitigation measures, in order to improve the level of service as well as facilitate logistics along the entire chain. Some initiatives that could be pursued are the following:

- Identify critical supply chains that are significant for studies and programs to improve security with support of customs intelligence processes;
- Categorize and monitor harmful factors and events that constrain the normal operation of supply chains on main and secondary roads or make it unsafe, with a special emphasis on routes used for shipping goods to international destinations;
- Develop specialized enterprise social networks to provide knowledge on the characteristics of criminal events a supply chain suffer, such as type of event, frequency of occurrence, degree of impact, estimated losses, type of cargo, modus operandi of the perpetrators, etc., and make this information available to all the supply chain members;
- Prepare an in-depth study on the role of physical infrastructure in security incidents along the significant supply chains: lack of unloading or rest areas for drivers, poor road maintenance, etc.;
- Design a pilot program to improve the traceability of cargo and carriers throughout a given group of critical supply chains;
- Develop a security program for small and medium-sized carriers and an expanded security program for cargo producers.

As further research, it is proposed to extend the current study to other countries in Latin America in order to generate a comprehensive benchmarking on the topic for the whole region. As a derivation of this study, a scenario analysis in terms of impacts when implementing SCSS could be performed, such that to derive recommendations and prioritize actions.

ACKNOWLEDGMENT

The authors thank Flora Hammer and anonymous reviewers for comments and suggestions that improved this paper.

NOTE

This document was based on the technical report of "The First National Exploratory Supply Chain Security Study in Mexico" which was awarded with the National Logistics Award granted by 10 institutions among them the Mexican Secretariat of Economy (SE), Mexican Secretariat of Transport (SCT), the Council of Supply Chain Management Professionals (CSCMP), the Association of Logistics, Distribution, and Freight Executives (ASELDyT), National Autonomous University of Mexico (UNAM), Ibero-American University, Anahuac University, National Polytechnic Institute (IPN), Confederation of Industrial Chambers (CONCAMIN), and Mexican Business Council for Foreign Trade, Investment, and Technology (COMCE).

FUNDING

The authors acknowledge all the support provided by the National Council of Science and Technology of Mexico (CONACYT) through the research program "Laboratorios Nacionales" and by the Bicentennial Program in Science and Technology (Chile). At the same time, the authors acknowledge the determination and effort performed by the Mexican Logistics and Supply Chain Association (AML) and the Mexican Institute of Transportation (IMT) for providing an internationally recognized collaboration platform, the International Congress on Logistics and Supply Chain [CiLOG].

REFERENCES

ACCM - American Chamber of Commerce of Mexico. (2011). *The Impact of Security in Mexico in the Private Sector.* Mexico: ACCM.

Autry, C., & Bobbitt, M. (2008). Supply chain security orientation: Conceptual development and a proposed framework. *International Journal of Logistics Management, 19*(1), 42–64. doi:10.1108/09574090810872596

Barbero, J. (2010). Freight Logistics in Latin America and the Caribbean: An Agenda to Improve Performance. Inter-American Development Bank, Infrastructure and Environment Department, Technical Notes No. IDB-TN-103, Washington D.C., USA.

Bertalanffy, L. (1968). *General systems theory foundations, development, applications.* New York: George Braziller.

Blackhurst, J., Craighead, C. W., Elkinsc, D., & Handfieldd, R. B. (2005). An empirically derived agenda of critical research issues for managing supply-chain disruptions. *International Journal of Production Research, 43*(19), 4067–4081. doi:10.1080/00207540500151549

Bueno, A., & Cedillo-Campos, M. (2014). Dynamic impact on global supply chains performance of disruptions propagation produced by terrorist acts. *Transportation Research Part E, Logistics and Transportation Review, 61*, 1–12. doi:10.1016/j.tre.2013.09.005

Carranza, O. (2007). *Global Perspectives-Mexico.* Council of Supply Chain Management Professionals.

Cedillo-Campos, M. (2010). Security risk level classification of products: A multi-criteria approach applied to warehouse management. *The 15th Annual International Conference on Industrial Engineering Theory, Applications & Practice.*

Cedillo-Campos, M. (2011). *The First National Exploratory Study in Mexico: Supply Chain Security in Export Operations.* Secretariat of Economy. Retrieved from http://www.elogistica.economia.gob.mx/swb/work/models/elogistica/Resource/3/1/images/2011_Evaluacion_del_Riesgo_en_las_Cadenas_de_Suministro.pdf

Cedillo-Campos, M.-G., & Cedillo-Campos, H. (2015). w@reRISK Method: A Tool for Security Risk Level Classification of Products. *Safety Science, 79*(November), 358–368.

Cedillo-Campos, M., & Sánchez, C. (2008). *Dynamic analysis of industrial systems*. Mexico: Trillas Publisher.

Cedillo-Campos, M., & Sanchez, C. (2013). Dynamic self-assessment of supply chains performance: An emerging market approach. *Journal of Applied Research and Technology*, *11*(3), 338–347. doi:10.1016/S1665-6423(13)71544-X

Cervantes, S. (2011). ¿Y dónde están los camiones? Revista T21. *Año*, *12*(Julio), 2223.

Chopra, S., & Meindl, P. (2013). *Supply chain management* (5th ed.). Prentice Hall.

Christopher, B. (2003). *Strategies for maximizing supply chain resilience: Learning from the past to prepare for the future*. (Master's Thesis). Massachusetts Institute of Technology, Cambridge, MA.

CIA – Central Intelligence Agency. (2012). *The World Factbook: Mexico*. Retrieved from https://www.cia.gov/library/publications/the-world-factbook/geos/mx.html

Cipoletta, G., Sánchez, R., & Pérez, G. (2010). *Políticas integradas de infraestructura, transporte y logística: experiencias internacionales y propuestas Iniciales*. Comisión Económica para América Latina ONU-CEPAL, División de Recursos Naturales e Infraestructura, Santiago de Chile. Retrieved from http://www.eclac.org/publicaciones/xml/7/39877/lcl3226e.pdf

Closs, D., & McGarrell, E. (2004). Enhancing security through the supply chain. IBM Centre for the Business of Government. Special Report Series (April).

Cruz, O. (2011). Robo al autotransporte más grave que el narco. Revista T21. *Año*, *12*(Julio), 25.

Dubois, A., & Gadde, L. (2002). Systematic combining: An abductive approach to case research. *Journal of Business Research*, *55*(7), 553560. doi:10.1016/S0148-2963(00)00195-8

Eisenhardt, K. M. (1989). Building theory from case study research. *Academy of Management Review*, *14*(4), 532–550.

Ellram, L. M. (1996). The use of the case study method in logistics research. *Journal of Business Logistics*, *17*(2), 93–138.

Financial Post. (2012). *Goldman Sachs economists: Mexico in the 2020 harvest the seventh largest economy*. Retrieved from http://www.financial-post.com/goldman-sachs-economists-mexico-in-the-2020-harvest-the-seventh-largest-economy.html

Finch, P. (2004). Supply chain risk management. *Supply Chain Management. International Journal (Toronto, Ont.)*, *9*(2), 183–196.

Friedman, T. (2012). How Mexico Got Back in the Game. *The New York Times*. Retrieved from http://www.nytimes.com/2013/02/24/opinion/sunday/friedman-how-mexico-got-back-in-the-game.html?_r=0

Gaonkar, R., & Viswanadham, N. (2009). A conceptual and analytical framework for management of integrated knowledge based logistics providers. *International Journal of Logistics Systems and Management*, *5*(1/2), 191209.

Ghauri, P., & Gronhaug, K. (2005). *Research methods in business studies: a practical guide* (3rd ed.). New York: FT Prentice Hall.

Giunipero, L., & Eltantawy, R. (2003). Securing the upstream supply chain: A risk management approach. *International Journal of Physical Distribution & Logistics Management*, *34*(9), 698713.

Hintsa, J., Ahokas, J., Zaghbour, K., Männistö, T., Hameri, A., & Holmström, J. (2010). A Conceptual model for assessing cost of security in global supply chains. EurOMA 2010, Porto, Portugal.

Host, M., & Runeson, P. (2008). *Case Studies on Software Engineering PhD Programmes*. Retrieved from http://citeseerx.ist.psu.edu/viewdoc/download?doi=10.1.1.98.4978&rep=rep1&type=pdf

INEGI – National Institute of Statistics and Geography of Mexico. (2012). *Estadísticas del comercio exterior de México*. INEGI.

Jackson, A. (2011). Trade facilitation and supply chain security a globally integrated approach. *World Customs Journal*, *5*(2), 123–128.

Jankowicz, A. D. (2000). *Business research projects* (3rd ed.). Thomson Business Press.

Jick, T. D. (1989). Mixing qualitative and quantitative methods: Triangulation in action. *Administrative Science Quarterly*, *21*(4), 602–611.

Jiménez, J., & De la Torre, E. (2011). *¿Cómo construir cadenas internacionales seguras?* Énfasis Logística. *Logistic Summit & Expo Edition*, *127*, 6267.

Jones, D. A., Farkas, J. L., Bernestein, O., Davis, Ch. E., Turk, A., Turnquist, M. A., & Sawaya, W. et al. (2011). U.S. import/export container flow modeling and disruption analysis. *Research in Transportation Economics*, *32*(1), 3–11. doi:10.1016/j.retrec.2011.06.003

Jüttner, U., Peck, H., & Christopher, M. (2003). Supply chain risk management: Outlining an agenda for future research. *International Journal of Logistics: Research and Applications*, *6*(4), 197–210. doi: 10.1080/13675560310001627016

Knemeyer, M., Zinn, W., & Eroglu, C. (2009). Proactive planning for catastrophic events in supply chains. *Journal of Operations Management*, *27*(2), 141153. doi:10.1016/j.jom.2008.06.002

Kommerskollegium. (2008). *Supply chain security initiatives: a trade facilitation perspective*. The National Board of Trade, Sweden.

Kovács, G., & Spens, K. (2005). Adductive reasoning in logistics research. *International Journal of Physical Distribution & Logistics Management*, *35*(2), 132144.

Lawrence, B., & Leon, J. (2010). *Importance of the North American regional manufacturing*. Unpublished manuscript, Supply Chain Systems Laboratory, Texas A&M University, College Station.

Lee, H., & Wolfe, M., (2003). Supply Chain Security Without Tears. *Supply Chain Management Review*, *7*(1), 12-20.

Leedy, P., & Ormrod, J. (2005). *A Handbook for Teacher Research from Design to Implementation*. Pearson Education.

Manuj, I., & Mentzer, J. (2008). Global supply chain risk management strategies. *International Journal of Physical Distribution & Logistics Management*, *38*(3), 192–223. doi:10.1108/09600030810866986

Marsden, G., & Docherty, I. (2013). Insights on disruptions as opportunities for transport policy change. *Transportation Research Part A, Policy and Practice*, *51*, 46–55. doi:10.1016/j.tra.2013.03.004

Maxwell, J. (1998). Designing a qualitative study. In L. Bickman & D. Rog (Eds.), *Handbook of applied social research methods*. Thousand Oaks, CA: Sage Publications.

Meredith, J. (1998). Building operations management theory through case and field research. *Journal of Operations Management*, *16*(4), 441–454. doi:10.1016/S0272-6963(98)00023-0

Peck, H. (2006). Reconciling supply chain vulnerability, risk and supply chain management. *International Journal of Logistics: Research and Applications*, *9*(2), 127–142. doi:10.1080/13675560600673578

Pfohl, H., Köhler, H., & Thomas, D. (2010). State of the art in supply chain risk management research: Empirical and conceptual findings and a roadmap for the implementation in practice. *Logistics Research*, *2*(1), 3344. doi:10.1007/s12159-010-0023-8

Pickett, C. (2003). *Strategies for maximizing supply chain resilience: Learning from the past to prepare for the future*. (Master's Thesis). Massachusetts Institute of Technology, Cambridge, MA.

PROMEXICO. (2012). *Mexico's Strengths*. Retrieved from http://pdf.aigroup.asn.au/trade/Mexicos_Strengths.pdf

Rao, S., & Goldsby, T. (2009). Supply chain risks: A review and typology. *International Journal of Logistics Management*, *20*(1), 97–123. doi:10.1108/09574090910954864

Rice, J., (2007). Rethinking security. *Logistics Management*, *46*(5), 28-32.

Rice, J. B., Caniato, J., Fleck, D., Disraelly, D., Lowtan, R., Lensing, & Pickett. (2003). *Supply Chain Response to Terrorism: Creating Resilient and Secure Supply Chains*. MIT Center for Transportation and Logistics Report.

Rosoff, H., & Von Winterfeldt, D. (2007). A Risk and economic analysis of dirty bomb attacks on the ports of Los Angeles and Long Beach. *Risk Analysis*, *23*(3), 533546. PMID:17640206

Runeson, P., & Höst, M. (2009). Guidelines for conducting and reporting case study research in software engineering. *Empirical Software Engineering*, *2*(14), 131–164. doi:10.1007/s10664-008-9102-8

Sheffi, Y., & Rice, J. (2005). A supply chain view of the resilient enterprise. *MIT Sloan Management Review*, *47*(1), 4148.

Sheffi, Y., Rice, J. B., Fleck, J. M., & Caniato, F. (2003). Supply chain response to global terrorism: A situation scan. *Proceedings of the First EurOMA-POMS Joint International Conference*.

Shook, C., Adams, G., & Ketchen, D. (2009). Towards a 'theoretical toolbox' for strategic sourcing, *Supply Chain Management. International Journal (Toronto, Ont.)*, *14*(1), 3–10.

Sodhi, M., & Tang, C., (2012). *Managing Supply Chain Risks*. Springer.

Stecke, K., & Kumar, S. (2006). *Sources of supply chain disruptions: factors that breed vulnerability, and mitigating strategies*. Richardson: School of Management, University of Texas at Dallas.

Tang, C. (2006). Perspectives in supply chain risk management. *International Journal of Production Economics*, *103*(2), 451–488. doi:10.1016/j.ijpe.2005.12.006

The White House. (2012). *National strategy for global supply chain security*. Retrieved from http://www.whitehouse.gov/sites/default/files/national_strategy_for_global_supply_chain_security.pdf

Thiétart, R. A. (1999). Méthodes de recherche en management. Ed. Dunod.

Verner, J., Sampson, J., Tosic, V., Bakar, N., & Kitchenham, B. (2009). *Guidelines for industrially-based multiple case studies in software engineering*. Fes, Morocco: IEEE Research Challenges in Information Systems. doi:10.1109/RCIS.2009.5089295

Voss, C., Tsikriktsis, N., & Frohlich, M. (2002). Case research in operations management. *International Journal of Operations & Production Management*, *22*(2), 195–219. doi:10.1108/01443570210414329

Wagner, S., & Bode, C. (2008). An empirical examination of supply chain performance along several dimensions of risk. *Journal of Business Logistics*, *29*(1), 307–325. doi:10.1002/j.2158-1592.2008.tb00081.x

Waters, D., (2007). *Supply Chain risk management: vulnerability and resilience in logistics*. Kogan Page.

Williams, Z., Lueg, J., & LeMay, S. (2008). Supply chain security: An overview and re-search agenda. *International Journal of Logistics Management*, *19*(2), 254–281. doi:10.1108/09574090810895988

Wilson, M. (2007). The impact of transportation disruptions on supply chain performance. *Transportation Research Part E, Logistics and Transportation Review*, *43*(4), 295320. doi:10.1016/j.tre.2005.09.008

Wu, T., Blackhurst, J., & O'grady, P. (2007). Methodology for supply chain disruption analysis. *International Journal of Production Research*, *45*(7), 16651682. doi:10.1080/00207540500362138

Yang, Z., Ng, A., & Wang, J. (2014). A new risk quantification approach in port facility security assessment. *Transportation Research Part A, Policy and Practice*, *59*, 72–90. doi:10.1016/j.tra.2013.10.025

Yin, R. (2003). *Case study research: design and methods* (3rd ed.). Thousand Oaks, CA: Sage Publications.

APPENDIX

Table 1. Supply Chain Security Standards

STANDARD	FOCUS	INITIATOR AND DRIVING FORCE
ISO 28001	Standards related to supply chain security.	It is a program supported by an international organization (International Organization for Standardization, ISO). The program is voluntary and applies for any type of industry.
ISO 31000	Standards related to risk management.	It is a program supported by an international organization (International Organization for Standardization, ISO). The program is voluntary and applies for any type of industry.
C-TPAT (Customs-Trade Partnership Against Terrorism)	Certification program in supply chain security that aim to minimize terrorism impacts and facilitate customs procedures for those companies certified.	It is a program initiated by the government of United States, specifically the Customs Authority. This is a voluntary program and the companies that participate are: manufacturing companies, transportation companies, and logistic operators. The initial focus was the countries of Canada and Mexico, but currently considers any country.
AEO (Authorised Economic Operator)	Supply chain security standards based on Customs-to-Business partnerships.	It is a core part of the WCO (World Customs Organization) framework of Standards to Secure and Facilitate global trade (SAFE). Operators can be accredited by Customs as AEOs when they prove to have high quality internal processes that will prevent goods in international transport to be tampered. All the companies that participate in global supply chains can be certified.
IRU (International Road Transport Union)	World road transport organization.	It is a program supported by a global organization that seeks to uphold the interest from large fleet to individual owner-operators.
BASC (Business Alliance for Secure Commerce)	Secure international trade.	It is a program supported by an international business alliance, created as a initiative of the private sector to promote secure international trade in cooperation with governments and international organizations, as a voluntary cooperation program.
CASCEM (Security Supply Chain of Mexico Foreign Trade)	Increase the reliability of Mexican companies and the fulfilment of secure international trade programs of United States and Mexico and the World Customs Organization.	It is an entrepreneurial Program to certify Mexican companies in supply chain security standards. The program is focused to all the companies of a global supply chain and is voluntary.
NEEC (The New Certified Companies Scheme)	Security and quality assurance for logistics chain procedures of Mexican companies.	It is a program entrusted to the Customs Authority in Mexico (Administración General de Aduanas) that was designed considering the criteria of the SAFE framework and C-TPAT. This is a voluntary program intended for manufacturing and commercialization companies that have conducted foreign trade operations for the past 5 years or more.
PIP (Partners In Protection), Golden List	Security of the supply chain and the facilitation of legitimate trade.	It is a voluntary program of the Canada Border Services Agency (CBSA) with no membership fee. PIP members agree to implement and adhere to high security standards and are recognized as being trusted traders.
TAPA (Transported Asset Protection Association)	Reduction of losses in the manufacture, transportation and distribution of high value and theft attractive products.	It is a program supported by the private sector that seeks to reducing losses from international supply chains. Is a voluntary program based on memberships and partners.

continued on following page

Table 1. Continued

STANDARD	FOCUS	INITIATOR AND DRIVING FORCE
ISA (Internet Security Alliance)	Standards, practices and technologies of cyber security.	It is a program supported by a multi-sector trade association that was founded in 2000 in collaboration with Carnegie Mellon University. ISA membership is open to public and privately held entities and currently has substantial participation from the aviation, banking, communications, defense, education, financial services, health care, insurance, manufacturing, security and technology industries.
CSI (Container Security Initiative)	Security of maritime containerized cargo shipped to the United States from around the world.	It is a program supported by the U.S. Customs and Border Protection (CBP) to examine high-risk maritime containerized cargo at foreign ports. The program considers port maritime terminals with traffic flows to United States.
APEC-STAR (Secure Trade in the APEC Region)	Security and efficiency in the APEC region's access points (ports, airports, etc.) and coordination between private and public sector to counteract terrorism threats.	It is a program supported by Secure Trade initiative in the Asia-Pacific Economic Cooperation (APEC) Region that considers the 21 APEC member economies, senior representatives from private sector and international organizations.
CSP (Customs Security Program)	Secure the international supply chain and protect European Union member states and markets.	It is a program implemented by the European Commission that impacts all importers, exporters, logistic service providers and other companies involved in moving goods in international trade.
ISPS (International Ship and Port Facility Security)	Establishment of an international cooperation framework among governmental agencies, local administrators and the shipping and port industry.	It is a program supported by the International Maritime Organization (IMO) as a comprehensive set of measures to enhance the security of ships and port facilities as defined by the International Maritime Organization (IMO).

Chapter 4
Intentional Food Contamination in the Food Supply Chain:
Proposal of a Management System for its Prevention

Ramón Navarrete Reynoso
Universidad de Guanajuato, Mexico

Cecilia Ramos-Estrada
Universidad de Guanajuato, Mexico

Omar J. Purata
Universidad de Guanajuato, Mexico

ABSTRACT

As globalization increases, supply chains are becoming increasingly complex and vulnerable, making the management of their security an important part of corporate management. The need to adopt control and prevention measures that allow for guaranteeing security in international commercial operations is essential to prevent becoming a target of these threats. Some of these threats are terrorism, piracy and theft. Within the food industry, food terrorism has gained relevance due to the extreme consequences it may have on both public health and the market. Food terrorism refers to the intentional contamination of food in any link of the supply chain, which covers everything from production to consumption. This chapter deals with the problems of food terrorism and makes a proposal that allows to establish a management system to prevent intentional food contamination in the supply chain through managing risks and improving security in the supply chain on this aspect.

INTRODUCTION

In today's globalized world the number of risks in supply chains are increasing and the threats are becoming more complex. All the risks and threats can have a severely negative impact on markets. For this reason it is necessary to have con-trolled conditions and preventive measures that guarantee that international commerce operations are carried out safely. Such measures help protect against being the

DOI: 10.4018/978-1-4666-9779-9.ch004

target of delinquent activities like drug-trafficking, patent violations and terrorism (Kleindorfer & Van Wassenhove, 2004; Kleindorfer & Saad, 2005; Rao & Goldsby, 2009).

Since the terrorist attacks of 2001, there has been greater emphasis on supply chain security. The United States was the first country to adopt and promulgate new heightened security laws due to the terrorist attacks. The United States Customs Agency began enforcing the Container Security Initiative (CSI) in the beginning of 2002. The Customs-Trade Partnership Against Terrorism (C-TPAT) was also founded in 2002. Today many countries are working in order to have a global framework to ensure that their supply chains are safe (Closs & Mcgarrell, 2004).

Intentional food contamination represents a real and potentially catastrophic threat for the society, given that it can produce long reach disastrous effects, including direct morbidity and/or mortality, interruption of food distribution, loss of confidence from the consumers in the Government and the food supply's responsible parties, company bankruptcies, trade restrictions and serious economic effects (Busta & Kennedy, 2011).

In this context, terrorist attacks on water and food supply is a risk that must be controlled. The food supply chain represents an attractive infrastructure that serves as a target for terrorist attacks. In order to have an idea of the magnitude of the food sector consider this: only in the United States, consumers spend more than $617 billion a year on food, of which $511 billion are spent on food within the agriculture sector (DHHS, 2005).

Food terrorism has been defined by the World Health Organization as "an act or deliberately try of food contamination for human consumption with chemical, physical or microbiological agents, with the purpose of causing damage or death to civil populations or to interrupt the social, politic or economic stability" (WHO, 2008; Veiga, 2011).

The Food Defense refers to the analysis, control and improvement of prevention mechanisms of those attacks; that is, the Food Defense involves a Risks Management. This management is based on the premise that absolute safety does not exist and that the reliability in each of the components, even the highest, does not imply reliability equivalent to the whole system. Risk management consists of recognizing the risks, evaluating them and regulating some in relation to others, leaving aside the attempt to restore situations in which the risk would be completely excluded (Dourlens et al. 1991). The oversight in this matter can lead to an increase in the failure probability in the security of the food supply chain (Food Supply Chain: FSC); and therefore, generate potentially high costs for its constituents.

Hence, the food supply chain must ensure that its activities follow a preventive approach, so the risks are as low as possible. This need for protection must be addressed from different points of view. One of the most important standpoints is related to the adaptation of the business processes and organizational structures that are involved in food exchange and handling.

The business process approach has gained importance in enterprises since the 90s compared to the traditional hierarchical departmental point of view (Aguilar-Savén, 2004). Some examples of business processes are described in Kettinger et al. (1997), Swanson (2003), Gaitanides (2007), Damij, et al. (2008) and Vanderhaeghen, et al. (2010). The process approach consists of compiling the enterprise activities in a logical sequence, thereby creating a clearer vision of the company's activities. One objective is to attain better understanding, control and productivity in the activities that generate value to the enterprises.

The government of the United States of America has developed ORM (Operational Risk Management) and CARVER+shock for the analysis of Food Defense, following a traditional hierarchical departmental approach (DD, 2000; FAA, 2000; DHHS, 2001; Rasco & Bledsoe, 2006; Kleter & Marvin, 2009). The business process approach would enable the integration of prevention, control and protection mechanisms to face food terrorism in the activities of food supply. Thus, this approach would be a clear and decisive contribution to Food Defense in a company or in its FSC.

However, despite the obvious benefit offered by the business process approach in the field of food defense, the consulted literature ignores the business processes point of view and offers a partial understanding of the vulnerability in each link of the FSC.

Based on the above, the aim of this chapter focuses on a proposal formulation that allows establishing a management system for preventing intentional food contamination and the improvement of the security of the supply chain. The aim is to manage the intentional contamination risks by means of a methodology that encompasses a risk analysis.

This proposal adopts a process approach for establishing, implementing, operating, monitoring, reviewing, maintaining and improving a system for preventing intentional food contamination. It establishes a system that is able to promote the security of each link, causing the whole supply chain to benefit from it, from the farm to the final consumers. The initial study covers only the theoretical methodological proposal and focusses only on the companies of the food sector of Guanajuato State in Mexico. However, it may extend into a research plan security of food supply chain.

Finally, this paper includes results of research in the doctoral thesis of Navarrete & Lario (2012), and previous results published in Navarrete & Lario (2009); Navarrete & Lario (2010) and Navarrete & Lario (2011), given the general importance of the topic within the trade relations between US-Mexico.

BACKGROUND

The Security in the Supply Chain is defined as the enforcement of policies, procedures and technologies to protect goods of the supply chains from robbery, damage or terrorism and to prevent the introduction of smuggled goods, people or weapons of mass destruction along the supply chain (Closs et al. 2004). As a way to make reference to the multiplicity of players and spheres of action, the term Supply Chain Security Management (SCSM) (Bowersox et al. 2007) has been coined. Hintsa, et al. (2009) later incorporated to this definition the dimension of the systemic solution pointing out that each SCSM measure should be oriented towards prevention, detection or recovery from a criminal act as quickly as possible.

For the World Health Organization, food terrorism has become one of the biggest global public health threats in the 21st Century. It expresses concerns about the possibility that physical, chemical or biological agents might deliberately be used to harm civilian populations. In this regard, food is recognized as a potential vehicle for disseminating such agents to a broad population (WHO, 2008; WHO, 2003).

Terrorism directed towards the food sector could have extreme consequences in civil populations, affecting their physical and psychological health, causing distrust concerning the safety of the affected food group (Peters, 2003; Onyango et al., 2005; Pistole, 2006). Deliberate contamination of food supply could have a devastating impact in public health and the global economy (DHSS, 2005; Boyle, 2005; Lyonga et al., 2006; Qing et al., 2008)

Brummer (2003) considers the following consequences: (i) physical consequences: an insufficient or non-edible food supply. The direct results could be diseases and mortality, and indirectly hunger and malnutrition of the affected populations; (ii) psychological consequences: there may be problems in the consumers' behavior, which could include the perception of an unsafe and vulnerable food supply; (iii) political consequences that could end up in civilian discord due to the food shortage, decline in confidence in governments, outbreaks of destabilization and anarchy (iv) economic consequences of variable length with loss of consumer trust and deterioration of the image in the market of the companies involved.

Some studies concentrate on possible consequences of food terrorist attacks through diverse estimates, e.g. Wein & Liu (2005), or response simulation models in cases of terrorist activity in food supply, e.g. Hartnett et al. (2009).

In the food supply chain some links are more vulnerable than others. Consequences of the contamination could vary according to the type of food and the specific link in the chain that is targeted (Alvarez, et al. 2010). An attack that targets a step closer to the consumer has a greater probability of success but affects fewer people. On the other hand, an attack in the early steps of the supply chain affects many more people, but has to evade many controls and countermeasures to be successful. The World Health Organization indicates that the potential of intentional contamination of products is probably greater near distribution points, and the potential of mortality becomes greater the closer the agent is introduced to the point of consumption (WHO, 2008).

Nevertheless, it does not happen when the products are consumed fresh as in the agricultural industry. Crutchley et al. (2007) indicates that the agricultural infrastructure is, due to its characteristics, extremely vulnerable to a terrorist attack, so it needs special attention. For Monke (2004), the farm is the most vulnerable link and prone to food terrorism (agroterrorism), because of its large expanses of land, little surveillance, and bulk storage.

Setola & de Maggio (2009) believes that when the contaminant agent is dispersed in the first links of the food supply chain it becomes more difficult to identify and detect the origin of the contamination. Above all chemical agents that when mixed can be imperceptible to quality control (Alvarez, et al. 2010).

Also, the transportation and storage steps are, in general, more vulnerable that the manufacturing step (Alvarez, et al. 2010). Packaged products are more susceptible to contamination during transportation and storage. However, contamination of bulk products would affect more people (Alvarez, et al. 2010).

Dalziel (2009) has conducted a systematic examination of incidents involving the intentional and malicious contamination of food from 1950 to 2008. The analysis reveals that almost 98% of the incidents occurred downstream in the food supply chain (e.g., at retail outlets, food service establishments, homes and the work-place). Typically, the incidents involved commonly-available household, agricultural or industrial chemicals. When more esoteric chemicals were used, the perpetrators often had access to these agents at work and also possessed the knowledge to use them. Incidents involving biological or radiological agents typically occurred at the retailer or at the consumer and had little impact on public health.

Preventive tools have been developed due to the complexity of terrorist activities and their potential effects on food supply. One of these tools is the ORM that originated in the United States Defense Department as a system for operational risk management to improve security in military installations and for their personnel. Since this institution supports the preservation of the security in the United States, the US government, through its Department of Health and Human Service (DHHS), the US Food and Drug Administration (FDA) and the Center for Food Safety and Applied Nutrition (FDA), has established the ORM tool, which is part of their vision and strategy in the prevention and protection against food terrorism (DHHS, 2001).

ORM shares similarities with other tools applied in the area of Risk Manage-ment. For instance, the Hazard Analysis and Critical Control Points (HACCP) is used in the food supply sector for the analysis of accidental food adulteration risks and to preserve their harmlessness (Arvanitoyannis & Varzakas, 2009). ORM also shares similarities with the Hazard and Operation (HAZOP), which is used in operational risk analysis to prevent industrial accidents (Cagno, et al. 2002) The aforementioned risk analysis tools are scientifically valid and widely accepted throughout the world (Rasco & Bledsoe, 2006). However, the HACCP concept is over 50 years old! It was developed as a common project between Pillsbury Company, U.S. Army and NASA to ensure the safe production of food for the manned space program. The original concept was based in three principles: identification and assessment of hazards, determination of critical control points and establishment of a system to monitor the critical control points (Surak, 2005).

Other of these food defense tools is the CARVER + Shock. It's a method is an offensive targeting prioritization tool that has been adapted for use in the food sector. This tool can be used to assess the vulnerabilities within a system or infrastructure to an attack. It allows you to think like an attacker by identifying the most attractive targets for attack (FDA, 2007). By conducting such a vulnerability assessment and determining the most vulnerable points in your infrastructure, you can then focus your resources on protecting your most vulnerable points. CARVER is an acronym for the following six attributes (discussed in further detail later) used to evaluate the attractiveness of a target for attack (FDA, 2007):

- Criticality: measure of public health and economic impacts of an attack.
- Accessibility: ability to physically access and egress from target.
- Recuperability: ability of system to recover from an attack.
- Vulnerability: ease of accomplishing attack.
- Effect: amount of direct loss from an attack as measured by loss in production.
- Recognizability: ease of identifying target.

In addition, the modified CARVER tool evaluates a seventh attribute, the combined health, economic, and psychological impacts of an attack, or the SHOCK attributes of a target.

Some countries are making efforts to raise awareness in preventing food terrorism. United States government provides training in preparedness against an intentional attack against food supply. The courses provide an understanding of and guidance for developing a food defense prevention based on a common sense approach. One of these courses is the ALERT course. This course was developed by the FDA to help stakeholders better understand food defense and how food defense applies to the food industry (FDA, 2015)

According to the above, obviously the food defense is one aspect which should be properly managed in the food industry, as well as, for many years it has been el food safety. The international standard for food safety is ISO 22000 – Food Safety Management Systems –Requirements for Any Organization in the Food Chain (ISO 22000: 2005), which can be used by all organizations in the food chain, incorporates the principles of HACCP, is an auditable standard and focuses only in food safety. The standard is composed of three major parts: requirements for HACCP, requirements for prerequisite programs (PRPs) and requirements for implementing a management system. The major parts can be seen as a three legged stool where all of the three are required for a food safety management system (FSMS). If one part is missing or lacking, the stool becomes unstable, which means the FSMS becomes ineffective in preventing food safety incidents (Surak & Lorca, 2007).

On the other hand, the most popular management systems are turning around risk management in its new versions. This year revisions of ISO 9001 and ISO 14001 standards are now taking into account the risk-based approach. There are several elements of the risk-based approach in the draft international standard (DIS) of ISO 9001:2015. For example: clauses 6.1.1 and 6.1.2 state that the organization must determine the risks and opportunities to ensure that the equality management system can achieve its intended results, prevent undesired effects and reach the continuous improvement (ISO/DIS 9001:2015). Specially related with the food supply can be seen what clause 8.5.5 says: (if applicable) the organization must determine and meet requirements for post-delivery activities associated with the nature and intended lifetime of the goods. In this sense, the organization is responsible for the risks associated with the goods and services, their use and lifetime.

Even when the ISO 22000 standard is fully compatible with current ISO 9001:2008 standard, there are important differences to point out: FSMS does not allow exclusions in the requirements, FSMS must conform to legal requirements, food safety must be included between the business objectives, there is an assumption that it is possible to enter unsafe product in the food chain, so the organization must have product withdrawal procedures and, concessions cannot be made if the product is unsafe or contains a food safety hazard, between another differences (Surak, 2007).

MAIN FOCUS OF THE CHAPTER (METHODOLOGY)

The development of this proposal has been carried out under an approach that results from adaptation (and/or combination) of different management systems that cover a risk analysis, such as the *Management System of Information Security (MSIS),* based on the standard *ISO/IEC 27001*[1]. The proposed procedure has been prepared in compliance with the Good Manufacturing Practices (GMPs) and the recommendations of the Food Defense proposed by experts in the government, academic and private fields of the food sector. A risk analysis approach was taken into account during the evaluations of the intentional contamination dangers of the supply chain links.

To establish and manage the *Management System for the Prevention of Intentional Contamination (MSPIC),* and the security improvement to the Supply Chain, the continuous *PDCA cycle* is used (Plan, Do, Check, Act), which is traditional in quality management systems (see Figure 1).

Figure 1. Continuous PDCA cycle (ISO/IEC 27001, 2005)
Source: ISO/IEC 27001, (2005).

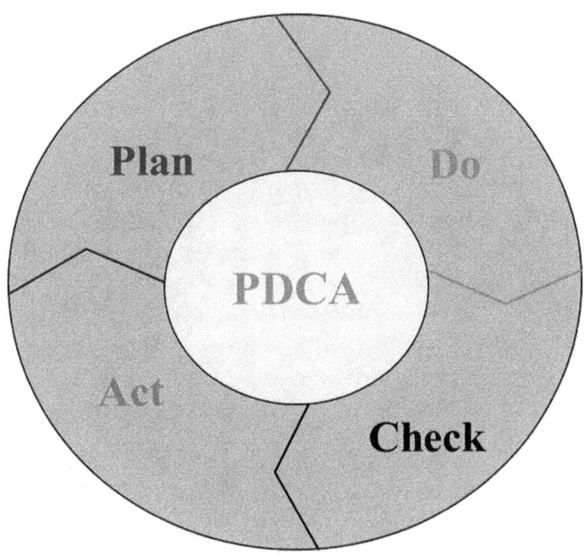

- **Plan:** establishes the MSPIC.
- **Do:** implements and uses the MSPIC.
- **Check:** monitors and reviews the MSPIC.
- **Act:** maintains and improves the MSPIC.

The adoption of an MSPIC should be a strategic decision for an organization. The MSPIC implementation depends on the objectives and the security requirements, the processes, the size and the structure of each link in the FSC.

This system adopts a process approach. An organization needs to identify and manage many activities in order to function effectively. A process is a sequence of activities that use resources (inputs) and transforms these inputs into products (outputs). Often the output from one process directly forms the input to the next process. The Figure 2 illustrates how an MSPIC takes as input *the Good Manufacturing Practices (GMP's)* and *food defense recommendations* and through the necessary actions and processes produces *food defense controls* that meets those requirements.

The GMP's have been developed for many years to prevent accidental contamination of food (FDA, 1999; CA, 2003), compared to the recommendations of the food defense which they are very recent and are used to prevent intentional contamination.

In the reviewed literature from various sources, including AIB (2015), C-TPAT (2015), FDA (2012), USDA (2004, 2005, 2008a, 2008b, 2009a y 2009b), DHSS (2001, 2005, 2007a, 2007b, 2007c), Rasco and Bledsoe, (2006), Valle et al. (2007), WHO (2008); described the general defense food recommendations to follow in order to prevent food terrorism. In some references, such as those presented by the USDA, they are grouped to have certain common characteristics. According to their scope, it is proposed to classify the recommendations of defense food used in six types:

1. Food defense programs
2. External physical infrastructure installation
3. Staff and visitors program

Figure 2. Continuous PDCA cycle applied to MSPIC processes
Source: adapted from ISO/IEC 27001, (2005).

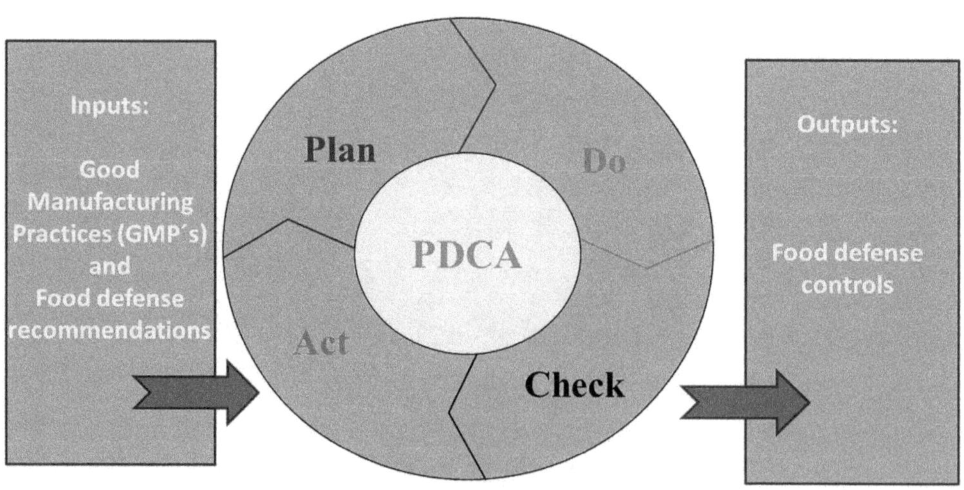

4. Reception of raw materials and supplies
5. Facility operations
6. Storage and shipment of the finished product

In the following table, there is a concrete compilation of some of the good practices of food defense, they could be considered as indispensable in any link of the FSC (see table 1):

Table 1. Best practices in food defense (compiled by the authors)

Typology	General Content	Good Practices
Food defense programs	Measures applied on the company strategy	Setting a food defense policy.
		A procedure for compliance with national and international legal requirements.
		Operational Risk Management (ORM) completed for the installation.
		Integrating a crisis management team of food terrorism (food defense equipment).
		Establishing food defense responsibilities to specific people or the team.
		Creating a list of key contacts of regulatory and enforcement agencies.
		Develop procedures for monitoring all food and ingredients from the manufacturer to the consumer's table.
		Developing a product (food) recall in the company.
		Recall practices of a product in the market led by the crisis management team with a frequency at least twice a year.
		Performing self-inspections of food defense on the premises planned and conducted by internal staff, at least quarterly.
		A program to ensure the safety of incoming mail and packages.
		Programs to protect and strengthen critical information for food defense and information technology (databases, computers, etc.).
		Developing procedures for the management and safe disposal of contaminated food.
		Written policies and procedures established for security contracted services or internal ones.
		A customer/consumer complaints program established and procedures implemented to investigate situations of product tampering.
External physical infrastructure installation	Implemented measures to improve the external physical security of the facilities.	External restriction perimeters to limit access to the facility. (E.g. with walls)
		Using security cameras on key points around the facility.
		Regular patrolling on outdoor areas and ceilings.
		Having a restricted Access to the buildings, silos, products storage, receiving stations, and lockable roofs, etc.
		Minimize potential "hiding" sites for aggressors (e.g. weeds in the garden, trees, etc.)
		Adequate lighting in the exterior, including: patios and gardens, parking lots, gates, loading platforms, etc.
		Establishing a system in place to control and identify authorized vehicles to enter and park nearby.
		Establishing a program to address any security problems detected outdoors
		The access points to the facility have been minimized to the barely minimum for an adequate operation and are being monitored.
		Solid doors at the entrances (e.g. reinforced metal doors).

continued on following page

Table 1. Continued

Typology	General Content	Good Practices
Staff and visitors program	Related measures to the programs and procedures for preventing the infiltration of malicious people (terrorists, criminals, etc.) and the mechanisms for the integration of the new staff (recruitment, selection and introduction).	Establishing a recruitment policy covering aspects of food defense prior to recruitment, for all employees and contractors.
		Prohibiting the staff or contracted individuals to begin work without completing the program of screening and approval.
		Implementing an identification system and staff recognition.
		Establishing a system to restrict the access to the staff to controlled areas both inside and outside the facilities.
		Creating a training program for employees to identify signs and evidence of any potential violation of the product.
		Having areas with restricted traffic for new employees. They can only move freely to resting areas.
		Having designated areas for the staff to store their belongings and implementing a procedure for the control of this aspect.
		The staff lockers in the locker rooms area assigned to store personal items are inspected according to a given frequency.
		Creating an installation policy for visitors.
		Visitors (contractors, guests, inspectors, sales representatives, etc.) report in a designated entrance and sign the log entries according to a procedure.
		Having a formal program to accompany visitors to the facility and verify access to vulnerable areas.
		Having a formal program of uniforms, for both staff and visitors.
Reception of raw materials and supplies	It refers to the guidelines for preventing pollution in the process of receiving materials (mainly raw materials).	Purchasing of all ingredients, food products, packaging materials and other supplies food service only to approved providers who have the appropriate permits or licenses.
		Keeping a list of phone numbers of suppliers and alternative suppliers (for situations when regular suppliers are unable to provide the product).
		Requiring suppliers to provide prior notice (by telephone, email, fax, or delivery schedule) for any delivery, especially out the established schedule.
		Assigning authorized and trained staff to inspect and receive deliveries in the facility.
		Having the suppliers providing evidence of their food defense programs.
		Having filed continuous guarantee letters from raw materials suppliers, including packaging.
		Having formally established programs for the analysis of raw materials and packaging (internal analysis, external analysis or certificates of analysis).
		Having established written procedures covering food reception.
		Having shipping checked with a scheduled delivery list and approved suppliers.
		Checking the transport arrival and the driver identification of the raw materials deliveries.
		Checking the reception and documentation documents.
		Documenting properly the results of inspections of incoming raw materials.
		Ensuring security and monitoring measures for the download process of food all the time.
		Having implemented procedures to dispose damaged or rejected materials.
		Having procedures to handle quarantines and releases, irregularities in amounts out of a predetermined range, evidence of violation or limitations of incoming materials.
		Establishing a policy for packaging of the received materials with systems that prevent their violation or visible possible tampering, when it may occur.

continued on following page

Table 1. Continued

Typology	General Content	Good Practices
Facility operations	It is about all the control measures over the physical infrastructure of the internal areas of the facilities and the business processes related to manufacturing operations of the product. .	It takes at least an external evaluation to indicate vulnerable areas of food terrorism: warehouses, water supply areas, compressed air supply areas, etc.
		Procedures for internal control and restriction of access to sensitive areas.
		Safety inspection in water supply and in related critical components (storage tanks, filters, backflow valves, etc.).
		Regular analysis of the potability of the water scheduled and performed following a scientific sampling procedure.
		Scheduled periodic monitoring for water treatment and / or filtration systems.
		Developing procedures to suspend the use of contaminated water.
		Inspecting the air supply system of the facilities and have it analyzed on a regular basis.
		Developing protocols to follow if there is a suspicion that the air flow (e.g. air conditioning, air cleaning, etc.) is contaminated.
		Having an appropriate access control, CCTV monitoring (Closed Circuit Television) and / or having a critical supervision on manufacturing and storage points.
		Having a restricted access to bulk commodities containers (e.g. silos)
		Having a restricted access to areas with dangerous chemicals and lubricants.
		Setting an established controlled access to the production or storage area. (e.g. maintenance contractors, pest control, or sanitation crews).
		Having programs to identify if a product package has been opened.
		Having a proven traceability for all raw materials, direct contact with the food product and product rework process.
		Controlling the access to areas that have the manufacturing equipment instrumentation used to ensure the product safety (autoclaves controls, pasteurizers controls, temperature controls, etc.)
		Having a program to prevent a deliberated combination of the rejected or suspicious products.
		Using packaging designed to prevent violations or using security seals on finished products.
		All the finished products are in lots properly identified.
		Getting the safety data sheets for hazardous chemicals and having them available to staff.
		Keeping a daily inventory of hazardous chemicals and investigate immediately any discrepancies.
		Having a program to manage product labels.
		Monitoring the use of chemicals to prevent a possible deliberate food contamination.
		Establishing procedures to monitor the operation of critical equipment (such as water storage systems, mixers and packing equipment) to prevent someone intentionally contaminating food during manufacturing.
		Creating a chart or a map which defines the limits of all service areas where visitors can stay and specific places for preparing and storing food.
		Limiting and controlling access to storage areas or other vulnerable areas.
		Securing the laboratories and access can only be granted to authorized personnel.
		Keeping locked the positive crop pathogens area.

continued on following page

Table 1. Continued

Typology	General Content	Good Practices
Storage and shipment of the finished product.	Safety measures on business processes of product storage and shipping.	Inspecting the safety in all the storing places (including temporary storage vehicles) regularly and keeping a record of the results.
		Maintaining accurate inventories of all products so unexplained additions or withdrawals decreases can be detected and investigated.
		Storing all surplus products in hermetically sealed containers (except during cooling), clearly labeled, and dated.
		Establishing policies and procedures to rule out products that are not properly sealed and labeled.
		Separating properly the finished products from fundamental and chemical substances.
		Having written procedures for inspecting all vehicles prior to be loaded.
		Handling adequately the removed waste or any other material from the vehicles after they are cleaned.
		Verifying the amount and number of lots of products during the transportation loading.
		Checking the documented identification of the driver.
		Having a washing certificate and or checking the stamps of the transportation vehicles (trucks, tankers, etc.)

The MSPIC is designed to ensure the selection of adequate and proportionate controls that protect against food terrorism. Controls are based on the results and conclusions of the risk assessment, legal or regulatory requirements, contractual obligations and the organization's business requirements for food defense.

The requirements set out in this system are generic and are intended to be applicable to all organizations of the food sector, regardless of type and size.

Plan

In this phase the MSPIC creation takes place, with the *scope* definition and the *Security Policy.* The scope refers to the limits the MSPIC will have in terms of the characteristics of the FSC chain, its organizational structure, its location, its activities, its technology and the details of and justification for anything excluded from its range. The MSPIC policy as defined and approved by the management will include: i) a key framework for objectives setting out a general sense of direction and principals for action relating to the prevention of intentional contamination; ii) the legal and commercial requirements; and iii) the definition of a criterion for evaluating risk that is in line with the organization's strategic risk management framework.

The fundamental core of this phase as well as of the MSPIC is the *risk analysis* performance that reflects the current situation of the entity. For the risks analysis it is necessary to define an adequate risk evaluation methodology, besides establishing the *risk acceptance criteria* and specifying the *acceptable risk levels.* It is of utmost importance to identify the threats within the business processes and infrastructure of the analyzed supply chain link. Afterwards, the risks related to the identified threats are being evaluated. Some of the methodologies, which could be adapted for this purpose, although currently used in the field of information security, are Magerit (Methodology of analysis and risk management information systems); the methodology detailed in the ISO/IEC 27005 standard, OCTAVE (Operationally

Critical Threat, Asset and Vulnerability Evaluation), etc. In the field of food safety, methodologies that could be adapted are ORM (Operational Risk Management), CARVER + Shock, or the methodology suggested for business processes in Navarrete et al. (2011).

Finally, the different risk treatment options are identified and evaluated. In this methodology, controls are selected to improve food defense performance. The Table 1 contains a comprehensive list of controls that have been found to be commonly relevant in food companies. Users are directed to Table 1 as a starting point for control selection to ensure that no important control options are overlooked.

Prepare a *Statement of Applicability*. It's a documented statement describing the control objectives and food defense controls that are relevant and applicable to the organization's MSPIC. It provides a summary of decisions concerning risk treatment. Justifying exclusions provides a cross-check that no controls have been inadvertently omitted (ISO/IEC 27001, 2013).

A Statement of Applicability shall be prepared that includes the following (ISO/IEC 27001, 2013):

1. The controls selected and the reasons for their selection.
2. The control objectives and controls currently implemented.
3. The exclusion of any controls and the justification for their exclusion.

Do

In this phase the *MSPIC' policy* and *controls* are implemented and operate with the *risks treatment plan* and its execution. The plan should identify the actions, resources, responsibilities and priorities in the risks management; with the purpose to achieve the identified control objectives. The possible actions for dealing with risk include: i) applying the controls; ii) consciously accepting the risk (doing nothing), as long as the criteria for risk acceptance are met; iii) avoiding risk (not carrying out the activity which entails risk); or iv) transferring the risk (for example by insuring or subcontracting). The *controls* have been prepared in compliance with the Good Manufacturing Practices (GMPs) and the Food Defense recommendations proposed by experts in the government, academic and private fields of the food sector. The definition of metrics and indicators is important to evaluate the efficiency of the implanted controls.

Training programs must also be established to ensure that the personnel who have responsibilities defined in the MSPIC are competent to perform the relevant tasks by (ISO/IEC 27001, 2013):

- Determining the necessary competencies for personnel performing work effecting the MSPIC.
- Providing training or taking other actions (e.g. employing competent personnel) to satisfy these needs.
- Evaluating the effectiveness of the actions taken
- Maintaining records of education, training, skills, experience and qualifications.

The personnel in charge must be aware of the relevance and importance of all the prevention activities to contribute to the food defense goals.

Also, you have to provide the necessary resources for MSPIC in order to (ISO/IEC 27001, 2013): a) establish, implement, operate, monitor, review, maintain and improve an MSPIC; b) ensure that procedures support the food defense requirements; c) identify legal regulatory requirements; d) maintain adequate prevention by correct application of all implemented controls; e) carry out reviews when necessary, and to react appropriately to the results of these reviews; and f) where required, improve the effectiveness of the MSPIC.

Check

During this phase different types of revisions are performed to check the correct system implantation. Among those, an independent and objective *internal audit* takes place; as well as a *global review* of the MSPIC by the Board.

Internal audits should be carried out at planned intervals to determine whether the controls, control objectives and system procedures: i) comply with legislation or international standards; ii) comply with good manufacturing practices (GMPs) and Food Defense recommendations; iii) are implemented and maintained effectively; and iv) are performed as anticipated.

The management must check the MSPIC at planned intervals to ensure continuity, effectivity and suitability. The results of both checks must be clearly recorded and the records must be maintained.

Some of the general objectives of these checks are: to identify the security gaps, to guarantee the effectiveness of the controls, to update the risks analysis or review the compliance with the MSPIC's policy and objectives.

Act

The result of the reviews should be reflected in the definition and implantation of corrective, preventive and improved actions to advance in the achievement of an effective and efficient MSPIC. At this stage, the FSC chain must do the following: i) implement the identified improvements; ii) establish preventative and corrective actions; iii) communicate results and actions to those effected at an appropriate level of detail; and iv) ensure that improvements achieve their stated objectives.

The organization continually improve the effectiveness of the MSPIC through the use of the food defense policy, food defense objectives, audit results, corrective and preventive actions and management review. Actions to eliminate the cause of nonconformities with the MSPIC requirements in order to prevent recurrence (ISO/IEC 27001, 2013).

The documented procedure for prevent and corrective action define requirements for (ISO/IEC 27001, 2013): i) identifying nonconformities; ii) determining the causes of nonconformities; iii) evaluating the need for actions to ensure that nonconformities do not recur; iv) determining and implementing the corrective action needed; v) recording results of action taken; and reviewing of corrective action taken.

Documentation Requirements

It is important that the documentation clearly reflects the established food defense commitments by management, where there is a correlation between the implemented controls and the results of the risk assessment. The MSPIC documentation include (ISO/IEC 27001, 2013): a) documented statements of the MSPIC policy and objectives; b) the scope of the MSPIC; c) controls in support of the MSPIC; d) a description of the risk assessment methodology; e) the risk assessment report; f) the risk treatment plan; g) documented procedures needed by the organization to ensure the effective application of its food defense recommendations and describe how to measure the effectiveness of controls; h) records required; and i) the Statement of Applicability.

All documents are protected and controlled. A methodology is provided in order to (ISO/IEC 27001, 2013): a) approve documents prior to issue; b) review and update documents as necessary; c) ensure that changes and the current revision status of documents are identified; d) ensure that relevant versions of applicable documents are available at points of use; e) ensure that documents remain legible and readily identifiable; f) ensure that documents are available to those who need them, and are transferred, stored and ultimately disposed of in accordance with the procedures applicable to their classification; g) ensure that documents of external origin are identified; h) ensure that the distribution of documents is controlled; i) prevent the unintended use of obsolete documents; and j) apply suitable identification to them if they are retained for any purpose.

The records must be kept; protected and controlled, because they provide evidence of effective MSPIC operation. Examples of records are a visitors' book, audit reports and completed access authorization forms. The records must be kept legible, identifiable and retrievable. Procedures for the identification, storage, protection, retrieval, retention time and disposition of records are established.

SOLUTIONS AND RECOMMENDATIONS

To address the issues raised in the introduction, a proposal for a management system for the prevention of intentional contamination (MSPIC) of food, considering a vision of processes, is performed.

The system uses the continuous PDCA cycle (Planning, Doing, Checking, Acting), which are traditional systems of quality management. In the planning phase, the MSPIC is established, i.e., the risk analysis is performed and the policies, objectives and procedures are established to be managed. For the doing phase, the MSPIC is implemented, in other words, executing the policy, the controls and the procedures. In the checking phase, the MSPIC evaluates the performance through internal audits and management reviews. Finally, on the acting stage, the MSPIC is maintained and improved. i.e., there are preventive and corrective actions performed based on the previous step.

The results from this research can be listed as follows:

1. A methodological option is provided: the improvement of the prevention of the intentional contamination based on a risks analysis.
2. The proposal establishes the use of metrics for the prevention of intentional food contamination and its possible impact on other metrics (cost and time) as a link of the supply chain.
3. According to the research reference frame, the proposal has the potential to improve the security of the supply chain in the food sector of Guanajuato State, Mexico.
4. In accordance to the research scope, the proposal has the potential to improve the food security (food safety and food defense) in the food sector links in Guanajuato.

According to the analysis of the theoretical framework of the present research, there is clearly a few research evidence on the food defense in the context of sup-ply chain management and business process management, having a vision of business processes. Regarding the review of the literature on the subject in Mexico, virtually almost no research has been done on these approaches.

Specifically, there is a gap of knowledge to establish the way to manage food defense on the links in the food supply chain, according to the market requirements, the academics and what experts recommend as suitable for preventing food terrorism.

The vision of business processes revised in this chapter is one of the differential approach among this investigation and others. Instead of studying the food defense simply using a traditional departmental perspective (based on analyzing functional areas, stages of production, equipment or specific infrastructure), using a business process approach.

Indeed, the study of food defense from the point of view of business processes is considered novel because risk analysis is performed on both physical infrastructure, activities and business processes.

Although there are many research areas untouched, there is a need to look especially for new ways of managing the defense in terms of food supply chain, where it is used to strengthen the collaboration and a more effective coordination among some entities for risk prevention.

From here, some areas of interest need to be studied; because they have been touched very superficially by previous investigators, which are shown in the next section.

FUTURE RESEARCH DIRECTIONS

1. From the research process and the solution to the given problem in this chapter, and due to the limitations, a series of future research has been recognized. This could extend the scope of the chapter. Among the future research directions, the following are identified:
2. The proposal of this chapter helps to develop a deeper investigation on food defense within the context of supply chain management and business process management, and according to the literature review, there is clear evidence about the lack of evidence on food defense in this context. It could be considered to continue filling this gap in this area of knowledge, using the business process approach.
3. Some benefits through the implementation of MSPIC in the food industry have been proposed. The immediate step is to evaluate the proposal in a food business in Guanajuato, Mexico. A possible extension of the research problem would contemplate more links in the supply chain, that is, taking in consideration producers and distributors into an 'upstream' and 'downstream' scenario of the customer and the supplier, located in different places, they may also be featured as multi-linked (multiple clients), so that various types of supply chain could contemplated.
4. An auspicious extension of the research problem would be to change the application to other geographic regions in other countries or continents. It could also be applied in various subsectors of the food industry, such as the meat or milk markets.
5. The theoretical review showed the possibility of integrating the various areas of industrial activity for getting a more comprehensive and integrated view of the enterprise system. Integrated business models may consist of the following (partial) models: functional model, information model, decisional model, organizational model, resource model or economic model. The research could lead to the development of enterprise modeling to analyze the variable food defense from other points of view, as an informational or decisional one, not only from a functional vision.
6. Once improved the estimated food defense it could be correlated to other parameters such as (cost and time). For a deeper analysis, it could be considered the use of a Key Performance Indicator (KPI) of food defense systems for comprehensive performance measurement, such as balance scorecard

(BSC), the performance prism (PRISM), integrated dynamic performance measurement systems (IDPMS), dynamic performance measurement system model (DPMSM) measurement system performance IE-GIP (PSM IE-GIP) or performance measurement system for Business Processes (PMS-BP) (Alfaro et al., 2007).

7. In regard of the MSPIC proposal, some alternatives could be developed for managing a food defense system that could be innovative. Also, the development and implementation of new techniques and risk analysis that could be used given the diversity of existing tools. A line of future work would be testing different combinations of risk analysis techniques and tools. The possible benefits or difficulties that may arise can be studied from the different combinations, and how that would affect the analysis and improvement of food defense.
8. Future research direction would be the development of new alternatives for the development of a food defense KPI food that could be helpful to measure improvement.
9. The development of alternatives to improve the food defense outside the scope of the processes can also be deepened. The tools that have been developed in the food terrorism prevention systems have been done without a vision of business processes, such as ORM and the CARVER + Shock. This line of investigation is ongoing to develop new options.
10. A possible future line of research includes the development of software that helps food defense management. It is vital that the exchange of information about the threats of food terrorism between government institutions and the links in the food supply chain is made quickly and effectively. Hence the importance of research for the development of common IT-based platforms, for the communication of truthful and confidential information. The shared information may serve to adjust prevention measures in the links in the food supply chain. In this line of work, the development of tools (software) that could support the implementation of the proposals is also desirable.
11. It can be observed in the research by Nganje et al (2007, 2008) and Zhuan, et al (2007).; that some issues of coordination among some links for controlling the threats of food terrorism and the implications on the market by applying alternative for traceability in food supply chains in the dairy industry are explored. A line of future research could include new approaches to coordination and collaboration among the supply chain to solve these issues for different types of supply chains in the food industry.
12. Another line of research can refer to calculate the implementation costs for mitigating the threats of terrorism in food supply chains. Which can take various approaches of financial or mathematical type.
13. Several studies have been focused on the possible consequences of a food terrorism attack through various scenarios, such as the ones shown by Wein et al. (2005) or there have been some simulation models for responding to terrorist events in the food supply, Hartnett et al. (2009). An interesting future line of research lies in the construction of new simulation models to estimate consequences for different scenarios for planning a response.
14. One research path in the area of science and food technology involves the need to develop portable detection systems to validate intentionally contaminated food to be very sensitive and accurate, also having rapid detection systems that can be used to study the pollutant and the interaction with the food, and developing appropriate decontamination methods and technologies for the disposal of the affected products.

CONCLUSION

The PDCA is a continuous life cycle, which means that the Acting phase brings back to the Planning stage to start a new cycle of the four phases. There-fore, the system enables continuous improvement in risk prevention of intentional contamination and increased security of the food supply chain, not only in the food sector in the state of Guanajuato, but it can be extended into a security re-search plan of the food supply chain.

The proposals given in this chapter, allow giving the business companies and the managers of the supply chain, the tools for implementing the concept of defense from the perspective of food processes. This implies a great advantage, since the necessary activities for the fulfillment of the recommendations can be implemented on an adequate food defense, in an easy and systematic way.

On the other hand, managers can also provide a management system and continuous improvement of food defense that can be applied in any food supply chain, which has a potential market value to use in consulting companies.

The contributions of the chapter also have potential to be used both by government and private institutions, to assist in meeting the food defense strategies. However, a major difficulty may be that not all companies or links in the food supply chain are interested in investing resources (time and money) in a pre-ventive procedure. Therefore it is useful to consider how to 'sell' the proposal to managers, giving a great importance to the potential consequences of an attack of food terrorism in contrast to the investment is actually negligible.

It is important to note that a second phase of the research is scheduled for implementation on any particular company in the food sector. By proposing the possible application of MSPIC to a real case, the validation of the proposal is made contrasting it with the business reality.

REFERENCES

Aguilar-Savén, R. S. (2004). Business Process Modeling: Review and Framework. *International Journal of Production Economics*, *90*(2), 129–149. doi:10.1016/S0925-5273(03)00102-6

Alfaro, J., Ortíz, A., & Poler, R. (2007). Performance measurement system for business processes. *Journal of the Production Planning & Control*, *18*(8), 641–654. doi:10.1080/09537280701599772

Alvarez, M., Alvarez, A., de Maggio, M. C., Oses, A., Trombetta, M., & Setola, R. (2010) Protecting the Food Supply Chain from Terrorist Attack. In Critical Infrastructure Protection IV, IFIP Advances in Information and Communication Technology. doi:10.1007/978-3-642-16806-2_11

American Institute of Baking (AIB). (2015). *Guía para la Seguridad de la Planta.* Retrieved June 4, 2015, from http://www.aibonline.org/aibOnline_/secure.aibonline.org/php/ecomm-catalog.aspx?catalogNbr=56-1460-S&l=1

Arvanitoyannis, I., & Varzakas, T. H. (2009). Application of ISO 22000 and comparison with HACCP on industrial processing of common octopus. *International Journal of Food Science & Technology*, *44*(1), 58–78. doi:10.1111/j.1365-2621.2007.01666.x

Bowersox, D. J., Closs, D. J., & Cooper, M. B. (2007). *Supply Chain Logistics Management* (2nd ed.). McGraw-Hill.

Boyle M. (2005). A Recipe for Disaster. *Fortune International Journal, 152*(9).

Brummer, B. (2003). Food Biosecurity. *Journal of the American Dietetic Association, 103*(6), 697–691. PMID:12778038

Busta, F., & Kennedy, S. (2011). Defending the Safety of the Global Food System from Intentional Contamination in a Changing Market. In *NATO Science for Peace and Security Series A* (pp. 119–135). Chemistry and Biology, Advances of Food Protection. doi:10.1007/978-94-007-1100-6_7

Cagno, E., Caron, F., & Mancini, M. (2002). Risk Analysis in plant commissioning: The multilevel HAZOP. *Reliability Engineering & System Safety, 77*(3), 309–323. doi:10.1016/S0951-8320(02)00064-9

Closs, D. & Mcgarrell, E. (2004). *Enhancing Security Throughout the Supply Chain*. Special Report Series, IBM Center for The Business of Government.

Codex Alimentarius, CA (2003). *Código Internacional Recomendado de Prácticas-Principios generales de higiene de los alimentos*. (CAC/RCP 1-1969, Rev. 4-2003).

Crutchley, T. M., Rodgers, H. P., Whiteside, J., Vanier, M., & Terndrup, T. E. (2007). Agroterrorism: Where Are We in the Ongoing War on Terrorism. *Journal of Food Protection, 70*(370), 791–804. PMID:17388078

Customs-Trade Partnership Against Terrorism (C-TPAT). (2014). *C-TPAT's Five Step Risk Assessment*. U.S. Customs and Border Protection, Office of Field Operations. Retrieved June 2, 2015, from http://www.cbp.gov/sites/default/files/documents/

Dalziel G.R. (2009). *Food Defense Incidents 1950-2008: a Chronology and analysis of Incidents involving the malicious Contamination of the Food Supply Chain*, Centre of Excellence for National Security (CENS) of the S. Rajaratnam School of International Studies (RSIS) at Nanyang technological University, Singapore.

Damij, N., Damij, T., Grand, J., & Jelenc, F. (2008). A Methodology for Business Process Emprovement and IS Development. *Information and Software Technology, 50*(11), 1127–1141. doi:10.1016/j.infsof.2007.11.004

Department of Defense (DD). (2000). *Standard Practice for System Safety*. United Estates of America Department of Defense (MIL-STD-882D). Retrieved May 14, 2015, from http://www.system-safety.org/Documents/MIL-STD-882D.pdf

Department of Health and Human Services (DHHS). (2001). *Food safety and security: operational risk management approach*. United States of America, US Food and Drug Administration, Center for Food Safety and Applied Nutrition. Retrieved May 11, 2015, from https://www.cdph.ca.gov/pubsforms/Guidelines/Documents/fdb%20ORM%202001.pdf

Department of Health and Human Services (DHHS). (2005), *Terrorism and Food Supply. Terrorism and other Public Health Emergencies: a reference guide.* United States of America, US Food and Drug Administration, Center for Food Safety and Applied Nutrition. Retrieved April 22, 2015, from http://www.achd.net/emergprep/pubs/pdf/HHSMedisReferenceGuideFinal.pdf

Department of Health and Human Services (DHHS). (2007a). *Cosmetics Processors and Transporters: Cosmetics Security Preventive Measures Guidance*. Estados Unidos. Retrieved May 14, 2015, from http://www.fda.gov/Food/GuidanceRegulation/GuidanceDocumentsRegulatoryInformation/FoodDefense/ucm082716.htm

Department of Health and Human Services (DHHS). (2007b). *Retail Food Stores and Food Service Establishments: Food Security Preventive Measures Guidance*. Estados Unidos. Retrieved May 25, 2015, from http://www.fda.gov/Food/GuidanceRegulation/GuidanceDocumentsRegulatoryInformation/FoodDefense/ucm082751.htm

Department of Health and Human Services (DHHS). (2007c). *Food Producers, Processors and Transporters: Food Security Preventive Measures Guidance*. Estados Unidos. Retrieved May 25, 2015, from https://www.hsdl.org/?view&did=451052

Dourlens, C., Galland, J. P., & Vidal-Naquet, P. A. (1991). Introduction. In Conquête de la sécurité, gestion des risques. París: L'Harmattan.

Federal Aviation Administration (FAA). (2000). Operational Risk Management (ORM). In System Safety Handbook: Practices and Guidelines for Conducting System Safety Engineering and Management. Author.

Food and Drug Administration (FDA). (1999). *Buenas prácticas de manufactura en la fabricación, empaque y manejo de alimentos para consumo humano.* (Parte 110 -21 CFR 110.1 - 110.110 -Código de Regulaciones Federales).

Food and Drug Administration (FDA). (2015). *Tools and educational materials of food defense*. Estados Unidos. Retrieved June 1, 2015, from http://www.fsis.usda.gov/wps/portal/fsis/topics/food-defense-defense-and-emergency-response/preparation-and-prevention/guidance-documents

Food and Drug Administration (FDA). (2007). *An overview of the carver plus shock method for food sector vulnerability assessments*. Estados Unidos. Retrieved June 1, 2015, from http://www.fsis.usda.gov/wps/wcm/connect/483f86d5-a566-44f8-90d5-05a16dbe3f78/CARVER.pdf?MOD=AJPERES

Food and Drug Administration (FDA). (2012). *Vulnerability Assessments of Food Systems. Final Summary Report*. Estados Unidos. Retrieved June 1, 2015, from http://www.fda.gov/downloads/Food/FoodDefense/UCM317547.pdf

Gaitanides, M. (2007). *Prozessorganisation –Entwicklung, Ansätze und Programme des Managements von Geschäftsprozessen* (2nd ed.). Munich: Vahlen.

Hartnett, E., Paoli, G. M., & Schaffner, D. W. (2009). Modeling the Public Health System Response to a Terrorist Event in the Food Supply. *Society for Risk Analysis, Journal of Risk Analysis*, *29*(11).

Hintsa, J., Gutiérrez, X., Hameri, A. P., & Wieser, P. (2009). Supply Chain Security Management: An overview. *International Journal of Logistics Systems and Management*, *5*(3/4), 344. doi:10.1504/IJLSM.2009.022501

ISO/IEC 27001. (2005). *Estándar Internacional: Tecnología de la Información-Técnicas de Seguridad-Sistemas de Gestión de la Seguridad de la Información*. ISO.

ISO/IEC 27001. (2013). *Estándar Internacional: Tecnología de la Información-Técnicas de Seguridad-Sistemas de Gestión de la Seguridad de la Información*. ISO.

ISO 22000. (2005). *Food safety management systems – Requirements for any organization in the food chain*. Geneva. Switzerland: ISO.

ISO/DIS 9001. (2015). *Draft International Standard – Quality management systems – Requirements*. Geneva. Switzerland: ISO.

Kleindorfer, P. R., & Saad, G. H. (2005). Managing Disruption Risks in Supply Chains. *Production and Operations Management*, *14*(1), 53–68. doi:10.1111/j.1937-5956.2005.tb00009.x

Kleindorfer, P. R., & Van Wassenhove, L. (2004). Risk Management in Global Supply Chains. In H. Gatignon & J. Kimberly (Eds.), *The Alliance on Globalization*. Cambridge University Press.

Kleter, G. A., & Marvin, H. J. P. (2009). Indicators of emerging hazards and risk to food safety. *Food and Chemical Toxicology*, *47*(5), 1022–1039. doi:10.1016/j.fct.2008.07.028 PMID:18761390

Lyonga, A.N., Sellnow T.L., Nganje, W.E. & Venette S. (2006). Insider Agroterrorism Threat and High Reliability in Turkey Processing. *Journal of Food Protection Trends*.

Monke, J. (2004). *Agroterrorism: Threats and Preparedness*. Paper presented at Congressional Research Service; CRS Report for Congress.

Navarrete, R. (2012). *Modelo de Bioseguridad en la Cadena de Suministros de Productos Alimenticios, teniendo en cuenta la Gestión de la Cadena de Suministros y la Visión de Procesos de Negocio. Aplicación a la Industria Alimenticia, de la Zona del Bajío (México)*. (Unpublished doctoral dissertation). Universidad Politécnica de Valencia, España.

Navarrete, R., & Lario, F. C. (2009). Modelo de gestión de la bioseguridad, basado en procesos de negocio de la cadena de suministro. Aplicación en la industria mexicana de los alimentos. In *Proceedings of XIII Congreso de Ingeniería de Organización, 3rd International Conference on Industrial Engineering and Industrial Management*.

Navarrete, R., & Lario, F. C. (2010). Propuesta de una Metodología para el Modelado AS IS y TO BE de Procesos de Negocio de Bioseguridad (Terrorismo Alimentario), dentro del Contexto de la Cadena de Suministro. Aplicación en la Industria Mexicana Alimentaria. In *Proceedings of XIV Congreso de Ingeniería de Organización, 4th International Conference on Industrial Engineering and Industrial Management*.

Navarrete, R., & Lario, F. C. (2011). Propuesta de un Procedimiento para la Elaboración de un KPI para la Medición de la Bioseguridad en Procesos de Negocio de la Cadena de Suministro Alimenticia. Aplicación en la Industria Mexicana Alimenticia. In *Proceedings of 5th International Conference on Industrial Engineering and Industrial Management, XV Congreso de Ingeniería de Organización.*

Nganje, W., Bier, B. V., Han, H., & Zack, L. (2008). Models of Interdependent Security along the Milk Supply Chain. *American Journal of Agricultural Economics*, *90*(5), 1265–1271. doi:10.1111/j.1467-8276.2008.01215.x

Nganje, W., Dahl, B., Wilson, W., Mounir, S., & Lewis, A. (2007). Valuing Private Sector Incentives to Invest in Food Security Measures: Quantifying the Risk Premium for RFEM. *Journal of International Agricultural Trade and Development*, *3*(2), 199–216.

Onyango, B., Turvey, C., & Hallman, W. (2005). *Public Attitudes and Perceptions of the Vulnerability of the U.S. Food Chain to Agroterrorism*. Paper presented at American Agricultural Economics Association annual meeting, Providence, RI.

Pennsylvania State University (PSU). (2010). *The basics of food defense*. Retrieved May 18, from http://www.fsis.usda.gov/shared/PDF/7_Food_Defense.pdf?redirecthttp=true

Peters, K. (2003). *Officials Fear Terrorist Attack on U.S. Food Supply*. Retrieved April 2015, from http://www. govexec.com/ dailyfed/0603/061003kp1.htm

Pistole J. (2006). Preparing for the Unthinkable: Agroterrorism. *Food Business News*, (1), 23-26.

Qing, L., Shao-ping G., Ming, Z., Shi-zhou, C., & Qian Y. (2008). American Food Defense Plan and Hazard Analysis and Critical Control Point. *Chinese Journal of Food Hygiene*.

Rao, S., & Goldsby, T. J. (2009). Supply Chain Risks: A Review and Typology. *International Journal of Logistics Management*, *20*(1), 97–123. doi:10.1108/09574090910954864

Rasco, B., & Beldsoe, G. (2006). *Bioterrorism and Food Safety*. Boca Ratón, FL: CRC Press.

Setola, R., & de Maggio, M. C. (2009). *Security of the Food Supply Chain*. Paper presented at 31st International conference of the IEEE EMBS.

Surak, J. G. (2005). ISO 22000: *Requirements for food safety management systems*. Paper presented at World Conference on Quality and Improvement.

Surak, J. G. (2007). A Recipe for Safe Food: ISO 22000 and HACCP. *Quality Progress*, *40*(10), 21–27.

Surak, J. G., & Lorca, T. A. (2007). *Moving from HACCP to ISO 22000*. Paper presented at World Conference on Quality and Improvement.

Swanson, L. (2003). An information-processing model of maintenance managemen. *International Journal of Production Economics*, *83*(1), 45–64. doi:10.1016/S0925-5273(02)00266-9

United States Department of Agriculture (USDA). (2004). *A Biosecurity Checklist for School Foodservice Programs*. Retrieved May 24, 2015, from http://education.ky.gov/federal/scn/documents/biosecurity%20 checklist.pdf

United States Department of Agriculture (USDA). (2008). *Guide to Developing a Food Defense Plan for Warehouse and Distribution Centers*. Retrieved May 20 2015, from http://www.fsis.usda.gov/shared/PDF/Guidance_Document_Warehouses.pdf

United States Department of Agriculture (USDA). (2009a). *Food Defense Guidelines for Slaughter and Processing Establishments*. Retrieved May 26, 2015, from http://www.fsis.usda.gov/wps/wcm/connect/cd18dfb5-9443-42f8-b8c5-cadf862fcbc4/SecurityGuide.pdf?MOD=AJPERES

United States Department of Agriculture (USDA). (2013). *Food Defense Guidelines for the Transportation and Distribution of Meat, Poultry, and Processed Egg Products*. Biosecurity Guidelines. Retrieved May 24, 2015, from http://www.fsis.usda.gov/wps/wcm/connect/4f9d737a-1f3e-49ff-851b-74884fa946bd/Transportation_Security_Guidelines.pdf?MOD=AJPERES

United States Department of Agriculture (USDA). (2009b). *General Food Defense Plan, 2009*. Retrieved June 2, 2015, from http://www.fsis.usda.gov/wps/wcm/connect/99f95182-0c9e-4214-9762-e98197f54ebf/General-Food-Defense-Plan.pdf?MOD=AJPERES

Valle, P., Girard, A. & Saldate, O. (2007). Defensa Alimentaria "Food Defense". *Revista Mundo Lácteo y Cárnico*.

Vanderhaeghen, D., Fettke, P. & Loos, P. (2010). Organizational and Technological Options for Business Process Management from the Perspective of Web 2.0. *Business & Information Systems Engineering*, (1), 15-28.

Veiga, A. (2011). Food Defense and Security: A New Reality. In NATO Science for Peace and Security Series C: Environmental Security, Environmental Security and Ecoterrorism (pp. 39-54). Academic Press.

Wein, L. M., & Liu, Y. (2005). Analyzing a Bioterrorism Attack on the Food Supply: The Case of Botulinum Toxin in Milk. *Proceedings of the National Academy of Sciences of the United States of America, 102*(99), 84–89.

World Health Organization (WHO). (2003). *Bioterrorismo: la amenaza en el continente americano*. Paper presented at Reunión Interamericana a Nivel Ministerial en Salud y Agricultura.

World Health Organization (WHO). (2008). *Terrorist Threats to Food: Guidance for Establishing and Strengthening Prevention and Response Systems*. Department of Food Safety, Zoonoses and Foodborne Disease, Cluster on Health Security and Environment.

Zhuang, J., Bier, V. M., & Gupta, A. (2007). Subsidies in Interdependent Security with Heterogeneous Discount Rates. *The Engineering Economist, 52*(1), 1–19. doi:10.1080/00137910601159722

KEY TERMS AND DEFINITIONS

Food Defense: Is the effort to protect the food supply against intentional contamination due to sabotage, terrorism, counterfeiting, or other illegal, inten-tionally harmful means. Potential contaminants include biological, chemical and radiological hazards that are generally not found in foods or their production environment.

Food Terrorism: An act or threat of deliberate contamination of food for human consumption with chemical, physical or microbiological agents for the purpose of causing injury or death to civilian populations and/or disrupting social, economic or political stability.

ISO/IEC 27001: Is an international standard that provides the basis for effective management of confidential and sensitive information, and for the application of information security controls.

Management System for the Prevention of Intentional Contamination (MSPIC): Is an overarching management framework through which the organization identifies, analyzes and addresses its food defense risks.

Risk Management: Is the identification, analysis, assessment, control, and avoidance, minimization, or elimination of unacceptable risks. An organization may use risk assumption, risk avoidance, risk retention, risk transfer, or any other strategy (or combination of strategies) in proper management of future events.

PDCA Cycle: Is an iterative four-step management method used in business for the control and continuous improvement of processes and products. It is also known as the Deming circle/cycle.

Statement of Applicability: Documented statement describing the control objectives and controls that are relevant and applicable to the organization's MSPIC.

Supply Chain Management: As a management philosophy has the following characteristics: i) a systems approach to viewing the supply chain as a whole, and to managing the total flow of goods inventory from the supplier to the ultimate customer; ii) A strategic orientation toward cooperative efforts to synchronize and converge intraorganizational and interorganizational operational and strategic capabilities into a unified whole; and iii) A customer focus to create unique and individualized sources of customer value, leading to customer satisfaction.

Supply Chain Risks: Any risks for the information, material and product flows from original supplier to the delivery of the final product for the end user.

Supply Chain Security: The application of policies, procedures and technology in order to protect the supply chains goods from theft, damage or terrorism and to prevent the introduction of unauthorized contraband, people or weapons of mass destruction along the entire supply chain.

ENDNOTE

1 Some concepts of this proposal are taken and adapted from ISO/IEC 27001 (2013).

Chapter 5
Clio-Combinatorics: A Novel Framework to Analyze Military Logistics Choices Using Operations Research Techniques

Jesus Gonzalez-Feliu
Ecole des Mines de Saint-Etienne, France

Antoine Parent
Sciences Po Lyon, France

ABSTRACT

This chapter proposes to apply combinatorial optimization to past military conflicts with the aim of producing quantitative data that help explaining history. To do this, we can go beyond the classical "problem solving" vision of operations research that focuses on algorithmic development and computation analysis to privilege solution analysis and the needs of matching the obtained solution to the reality we aim to represent, study and analyze. In particular, we propose an iterative logic search method that aims to identify and analyze military strategic logic in terms of logistics is presented. Then, to illustrate it, an application to the French troop assignment plan (Plan XVII of Joffre, 1932) is made to analyze which could be the subjacent logic behind the defense plan of French troops and state on the consequences of the optimization choices in terms of regional distribution of troops. A discussion of the proposed framework and the directions to generalize it will be presented as a conclusion.

1. INTRODUCTION

Operations research and combinatorial optimization are often used in many business and management fields, and are one of the main groups of techniques used to assess performance and efficiency in enterprise and organization planning and management, at different horizons (strategic, tactical, operational and executive). Moreover, the application of operations research techniques is mainly related to logistics and resource management, since they derive from military uses (although the discipline is said to be formalized during the Second World War for the optimal location of radars in the English armies, many

DOI: 10.4018/978-1-4666-9779-9.ch005

documents show that such techniques were already used for military logistics and supply optimization in the precedent conflict, i.e. the first world war). Coinciding with the celebration of the first century of the First World War, and the increasing phenomenon of quantitative history, we aim to propose a new vision of operations research and statistics to retrospectively apply them to past war conflicts, completing the existing field constituted by cliometrics and cliodynamics by introducing clio-combinatorics, i.e. the deployment of combinatorial optimization to identify and understand military strategic logistics in past war conflicts.

The aim of this chapter is twofold. The first contribution is to introduce the need and opportunity of applying combinatorial optimization to past military conflicts in order to produce data that help explaining history, and present a systematic framework to this purpose. But to do this, we can go beyond the classical "problem solving" vision of operations research that focuses on algorithmic development and computation analysis to privilege solution analysis. For that reason, we propose here a framework deriving from the "solution probleming" vision (in the sense of Ackoff, 1976) to identify and analyze military strategic logic in terms of logistics is presented. Then, to illustrate it, an application to the French troop assignment plan (Plan XVII of Joffre, 1932) is made to analyze which could be the subjacent logic behind the defense plan of French troops and state on the consequences of the optimization choices in terms of regional distribution of troops. A discussion of the proposed framework and the directions to generalize it will be presented as a conclusion.

2. EXPLAINING HISTORICAL FACTS BY QUANTITATIVE METHODS: FROM CLIOMETRICS TO CLIO-SCIENCES

A good definition of cliometrics to start with is the one by Avner Greif, 'Cliometrics after 40 years' (1997): "As an intellectual movement, cliometrics aspired to enhance the study of past economies by subjecting them to the rigor of economic theory and quantitative analysis, while utilizing the richness of history to evaluate and stimulate economic theory and improve our comprehension of long-run economic processes. The contribution of this approach is immeasurable: it has altered and enriched our perceptions regarding numerous issues in economic history (p. 400)".

It has to be noted that the topic of economics of transportation became the heart of the development of cliometric, since Fogel's (1964) seminal book *Railroads and American economic growth*. Curiously, the usual methodology in this field of research, namely operational research which currently applies to spatial economics and economics of transportation today, is totally absent of Fogel's approach. We want here to fill a gap in the cliometric literature by emphasizing the necessity to revisit standard cliometrics of transportion by introducing in this approach the means of operation research. By doing so, we are in line with the original cliometric purpose as developed by Fogel and North, since we wish to innovate by developing a new branch of cliometrics of transports, that we call clio-combinatorics. We shall explicit below what makes the originality and continuity of our research program. Fundamentally, we rely on the distinction initially established by Fogel, between theory and modeling, and argue that the operation research delivers a good methodological framework to assess the behavior of the agent in the past, whatever the type of action he undertakes.

Clearly, cliometrics is indebted to the economics of transportation. Goldin (1994) recalls that cliometrics was partly born from questioning conventional wisdom that the railroad was the engine of economic growth in 19th century America and caused it. Implicitly, this supposes that the growth would have been considerably less in the absence of the railroads. Fogel (1964) restated these claims as a "counterfactual". According to Fogel, the historians advocating the theorem of indispensability of the railroads were implicitly suggesting this experiment. Fogel became the one who made the claim explicit by submitting it to counterfactual analysis. One can say that the birth of cliometrics partly coincided with Fogel's early work on transportation.

As stated by Goldin (1994), "Fogel's *Railroads and American Economic Growth* was an immediate classic. Derived from his doctoral dissertation at John Hopkins University, it is a volume in search of a single number, the social savings of the American railroads in 1890. By the term 'social savings', Fogel meant the increase in social surplus from the decrease in transport costs". Fogel distinguished the "interregional" social savings from the "intraregional" one. Interregional transport referred mainly at the end of the 19th century to the transport between the food exporting cities of the Midwest to the food importing cities of the East. Fogel calculated that if in 1890 the railroads had not existed and if the four major agricultural goods (wheat, corn, pork and beef) were transported on existing water routes, gross national product would have declined by less than 0.6%. These preliminary results from the interregional social savings calculation were presented by Fogel to the first Cliometrics meetings held at Purdue University in December 1960. We are not going to discuss here the relevancy of such outcomes which have been source of lots of debates in cliometrics. It is clear that the robustness of the outcomes depends on the robustness of the modeling. Fogel used a partial equilibrium model that Williamson (1974) and O'Rourke and Willamson (1999) disputed: using computable general equilibrium model, these authors found, conversely to Fogel, that the sector of railroads had been crucial to the 19th century US economic development. We just want here to recall the genesis of the economics of transportation in cliometrics, not calling into question the validity of Fogel's outcomes which seem very dubious today. *Per se*, Fogel's result is not surprising that much in the case of US: the US had plenty of navigable waterways and the cost of water transportation was at that time much lower than for railroads, even if, most of the time, it required also some land transport. Fogel gives evidence in his book that for instance Mexico lacked navigable waterways and consequently interregional social savings were more important in this country.

As noted by Goldin (1994), "in some sense the real controversy, and the more difficult calculation, concerned the social savings of the railroads within regions, termed the 'intraregional' social savings. The intraregional calculation involved the tough-experiment of removing all the railroads that transported goods from farms to the market of the Midwest and computing how much gross national product would have declined. The railroads had more costly alternatives here than they did for the long-haul routes". Fogel (1964) made four calculations of the within-region social savings, depending upon various assumptions concerning the alternative mode of transportation and the area cultivated. The alternative modes involved either existing water and land routes, along with a set of canals that would probably have been built had the railroads not existed. Fogel contends that the higher the costs of transportation, the bigger the cultivated areas of the country. He estimated the intraregional social savings to be around 1% of 1890 GNP. Globally, the total social savings from shipping agricultural commodities both between and inside regions is supposed to represent less than 2% of GNP in 1890. Agricultural commodities, by ton-miles shipped, represented roughly one fourth of all goods shipped by rail. According to Fogel, multiplying the agricultural social savings figure by four to get an exact idea of the total productivity gains generated by railroads in 19th century economic growth America would be incorrect.

Rather than the robustness of these peculiar outcomes, the intellectual approach of Fogel deserves to be examined more carefully: the goal is mainly to kill a myth (here, the myth of the American frontier) thanks to the tool of counterfactual analysis. Let us now refer to Fogel's own analysis of cliometrics objectives.

Assessing his own contribution to economic history, Fogel (1970) first noted that "the introduction of mathematical methods has not resulted in a narrowing of the range of issues treated by the discipline" (p. 246). In one of his most famous quotations, he recalls that: "The most distinctive feature of the new economic history is its introduction of mathematics (equations) into what used to be a literary discipline. Mathematics has always been present in economic history. In the past, however, it was implicit, covert and subliminal. The contribution of cliometricians is that we have made implicit mathematics, explicit (p. 252)". He devoted a full paragraph to "The problems of counterfactuals" (p. 256). For most of the critics, the reprehensible objective of cliometrics is the importance given to counterfactual analysis. Many historians assimilate this approach to "quasi-history" (false history). Fogel's answer is straightforward: "Historians who write that Wilson or Johnson made mistakes are implicitly asserting that the particular course of action that each man followed was inferior to an alternative that was available to him. In branding actual behavior as an error, one presumes knowledge of the course of events in situations that never occurred… The crucial step in the verification of a counterfactual statement is the determination of the empirical validity of the explicit or implicit equation (p. 258)".

Let us consider now Douglass C. North, the other Nobel prize of the field. This author underlines that "The great advantage of the new economic history (NEH hereafter) is that it uses simple theory which can produce predictable and specific results rather than indeterminate consequences" (North, 1970, p. 188). Should we fear to rest on a behavioral assumption of rational, calculating human beings who make explicit choices, at any period of history? Is this methodological break with the old economic history plausible? North's answer is straightforward, as Fogel's one was: "Systematic advances to our understanding of the past require explicit theory and systematic testing procedures… The NEH is a beginning, even if a very modest beginning, towards such an approach. Its fundamental contribution is to offer a consistent theoretical analysis of human behavior in the past, which offers the promise of advancing our knowledge about man's economic past" (North, 1970, p. 192).

Let us turn now towards another text of Douglas C. North (co-authored with John V.C. Nye), written forty years after the advent of the cliometric revolution. Those passages (North & Nye, 1997) do not reveal any change in spirit with the quotations mentioned above but show the invariance in the goal and tools of Cliometrics:

Although there was a surprisingly large amount of quantitative history before cliometrics much of it was a-theoretical. Four decades after the advent of the cliometric revolution, we should be adding a missing dimension to the two fields from which we have evolved. To history we should provide the analytical rigor of sound theory and quantitative methods. To economics we must add the dimension of time and an analytical account of the way an economy evolves through time that is fully cognizant of the evolution of political as well as economic institutions (North & Nye, 1997, p. 1).

Forty years after the birth of the discipline, the objectives of cliometrics remain the same. Cliometrics, asserted the Nobel prize, "is not a radical new methodology" (p. 9)… "Our work combines the strengths of both disciplines, [ie. economics and history]" (p. 10). We want to stress that our research project on "clio-combinatorics" is perfectly in line with what promotes DC. North. We are not inventing anything new, but simply advocate for using formal theory to illuminate the study of war and economics of transportation thanks to operational research.

North and Nye (1997, p. 13) established a second distinction that we intend to make ours as well: "We need to understand that modeling is not the same thing as theorizing. The profession has so distorted the idea of theory that it has become synonymous with mathematical modeling… But in fact, modeling makes sense only after the process of understanding and developing hypotheses into coherent, though not formal theories. Mathematical modeling is only one way of generating useful claims about the economy. According to Drobak and Nye (1997), Fogel said: *"30 years ago it was common to call courses in economic theory 'economic theory' and to call courses that dealt with mathematical models of economic behavior 'mathematical economics"* (in Drobak & Nye, 1997, p. 13). In the 1950s and 1960s, course in economic theory was presumed to summarize the body of generalizations about the operation of the economy that economists thought were especially valid'. History needs to be seen as an important source of such generalizations… But since the cliometrics revolution, there has been a tendency to cede the theoretical ground to modelers". Then, this author supports his own contribution to cliometrics as follows: *"The institutional narrative is a method of reasserting our rightful role as creators of new hypotheses and new theories"*. We invite the reader to replace "institutional narrative" (the core of the research and contribution to economic history of North) by "clio-combinatorics" to understand that our research project fully matches the original goal enacted to Cliometrics by North. Indeed, our project is threefold:

- Firstly, as professed by North, we take into account the contemporary institutions to assess the rationality of high command during the Great War, namely the existing railroads network and the military forces (regiments) in presence. Our research program cannot be characterized as ahistorical.
- Secondly, operation research is a powerful way to systematize the hypothesis of economic rationality, without imposing any form of modeling. Using optimization techniques is an appropriate (and entirely new) tool to assess decision making and rational choices taken in the past by the high military command. Notably, did the high command take correct decisions regarding the existing railroads network during WW1? We need to mention here that operation research stands out from a pure neo-classical modelling (implying the institution of the market), since it purely relies on a very simple hypothesis of rational choice.
- Thirdly, we are part of the cliometric tradition as enacted by North since, in order to reinforce our knowledge of the past, we use operational research, which is an unprecedented tool in cliometrics. In this project, we believe that nothing is more central and more likely to reinvigorate the field of cliometrics of transportation than "clio-combinatorics". The main contribution of the New Economic History (NEH) has been the systematic use of theory and quantitative methods to history. In the words of North (1974) in his presidential address *"The NEH has made a significant contribution to revitalizing the field and advancing the frontiers of knowledge"* (p. 1). We believe that operational research applied to history (most notably of war and transportation) will be a powerful tool of analysis to shed light upon our economic past (as other economic theoretical tools have already repeatedly demonstrated when applied to history). Accordingly to North (1974) in his Presidential address, we deeply believe that *"few of man's economic problems are new, that*

most have recurred endlessly in the past" (p. 5). According to us, this is a sufficient motive to give a new impulse to the cliometrics of transportation and war. Nevertheless, by doing so, we are well aware that the critics of historians against rational economic motivation as a behavioral assumption still apply.

In his presidential address, Douglass C. North (1974) surveys some usual shortcomings in cliometrics:

How can one talk seriously about the economic past without an explanation for non-market decision making? Neo-classical economic theory has two major shortcomings for the economic historian. One, it was not designed to explain long-run economic change; and two, it provides quite limited answers since it is immediately relevant to a world of perfect markets (that is perfect in the sense of zero transaction costs): the costs of specifying and enforcing property rights. Accordingly, a theoretical analysis of the changing rules of the game is at the very core of the subject matter of economic theory. (North, 1974, p. 3)

Thus, North paves the way for what cliometrics should be ideally, i.e. a theory of social change. Nevertheless, he remains well aware of a certain number of limits inherent to the use of standard economics to history. The same reservations about NEH can be found in his *New Economic History After 20 Years* (North, 1977):

The neoclassical economic theory upon which most of the new economic history is based offers a sophisticated and powerful set of tools to analyze the operations of markets under the conditions of zero transaction costs. It assumes a frictionless world... The problem for the economic historian is that transaction costs are precisely at the center of the economic historian's problems... The rules of the game that specify the conditions under which exchange takes place (in effect the costs of specifying and enforcing a body of property rights) are never neutral.l (p. 193)

Neoclassical theory deals with the operation of markets. Yet, throughout history, and even in our society today, economic resources to a large extent have not been allocated through markets. As a result, neoclassical theory has had very little to say about many important forms of resource allocation; specifically allocation of resources by households, by a vast array of voluntary organizations, including corporations and other kinds of firms, guilds, and trade unions that have characterized economic activity throughout history... the failure to theorize about resource allocation outside organized markets precludes the serious study of most economic history. (p. 194)

It seems that that nothing has changed from this seminal article regarding the objectives of Cliometrics. What should we keep and reject for the New Cliometrics we intend to promote with Clio-Combinatorics?

We recognize ourselves in the definition given by Meiners and Nardinelli (1986), who characterize cliometrics as follows:

The method used by cliometricians reflects the belief that the effects of cultural factors should be measured as a residual after the effect of changes in economic variables have been estimated... The history of cliometrics can be described as he spreading use of economic reasoning and methods in the study of history. Indeed, the history of cliometrics could well be described as the continual reduction of the "unexplained residual. (Meiners & Nardienelli, 1986, p. 513)

This movement continues: what we want to do is extending explicit economic modeling relying on operation research between the two traditional branches of Cliometrics (retrospective econometrics and institutional cliometrics focusing on institutional change, property rights, and law and economics). We are well aware that the reliance on neo-classical economics may have constrained economic history analysis by limiting the issues that could have been examined. We accept D. North view when he wrote: "it is the systematic use of standard neoclassical economic theory which both provided the incisive new insights into man's economic past and also serves to limit the range of enquiry" (1974, p. 1). But, as mentioned above, using operational research is a useful tool to assess the decision making of high command during the war. In this case, the usual criticism according to which the ahistorical nature of cliometrics reflects theoretical assumptions rather than empirical observations does not apply. Indeed, regarding the relevance of decision making of military authorities, their correct (or wrong) use of contemporary institutions (such as the railroads network for troupes transport), using operation research appears as a prerequisite to better evaluate the rationality of military choices at war.

We are deeply aware that the use of a single body of theory to direct all economic history analyses for all issues in all past and present economies would be misleading. But this is not our purpose here. Our will is to fill a gap in the literature, by introducing operation research, in the field of cliometrics of transport and war, two areas where, inexplicably, they were still absent. By that way, we manage to provide useful knowledge in economic history, in the sense given by James J. Heckman in *The value of quantitative evidence on the effect of the past on the present*, (Heckman, 1997): "*What is the useful knowledge provided in economic history? To place current development in the economy in historical context*". To do this in the case of World War 1 and using combinatorial optimization techniques, it is important to set the vision we aim to give to operations research applications and to set the contextualization bases that will place the clio-combinatorics analysis in its historical context, as will be explained below.

3. OPERATIONS RESEARCH, OBJECTIVENESS, APPLICABILITY AND NEW OPPORTUNITIES

Operations research is nowadays mainly related to what most scholars call "problem solving", i.e. the identification then the formalization of a problem and the search of the optimal solution to the given problem formulation. However, and as already signaled by Ackoff (1977), problem solving is associated to a quest of objectiveness that often increases the gap between the mathematical vision of the problem and the represented reality, sometimes for computational needs, other times as a limitation of this objectiveness' needs. Moreover, the classical problem solving approaches are focused on the quality of the algorithms in finding an optimum with respect to a mathematical formulation without questioning its real meaning. In other words, problem solving approaches focus in formalizing then finding optima to those formulations without a return to reality to examine if the chosen representation of the reality is relevant and coherent and if the solution is satisfactory in meeting the modelling and probleming aims (Ackoff, 1977).

Ackoff (1977) proposes an alternative to problem solving, that of "solution probleming", which is not in opposition to the vision above but completes it by relating the obtained solutions to both elements: the first is the model itself and its capacity to represent the observed reality (i.e. the pertinence of the chosen model and its representativeness); the second is the capacity of the solution to meet the objectives of the analysis. In other words, in solution probleming, once a solution is found the problem is not "solved", but the real analysis starts, aiming to better represent reality and giving relevance and robustness with respect to both the empirical data collected to its construction and the theory that the model will refer to (Donnelly et al., 2012). Although this can mean being ease of use while respecting a minimum computation performance (Cordeau et al., 2002), other aspects would be considered, such as the capability of the model to react in a similar way we interpret the reality (i.e. the behavior of the model reflects the reality we observe, c.f. Barlas, 1996).

Opposing to problem solving, "solution probleming" as expression is not very used in literature (less than 10 papers using the expression resulted from a Google Scholar search) but the idea of analyzing the potential of a solution to well fit the users' needs and relating the model (and the solution) to reality by examining its capacity to represent this reality have been treated by a few hundreds of works[1]. Those works (which a significant illustration can be found in Zbilut and Giuliani, 2007) can be then put under the banner of solution probleming and constitute its main families and variants[2]. However, the general methodological patterns of "solution probleming" can be related to a general macro-structure. We aim to present it summarized in the following 8 steps:

1. Definition of the aims and scope of the analysis.
2. Representation of the reality and subsequent modelling.
3. Data production and definition of situations to assess using the chosen model.
4. Choice of a first solving method for the proposed model applied to the produced data and situations.
5. Robustness analysis of the algorithm.
6. Pertinence, relevance, coherence and robustness analysis of the solution with respect to the aims and scope of the analysis.
7. Representativeness analysis of the model with respect to the observed reality.
8. Eventually, reformulation of the problem, new mathematical formation of the model or revision/change of the solving method and iterative analyses until converging to a satisfactory combination (problem, model, solving method and solution).

This explicit description of solution probleming methodological patterns leads to question the pertinence of operations research as well as several potential applications. Although currently used mainly on a "total optimization" vision, for example in freight transport (Taniguchi et al., 2001, Barnhart & Laporte, 2007), operations research (which was born for military tactics reasons) can also be applied with other aims to different fields, like in logistics for simulation and evaluation purposes (Gonzalez-Feliu et al., 2014; Macharis et al., 2014; Taniguchi & Thompson, 2015). Most of current applications are however of forecasting and prevision nature (for both total optimization or simulation/assessment applications).

Here, we think that a new use of operations research can be found, not for forecasting goals but with a retrospective perspective. Since other quantitative methods and frameworks have already been applied to explain history (and more precisely economic history), why not apply operations research to explain past decision making processes?

In this context, and since operations research was born for military reasons, we think that the main focus of this retrospective vision has to be related to military actions. First because military choices and reasoning related to war tactics uses since WWII operations research methods to support decisions, but also because the date of birth of operations research (which can be considered as being in 1937 according to Ackoff and Sasinieni, 1968) is that of the formal definition of the discipline and the moment were a name was given to the set of methods and frameworks that were already used, in the same form or in preliminary declinations, in past actions.

We need to recall that History counts several "military tactics geniuses" before World War II, like Alexander the Great, Attila, Julius Caesar or Napoleon Bonaparte among others. And that the Ancient Greece was known for its "mathematics and physics erudite personalities" like Archimedes and Pythagoras, that make us think that mathematics were already used in military tactics even in the Ancient World. Moreover, and focusing on World War I, the Joffre/Schlieffen opposition (named like this for the names given respectively to the French and German troop allocation plans deployed at the beginning of the conflict) is one of the main research subject in military-based war history (Audoin-Rouzeau & Becker, 2004; Joll & Martell, 2007; Sondhaus, 2011; Ehlert et al., 2014). Following the cliometric approach as developed in section 1, it seems relevant to extend mathematical and computational methods (and more precisely those associated later to operations research) to the study of the understanding of World War 1. This is a way to assess retrospectively the depth of high military command choices and strategy.

To what extent can we apply operations research to historical analysis purpose? Actually, techniques nowadays included inside operations research compendiums were already used for military reasons before WW1, like assignment and matching algorithms, supply problems, base location or scheduling-based applications. However, and since operations research was born for military logistics support then widely extended to civil logistics questions, we think that most military logistics decisions in history can be assessed using those methods. The main actions that would concern WW1 for a possible retrospective decision analysis are the following:

- Troop allocation: the choice of the best troop allocation plan that meets the Military Command aims (in a solution probleming vision).
- Military transport: design and development of the military transport network (essentially rail-based), troop transport routing and scheduling.
- Supply network: estimation of logistics needs and allocation to the different concentration points, military supply chain design, definition of supply plans and stock management actions.

In this chapter, we will focus only in troop allocation and transport questions. However, we aim to propose a generic framework (clio-combinatorics) that would be applied to the different fields of military history. This framework can be defined as the "iterative methodical logic search" method and is presented in next section.

4. THE "ITERATIVE METHODICAL LOGIC SEARCH" METHOD: PRINCIPLES AND APPLICATIONS

As said in section 2, clio-sciences (i.e. all subdisciplines and research fields that derive from cliometrics) have seen in the last years important evolutions (mainly with the renewal of classic cliometrics and their application to microeconomic and sectorial fields) and the development of cliodynamics.

In this context, and when dealing with the history of military economics and management, operations research can play an important role (as also shown above, in section 3). More precisely, combinatorial optimization and combinatorics in general seem useful to identify and analyze several military choices that have been taken from a large set of alternatives. In a way, the principle is the same as that of current combinatorial optimization (for present problems or forecasting issues), but the objectives are different. Indeed, the main aim of clio-combinatorics is not to fully optimize an organization (here, a military schema) but to find the most suitable choice taken into account in a given historical context.

In our aim of setting the bases of clio-combinatorics, we will start to present the method we propose for analyzing the military choices behind the French Troop Allocation Plan (Plan XVII of Joffre, 1932) and illustrate it via an example[3]. Then, in next section, we will try to theorize (as on classical case study research, Eisenhardt, 1982) in order to propose a first overview and set the bases of clio-combinatorics

4.1. Goals, Hypotheses, and Method Description

The goals of the "iterative methodical logic search" method (IMLS) are to reproduce the most literally a historical choice, using simple algorithms that could be deployed in the historical period where the choice was made. More precisely here, we aim to reproduce the results of the Plan XVII of Joffre (1932), i.e. the French troop allocation plan, using an algorithm that could be done without the help of complex computation infrastructures such as is nowadays. The resulting algorithm will then be linked with a logic, and more precisely, an optimization objective, model and development pattern, which will be extrapolated to define the logic of the algorithm and then the logic behind the troop allocation plan of Joffre (1932).

This method supposes that the choice and its consequences are known, i.e. we know the troop allocation plan and can reproduce it in any mathematical representation

The method is articulated as follows:

1. Setup of the hypotheses and assumptions
2. Mathematical formulation of the problem: decision variables, objective function and constraints
3. Problem resolution using an exact method. Comparison of the solution obtained with the exact method and the real troop allocation plan and gap estimation.
4. Solution probleming phase 1: Interpretative analysis to find the causes of the gap and identification of the main variables/procedures that need to be adapted or change
5. Solution probleming phase 2: Development of a heuristic algorithm able to implement the changes needed as found in step 4.
6. Comparison of the solution obtained by the new algorithm to the real troop allocation plan and gap estimation.
7. If the gap is considered satisfactory (0 gap or a small value able to be considered as the best possible with the given resources), the solution of the final version of the algorithm is retained. Elsewhere, steps 5 to 7 are repeated until finding a solution considered satisfactory (i.e. that reproduces in a realistic way the real troop allocation plan).
8. Analysis of the final gap, identification and definition of the chosen logic and interpretative-based search of possible causes behind the logic chosen as the more realistic.

4.2. An Example of Application: Retrospective Analysis of the Troop Assignment in the French "Plan XVII" of Joffre (First World War)

To illustrate the proposed method, we aim to apply it to the troop allocation plan of Joffre (1932) in World War 1 (also called Plan XVII). The execution of this plan and the controversial choices of troop assignment had a direct consequence on the fatalities during the war, notably on the inequalities of treatment of some regions in the allocation and assignment of troops (Férier, 1984; Forestier, 2014; Loez, 2014). However, all those analyses are narrative or interpretative, without providing any clear cut evidence sustaining the given purposes. In order to give to interpretative history the elements to feed with theory and discussion, clio-combinatorics delivers an appropriate framework to produce quantitative data and identify different logic patterns in historical choices. More precisely, in WW1 and related to Plan XVII, clio-combinatorics will be used here to reproduce the troop allocation plan using general logic patterns (i.e. an algorithm that follows an objective logic) and analyze the gap of the "objective logic" to that of Plan XVII. By comparing the two plans (the "objective logic" one and the real one) we can conclude on the objectiveness (Ackoff, 1977) of the plan and discuss if those inequalities stated by several historians are a consequence of optimization choices or have been influenced by subjective reasons.

In this context, the main aim of the proposed framework is to reproduce in the most accurate way the models and choices made at the triggering of the war (the big lines of this plan can be found in Le Hénnaff, 1922; Joffre, 1932). Moreover, the researches, studies and analyses made to this purpose have been developed under the military secret, so it is nowadays very difficult to get an extensive detail of data related to troop assignment (except some generalities of the plan and several details on the geography, territory and transport network characteristics). For those reasons, the IMLS method would be an appropriate tool to reproduce what cannot be obtained by real data and compare the outputs of our method with the results of the real Plan XVII, and then induce the main logic choices behind troop allocation of French armies in World War I.

The first step in deploying the IMLS clio-combinatorial method to the subjacent logic of Plan XVII is to set the historical context and the hypotheses under which the "solution probleming" analysis will be carried out. The French troop allocation plan being essentially defensive, (Bourget & Anthoine, 1989) the first hypothesis is related to forecasts of the services of intelligence about the attacks by the German army[4]. The anticipation and the forecasting of German attacks made by the French Military decision makers changed seventeen times since the initial troop allocation plan decided after the Franco-Prussian war of 1870 to Plan XVII[5], coordinated by Maréchal Joseph Joffre and approved in 1913.

In the Plan XVII, the main military operations were expected on the Franco-German border, i.e. the forecasting of German attacks excluded any other confrontation line (Audoin-Rouzeau & Becker, 2004). Nevertheless, an army is positioned in the border with the Italy in the Alps, which constitute a scenario, bound to the supposed neutrality of this neighboring country (Le Hénnaff, 1922). As a result, the main destinations would be located between Belfort and Charleville-Mézières (Le Hénnaff, 1922).

Then, a set of hypotheses need to be made regarding the context in which the transport of troops could be realized. Indeed, taking into account the geography of France in 1914 (Maurette, 1927) and the development of transport networks in that period, the only possible transport mode to deploy troops efficiently was the railway (for long distances), of course with a small part of road transport (a few miles) for reaching the railway network at the origin and the concentration locations near the front at destination. Moreover, the railway network in France was of type hub-and-spoke (Beuthe & Kreuzberger, 2001; Gonzalez-Feliu, 2006), i.e. with a center in Paris where most main lines converged. So to go from any origin to any destination (except short-distances lines), the French capital was a compulsory stage in the transport chain. This network can be modelled precisely, mainly with current techniques that take into account the complexity of such transport schemes (Kreutzberger, 2008, Gonzalez-Feliu, 2008). However, applying the "solution probleming" vision of Ackoff (1977) legitimates to wonder whether it is relevant to model the network in a complex way or in a simplified vision (Zbilut & Giuliani, 2007). The centralization of all troops in Paris, then redistributed into a system optimum viewpoint seems to be an interesting option today, but the Plan XVII assigned a priori each military region to a destination, without reorganization at the Paris node (Le Hénnaff, 1922). Moreover, simplifications of the network were planned to facilitate the assignment operations (Le Hénnaff, 1922):

1. The military authority took precedence to organize and manage the network for the sake of the distribution of troops (military convoys to civilian traffic priority). Nevertheless, the high command did not have in 1913 a global vision together with a legitimacy for completing the plan of deployment of troops, that it will acquire later on (Joffre, 1932). .
2. Links from Paris to the East and North were excellent, the rest of the links being considered as sufficient by Le Hénnaff, (1922). Several schemes have been foreseen in the years before the war to improve certain corridors.
3. Since repositioning troops once they have been moved to initial destination requires the use of the route in detriment of the railway, troops are not interchangeable: they are assigned to a destination and are not supposed to move to any other once assigned and moved to destination (Le Hénnaff, 1922). For that reason, plan allows little flexibility and few possible changes. Indeed, in addition to the army component number 19 (divisions of Algeria) and Alpine divisions (assigned ex officio at the Franco-Italian border), the main infantry components have a firm destination. However, four army components are designed as mobile: the 12e and 17e army components are the adjustment variable and can be moved if the German army attacks from Belgium (Le Hénnaff, 1922), which would, in a more limited way, provoke a contingent displacement to the north of the 10e and 11e army components to cover the positions initially expected for the 12th and 17th components (Le Hénnaff, 1922, p. 69).
4. The main intention of the plan is to "give preference to the speed at the expense of flexibility" (Le Hénnaff, 1922, p. 69). It is therefore an optimization in terms of time of mobilization that seems to have been considered priority by the French high command. It is this assumption that should be tested.

In Figure 1 we can see a first simplification of the network. However, in a first analysis, we will assign origins to destinations on the basis of direct Euclidean distances (i.e. the distance "as the crow flies" between two points); then, we make the assumption that the higher the Euclidean distance, the longer the time to reach a destination from a given origin.

From those assumptions, we can define the optimization problem that meets the requirements stated previously, within the constraints imposed by the railway network. It is important to begin to clarify the decision variables. The main goal is to assign troops on a given front. The decision variable refers to the troops of each region. In this study, the selected variable is the army components: a component will be, in our model, assigned in its entirety to a single front, without the ability to split it on several destinations. This corresponds to the initial choice of the plan XVII (Le Hénnaff, 1922).

Figure 1. Presentation of the modeled network

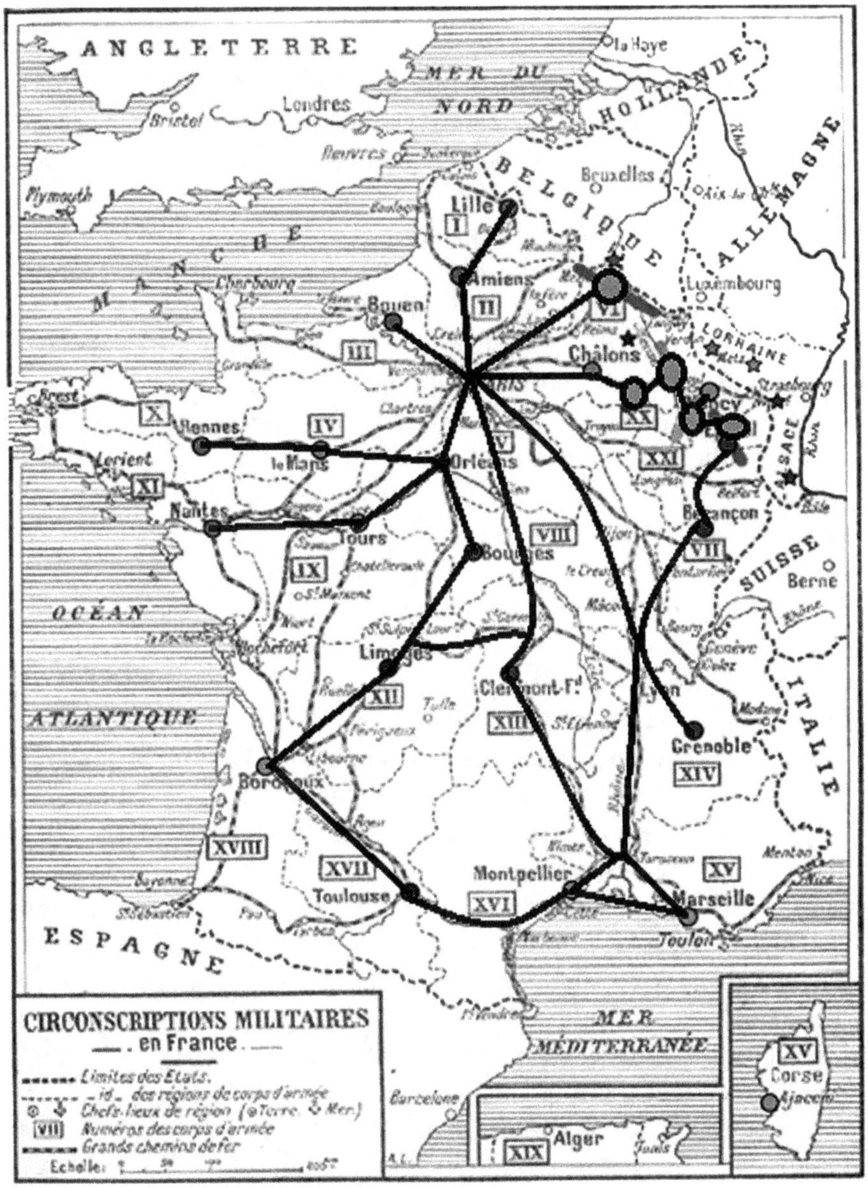

The central element of any model of optimization is known as the function objective test. Le Hénnaff (1922) and Joffre (1932) point out that the essential criterion was the speed to join the front. Thus, two kinds of optimization are possible:

1. Minimize the total time to browse (i.e., with a system optimum viewpoint, find the solution which minimizes the sum of all the journey time).
2. Minimize the time of each route, without getting a mathematical minimum at the global level. In other words, find the minimum cost for each assignment (here the time of course), and then infer

the total time. Wardrop (1952) shows that this approach, called balance the user gives a result of cost (or time) equal to (or greater) than the system optimum.

We will test in this study these two visions. Once defined the function goal, it is important to set the constraints. We will not detail here mathematical formulas but explain in clear language how these constraints are reflected in an optimization model in the sense of Jacobsen and Madsen (1980).

The first and main constraint is that the assignment of troops needs to be respected. We assume that the troops for each front needs are fully used as input.

For this, we have assumed that the front was divided into 5 main areas (taking into account the proximity of the various destinations): Belgium-Ardennes, Argonne [5], Mulhouse-Sarrebourg, Morhange and Metz. This choice reflects the decision of the staff to strengthen the border with the Germany and therefore concentrate troops in this area. In addition, troops assigned to the Italian border are not considered in this problem because they have been assigned automatically according to a criterion of proximity. So, for each of the 5 fronts identified, troops from the staff requirements are as follows: 5 components in Belgium-Ardennes, 2 in Argonne, 5 in Mulhouse-Sarrebourg, 5 in Morhange area, and 3 in Metz, making a total of 20 components (the region 19, Algeria, because of its remoteness, is not taken into account in the constitution of this assignment of the troops).

After this, it is necessary to build the data for the simulation of the logic of distribution of troops. For this, we have built the input of our optimization models, i.e., the needs of troops to the five front lines and the availability of troops in each military region, from the archives of the war, Series N 1872 inventories-1919, enriched by the assumptions of work extracted from the memoirs of Marshal Joffre (1932), of Colonel Le Hénnaff analysis (1922) and rail supply in 1914 from Maurette (1927). To model the constitution of the transport network for the concentration of troops, we define a graph (i.e., a set of vertices connected by arches) which represents the starting and arrival areas, and the rail network that connects. For this, each zone is represented by a Summit, which corresponds to its main railway station. For points of departure, it is relatively simple to identify the main train station (in terms of traffic and capacity) for each zone (see table 1); it corresponds in fact to the regional capital where troupes were gathered. For the arrival areas, 5 areas of the front have been distinguished; note that these lines were relatively extensive, sometimes of about 200 km, and the troops distributed along these lines, falling in various stations. To account for this, we make the assumption that the starting points correspond to a station (hence a Summit) but that arrival points correspond to a set of vertices (this also has its representation in the theory of graphs, see Gonzalez-Feliu, 2013). Every possible route (between a starting point and a finish line) is therefore defined by its total journey time taking into account the performance of the rail network at the time.

In addition, each region has its own army components. We have not detailed the composition of these components which are assumed to be homogeneous. Furthermore, an army components is always assigned in its entirety at one of the 5 lines of destination.

It therefore consists in summing up 20 army components (one per military region under the assumption that region 19, Algeria, is withdrawn for the reasons given above) and 5 points of arrival. The problem is to minimize the time to deploy these 19 army components and optimally allocate them to the 5 areas of destination in accordance with the needs of the front.

Table 1. Composition of armies (only infantry), assignment according to the plan XVII and stations of concentration on the front (developments of the authors according to Joffre, 1932)

	Commander	Army Components	Staff	Final Destination	Stations of Concentration
1st armey (Dole)	Auguste Dubail	7, 8, 13, 14, 21	266,452 men	Mulhouse and Sarrebourg	Remiremont to Charmes
2nd army (Dijon)	Édouard de Castelnau	9.15, 16, 18, 20	323,445 men	Morhange	from Pont-Saint-Vincent to Neufchâteau
3th army (Châlons)	Pierre Ruffey	4, 5, 6	237,257 men	Metz	from Saint-Mihiel to Verdun
4th army (Fontainebleau)	Fernand Langle Cary	12, 17	159,588 men	Argonne (only mobile army)	From Saint-Dizier to Bar-le-Duc
5th army (Paris)	Charles Lanrezac	1, 2, 3, 10, 11	299,350 men	Belgian and massive border Ardennes	From Hirson to Dun-sur-Meuse

The optimization model can be formulated as follows:

$$Min \sum_{i=1}^{n_o} \sum_{j=1}^{n_d} t_{ij}.x_{ij} \tag{1}$$

s.t.

$$\sum_{i=1}^{n_o} x_{ij} = b_j \ ; \forall j = \{1; 2; \ldots; n_d\} \tag{2}$$

$$\sum_{j=1}^{n_d} x_{ij} = 1 \ ; \forall i = \{1; 2; \ldots; n_o\} \tag{3}$$

$$x_{ij} \in \{0; 1\} \ \forall i = \{1; 2; \ldots; n_o\}; \forall j = \{1; 2; \ldots; n_d\} \tag{4}$$

The objective function (1) aims to minimize the total travel time of troops from each origin ot each destination. Constraints (2) and (3) ensure respectively that the good number of troops is allocated to each destination (2) and that each region sends one army components (3). Equation (4) indicates the nature of the variables (binary). This problem, formulated as above, seems to be an adaptation of the Hitchcock's (1942) Transportation problem, which has been formulated for the transport of goods with trucks from several origins to several destinations. The main difference arises on the number of troops at the origin (one components per region) which simplifies slightly the problem with respect to classical

Hitchcock problem (there is not the possibility to having one origin sending troops to more than one destination, so the problem is close to a matching problem). However, we can see from this formulation that the logic behind troop allocation can be modelled with a linear problem formulation, so the use of combinatorial optimization seems adapted to the issue of troop allocation.

This problem can be optimized using an exact or a heuristic method. The IMLS method will then focus on the possible way the military decision makers would solve the problem with the available resources in 1913 (so without a personal computer and in a given time that seems realistic, i.e. methods able to be solved mathematically "by hand").

The IMLS method is articulated for this application in the following steps:

1. We transform the problem of Hitchcock by modeling it as a decision tree. Then, to solve the problem, a Minimum Spanning Tree problem is applied using Kruskal's (1956) algorithm. This problem is applied by converting Euclidean distances into average journey times between O-D couples using conversion coefficients.
2. Once the Kruskal solution to the simplified decision problem is found, a comparison to the real plan is made. We observe that the correspondence between the Kruskal solution and the real Plan XVII is not total. To overcome this gap, we propose to deploy a Greedy heuristic, i.e. a nearest-neighbor algorithm that needs an order of examination to work. In other words, a sequence of assignment is given (i.e. of destinations). The algorithm depends on the order we consider for the potential assignments, so in this step the choice of the best order to be applied is made by iteration of possible sequences.
3. After analyzing the results of this algorithm, we observe that although the Greedy approach gives better results with respect to the Kruskal one, there is still a gap between the Greedy results and those of the real Plan XVII, even for the best sequence explored. For that reason we propose to make a hybrid algorithm mixing the Kruskal vision and the greedy vision.
4. After examining several possibilities (in an iterative way), the sequence that leads to the troop allocation plan closer to the real Plan XVII is chosen.
5. Finally, an analysis and a discussion of the 3 proposed plans and the gaps to the real Plan XVII are made.

We present in Table 2 the main results of the IMLS method.

Table 2 shows that most of troops seem having been assigned using a rational logic: that of nearest-neighbor (or Greedy) algorithm following travel times obtained via the conversion of Euclidean distances into times using a speed coefficient. However, the classical Greedy algorithm seems to be combined with a Kruskal algorithm, in the following way. First, we affect to each origin the closest destination (even if we do not respect the number of components at each destination). For each army (except army 4), we validate all assignments that are optimal (i.e., locally the least costly). If the number of assignments selected in this way is lower than the number of components of the army, we validate all of them; if not, we validate the n best assignments, n being the number of components of the selected army. In the analysis of the different schemes tested, we observed that army 4 is one of the closest to most destinations but the chosen assignments for this army are not the closest ones. We decided then not to include it in the first iterations and assign the remaining troops to army 4 to test the hypothesis of assigning to the mobile army the two components that are far from all assignments (the hypothesis of assigning the two closest ones is not verified).

Table 2. Main results of clio-combinatorics assessment

Region	Main City	Kruskal-Based Algorithm	Pure Greedy Algorithm	Mixed Kruskal-Greedy Algorithm	Real Assignment
1	Lille	5	5	5	5
2	Amiens	5	5	5	5
3	Rouen	5	5	5	5
4	Le Mans	4	3	3	3
5	Orléans	4	4	3	3
6	Châlons	3	4	3	3
7	Besançon	3	1	1	1
8	Bourges	3	3	1	1
9	Tours	2	3	2	2
10	Rennes	5	5	5	5
11	Nantes	5	5	5	5
12	Limoges	2	1	2	4
13	Clermont-Ferrand	1	1	1	1
14	Lyon	1	1	1	1
15	Marseille	2	2	4	2
16	Montpellier	2	2	2	2
17	Toulouse	1	2	1	4
18	Bordeaux	2	2	4	2
20	Nancy	2	2	2	2
21	Épinal	1	1	1	1
Gap (in travel time) to theoretical optimum (%)		-	+9%	+10%	+11%
Number of assignments coinciding with Plan XVII		13 (65%)	14 (70%)	16 (80%)	-

Let us illustrate it. While assigning to each origin the closest destination, we observe that army 4 is the closest one to the high number of origins (10), then army 5 (5), army 1 (4) and army 2 (1). Army 3 is not the closest destination for any origin. So according to what said above, we assign all origins to army 5 (that correspond to the real assignments), 4 of 5 to army 1 (three of them corresponding to the real assignment) and that of army 2 (also corresponding to the real assignment). In this way, 10 of 20 assignments are realized (9 of them corresponding to those of real Plan XVII of Joffre).

After that, we assign the 10 remaining army components following a Greedy algorithm. To do that, we need to define a sequence to examine the different O-D couples. Since Army 5 is entirely assigned, we focus on the remaining three (we remember that army 4 has been excluded from the analysis). We make then the choice to assign one entire army after another, in other words, once one army is chosen we complete the remaining non-assigned army components until all needed troops to compose it are assigned. Then we can choose another army and repeat the procedure until all three armies are completed. Then, the two remaining military regions are assigned to the mobile army (i.e. army 4). As only three armies remain, the possible permutations (in this case the order of the sequence is important) remain contained (only 6 possibilities) so all of them can be analyzed. By analyzing them, we observe that the sequence 3-1-2 (i.e. first assigning all components of army 3, then completing army 1 then finishing with army 2) give the best results in terms of gap (in number of coinciding assignments) with respect to the real Plan XVII assignment.

We observe that in the best case (the procedure presented here above) the coincidence with the real Plan XVII is of 80% (16 out of 20). That means than in 80% of the cases, the proposed procedure assigns troops in the same way they have been assigned in Joffre's (1932) Plan XVII. In a solution probleming viewpoint, it is however important to see the difference between total travel times between the different possibilities to examine the gap to a theoretical optimum (the solution obtained with the Kruskal algorithm) in terms of the optimization criteria (or objective function), i.e. the total travel time. We observe that the Greedy algorithm is (as appears often in combinatorial optimization literature) at 10% of theoretical optimum (i.e., travel times obtained with the greedy algorithm are 10% higher than those of the theoretical optimum). However, changes implemented with the other two algorithms increase travel times of 2% maximum (1% for the pure greedy algorithm and 2% for the mixed algorithm), which can be considered as negligible if we take into account incertitude and eventual bias in the estimation of travel times and those related to the approximation made in estimating distances between O-D couples. So we can consider the two greedy-based procedures as having similar results in terms of total travel times, so as being equivalent. That is valid also for the real assignment (the gap between the two greedy-based algorithms and the real assignment is about 1-2%).

This allows us to conclude that in general, the Plan XVII of Joffre followed an optimization logic for troop assignment: that of minimizing the total travel time, based on the following procedure: first, the O-D couples that have a minimum travel time are selected. Then, excluding the O-D couples for which the army 4 would have the minimum travel time, assignments are made to the selected army. In this way 10 assignments are ensured, including all components of army 5. After that, army 3 is completed using a Greedy algorithm, then army 1 and 2.

Concerning army 4, we cannot conclude this logic was applied to army 4 (mobile army). This led us to conclude that the mobile army seems not to have followed the same logic. It seems then that a pre-assignment of army 4 would have taken place. However, to analyze the motives of this pre-assignment and find the appropriate underlying logic requires further computations that will be part of future research.

5. CONCLUDING REMARKS: TOWARDS A SYSTEMATIC VISION OF CLIO-COMBINATORICS

After having illustrated the IMLS method and seen a few analyses of the results, we can state that clio-combinatorics can add new information and quantitative elements for historical analyses and discussion. However, it is important to formalize clio-combinatorics in order to make potential users having the necessary tools and elements to carry out their analyses. This final section aims to present a first systematic vision of clio-combinatorics and set its theoretical bases from the example above, as previously developed in case study research (Eisendhart, 1982).

First, this case study illustrates correctly the objectives of clio-combinatorics analyses and the link of this framework to solution probleming: the aim of clio-combinatorics is not to optimize "in absolute way" or with respect to a cost, distance or time-related objective function, but to try to reproduce a given reality. In this way, the aim of clio-combinatorics is twofold:

1. To find the modelling framework that meets the best an optimization problem in a past situation.
2. To find an algorithm that minimizes the gap (in terms of solution composition of the combinatorial sequence) with respect to the real solution used in past.

In this way, the advantage that clio-combinatorics offers with respect to forecasting combinatorial optimization is that the "target solution" is known (we will not use the term "optimum" since we are aware that the choice made in the past is not the theoretical optimum, mainly when computation methods were still primary). In this context, we can develop an optimization method which criteria is to minimize the gap between the given solution and the real solution, and revise it until the obtained solution gives a result that is considered satisfactory. This way of proceeding is close to that described by Ackoff (1977) in the solution probleming founding paper.

To proceed, we have started first by defining the optimization problem (i.e. the mathematical formulation) then solving it when possible. If the obtained solution is far (in terms of correspondence) to the real solution, two alternatives are possible: the first is to revise the solution method; the second is to question the mathematical model itself.

If we consider the mathematical formulation as corresponding to the logic used in the historical decision, to approach the real solution we need to modify the solving method and algorithm. This alternative supposes that the aim of the decision was really to optimize the modelled problem but the techniques available in the historical period would not be ready to produce an exact or advanced solution. The target of the clio-combinatorial analysis will then be to reproduce the sequence of the algorithm (which will be a heuristic) potentially used in the context. This is the methodological framework followed in the proposed example.

If we suppose that the gap comes from inappropriate (or inaccurate) modelling, the second way to proceed is to question then modify the decision problem itself. This methodological framework is not considered here. However, in some cases it seems interesting to examine various mathematical decision models and compare them to examine the validity and robustness of the chosen one. This could become a major research edge, more related to conceptual and mathematical analysis.

REFERENCES

Ackoff, R. L. (1970). Some ideas on education in the management sciences. *Management Science*, *17*(2), B-2–B-4. doi:10.1287/mnsc.17.2.B2

Ackoff, R. L. (1976). Optimization+ objectivity= optout. *European Journal of Operational Research*, *1*(1), 1–7. doi:10.1016/S0377-2217(77)81003-5

Ackoff, R. L. (1979). Resurrecting the future of operational research. *The Journal of the Operational Research Society*, *30*(3), 189–199. doi:10.1057/jors.1979.41

Ackoff, R. L. (1983). An interactive view of rationality. *The Journal of the Operational Research Society*, *34*(8), 719–722. doi:10.1057/jors.1983.165

Ackoff, R. L., & Sasinieni, M. W. (1968). *Fundamentals of operations research*. New York: John Wiley & Sons.

Audoin-Rouzeau, S., & Becker, J. J. (Eds.). (2004). *Encyclopédie de la Grande Guerre, 1914-1918: histoire et culture*. Paris: Bayard Centurion.

Greif Avner, (1997). Cliometrics after 40 years. *American Economic Review, Papers and Proceedings, 87*(2).

Barlas, Y. (1996). Formal aspects of model validity and validation in system dynamics. *System Dynamics Review, 12*(3), 183–210. doi:10.1002/(SICI)1099-1727(199623)12:3<183::AID-SDR103>3.0.CO;2-4

Barnhart, C., & Laporte, G. (2007). *Transportation, handbooks in operations research and management science*. Amsterdam: Elsevier.

Becker, J. J. (1988). *La France en guerre, 1914 – 1918, La grande mutation*. Editions Complexe.

Berstein, S., & Milza, P. (1999). *Histoire de la France au XXe siècle: 1900-1930 1*. Editions complexe.

Bourget, P., & Anthoine, P. (1989). La «déposition» du général anthoine sur la guerre de 1914-1918. *Guerres Mondiales et Conflits Contemporains*, (156), 105–108.

Cochet, F., & Garde, R. (Eds.). (2008). Dictionnaire de la Grande Guerre 1914-1918. Paris: Robert Laffont.

Cordeau, J. F., Gendreau, M., Laporte, G., Potvin, J. Y., & Semet, F. (2002). A guide to vehicle routing heuristics. *The Journal of the Operational Research Society*, *53*(5), 512–522. doi:10.1057/palgrave.jors.2601319

Daviet, J. P. (1997). *La société industrielle en France 1814-1914*. Paris: Seuil.

de Rham, C. (1980). La classification hiérarchique ascendante selon la méthode des voisins réciproques. *Les Cahiers de l'Analyse des Donnees*, *5*(2), 135–144.

Delhez, J. C. (2013), La bataille des frontières. Joffre attaque au centre. 22-26 août 1914. Economica.

DeVore, R. A., & Temlyakov, V. N. (1996). Some remarks on greedy algorithms. *Advances in Computational Mathematics*, *5*(1), 173–187. doi:10.1007/BF02124742

Donnelly, R., Thompson, R. G., & Wigan, M. (2012). Process validation of urban freight and logistics models. *Procedia: Social and Behavioral Sciences*, *39*, 400–408. doi:10.1016/j.sbspro.2012.03.117

Ehlert, H., Epkenhans, M., Gross, G. P., & Zabecki, D. T. (Eds.). (2014). *The Schlieffen Plan: International Perspectives on the German Strategy for World War I*. University Press of Kentucky.

Faure, R., Guillat-Le Garff, N. S., & Bloch, M. (1979). *Précis de recherche opérationnelle*. Paris: Dunod.

Férier, G. (1987). Les conséquences démographiques de la grande guerre en Auvergne. *Guerres Mondiales et Conflits Contemporains*, (145): 81–94. PMID:11634995

Fogel R.W. (1964), *Railroads and American economic growth*. Academic Press.

Fogel, R. W. (1970). Historiography and Retrospective Econometrics. *History and Theory*, *9*(3), 245–264. doi:10.2307/2504408

Forestier, P. (2014). La citoyenneté en guerre? Les soldats guadeloupéens sur le front pendant la Première Guerre mondiale. *Bulletin de la Société d'Histoire de la Guadeloupe*, *168*(168), 3–20. doi:10.7202/1026842ar

Gilles, H., Guironnet, J. P., & Parent, A. (2014a). Géographie économique des morts de 14-18 en France. *Revue Économique, 65*(3), 519-532.

Gilles, H., Guironnet, J. P., & Parent, A. (2014b, May 13). La géographie des Morts pour la France. *Le Monde*.

Goldin, C. (1994). *Cliometrics and the Nobel.* Historical paper N° 65, NBER Working Paper.

Gonzalez-Feliu, J. (2013). Multi-Stage LTL Transport Systems in Supply Chain Management. In J. Cheung & H. Song (Eds.), *Logistics: Perspectives, Approaches and Challenges* (pp. 65–86). New York: Nova Science Publishers.

Gonzalez-Feliu, J., Ovtracht, N., & Parent, A. (2015). *Did Joffre's Plan XVII follow an economics logic?* 4th seminar Clio – Sciences Po Lyon-LET, Lyon, France.

Gonzalez-Feliu, J., Ovtracht, N., & Parent, A. (2014). Gare et Front: la logique de répartition des troupes dans le plan XVII de Joffre. *Colloque « Gares en Guerre, 1914-1918 »*, Paris, 3-5 septembre 2014.

Gonzalez-Feliu, J., Semet, F., & Routhier, J. L. (Eds.). (2014). *Sustainable urban logistics: Concepts, methods and information systems*. Heidelberg, Germany: Springer. doi:10.1007/978-3-642-31788-0

Hanotaux, G. (1916). *Histoire illustrée de la guerre de 1914* (Vol. 3). Paris: Gounouilhou.

Heckman, J.J. (1997). The value of quantitative evidence on the effect of the past on the present, *AER, Papers and Proceedings*, *87*(2).

Hitchcock, F. L. (1941). The distribution of a product from several sources to numerous localities. *Journal of Mathematical Physics*, *20*(2), 224–230. doi:10.1002/sapm1941201224

Jacobsen, S. K., & Madsen, O. B. (1980). A comparative study of heuristics for a two-level routing-location problem. *European Journal of Operational Research*, *5*(6), 378–387. doi:10.1016/0377-2217(80)90124-1

Joffre, J. (1932). *Mémoires du Maréchal Joffre (1910-1917). Tome premier*. Paris: Librairie Plon.

Joll, J., & Martel, G. (2007). *The origins of the first world war*. Pearson Education.

Le Hénaff, J.H.F. (1922), La préparation et l'exécution d'un plan de transport de concentration (août-octobre 1914). *Revue Militaire Française*, (10), 62-84.

Loez, A. (2014). Les expériences combattantes. La Grande Guerre, 45-60.

Macharis, C., Melo, S., Woxenius, J., & Van Lier, T. (Eds.). (2014). Sustainable Logistics. Transport and Sustainability, 6.

Mallouy, J. (1914). Les Transport stratégiques. *Armée et Marine: revue hebdomadaire illustrée des armées de terre et de mer*, (24), 615.

Maurette, F. (1927). *Géographie de la France*. Paris: Editions Hachette.

Meiners, R. E., & Nardinelli, C. (1986). What Has Happened to the New Economic History? *Journal of Institutional and Theoretical Economics*, *14*(2/3), 510–527.

North, D.C., (1974). Beyond the new economic history. *Journal of Economic History*, *34*(1).

North, D. C. (1977). The New Economic History After 20 Years. *The American Behavioral Scientist*, *21*(2), 187–200. doi:10.1177/000276427702100203

North, D. C., & Nye, J. V. C. (1997). *Cliometrics, the new institutional economics and the future of economic history*. Economic History Association, 62nd Annual Meeting, "Private versus Public Institutions".

Sondhaus, L. (2011). *World War One: The Global Revolution*. Cambridge University Press.

Taniguchi, E., & Thompson, R. G. (2015). *City Logistics. Mapping the future*. Boca Raton, FL: CRC Press.

Taniguchi, E., Thompson, R. G., Yamada, T., & Van Duin, R. (2001). *City Logistics. Network modelling and Intelligent Transport Systems*. Amsterdam: Elsevier.

Union Internationale des Chemins de fer. (2014). *Railisia. The online UIC Railway Database*. Retrieved from http://www.uic.org/spip.php?article1352

Wardrop, J. G. (1952). Road paper. Some theoretical aspects of road traffic research. *ICE Proceedings: Engineering Divisions, 1*(3), 325-362.

Williamson, J. (1974). *Late 19th Century American Development, A General Equilibrium History*. Cambridge University Press.

Williamson, J., & O'Rourke, K. (1999). *Globalization and History. The evolution of a 19th Atlantic Economy*. MIT Press.

Winston, W. L. (2003). *Operations Research: Applications And Algorithms*. Andover: Duxbury Press.

Winter, J., & Prost, A. (2005). *The Great War in history: debates and controversies, 1914 to the present (No. 21)*. Cambridge University Press. doi:10.1017/CBO9780511614811

Zbilut, J. P., & Giuliani, A. (Eds.). (2007). *Simplicity: the latent order of complexity*. New York: Nova Science.

ENDNOTES

1 Those conclusions are obtained after a Google Scholar search of citations and citations of citations of Ackoff (1977) and Ackoff (1979) dealing with the limits of current « problem solving » visions of operations research as well as proposing alternative visions and methodological frameworks.

2 The classification and typology of solution probleming methods is out of the scope of this paper and would need a specific systematic literature analysis which will not be held here.

3 Joffre details in his Memoirs published in 1932 the logics of French troops allocation that prevailed at the begining of the war. This allocation of troops was effective the 2 August 1914.

4 Indeed, in the context of a defensive plan, the main issue when constructing it is to anticipate the attack plans of the enemy, so the intelligence services' contribution and efficiency is essential to the good deployment of those plans.

5 Actually, 18 plans have been deployed in France, since a a Plan XV bis had existed, the last one being the Plan XVII which was applied at the very beginning of World War I (Le Hénnaff, 1922)

Chapter 6
Optimization of Utility Functions in an Admissible Space of Higher Dimension

German Almanza
University Autonomous of Juarez City, Mexico

Victor M. Carrillo
University Autonomous of Juarez City, Mexico

Cely C. Ronquillo
University Autonomous of Juarez City, Mexico

ABSTRACT

S. Smale published a paper where announce a theorem which optimize a several utility functions at once (cf. Smale, 1975) using Morse Theory, this is a very abstract subject that require high skills in Differential Topology and Algebraic Topology. Our goal in this paper is announce the same theorems in terms of Calculus of Manifolds and Linear Algebra, those subjects are more reachable to engineers and economists whom are concern with maximizing functions in several variables. Moreover, the elements involved in our theorems are accessible to graduate students, also we putting forward the results we consider economically relevant.

1. INTRODUCTION

We introduce four definitions that are fundamental in the study of optimization of several utility functions into a pure exchange economy. The first of them is a *convex cone* defined in terms of the utility function; more precisely, we defined the convex cone like a subset in the tangent space of the commodity space. The definitions complete the assumptions of a pure exchange economy. *Curve that maximize the utility*, mathematically this is an argument used in differential topology and we introduced into the approach of the main problem, because is a strong argument in the proof of the main theorems. Moreover, the curve that maximize the utility function introduces an idea of a path that converge into the optimum point in

DOI: 10.4018/978-1-4666-9779-9.ch006

the commodity space and we may thought this curve as a trade that lead the Pareto optimum point to our agents whose are playing in this economy. The *Pareto critical set*, is a more general definition that the one used in the classical theory, our definition have the needs to deals with the main problem, we also prove two statements one is a topological characterization of this set and the other is an analytical characterization of this set, both of them are very important in the proof of our theorems. The *Generalized Hessian* is the last definition we introduced, actually this is a framework in the theory of singularities is studied like a generalization of the classical definition and is applied into high dimensional manifolds, but we used to give an analogous condition of the second derivative used in elemental calculus. Finally we introduce two theorems, *Necessary conditions theorem* which is the analogous to find a critical point for a one variable function, up to in this case we are concerning with the Pareto critical set. The last statement is the *sufficient condition theorem* involves the generalized Hessian.

2. METHODS

2.1 Utility Function

We assume an economy with l numbers of commodities and the market of each commodity have defined a unity of weight for each good, i.e. each good is quantified and is related with a real number, we will denote the real number set with $\mathbb{R}$. We are considering only those commodities that have in the market, that mean we use only positive real numbers. Then we denote $\mathbb{R}^l_+ := \{x \in \mathbb{R}^l \mid x > 0\}$ the orthant positive of $\mathbb{R}^l$ (the real space of dimension l) and P to the commodity space, then there exist a homeomorphism of P with an open subset of $\mathbb{R}^l_+$ indeed there is a map

$i : P \to \mathbb{R}^l_+$, which is an *inclusion.*

The coordinate $x \in P$ is a commodity bundle that belongs to a consumer, or the consumer wish to choose between all the l commodities. Also assume m agents in the market, we denoted x_i to the commodity bundle of the *i*-agent, then exist $x_i \in P$ bundles for $i = 1, 2, ..., m$ agents. The status of the economy assuming the commodities of the m agents is denoted with the point $x \in P^m$; i.e. the complete economy of the m agents is represented with a point in the commodity space of dimension m. In our economic model we assume exhausted resources and those are quantified, then the next space is a natural definitions.

Definition 1: Let $p \in P$ the total commodity that agents are allow to choose, we named *attainable space* to the set

$$W = \left\{ x \in P^n \,\middle|\, \sum_{i=1}^{n} x_i = p \text{ with } x = (x_1, x_2, ..., x_n) \right\}.$$

In the above space there are interacting the m agents and it represent the whole space of consumption or choice possibilities.

Remark 1: The space W is identified with an open subset of $\mathbb{R}^n$, with $n = l \cdot m$ then we assume to W is a n-manifold. The manifold W has a topological property of compact closed. The Heine-Borel theorem (cf. Apostol, 1983) said: *Every compact subset in $\mathbb{R}^n$ is closed and bounded.* The above theorem has an economic meaning, the relation with the W space is: *any agent can increase or decrease fairly his commodities, until reach or exhaust any combination of commodity desired.* This property is economically relevant, because is important that any agent in the market exhaust or reach their preferences.

Definition 2: *The utility function* of *i-agent* is defined as

$u_i : P \to \mathbb{R}$ *where*

$u_i(x') > u_i(x)$ means: absolute preference of x' over x, and
$u_i(x') \geq u_i(x)$ means: that x' comply equally or more than x

2.2. Pure Exchange Economy

$$u : W \to \mathbb{R} \text{ is at least } C^2. \quad (1)$$

$$S^n = \left\{ (x_1, x_2, \ldots, x_{n+1}) \in \mathbb{R}^{n+1} \middle| x_1^2 + x_2^2 + \ldots + x_{n+1}^2 = 1 \right\},$$

let be $g(x)$ the *normal unitary vector* oriented with the indifference surface $u^{-1}(c)$, such that $u(x) = c$. Observe that $g(x)$ is the map defined like

$$g : W \to S^{n-1}, \text{ such that } g(x) = \frac{gradu(x)}{\|gradu(x)\|},$$

where *grad* $u(x) = \left(\frac{\partial u}{\partial x_1}, \frac{\partial u}{\partial x_2}, \ldots, \frac{\partial u}{\partial x_n} \right)$ is the gradient of the map $u(x)$.

The map $g(x)$ is an important element in the *supply theory* and the *consumer preference* (cf. Debreu, 1972, pp. 603-615), and is related with the price vector of *the general equilibrium theory.*

Now we will announce a "strong version" of differentiability, it is also named *monotonicity,* which in economic theory means free purchase "more commodities, are mayor satisfaction", mathematically is

$$g(x) \in W \cap S^{n-1} = \operatorname{int}(S^{n-1}) \text{ for each } x \in W, \quad (2)$$

the notation $\operatorname{int}(S^{-1})$ means that $\frac{\partial u(x)}{\partial x_i} > 0$ for all $i = \{1, 2, \ldots, n\}$. We denoted $g(x)^{\perp}$ to the *orthogonal complement* of the function $g(x)$ considering the map $Dg(x) : \mathbb{R}^n \to g(x)^{\perp}$ which can be relate with the tangent space $T_{g(x)}(S^{n-1})$ or the hypersurface tangent to the indifference surface in *x*. The restriction $Dg(x): g(x)^{\perp} \to g(x)^{\perp}$ is a symmetric linear map. This property allow permutations of the coordinates of *x*.

Convexity of the map is a necessary condition into the optimization theory, the next is a strong version of convexity in terms of the differentiability of *g(x)*.

$Dg(x): g(x)^{\perp} \to g(x)^{\perp}$, has negative eigenvalues. (3)

The above condition is well know in the classic literature named *second order differentiability of convex function*, when this is announced in terms of the Hessian or the second order differential, i.e.

$D^2u(x): W \to \mathbb{R}$ is a matrix negative defined. (3')

The last condition on *u* is over the boundary of $\mathbb{R}^n_+$.

The indifference curve $u^{-1}(c)$ is closed in $\mathbb{R}^n$ for each *c*. (4)

The above condition have an economic interpretation, it mean any agent wish to keep at least few of each commodity allowed in *W* (cf. Debreu, 1972; Debreu, 1972, pp. 603-615; Smale, 1981). The figure 1, represent the action between these maps.

Definition 3: A *pure exchange economy* consist of the next elements:

- Exist *m* agents which are traders and they deal in the attainable space *W*,
- The *i*-agent for $i\{1,2,\ldots,m\}$ has a preference represented for the utility function $u_i: P \to \mathbb{R}$ which satisfied the conditions (1) to (4),
- The *i*-agent has an aid $e_i \in P$, then for the price $p \in \mathbb{R}^n_+$ the income or welfare of the *i*-agent is $p \cdot e_i$

Figure 1. The maps $g(x)$ *y* $g(x)^{\perp}$ *on* W *space*

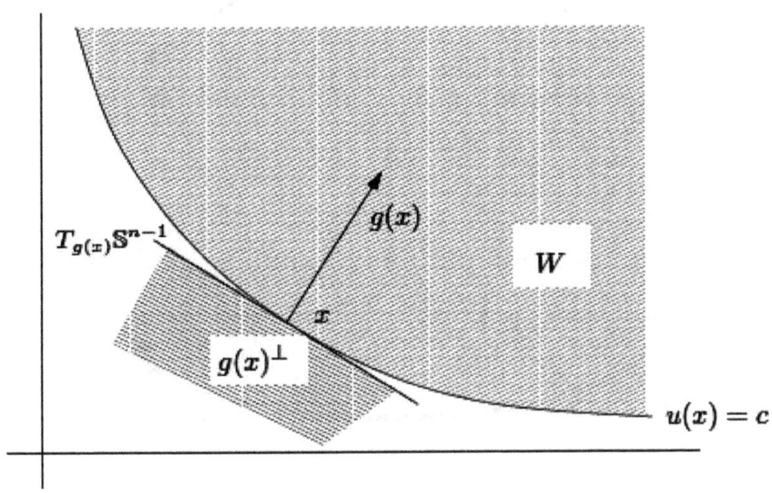

Remark 2: In this model we can interpret like an economy where any trade his own aid reaching for a better aid or even make an optimum satisfaction (opposite with the restrictions models).

Exist a homeomorphism $h: W \to U$, where U is an open subset of the positive orthant in $\mathbb{R}^n$. Then, the homeomorphism h induce the topology of $\mathbb{R}^n$ in *W*. Now suppose a welfare economy and *m* agents which trade on this space and any agent have an aid P of commodities, the Cartesian product $P^m = P \times P \times \ldots \times P$ denote the collection of commodities of the *m* agents, then exist a *natural inclusion* $i: W \to P^m$ and the homeomorphisms $\varphi_n : P^m \to U$ y $\varphi : P \to V$ where U and V are open sets of the orthant $\mathbb{R}^n$ y $\mathbb{R}^l$ respectively. We denoted by $\pi_i : P^m \to P$ the canonical projections that are represented the commodity space of the *i*-agent and $p_i : \mathbb{R}^m \to \mathbb{R}$ denote the preference of the *i*-agent. Then, the maps for concern are

$$u(x) = \tilde{u} \circ h(x) \text{ y } u_i(x) = \tilde{u}_i \circ \tilde{\pi} \circ h(x),$$

as is shown in Figure 2.

2.3. Multi-Objective Function

The space W is an n-manifold, we expect to get the Pareto optimum of the multi-objective function

u: $W \to \mathbb{R}^m$, where $m \leq \dim(W)$, with *u* at least C^2. (6)

Figure 2.

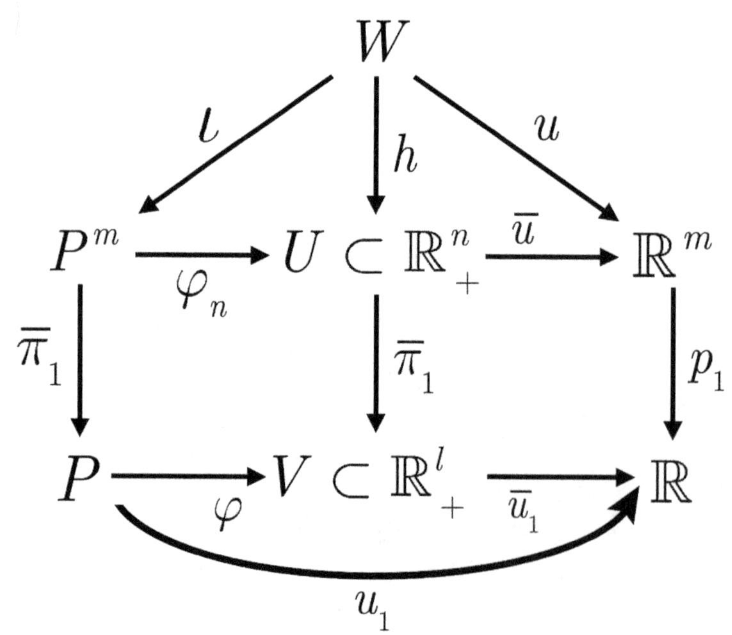

The function $u_i : P \to \mathbb{R}$ can split like a vectorial function

$u_i : P \to \mathbb{R}$ where $u(x) = (u_1, u_2, \ldots, u_m)(x)$,

with u_i at least C^2 for each $i = 1, 2, \ldots, m$. Assume $u(x)$ satisfy the properties (1) to (4) in the section §2.2.

Definition 4: The point $x \in W$ is Pareto optimum, if not exist a $x' \in W$ such that $u_i(x) \geq u_i(x)$ for all $1 \leq i \leq m$ y $u_j(x') > u_j(x)$ for some j. If exist a neighborhood $U(x) \subset W$ for the point x, where x is Pareto optimum, then we called local Pareto optimum local. We denote those set PO and LPO respectively.

In a multi-objetive function like the equation (6), the Pareto optimum point is not unique, It is usually a not convex set, then the required convexity assumptions in the utility function are not always satisfied for the multi-objective functions, then the PO set neither satisfied. The set PO is usually disconnected with lower dimension than dim(W).

2.4. Convex Cone

Let $Du(x)$ a derivate in the point $x \in W$ and the operator differential in *x* is denoted

$Du(x) : T_x W \to \mathbb{R}^m$.

Definition 5: Let *V* a vectorial space, a subset $C \subset V$ is a *cone* if it is closet by a positive scalar product, in other words *C* is a cone if

$\lambda x \in C$ for all $x \in C$ and all $\lambda \geq 0$.

We denoted *U* to a open set and U_x to a neighborhood in the point *x*.

More over, if the cone C is closet under addition, i.e.

$x + y \in C$ for all x, y $\in C$,

then we called convex cone to C.

The utility function has codomine in the positive cone Pos$\subset \mathbb{R}^m$, i.e. we denoted

$Pos := \{ y \in \mathbb{R}^m \,|\, y_i > 0 \text{ for all } i = 1, 2, \ldots, m \}$

Definition 6: For any $x \in W$ is defined the set

$$C_x := \{ v \in T_x W \,|\, Du(x) \cdot v \in Pos \},$$

in other words $C_x := Du(Pos)^{-1}$ for each $x \in W$. The operator *Du(x)* is linear and C_x is the preimage of the cone *Pos* which is named *convex cone related* to x.

Definition 7: In the vetorial space *V* of finite dimension, we said that the set of vectors $\{v_1, v_2, \ldots, v_n\} \in V$ belong to the same *hemi-space,* if exist a linear application $p : V \to \mathbb{R}$ such that $p(v_i) > 0$ for all *I,* where the hemi-space is $p^{-1}(0, \infty)$ (see Figure 3).

2.5. Utility Curve

The next definition is a very common tool in differential geometry.

Definición 8: Let $\varphi : [0,1) \to W$ be an injective function, we called *admissible curve to* φ if

$$\frac{d}{dt} u_i(\varphi(t)) > 0 \text{ for each } t \in [0,1) \text{ and all } 1 \le i \le m.$$

Notice that the admissible curve start in x and ended in x_0 if

$$\varphi(0) = x \text{ and } \lim_{t \to 1} \varphi(t) = x_0$$

Figure 3. The convex cone in the tangent space $T_{x_i} W$

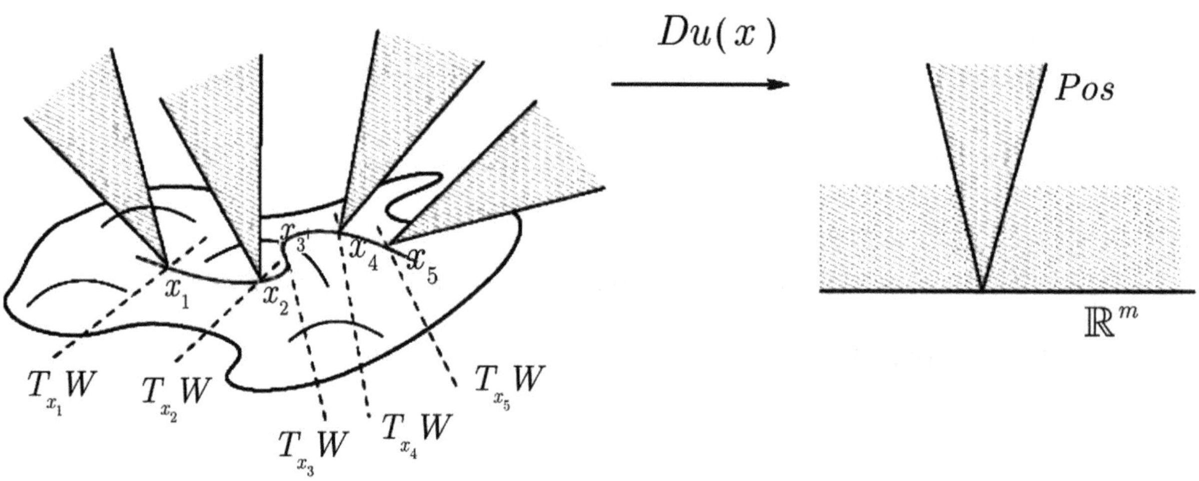

Remark 3: Any admissible curve define a path where the utility function is strictly increasing, in terms of convex cone we said the curve $\varphi : [0,1) \to W$ is admissible if the tangent vector to $\varphi(t)$ belong to the cone field corresponding to $\varphi(t)$, i.e. $\varphi'(t) \in C_{\varphi(t)}$. The fact to $\varphi(t)$ is injective; exclude the possibility that the curve have self-intersection, i.e. $\varphi(s) \neq \varphi(t)$ for all $s,t \in [0,1]$.

The admissible curve explain a trade starting with the commodity $x \in W$ it reach the commodity x_0, which gives a better utility than *x;* i.e., the limit $\lim_{t\to 1} \varphi(t) = x_0$ with $\varphi(0) = x$, is a continuous succession such that the commodity *x* converge to the local Pareto optimum x_0, then $u(x_0) > u(x)$ for all $x \in U(x_0)$.

Remark 4: If x_0 is a Pareto optimum point, then not exist an admissible curve crossing through x_0, i.e. $\varphi(t) \neq x_0$ for all $t \in [0,1)$.

2.6. Pareto Critical Set

Definition 9: We named *Pareto critical set* to the set

$$\theta := \{x \in W \mid C_x = 0\}.$$

Exist a natural contention in the Pareto sets we have defined. If $x \in PO$, then not exist $x' \in U(x)$, such that $u_j(x') > u_j(x)$ for some j, then $x \in LPO$. Now, if $x \in POL$ we can see there not exist an admissible curve goes through x, then $x \in \theta$; this mean

$$PO \subset POL \subset \theta$$

By other hand, a point $x \in POL$ is not necessary Pareto optimum, more over a point $x \in \theta$ is neither Pareto optimum. The next result characterize the topology of the set θ.

Proposition 1: *The set* θ is closet in W.

Proof: If is not, then we have $u_i(x) = c$ is a constant for all $1 \leq i \leq m$, which is a contradiction. ⌘

Corollary 1: *If* $x \in \theta$ then $\mathrm{Im}(Du(x)) \cap Pos = 0$.

Definition 10: The point $x_0 \in \theta$ is *critical stable* if each neighborhood $U(x_0) \subset W$ of x_0, there is another neighborhood $V(x_0) \subset U(x_0)$ also of the point x_0, such that any admissible curve $\varphi : [0,1) \to W$ with $\varphi(0) \in V(x_0)$ satisfy $\varphi(t) \in U(x_0)$ for all $t \in [0,1)$; more over $\varphi(t) \to PO$ when $t \to 1$. We denoted the critical stable set like θ_E.

Remark 5: When $\dim(W) = 1$ is verified that the set θ is the critical points set of the function $u : W \to \mathbb{R}$, and the set θ_E is the set of local maximum points of *u,* because $u'(x) > 0$.

2.7. Necessary Condition

In the next theorem we assume a smooth function $u : W \to \mathbb{R}^m$, such that $u = (u_1, u_2, ..., u_m)$, with $u_i : W \to \mathbb{R}$ and W is an n-manifold differentiable with $n \geq m$.

Theorem (Necessary Condition): $x \in \theta$ *if and only if*

1. *The set* $\{Du_i(x)\}_{i=1}^m$ do not belong to the same hemi-space of the cotangent space

$T_X^* W$.

2. Exists λ_i for each i = {1,2,..., m} not all null, such that

$$\sum_{i=1}^{m} \lambda_i Du_i(x) = 0 \tag{7}$$

Remark 6: The part (ii) of the theorem implies the set $\{Du_i(x)\}$ is linearly dependent, then the vectors $\{Du_i(x)\}$ are opposite between them. An immediate consequence of linear algebra is the operator $Du(x)$ is a linear transformation not surjective.

Remark 7: The *Riemannian metric* is a bilinear form of the tangent space, then for any Riemannian metric in *W* we have

$$\sum_{i=1}^{m} \langle grad(u_i), v \rangle = \sum_{i=1}^{m} Du_i(x) \cdot v = \sum_{i=1}^{m} \lambda_i Du_i(x),$$

where $\langle , \rangle$ is the inner product induced for the Riemannian metric, $v \in T_x W$ with $v = (\lambda_1, \lambda_2, ..., \lambda_m)$. This give us a geometric meaning of the theorem above, inducing the set $\{grad(u_i)\}_{i=1}^m$ is oriented in opposite directions.

2.8. Multidimensional Hessian

The next theorem is related with the *Second Intrinsic Derivative,* this is a very useful tool in the singularity theory introduced by Porteous (1971) and Mather (1971). Before to introduce this concept, we begin with a corollary of the Implicit Function Theorem (cf. Spivak, 1971).

Teorema: *Let* $u : W \to \mathbb{R}^m$ be a function of type at least C^1, with $\dim(W) = n \geq m$ for $n \geq 2$, with range of $Du(x) : T_x W \to \mathbb{R}^m$ equal to m. *Then, exist a neighborhood* $U_x \subset W$ of x such that $U_X \cap u^{-1}(u(x))$ is a smooth manifold W of codimension m; *i.e.* $M = U_x \cap u^{-1}(u(x))$ have dimension $(n-m)$.

We denoted K_x to the kernel of the map $Du(x): T_x W \to \mathbb{R}^m$, i.e.

$$K_X := \bigcap_{I=1}^{M} \ker\left\{Du_i(x) : T_x W \to \mathbb{R}\right\} \subset T_x W .$$

By other hand, the first isomorphism theorem (cf. Roman, 2005) allow us defined the next map

$$D^2u(x): K_x \times K_x \to \mathbb{R}^2 / \operatorname{Im}(Du(x)) . \tag{8}$$

An important condition of the sufficient theorem is when the range of $Du(x)$ is $m-1$(i.e. the $corank(Du(x)) = 1)$); then, if $x \in \theta$ implies $\operatorname{Im}(Du(x)) \cap Pos = 0)$).

The isomorphism theorem suggest a canonical positive ray on the space $\mathbb{R}^m / \operatorname{Im}(Du(x))$; moreover, if the range of *Du(x)* is $m-1$, the theorem inverse image (cf. Mather, 1971) suggest that $\dim(\mathbb{R}^m / \operatorname{Im} Du(x)) = 1$. Then we defined the next bilinear form.

Definition 11: We called *Multidimensional Hessian* to the symmetric bilinear form induced for the equation (8) and we denoted

$$H_x : K_x \times K_x \to \mathbb{R}$$

2.9. Sufficient Condition

The sufficient condition theorem require the next definition, we considered the boundary of the Pareto critical set

$$\partial\theta\left\{x \in \theta \middle| \operatorname{Im} Du(x) \cap \left\{cl(\bar{Pos}) / 0\right\} \neq 0\right\},$$

where $cl(Pos)$ denoted the closure of the set *Pos.*

Theorem (Sufficient Condition): Let be $x \in \theta$ and $c \notin \partial\theta$ the $corank(Du(x)) = 1$, then if the hessian H_x is negative defined, then $x \in \theta_E$. Let be $\lambda_i \geq 0$, for all $i = 1, 2, ..., m$ similar than hypothesis of the theorem of necessary conditions, then (up to positive scalar multiplication)

$$H_x = \sum_{i=1}^{m} \lambda_i D^2 u_i(x), \text{ on the ker Du(x).}$$

Remark 8: *The multidimensional Hessian or* Second intrinsic derivative *it was introduced by Porteous (1971) and Mather (1971) they used like a differential invariant, but Smale used to get dynamical systems properties i.e.* H_x *describe the notion of attractor/repulsor which have an admissible curve in the neighborhood of the critical point, we can consult these applications in Smale (1967).*

3. RESULTS

In the above section we developed a theorems that resolves the problem of optimizing several utility function at once, only using Calculus on Manifolds and Linear Algebra, between other mathematical tool that concern with economic theory and business administration. Next we summarize the applications of the theorems.

- **The Attainable Space:** In the definition 1, is the space of all commodities that any agent in the market is allow to choose. This space is equipped with the topology of real numbers, this is mean that any agent can increase or decrease fairly his commodities, until reach or exhaust any combination of commodity desired available. With this topology in the attainable space, we can model any kind of preference in order to optimizing the utility of the agent.
- **The Indifference Surface:** $u^{-1}(c) \subset W$, is the generalization of the indifference curve with a multivariable function (the inverse image of *u* map), we assume the map $u \in C^2$ then we can use all tools of Calculus in Manifolds. The convexity of this map, imply that any agent wish to keep at least few of each commodity allowed in *W*. This is a very common situation in our real economy

The *admissible curve* in definition 8, define a path where the utility function is strictly increasing, explain a trade starting with the commodity $x \in W$ then reach the commodity x_0, which gives a better utility than *x*. The map $\varphi : [0,1) \to W$ models a trade which is better for the economy.

- **The Pareto Critical Set:** In definition 9, actually have not economical interpretation, but is a stronger necessary element in the main theorem we formulate this definition only in terms of Linear Algebra and Multivariable Calculus which is reachable for any graduate student.
- **The Two Theorems: Sufficient and Necessary Conditions:** Solve the optimization problem of several utility functions at once. The importance of these theorems is our statement via Multivariable Calculus, an unsolved problem until S. Smale but he's proof is via Morse Theory a higher nontrivial subject.

REFERENCES

Apostol, T. M. (1983). *Mathematical analysis*. Addison-Wesley.

Debreu, G. (1972). *Theory of value: An axiomatic analysis of economic equilibrium. Cowles Foundation Monograph*. University Press.

Debreu, G. (1972). Smooth preference. *Econometrica*, *40*(4), 603–615. doi:10.2307/1912956

Guillemin, V., & Pollack, A. (2010). *Differential topology*. AMS Chelsea Publishing.

Mather, J. (1971). Stability of C°° mappings VI: The nice dimensions. In *Proceedings of Liverpool Singularities Symposium (1969/70)* (LNIM), (vol. 192, pp. 207-253). Berlin: Springer. doi:10.1007/BFb0066824

Moore, J. C. (2007). *General equilibrium and welfare economics: An introduction*. Springer.

Porteous, I. (1971). Simple singularities of maps. In *Proceedings of Liverpool Singularities Symposium (1969/70)* (LNIM), (vol., 192, pp. 286-307). Berlin: Springer. doi:10.1007/BFb0066829

Roman, S. (2005). *Advanced linear algebra. In Graduate texts in mathematics* (Vol. 135). Springer.

Smale, S. (1967). Differentiable dynamical systems. *Bulletin of the American Mathematical Society, 73*(6), 747–817. doi:10.1090/S0002-9904-1967-11798-1

Smale, S. (1975), Optimizing Several Functions. In *Proceedings of the International Conference on Manifolds and Related Topics in Topology* (pp. 69-75). Tokyo: University of Tokyo Press.

Smale, S. (1981). Global Analysis and Economics. In K. J. Arrow & M. D. Intrilligator (Eds.), *Handbook of mathematical economics* (Vol. 1, pp. 331–370). Amsterdam: North – Holland.

Spivak, M. (1971). *Calculus on Manifolds*. New York: Westview Press.

Varian, H. R. (1992). *Microeconomic Analysis* (3rd ed.). New York: W. W. Norton & Company.

Section 2
Implementations of Logistics

Featuring seven chapters related with different comparatives of logistics in the search to improve resources in diverse aspects of our transportations or improve process to this.

Chapter 7
Analyzing Airport Capacity by Simulation:
A Mexican Case Study

Catya Zuniga
Aeronautical University in Queretaro, Mexico

Miguel Mujica Mota
Amsterdam University of Applied Sciences, The Netherlands

Alfonso Herrera García
Instituto Mexicano del Transporte, Mexico

ABSTRACT

Air transportation has grown in an unexpected way during last decades and is expected to increase even more in the next years. Traffic growth tendencies forecast an expansion in the demand and greater aviation connectivity, but also higher workload to the different airspace users, especially for airport and services. Therefore, it is essential to employ strategies designed to use efficiently valuable corporate resource. Airport authorities around the world are investing in large capital projects, including new or improved runways, terminal expansions, and entirely new airports. However, this effort is sometimes limited due to their geographic location. In this work, two main objectives are pursued: first, to highlight the importance of the industry by exposing the current situation and future trends all over the world focusing in the Mexican industry; and second, to introduce a simulation model which can be used as a decision making tool for the upcoming demand. The analysis of the scenarios illustrates how to develop strategies to cope with the different airspace user's needs.

INTRODUCTION

The air transportation industry plays an essential role for global business and tourism but also in the economic growth worldwide. It enables countries to participate in the global economy by increasing access to international markets and, at the same time, allowing globalization of production, among many other benefits. The air transportation industry is a team effort, each time an airplane is in the air, there

DOI: 10.4018/978-1-4666-9779-9.ch007

is needed a coordination and interaction across a complex chain of partners, from airports, airlines, air navigation service providers (ANSPs), fuel suppliers, ground handlers, systems providers, catering companies, travel agencies, security screeners, and many more.

The air transportation has grown in an unexpected way during last decades and is expected to increase even more in the next years. The industry is wide, and with an innovative and excellent increasing perspective. Investments, business operations and efficiency, around the world are encouraged by air connectivity. The industry is evolving with several organizational and technical restructuring created by the globalization itself, the constant growing economic environment, and the development of new technologies, among other factors.

The first commercial passenger flight took place on January 1st, 1914 with a single passenger on it. Since then, the industry has not stopped growing. Twenty years later, 652 000 passengers were transported on commercial flights, reaching almost 40 million in 1952. Nowadays, around 3.3 billon people are safely transported, see Figure 1. The year of 2014 was outstanding for the transportation industry; passenger traffic as measured by revenue passenger kilometers (RPK) was up nearly 6% than 2013, and capacity was up nearly 5.8%. Lower oil prices were one of the key factors for airlines profits; they accounted around 20 billion USD; but also 2014 was a record year for manufacturers such as Boeing and Airbus; over 1490 airplanes were delivered and 3680 were ordered by airlines. The air transportation business activity generated around 2.4 trillion USD.

In 2014, the air transportation industry generated around of 58 million jobs in the world; around 9 million direct jobs: airlines, air navigation service providers and airports directly employed 7.6 million people. The civil aerospace manufacture sector employed more than 1 million people. It has been

Figure 1. Evolution of Passengers carried by airlines Source (IATA, 2014)

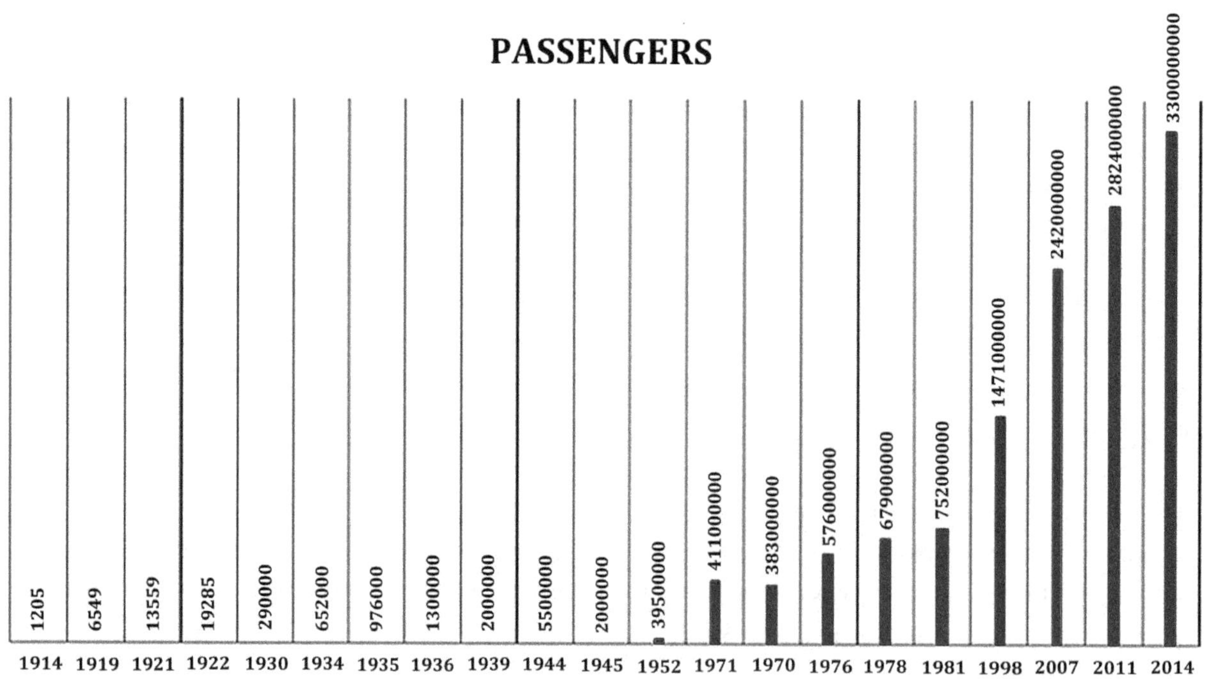

accounted more than 9 million indirect jobs generated through purchases of goods and services from companies in their associated supply chain. Furthermore, the tourism is enabled by air transport and it generated around 35 million jobs globally (ATAG, 2012).

During the past decade, 2001-2011, the world economy grew at an average annual rate of 3.6%, measured in terms of Gross Domestic Product (GDP). In the same period, the total scheduled airline traffic grew at an average annual rate of 5%, reaching around $2.6 billion USD annually in RPK, totaling nearly 5 trillion USD in 2010, and nearly 48 million tons of freight reaching to 172 billion freight ton kilometers (FTK). The total value of goods transported by air in 2011 represented the 35% of all international trade, (ATAG, 2012).

The air transport industry is a complex system, along with its different airspace users, airports, ANSPs, airlines, manufacturers, regulators, and multiple international organizations worldwide. Traffic growth tendencies forecast an expansion in the demand and greater aviation connectivity, even with multiple crises over the years. Unfortunately, higher demand means higher workload to the different airspace users, especially for airport and ANSPs. They are facing new challenges which require that government and industry work together, but with excellent opportunities to share, develop or enhance their capacity (Belobaba, Odoni & Barnhart, 2009), (De Neufville & Odoni, 2013).

Therefore, it is essential to employ airport management strategies designed to use efficiently valuable corporate resource within airport in order to meet the demands of the diverse airspace users. Airport authorities around the world are investing in large capital projects, including new or improved runways, terminal expansions, and entirely new airports. However, some airports are limited for expansions because of their geographic location, meaning that it is very complicated to face the increasing passengers' traffic demand by projects that involve infrastructure growth. On the other hand, airlines have implemented diverse approaches to manage traffic growth. They have been replacing smaller airplane by bigger ones to increase the cost/efficiency ratio. Code sharing, the development of alliances and joint ventures are all also different ways to cope with the increasing demand (Belobaba et al., 2009), (De Neufville et al., 2013). Secondary hubs, expanding service (route decentralization) to secondary airports have also been proposed to ease congestion at the busiest airports. Air Traffic Management promotes collaboration among all subsystems in order to improve decision making.

Diverse modernization projects such as The Single European Sky ATM Research (SESAR) launched by the European Community and the Next Generation Air Transportation System (NextGen) launched by US government, among others, are aimed to ensure the safety and fluidity of Air Transport over the next thirty years. These projects are revisiting some key aspects to overhaul airspace systems (EUROCONTROL, 2013).

In consequence, it is necessary to exploit different techniques for analyzing the use of the different system resources. Simulation answers the questions; What if?, through the display and analysis of different scenarios in which many variables (deterministic and stochastic) can be investigated. A simulation approach can provide the means to incorporate different stakeholder issues and provide effective 'what-if' analysis based on different scenarios. The simulation technique allows analyzing diverse activities and processes for managing in a sustainable manner airport and air traffic infrastructure. Specially, Discrete Event Simulation (DES) involves events (arrival, departure, etc.) that occur at discrete points in time. Hence, DES models are dynamic, stochastic and discrete by nature, i.e., the passage of time plays a crucial role while some uncertainties can be modeled in certain and known time periods (Banks, Carson & Nelson, 2010).

Over the last years, there have been several studies which exploited the benefits of a DES approach. These studies have modeled a wide range of systems, from human systems to underground haulage systems, and more. Moreover, the models have proved value to this technique. In the air transport industry, it has been also widely used to support airport stakeholder decision-making process (De Neufville et al., 2013).

In this work, two main objectives are pursued: first, to highlight the importance of the industry by exposing the current situation and future trends all over the world focusing in the Mexican industry; and second, to introduce a simulation model which can be used as a decision making tool for the upcoming demand. The departure process in a medium size airport terminal building in Mexico is analyzed using two different scenarios. The analysis of the scenarios illustrates how to develop strategies to cope with the different airspace user's needs.

This work is organized as follows. Section 1 introduces the importance of the industry by means of its current and future growth tendencies around the world and particularly in Mexico. In section 2 is examined the typical design process of a passenger terminal building and then, later on, in section 3, it is referred to some of the works published on analysis and design of the process within terminal buildings to enhance capacity. Section 4 is focused on modeling the departure process using a Discrete Event Simulation approach. Finally, Section 5 shows some preliminary results to highlight the benefits of the simulation approach. At the end of the work, conclusion and further work are presented.

THE AIR TRANSPORTATION INDUSTRY IN NUMBERS

The economic impact of the air transportation industry, particularly the commercial sector, is crucial around the world. As previously said, in 2014, 3.3 billion people were transported safely all over the world, not forgetting the cargo sector with 50 million metric tons moved during the same year, (IATA, 2014). The aviation's global economic benefits are quite big; it is estimated at $2.2 trillion USD in direct, indirect, induced and tourism-catalytic included (ATAG, 2012). For instance, in the United States, one of the leaders of the sector, the commercial aviation represents around 8% of their GDP. The annual investment in American airport infrastructure was nearly $10 billion USD; in 2010 airports invested $26 billion USD in construction projects. The number of passengers has risen and cargo volumes have grown fourteen fold, despite repeated shocks from recessions, wars, terrorism and disease (ICAO, 2013), (IATA, 2014).

In 2013, over 1500 airlines operate a total fleet of nearly 20 310 airplanes serving almost 4000 airports and managed by around 190 air navigation service providers, (ATAG, 2012). Airlines all around the globe have put on new services for the network connectivity, with a seasonal peak to over 40 000 pairs, an increase of 2.5% over the same month a year earlier (IATA, 2011). In 2014 a total of 50 000 scheduled airport pairs registered a $6.6 trillion USD as result to the connectivity (IATA, 2014). In 2012, the airlines registered an impressive 11% growth, while during 2011 the growth was about 7.5%. It has been forecasting for the year 2031 around 30.17 million passengers with an average annual growth of 8.9% and 400 million tons of cargo (ICAO, 2013). In 2012, CAR-SAM region registered 119 commercial airlines from which it can be pointed out in two largest markets: Brazil and Mexico, representing the 17% and 19%, respectively.

Regarding the air traffic passengers' distribution in the world in 2010, the Asian-Pacific region had over the 36% of passengers while North American region and Europe, obtained the 27% of the total. Latin America and the Central America and South America Region (CAR-SAM) transported the 6% of the world passengers.

For instance, Europe accounted between 25 000 and 30 000 flights daily, using 5 000 airplanes which land and depart from around 1000 airports. The economic benefits generated by the European air transport industry are approximately of 20 000 million USD per year. The contribution of the European Air Traffic Management is over 9 000 million dollars per year, representing approximately the 45% of the total obtained benefits. On the other hand, in the United Sates, the airline company market was conformed by over 2000 airline companies with approximately 23 000 commercial airplanes, that provided services to over 3700 airports (IATA, 2014).

According to AIRBUS (2014), in 2013, the biggest fleet was owned by United States, with a total of 6590 airplanes, followed by Asia Pacific region with 5090 and Europe with 4390. Latin America and Middle East account for 1280 and 1140 respectively. New airplanes replaced older, less efficient airplanes, reducing the cost of air travel and decreasing carbon emissions. Nowadays, United States still owns the biggest fleet, with 6700 airplanes, but not for much, the Asia Pacific region keeps growing with a total of 5850 airplanes, i.e. new 760 airplanes were received in around 2 years with a value of 2200 billion USD. Most of Asia Pacific fleet is composed by single aisle airplanes, around 70%, other 13% is represented by small widebody airplanes and almost 10% is medium widebody.

The Asia Pacific region has demanded 12 820 new airplanes with a total value of 1890 billion USD. In the case of Europe and United States, 7460 and 7250 airplanes with a value of 1020 billion USD and 810 USD, respectively, have been demanded. Latin America has also asked for 2900 airplanes with a value of 300 billion USD. It has been predicted that in 2032 the Asia Pacific region will account with the biggest fleet, 14750 airplanes in total, followed by United States with 8810 and Europe with over 7000. Latin America will improve its fleet with a total of 3790 in the same year. But not only more airplanes have been requested also the seat occupancy have become more productive, i.e. the load factors has incremented to an 80%, see Figure 2.

Figure 2. Passengers carried by airlines in 2010 Source (ATAG, 2012)

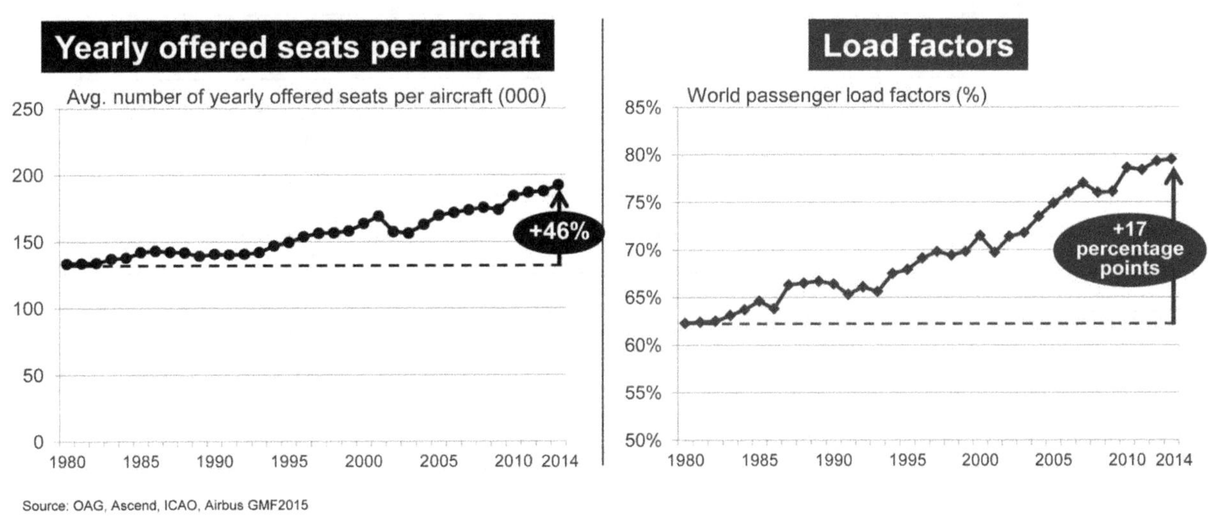

In 2013, there were nearly 7 780 airports in the world with IATA/ICAO identification code, 34% of them belong to United States of America, followed by the 25.64% of Europe, and the 12.57% corresponded to Latin America. In the same year, the world registered over 82 million aircraft movements, from which North America region made the 35.8% of the total, followed by Europe with the 25.3% and Asia Pacific with the 22%. Latin America and CAR/SAM accounted a 10% of the total of movements and Africa and Middle East represented each one around the 3% of the total (ACI, 2012). In 2013, Atlanta was the busiest airport on the globe both in movements and passenger transported, with 911 074 movements and 94.4 million passengers. Chicago airport takes the second place with 883 287 movements and Los Angeles airport is in the third place, 696 443 movements.

According to AIRBUS (2014), 47 mega-cities, airports, possess the mayor air traffic in the world moving around 0.9 million passenger per day. Table 1 provides detailed information about top 10 busiest airports in the world. It has been forecasted that in 2034 there will be 91 mega-cities moving daily around 2.3 million passengers. Furthermore, 39 out of 47 airports are already constrained, see Figure 3.

Regarding cargo operations, Table 2 introduces the top 10 cargo airports in 2013, where Hong Kong (HKG) and Memphis (MEM) took the first places as the busiest airports in terms of air cargo traffic. Each airport handled over 4 million metric tons in 2013. Meanwhile, as a region, Asia-Pacific region handled the largest amount of air cargo during the same year, moving 37 million metric tons, followed by North America with 27.9 million metric tons. Europe also represents a good player within the industry, moving 17.8 million metric tons. Middle East and Latin America-Caribbean moved 6.5 million and 5.1 million metric tons, respectively (ACI, 2013).

Today, there are nearly 1,650 dedicated freighter aircraft with a cargo hold of at least 10 tons and more than 200 airlines. For the period 2013-2032, worldwide air freight is expected to grow at 4.8% per year; a demand for 1,859 converted aircraft and 871 new aircraft is also predicted.

The airports in Latin America with mayor passengers' traffic are the International Airport of São Paulo-Guarulhos (São Paulo, Brazil), the Mexico City International Airport (MCIA) (Mexico City, Mexico) and the International Airport El Dorado (Bogotá, Colombia) (ICAO, 2013). México and Brazil are the principal representatives of the low cost carrier sector in Latin America, see Figure 4.

Table 1. Top 10 busiest airports by passengers in the world Source: (ACI, 2013)

Rank 2012	City	Code	Total passengers (in million)
1	Atlanta G.A., USA	ATL	94.4
2	Beijing, China	PEK	83.7
3	London, UK	LHR	72.3
4	Tokyo, Japan	HND	68.9
5	Chicago IL, USA	ORD	66.7
6	Los Angeles CA, USA	LAX	66.6
7	Dubai, UAE	DXB	66.4
8	Paris, France	CDG	62.0
9	Dallas/Fort Worth TX, USA	DFW	60.4
10	Jakarta, Indonesia	CGK	60.1

Figure 3. Current and future mega-cities in the world Source (AIRBUS, 2014)

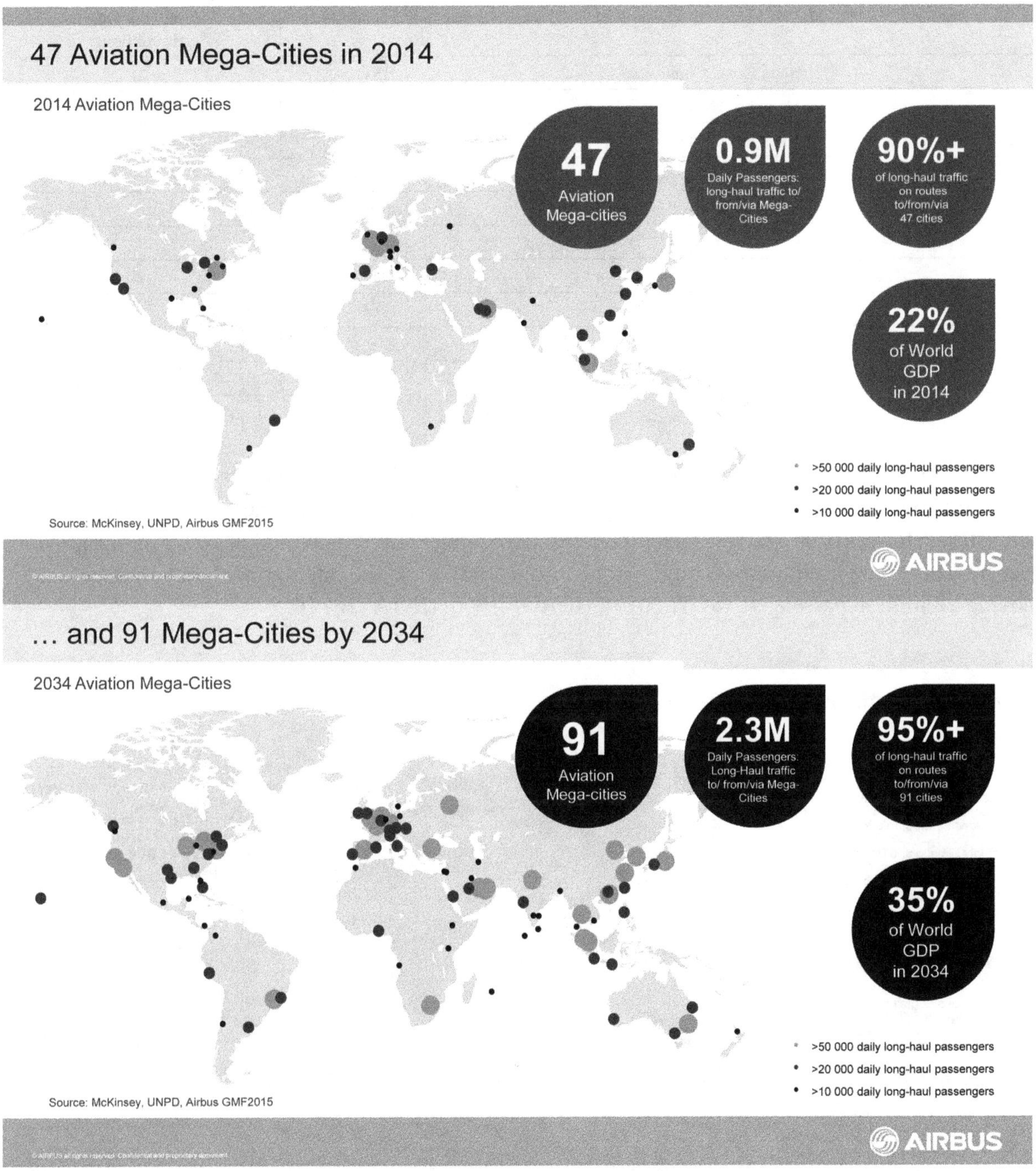

Mexico transported in 2012 over 56.8 million passengers through 618 regular routes, 355 international and 263 domestic. The Mexican Airline industry operated more than 9369 airplanes, 2166 commercial service, 6632 private and 569 from government service during that year. Passenger traffic grew around 8% (RPK), it moved 605.4 mil tons of cargo in the same year, which meant a 7% reduction in compari-

Table 2. Top 10 busiest airports by cargo in the world Source: (ACI, 2013)

Rank 2012	City	Code	Total passengers (in million)
1	Hong Kong, China	HKG	4.16
2	Memphis TN, USA	MEM	4.13
3	Shanghai, China	PVG	2.92
4	Anchorage, OK, USA	ANC	2.42
5	Incheon, Korea	ICN	2.46
6	Dubai, UAE	DXB	2.43
7	Louisville KY, USA	SDF	2.21
8	Frankfurt, Germany	FRA	2.09
9	Paris, France	CDG	2.06
10	Tokyo, Japan	NRT	2.01

son with 2011 (651 200 tons). The total number of operations have reached more than 1 million, 748 000 of the total correspond to domestic flights and 281000 to international ones. On the other hand the domestic sector has been growing faster than the international one, it increased in 10% over the previous year transporting 34 million passengers (60% of the total) while the international only increased a 7% moving 22 million passengers (SCT, 2015), (ICAO, 2013), (IATA, 2014).

Figure 4. Latin America's top 10 airports Source: (ICAO, 2013)

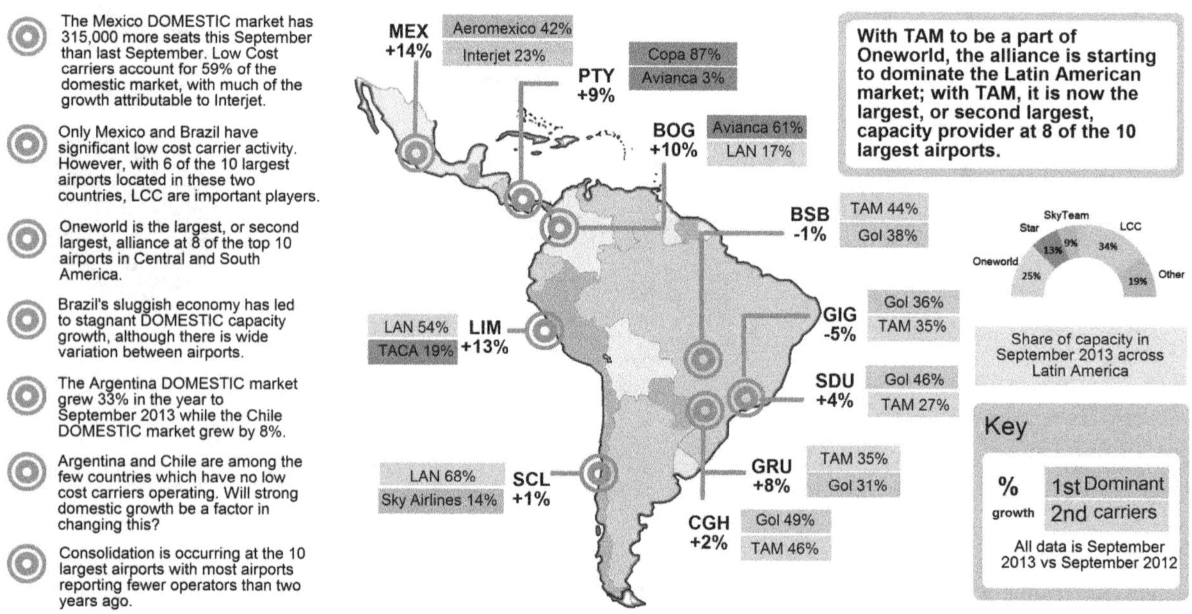

The Mexican Transport System

Mexico transported in 2014 over 65 million passengers, an increase of 8.5% compare with previous year. The total number of operations have reached more than 1 million, 800 mil of the total correspond to domestic flights and over 290 000 to international ones. The domestic airline sector has been growing faster than the international one, it were transported 40.6 million passengers (62.5% of the total) while the international only increased a 7% moving 24.4 million passengers. Figure 5 shows 7 out of the 9 regular passenger commercial airlines in México which serve domestic and international routes in 2014. It can be noticed that the biggest domestic airlines in terms of transported passengers are Aeromexico, Volaris, Interjet and Aeromexico Connect which moved 9510, 9363, 8694 and 7488 million passengers, respectively. The rest of passengers, i.e. 298 000, where transported by 8 charter airlines (SCT, 2015).

Viva Aerobus, which started in 2006 its services is growing quite fast and is forecasted to be one of the leaders in the low cost sector. In fact, as it can be seen in Figure 6, the low cost sector is gaining force since 2005, and in 2013 it accounted with the 60% of the market. Volaris and Interjet together with Viva Aerobus are categorized as the current Mexican low cost carriers.

It has been accounted 750 regular routes, 402 international and 348 domestic routes. Table 3. and Table 4 introduce the top 10 domestic routes. Within the domestic routes, 47 concentrate the 80.2% of the total passengers; while the 80% of the international tourism uses 94 routes (SCT, 2015).

Figure 5. Passenger transported by domestic airlines in domestic and international routes in 2014 Source: (SCT, 2015)

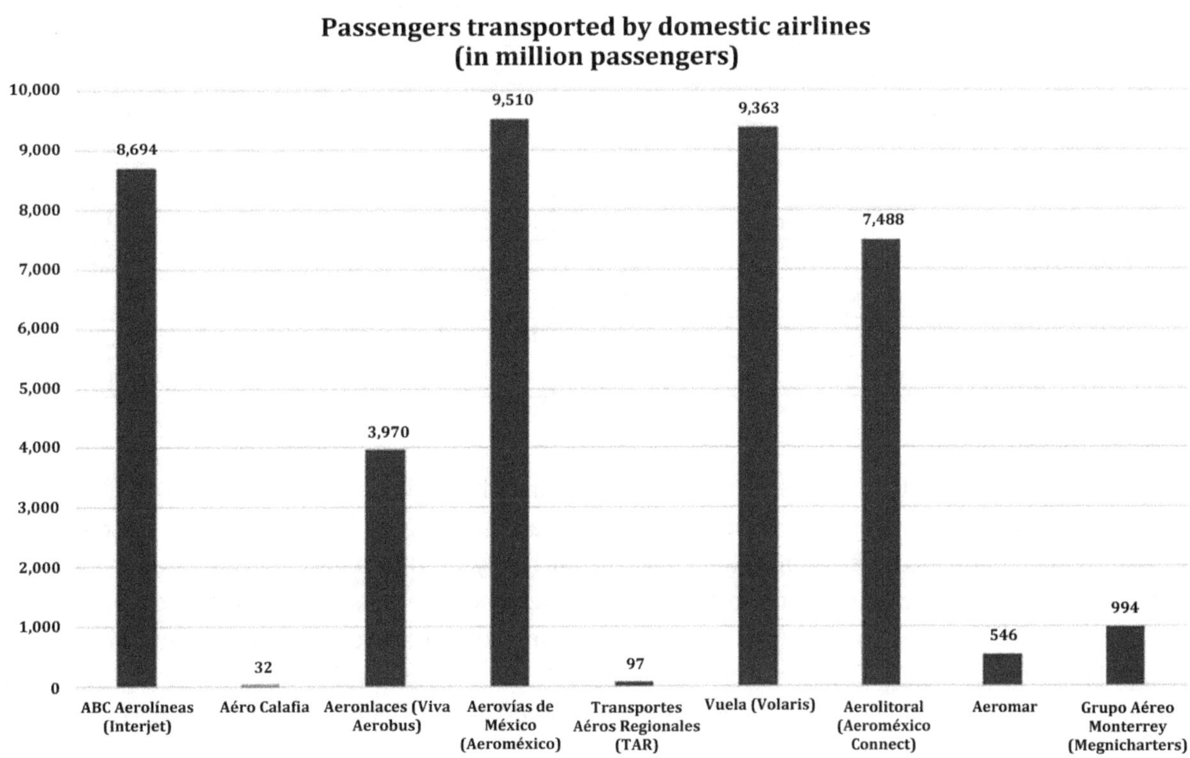

Figure 6. Main development of Mexican airlines since 2005 Source: (SCT, 2013)

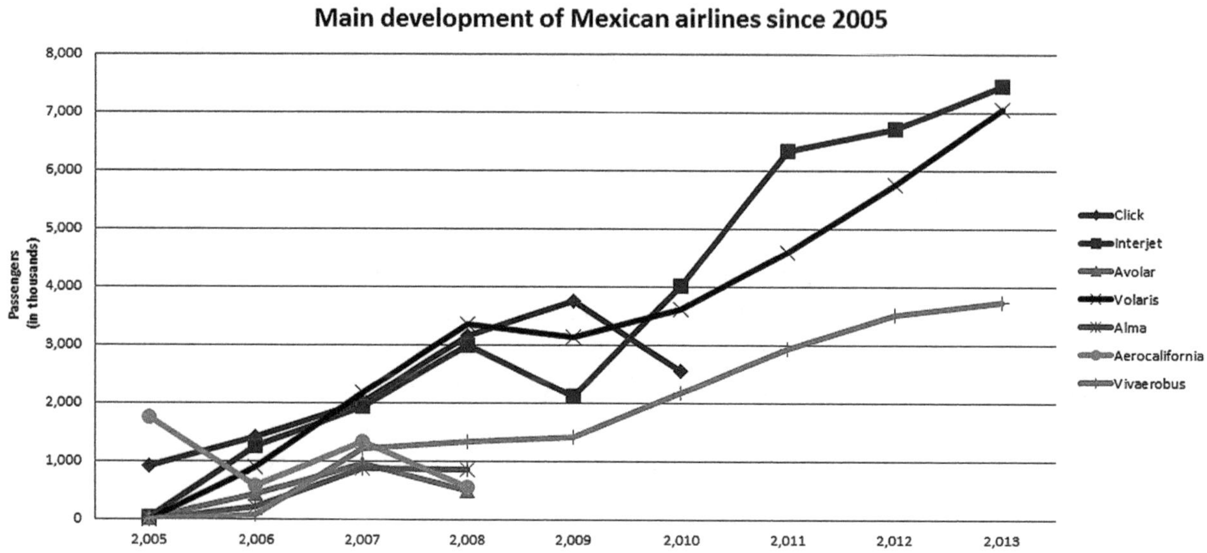

Table 3. Top 10 domestic routes in México Source: (SCT, 2015)

	Origin	Destination	Transported passengers (thousands)		Growing	Origin-Destination vs. Total %	
			2013	2014	2013/2014	2013	2014
1	Mexico	Cancun	3,295	3,524	7.0%	10.8%	10.7%
2	Monterrey	Mexico	2,460	2,736	11.2%	8.1%	8.3%
3	Mexico	Guadalajara	2,278	2,379	4.4%	7.5%	7.2%
4	Tijuana	Mexico	1,241	1,266	2.0%	4.1%	3.8%
5	Mexico	Merida	1,050	1,131	7.8%	3.4%	3.4%
6	Tijuana	Guadalajara	941	1,025	9.0%	3.1%	3.1%
7	Villahermosa	Mexico	700	776	11.0%	2.3%	2.4%
8	Tuxtla Gutierrez	Mexico	684	728	6.5%	2.2%	2.2%
9	Monterrey	Cancun	673	712	5.9%	2.2%	2.2%
10	Puerto Vallarta	Mexico	527	606	14.9%	1.7%	1.8%

Figure 7 shows the international traffic by region carried by Mexican airlines in 2014. It can be noticed that most of the passengers come from United States, this could be linked to two main factors, the first one is the alliances between airlines and the currently used hub and spoke model of airports; and the big amount of Mexican living in this country. Aeromexico transported the biggest amount of passengers to United States with a total of 2.8 million in 2014, followed by Volaris with almost 1.7 million. To Europe, Asia and Canada, Aeromexico was the only airline which transported passengers, a total of 384, 120 and 83 thousand passengers were accounted, respectively. To Central America and the

Table 4. Top 10 international routes in México Source: (SCT, 2015)

	Origin	Destination	Transported passengers (thousands)		Growing	Origin-Destination vs. Total %	
			2013	2014	2013/2014	2013	2014
1	Mexico	Los Angeles	783	813	3.8%	2.7%	2.5%
2	New York	Cancun	731	803	9.8%	2.5%	2.5%
3	Los Angeles	Guadalajara	746	781	4.7%	2.5%	2.4%
4	New York	Mexico	710	760	7.2%	2.4%	2.4%
5	Cancun	Atlanta	661	704	6.6%	2.2%	2.2%
6	Miami	Mexico	718	694	-3.4%	2.4%	2.2%
7	Mexico	Houston	620	693	11.7%	2.1%	2.1%
8	Dallas	Cancun	630	678	7.7%	2.1%	2.1%
9	Houston	Cancun	561	585	4.3%	1.9%	1.8%
10	Mexico	Bogota	469	572	21.9%	1.6%	1.8%

Figure 7. International traffic by region Source: (SCT, 2015)

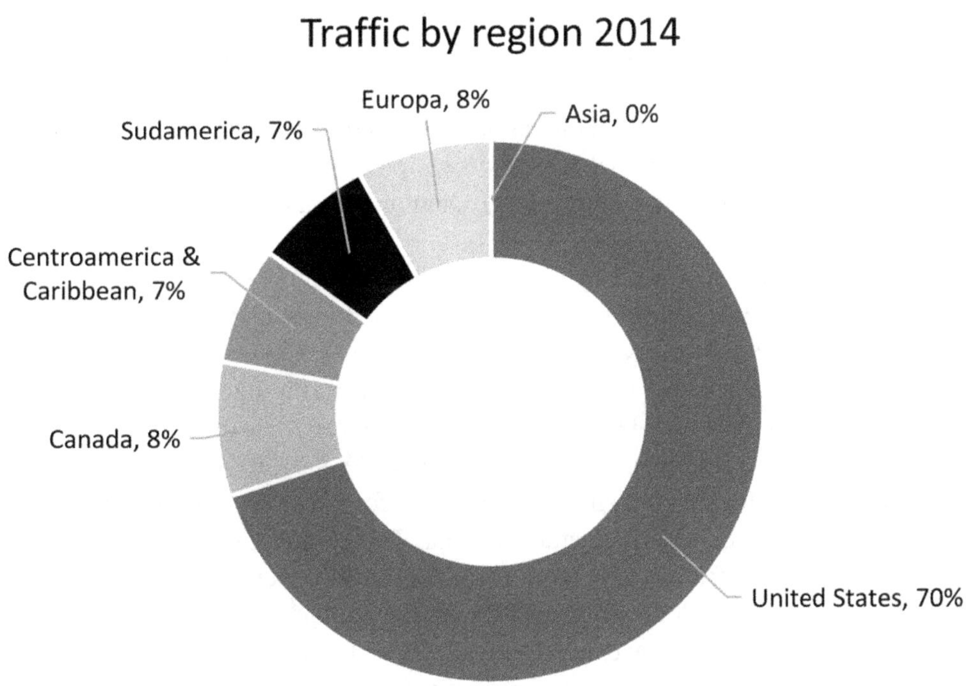

Caribbean, Aeromexico, Aeromexico Connect and Interjet transported 196 000, 235 000 and 270 000 passengers, respectively. To South America, there were transported by Aeromexico and Interjet 881 and 76 thousand passengers, respectively (SCT, 2015).

617.5 million tons of cargo were moved in 2014, which meant a 5.6% increase in comparison with 2013. From the total cargo moved, 43.1% corresponded to domestic carriers and 56.9% to international ones. The Mexican carriers moved almost 285 000 tons. Within the commercial carriers, the most prominent were Aeromexico which moved 57.5 tons, followed by Volaris, Aeromexico Connect and Interjet moving 16.4, 15.1 and 10.9 tons, respectively. The specialized cargo carriers moved 183 000 tons, Mas Air, Aero Unión and Estafeta were the top 3 carriers moving 71 100, 62 800 and 30 800 tons, respectively. International cargo was guided by United States and Europe whose airlines moved 160 000 and 123 000 tons, followed by Central and south American carriers with almost 41 000 tons. The Asiatic carriers movement accounted for 26 500 tons. Figure 8 introduces the flight concentration regarding the cargo in the airport system. It can be noticed that Mexico City International Airport moves the 65.4% of the total international cargo in Mexico and 38% in the domestic sector. Guadalajara is a prominent competitor with 19.2% of the international cargo and 15% of the domestic cargo.

The Mexican Airline industry operated 8961 aircraft, from which 2011 were for commercial services, 6509 were private and 441 were for government service. The 2011 aircraft registered in the commercial services included commercial and cargo airlines and aerotaxis. Commercial airlines owned 289 aircraft with the following distribution: Aeromexico and Aeromexico Connect accounted 64 aircraft each one, followed by Interjet and Volaris with 52 and 44 aircraft, respectively. Viva Aerobus owned 21 aircraft, Aeromar 19 and Magnicharters11. There were 54 aircraft owned by cargo airlines, Aeronaves TSM had the biggest amount of aircraft 31 in total, Aeroservicios de la Costa, Aero Union, Estafeta Carga Aerea and Aerotransportes MAS de Carga (MAS AIR) had 7, 6, 6 and 4 aircraft respectively.

Mexico counts with 76 airports, 63 of them are international airports and 13 domestic, in addition there are 1431 aerodromes register in the country. This places Mexico as one of the first countries in Latin America with the major airport network. The states of Coahuila, Sonora and Tamaulipas, in the north part of the country, register 5 international airports each one, followed by Baja California and Baja California South with 4 international airports each, placing them as the states with the biggest concentration of international airports in the country. Quintana Roo, Oaxaca and Sinaloa have each one 3 international airports.

Figure 9 introduces the 10 top airports by passenger traffic within Mexico in 2015. It can be noticed that Mexico City International Airport moves the 35% the total domestic traffic of the country, followed by far for four other airports: Monterrey (10%), Guadalajara (9%), Cancun (8%) and Tijuana (6%), respectively. In the international context, Cancun International airport is a good opponent to Mexico City International airport moving 34% and 33% of the total, respectively.

It can be said that the busiest airport in Mexico is called Mexico City International Airport (ICAO code: MMMX), located in Mexico city, and which also conforms, since 2003 the pillar of the Metropolitan Airport system, together with Queretaro, Puebla, Toluca and Cuernavaca. In 2014, it moved almost 34.2 million passenger, from which more than 22.7 million were domestic and around 11 million international. The amount of domestic and international passengers transported, located MMX as the first and second top 10 airports in Mexico. In the cargo sector, MMMX moved almost 33 500 tons from which 27 600 tons where from international destinations and 5700 domestic ones. This located MMMX as the main airport for international and domestic cargo controlling the 65% and 38% of the total, respectively, see Figure 8. The airport handled over 410 000 flight operations, most of them were commercial flights: 65% were domestic and 23% international; the cargo carriers performed 11 252 operations which represented the 3% of the movements. However, the domestic general aviation sector accounted with the 8.5% of the total movements.

The first international airport is Cancun, located in the state of Quintana Roo in Southeastern Mexico. Quintana Roo is also the home of the islands of Cozumel and Isla Mujeres, which together with Cancun

Figure 8. Domestic and International cargo by main airports in Mexico Source: SCT (2014)

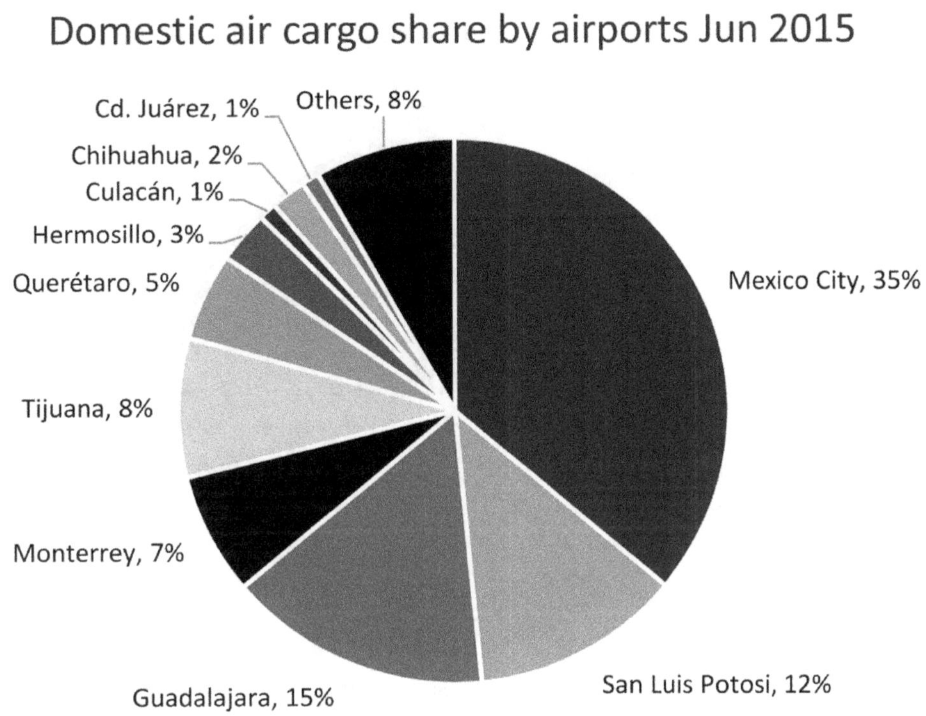

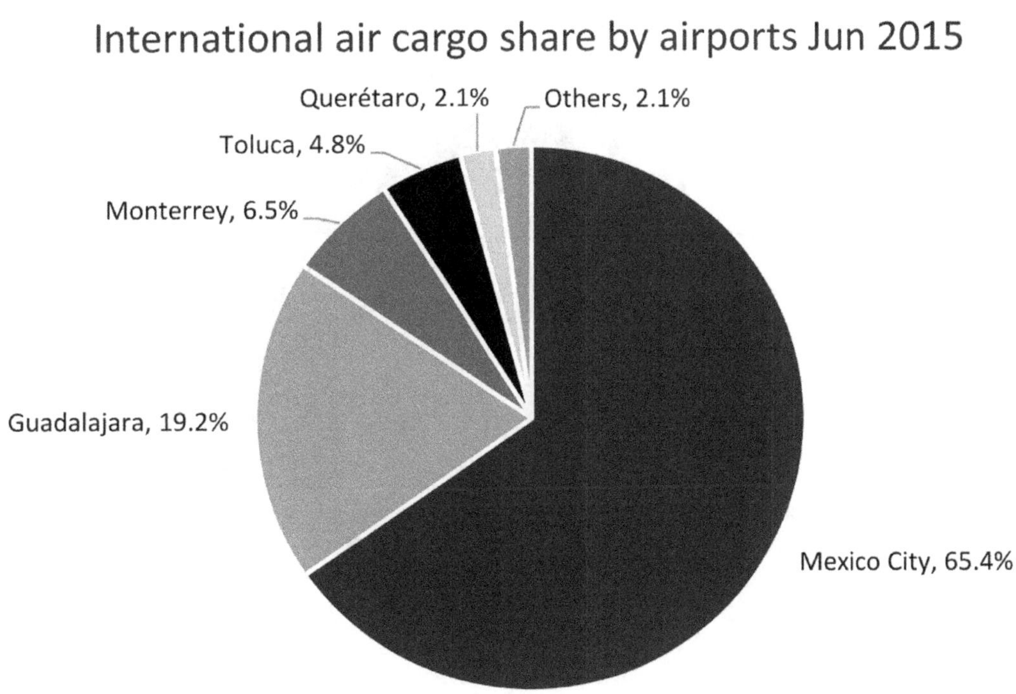

Figure 9. Domestic and International passenger traffic by main airports in Mexico Source: (SCT, 2015)

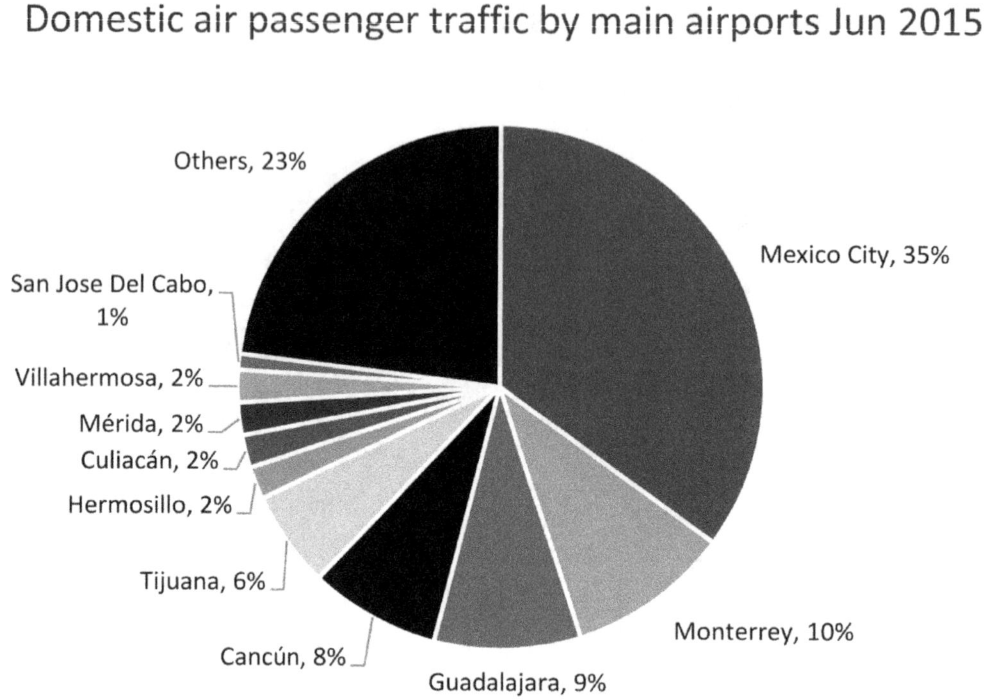

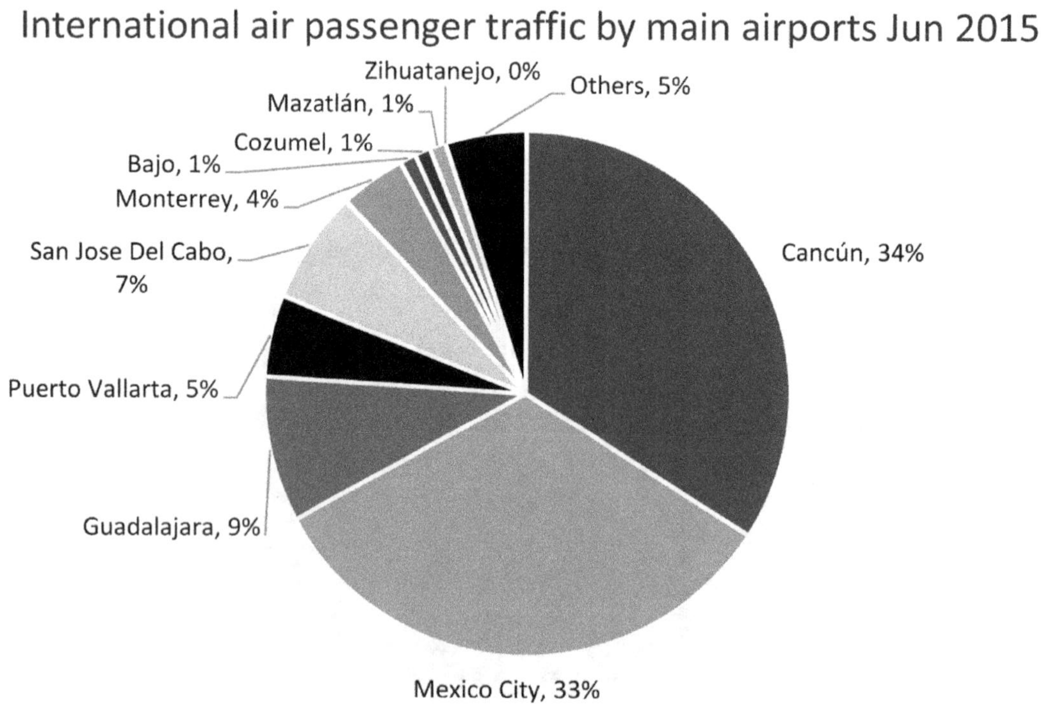

are one of the first tourist destinations in Mexico and in the world. Cancun international airport (ICAO code: MMUN) is one of the busiest airports in the Caribbean and in Mexico. In 2014 it transported almost 17.5 million passengers from which 12 million were international and 5.4 million were domestic travelers.

ANALYZING PROCESSES WITHIN PASSENGER TERMINAL BUILDINGS

Nowadays, an efficient design and operation of terminals buildings is essential, especially in view of the increasing demand all over the world. Terminal buildings are intermodal facilities that help to transfer passengers and goods between air and land transportation. Given the projected traffic increase, it is critical that new or existing terminals move efficiently large volumes of air passengers while providing a pleasant atmosphere for traveling and a harmonious operation within it. Terminals should be designed to naturally direct air passengers from the terminal curb to the gate, passing through each of the processes smoothly.

Those processes include ticket counters, check-in counters, security controls, passport controls, baggage carousels, customs counters, holding areas (e.g. lobbies, atria, gate lounges, etc.) and passageways (e.g. corridors, escalators, moving sidewalks, etc. plus amenities, such as restrooms, restaurants, shop, banks and other facilities required. In addition to move efficiently passengers, terminal buildings also should provide passengers, employees and other airport user's services to make their stay at the airports pleasant.

Analyzing the different process that occur within the passenger terminal buildings could be somehow quite similar than the design process. Normally, a process analysis can be used to improve the understanding of how the process operates, and to determine potential targets for process improvement through removing waste and increasing efficiency. It is needed a 'as-is' process mapping and measurement of the different activities within it but also as in a design process, it is needed start from a green field and map out the process together with the employees as it 'should-be'. The process analysis 'as-is' is the assessment of the current situation in diverse aspects: a) it answers the questions What is the problem?; How are things done?; and Where are the root causes of the problems?. The process analysis consists on the following steps: define objective and scopes, locate and gather data, map actors, functions and relationships of the process, verified, validate and analyze the current behavior of the system. Following this, an identification of possible improvements can be performed, see *Figure 10.*

Figure 10. Process Analysis

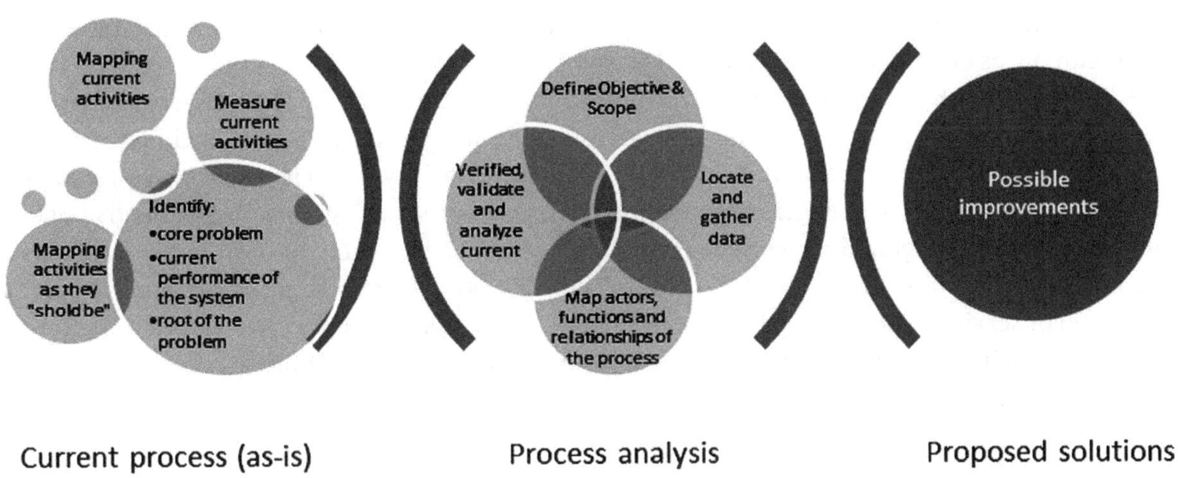

To understand the processes, facilities and services provided for passengers travelling through airports, and how to monitor and maintain efficient passenger flow are needed. Particularly, for the design and analysis of terminal, it is needed *forecasting, specification of level-of-service, flow analysis* and *design/ analysis of spaces* (De Neufville et al., 2013), (Odoni & De Neufville, 1992a).

Forecasting traffic levels for peak hours: A traffic forecast is the basis not only for capacity and resource planning at an airport, but also provides the foundation for an airport's business development and overall strategy. The objective of this exercise is to produce highly detailed, peak-hour demand scenarios for the design day many years ahead. Therefore, the definition of the passenger peak-hour is fundamental for a facility's design purposes. There are many definitions for the "peak-hour" concept at an airport, but most of them encompass a key concept: planning should aim to satisfy demand at some level below this peak, i.e. the majority of passengers receive adequate service levels and only an acceptably small proportion experience the impact of congestion, with just a few hours of operation during the year in which this passenger throughput might be surpassed (Wang & Pitfield, 1999).

The forecast, in turn, is used to develop hour-by-hour traffic scenarios, down to the level of a specific schedule of flights, for which assumptions must be made concerning the type of aircraft involved, their origin or destination, load factors, percentage of transfer or transit passengers, etc. However, according to Odoni et al., (1992a), forecasts are demonstrably inaccurate but extremely helpful and necessary to develop strategic and operational plans. Even though, frequent revisits to key operations are needed to ensure the level-of-service standards and the proper airport performance. Forecasts must be done for both airside and landside to have estimation on future activity in each area. The level and type of aviation activity expected at the airport are some of the key factors that determine the parameters that must be forecasted. Generally, the most important activity forecast for airfield planning is the level and type of aviation demand generated at the airport (as measured by aircraft operations), because this demand that defines the runway and taxiway requirements. Some of the key airfield typical design parameters are depicted in Figure 11. For the landside, the most important activity to be forecasted corresponds to ground handling services, traffic flows and type and destination of passengers (FAA, 2001), see Figure 12.

It is fundamental to define certain magnitude of *level-of-service* (LOS) standards that the capacity of the airport can handle with an acceptable quality. The specification of level-of-service standards aim to specify minimum standards for waiting times and space allocation, i.e. area per occupant at the processing facilities, the holding areas, and the passageways of the terminal. Hence, the LOS for space are usually defined in terms of "space conversion factors" giving the appropriate space per simultaneous occupant; it should be pointed out that higher standards imply more space and cost, which has to be compatible with the financial objectives of the airport manager. Each institution has its own standards, Table 5 and Table 6 introduces common LOS standards, (De Neufville & Odoni, 2013).

Recent works, such as the one in (FAA, 2001) have evaluated appropriate level-of-service standards applied in the terminal planning and design process while testing the continued validity of historic space allocation parameters that have been in use for more than 30 years. The researchers used a new approach to measure how passengers perceive the sufficiency of space, relying on quantitative data in combination with ethnographic interviews. Interviews were conducted on site within the terminals at seven case-study airports. The research also examined what other factors might affect positive perception of level of service, such as availability of extended information resources plus opportunities for use of technology (wireless connectivity, power connections for computers and other electronic equipment, and other innovations).

Figure 11. Airside typical design parameters

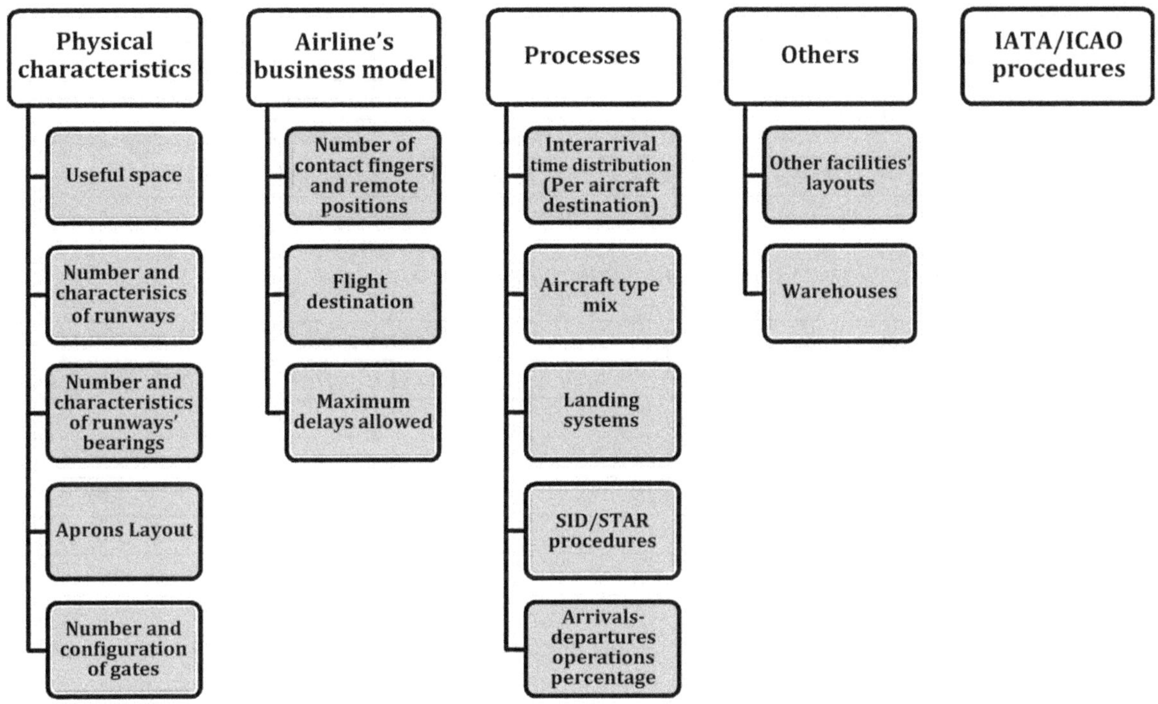

Figure 12. Landside typical design parameters

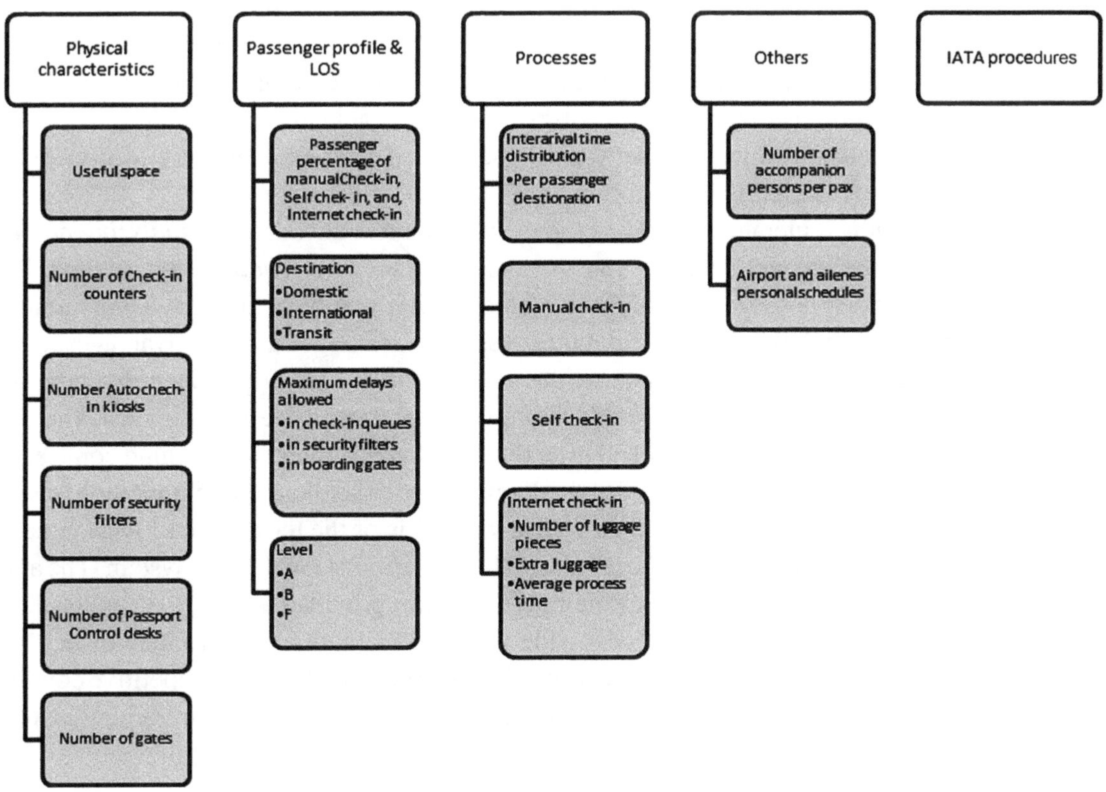

Table 5. Categories for Level of Service Indicators

Subsystem	Level of Service				
LOS	G- Excellent	H- High	I- Good	J- Adequate	K- Inadequate	L- Unacceptable
Flows	Free	Stable	Stable	Unstable	Unstable	
Delays	None	Very Few	Acceptable	Passable	Unacceptable	System Breakdown
Comfort	Excellent	High	Good	Adequate	Inadequate	

Table 6. Typical Values for Level of Service Indicators for Space Standard

Subsystem	Level of Service				
	A	B	C	D	E
Wait/circulate	2.7	2.3	1.9	1.5	1.0
Bag Claim	2.0	1.8	1.6	1.4	1.2
Check-in Queue	1.8	1.6	1.4	1.2	1.0
Hold-room Inspection	1.4	1.2	1.9	0.8	0.6

Even though the LOS can be converted into space indicators, there should be no confusion about allocating space in proportion to the number of people simultaneously in any particular part of the terminal. There is a key concept when designing which is called *dwell time*, according to Odoni (1992a) it is the amount of time spent in a particular area, which is central when determining the number of simultaneous occupants per area. The overall objective is not to provide facilities that result in a perfect match of unrestrained service levels for the peak hours but to provide a robust design that may not result in a waste of resources at other times. One of the most common mistake is to design spaces not considering the dwell time but only the number of peak hour passengers (also called typical peak hour passengers, TPHP). Therefore, a design must incorporate the time passengers actually spend in a space, i.e. their dwell time. It should be also distinguished between domestic and international traffic, for passengers, cargo and airplanes, if they are in transit or not.

In the following section, a literature review of the most prominent advances on *analyzing passenger flows* and *to determine the amount of space and the number of servers required* is presented. Essentially there are three popular approaches to do it: queuing theory; graphical analysis using cumulative diagrams; and computer simulations. Each technique presents advantages and disadvantages, queuing theory have not provided efficient designs mainly due to the processes in airports are dynamic by nature and the analytical solutions do not capture the complexity of those systems. Graphical analyses of the cumulative arrivals and service have worked effective when analyzing and designing some specific elements of the terminal such as departure and arrival lounges or ticket-counters. Simulation technique is perhaps the most robust approach to deal with the flow analysis of the terminal building. Simulation allows concentrating effort in investigating the effects of the uncertainties within the system. The appearance and development of computers as well as their systems have provided access to an endless source of tools to find alternatives for the solution of a wide variety of situations in all kinds of fields with high complexity. Thus, this resulted in the application of simulation which implies repetitive continuous experimentation without high costs, under safety environments and helpful for the training of people.

Simulation approaches allows the thorough analysis and the obtainment of an individual solution for a particular problem while analytical methods only provide general ones. The application of computer simulation has allowed an easy experimentation with virtual environments with such an important level of detail even in complex problems (Banks et al, 2010).

The approach and the methodology that is embraced within this work, is the use of simulation modeling and analysis, which is quite suitable for studying passenger flow in an airport terminal. Simulation can be considered as one of the best-fit solutions for modeling and analyzing a passenger terminal for capacity analysis under varying scenarios, taking into consideration LOS and other stochastic variables since simulation can, by nature, model deterministic as well as stochastic systems. It considers the variability and the dynamic features of the system which allows the end-to-end passenger experience analysis under a variety of scenarios.

LITERATURE REVIEW

Authors have tackled isolated each of the elements of the land and airside to enhance capacity and other as integrated systems. These works can be classified respect to three main aspects: level of detail (macroscopic and microscopic), methodology (commonly analytical and simulation models), and coverage (a single element or integrated system).

Macroscopic approaches lead to approximate answers mainly for planning purposes and some design issues, with an emphasis on assessing the relative performance of a wide range of alternatives. These kinds of approaches allow a strategic perspective, which is crucial in the current context for the major airports around the world, especially when there is not enough financial resources for taking the decision of investing in the construction of more facilities. There are few works that address the airport terminal capacity problem in a macroscopic approach. Furthermore, according to Solak, Clarke & Johnson (2009), there are no works which actually address the airport terminal capacity problem in a truly holistic fashion, because of the difficulty of modeling passenger flow in a complex terminal structure with transient demand patterns.

It can also be found diverse literature that cover the check-in allocation, security and immigration check points, baggage carousels, customs counters, holding areas and even amenities, i.e. restrooms, restaurants, shop, banks and other facilities.

Some of the first studies (Cox & Smith, 1961) and (Lee, 1966) use queuing models for enplanement passenger check-in to determine the average service and wait times. This approach was used to design or redesign efficient ticketing areas, including optimum capacity and space allocation. In the work of Odoni and Neufville (1992b) it has been analyzed the entire departure lounge in an airport terminal. Other functional areas such as corridors, seating, and immigration points have been included into the analysis (De Neufville and Grillot, 1982).

Some other developed methodologies to evaluate capacity at airport terminal facilities as in Mumayiz & Ashford (1986). The authors examined 6 airport facilities check-in, security control, immigration, baggage recovery and customs control. It was constructed a model of service perception and response based on processing time using 3 LOS. It was possible to establish the relationship between passenger demand and LOS provided. The FAA (2010) has also produced techniques to evaluate capacity and formulate recommendations to measure capacity. For each facility it is offered description and demand for components, operational characteristics, and diverse methods to evaluate capacity. Within this document, the FAA provides well defined requirements and data for airport planning, design, and modeling. Other works can be the ones from Tosic (1992) and Martel & Seneviratne (1990) where surveys of researches can be found on the area of airport passenger terminal operations analysis and modeling, including applications, methods and techniques.

Other authors have focus their attention on developing analytical models and simulation tools to monitor and analyze airport operations and optimizing the airport system and network with respect to capacity and delays, as well as cost/benefit analysis see (Odoni, 1991), (OPAL, 2003), (Andreatta et al., 1999), (Zografos & Madas, 2004), and more recently (Delahaye & Puechmorel, 2013), to mention some.

In relation to simulation of Mexican airports very few studies have been performed. Generally, these have been focused on Mexico City International Airport mainly modeling airside to reduce operating costs and passenger delays at runways (Herrera, 2001) and (Herrera, 2012) and recently (Mendoza et al., 2015). However, this work does not consider the landside and terminal area.

The check–in allocation problem, for example, has been studied by Parlar, Rodrigues, & Sharafali, (2013) using a dynamic programming approach. The authors find solution under some assumptions as closing or opening counters depending on the demand. Littler and Whitaker (1997) aim to estimate staffing requirements to meet a pre–set processing time target by means of simulation. Their approach was used as a decision support tool in the design. Park & Ahn, (2003), focus on the check-in problem at Gimpo airport to determine the most appropriate number of check–in counters. Mujica & Zuñiga (2015) employ a hybrid methodology to simulate the check-in problems using Genetic Algorithms and Discrete Event Simulation techniques. They propose an interesting methodology to address stochasticity in some variables.

Other works have focused their attention on the simulation of passenger security system as in (Penergraft, Robertson, & Shrader 2004). Their work applied the business process re-engineering approach to the passenger process. Their approach was used to provide analytical support for operational policy development of the airport. In the work of Wilson, RoSe, and So (2006) the impact of new or modified airport procedures and facilities on the security effectiveness, operational costs were evaluated to provide planning and operational support.

MODELING APPROACH

Simulation represents a fruitful technique since it generates an artificial history of a system, which provides the base for inferences with regards of the operating characteristics of the real system. With that, it gives certainty on the decisions due to its correct course of action. It is broad and flexible, providing a basis for continuous improvement according to the initial results and their performance.

Simulation can be defined as a numerical technique for producing experiments in a digital computer, using graphics, animation and others technological devices; which involve some mathematic and logical models, which describe the behavior of a system (Flores et al., 2013). Simulation involves the possibility to explore new policies and procedures without disrupting ongoing operations of the real system; also new systems can be tested without committing resources for their acquisition. In that sense, the time (a very valuable resource) can be compressed or expanded with the simulation. Also the hypothesis of the researching can be tested for feasibility; and insight can be obtained about the interaction of variables.

A simulation study can help in understanding how the system operates rather than how individuals think the system operates. In brief, the simulation can answer the question What if? This is useful in the design of new systems.

Simulation is used to model systems (or real-world process), which can be defined as a set of objects linked by some regular interaction or interdependence over time, concerning the consecution of some purpose. It was also established that a system is more than the sum of its parts; it is an indivisible whole. It loses its essential properties when it is taken apart. The elements of a system may themselves be systems, and every system may be part of a larger system. Every system includes other systems (called sub-systems), but at the same time the whole system is a part of a superior system (supra-system). Because of that, the interaction and interrelation between each component of a system is fundamental to understand the whole system (Banks et al, 2010), (Weiner, 2009).

On the other hand, a system is made up of the following elements*: entity, attributes, activities, state of a system* and *event*. The entity is an object or component in the system which requires explicit representation in the model. An entity can be dynamic in that it "moves" through the system, or it can be static in that it serves other entities; the attributes are properties of that entity; whereas an activity represents a time period of specified length. Other important concepts are called the states and events. A state is the collection of variables necessary to describe the system at any time, relative to the objectives of the study; these variables are called state variables; while an event is an instantaneous occurrence that may change the state of the system (Banks et al, 2010).

According to Weiner (2009), systems can be classified depending on how the states variables and time are represented in the model as *continuous* and *discrete*. According to time base, there are continuous time paradigms, where time evolves continuously, and discrete time paradigms, where time evolves by advancing in discrete portions. On the other hand, if the values of the state variables are considered, there are continuous models, where the variables take their values from continuous set represented as a real number, and discrete models, where the variables are discrete and can be represented as a finite set of integer numbers.

Discrete Event Simulation (DES) consists of modeling a system over time with the development of changes on its state at discrete points in time. It develops after producing a sequence of system images, representing the evolution of the system through time (Banks et al, 2010). There is a large literature in the Discrete Event System (DES) field such as in (Cassandras & Lafortune, 2009), (Banks, 2010) and (Weiner, 2009), among others.

The Departure Process

The approach proposed within this work provides a model that mimics departure operations within a medium size Mexican airport, including all the departure functional areas. The departure process is illustrated through the flowchart presented in *Figure 13*. It starts when passengers arrive to the airport departure hall which is used to serve both, domestic and international passengers, as well as accompanying. The first process to be modeled is the check-in for both domestic and international flights. This process models passenger arrival before flight until the check-in is performed.

The proposed model can be adapted to any airport and can be filled with data for any period of time (i.e. peak or non-peak hours). Figure 14 introduces the daily arrival distribution used within this work. The simulation is executed for 1 day period but it can also be simulated for different periods i.e. one

Figure 13. Passenger arrival procedure for international and domestic flights

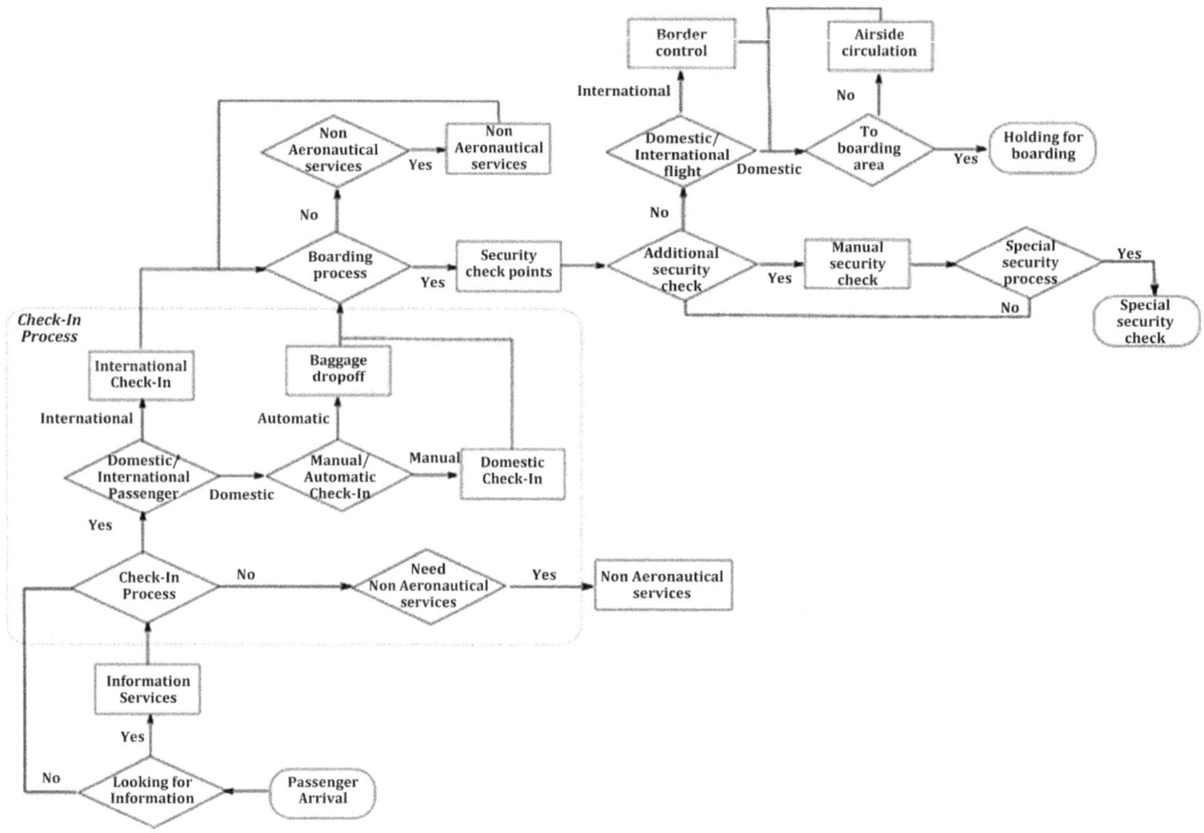

Figure 14. Typical daily passenger arrival distribution in a medium size Mexican airport

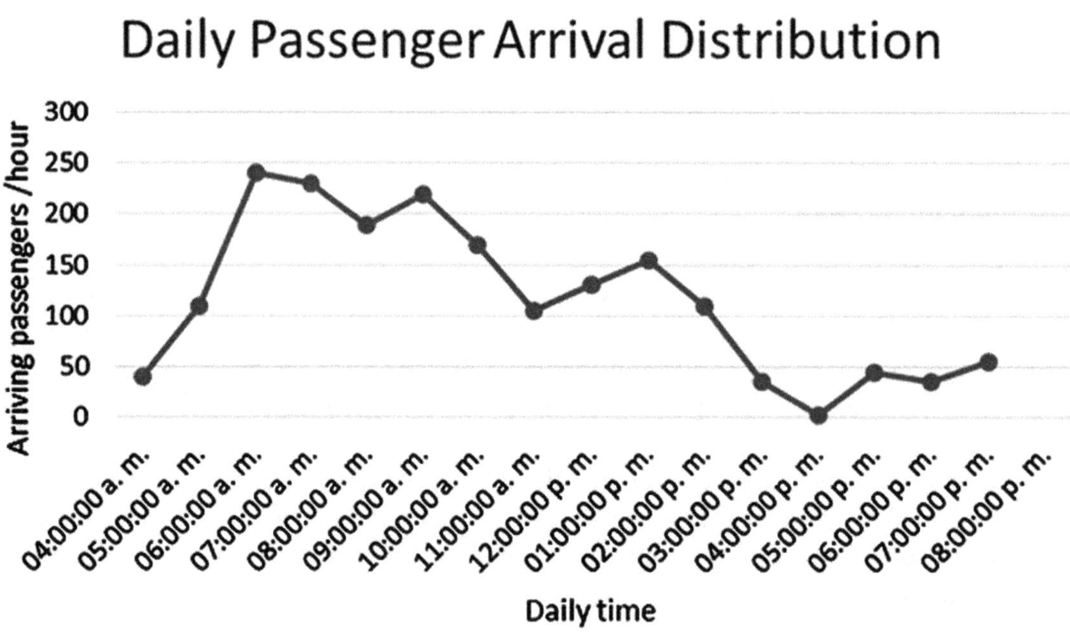

or more days, weeks, months or years of operation, depending of the planning purpose. The model of the boarding process has been developed using a general purpose simulation software called SIMIO (SIMIO, 2015).

Normally, passengers and their companions arrive from a time-based arrival schedule which depends on their trip (see Figure 15 as an example). Each passenger has different characteristics such as group size, international/domestic trip destination, number of bags, and servicing airline and flight, among others.

The arrival profiles of the passengers have been taken as three different types: business, tourist and special passengers since the objective of the work only is to present the approach for supporting decisions in operation planning. Each type of passenger arrival has been modeled with a discrete distribution of 30% for business arrival, 50% of tourist and 15% for special passengers. According to (Schultz & Fricke, 2011), the movement behavior mainly depends on passengers characteristics such as baggage (e.g. trolley, baggage cart, or rucksack), gender, group size and travel purpose (business or tourist), among others. Hence an average normal distribution has been taken for each kind of the three passengers modeled, see Table 7.

To locate the corresponding check-in counter it can be used the flight information desk or display screens. The check-in area might be composed by manual check-in or automatic (self-service check-in/ online check-in) and baggage drop to complete the check-in process. 14 manual ticketing counters and 3 automatic ones (kiosks) have been modeled. Passengers are routed to their respective desks based on their characteristics, i.e. servicing airline, destination, and corresponding class and flight desk. Furthermore, passengers can choose between manual and automatic ticketing to perform their check-in process. For domestic flights, both options are normally available, but for international ones, it should be performed normally a manual check-in. The manual check-in includes baggage acceptance, security

Figure 15. Example of passenger arrival distribution in an international flight

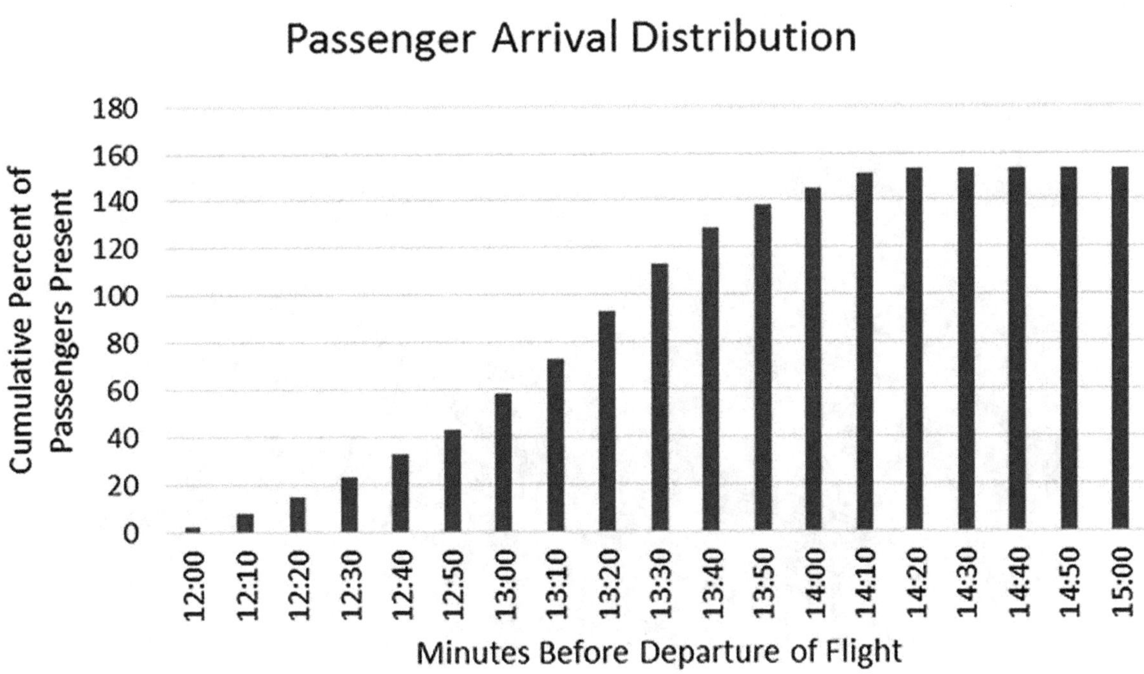

Table 7. Speed profiles for different passenger types

Passenger Type	Speed (*m/s*)
1: Business	Uniform (0.4,0.5)
2: Tourist	Uniform (0.3,0.9)
3: Special passenger	Uniform (0.7,1.2)

questions, documentation checks, boarding card issue, and gate information. In most of the flights, there is a preferential desks(s), for first class, business or special travelers. The automatic check-in includes documentation checks, boarding card issue, and gate information, the baggage drop-off has to be performed in a special desk or machine.

The check-in area might be composed by manual check-in or automatic (self-service check-in/online check-in) and baggage drop to complete the check-in process, see Figure 16. Passengers are routed to their respective desks based on its characteristics, i.e. servicing airline, destination, and corresponding class and flight desk. Furthermore, passengers can choose between manual and automatic ticketing to perform their check-in process if they are taking a domestic flight, but for international passengers, the process should be performed within manual check-in. The manual check-in includes baggage acceptance, security questions, documentation checks, boarding card issue, and gate information. In most of the flights, there is a preferential desks(s), for first class, business or special travelers. The automatic check-in includes documentation checks, boarding card issue, and gate information, the baggage drop-off has to be performed in a special desk or machine.

Table 8 show some simulation parameters employed in the model. Data was taken from the work of (Siroky, 2014). It is introduced the service time and probability that passengers use the different services available, i.e. information desk, leisure services or passengers to direct to the check-in process and the arrival distribution expression of time spend within each service.

Figure 16. Snapshot or check-in process

Table 8. Simulation parameters

Probability to use a Specific Service		Service Time (min)		
		Business	**Tourist**	**Special**
Information desk	35	Triangular(0.4,1,2)	Triangular(0.8,3,5)	Triangular(1,4,8)
Shops	10	Triangular(1,3,5)	Triangular(3,6,10)	Triangular(10,15,25)
Restrooms & phones	10	Triangular(2,3,4)	Triangular(2.5,3,5)	Triangular(3,5,10)
Restaurants	10	Triangular(0,7,15)	Triangular(5,30,60)	Triangular(20,30,60)
Waiting area	50	Triangular(15,25,40)	Triangular(20,45,60)	Triangular(30,45,60)
Manual check-in	80	Lognormal(3,0.25)	Lognormal(4.05,0.45)	Lognormal(6.058,0.50)
Automatic check-in	20	Gamma(5,2)	Gamma(14.37,2.29)	Gamma(19.7,3)
Baggage drop-off		Lognormal(2,0.2)	Lognormal(2.39,0.25)	Lognormal(4,0.5)
Security check	100	Gamma(5,1)	Gamma(14.37,2.29)	Gamma(20,2)
X-ray screening	100	Exponential(0.02)	Exponential(0.0464)	Exponential(0.06)
Immigration checkpoint	30	Triangular(10,15,20)	Triangular(15,20,25)	Triangular(15,20,25)

At any moment before the security points, passengers can use other supplementary facilities and services such as restrooms, banks, shops, telephones, catering facilities (restaurants, cafe, bars, etc.), children's play areas, leisure areas, baggage trolleys, information counters, airline customer service, sales desk, etc. which are located in the public common area, see Figure 17. There are also special passenger-handling services and facilities which can be used for passengers with reduced mobility and unaccompanied minors.

After check-in, passengers should go to security screening checkpoints to enter the area for boarding. The security checkpoint is typically composed of a combination of regular lanes and employee lanes, see Figure 18. In our case, two regular lanes and two scanning devices have been modeled. This area also

Figure 17. General services provided for passengers

Figure 18. Security control points

includes preceding queuing area, individual queues, carry-on luggage drop off, scanning devices, and bag pick up. One of the main objectives of the luggage inspections and physical screening is to prevent airplane hijacking, therefore there are some prohibited or restricted items and liquids in the carry-on luggage. It is normally asked to passengers to show their boarding pass and passport to the security staff and put them onto the carry-on luggage or bins. Within this process, passengers are asked to leave their carry-on luggage at the special counter lines and to remove small objects from them (keys, wallet, cell phone, scarf, etc.) and place them either in their carry-on luggage or in a tray to be screening. Once all objects are removed, the passenger can join the line and are asked to walk through the metal detector, if it is necessary, a physical or manual screening could be performed.

In the case that the passenger will perform an international flight, an immigration control procedure must be carried out by the immigration border control police. International passengers are asked to show their proper immigration documents and to answer some questions to verify their destination, among other tasks. If the passenger is consider "clean" he can continue his path to the waiting area. Figure 19 depicts the four border control counters modeled within this work.

Regarding the terminal building for commercial flights, it has an area of 200 m2. The layout, it is depicted in Figure 20. It can be noticed that each area has been label with the proper process which is performed. At the right side of the entrance, it is located the check in area which is 14x6 m; the general non-aeronautical services are located straight ahead of entrance in an area of 7x6 m. The security area is an area of 6x4 m and immigration is 6x5 m.

The first scenario to be tested corresponds to data obtain from a sample day at a Mexican airport, which includes 5 airlines and 17 flights in a typical day. Table 9 introduces the flights considered as well as destination and schedule date. A second scenario has been introduced modeling 3 more flights with a total of 250 passengers to test the behavior of the system under the growing demand predicted by the Mexican aeronautical organization, see Table 10. It has been marked in red the inclusion of 3 flights in a single day. In both scenarios, check-in desk have been assigned in the following way: Aero-

Figure 19. Immigration control points

Figure 20. Airport layout.

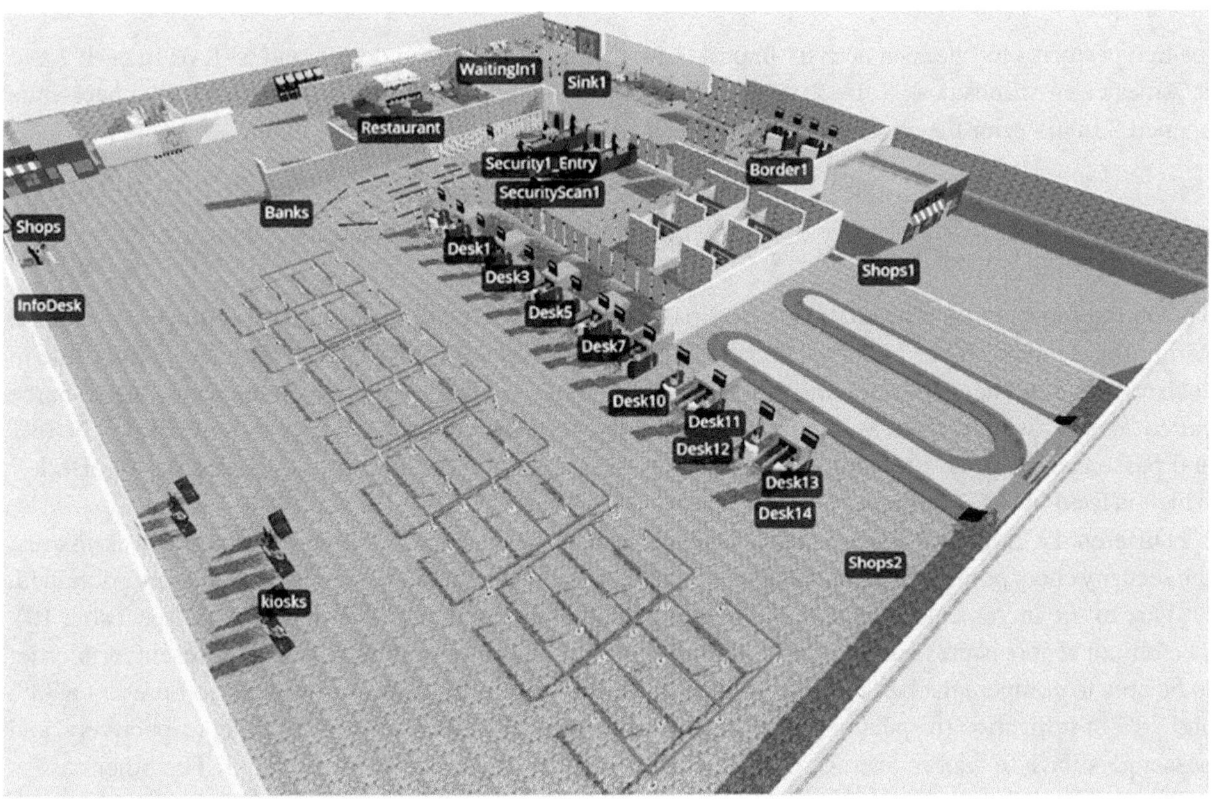

Table 9. Scenario I

Airlines	Flight	From	To	Scheduled	Passengers per Flight
Aeromexico	2237	Querétaro	Monterrey	06:10:00 a. m.	60
TAR	LCT-764	Querétaro	Monterrey	06:45:00 a. m.	50
United Airlines	5216	Querétaro	Houston	06:50:00 a. m.	90
American Airlines	5788	Querétaro	Dallas	07:15:00 a. m.	60
Aeromexico	2264	Querétaro	Monterrey	08:02:00 a. m.	50
Volaris	207	Querétaro	Monterrey	09:45:00 a. m.	50
TAR	LCT-432	Querétaro	Guadalajara	10:10:00 a. m.	50
United Airlines	4483	Querétaro	Houston	11:43:00 a. m.	100
Volaris	847	Querétaro	Tijuana	12:20:00 p. m.	50
American Airlines	5780	Querétaro	Dallas	01:49:00 p. m.	100
Volaris	112	Querétaro	Cancun	01:51:00 p. m.	60
United Airlines	5217	Querétaro	Houston	02:19:00 p. m.	80
TAR	LCT-762	Querétaro	Monterrey	02:30:00 p. m.	50
Aeromexico	2226	Querétaro	Monterrey	03:15:00 p. m.	60
TAR	LCT-500	Querétaro	Acapulco	04:15:00 p. m.	50
TAR	LCT-766	Querétaro	Monterrey	06:55:00 p. m.	50
Aeromexico	2268	Querétaro	Monterrey	08:10:00 p. m.	50

mexico performs its check-in activity in desk 1 to 4; United Airlines in desk 5 and 6; TAR in desk 7 and 8; American Airlines desk 9 to 12; and Volaris desks 13 and14. For domestic flights each check-in is opened 2 hours prior the departure and for international ones three hours.

RESULTS

The output report of both simulation scenarios (I and II) presented in the previous section are presented within this section. Table 11 shows diverse key performance indicators (KPIs) such as waiting time in each resource and percentage of utilization. As an example, the reader must pay attention at the resource called "desk1" which refers, as it names indicates, to check-in counter #1. It reports a utilization of 73% and passengers that have to wait in it in average spend 0.056 hrs. The most congested desk is Desk 7 which is used by TAR airlines.

Different LOS indicators have also been evaluated using the simulation model for the check-in area, the security checkpoint, immigration checkpoint and waiting areas. Table 12 presents the results obtained.

Due to the increment forecasted by authorities, other scenario has been introduced (see Table 10). The output report is presented in Table 13 and Table 14 . The same KPIs have been presented in order to be able to compere the behavior of the model. Meanwhile desk 13 and 14 in scenario I present a 73% and 78% of utilization (respectively), in scenario II it is used 75% and 85% of the time, respectively; and passengers have to wait in average 0.049 and 0.055 hrs, respectively, to be serviced. The other case is

Table 10. Scenarios II

Airlines	Flight	From	To	Scheduled	Passengers per Flight
Aeromexico	2237	Querétaro	Monterrey	06:10:00 a. m.	60
TAR	LCT-764	Querétaro	Monterrey	06:45:00 a. m.	50
United Airlines	5216	Querétaro	Houston	06:50:00 a. m.	90
American Airlines	5788	Querétaro	Dallas	07:15:00 a. m.	60
Aeromexico	2264	Querétaro	Monterrey	08:02:00 a. m.	50
Volaris	111	Querétaro	Mexico city	08:20:00 p. m.	80
Volaris	207	Querétaro	Monterrey	09:45:00 a. m.	50
TAR	LCT-432	Querétaro	Guadalajara	10:10:00 a. m.	50
United Airlines	4483	Querétaro	Houston	11:43:00 a. m.	100
Volaris	847	Querétaro	Tijuana	12:20:00 p. m.	50
American Airlines	5780	Querétaro	Dallas	01:49:00 p. m.	100
Volaris	112	Querétaro	Cancun	01:51:00 p. m.	60
United Airlines	5217	Querétaro	Houston	02:19:00 p. m.	80
TAR	LCT-762	Querétaro	Monterrey	02:30:00 p. m.	50
Aeromexico	2226	Querétaro	Monterrey	03:15:00 p. m.	60
TAR	LCT-500	Querétaro	Acapulco	04:15:00 p. m.	50
TAR	LCT-766	Querétaro	Monterrey	06:55:00 p. m.	50
Aeromexico	2268	Querétaro	Monterrey	08:10:00 p. m.	50
United Airlines	5217	Querétaro	Miami	08:50:00 p. m.	90
Volaris	111	Querétaro	Mexico city	08:20:00 p. m.	80

presented due to the increment to this flights corresponds to desk #5 and #6, belonging to United airline which, due to the new international flights, has been used 89% and 85% of time (respectively) and passengers have to wait 0.064 and 0.063 hrs, respectively, on average.

The LOS values help to decide if space utilization in peak hours is adequate. In our example, within the check-in area, in Scenario I, a minimum of 9.5 m2/pax has been accounted meanwhile in Scenario II the space is reduced to 8.87 m2/pax. It indicates that the space in the check-in area will decrease around 0.63 m2/pax in peak hours. According to Table 5., it places the service of both scenarios as level "A" meaning that the comfort and level of service is excellent, delays are not expected and the flow of passengers will be excellent within this area.

Analogous, the security checkpoints area has also been affected changing the available space from 19.65 m2/pax to 17.56 m2/pax. However, this decrement of space does not affect passenger's LOS, comfort and flows. Immigration checkpoints space decrement is not really meaningful respect to the LOS, mainly due to the nature of the new flights, i.e. only one out of three new flights is international.

In our example, the decision makers will approve the expected incoming flights because they will not decrease the level of service at the airport. Furthermore, it could be inspected new scenarios with bigger flight demand or new incoming flights. Nevertheless, the common areas have to be revisited integrating the departure process to have a complete overview of the behavior of the system.

Table 11. KPIs numerical results from scenario I

Resource	Utilization (%)	Pax Time (hrs)		
Check-in Desks				
	Average	*Average*	*Minimum*	*Maximum*
Desk1	73	0.056	0.055	0.056
Desk2	72	0.053	0.054	0.056
Desk3	79	0.057	0.056	0.059
Desk4	75	0.056	0.056	0.058
Desk5	80	0.061	0.063	0.068
Desk6	85	0.065	0.066	0.069
Desk7	87	0.066	0.068	0.072
Desk8	86	0.065	0.068	0.070
Desk9	65	0.044	0.047	0.048
Desk10	52	0.040	0.041	0.042
Desk11	57	0.041	0.042	0.043
Desk12	61	0.042	0.046	0.0047
Desk13	73	0.056	0.056	0.058
Desk14	78	0.047	0.051	0.053
Security Checkpoints				
Security Counter 1	67	0.083	0.083	0.083
Security Counter 2	65	0.086	0.086	0.088
Security Scan 1	25	0.0030	0.0030	0.0031
Security Scan 2	29	0.0041	0.0041	0.0042
Border Control Checkpoints				
Border Counter 1	53	0.044	0.044	0.044
Border Counter 2	42	0.043	0.044	0.046
Border Counter 3	49	0.042	0.045	0.045
Border Counter 4	45	0.041	0.043	0.043
Inside Waiting Area				
Waiting area 1	85	0.45	0.45	0.49
Waiting area 2	87	0.56	0.58	0.52

Table 12. LOS numerical results from scenario I

	Average	Minimum	Maximum
Check-in LOS	11.81	9.5	15.8
Security LOS	19.65	16.2	24.21
Border LOS	22.15	19.54	27.19

Table 13. KPIs numerical results from scenarios II

Resource	Utilization (%)	Pax Time (hrs)		
Check-In Desks				
		Average	Minimum	Maximum
	Average	*Average*	*Minimum*	*Maximum*
Desk1	73	0.056	0.055	0.056
Desk2	72	0.053	0.054	0.056
Desk3	79	0.057	0.056	0.059
Desk4	75	0.056	0.056	0.058
Desk5	89	0.064	0.066	0.075
Desk6	85	0.063	0.071	0.073
Desk7	87	0.066	0.068	0.072
Desk8	86	0.065	0.068	0.070
Desk9	65	0.044	0.047	0.048
Desk10	52	0.040	0.041	0.042
Desk11	57	0.041	0.042	0.043
Desk12	75	0.049	0.051	0.0055
Desk13	81	0.057	0.059	0.062
Desk14	85	0.054	0.055	0.065
Security Checkpoints				
Security Counter 1	67	0.083	0.083	0.083
Security Counter 2	65	0.086	0.086	0.088
Security Scan 1	25	0.0030	0.0030	0.0031
Security Scan 2	29	0.0041	0.0041	0.0042
Border Control Checkpoints				
Border Counter 1	53	0.044	0.044	0.044
Border Counter 2	42	0.043	0.044	0.046
Border Counter 3	49	0.042	0.045	0.045
Border Counter 4	45	0.041	0.043	0.043
Inside Waiting Area				
Waiting area 1	85	0.45	0.45	0.49
Waiting area 2	87	0.56	0.58	0.52

Table 14. LOS numerical results from scenarios II

	Average	Minimum	Maximum
Check-in LOS	10.1	8.87	13.8
Security LOS	17.56	14.58	21.59
Border LOS	19.89	17.58	24.58

CONCLUSION

Air transport contributes to sustainable development worldwide not only due to the transportation of tourists and freight which moved 3.3 billion people and 50 million metric tons in 2014, but also because it provides direct and indirect jobs which in turn generate economic growth in a regional and domestic scale, to be precise $2.2 trillion USD in 2012. In the same way, it develops social and economic networks that will have long-lasting effects and closer relationships between countries, both from an individual perspective and at a country level. It improves living standards and facilitates the international trade as it has been demonstrated. All over the world, it offers a vital lifeline to communities that lack adequate road or rail networks. Summarizing, air transport is a vital component of modern life.

The traffic growth tendencies are quite favorable to the industry and all its different airspace users all over the world. As an example it can be pointed out the Asian-Pacific region that in 2032 is forecasted to account half of the fleet in the world. Nevertheless, the bigger demand and greater connectivity have to properly address to bring its benefits for the industry. Specially, in the commercial sector, the use of different and innovative airport management strategies is essential to use efficiently valuable corporate resource while meeting the growing demand of the diverse airspace users.

It has been said that all the airport network will grow doubling the number of mega-cities around the world and moving daily around 2.3 million passengers. In Mexico, the principal airport, Mexico City International Airport, is already suffering the consequences of this non stopping increasing demand. The government recently announced a project to build a new one to replace the older airport, it is and will be needed to provide decentralized strategies together with a deep analysis of the rest of the network.

Our proposed approach provides a model that generically mimics the departure process in a medium size airport, which is the case of more than the 90% of the Mexican airports. Most of the functional areas and operations within the departure process have been included. The most important KPIs together with the LOS indicators have been analyzed to highlight the benefits of this approach. The variable nature of airport operations make suitable the used of discrete event simulation to perform the analysis. It provides meaningful information in a timely and efficient manner.

As the purpose of this work was to show the benefits of the approach, lot of work can be done as future research, starting by applying real data from a case study, and analyzing more deeply the results obtained. A broader forecast can be employed in order to design the scenarios and different KPIs can be implemented.

REFERENCES

Air Transportat Action Group. (2012). *Aviation benefits beyond borders.* Retrieved from http://www.atag.org/

AIRBUS. (2014). *Global Market Forecast 2015-2034*. Flying on Demand Deutsche Bank Toulouse.

Airport Council International. (2012). *World Airport Traffic Report.* Author.

Airport Council International. (2013). *World Airport Traffic Report.* Author.

Andreatta, G., Brunetta, L., Odoni, A. R., Righi, L., Stamatopoulos, M. A., & Zografos, K. G. (1999). A Set of Approximate and Compatible Models for Airport Strategic Planning on Airside and Landside. *Air Traffic Control Quarterly*, *7*(4), 291–319.

Banks, J., Carson, J. S., & Nelson, B. L. (2010). *Discrete Event System Simulation* (5th ed.). United States: Pretince Hall.

Belobaba, P., Odoni, A., & Barnhart, C. (Eds.). (2009). *The global airline industry*. John Wiley & Sons. doi:10.1002/9780470744734

Cassandras, G. C., & Lafortune, S. (2009). *Introduction to discrete event systems*. Springer.

Cox, D. R., & Smith, W. L. (1961). *Queues*. New York: John Wiley & Sons.

De Neufville, R., & Grillot, M. (1982). Design of pedestrian space in airport terminals. *Journal of Transportation Engineering*, *108*, 87–102.

De Neufville, R., & Odoni, A. (2013). *Airport Systems. Planning, Design and Management*. McGraw-Hill Education LLC.

Delahaye, D., & Puechmorel, S. (2013). *Modeling and Optimization of Air Traffic*. Wiley-ISTE. doi:10.1002/9781118743805

EUROCONTROL. (2013). *Challenges of Growth 2013 Task 4: European Air Traffic in 2035*. Technical Report.

Federal Aviation Administration. (2001). *Forecasting aviation activity by airport*. Federal Aviation Administration, Technical Report.

Federal Aviation Administration (FAA). (2010). Airport Passenger Terminal Planning and Design. In Guidebook, Office of Oper. Research (vol. 1). Federal Aviation Administration.

Flores, I., Figueras, J., Guasch, A., Mujica, M., Narciso, M., & Piera, M. (2013). *Simulation Models: Using Simio and Preti Nets*. Mexico: Universidad Nacional Autonoma de Mexico.

Herrera, G. A. (2001). *Simulación de operaciones aeroportuarias. El caso de despegues y aterrizajes en el Aeropuerto Internacional de la Ciudad de México. Publicación Técnica No. 180*. México: Instituto Mexicano del Transporte.

Herrera, G. A. (2012). *Modelo de simulación de operaciones aéreas en aeropuertos saturados. El caso del Aeropuerto Internacional de la Ciudad de México. Publicación Técnica No. 365*. México: Instituto Mexicano del Transporte.

International Air Transport Association. (2011). Vision 2050. Technical Report. Singapore.

International Air Transport Association. (2014). *Annual Review 2014*. Technical report 69th Annual General Meeting Cape Town.

International Civil Aviation Organization. (2013). *Caribbean/South American Regional Traffic Forecasts 2011-2031*. Report of the Ninth Meeting. Of the CAR/SAM Traffic. Forecasting Group (CAR/SAM TFG).

International Civil Aviation Organization. (2013b). *2013–2028 Global Air navigation Plan. ICAO 9750.* Canada: Published in Montréal.

Lee, A. M. (1966). *Applied Queuing Theory*. New York: St. Martin's Press Company.

Littler, R. A., & Whitaker, D. (1997). Estimating staffing requirements at an airport terminal. *The Journal of the Operational Research Society*, *48*(2), 124131. doi:10.1057/palgrave.jors.2600344

Martel, N., & Seneviratne, N. P. (1990). *Analysis of factors influencing quality of service in passenger terminal buildings. In Airport Terminal and Landside Design and Operation 1990, Transportation Research Record 1273* (pp. 1–10). Washington, DC: Transportation Research Board, National Research Council.

Mendoza, E., Zuñiga, C. A., Mujica, M. A., & Delahaye, D. 2015, *Simulating Airport Capacity: Mexico City Airport Case.* Conferences in Air Transport & Operations. in Proceeding of the 3rd International Conference on Interdisciplinary Science for Innovative Air Traffic Management (ISIATM2015), Netherlands.

Mujica, M. A., & Zuñiga, C. A. (2015). Allocation of Airport Check-in Counters Using a Simulation-Optimization Approach. In M. A. Mujica, I. De La Mota-Flores, & D. Guimarans-Serrano (Eds.), *Applied Simulation and Optimization: In Logistics, Industrial and Aeronautical Practice* (pp. 203–229). Springer International Publishing. doi:10.1007/978-3-319-15033-8_7

Mumayiz, S., & Ashford, N. (1986). Methodology for planning and operations management af airport terminal facilities. *Transportation Research Record*, *1094*, 24–35.

Odoni, A. R. (1991). Transportation modeling needs: airports and airspace. Volpe domestic Transportation Systems Center, Cambridge, MA.

Odoni, A. R., & De Neufville, R. (1992a). Passenger terminal design. *Transportation Research Part A, Policy and Practice*, *26*(1), 27–35. doi:10.1016/0965-8564(92)90042-6

Odoni, A. R., & De Neufville, R. (1992b). Walking distance minimization for airport terminal configurations. *Transportation Research Part A, Policy and Practice*, *26*(1), 59–74. doi:10.1016/0965-8564(92)90045-9

OPAL Consortium. (2003). *Inventory of Airport Performance Models. Technical WP1 Report - Public Version, Optimisation Platform for Airports including Landside (OPAL) Research Project funded by the European Commission (Directorate-General Energy and Transport - DG-TREN).* Brussels, Belgium: European Commission.

Park, Y., & Ahn, S. B. (2003). Optimal assignment for check-in counters based on passenger arrival behaviour at an airport. *Transportation Planning and Technology*, *26*(5), 397–416. doi:10.1080/03081060310001635887

Parlar, M., Rodrigues, B., & Sharafali, M. (2013). On the allocation of exclusive-use counters for airport check-in queues: Static vs. dynamic policies. *OPSEARCH*, *50*(3), 433–453. doi:10.1007/s12597-012-0110-5

Parlar, M., & Sharafali, M. (2008). *Dynamic allocation of airline check-in counters: A queueing optimization approach*. *Management Science*, *54*(8), 1410–1424. doi:10.1287/mnsc.1070.0842

Pendergraft, D. R., Robertson, C. V., & Shrader, S. (2004). Simulation of an airport passenger security system. In *Proceedings of the 36th conference on Winter simulation (WSC '04). Winter Simulation Conference* (pp. 874-878). doi:10.1109/WSC.2004.1371402

Schultz, M., & Fricke, H. (2011). *Managing Passenger Handling at Airport Terminals. Individual-based Approach for Modeling the Stochastic Passenger Behavior.* Ninth USA/Europe Air Traffic Management Research and Development Seminar (ATM2011).

Secretaria de Comunicaciones y Transportes. (2015). *Estadística histórica (1992-2014) / Historical statistics (1992-2014).* Retrieved from http://www.simio.com

Siroky, J. & Hlavsová, P. (2014). *Optimizing process of check-in and security check at airport terminals.* Academic Press.

Solak, S., Clarke, J. P. B., & Johnson, E. L. (2009). *Airport terminal capacity planning. Transportation Research Part B: Methodological, 43-6*(6), 659–676. doi:10.1016/j.trb.2009.01.002

Tosic, V. (1992). A review of airport passenger terminal operations analysis and modeling. *Transportation Research Part A, Policy and Practice, 26*(1), 3–26. doi:10.1016/0965-8564(92)90041-5

Wang, P. T., & Pitfield, D. E. (1999). The derivation and analysis of the passenger peak hour: An empirical application to Brazil. *Journal of Air Transport Management, 5*(3), 135–141. doi:10.1016/S0969-6997(99)00007-1

Weiner, G. A. (2009). *Modeling and Simulation: A practitioner approach.* CR Press. doi:10.1201/9781420053371

Wilson, D., Rose, E. K., & So, A. A. (2006). Security checkpoint optimizer (SCO): An application for simulating the operations of airport security checkpoints. In *Proceedings of the 2006 inter Simulation Conference.* doi:10.1109/WSC.2006.323126

Zografos, K. G., & Madas, M. A. (2004). Design and Development of an Integrated Computational Platform for Total Airport Performance Analysis. In *Proceedings of the 2 International Congress on Transportation Research in Greece, Session on Air Transport.*

KEY TERMS AND DEFINITIONS

Airport Planning and Capacity: Airport planning encompasses capacity, master and regional planning, aviation forecasting, and airspace planning. Activities include providing planning guidance, producing the biennial National Plan of Integrated Airport Systems, developing capacity studies, collecting boarding and cargo data to determine airport categories.

Capacity: The airport operating level, expressed as the number of aircraft movements that can occur at an airport over a specified time period.

Commercial Aviation: Aircraft activity licensed by state or federal authority to transport passengers and/or cargo on a scheduled or non-scheduled basis.

Discrete Event Simulation: The process of codifying the behavior of a complex system as an ordered sequence of well-defined events. In this context, an event comprises a specific change in the system's state at a specific point in time.

Facility Design: Ensure a smooth flow of resources such as work, material, or information through a specific space within the system.

Peak Hour: Represents that highest number of operations or passengers during the busiest hour of an average day of a peak month.

Terminal Building: That building on an airport which is used in making the transition between surface and air transportation.

Chapter 8
VisTHAA:
A Statistical Tool for Comparison of Heuristics

Laura Cruz-Reyes
National Mexican Institute of Technology, Mexico

Claudia Gómez
National Mexican Institute of Technology, Mexico

Mercedes Pérez Villafuerte
National Mexican Institute of Technology, Mexico

Nelson Rangel Valdez
National Mexican Institute of Technology, Mexico

Marcela Quiroz-Castellanos
National Mexican Institute of Technology, Mexico

César Medina Trejo
National Mexican Institute of Technology, Mexico

Enith Martínez-Cruz
National Mexican Institute of Technology, Mexico

ABSTRACT

In this chapter, a scientific tool designed to facilitate fair comparisons of heuristics is introduced. Making a fair comparison of the performance of different algorithms is a general problem for the heuristic community. Most of the works on experimental analysis of heuristic algorithms have been focused on tabular comparisons of experimental results over standard sets of benchmark instances. However, from a statistical point of view, and according to the experimental design theory, a minimum requirement to compare heuristic algorithms is the use of non-parametric tests. Non-parametric tests can be used for comparing algorithms whose results represent average values, in spite of the inexistence of relationships between them, and explicit conditions of normality, among others. The proposed tool, referred to as VisTHAA, incorporates four non-parametric statistical tests to facilitate the comparative analysis of heuristics. As a case study, VisTHAA is applied to analyze the published results for the best state-of-the-art algorithms that solve the one-dimensional Bin Packing Problem.

DOI: 10.4018/978-1-4666-9779-9.ch008

INTRODUCTION

Many real-world problems are NP-hard, i.e. it is believed that there is no exact algorithm that solves them, and whose running time does not increase exponentially with the problem size (Garey & Johnson, 1979). There are two ways to attack the NP-hard problems. The first one is using *exact methods* that require exponential computational time. The second way, the most used in practice for large problems, corresponds to the use of *heuristic algorithms*, commonly called *heuristics*. Heuristics are procedures for which it is not possible to prove that the optimal solution will be obtained in finite time. Hence, a heuristic can be considered as a rule of thumb for guiding the solution search, achieving a good performance in terms of runtime and solution quality.

Heuristic algorithms include strategies based on common sense to provide approximate solutions in a reasonable time, but do not guarantee the optimal solution. Throughout the search for the best possible solutions for NP-hard problems, a wide variety of heuristic algorithms have been proposed. However, the "No free lunch" theorem (Wolpert & Macready, 1997) states that there is no efficient algorithm being better in behavior for all possible situations.

Metaheuristic algorithms are a special class of heuristics that include general algorithmic frameworks that can be applied to an extensive range of problems for producing heuristics of specific purpose. Generally, the principle of guided search of a metaheuristic is inspired by some natural phenomena. For example, in the case of evolutionary algorithms, they are based on genetic inheritance, and the Darwinian metaphor of "Natural Selection".

The behavior of a metaheuristic is controlled by a set of components and parameters that are set to induce a desirable performance on a given optimization problem. When a metaheuristic is applied to a particular NP-hard problem, some modules need to be developed and some parameters need to be tuned. These two problems are called *structural* and *parametric* setting, respectively; the combination of these two problems is called *tuning*. The tuning process is crucial for optimizing a metaheuristic; the choice of values for the search parameters has a substantial impact on the performance of the procedure (Halim, 2009).

Heuristic algorithms are too complex to be studied analytically, but become accessible through an experimental analysis. In many cases, the design and tuning of metaheuristics are made manually and without theoretical support, hindering the understanding of their good or bad performance (Snodgrass, 2010). From a scientific perspective, a new algorithmic proposal must demonstrate an improvement in performance when comparing with the state-of-the-art algorithms, making a series of computational experiments and demonstrating statistical validity (Adenso-Diaz & Corral, 2007). Several situations have hampered the fair comparison between algorithms, including: the lack of standard test cases; the lack of standards for reporting computational experiments; incomplete and heterogeneous experimental results addressing one problem, presenting a single result, or using different equipment with unstandardized processing time, among others; and the ignorance of the most appropriate statistical test to validate results. Currently there are many attempts to rectify this situation. Guides to properly report computational experiments are presented in (Crowder et al., 1979; McGeoch, 2007; Preuss, 2007). A simple methodology for scaling different computer equipments is reported in (Quiroz-Castellanos et al., 2015). Some guides to identify the most appropriate statistical test can be found in (Adenso-Diaz & Corral, 2007; Derrac et al., 2011).

The main objective of this research is to contribute to the automation of experimental analysis of heuristic algorithms, by means of a statistical tool that allows the comparison among different algorithms, or different configurations for a single algorithm. The proposed scientific tool is referred to as VisTHAA (Visualization Tool for Heuristic Algorithm Analysis). The first version of VisTHAA was reported in a previous work (Castillo et al., 2011); the architecture of the tool has been redesigned to allow statistical and visual diagnosis. It is expected that VisTHAA contributes to the area of experimental algorithms with an architecture that provides several techniques of analysis and design.

The remainder of this chapter is structured as follows. The Section of *Background* presents a brief introduction to the statistical analysis using hypothesis testing methods. The Section *Statistical Analysis of Heuristic Algorithms* discusses related work about the use of nonparametric tests for heuristic comparison and describes some relevant nonparametric tests. The Section *A Visualization Tool for Heuristic Algorithm Analysis* presents the architecture of VisTHAA and the incorporation of nonparametric statistical tests. The Section *Case of Study: Performance of Bin Packing Algorithms* presents a statistical analysis, through the nonparametric tests incorporated in VisTHAA, of the best state-of-the-art algorithms that solve the Bin Packing Problem. Finally, the Section *Conclusions* concludes with final remarks, highlighting open problems and proposing future work.

Bin Packing Problem

The Bin Packing Problem (BPP) is an NP-hard problem (Garey & Johnson, 1979) and is considered intractable because they demand many resources for their solution, it consists in packing a set of n items of different sizes $W = \{w_1, \ldots, w_n\}$ in the minor number of fixed size bins without violating the capacity c of any bin.

The importance of BPP is that it has a large number of industrial and logistics applications, and many real practical problems can be modeled as packing problems. BPP is of great interest in: planning of telecommunication, backups in media, transportation, production and logistics/supply-chain systems, among others (Crainic et al., 2011). In military and maritime areas it can be applied in the transportation and logistics problems such as filling up containers and loading trucks with weight capacity constraints.

For finding good and better solutions for BPP, a variety of algorithms have been designed. The most significant results were obtained using metaheuristics algorithms. The performance of one of them is the case study presented in this chapter.

BACKGROUND

The design of experiments is a task of great importance in the development of new algorithms, and it is commonly supported by statistical analysis. The validation of new algorithms requires an experimental framework that allows the inclusion of a wide range of test problems and state-of-the-art algorithms. The critical part of these comparisons lies with the statistical validation of the results, contrasting the differences between methods. This section presents a brief introduction to the statistical analysis using hypothesis testing methods.

Statistical Analysis

Researchers from different areas of knowledge use different resources for multivariate data analysis based on hypothesis testing. For this purpose there are commercial statistics tools (SPSS®, SAS®, Minitab®, Matlab, etc.), as well as free versions (e.g., DataPlot, R and Octave). However, due to the particular features that distinguished each field of application, and the nature of the analyzed data, researchers generally use more than one of these software tools to complement the analysis.

The scientific literature provides recommendations over hypothesis testing that are appropriate for performance analysis of heuristic algorithms. However, there is a lack of accessible documentation over this topic. Because of this situation, this chapter presents a review of statistical test following an approach oriented to their implementation, and based on the proposed architecture for the analysis of heuristic algorithms.

Statistical Test

The verb *infer* means "*to draw conclusions as a consequence or a probability*" (Siegel et al., 1995). A key feature of modern statistics is statistical inference. Inferential statistics is an area focused in two types of problems: the estimation of population parameters, and hypothesis testing. The statistical inference is oriented to draw conclusions about large groups of individuals or events, based on observations of a few subjects, or in past situations; the statistics provide tools that formalize and standardize procedures for such conclusions.

A common problem in statistical inference is to determine, in terms of probability, whether the observed differences between two samples represents that the sampled populations are in fact different. Another common problem is to determine if a sample from scores belongs to a specific population. An additional problem the statistical inference deal with is to answer properly the question whether several groups differ from each other or not. The procedures of statistical inference for hypothesis testing can determine if the results obtained by observation, for problems such as the previously described, were within the extent in which they might have occurred, or were just random. The elements involved in hypothesis testing are described below (Mendenhall & Sincich, 1997; Berenson, Levine & Krehbiel, 2001):

- **Hypothesis:** It is a proposition, or set of propositions, set forth as an explanation for the occurrence of one or more characteristics of a population. Features may be the parameters of a predetermined probability distribution, which has been chosen for the population.
- **Hypothesis Testing:** It is a method of statistical inference used for accepting or rejecting a hypothesis. There are generally two hypotheses in contrast: a) the null hypothesis (H_0), which is the hypothesis to be validated; and, b) the alternative hypothesis (H_1), which is the hypothesis that is accepted when the null one is rejected. In a one-sided test the critical area of a distribution is one-sided, so that either it is greater than or less than a certain value, but not both. The two-sided test requires testing the area under both of the sides (tails) of a distribution.
- **Test Statistic:** It is a random variable used to estimate a parameter of a population (e.g. a mean) from some attribute of a sample (e.g. its arithmetic mean value). The test statistic distribution is known when H_0 is true.

- **Significance Level** (α): It is a probability threshold below which the null hypothesis H_0 will be rejected. Common values are 5% and 1%. The criterion is based on the probability of obtaining a statistic measured in a sample if the value stated in the null hypothesis were true.
- **Region of Acceptance and Critical Region:** Statistical set of values that lead to acceptance or rejection of H_0, respectively.
- ***p*-Value:** When H_0 is true, it is the probability of obtaining the observed results in the sample, or more extreme results. The extreme values depend on the way (one-sided or two-sided) that the hypothesis is tested. H_0 is accepted if p-value $> \alpha$, otherwise, it is rejected and the evidence suggests that the observed data are inconsistent with the assumption that H_0 is true. In other words, it is the least value that α can take such that the test would reject the null hypothesis.
- **Decision Errors:** Two types of errors can result from a hypothesis testing. A Type I error defines the cases in which a null hypothesis is rejected when it is actually true; the probability of commiting this type of error is controlled by the significance level α. A type II error (or β error) occurs when a null hypothesis is accepted when it is false; β is the probability of retaining a null hypothesis that is actually false. The power of the test of the null hypothesis H_0 versus the alternative H_1 is the probability of not commiting a Type II error (the complement of β), so is the probability that the test will reject the null hypothesis when in fact the alternative hypothesis is true. Decreasing α generally causes β to increase and vice versa, while increasing the sample size will decrease both.

General Procedure for Hypothesis Testing

In general, the method for a hypothesis test consists in taking a random sample of the population under study and formulating both, null and alternative hypothesis, about the population. If the observations do not support the assertion postulated in the null hypothesis then it is rejected, otherwise it is accepted. The significance level α is associated with the decision, which can be interpreted as the rate of falsely rejecting the null hypothesis (or type I error). A more detailed description of the procedure for hypothesis testing can be found in (Bartz-Beielstein & Preuss, 2010), which involves the following steps, assuming that we have a sample *X*:

1. **State a null hypothesis H_0 and an alternate hypothesis H_1:** The hypotheses are formulated on a parameter Θ related to the statistical distribution of the population.
2. **Specify a test statistic and its distribution:** A test statistic *T*(*X*) is considered as a numerical summary of a sample data *X* that can be used to evaluate the null hypothesis. The sampling distribution, with respect to a parameter Θ, $F_\Theta(t)$ of a test statistic under the null hypothesis must be known and calculable, either exactly or approximately, which allows p-values to be calculated.
3. **Set the criteria for a decision:** Define a rejection region (the critical region, *R*) for the expected distribution of the test statistic and based on a pre-assigned significance level α (generally $\alpha =$ 0.05). In a 2-sided test, with $\alpha = 0.05$, for example, the critical region would have 0.025 in either tail.
4. **Compute the test statistic from a sample (observed data):** It allows determining the likelihood of obtaining sample outcomes if the null hypothesis were true.

5. **Take a decision:** In classical hypothesis testing, the null hypothesis is rejected if the test statistic is within the critical region; the decision is based on the probability of obtaining a value of the statistic used at least as extreme as the observed one, given that the value stated in the null hypothesis is true, stated by the p-value. H_0 is rejected if the p-value is smaller than α. If H_0 is not rejected, the p-value may be very close to α, allowing the option of rejecting at a slightly reduced level. The p-value provides more information than just that the hypothesis was rejected or not.

STATISTICAL ANALYSIS OF HEURISTIC ALGORITHMS

Experimental analysis of the performance of heuristics is a crucial and necessary task in an investigation. There are two main types of statistical test for analyzing the results obtained by heuristic algorithms: parametric tests and nonparametric tests. The decision to use the former or the latter may depend on the properties of the sample of results to be analyzed.

A parametric statistical test specifies certain conditions on the response distribution of the population from which the investigated sample was obtained: independence, normality, and homoscedasticity. These conditions are usually assumed. The significance of the parametric test results depends on the validity of these assumptions. A proper interpretation of the parametric tests based on the normal distribution supposes that the scores that are being analyzed result from measurements of at least one interval scale. Parametric tests have been commonly used in the analysis of experiments with heuristic algorithms. However, most of the experimental data do not satisfy the required properties for a proper application of parametric procedures.

The lack of the required properties for a proper application of parametric procedures yields to nonparametric ones the task of performing a rigorous comparison among algorithms (Derrac et al., 2011). A nonparametric statistical test is based on a model that specifies only very general conditions and none about the specific form of the distribution of which was obtained the sample. Certain assumptions are associated with most nonparametric tests, e.g., that the observations are independent and perhaps the variable under study is continuous; but these assumptions are fewer and weaker than those associated with the parametric tests. The nonparametric procedures test different hypotheses about the population, unlike the parametric methods. In contrast to the parametric tests, there are nonparametric tests that can properly be applied to data measured on an ordinal scale, and other tests to data measured on a nominal or categorical scale.

If the sample size is very small, then it might be possible that there is no other option than using a nonparametric statistical test, unless the nature of the distribution of the population is known with accuracy. Nonparametric tests typically make less assumptions over the numerical data, and might be more relevant in a particular situation. Also, nonparametric tests are typically easier to learn and apply than parametric tests. In addition, its interpretation is often more direct than the interpretation of the parametric tests.

Given the randomness involved in a metaheuristic, it requires a particular ensemble formed by several runs to perform a statistical estimation. The estimation is generally an analysis of the mean of the solutions, errors, or other indices obtained in each run, which allows a comparison in the performance of different algorithms. Since the distribution of the data derived from a metaheuristic is unknown, it is critical the use of nonparametric statistical tests, as they do not assume a probability distribution for the data. Finally, using a proper data set, these tests can determine for two different samples the following task: a) if there is a relationship among them; b) if they are independent; c) if there are significant differences, among others. According to the literature, nonparametric tests can be classified considering the type of comparisons they made:

1. **Pairwise comparisons (P):** It includes tests for comparisons of two algorithms over multiple cases of problems. The most recommended method of this type is the Wilcoxon signed-rank test.
2. **Multiple comparisons with a control algorithm 1×N (a control algorithm x N algorithms):** It includes tests to compare all of the other algorithms with a control algorithm. Note that the p-value in a pairwise comparison is independent from another one, so if we try to make conclusions involving more than one pairwise comparison, we will obtain an accumulated error coming from the combination of comparisons; for this reason, the multiple comparison tests should be used instead.
3. **Multiple comparisons among all algorithms N×N (N control algorithms x N algorithms):** It is a test to detect differences considering the global set of algorithms. Friedman's test can be used to perform these types of comparisons.

Using Nonparametric Tests to Analyze Heuristic Algorithms

An extensive study made by Demsar (2006) examined over 120 papers in the field of machine learning. He found that less than half of the papers contained some statistical procedure for comparing the performances of their algorithms. His work generated theoretical and experimental studies aimed to establish tutorials and guides for comparing the performance of heuristics. This section reviews some of these studies.

Taillard (2003) presents the U test, a nonparametric test that aims to answer the question commonly ask by the researchers: Is a success rate of $a = n$ significantly higher than the success rate $b = m$?, this is of interest for those who wish to compare different heuristic algorithms that do not necessarily finish with feasible (or satisfactory) solutions. This test has been designed to work with very small sample sizes, which means that it is possible to reduce the computational effort when numerical experiments are conducted.

García et al., (2009) focus on the study of the use of statistical techniques to analyze the behavior of evolutionary algorithms in optimization problems. The study is conducted in two ways: one problem analysis, and the analysis of multiple problems. The results establish that a parametric statistical test may not be appropriate especially when dealing with results from multiple problems. In the analysis of multiple problems, it proposes the use of nonparametric statistical tests because they are less restrictive than the parametric ones, and the results can be used in small samples. The parametric tests require the verification of three conditions: a) independence (which can be assumed in the case study analyzed); b) normality, verified with tests of normality Kolmogorov-Smirnov, Shapiro-Wilk and D'Agostino-Pearson; c) heteroscedasticity, verified with the Levene test. For the analysis of a single problem both, the parametric paired t-test, and the nonparametric Wilcoxon test, were applied; in the case of the analysis of multiple problems, Friedman test, Iman Davenport, Bonferroni-Dunn, Holm, Hochberg and Wilcoxon were applied.

Derrac et al., (2011) made a review of the most representative nonparametric methods for multiple comparisons over a case study. The authors considered in the experimental design a a comparison based on the 25 issues presented at the Special Session of optimization parameters Royal IEEE Congress on Evolutionary Computation 2005 (CEC'2005). The algorithm used in the comparison were: IPOP-CMA-ES, CHC, SS-arit, SS-BLX, DE-Exp, DE-Bin, SaDE and PSO. The authors used the Friedman test as a procedure for multiple comparisons between different algorithms. In addition, the tests Friedman Aligned and Quade, as advanced versions of the same, were applied. They also applied tests for pairwise comparisons like Sign test, Wilcoxon Test, Contrast Estimation Test, Multiple Sign Test.

Knowledge Extraction based on Evolutionary Learning (KEEL) is a software tool to evaluate evolutionary algorithms applied to the solution of data mining problems, which includes regression, classification, clustering, and pattern mining, among others. Some of the Statistical procedures available in KEEL are: T test based on the Student's t distribution, F test based on the Snedecor's F distribution, Shapiro-Wilk test, Mann-Whitney U test, Wilcoxon test, Friedman test, among others Post-Hoc procedures. The description of this tool is available in Fernández et al. (2014).

Table 1 describes the use of some nonparametric tests, the type of comparisons made with them, and also presents related work that use them.

VisTHAA incorporates four nonparametric statistical tests to perform the three types of comparisons: multiple sign test, contrast estimation test, Wilcoxon signed-rank test and Friedman test. These tests are described in the next subsections. The description of the statistical tests considers some works focused on statistical analysis of algorithms. Among them are: Derrac et al. (2011), García et al. (2009), Luengo et al. (2009) and Sheskin (2006).

Table 1. Works related to the statistical analysis of heuristic algorithms; type of comparison: (P) pairwise comparison, (1 × N) multiple comparisons with control method, (N × N) multiple comparisons

Statistical Test	Purpose	Type of Comparison	Related Work
Wilcoxon signed-rank test	This test responds to the following question: are two samples representing two different populations?	P	García et al. (2009) Derrac et al. (2011)
U test	It defines whether a rate of success is really better than another.	P	Taillard (2003)
Multiple sign test	This test serves as a filter to remove samples that are easily distinguishable as worse than the control method.	1×N	Derrac et al. (2011)
Contrast Estimation test	It calculates the estimation of the difference among the performance of two algorithms.	1×N	Derrac et al. (2011)
Friedman test	This test considers that all the problems have equal importance.	1×N, N×N	García et al. (2009) Derrac et al. (2011)
Friedman aligned ranks test	It makes comparisons between performances of algorithms as a single whole, without considering the interrelationships that exist between the sets of the complete test.	1×N	Derrac et al. (2011)
Quade test	It defines whether the differences between the observations and recorded data, are significantly larger or not, i.e., if they are more difficult.	1×N	Derrac et al. (2011)

Multiple Sign Test

The multiple sign test is a non-complex procedure to do 1 x *N* comparisons (one control method against N methods). This procedure can be useful in such cases where we want to perform a first preliminary study of results; it can also be used as a first filter for removing samples that are easily distinguishable as worse than the control method. This procedure was proposed in (Steel 1959; Rhyne & Steel, 1965) and discussed in (Hochberg & Tamhane, 1987; Derrac et al. 2011).

This test is an extension of the conventional sign test that is capable of simultaneous comparison of *k* methods against a control method; it consists of the following steps:

1. Denote by x_{i1} the performance values of the control method on the *i*th set. Represent by x_{ij} the performance values of the remaining methods (*j*th method on the *i*th set).
2. Calculate the signs of the resulting difference of $d_{ij} = xi_j - x_{i1}$. If the performance is measured with the success rate, assign (+) when the control method yields the worst performance, or assign (-) when it is the best, or discard the result when the difference is 0. The signs are inverted when the performance is measured in error rate.
3. Calculate r_j as the number of differences d_{ij} with the less frequent sign, being (+) or (-), between a pairing of the *j* algorithm with the control algorithm.
4. Assume medians for testing, being M_1 the median of one sample of results from the control algorithm and M_j the median of one sample of results from the *j*th algorithm.
5. Make the application of one of the following rules. Note that the two first two rules are for a one-tailed critical region (the first is for success rate and the second is for error rate), and the third is for a two-tailed critical region.
   - For Testing $H_0 : M_j \geq M_1$ against$_1$: $M_j < M_1$, H_0 is rejected if the number of (+) signs is equal or less than the critical value of r_j from the Table 769 in (Steel 1959), for $k-1$ (the number of algorithms excluding the control algorithm), *n* (number of problems) and the chosen significance level.
   - For testing $H_0 : M_j \leq M_1$ against $H_1 : M_j > M_1$, H_0 is rejected if the number of (-) signs is equal or less than the critical value of r_j from the Table 769 in (Steel 1959), for $k-1$ (the number of algorithms excluding the control algorithm), *n* (number of problems) and the chosen significance level.
   - For testing $H_0 : M_j \leq M_1$ against $H_1 : M_j \neq M_1$, H_0 is rejected if the number of signs in r_j is equal or less than its critical value from the Table 770 in (Steel 1959), for $k-1$ (the number of algorithms excluding the control algorithm), *n* (number of problems) and the chosen significance level.

Tables 769 and 770, which report approximate critical value of r_j, were obtained by taking the integral part of the number computed through the following equation, with *t* from Dunnett's tables (Dunnett 1955).

$$r = \frac{n-1}{2} - t\frac{\sqrt{n}}{2}$$

A better approximation for the critical values of R_j can be found in Tables 1 and 2 in (Rhyne & Steel 1965), instead of Tables 769 and 770, respectively.

Contrast Estimation Test

The contrast estimation test was proposed in (Doksum 1957) and described in (Derrac et al. 2011; Wolfe 1998). It is a procedure to make 1 x *N* comparisons (one control method x *N* methods). Given the data resulting from the runs of several algorithms on multiple problems, a researcher could be interested in the estimation of the differences between the performances of two algorithms.

A procedure to achieve this purpose is the contrast estimation, which supposes that the expected differences of the algorithmic performances are the same among problems. We assume that every measurement of performance is reflected as the difference in the performance of the algorithms. We are interested in estimating the contrast between medians of the results derived from the samples considering, all the pair combinations. For this purpose, the contrast estimation obtains a quantitative difference calculated using the means between two algorithms. The procedure is as follow:

1. For each pair (*uv*) of *k* algorithms in the experiment, compute the difference between the performances of both algorithms in each one of the *n* problems, $D_i(uv) = x_{iu} - x_{iv}$ where $i = 1 \ldots n$, $u = 1 \ldots k$, $v = 1 \ldots k$, only the differences where $u < v$ are considerate.
2. Compute the median of each set of differences, Z_{uv}. This is an unadjusted estimator of the difference between the medians of *u* and *v*. Note that $Z_{uu} = 0$.
3. Calculate the mean of each set of unadjusted medians that have the same prefix *u*, m_u:

$$m_u = \frac{\sum_{j=1}^{k} Z_{uj}}{k}, u = 1, \ldots, k$$

Table 2. Nonparametric statistical tests included in VisTHAA; type of comparison: (P) pairwise comparison, (1 × N) multiple comparisons with control method, (N × N) multiple comparisons

Statistical Test	Purpose	Comparison Type
Wilcoxon signed rank test	This test aims to detect if two samples are representing two different populations, i.e., if exist significant differences between the behavior of two algorithms based on the median of its difference scores.	P
Multiple sign test	This test is used to compare a set of algorithms with a control one. It serves as a quick filter to remove samples that are easily distinguishable as worse than the control method and take a first glance at the results.	1×N
Contrast Estimation	In this test, the performance is reflected by the differences between performances of the algorithms. So, the contrast between medians of differences is estimated considering all pairwise comparisons.	1×N
Friedman Test	The test aims to detect significant differences between the behaviors of two or more algorithms. This test considers that all the problems have equal importance.	1×N, N×N

4. Define the estimator $M_u - M_v$ as $m_u - m_v$, where u and v are in the range from 1 to k, e.g., the difference between M_1 and M_2 is estimated by $m_1 - m_2$.

The $M_u - M_v$ estimators can be assumed as a measure of overall performance. When the performance measure is the error rate, algorithms with negative values in the estimators are those with a lower percentage of error. Moreover, positive values confirm a better performance when it is measured in success rate. Although the contrast estimation test does not provide an error probability related with the rejection of the null hypothesis of equal performance, it can be used to estimate how much an algorithm outperforms another.

Wilcoxon Signed-Rank Test

Wilcoxon´s test is a nonparametric pairwise test for repeated measurements of a single sample. The Wilcoxon signed rank test is a nonparametric statistical test for repeated measures of a single sample. This statistical test is used for samples that are not normally distributed. With this test we can know if the result of using an algorithm is significantly different from that obtained when using another algorithm to solve the same sample. This is a direct comparison test, i.e., it only studies a pair of methods.

This test is used to answer the question whether two samples represent two different populations or not. It is a nonparametric procedure used in hypothesis testing in situations involving a design with two samples. This test is analogous to the paired *t*-test for nonparametric statistical procedures; therefore, it is a test of pairs to identify significant differences between two sample means, which is the behavior of two algorithms.

The Wilcoxon test is defined as follows.

1. Let d_i be the difference in performance given by the results of two algorithms in the *i*th of the *n* problems (if these performances are known to be represented in different ranks, they can be normalized in the range [0,1], in order to not prioritize any problem).
2. Hierarchize the differences according to their absolute values. In case of ties (the difference score is zero because the score is the same in both algorithms), an alternative is the use of the average of the ranks with ties (for example, if two differences are tied for the allocation of ranks 1 and 2, then assign the range 1.5 to both differences). Another alternative is to discard the score from the analysis and reduce the sample size; if exists a lot of ties, this procedure will undermine the reliability of the test.
3. Let W^+ be the sum of the ranges for the problems in which the first algorithm outperformed the second one, and W^- be the sum of the ranges for the opposite case. In case of inclusion of ties ($d_i = 0$), the ranges are equally divided between W^+ and W^-; if the number of these ranges is odd, one is ignored.

$$W^+ = \sum_{d_i>0} rank(d_i) + \frac{1}{2}\sum_{d_i=0} rank(d_i)$$

$$W^{-} = \sum_{d_i < 0} rank(d_i) + \frac{1}{2} \sum_{d_i = 0} rank(d_i)$$

4. Compute *W* statistic, defined as $W = \min(W^{+}, W^{-})$.
5. Compare *W* with the critical value of the distribution of Wilcoxon for *n* degrees of freedom (see Table A5 in Sheskin 2006), If *W* is less than or equal to the critical value, the null hypothesis of equal means is rejected; this means that a given algorithm outperform the other, with the associated *p*-value, which is obtained through the normal approximation for the Wilcoxon *W* statistic with the next equation (see Section VI, Test 18 in Sheskin 2006), and considering if the critical region is of one-tailed or two-tailed.

$$z = \frac{W - \frac{n(n+1)}{4}}{n(n+1)(2n+1)}$$

The Wilcoxon signed rank test is more sensitive than the *t*-test. It assumes commensurability of differences, but only qualitatively: biggest differences count more, which is probably desired, but the absolute magnitudes are ignored.

From a statistical point of view, the test is safe as it not assumes normal distributions. In addition, outliers (exceptionally good/bad performances of some problems) have less effect on the Wilcoxon test than in the *t*-test. Wilcoxon test assumes d_i continuous differences; therefore, there should not be rounded to one or two decimal places, since this would reduce the power of the test in the case of a high number of ties.

Friedman Test

In 1937 Milton Friedman proposed a rank order in the analysis of variance in a work that even today is remembered as the basic contribution in the development of nonparametric methods for analysis of variance.

The years of dedication to the statistics at Columbia University during the wars turned out into another important work (1948) on sequential sampling. Friedman proved that in the exploration of a sample, the same verification process provided information on the level of confidence reached, so that it could interrupt the sample before reaching the predetermined sample size.

Friedman test is equivalent to an ANOVA (analysis of variance) test for two factors and allows determining whether there are significant differences between a set of tests, and whether they are interrelated. Friedman test works as follows:

1. Assign ranks r_{ij} to the results obtained by each *j* algorithm on each problem *i*. That is, for every problem, a rank $1 \leq r_{ij} \leq k$ is assigned, where *k* is the number of algorithms to compare. These ranks are assigned in ascending order, i.e., 1 to the best result, 2 to the second one, etc. (in case of ties, mid ranks are assigned).

2. Compute the average rank of the algorithms on the n problems, using this equation:

$$R_j = \frac{\sum_{i=1}^{n} r_{ij}}{n}$$

3. Compute the Friedman statistic with the equation given below. The null hypothesis asserts a similarity in the behavior of all the algorithms, so their R_j rankings should be similar. Following this hypothesis Friedman statistic is distributed according to a χ^2 distribution with $k-1$ degrees of freedom.

$$F_F = \frac{12n}{k(k+1)}\left[\sum_j R_j^2 - \frac{k(k+1)^2}{4}\right]$$

4. Evaluate the computed Friedman statistic value with Table A4 (Table of the Chi-square Distribution) in Sheskin (2006). In order to reject the null hypothesis, the computed value F_F must be equal to or greater than the tabulated critical chi-square value at the pre-specified level of significance.

Because Friedman´s method is a *N*×*N* test, once this test rejects the null hypothesis, we can proceed with a post-hoc test in order to find the specific pairwise comparison, which produces the differences among them. A *post-hoc* analysis consists of looking in the data, after the experiment has concluded, for patterns that were not specified a priori. For more information on post-hoc tests, see Trawinski et al., (2012).

A VISUALIZATION TOOL FOR HEURISTIC ALGORITHM ANALYSIS

This section describes the functional characteristics included in VisTHAA, and its current development state, emphasizing the statistical module for the comparative analysis of algorithms. The visual diagnostic tool VisTHAA requires the definition and planning of functional modules. Due to the development of a tool like VisTHAA is vast and complex, this chapter only addressed a few of these modules.

Specifically, the module for the comparative analysis of algorithms which is supported by nonparametric statistical tests is reported. Figure 1 shows a scheme of VisTHAA, and the modules that it will provide to the final user; in this scheme it is enclosed in a red box the module that is presented in this work. It is worth of noting that in a first stage, the VisTHAA tool is planned as an offline analysis tool, i.e., it does not have a direct interaction with executions of algorithms, because it only needs the results derived from each execution. The software VisTHAA and all files of the case of study are available at https://sth-se.diino.com/lauracruzreyes/VisTHAA2.

Module for Structural Characterization of Instances and Problems

This module is planned to help the researcher to obtain the relevant characteristics of NP-hard optimization problems. The features of interest are obtained and displayed in terms of: a) new metrics, or metrics

Figure 1. Modular scheme of VisTHAA

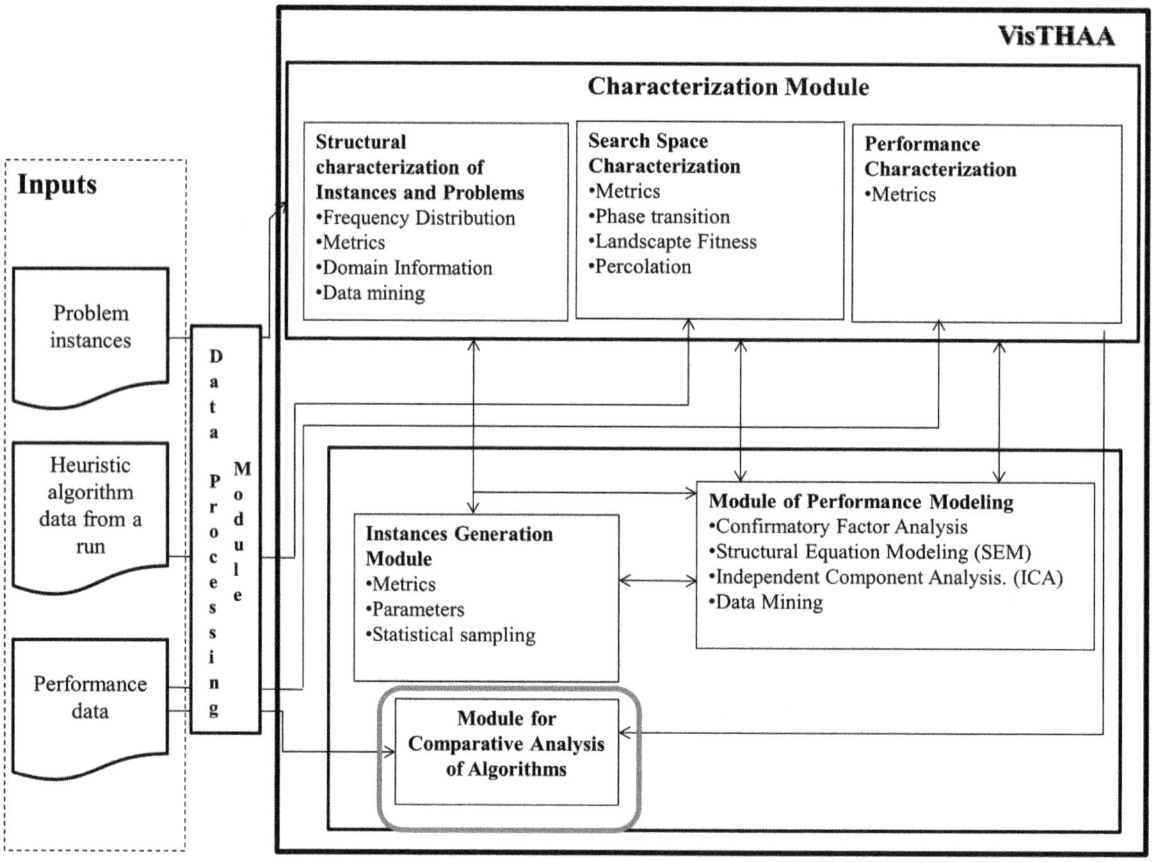

taken from the literature; b) the frequency distribution of parameters of the problem, e.g., it may be of interest to know and display the distribution of the weights of the objects in a particular instance of the well-known Bin Packing Problem (BPP); c) the problem domain information provided by the researcher; and, d) the ability to associate related instances using data mining technique or some other grouping.

Currently, the module allows the user to load a file which must contain the name of the set of instances to be analyzed. Once that the instances has been specified, the user can select the indices or metrics that are available for them.

Module for Search Space Characterization

This module aims to characterize the searching process of a heuristic algorithm using metrics derived from two types of studies: a) the analysis of phase transition, to determine that an instance is difficult if the metric is located in the transition zone, i.e. the point where an abrupt change in performance occurs; and b) the fitness surface analysis to measure the variability of solutions generated during the search. In addition, there will be a simulated visualization of the solution process; this simulation will show in time and space: a) the line of the search path, e.g., the performance indicator; b) the basic statistics on the solution values, such as frequency, minimum, and maximum values, and proximity to the best found; and c) the value of the metrics.

Currently, this module allows the user to load a file containing information about the search process. Another feature incorporated in this module is the visualization of the behavior of the algorithm in the search space; with this option, a researcher can see how good the diversification and/or intensification processes are in the algorithm. The manipulation allows the rotation of the displayed graphic to show different points of the process search.

Module for Performance Characterization

The characterization of the performance of heuristic algorithms incorporates metrics that express three types of information, which are: 1) the quality of the solution (the usual metric for BPP is the ratio obtained from the solution and the best known solution); 2) the consistency of the results (usually estimated by measuring the standard deviation of the results); and, 3) the time spent by the algorithm to find the best solution. Currently, the characterization module has a greater flexibility due to a component that allows the incorporation of new metrics.

Module for Performance Modeling

Building models for the performance of metaheuristics is a complex task. Although these models have the potential to generate explanations about the performance of metaheuristics, the experimental studies rarely have focused on analyzing the contributions and interactions of the strategies involved, something that commonly is of greater interest when comparing their efficiency. In the module described in this section, it is planned the incorporation of exploratory techniques, such as Confirmatory Factor Analysis, Structural Equations, Independent Component Analysis, Causal Analysis and Data Mining. According to the scientific literature, it is possible to visualize, summarize, and build models for algorithmic performance with the use of these techniques (Cohen, 1995; Liu, 1996; Feitelson, 2006).

Module for Instances Generation

Problems that have been widely studied as the BPP require the replacement of old instances for new ones that are representative of the problem, and of large scale. Some techniques that have been explored are: a) the use of metric for characterization; b) the specification of parameter ranges; and, c) statistical sampling (Cruz, 2004). These techniques are contemplated to be incorporated into the VisTHAA tool.

Module for Data Processing

VisTHAA has been planned for the analysis of different instances of a problem. Also, it is desired the possibility of reading the results of the analysis of metaheuristic algorithms without the modification of the data reading process.

Because there is no standard for the representation of instances of optimization problems, nor for the output format and options from the solvers (Fourer & Martin, 2010), it has been designed a metadata to describe in plain text the structure of the content derived from instances, and their data, that will serve for the analysis process. Metadata is a term used to describe data that give type and class of the information, i.e. it is data about data, which provide the necessary information so that it can be used nimbly in different applications (López, 2000).

The proposed design helps the users to describe the contents of their instances, giving the possibility of introducing only one instance at a time, or a list of all the instances to be evaluated. This idea can be extended to read data from other entries, e.g., the data resulting from a run of a metaheuristic algorithm.

The metadata design created the design of a meta-language to handle data processing. Through a meta-language, also referred as formal system, encoding is added to the digital form of a document, either to control processing or to represent their meaning (Lamarca, 2009). Based on this idea, it is proposed a data structure for the efficient storage of instances derived from generic optimization problems, which is used by the meta-language.

Figure 2. shows an example with the files involved in the data processing required for subsequent solution of BPP. Figure 2.a corresponds to the batch file, or just *batch*. This file contains the number of instances that will be processed with the metadata file described in Figure 2.b. The metadata file describes the format necessary to read each file instance Figure 2.c.

Module for Comparative Analysis of Algorithms

This module uses nonparametric statistical tests (described in the background section), in order to do the comparative analysis of the performance of the algorithms. The organization of this module is shown in Figure 3, and it shows the nonparametric statistical added to VisTHAA. These tests supports the module for comparative analysis of heuristic algorithm. Let's point out that the tests are implemented such that the algorithm analysis is not exclusive.

For each test, a file containing the results of the performance of the algorithms, and an appropriate solicitude of specific information made to the interface, are required. The end user can see the results of the test and, if needed, will also be displayed intermediate test procedures.

Table 2 concentrates information about nonparametric statistical tests that are included in VisTHAA. This table provides a summary of what has been previously described on these tests in the background section.

Figure 2. Files involved in the data processing

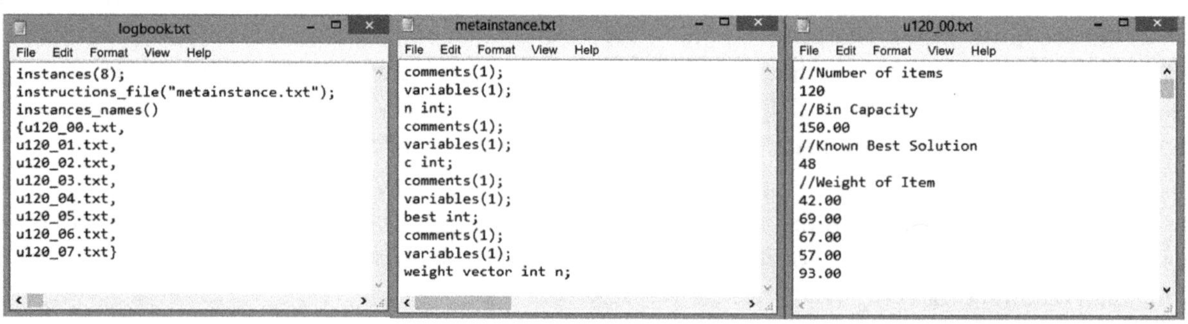

a) Batch file b) Metadata file c) BPP instance

Figure 3. Module for algorithm comparative analysis

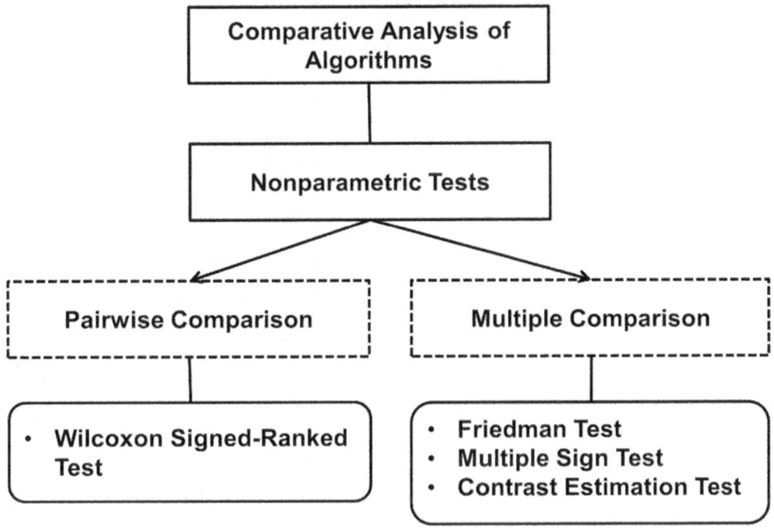

CASE OF STUDY: PERFORMANCE OF BIN PACKING ALGORITHMS

To evaluate the contribution of the proposed tool we performed a study of the performance of the best state-of-the-art algorithms for BPP. The case study confirms the importance of applying nonparametric statistical tests to facilitate comparative analysis of the performance of algorithms.

An experimental analysis of the global performance of eight BPP algorithms over the BPP standard instances was performed, comparing the results reported for the heuristics GGA-CGT (Quiroz-Castellanos et al., 2015), Perturbation-SAWMBS (Fleszar et al., 2011), HI_BP (Alvim et al., 2004), HGGA-BP2 (Quiroz et al., 2013), WABP (Loh et al. 2008) and the well-known heuristics FFD, BFD and WFD (Johnson, 1974). Table 3 shows the fraction of optimal solutions found in each class of instances for each algorithm (success rate). From the table, it can be observed that the classes that include instances that match to be difficult for all algorithms are the sets: Gau 1 and Hard28. The statistical tests included in VisTHAA, allow investigating the effectiveness of the GGA-CGT heuristic.

Multiple Sign Test

Figures 4 and 5 show the use and the results obtained when applying the Multiple Sign Test included in VisTHAA to the performance of the BPP algorithms included in Table 3. From Figure 4, it can be observed that we choose a level of significance $\alpha = 0.05$, indicate that the control method is GGA-CGT, and establish our hypotheses as $H_0 : M_j \leq M_1$ and $H_1 : M_j < M_1$ to reflect the assumption that our control algorithm GGA-CGT is significantly better than the remaining algorithms in terms of success error. In Figure 5, VisTHAA reports the number of pluses, the number of minuses, and for m=7 ($m = k - 1$) and $n = 9$ that the critical value of r_j is 0. Since the number of pluses in the pairwise comparison between the control and the other algorithms is equal to 0, satisfying the condition to reject H_0, we may conclude that GGA-CGT has a significantly better performance than them.

Table 3. Result of the state-of-the-art BPP algorithms (k=8)

Class	GGA-CGT	Pert.-SAWMBS	HI_BP	HGGA-BP2	WABP	FFD	BFD	WFD
Uniform	1	0.98	1	0.98	0.88	0.07	0.07	0.02
Triplets	1	1	1	0.75	0	0	0	0
Data Set 1	1	1	1	0.99	0.97	0.75	0.75	0.68
Data Set 2	1	1	1	1	0.97	0.49	0.49	0.46
Data Set 3	1	1	1	0.9	0.9	0	0	0
Was 1	1	1	1	1	1	0	0	0
Was 2	1	1	1	1	1	0	0	0
Gau 1	0.94	0.94	0.7	0.94	0.7	0.11	0.11	0.11
Hard28	0.57	0.17	0.17	0.35	0.17	0.17	0.17	0.07

Figure 4. Multiple Sign Test in VisTHAA, selection of control algorithm

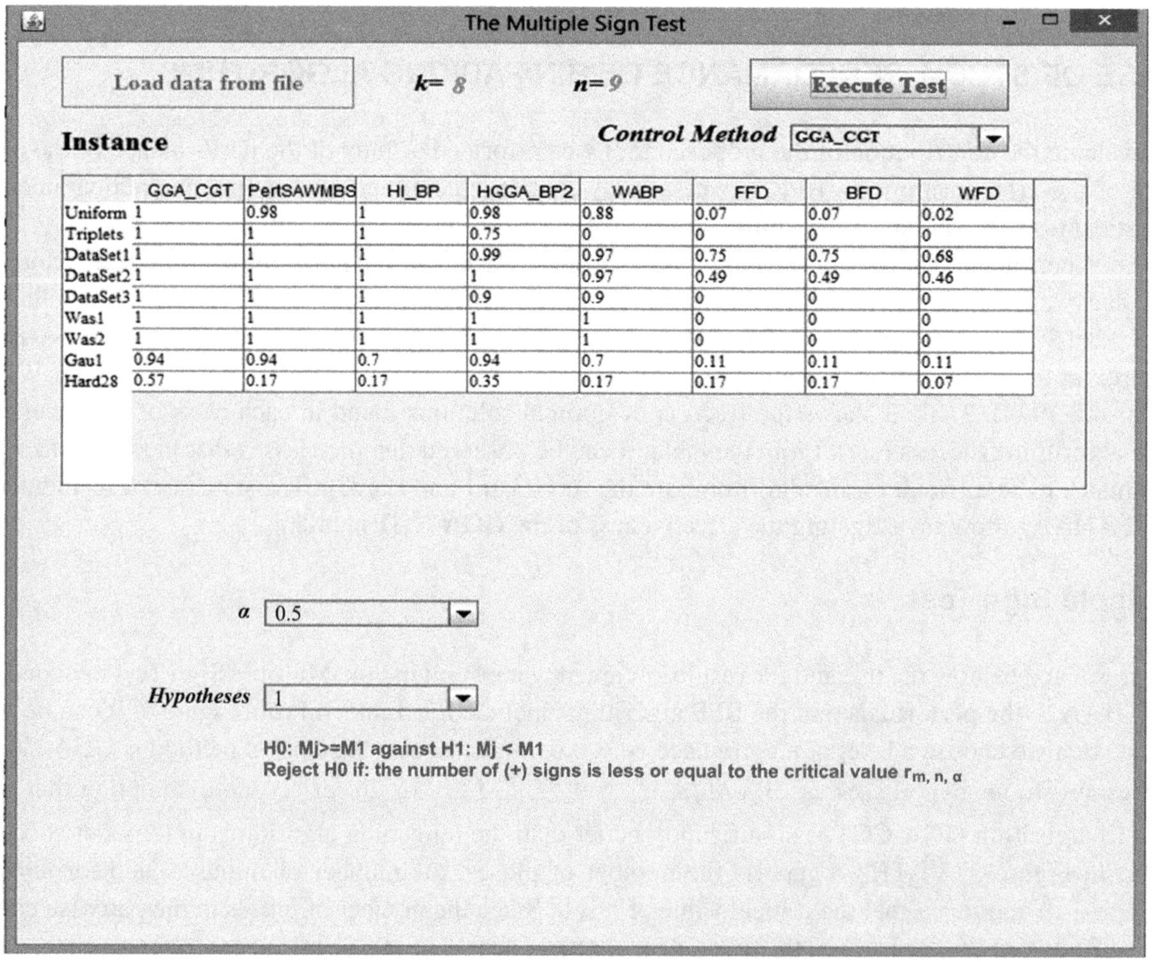

Figure 5. Results obtained for the Multiple Sign Test in VisTHAA

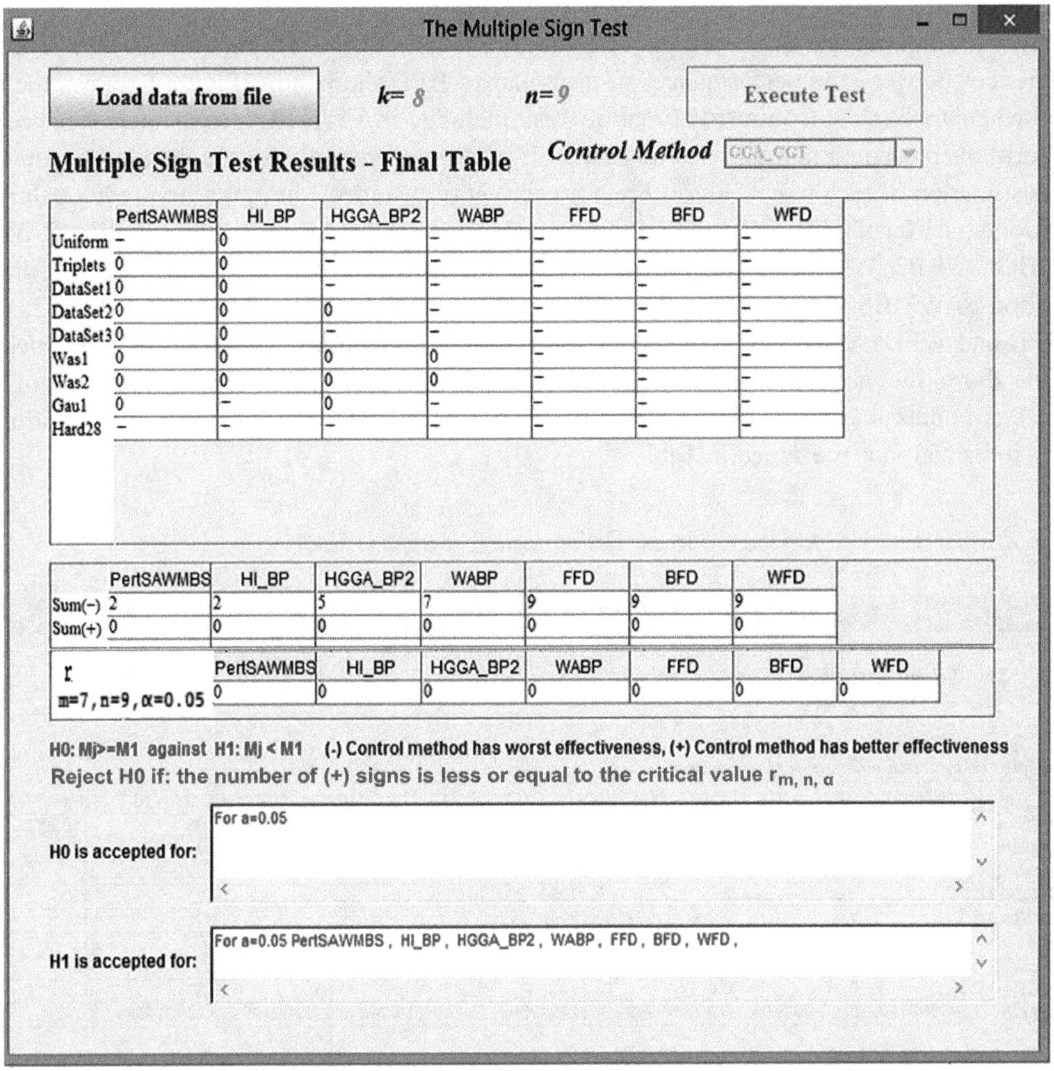

Figure 5. also shows the comparison of the efficiency of GGA-CGT against the rest of the BPP algorithms in each set of instances. Because the performance is measured in terms of success rate, a positive sign (+) indicates that GGA-CGT shows a lower performance that the algorithm in the corresponding column, a negative sign (-) indicates that GGA-CGT shows a higher performance, and zero (0) indicates that there are no differences in the performance of the algorithms. As it can be seen, none of the algorithms improves the GGA-CGT performance in any of the sets of instances (there are no positive signs). The first column includes the comparison of CGT-GGA with Perturbation-SAWMBS; a negative sign (-) was assigned in the Uniform and Hard28 sets; this indicates that GGA-CGT shows a better performance than Perturbation-SAWMBS in these two sets. The rest of the columns include the comparison of GGA-CGT with the other six algorithms in Table 3. It is also seen that for the Hard28 set, when GGA-CGT is compared against the BPP algorithms, it always obtain a negative sign (-), these results confirm quickly the superiority of the GGA-CGT while considering the class of Hard28 instances.

Contrast Estimation Test

The contrast estimation test allow proving the superiority of GGA-CGT; this test is used to estimate the differences between the performances of each pair of BPP algorithms. Figure 6 shows the results obtained when applied the Contrast Estimation Test, included in VisTHAA, to the performance of the BPP algorithms presented in Table 3. The first table of this figure includes the detailed results for the contrast estimation of each pair of algorithm, for each set of instances; in it, the first seven columns include the comparison of CGT-GGA with the other state-of-the-art algorithms (Perturbation-SAWMBS, HI_BP, HGGA-BP2, WABP, FFD, BFD and WFD), the next six columns include the comparison of Perturbation-SAWMBS with the other algorithms excluding GGA-CGT (HI_BP, HGGA-BP2, WABP, FFD, BFD and WFD), and so on. The second table shows the median of each set of differences. The third table shows the mean of each set of medians for each algorithm. Finally, the last table of Figure 6 includes the contrast estimator considering the global performance of each pair of algorithms; the complete information is presented in Table 4.

Figure 6. Results obtained for the Contrast Estimation Test in VisTHAA

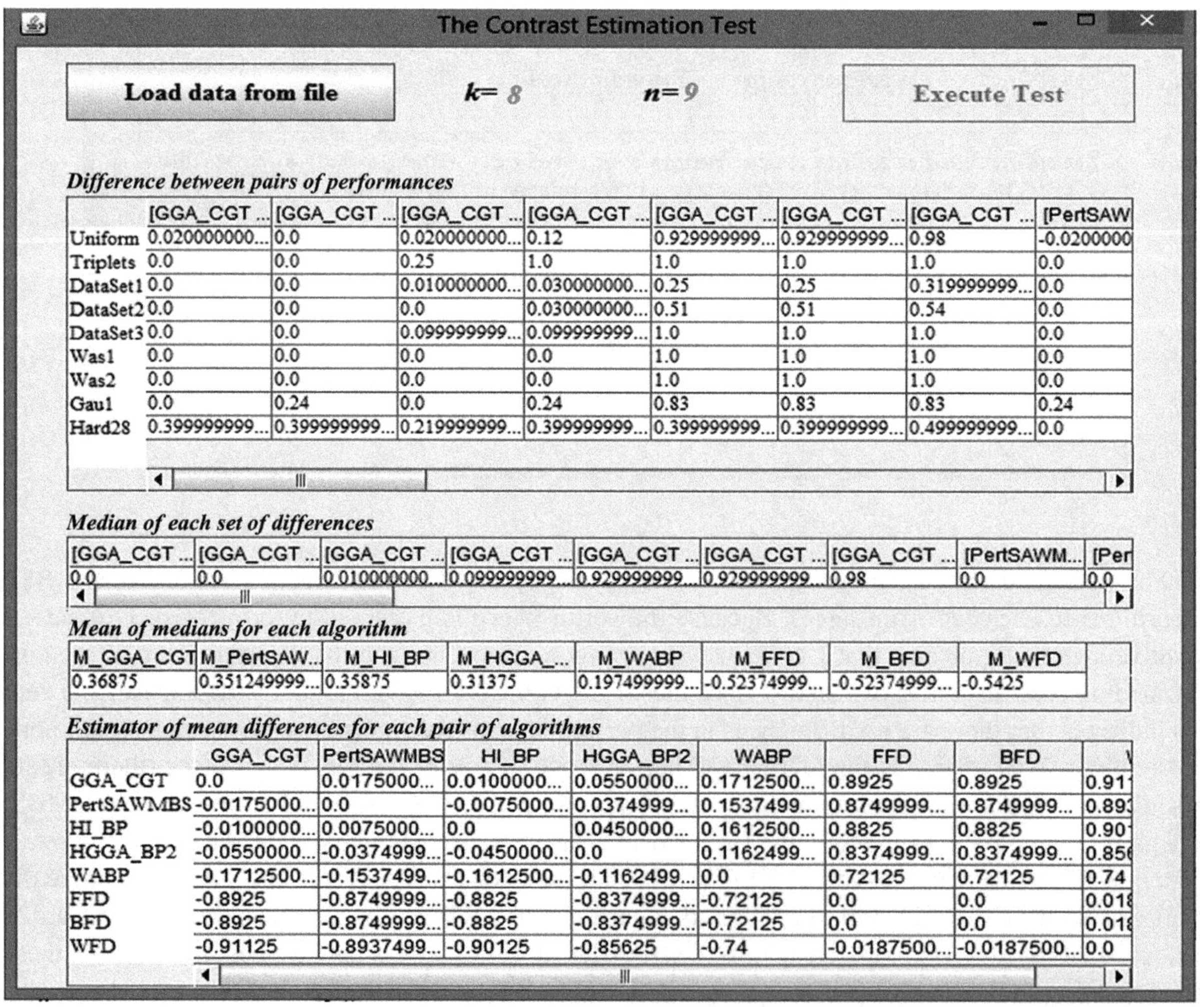

Table 4. Table of Contrast Estimation Test for the BPP algorithms

	GGA-CGT	Pert.-SAWMBS	HI_BP	HGGA-BP2	WABP	FFD	BFD	WFD
GGA-CGT	0	0.0175	0.01	0.055	0.17125	0.8925	0.8925	0.91125
Pert.-SAWMBS	-0.0175	0	-0.0075	0.0375	0.15375	0.875	0.875	0.89375
HI_BP	-0.01	0.0075	0	0.045	0.16125	0.8825	0.8825	0.90125
HGGA-BP2	-0.055	-0.0375	-0.045	0	0.11625	0.8375	0.8375	0.85625
WABP	-0.17125	-0.15375	-0.16125	-0.11625	0	0.72125	0.72125	0.74
FFD	-0.8925	-0.875	-0.8825	-0.8375	-0.72125	0	0	0.01875
BFD	-0.8925	-0.875	-0.8825	-0.8375	-0.72125	0	0	0.01875
WFD	-0.91125	-0.89375	-0.90125	-0.85625	-0.74	-0.01875	-0.01875	0

From Table 4 it can be seen that GGA-CGT shows the best performance, since all their row estimators have positive values; this means that it has a higher level of efficiency, followed by HI_BP, which have only one negative estimator. It can also be seen that the algorithm showing the worse performance is WFD, this algorithm is outperformed for the other seven BPP algorithms.

Wilcoxon Signed-Rank Test

The results of the previous tests showed the superiority of the GGA-CGT metaheuristic while considering the class of Hard28 instances, which contain the most difficult instances for the known algorithms. Table 5 presents the detailed results of the GGA-CGT algorithm and some of the best state-of-the-art algorithms on the Hard28 set; only Perturbation-SAWMBS and HI_BP are included because they have the same performance for this set of instances and HI-BP had the second position in the contrast estimation test. For each instance, it is included, the error obtained by each algorithm, which is defined as $(y - x) / x$, where x is the optimal solution value, and y is the number of bins obtained by the algorithm.

The Wilcoxon Signed-Rank Test allows the comparison of the two samples in Table 5 (GGA-CGT Error with Pert.-SAWMBS and HI-BP Error) and to assess if there is a difference in the performance of the algorithms. Figure 7 shows the results obtained when applied the Wilcoxon Signed-Rank Test included in VisTHAA to the data in Table 5. Considering that the hypothesis H_0 is that there is no significant difference in the performance of the algorithms. From Figure 6, it can be seen that H_0 is rejected; this result demonstrates that the performances are significantly different and GGA-CGT is the best strategy to resolve the Hard28 set.

Friedman Test

The results of the above statistical tests indicate that GGA-CGT outperforms the results of the best state-of-the-art algorithms on the BPP standard instances. The Friedman test was useful for the experimental design, it was applied to study the impact of some of the key strategies that are included in GGA-CGT. The test was applied to compare the results obtained by different configurations of the algorithm, with the objective of analyzing the impact of the crossover and mutation operators, and the influence of the

Table 5. Results of GGA-CGT and the best state-of-the-art algorithms on the set Hard28

Instance	GGA-CGT Error	Pert.-SAWMBS and HI-BP Error	Instance	GGA-CGT Error	Pert.-SAWMBS and HI-BP Error
hBPP14	0	0	**hBPP359**	0	0
hBPP832	0.01666667	0.01666667	**hBPP485**	0.01408451	0.01408451
hBPP40	0.01694915	0.01694915	**hBPP640**	0	0.01351351
hBPP360	0	0.01612903	**hBPP716**	0	0
hBPP645	0	0.01724138	**hBPP119**	0	0
hBPP742	0	0.015625	**hBPP144**	0.01369863	0.01369863
hBPP766	0.01612903	0.01612903	**hBPP561**	0	0.01388889
hBPP60	0.01587302	0.01587302	**hBPP781**	0	0.01408451
hBPP13	0.01492537	0.01492537	**hBPP900**	0.01333333	0.01333333
hBPP195	0	0.015625	**hBPP175**	0	0
hBPP709	0.01492537	0.01492537	**hBPP178**	0.0125	0.0125
hBPP785	0	0.01470588	**hBPP419**	0.0125	0.0125
hBPP47	0	0.01408451	**hBPP531**	0	0.01204819
hBPP181	0.01388889	0.01388889	**hBPP814**	0	0.01234568

Figure 7. Results obtained for the Wilcoxon Signed-Rank Test in VisTHAA

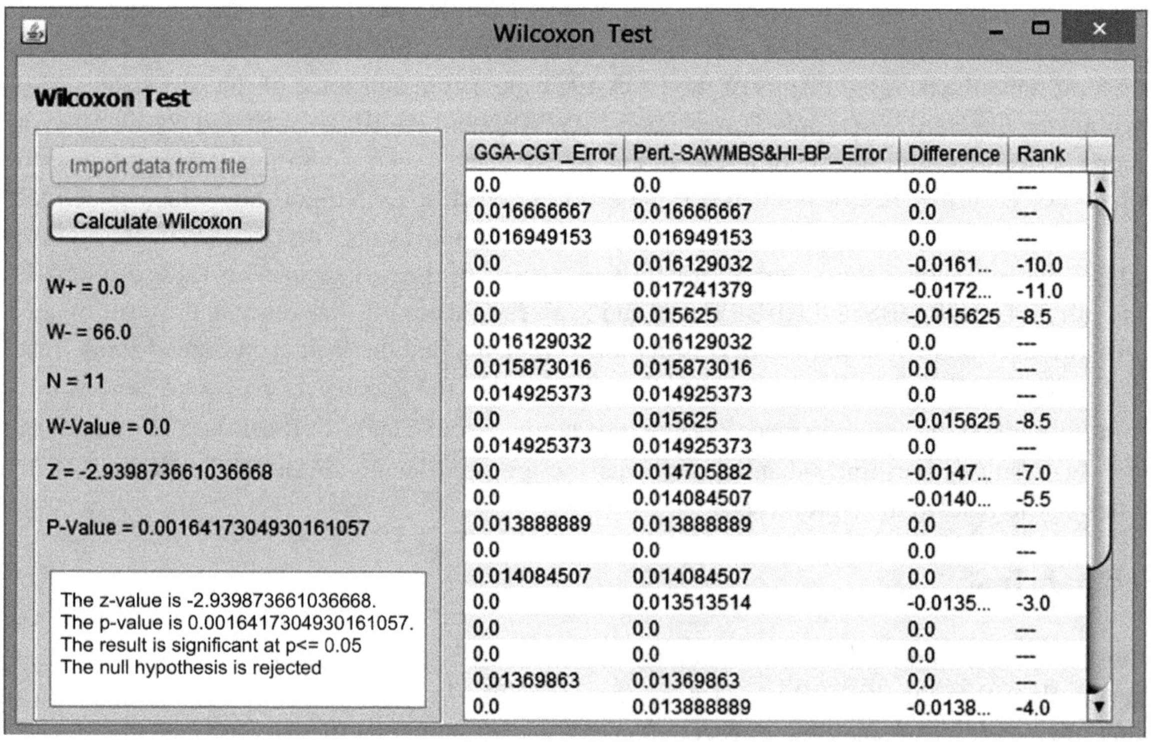

elite group B. In all of the cases when the value for a parameter is not specified, the parameter will have the value given by Quiroz-Castellanos et al. (2015).

Table 6 shows the performance of GGA-CGT algorithm and the performance of three versions of this algorithm where the values of the parameters $n_m = 0$, $n_c = 0$ and $|B| = 0$ are set to zero, respectively. The table includes the fraction of optimal solutions found in each class of instances for each configuration of the algorithm. The Friedman test was used to detect significant differences in the behavior of the algorithm with the different configurations.

Figure 8 shows the results obtained when applied the Friedman Test included in VisTHAA to the data in Table 6. Considering that the hypothesis H_0 is that there is no significant difference in the performance of the algorithm with the different configurations, the rankings R_i should be similar. From Figure 8, it can be seen that H_0 is rejected because the Friedman statistic is greater than its critical value (18.731>7.815) and the p-value is less than α (0.0003108<0.05). This figure also shows that the full GGA-CGT has the best ranking (31.5), followed by $n_c = 0$ (24) and $|B| = 0$ (24.5) that show similar efficiency, and the worst efficiency is obtained when $n_m = 0$ (10). This result demonstrates the importance of the integration of these three strategies to obtain a high performance of the algorithm.

CONCLUSION

A new tool was developed for the design and study of heuristic algorithms; this tool is referred to as VisTHAA (Visualization Tool for Heuristic Algorithm Analysis). The proposed tool includes modules that guide the scientific design, visual diagnosis, and evaluation of a heuristic algorithm and that allow for the statistical analysis and comparison of experimental results.

VisTHAA can be a helpful tool to confirm that a new heuristic shows a better performance than the state-of-the-art algorithms. VisTHAA can also be useful to analyze metaheuristic algorithms, and for the identification of the features that impact on their performance, by analyzing if a method or a configuration of a metaheuristic is really providing a significant improvement.

Table 6. Results for different configurations of GGA-CGT

Class	GGA-CGT	$n_m = 0$	$n_c = 0$	$\|B\| = 0$
Uniform	1	0.95	1	0.9875
Triplets	1	0	1	0.5125
Data Set 1	1	0.986111111	0.983333333	0.997222222
Data Set 2	1	0.93125	1	0.9875
Data Set 3	1	0.7	0.9	0.9
Was 1	1	0.88	1	1
Was 2	1	0.94	1	1
Gau 1	0.94	0.705882353	0.882352941	0.941176471
Hard28	0.57	0.178571429	0.321428571	0.428571429

Figure 8. Results obtained for the Friedman Test in VisTHAA

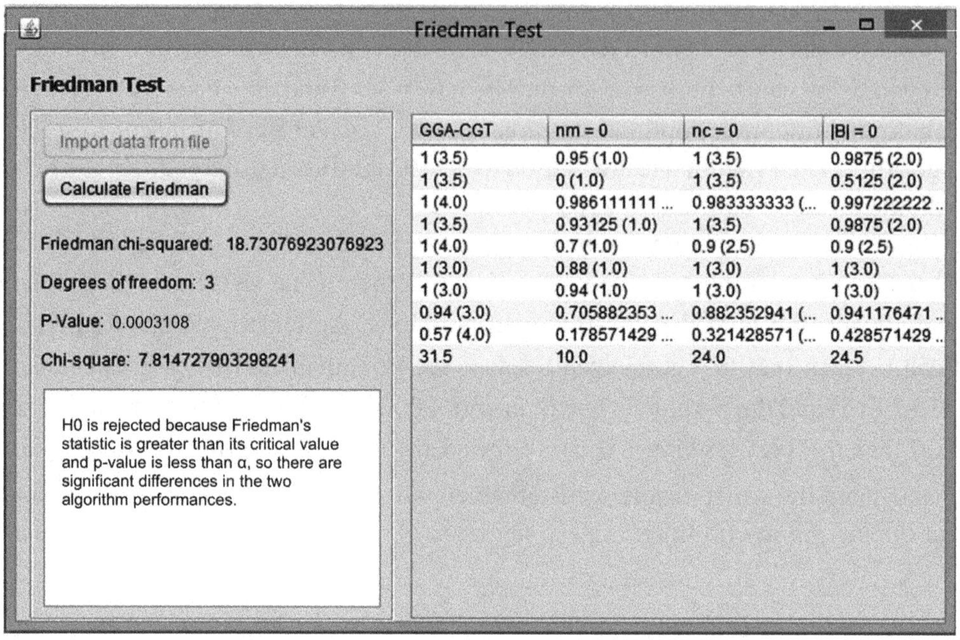

The study of the literature revealed that most of the experimental studies of heuristics do not meet the properties required to implement the well-known parametric tests, yielding to nonparametric tests the task of performing a rigorous comparison among algorithms. VisTHAA incorporates four nonparametric statistical tests to facilitate the comparative analysis of heuristics.

As a case study, VisTHAA was applied to analyze the published results for the best state-of-the-art algorithms that solve the one-dimensional Bin Packing Problem (BPP) and to evaluate the contribution of the methods included in this best state-of-the-art algorithm. The obtained results confirm the usefulness of the tool to validate and understand the performance of a heuristic algorithm.

Among the future works to enrich the tool is necessary to extend the metadata entry parameters in VisTHAA to more than one dimension. Another objective to be achieved in the future is to provide VisTHAA the ability of working online. It is also necessary to incorporate multivariate analysis techniques, implement parametric tests and extend the set of nonparametric test.

ACKNOWLEDGMENT

This work has been partially supported by CONACYT Project 254498 and PRODEP.

REFERENCES

Adenso-Diaz, B., & Corral, N. (2007). A guide to statistical analysis for single criteria heuristics evaluation. *International Journal of Innovative Computing and Applications*, *1*(2), 94–107. doi:10.1504/IJICA.2007.016791

Alvim, A. C., Ribeiro, C. C., Glover, F., & Aloise, D. J. (2004). A hybrid improvement heuristic for the one-dimensional bin packing problem. *Journal of Heuristics*, *10*(2), 205–229. doi:10.1023/B:HEUR.0000026267.44673.ed

Bartz-Beielstein, T., & Preuss, M. (2010). The future of experimental research. In *Experimental Methods for the Analysis of Optimization Algorithms* (pp. 17–49). Springer Berlin Heidelberg. doi:10.1007/978-3-642-02538-9_2

Berenson, M. L., Levine, D. M., & Krehbiel, T. C. (2001). *Estadística para administración*. Pearson Education.

Castillo, N. (2011). *Evaluación de Estrategias de Mejora del Desempeño de Metaheurísticos Aplicados a BPP Vía Diagnóstico Visual*. (Master's thesis). Ciudad Madero, Tamaulipas.

Cohen, P. R. (1995). *Empirical Methods for Artificial Intelligence*. Bradford Books.

Crainic, T. G., Perboli, G., Rei, W., & Tadei, R. (2011). Efficient lower bounds and heuristics for the variable cost and size bin packing problem. *Computers & Operations Research*, *38*(11), 1474–1482. doi:10.1016/j.cor.2011.01.001

Crowder, H., Dembo, R. S., & Mulvey, J. M. (1979). On reporting computational experiments with mathematical software. *ACM Transactions on Mathematical Software*, *5*(2), 193–203. doi:10.1145/355826.355833

Cruz, L. (2004). *Clasificación de Algoritmos Heurísticos Para la Solución de Problemas de Bin Packing*. (Doctoral dissertation, PhD thesis). CENIDET, Cuernavaca, Mexico.

Demšar, J. (2006). Statistical comparisons of classifiers over multiple data sets. *Journal of Machine Learning Research*, *7*, 1–30.

Derrac, J., García, S., Molina, D., & Herrera, F. (2011). A practical tutorial on the use of nonparametric statistical tests as a methodology for comparing evolutionary and swarm intelligence algorithms. *Swarm and Evolutionary Computation*, *1*(1), 3–18. doi:10.1016/j.swevo.2011.02.002

Doksum, K. (1967). Robust procedures for some linear models with one observation per cell. *Annals of Mathematical Statistics*, *38*(3), 878–883. doi:10.1214/aoms/1177698881

Dunnett, C. W. (1955). A multiple comparison procedure for comparing several treatments with a control. *Journal of the American Statistical Association*, *50*(272), 1096–1121. doi:10.1080/01621459.1955.10501294

Feitelson, D. G. (2006). *Experimental computer science: The need for a cultural change*. Retrieved from http://www.cs.huji.ac.il/~feit/papers/exp05.pdf

Fernández, J., García, S., & Derrac, J. (2014). KEEL: A software tool to assess evolutionary algorithms for Data Mining problems (regression, classification, clustering, pattern mining and so on). *Keel.es*. Retrieved 02 July 2015, from http://www.keel.es/

Fleszar, K., & Charalambous, C. (2011). Average-weight-controlled bin-oriented heuristics for the one-dimensional bin-packing problem. *European Journal of Operational Research*, *210*(2), 176–184. doi:10.1016/j.ejor.2010.11.004

Fourer, R., Ma, J., & Martin, K. (2010). Optimization services: A framework for distributed optimization. *Operations Research*, *58*(6), 1624–1636. doi:10.1287/opre.1100.0880

García, S., Molina, D., Lozano, M., & Herrera, F. (2009). A study on the use of nonparametric tests for analyzing the evolutionary algorithms' behaviour: A case study on the CEC'2005 special session on real parameter optimization. *Journal of Heuristics*, *15*(6), 617–644. doi:10.1007/s10732-008-9080-4

Garey, M. R., & Johnson, D. S. (1979). *Computers and intractability: a guide to the theory of NP-completeness*. San Francisco: WH Freeman & Co.

Halim, S. (2009). *An integrated White + Black box approach for designing and tuning stochastic local search algorithms*. (Doctoral dissertation).

Hochberg, Y., & Tamhane, A. C. (1987). *Multiple comparison procedures*. Wiley.

Johnson, D. S. (1974). Fast algorithms for bin packing. *Journal of Computer and System Sciences*, *8*(3), 272–314. doi:10.1016/S0022-0000(74)80026-7

Lamarca, M. (2009). *Hipertexto: El nuevo concepto de documento en la cultura de la imagen: Internet*. (Doctoral dissertation, PhD thesis). Complutense University of Madrid. Available at http://www.hipertexto.info

Liu, X. (1996). Intelligent data analysis: Issues and challenges. *The Knowledge Engineering Review*, *11*(04), 365–371. doi:10.1017/S0269888900008055

Loh, K. H., Golden, B., & Wasil, E. (2008). Solving the one-dimensional bin packing problem with a weight annealing heuristic. *Computers & Operations Research*, *35*(7), 2283–2291. doi:10.1016/j.cor.2006.10.021

López, C. (2000) *Modelo para el Desarrollo de Bibliotecas Digitales Especializadas*. (Master's thesis, ITAM). Retrieved on June 4, 2014, from http://www.bibliodgsca.unam.mx/tesis/tes7cllg/sec_10.htm

Luengo, J., García, S., & Herrera, F. (2009). A study on the use of statistical tests for experimentation with neural networks: Analysis of parametric test conditions and nonparametric tests. *Expert Systems with Applications*, *36*(4), 7798–7808. doi:10.1016/j.eswa.2008.11.041

McGeoch, C. C. (2007). Experimental algorithmics. *Communications of the ACM*, *50*(11), 27–31. doi:10.1145/1297797.1297818

Mendenhall, W., & Sincich, T. (1997). *Probabilidad y estadística para ingeniería y ciencias* (4th ed.). México: Prentice Hall.

Preuss, M. (2007). *Reporting on experiments in evolutionary computation*. Univ., Sekretariat des SFB 531.

Quiroz, M., Cruz-Reyes, L., Torres-Jiménez, J., & Melin, P. (2013). Improving the performance of heuristic algorithms based on exploratory data analysis. In *Recent Advances on Hybrid Intelligent Systems* (pp. 361–375). Springer Berlin Heidelberg. doi:10.1007/978-3-642-33021-6_29

Quiroz-Castellanos, M., Cruz-Reyes, L., Torres-Jimenez, J., Gómez, C., Huacuja, H. J. F., & Alvim, A. C. (2015). A grouping genetic algorithm with controlled gene transmission for the bin packing problem. *Computers & Operations Research*, *55*, 52–64. doi:10.1016/j.cor.2014.10.010

Rhyne, A. L., & Steel, R. G. D. (1965). Tables for a treatments versus control multiple comparisons sign test. *Technometrics*, *7*(3), 293–306. doi:10.1080/00401706.1965.10490264

Sheskin, D. J. (2006). *Handbook of Parametric and Nonparametric Statistical Procedures*. Boca Raton, FL: CRC Press.

Siegel, S. C., John, N., Aragón Borja, L. E., & Fierros Dávila, L. E. (1995). Estadística no paramétrica: aplicada a las ciencias de la conducta (4th ed.). México: Trillas.

Snodgrass, R. T. (2010). *Ergalics: A natural science of computation*. University of Arizona.

Steel, R. G. (1959). A multiple comparison sign test: Treatments versus control. *Journal of the American Statistical Association*, *54*(288), 767–775. doi:10.2307/2282500

Taillard, E. D. (2003). A statistical test for comparing success rates. In Metaheuristic international conference MIC'03.

Trawiński, B., Smętek, M., Telec, Z., & Lasota, T. (2012). Nonparametric statistical analysis for multiple comparison of machine learning regression algorithms. *International Journal of Applied Mathematics and Computer Science*, *22*(4), 867–881. doi:10.2478/v10006-012-0064-z

Wolfe, D. A. (1998). *Nonparametric Methods*. Wiley Encyclopedia of Clinical Trials.

Wolpert, D. H., & Macready, W. G. (1997). No free lunch theorems for optimization. *Evolutionary Computation. IEEE Transactions on*, *1*(1), 67–82.

Chapter 9
Modelling the Route Choice:
The Role of Volume-Delay Functions in Transport Planning

Eric Moreno-Quintero
Mexican Transportation Institute, Mexico

ABSTRACT

The issue of route choice is a key factor for the freight transport performance. Congestion at roads encourages hauliers to change routes to minimize the delays and keep lead times in a reliable range. In the context of transport planning, the route choice problem can be assessed by modelling the travel times needed to reach a destination through the different routes in a road network. Is in this point where the volume-delay functions become relevant. A Volume-Delay Function (VDF) is a mathematical representation of the increase of the travel time as more and more vehicles utilize the routes, causing congestion on the road networks. The related literature and practitioners report on the use of some known functional forms, as the BPR function, the Conical volume-delay function or the Akcelik's function, which are widely utilized in flow's assignation modelling in transport planning. A successful application of VDFs requires a proper fitting of the function's parameters. In a classical focus these parameters can be deduced from speed-flow surveys carried out at the routes or links of interest; these surveys generally require time money and personnel. As an alternative to this classical focus, particularly when facing scarcity of resources, this work carries out a mathematical analysis of the VDF functional forms, as well with an interpretation of their parameters in relation to road's operation. The results of these analyses clarifies the meaning of the functional forms for the VDFs and their parameters, and suggest other ways to assess those parameters which may be more practical for the purpose of modelling the choice of route in freight transport. Some considerations to put into practice this in Mexico are discussed at the end of this work.

DOI: 10.4018/978-1-4666-9779-9.ch009

1. INTRODUCTION

In any surface transportation system, a reasonable forecast of vehicle flows on the diverse routes available to connect origins and destinations is of interest for transport planners, road users and hauliers moving freight. For transport planners, these forecasts guide the design of control policies for traffic and reduction of impacts like congestion, accidents and environmental damage. For users and hauliers, forecasts give insight on congestion conditions, travel times and the alternative routes, in order to plan logistics issues.

Forecasts of vehicle flows are based on mathematical assignment models, which determine the expected vehicle volumes on each route in a transport network, once the network topology, an origin-destination matrix and the volume-delay functions on the network links are known.

A volume-delay function (VDF) is a mathematical function whose independent variable is the vehicle volume on a network's link, or alternatively the percentage of link's capacity that is used by traffic, and gives as response the expected time needed to cross that link.

The formulation of a VDF aims to represent the delay suffered by trips as a consequence of the growing number of vehicles on the road. Thus, VDFs are usually proposed looking for some common desired characteristics as: being continuous functions, monotonically increasing and bounded on their domain, and having derivatives at least of the first and second order. The latter is convenient for modelling marginal costs or when dealing with optimizing algorithms based on gradient methods. The role of the VDFs in the assignment model is to estimate the effect that occupancy of the road by vehicles has on the trip times. Based on this estimation, the assignment model uses route choice criteria to optimize a desired objective (e.g. time or cost) and calculates the vehicle flow split on the available routes.

In practice, modelling of route choice usually is done in reference to the equilibrium concepts proposed by J.G. Wardrop around the 1950s. These two equilibrium concepts are: user's equilibrium (UE) and system's Equilibrium (SE).

User's equilibrium occurs on a road network when users choose the more convenient route to reach their destinations freely on a self-interest basis. When the network begins to be occupied, early users choose usually the shortest routes to their destinations. As more users enter into the roads, the congestion effects motivate some users to change route in order to improve their travel times. These changes happen without any agreement among users, and finally they reach an equilibrium point where travel time on any used route is the same, so nobody gets improvement by changing route.

System's equilibrium appears on a road network when users choose their routes in such a way that the total time of their trips, or also the average trip time on all used routes is a minimum. This equilibrium does not happen spontaneously, and in general it is not easy that users agree among them to produce this effect. That is why results from a SE model have a normative focus, in the sense that the forecasted flows are those desirable from the planner's viewpoint, so in order this really happen the network has to be controlled, e.g. by fostering circulation on certain routes or inhibiting it on another parts of the network.

An additional criterion used in practice is the so called "All-or-Nothing". This criterion does not consider congestion effects on the network, and assumes that costs (times) are fixed on the routes; it is useful for networks having low traffic levels or with operating conditions very well controlled, as is the case of railroad networks.

Several functional forms for VDFs are found in the assignment modelling literature. Probably the most commonly used is that proposed by the US Bureau of Public Roads, known as BPR function; other typical formulations used in practice are the Conical function and Akcelik's function. These functions are usually available in transportation planning software packages like TransCAD from Caliper Corporation or VISUM from PTV Group. In this work this three VDFs are examined regarding their mathematical formulation, their behaviour on their domains, the meaning of their parameters and the possibilities of application into the context of surface transport planning in Mexico.

2. FLOWS' ASSIGNMENT ON TRANSPORTATION NETWORKS

An assignment model forecasts the vehicle flows circulating on a transportation network, given a set of assumptions considered for the desired planning scenarios (OD matrices used, vehicle operating costs, free-flow times, etc.). This is not the only objective in an assignment model. From the transport planning perspective, the main objectives for an assignment model are (Ortúzar and Willumsen, 2011):

- Get good global estimates of the network performance, e.g. total vehicle flowing into the network, fares collected at public transport, etc.
- Obtain estimates for the trip times between major zones in the network, for different demand levels.
- Generate reasonable for the vehicle flow on certain routes, and the resulting congestion.

Other secondary objectives of interest are:

- Identify the routes used for trips among any origin-destination pair.
- Analyse which origin-destination pairs are involved on a specific route.

Results from the assignment model allow then estimate the economic and environmental benefits and impacts, so giving useful information to policy designers who are interested in improvements for the transportation system analysed. Figure 1 shows a graph directly on a map of Mexico, of an assignment exercise developed with TransCAD.

The network correspond to the federal highway system, and the flows represent truck movements with origin-destination data 2010 (Moreno et al, 2014). The green line thickness is proportional to the traffic level. The route choice criterion is solely the average trip time on the road's links, and a BPR function represents congestion.

The flows in Figure 1 show a more intense traffic on the central region surrounding the Metropolitan Area of Mexico City and its connections to the Mexico-USA border by Monterrey and Guadalajara, and also for the flows to the Gulf coast by the oil zone in Tabasco.

The assignment of flows on a network is based on some criteria; the more common are the following.

2.1. All or Nothing Criterion

It is the typical shortest route model found in Graph Theory, that considers a network with n nodes, where any *arc (I, j)* connecting node i to node j has a fixed distance or cost Cij. The related optimisation problem is as follows, where Xij is the flow from node i to node j:

$$MinZ = \sum_{i=1}^{n} \sum_{j=1}^{n} C_{ij} X_{ij}$$

Figure 1. Tonnes assignment in road transport based on travel times (Source: own produced)

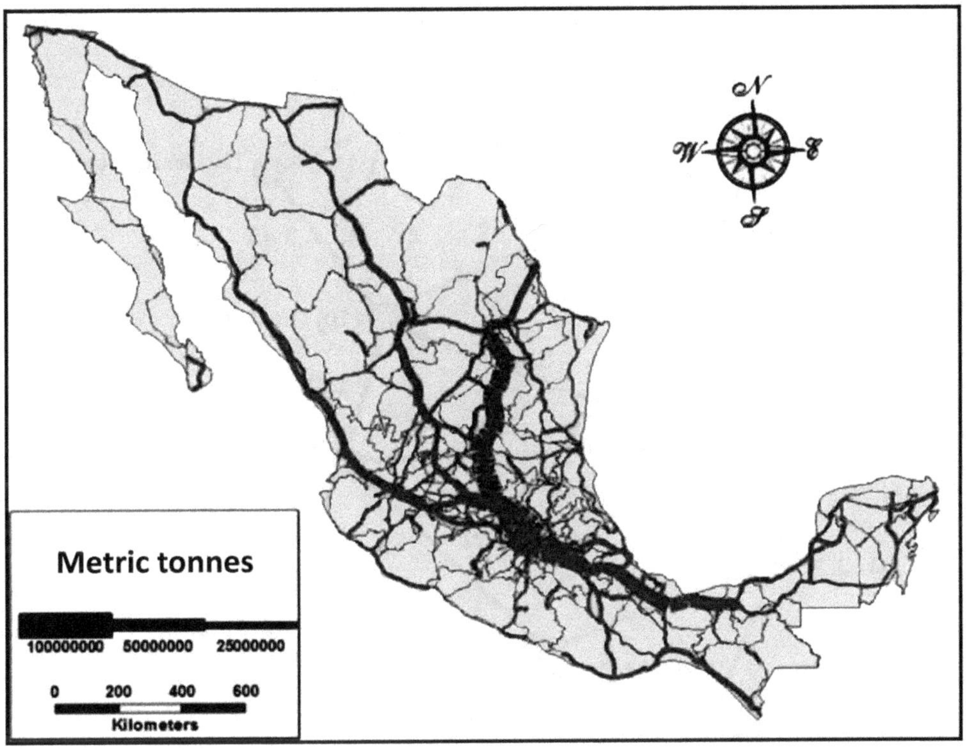

$$subject\ to: \sum_{j=1}^{n} X_{ij} - \sum_{k=1}^{n} X_{ki} = \begin{cases} 1 & if\ i=1 \\ 0 & if\ i \neq 1 \quad or\ i=n \\ -1 & if\ i=n \end{cases}$$

$$X_{ij} \in \{0,1\} \quad i,j \in \{1,2,\cdots,n\}$$

Problem constraints represent the flow equilibrium at nodes and decision variables are binary, with the value "1" when there is a flow through the arc and "0" when there is no flow. This criterion does not consider increasing travel times as more vehicles occupy the network, nor the existence of alternate routes, because the cost *Cij* for crossing any arc *(i, j)* it be distance or time, is fixed. This model is used for networks with low vehicle volumes without congestion effects, or in networks with controlled flows where vehicles follow the shortest routes in a natural way as is the case for rail movements.

The basic model for shortest route from an origin to a destination has been solved with quite efficient methods, being Dijkstra's and Floyd's algorithms the most used.

Dijkstra's algorithm finds all the shortest routes from an initial fixed node towards the rest of nodes in the network, whereas Floyd's algorithm finds the shortest route for any pair of nodes in the transport network.

Figure 2 shows a simplified scheme for the federal road network from Mexico City, Federal District, to the city of Nuevo Laredo at the Mexico-USA border, where the values on the arcs are travelling times.

Figure 2. Mexico City-N. Laredo highway scheme (Trip times estimates by Google Maps, May 2015)

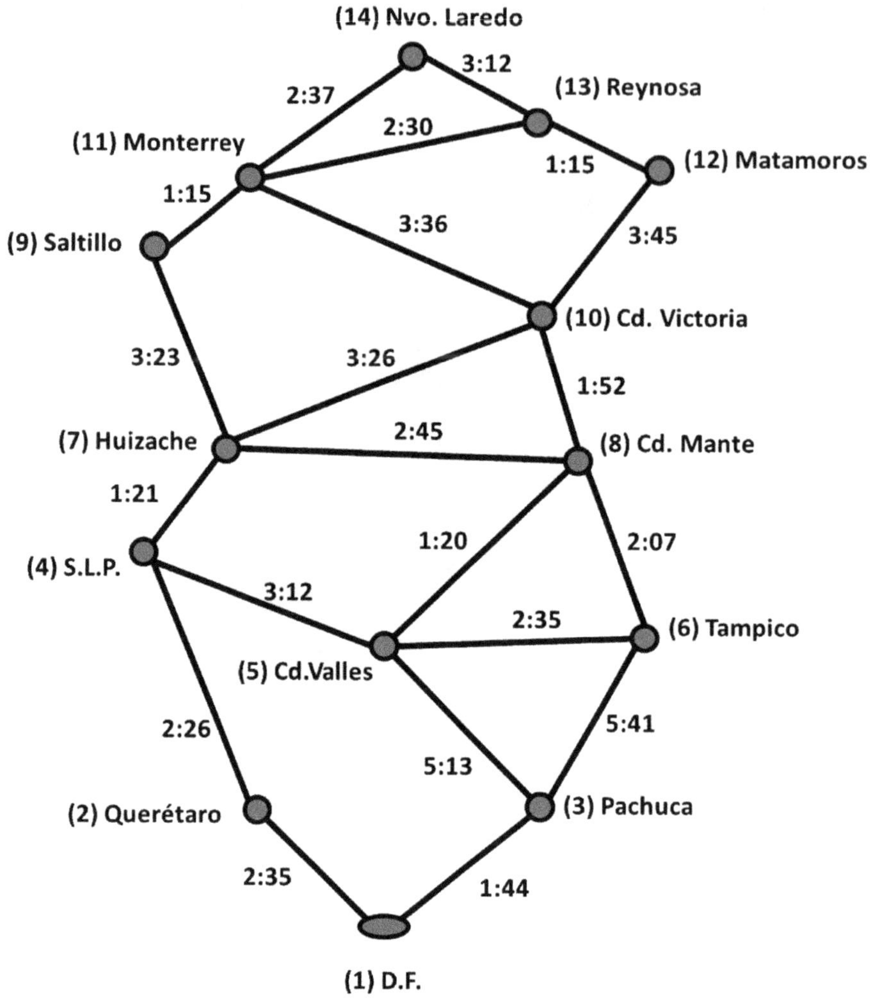

Thus in this case the shortest route for any pair of nodes results the fastest one.

Using Dijkstra's algorithm from the initial node (1) DF, the shortest routes tree (minimal trip time) to the rest of the nodes is obtained as is shown in Figure 3. This solution was obtained from TORA (Taha, 2011). The optimal route DF – NUEVO LAREDO is: 1 – 2 – 4 – 7 – 9 – 11 – 14, corresponding to route:

DF – QUERETARO – SLP – HUIZACHE – SALTILLO- MONTERREY – NUEVO LAREDO, with a total time of 14 hr. 4 min (14.06 hr).

Floyd algorithm results are shown in Figures 4 and 5.

Each matrix entry in Figure 4 is the shortest route time from row's node to column's node; e.g. the minimal time trip from Tampico (N6) to Monterrey (N11) is 7 hr. 35 min (7.59 hr.), while the minimal one for the pair Pachuca (N3) – Cd. Victoria (N10) is 8 hr. 25 min (8.42 hr).

Table in Figure 5 allows to find optimal routes for any origin-destination pair with the basic criterion of intermediate nodes (Taha, 2011). According to this criterion, k is an intermediate node for the pair (i, j) if entry (i, k) in the table equals "k" whereas entry (k, j) equals "j".

Figure 3. Optimal routes from node DF to the rest of nodes in the simplified network with Dijkstra's algorithm (using TORA, Taha, 2011)

From	To	Distance	Route
1-DF	2-QRO	2.58	1-2
1-DF	3-PACH	1.75	1-3
1-DF	4-SLP	5.01	1-2-4
1-DF	5-VALLES	6.97	1-3-5
1-DF	6-TAMPIC	7.43	1-3-6
1-DF	7-HUIZAC	6.36	1-2-4-7
1-DF	8-MANTE	8.30	1-3-5-8
1-DF	9-SALTIL	10.19	1-2-4-7-9
1-DF	10-VICTOR	9.79	1-2-4-7-10
1-DF	11-MTY	11.44	1-2-4-7-9-11
1-DF	12-MATAM	13.54	1-2-4-7-10-12
1-DF	13-REYN	13.94	1-2-4-7-9-11-13
1-DF	14-NLARED	14.06	1-2-4-7-9-11-14

Figure 4. Shortest times among OD pairs at the simplified network based on Floyd's algorithm (using TORA, Taha, 2011)

Iter 14	D 14												
	N 2	N 3	N 4	N 5	N 6	N 7	N 8	N 9	N 10	N 11	N 12	N 13	N 14
N1	2.58	1.75	5.01	6.97	7.43	6.36	8.30	10.19	9.79	11.44	13.54	13.94	14.06
N2		4.33	2.43	5.63	8.21	3.78	6.53	7.61	7.21	8.86	10.96	11.36	11.48
N3	4.33		6.76	5.22	5.68	8.11	6.55	11.94	8.42	12.02	12.17	13.42	14.64
N4	2.43	6.76		3.20	5.78	1.35	4.10	5.18	4.78	6.43	8.53	8.93	9.05
N5	5.63	5.22	3.20		2.58	4.08	1.33	7.91	3.20	6.80	6.95	8.20	9.42
N6	8.21	5.68	5.78	2.58		4.87	2.12	8.70	3.99	7.59	7.74	8.99	10.21
N7	3.78	8.11	1.35	4.08	4.87		2.75	3.83	3.43	5.08	7.18	7.58	7.70
N8	6.53	6.55	4.10	1.33	2.12	2.75		6.58	1.87	5.47	5.62	6.87	8.09
N9	7.61	11.94	5.18	7.91	8.70	3.83	6.58		4.85	1.25	5.00	3.75	3.87
N10	7.21	8.42	4.78	3.20	3.99	3.43	1.87	4.85		3.60	3.75	5.00	6.22
N11	8.86	12.02	6.43	6.80	7.59	5.08	5.47	1.25	3.60		3.75	2.50	2.62
N12	10.96	12.17	8.53	6.95	7.74	7.18	5.62	5.00	3.75	3.75		1.25	4.45
N13	11.36	13.42	8.93	8.20	8.99	7.58	6.87	3.75	5.00	2.50	1.25		3.20
N14	11.48	14.64	9.05	9.42	10.21	7.70	8.09	3.87	6.22	2.62	4.45	3.20	

Figure 5. Data to determine fastest routes among all origin-destination pairs of the simplified network, using Floyd's algorithm (using TORA, Taha, 2011)

	S 14													
	N 1	N 2	N 3	N 4	N 5	N 6	N 7	N 8	N 9	N 10	N 11	N 12	N 13	N 14
N1		2	3	2	3	3	4	5	7	7	9	10	11	11
N2	1		1	4	4	5	4	7	7	7	9	10	11	11
N3	1	1		2	5	6	4	5	7	8	10	10	12	11
N4	2	2	2		5	5	7	7	7	7	9	10	11	11
N5	3	4	3	4		6	8	8	8	8	10	10	12	11
N6	3	5	3	5	5		8	8	8	8	10	10	12	11
N7	4	4	4	4	8	8		8	9	10	9	10	11	11
N8	5	7	5	7	5	6	7		7	10	10	10	12	11
N9	7	7	7	7	8	8	7	7		11	11	13	11	11
N10	7	7	8	7	8	8	7	8	11		11	12	12	11
N11	9	9	10	9	10	10	9	10	9	10		13	13	14
N12	10	10	10	10	10	10	10	10	13	10	13		13	13
N13	11	11	12	11	12	12	11	12	11	12	11	12		14
N14	11	11	11	11	11	11	11	11	11	11	11	13	13	

By example, for the optimal route for pair Tampico (N6) and Monterrey (N11): the entry (N6, N8) equals 8, while entry (N8, N10) equals 10 so N8 is an intermediate node, and the first try for route is: 6 – 8 – 11. In the same way, entry (N8, N10) equals 10, and entry (N10, N11) equals 11 so N10 is intermediate for the pair (N8, N11). The final route is 6 – 8 – 10 – 11, corresponding to TAMPICO – CD.MANTE – CD. VICTORIA – MONTERREY with a total time of 7 hr. 35 min (7.59 hr.) as is shown in Figure 4.

To determine a flow using the All-or-Nothing criterion it suffices to find all the optimal routes adding flows on the arcs shared by those routes in order to total the flows crossing on each arc. The next numerical example illustrates this.

Figure 6 shows an OD matrix with the number of vehicles at the origins wishing to reach the destinations, and also the corresponding optimal routes. By examining these routes, the shared arcs are identified, and the corresponding flows of the routes crossing on the same arc are added (in both directions, in this example) to get the total flows on arcs.

Figure 7 shows the assignment result, where the numbers at the arcs indicate the total flow on these arcs, and the thickness of the arc-lines are proportional to these flows. None route uses the arcs shown with dotted lines.

In this example, arc SLP–HUIZACHE has the major flow with 201 vehicles. Four routes share this arc: DF–N.LAREDO; MATAMOROS–DF; MATAMOROS–SLP and MONTERREY-DF, as shown in Figure 6.

So, adding the flows over the routes equals: 80 + 60 + 20 + 50 = 210. The first flow value (80) goes from SLP to HUIZACHE and the rest (130) goes from HUIZACHE to SLP.

2.2. Equilibrium Criteria

These criteria consider that travelling times on arcs depend on the vehicle volumes circulating, so it is possible to model congestion. Two basic types of equilibrium proposed by J.G. Wardrop around the 1950's are usual in transportation planning: User equilibrium (UE) and System equilibrium (SE) (Ortúzar and Willumsen, 2011).

Figure 6. OD Matrix and optimal routes for the numerical example (using TORA, Taha, 2011)

		DESTINATION				
		NLAREDO	TAMPICO	SLP	DF	CDVALLES
ORIGIN	DF	80	20			
	MATAM			20	60	
	MTY				50	25

From	To	Distance	Route
1-DF	14-NLARED	16.87	1-2-4-7-9-11-14
1-DF	6-TAMPIC	7.50	1-3-6
12-MATAM	1-DF	15.25	12-10-7-4-2-1
12-MATAM	4-SLP	9.75	12-10-7-4
11-MTY	1-DF	13.25	11-9-7-4-2-1
11-MTY	5-VALLES	9.00	11-10-8-5

Figure 7. All-or-Nothing traffic assignment. Numerical example on the simplified network

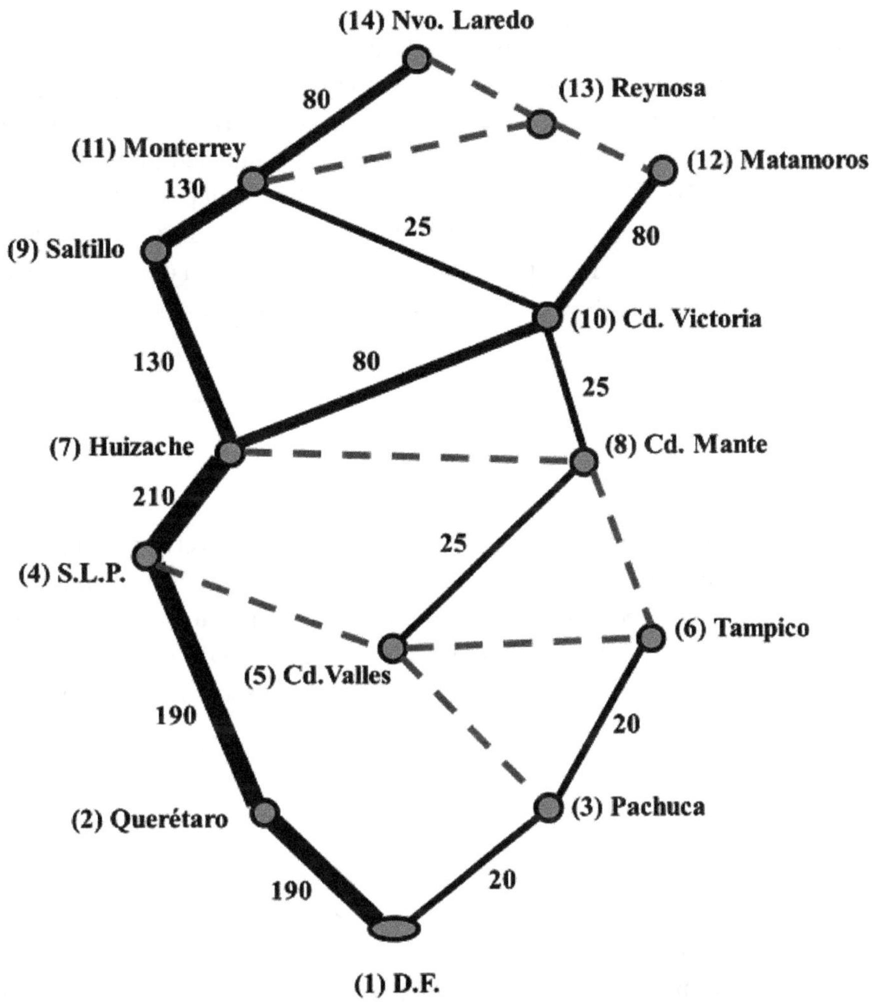

User Equilibrium (UE)

In this criterion, users choose freely and individually their best routes, without cooperation among them. By doing this, routes begin to be occupied by vehicle's traffic and an equilibrium point is reached, when any used route in a particular OD pair gets the same travel time, so nobody has advantage by changing route. The mathematical formulation is as follows:

$$Min\ C = \sum_{j} \int C_j(v_j)dv$$

$$s.t.\quad n_1 + n_2 + \cdots + n_k = N$$

where index j represents the arcs in the network, Cj is the cost function (volume-delay function) over the arc j, v_j is the vehicles volume on the arc j, n_k is the number of vehicles leaving at node k and N is the total number of vehicles into the network.

System Equilibrium (SE)

This equilibrium arises when the total travelling time or the average travel time of users is a minimum, thus implying a minimal use of resources. This equilibrium focuses on the planning viewpoint, because the fuel consumption and infrastructure wear, as a first estimation, are proportional to the total travelling time of the vehicles using the network. So transportation planners are interested in minimising this quantity.

The solution flows obtained by SE are usually different from those calculated by the User equilibrium, with some arcs where users have bigger travelling times than those in the corresponding UE solution. This means that individual interest can be affected in favour of the social interest of a rational use of resources. The mathematical formulation is as follows, with the same meaning of variables and indexes as before:

$$Min\ C = \sum_j C_j(v_j)v_j$$

$$s.\ t. \quad n_1 + n_2 + \cdots + n_k = N$$

The cost functions $Cj(v_j)$ into the objective functions C of the previous criteria represent the difficulty of traversing the arcs of the network. The independent variable is the volume of vehicles or alternatively its percentage of the capacity (veh/hr) at the arc, and they have parameters to be calibrated in order to model the target network. These functions are called volume-delay (VDF) by their intuitive meaning of impedance to traverse the network, that models the effect of speed reduction as more vehicles occupy the network.

A Numerical Example

This example shows the differences among the two equilibrium criteria previously explained. Assuming there are 10,000 vehicles at node DF with destination at node Monterrey, and two available routes as shown in Figure 8; the dotted-line arcs are not used.

On the first route: DF – QRO – SLP – HUIZACHE – SALTILLO – MONTERREY, the travel time is 13.25 h (13h 15 min). On the second route: DF – PACHUCA – TAMPICO – CD.MANTE – CD.VICTORIA – MONTERREY, the travel time is 16.25 h (16 h, 15 min). These are the free-flow times, without congestion.

In order to model congestion, it is assumed that every 100 vehicles entering in the first route, increase travel time in 12 minutes, whereas in the second route the corresponding increase is 15 minutes. In this way the cost functions for traversing the network are defined; these functions actually are volume-delay functions (VDFs). Thus, depending on the vehicle volumes circulating: V_1 at the first route and V_2 at the second one, the corresponding VDFs are:

First Route: $f_1(V_1) = 13.25 + 0.002V_1$

Second Route: $f_2(V_2) = 16.25 + 0.0025V_2$

Figure 8. Equilibrium flow assignment. Numerical example for simplified network

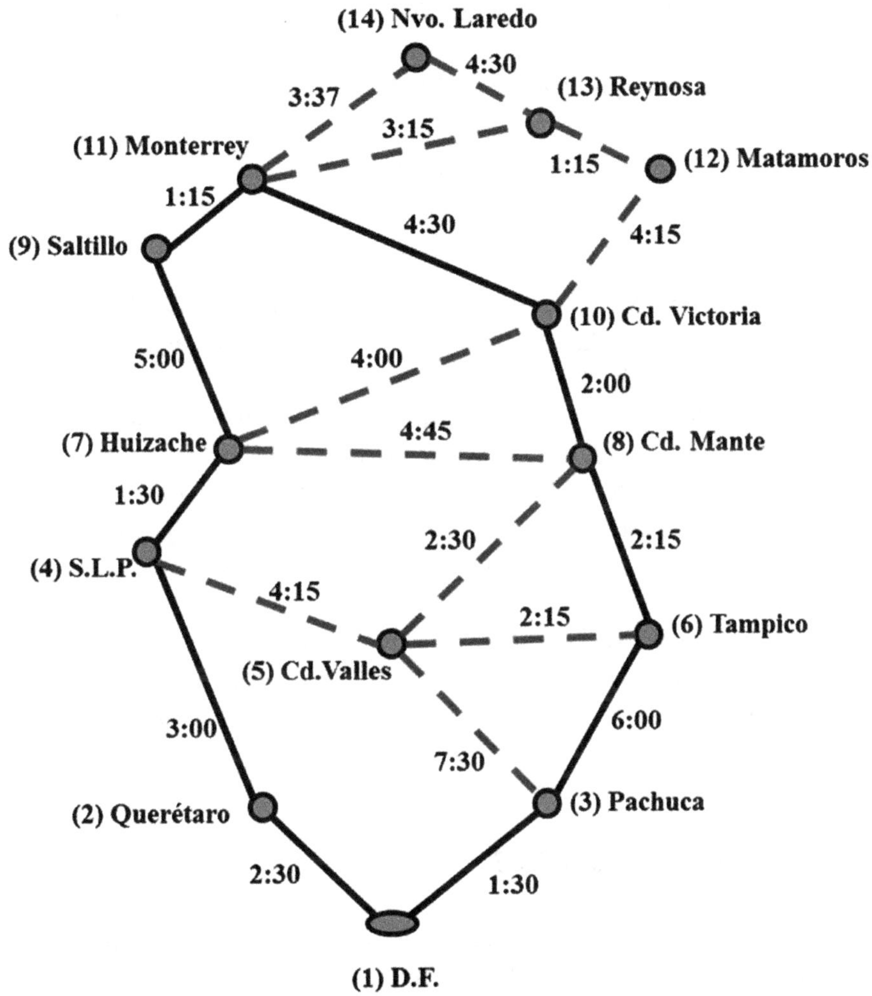

Then, the problem to solve is to find the number of vehicles moving on each route under the two equilibrium criteria and the resulting cost in each case.

To solve the UE, the following optimization problem is posed:

$$Min\ C = \int (13.25 + 0.002V_1)\, dV_1 + \int (16.25 + 0.0025V_2)\, dV_2$$
$$s.\ t. \quad V_1 + V_2 = 10000; \quad V_1, V_2 \geq 0; \quad V_1, V_2\ integer$$

Using *Excel's SOLVER*, the next result is obtained:

V_1 = 6,223 veh; V_2 = 3,777 veh; Total veh-hour: 256,946.8.

Each trip on the first route takes 13.25 + 0.002×6223 = 25.7 hr and each trip on the second route takes 16.25 + 0.0025×3777 = 25.7 hr. This is the expected equilibrium condition, where nobody improves by changing route.

To solve the SE, the associated optimisation problem is

$$Min\ C = \left(13.25 + 0.002V_1\right)V_1 + \left(16.25 + 0.0025V_2\right)V_2$$
$$s.\ t.\quad V_1 + V_2 = 10000;\quad V_1, V_2 \geq 0;\quad V_1, V_2\ integer$$

The corresponding solution is: V_1 = 5,889 veh; V_2 =4,111 veh; Total veh-hour:256,444.4.

Each trip on the first route takes 13.25 + 0.002×5889 = 25.0 hr; whereas each trip on the second one takes: 16.25 + 0.0025×4111 = 26.5 hr. The example shows how under the User equilibrium, veh-hr total is greater than the corresponding under the System Equilibrium, in this case a difference of502.3 veh-hr.

Under SE, the first route has a better travel time compared to the second route and also compared to those obtained under UE. But also this second route has the worst of the travel times. This issue poses the planner's challenge to induce users to adopt the SE flow patterns, in such a way that some of them accept greater travel times in order to benefit to the transportation system by reducing the total veh-hr involved in the movements.

The assignment solutions to determine the number of vehicles using the different routes depend directly from the volume-delay functions, which mathematically represent the congestion. For this reason, these functions have to be formulated with mathematical properties adequate to model the effects of congestion on the networks.

3. THE VOLUME-DELAY FUNCTIONS

Volume-delay functions (VDF) are a key element in the assignment model to achieve approximate representations of the flows circulating on the different routes in the transport network.

Since a VDF $f(x)$ aims to replicate the traffic behaviour under congestion, some mathematical properties usually desirable for $f(x)$ are as follows:

1. $f(x)$ represents the travel time on an arc; the argument x can be the volume of vehicles crossing the arc or the percentage of the arc capacity used; in the last case, $x \in [0, 1]$.
2. $f(x)$ is non-negative, continuous, monotonically increasing and bounded on its domain of definition. Thus there is a correspondence between the increase of vehicle volumes and the corresponding increase in travel times, as well as a maximal value for the travel time once saturation is reached.
3. $f(x)$ should have a straightforward analytical formula to do calculations, and facilitate transferability among different computer platforms.
4. $f(x)$ should have derivatives of first and second order at least, in order to get a soft curve and as well to calculate marginal costs and to allow the use of gradient-based optimisation methods.

These issues are examined for some typical VDFs in the following sections.

3.1. The Speed-Flow Relationship

The more directly noticed effect of congestion on a network is the speed reduction as vehicles volume increases. The mathematical representation of this effect is a speed-flow curve, which describes the variation in speed on the network arcs as a function of the vehicle volumes there.

An example of this type of curve is found in COBA program (Cost-Benefit Analysis) from the British Department for Transport. A general form of this type of curve is as follows (Lancashire County Council, 2010).

$$S(V) = \begin{cases} S_0 & for\ V \le F \\ S_0 + \dfrac{(S_1 - S_0)(V - F)}{(C - F)} & for\ F < V \le C \\ \dfrac{S_1}{1 + \dfrac{S_1 (V - C)}{8dC}} & for\ V > C \end{cases}$$

where

S = speed on the link (km/h); S = d/t.
t = travel time for the link (hours).
d = link's length (km).
V = flow at the link (PCU/h).
S_0 = free-flow speed.
S_1 = speed at link's capacity.
F = maximal flow at which free-flow conditions still hold (PCU/h).
C = link's capacity (PCU/h).

Table 1 shows some parameters for these curves calibrated for British highways. It can be noticed that in all cases the speed at the saturation point (flow = capacity) is the same (45 km/h). Critical flow *F* is the maximal value of flow at which free-flow conditions are still valid.

Figure 9 shows the speed curves (km/h) and travel time (min) for an interurban road (Rural D" Motorway) with the following parameters:

S_0 =free-flow speed = 112 km/h
S_1 = speed at capacity = 45 km/h
F = critical flow (limit for free-flow conditions) = 2400 pcu/h
C = link's capacity = 4860 pcu/h

Table 1. Parameters for speed-flow curves in the United Kingdom (Lancashire County Council, 2010).

Road Type	S0 Free-Flow Speed (km/h)	S1 Speed at Capacity (km/h)	F Critical Flow (pcu/h)	Capacity (pcu/h)
Rural D3 Motorway. Interurban, 3 lanes each direction	116	45	3600	7560
Rural D2 Motorway. Interurban, 2 lanes each direction	112	45	2400	4860
Rural D2/AP (A Trunk). Interurban, 2 lanes each direction. All purpose, trunk.	108	45	2160	4520
Rural D2/AP. Interurban, 2 lanes each direction. All purpose.	105	45	2160	4360
Rural D3/AP (A Trunk). Interurban, 3 lanes each direction. All purpose, trunk.	108	45	3240	6780
Rural D3/AP. Interurban, 3 lanes each direction.	105	45	3420	6540

Figure 9. Speed-flow curve example (based on COBA, Lancashire County Council, 2010)

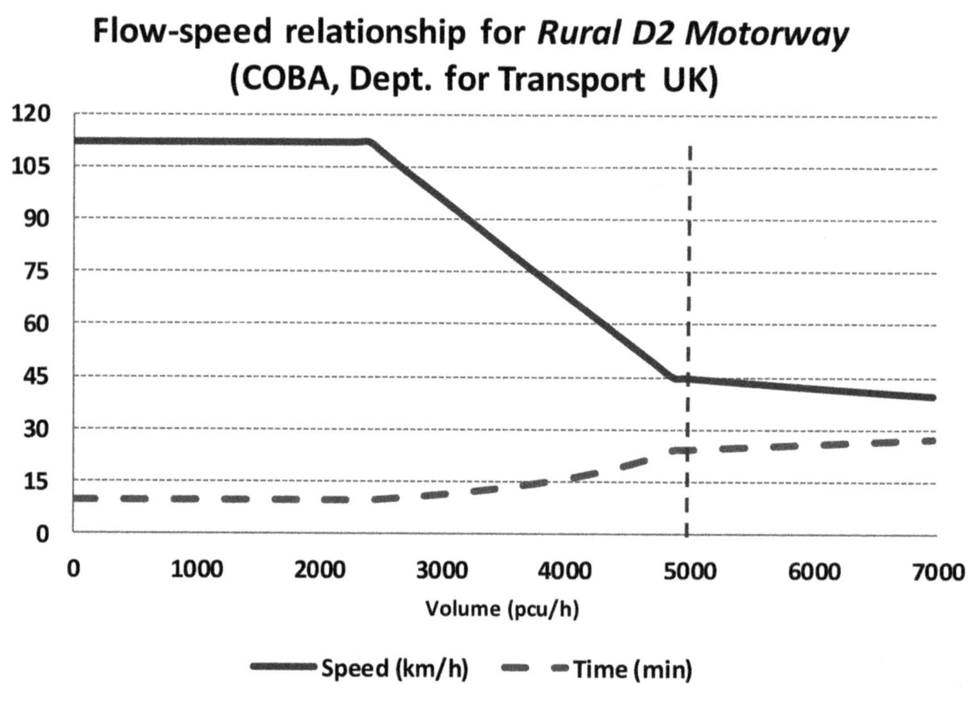

And assuming that d = link' length = 18 km. The corresponding time curve is obtained from $t = d/S$. Thus the free-flow travel time is $t = 18/112 = 0.161\ h \approx 9.6\ min$. At the saturation level (4860 PCU/h) the speed reduces to 45 km/h and the travel time is 24 minutes approximately.

The speed-flow curve at Figure 9 is a piecewise linear function in the most part of its domain, from cero to the saturation level, marked by the dotted vertical line on the graph. For volumes V higher than the saturation level C, the functional form is:

$$\frac{S_1}{1+\frac{S_1(V-C)}{8dC}}$$

The mathematical model on this part of the domain reduces speed to zero asymptotically, with the consequent raise of the travel time, in view that:

$$\lim_{V\to\infty}\left[\frac{S_1}{1+\frac{S_1(V-C)}{8dC}}\right]=0$$

Due to its piecewise linearity, the previous function has two points where the slope change is not continuous, i.e. at V = F and V = C; at these points a derivative does not exist. Where V = F (the critical flow) slope changes from zero to:

$$\frac{(S_1-S_0)}{(C-F)}=\frac{45-112}{4860-2400}=\frac{-67}{2460}$$

Corresponding to the slope of the decreasing line up to V = C. For values V > C, slope changes like the derivative of the corresponding formula, that is:

$$\frac{-S_1^{\,2}}{8dc\left[1+\frac{S_1(V-C)}{8dC}\right]^2}$$

Thus for V = C, slope changes from previous $\frac{-67}{2460}$ to:

$$\frac{-S_1^{\,2}}{8dc}=\frac{-45^2}{8\times18\times4860}=\frac{-5}{1728}$$

The characteristic shown by the function may to complicate optimisation of this type of functions under Wardrop principles when gradient-based techniques are used.

For flow values V > C it should be noted that actually, flow does not go further. This condition appears when additional flows trying to enter into the network, once saturation has been reached, generate instability and queues at networks entries. For modelling purposes, sometimes is convenient consider that for flows greater than capacity at link, speed keeps constant at the saturation level (Ortúzar and Willumsen, 2011); in the example shown it correspond to 45 km/h. Other volume-delay functions are found in the literature, with improved mathematical performance and giving reasonable results when modelling.

3.2. Volume-Delay Functions Commonly Used

Three of the most common VDF found in the literature are: the BPR (Bureau of Public Roads) function, the Conical function and Akcelik's function, described below.

Bureau of Public Roads Function (BPR Function)

$$t(x) = t_0\left[1 + \alpha\left(\frac{x}{C}\right)^{\beta}\right]$$

where t = travel time; t_0 = free-flow time; x = fow at the link; C = link's capacity; α, β calibration parameters (Ortúzar and Willumsen, 2011).

Conical Volume-Delay Function

$$t(x) = t_0\left[2 + \sqrt{\alpha^2\left(1 - \frac{x}{C}\right)^2 + \beta^2} - \alpha\left(1 - \frac{x}{C}\right) - \beta\right]$$

where t = travel time; t_0 = free-flow time; x = flow at the link; C = link's capacity; and $\alpha > 1$; $\beta = \frac{2\alpha - 1}{2\alpha - 2}$ are calibration parameters (Spiess, 1990).

Akcelik's Volume Delay Function

$$t(x) = t_0 + D_0 + 0.25T\left[(x-1) + \sqrt{(x-1)^2 + \frac{(16JL^2)x}{T^2}}\right]$$

where t = travel time; t_0 = free-flow time; D_0 = delay at node under free-flow; T =expected duration of demand; L = link's length; J = calibration parameter; x = flow to capacity rate. (Caliper, 2005)

Figure 10 shows examples of these functions with the more known parameters for BPR ($\alpha = 0.15$, $\beta = 4$) and other parameters for Conical and Akcelik's functions, as it is shown in the graph. The variable x at Figure 10 is the saturation level on the link, i.e. x = (vehicles volume) / (link's capacity). The ordinate axis represents the travel time, starting from unit value corresponding to the free-flow time.

The general behaviour of the curves is similar, though it can be noticed that BPR function has the slowest growing. In the three cases, for saturation levels from 0 to 0.125, all curves remain almost constant. The conical function has the biggest level at the saturation point; this function always double the free-flow time at this point.

Figure 10. Some volume-delay functions curves

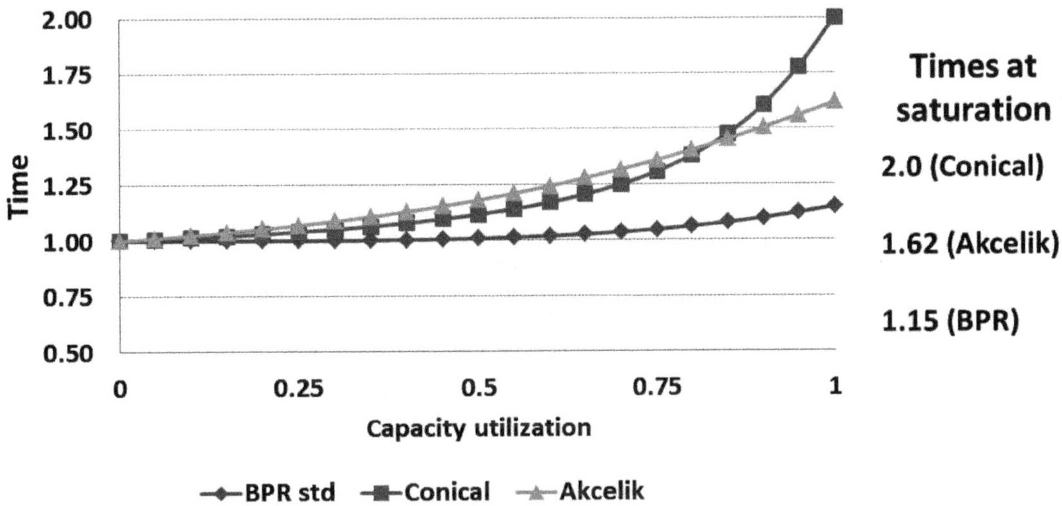

BPR	
alfa	0.15
beta	4

Conical	
alfa	5
beta	1.125

Akcelik	
J	0.000085
T (hrs)	3.5
L (km)	67

This example suggests that a realistic characterization of vehicle traffic with VDFs requires a good adjustment of the corresponding parameters in these functions. The problem of VDF parameter estimation is examined in the next section.

4. FUNCTION'S BEHAVIOUR AND ITS PARAMETERS

In principle, a proper use of VDFs requires calibrating their parameters using real data (field observations) of speeds and flows. When it is possible to do this, a classical curve speed-flow is obtained; this curve is widely used in the theory of traffic flow. An example is shown in Figure 11.

The curve in Figure 11 is based on observed data at an urban road in Melbourne. Australia, and is a good example of the relationship between speed and traffic intensity on a link without signalling. This curve has two parts clearly defined; the first one before the maximal flow (2500 veh/h, i.e. the road capacity) and the second one after this point.

Along the first part, the speed curve softly decreases as the traffic intensity grows up to the maximal flow, thus suggesting that drivers choose freely their speeds. When approaching the maximal flow, speed reduction intensifies, reflecting that drivers obstruct each other, thus limiting the choice of speed.

The first part of the curve corresponds to operating conditions where it is possible to define the five service levels (A to E) widely used in Traffic Engineering, and is the region where a stable flow's behaviour is found. The second part of the curve corresponds to level of service "F", where operating conditions are unstable and a forced flow occurs, with strong congestion effects. The operating speed at capacity, called sometimes "optimal speed" or "critical speed" (Kutz, 2004) is an important value for

Figure 11. Estimated and measured speed as a function of traffic flows at Eastern Freeway in Melbourne, Australia (including forced flow conditions). (Akcelik, R., 2003)

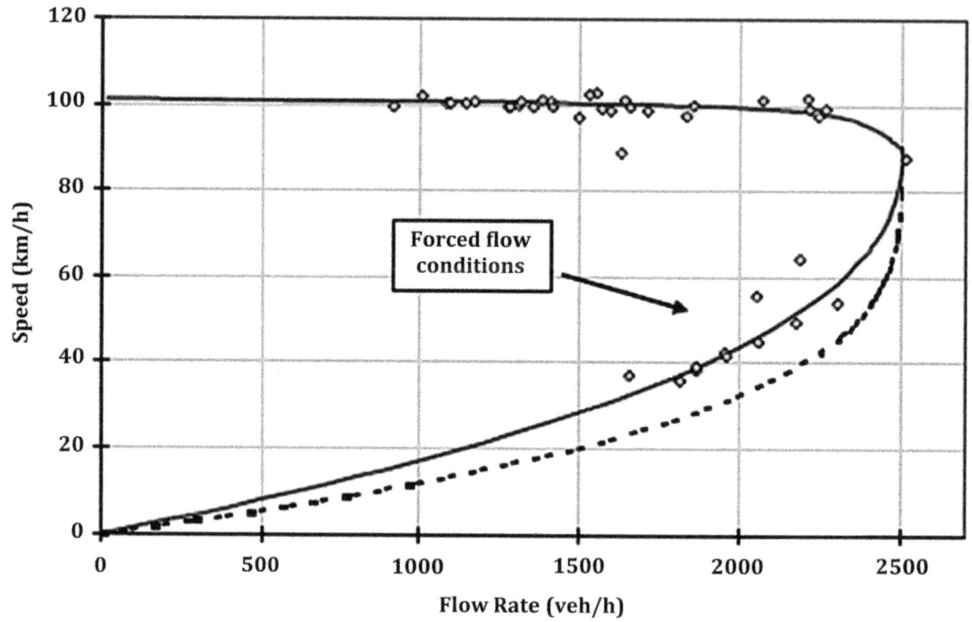

modelling purposes, as it will be shown forward. It should be noticed that observed behaviour in Figure 11 corresponds to an Australian road observed in particular, and in general each road section has a different behaviour, but similar in general.

Once the speed curve is determined, the travelling times are estimated for different volume levels, thus leading to a volume-delay function's calibration for the corresponding arc on a statistical basis.

However, gathering field data to determine speed-flow curves is not always an easy thing, because it requires resources, staff, measuring devices and mainly financing. These constraints to estimate VDFs can be diminished by analysing the functions' behaviour on a mathematical basis as ix explained below.

To start, the values of the VDFs are considered at the domain's extreme points, i.e. $x = 0$ (meaning time at free-flow) and $x =$ *Arc's Capacity* or $x = 1$ in the case x equals the (flow/capacity) rate. Results for the three analysed VDFs are as follows.

- **BPR:** $t(0)=t_0$; $t(C)=t_0(1+\alpha)$
- **Conical:** $t(0)=t_0$; $t(C)=2t_0$. For this function, saturation time is always twice the free-flow time.
- **Akcelik:** $t(0)=t_0$; $t(1)=t_0 + D_0 + \sqrt{J}$. Akcelik's function includes the D_0 term for delay at signalised nodes, and also the arc's length L. The saturation time is proportional to L.

As it can be seen, all the three VDFs assign consistently the free-flow time t_0 at the value $x = 0$ and the expected travelling time at saturation (maximal vehicle volume) when $x =$ *Arc's Capacity* or $x = 1$.

The three VDFs are non-negative, continuous, monotonically increasing and differentiable at their domains. These characteristics allow use the VDFs with Wardrop's equilibrium optimisation calculations based on gradient techniques and besides allow to obtain marginal cost functions on the basis of the corresponding derivatives.

4.1. The BPR Function

This is a widely used function in traffic modelling, and its simple formulation allows a useful interpretation of its parameters. The usual recommendation in basic references and software manuals is to take the "standard" BPR with $\alpha = 0.15$ and $\beta = 4$; parameters that have been used for urban networks. This "standard" BPR function was obtained through polynomial equations adjusted from American highways without congestion at the end of the 1950s, under the development of the analysis procedures of the *1965 Highway Capacity Manual* (Skabardonis and Dowling, 1997).

Other values have been calibrated in the United States for roads with different design speeds. Table 2 shows some values obtained with the procedures of the *1985 Highway Capacity Manual* (NHCRP, 1998); these values can be used as a general guide.

Figures 12 and 13 show BPR functions with data from Table 2 for American Freeways and Multilane roads, compared to the "standard" function with $\alpha = 0.15$ and $\beta = 4$.

These curves show the fact that the multilane roads, unlike freeways, have no controlled access, so vehicle flow increments have bigger effect on speed reduction than the corresponding at freeways.

Table 3 shows estimations for α and β from the 2000 Highway Capacity Manual. These values are estimated for a Level of Service C (TRB, 2000). At this level of service, the vehicle flow is stable, but

Table 2. α and β parameters at different American roads (NCHRP, 1998)

	Freeways			Multilane		
Coefficient	**70 mph 113 km/h**	**60 mph 97 km/h**	**50 mph 81 km/h**	**70 mph 113 km/h**	**60 mph 97 km/h**	**50 mph 81 km/h**
α	0.88	0.83	0.56	1	0.83	0.71
β	9.8	5.5	3.6	5.4	2.7	2.1

Figure 12. BPR curves for USA Freeways (based on data from NCHRP, 1998)

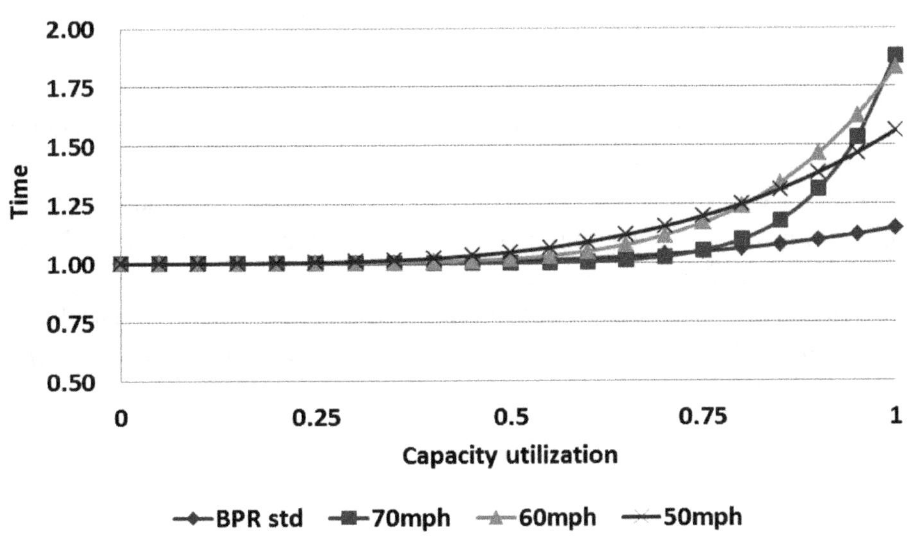

Figure 13. BPR curves for USA Multilane roads (based on data from NCHRP, 1998)

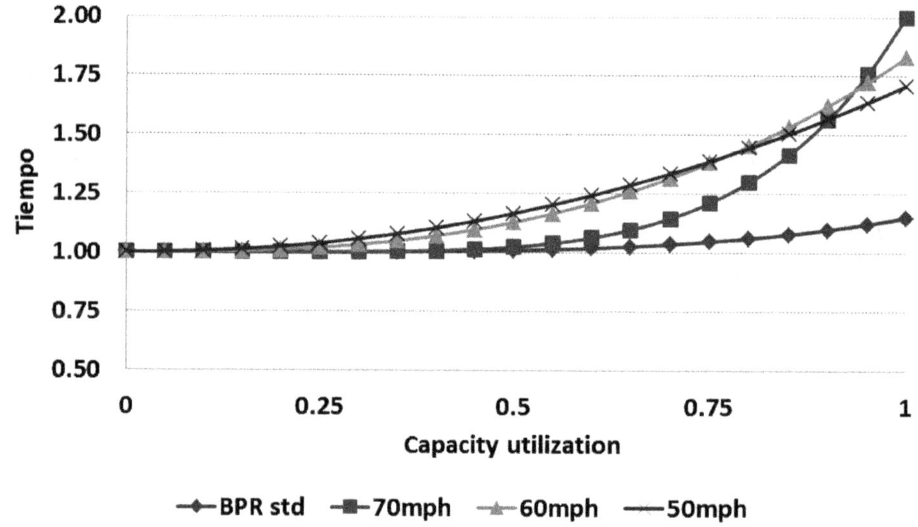

Table 3. and αβ parameters for different USA roads at Level of Service "C" (TRB, Highway Capacity Manual 2000)

		Free-Flow Speed (km/h)	Speed at Capacity (km/h)	α	β
Freeway	1	120	86	0.39	6.3
	2	112	85	0.32	7.0
	3	104	83	0.25	9.0
	4	96	82	0.18	8.5
	5	88	80	0.10	10.0
Multilane	6	96	88	0.09	6.0
	7	88	82	0.08	6.0
	8	80	75	0.07	6.0
	9	72	67	0.07	6.0

is the initial level of vehicles volume at which current users feel significantly affected by new users entering into the network (Papacostas, 1987).

Figure 14 shows some of the BPR functions from Table 3, i. e. freeways 1 and 4; and Multilane 6 and 8. On these graphs the increase in travelling time is softer as compared to Figures 13 and 14, because the estimated parameters are referred to level of service "C".

The BPR's functional form allows an interpretation of its parameters that can be useful for calibration purposes. An interpretation for parameter α is shown below, taking as the variable x = *(flow/capacity)* in order to simplify the presentation.

Figure 14. BPR curves for USA freeways and multilane roads at level of service "C" (based on data from TRB, HCM 2000)

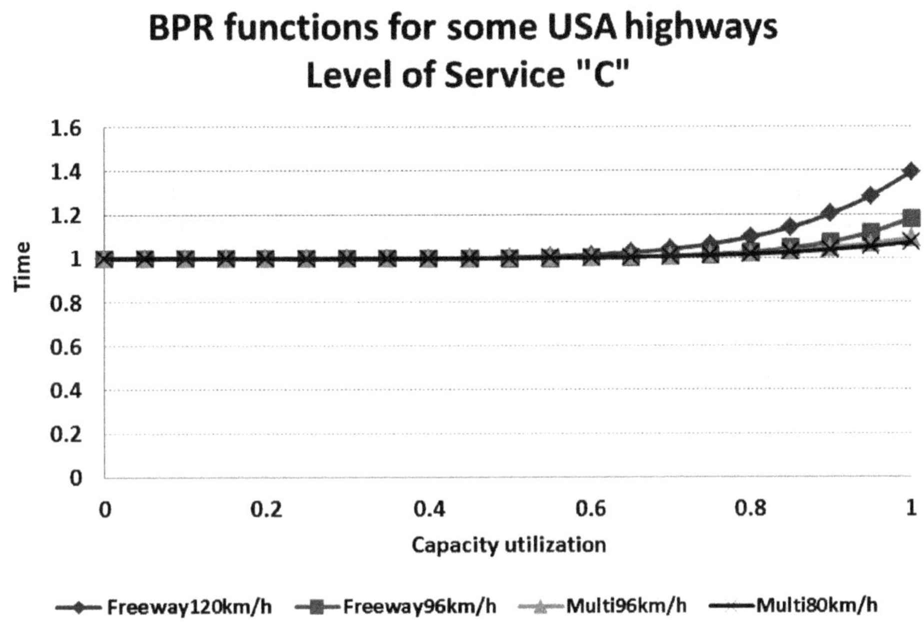

Interpretation of α in BPR Function

BPR's function writes as: $t(x) = t_0\left[1+\alpha x^{\beta}\right]$ where x = *saturation level, $0 \le x \le 1$.*

Since at capacity, $x = 1$ then: $t(1) = t_0\left[1+\alpha\right]$ therefore $\alpha = \dfrac{t(1)-t_0}{t_0}$

In this way, α can be interpreted as the percentage increase of t_0, the free-flow time when flow reach the saturation level. Table 4 gives an example of this interpretation of α with data from USA highways, assuming a free-flow time $t_0 = 45$ minutes.

By example, in a freeway with a design speed of 60 mph (97 km/h), α = 0.83 means 83% extra time due to delay at the saturation level in the arc, implying 37.4 extra minutes, giving a total time of 45min + 37.4 min = 1 h 22 min. Alternatively, α parameter can be expressed by a relationship between the

Table 4. Parameter interpretation for USA roads (based on data from αNCHRP, 1998).

Coefficient	Freeways			Multilane			BPR standard
	70 mph 113 km/h	60 mph 97 km/h	50 mph 81 km/h	70 mph 113 km/h	60 mph 97 km/h	50 mph 81 km/h	
α	0.88	0.83	0.56	1	0.83	0.71	0.15
Extra minutes	**39.6**	**37.4**	**25.2**	**45.0**	**37.4**	**32.0**	**6.8**
Total time	**1h 25 min**	**1h 22 min**	**1h 10 min**	**1h 30 min**	**1h 22 min**	**1h 17 min**	**51.8 min**

free-flow speed v_0 and the critical speed v_c (when saturation is reached). Based on the definition of speed, it is easily verified that:

$$\alpha = \frac{v_0}{v_C} - 1.$$

So α is directly proportional to free-flow speed, which in turn is higher for roads with better specifications. That is why Table 4 shows higher values for α at roads with the best specifications.

Interpretation of β in BPR Function

A direct approach to express BPR's β could give the following equation:

$$\beta = \frac{Ln\left[t(x) - t_0\right] - Ln\left[\alpha t_0\right]}{Ln(x)}$$

which in turn does not suggest a clear meaning for β. However, through the concept of elasticity it is possible to do this interpretation. Elasticity is an economic concept. It measures the variations of a function with respect to its variable, in terms of relative percentage changes.

If the variable x increases a specific percentage: $\Delta x/x$ the corresponding change in the function will be $\Delta f(x)/f(x)$. The elasticity at a point x in the domain is the proportion of percentage change of the function to the corresponding percentage change in the variable. So elasticity indicates the change in percentage points of the function $f(x)$ for each percentage point change of the variable x.

For BPR function, the elasticity delay-flow is the quotient of the percentage change in delay to the percentage change in flow which causes that delay. This can be expressed by the function E(x) as:

$$E(x) = \frac{\%\ delay\ change}{\%\ delay\ flow}$$

This expression varies as x changes. Using the mathematical definition of elasticity, E(x) is written as:

$$E_t(x) = \frac{x}{t}\frac{dt}{dx} = \beta\frac{\alpha x^{\beta}}{1+\alpha x^{\beta}}$$

Taking $x = 1$ at the last equation, it results:

$$E_t\left(1\right) = \beta\frac{\alpha}{1+\alpha} \text{ and therefore: } \beta = \frac{1+\alpha}{\alpha}E_t\left(1\right)$$

So β could be seen as the delay-volume elasticity at saturation level multiplied by the factor $\frac{1+\alpha}{\alpha}$.

Alternatively, β can be estimated considering the link's flow at Level of Service "C". According to the *Highway Capacity Manual 2000*, flow at Level of Service "C" corresponds to 80% of the link's capacity. Assuming this level of service is observable, and so the corresponding travel time, denoted here as $t_{0.8}$, using the BPR's definition it results:

$$t_{0.8} = t_0\left[1+\alpha\left(0.8\right)^{\beta}\right] \text{and then } \beta = \frac{Ln\left(\frac{t_{0.8}-t_0}{\alpha t_0}\right)}{Ln(0.8)}$$

Thus, β can be estimated by the last equation once the travel time at Level of Service "C", $t_{0.8}$ is known.

Table 5 shows the elasticities at saturation level based on the β values of the BPR functions in each road type.

Table 5 shows higher values for elasticity at the saturation point in roads with better specifications; this suggests that marginal percentage increments of travel time are bigger for roads with better operating conditions.

Some Pitfalls in the Use of BPR Functions

BPR function has been widely used in traffic assignment modelling and is usually found available in transportation planning software. However experiences from users and practitioners can be found in the literature revealing some technical difficulties as is described below.

Spiess (1990) points out that it is desirable that BPR's response at saturation level be twice the corresponding free-flow time, for any possible alpha value; a traffic condition seen by transport analysts as reasonable. This does not happen every time, as it was noticed before.

BPR function valued at saturation level multiplies free-flow time by (1 + α), and this value equals two only if α = 1. Moreover, solution algorithms for Wardrop's equilibria quite often predict flow levels exceeding saturation, three, five or more times the arc's capacity during the first iterations, which gives unrealistic travel times in some arcs.

Table 5. Elasticities at saturation level in some American roads (NCHRP 1998).

	Freeways			Multilane			BPR Standard
Coefficient	**70 mph 113 km/h**	**60 mph 97 km/h**	**50 mph 81 km/h**	**70 mph 113 km/h**	**60 mph 97 km/h**	**50 mph 81 km/h**	
α	0.88	0.83	0.56	1	0.83	0.71	0.15
β	9.8	5.5	3.6	5.4	2.7	2.1	4
Elasticity at saturation	**4.6**	**2.5**	**1.3**	**2.7**	**1.2**	**0.9**	**0.5**

In addition, for flows below the maximal, particularly for high β values, travel times become practically independent from volumes over a major part of the BPR's domain. To illustrate this, Spiess takes $\beta = 12$ on a link with capacity equals to 1000. Figure 15 shows the behaviour of three BPR curves for $\alpha = 2$, 4 and 8.

Figure 15 looks practically with no changes in travel time for flows under 600 veh/h. Close to the saturation point (1000 veh/h), travel times increase sharply with respect to the unit value for the free-flow time.

This behaviour of BPR functions when β has big values may cause that assignment procedures degenerate on a local region of the domain towards a "All-or-Nothing" solution in view of the quasi-linear behaviour for flow values below the maximal, this occurs apart from the fact that the function is not increasing on this range, and consequently the solution could be non-unique (Spiess, 1990).

These difficulties on the BPR's utilization, motivated the search of another VDFs, as is the case of the Conical function and the Akcelik's function, that are discussed below.

4.2. The Conical Volume-Delay Function

This function was proposed by Heinz Spiess in 1990, as an alternative to improve results obtained with the BPR function (Spiess, 1990). This function has two parameters α and β, though the latter is defined in terms of α. The formulation of the Conical VDF is as follows, where X equals the (flow/capacity) ratio.

$$t(X) = t_0 \left[2 + \sqrt{\alpha^2 \left(1 - X\right)^2 + \beta^2} - \alpha\left(1 - X\right) - \beta \right]$$

where

Figure 15. BPR Curves for β = 12 (Based on Data from Spiess, 1990)

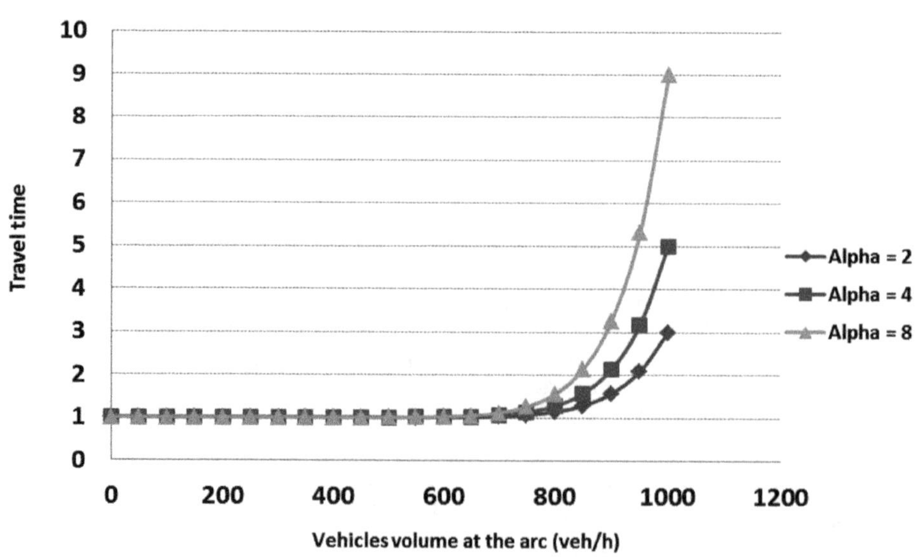

$$X = \frac{x}{C}; \quad \alpha > 1; \quad \beta = \frac{2\alpha - 1}{2\alpha - 2}; \quad x = \text{flow}; \quad C = \text{capacity}.$$

Values for parameters α and β have been calibrated for various road types at the USA, and can be used as a reference. Table 6 shows some of these values estimated for several highway types in the state of FLorida (Mtoi & Moses, 2014).

Figure 16 shows the curves for rural and urban freeways referred in Table 6. It can be noticed a practically linear behaviour and a quite closeness between these curves. At the saturation level, both curves predict a saturation time equal to twice the free-flow time. This characteristic comes from the very definition of the Conical function, as it was observed at the beginning of section 4.

Interpretation of α in the Conical Function

By using the definition of the Conical function, at the saturation level, $(1) = 2t_0$, so this function always assigns to saturation level a time twice the free-flow time. Calculating the delay-volume elasticity $E_t(X)$, according to the mathematical definition it results:

Table 6. α and β parameters for the Conical function estimated at Florida's roads, USA. (Mtoi and Moses, 2014).

	Freeways			Multilane		
Coefficient	**Urban**	**Residential**	**Rural**	**Urban**	**Residential**	**Rural**
α	1.028752	1.028752	1.035562	1.028752	1.03552	1.03552
β	18.390	18.390	15-060	18.390	15.064	15.064

Figure 16. Conical curves estimated for some roads in Florida, USA (Mtoi and Moses, 2014)

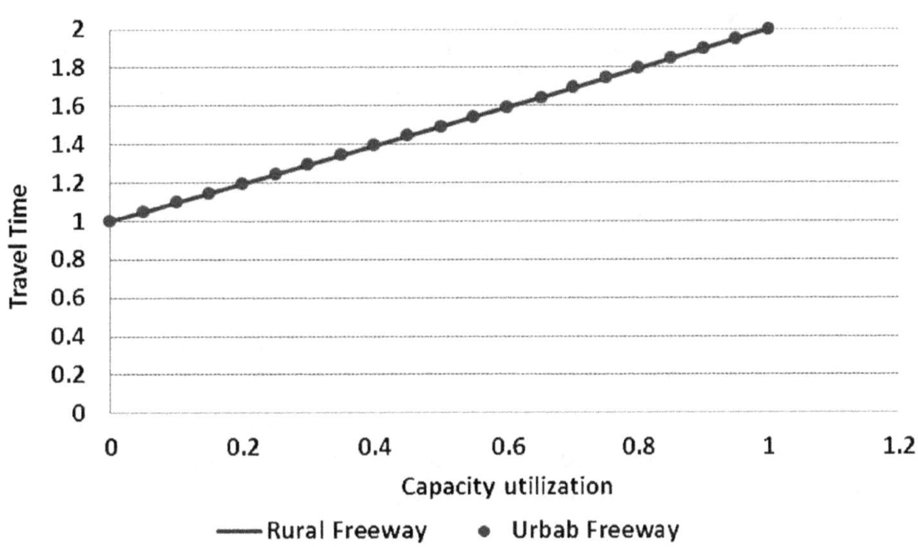

$$E_t(X) = \frac{\alpha X\left[\sqrt{\alpha^2(1-X)^2+\beta^2}-\alpha(1-X)\right]}{\sqrt{\alpha^2(1-X)^2+\beta^2}\left[2+\sqrt{\alpha^2(1-X)^2+\beta^2}-\alpha(1-X)-\beta\right]}$$

Evaluating $E_t(X)$ at $X = 0$ and $X = 1$ it results:

$$E(0) = 0; \quad E(1) = \frac{\alpha}{2}; \quad \textit{therefore} \quad \alpha = 2E(1)$$

In this way, parameter α can be interpreted as twice the delay-volume elasticity at the saturation level; β is obtained once α is known as:

$$\beta = \frac{2\alpha - 1}{2\alpha - 2}$$

4.3. The Akcelik's Function

This volume-delay function, proposed by Rahmi Akcelik and referred in the Highway Capacity Manual 2000 (HDM 2000), has a single parameter *J* to calibrate. The function includes besides of free-flow time, the estimated duration of demand, the length of the arc and optionally a measure of delay at intersections due to signalisations. The formulation is as follows (HDM 2000):

$$t(x) = t_0 + D_0 + 0.25T\left[(x-1) + \sqrt{(x-1)^2 + \frac{(16\,J\,L^2)\,x}{T^2}}\right]$$

where t = travel time; t_0 = free-flow time; D_0 = delay at intersection node by signalization; T = expected duration of demand; L = arc length; J = calibration parameter; x = (flow/capacity) rate. For the purpose of modelling interurban road freight flows, $D_0 = 0$ is assumed, so Akcelik's function becomes:

$$t(x) = t_0 + 0.25T\left[(x-1) + \sqrt{(x-1)^2 + \frac{(16\,J\,L^2)\,x}{T^2}}\right]$$

Some estimations of parameter *J* has been done in the USA. Table 7 shows this for some freeway and multilane roads referred in the HDM 2000.

Figure 17 show the corresponding Akcelik's curves for a freeway of 120 km/h free-flow speed and a multilane road of 100 km/h free-flow speed, as referred in Table 7. The assumed expected duration of demand was T = 1 h 15 min and the arc's length of 25 km.

Table 7. Akcelik's J parameter estimated for American freeways and multilane roads (Highway Capacity Manual 2000)

Type of Road	Free-Flow Speed (km/h)	Speed at Saturation (km/h)	Parameter J
Freeway	120	86	0.0000109
	110	84	0.0000079
	100	82	0.0000048
	90	80	0.0000019
Multilane	100	88	0.0000019
	90	81	0.0000015
	80	74	0.0000010
	70	68	0.0000002

Figure 17. Akcelik's curves estimated for American freeways and multilane roads. (based on Highway Capacity Manual 2000)

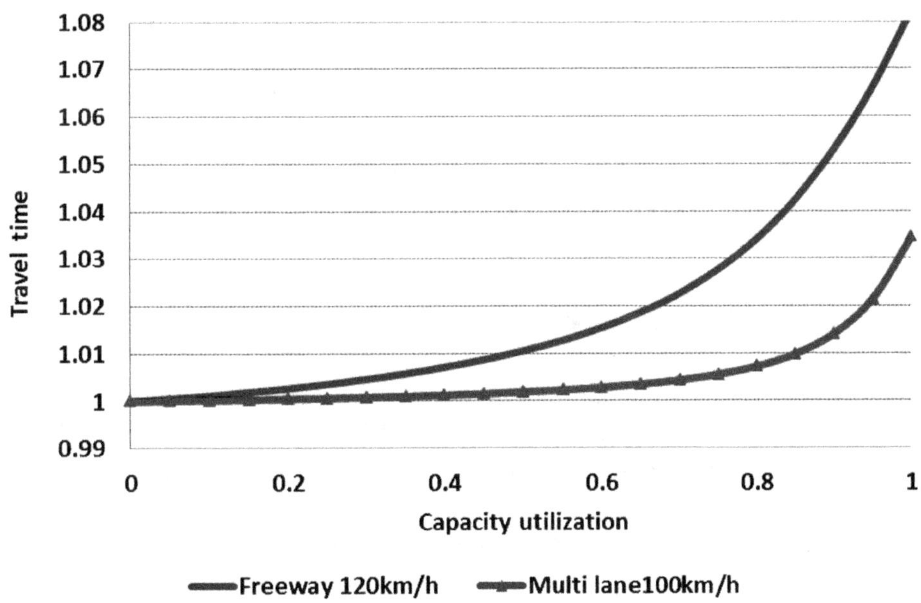

Interpretation of Akcelik's *J* parameter

At saturation level, $x = 1$, so travelling time is:

$$t(1) = t_0 + L\sqrt{J}$$

By solving for J it results:

$$J = \left(\frac{t(1) - t_0}{L} \right)^2 = \left(\frac{t(1)}{L} - \frac{t_0}{L} \right)^2 = \left(\frac{1}{L/_{t(1)}} - \frac{1}{L/_{t_0}} \right)^2$$

Last term in previous equation can be written as:

$$J = \left(\frac{1}{Speed\ at\ saturation} - \frac{1}{Free\ flow\ speed} \right)^2$$

Thus, J can be interpreted as the square of the difference of the inverses of the speed at saturation and the free flow speed.

The volume-delay functions discussed above are widely used in practice. Any of them have pros and cons in relation to its simplicity of formulation, their ability to model congestion or the response of travel time while vehicles flow is increasing. In order to apply any of these VDFs, the issue of parameter calibration has to be solved in each case.

In the literature there is no a general agreement about which VDFs is the best one; sometimes they are used in combination in order to improve the assignment results. By example, TransCAD software is able to model some links in a network using a BPR function and at the same time a Conical function in other links at the same assignment routine.

The next section explores some details in relation with the application of the VDFs discussed; the last section resumes the findings in this work and proposes suggestions for a practical use of the VDFs.

5. UTILIZING VDFS IN MODELLING

To utilize a VDF in an assignment model for a particular road network, good estimates for the corresponding parameters are needed. Examples of the use of BPR, Conical and Akcelik's functions have been reported in several countries, with the respective parameters assessed by calibration exercises at specific road networks.

The American TRB's report NHCRP 365 gives some α and β values for the BPR function for freeways and multilane roads as it was shown in Table 2. These values were obtained through the procedures described in the *Highway Capacity Manual 1985*. Similarly, the *Highway Capacity Manual 2000 (HCM 2000)* proposes α and β values for the BPR function assuming that the Level of Service "C" is represented, as it is shown in Table 3. Taking the original formulation for the BPR function:

$$t(x) = t_0 \left[1 + \alpha \left(\frac{x}{C} \right)^{\beta} \right]$$

where x corresponds to vehicles flow and C corresponds to link's capacity; the reference to Level of Service "C" is linked to the idea of *practical capacity* of the link, approximately 80% of the link's capacity, and is the flow level to consider for the BPR's use. That is why it is recommended to modellers not to use the BPR formulation with $x = C$, but with the practical capacity in order to have good representations of travel times at the link (Skabardonis and Dowling, 1997).

HCM 2000 also informs estimations for the Akcelik's J parameter; these are shown in Table 8, that extends information in Table 8.

Australian consulting Akcelik & Associates Pty Ltd, funded by Rahmi Akcelik who authored the VDF with his name, has informed estimations for J parameter, as is shown in Table 9. These values were estimated at Australian roads.

Considerations to Put into Practice VDFs in Mexican Roads

Previous discussion on VDFs gives a general view of the value ranges for the parameters used in these functions. These values, although referred for highways in other countries, can help the modelling process for Mexican roads, by considering similarities among the diverse interurban roads.

Thus, considering that Mexican roads have average operating conditions less favourable than the corresponding American highways, it is proposed to assimilate Mexican highways (at 100 km/h) to American

Table 8. Akcelik's J parameter estimated at American highways (Highway Capacity Manual 2000)

Type of Road	Free-Flow Speed (km/h)	Speed at Saturation (km/h)	Parameter J
Freeway	120	86	0.0000109
	110	84	0.0000079
	100	82	0.0000048
	90	80	0.0000019
Multilane	100	88	0.0000019
	90	81	0.0000015
	80	74	0.0000010
	70	68	0.0000002
Two-lane roads	100	60	0.0000444
	90	50	0.0000790
	80	40	0.0001560
	70	30	0.0003620

Table 9. Akcelik's J parameter estimated at Australian highways (Akcelik & Associates, 1991)

Type of Road	Description	Free-Flow Speed (km/h)	Capacity veh/h/ lane	Parameter J
1	Freway	120	2000	0.1
2	Arterial, uninterrupted	100	1800	0.2
3	Arterial, interrupted	80	1200	0.4
4	Secondary, interrupted	60	900	0.8
5	Secondary, high friction	40	600	1.6

freeways at 60 mi/h, and the Mexican multilane-roads (100 or 80 km/h) to the American freeways at 60 mi/h in some cases and 50 mi/h in other cases with lower operating conditions.

The tentative parameters to use in Mexican roads are shown in Table 10.

The corresponding BPR curves for Table 10 BPRs is shown in Figure 18, taking the value of free-flow time as $t_0 = 1$.

BPR curves at Figure 18 show increases of the travel time at the saturation point between 70% /t = 1.7) and 80% (t = 1.8), that seem a reasonable representation of time delay under congestion.

Curves for multilane roads, usually without a toll, start to increase their trip times at lower flow levels as compared to the highways curves. This reflects the fact that Mexican highways have better specifications as those of multilane roads, besides because highways have a number of limited accesses as opposed to multilane roads, where the traffic merges are more frequent. Travel times for the curves in Figure 18 are shown in Table 11, assuming the free-flow time as $t_0 = 1$.

The corresponding velocities at Level of Service "C" (80% of capacity) and flow at capacity are easily deduced from the relationship t =distance/time; these are shown in Table 12 for the tentative BPR functions.

Speed values in Table 12 for the Level of Service "C" seem a reasonable representation of the flow behaviour in Mexican roads, so these functions can be used as a first attempt in the modelling process for interurban traffic in Mexico.

In respect of the parameter interpretations commented in previous Section 4, some of these parameters could be assessed by direct observations on the roads of interest. Thus, in view that α is the percentage increase of free-flow time when saturation is reached, and this value is easily observed this is a first step

Table 10. Tentative parameters for BPR function in Mexican roads

Coefficient	Highways		Multilane	
	60 mph 97 km/h	50 mph 81 km/h	60 mph 97 km/h	50 mph 81 km/h
α	0.83	0.56	0.83	0.71
β	5.5	3.6	2.7	2.1

Table 11. Travel times for tentative BPR function t_0=1 in Mexican roads (own elaboration)

Saturation Level	Highway 100 km/h	Highway 80 km/h	Non Tolled 100 km/h	Non Tolled 80 km/h
0	1	1	1	1
0.8	1.24	1.25	1.45	1.44
1	1.83	1.56	1.83	1.71

Table 12. Speeds for BPR tentative functions (km/h) in Mexican roads (own elaboration)

Saturation Level	Highway 100 km/h	Highway 80 km/h	Non Tolled 100 km/h	Non Tolled 80 km/h
0	100	80	100	80
0.8	80.4	64.0	68.8	55.4
1	54.6	51.3	54.6	46.8

Figure 18. BPR functions. Proposal for Mexican roads (own elaboration)

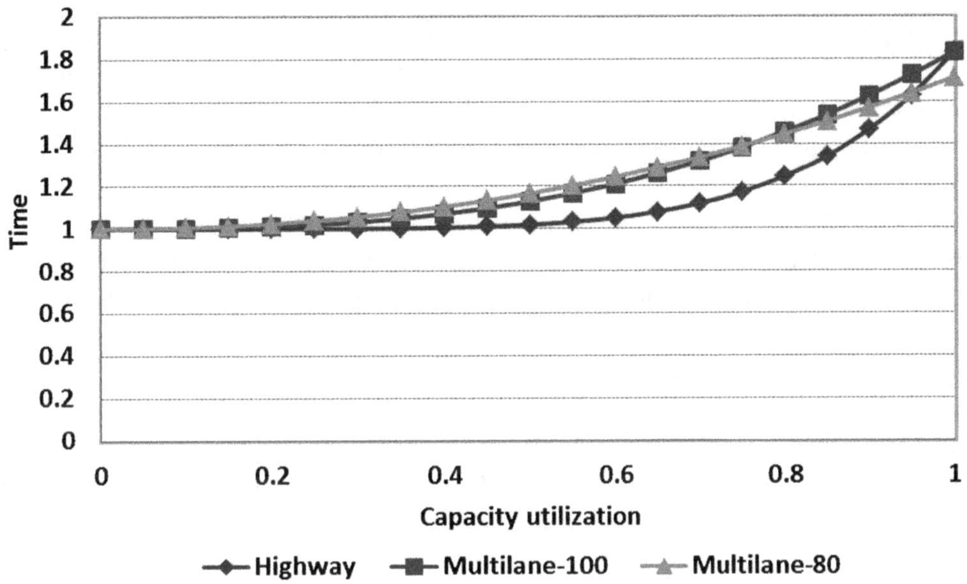

for the modelling process. Regarding the β value, a first try implies assessing the delay-volume elasticity at the saturation point, which may be not an easy task. However, work on the level of service "C", by observing travel times at flows 80% of capacity may be used to assess the value of β.

For the conical function, α parameter is equally linked to measures of elasticity at the saturation point, and presents the same difficulty than the BPR case.

In the case of Akcelik's function, parameter J is linked to measures of free-flow speed and the speed at saturation. These values are easily observable, so facilitating the modelling process.

Finally, once a VDF has been essayed, it is always convenient to compare the results of the assignment model with real observed flows at roads. In Mexico, the General Directorate of Technical Services, at the Secretariat of Communications and Transport publishes information on average annual daily traffic on most of the Mexican roads, a source that can be used as a reference to check assignment models on an annual basis, by now.

6. CONCLUDING REMARKS

Modelling of flows on road networks is a fundamental activity in the analysis of road transport systems, in the planning of transportation activities and finally in the economic and social planning. The quality of the model determines the corresponding quality of impact estimations that come from changes in transport policies or in the activity systems organization.

Vehicles' flow estimations are a basic input for design and evaluation of road infrastructure's modifications and expansions. These estimations, in turn depend on a realistic representation of vehicle's flow and its congestion effects by means of well calibrated volume-delay functions.

Three of the more used VDFs reported in the literature are: a) the BPR function; b) the Conical function and c) Akcelik's function. Due to its simple formulation, BPR function has been widely used in transportation planning projects. The "standard" values $\alpha = 0.15$, and $\beta = 4$ are usually referred as for general use at urban environments, but other values have been reported for different road types in some countries. In this way, the α and β estimations of BPR for American freeways and multilane roads can help the modelling process to assess Mexican highways on the basis of similarity among the roads in the two countries.

The classical procedure to estimate VDFs parameters, based on field observations of speeds and flows and the subsequent statistical analysis to obtained adjusted curves and values, even if usual in traffic engineering studies, requires availability of staff, time and money, not always available in particular for government agencies.

As an alternative to this problem, the parameter's interpretations for the VDFs discussed in this work can be a guide to collect more easily field's observations to estimate those parameters. Thus, the interpretations of: α and β for the BPR function; α for the Conical function and J for the Akcelik's function, suggest to collect information at the interest routes at the Level of Service "C" (80% of capacity) and at the saturation level (flow at capacity, Level of Service "E") in order to:

- Assess the percentage increase of free flow time at saturation level to estimateα and estimate the travel time at Level of Service "C" to estimate β for the BPR function.
- Assess the delay-volume elasticity at the saturation level as to estimate β for the BPR function, or α for the Conical function.
- Assess time or speed at saturation level as to estimate J for Akcelik's function.

These interpretations of VDFs parameters allow their estimation using a less number of observations as compared to the classical approach for speed-flow curves, thus simplifying the modelling process. In addition, these parameter's interpretations help to clarify their operational meaning into the VDF formulation and their role in the characterization of the roads that are modelled, and this general idea of mathematical analysis and interpretation of parameters could be extended to another formulations of VDFs in the existing literature.

REFERENCES

Akçelik, R. (1991). Travel time functions for transport planning purposes: Davidson's function, its time-dependent form and an alternative travel time function. *Road Research, 21*(3), 49–59.

Akcelik, R. (2003). *Speed-Flow Models for Uninterrupted Traffic Facilities. Technical Report*. Australia: Akcelik & Associates Pty Ltd.

Bazaraa, M. S., & Jarvis, J. J. (1977). *Linear programming and network flows*. New York: John Wiley and Sons.

Caliper Corporation. (2005). *Travel Demand Modeling with TransCAD 4.8*. USA: Caliper Corporation.

Centeno, A., & Mendoza, A. (2003). *Modelo de asignación intermodal multiproducto para las operaciones de carga por autotransporte y ferrocarril. Publicación Técnica 222. Instituto Mexicano del Transporte*. Querétaro, México: Spanish.

Gradilla, L. (2011). *Planeación de infraestructura de transporte: Identificación de tramos críticos para el funcionamiento de redes carreteras. Publicación Técnica 354. Instituto Mexicano del Transporte*. Querétaro, México: Spanish.

Kutz, M. (2004). *Handbook of Transportation Engineering*. Mc Graw-Hill.

Lancashire County Council. (2010). *Heysham – M6 Link Road Traffic Model. Local Model Validation Report*. Lancashire, UK: Appendix D.

Moreno, E., Rico, O., Bustos, A., Martner, C., & Montooya, R. (2014). *Reparto Modal Óptimo del Transporte Terrestre de Carga en México. 1ª. Etapa. Publicación Técnica No. 413. Instituto Mexicano del Transporte*. Querétaro, México: Spanish.

Mtoi, E., & Moses, R. (2014). Calibration and Evaluation of Link Congestion Functions: Applying Intrinsic Sensitivity of Link Speed as a Practical Consideration to Heterogeneous Facility Types within Urban Network. *Journal of Transportation Technologies*, *4*(02), 141–149. doi:10.4236/jtts.2014.42014

NCHRP. (1998). *Travel Estimation Techniques for Urban Planning. Report 365*. Washington, D.C.: TRB.

Ortúzar, J. D., & Willumsen, L. (2011). *Modelling Transport* (4th ed.). Chichester, UK: John Wiley. doi:10.1002/9781119993308

Papacostas, C. S. (1987). *Fundamentals of Transportation Engineering*. USA: Prentice-Hall Inc.

Sheffi, Y. (1985). *URBAN TRANSPORTATION NETWORKS: Equilibrium Analysis with Mathematical Programming Methods*. New Jersey, USA: Prentice-Hall Inc.

Skabardonis, A., & Dowling, R. (1997). *Improved Speed-Flow Relationships for Planning Applications. Transportation Research Record No. 1572*. Washington, DC: Transportation Research Board.

Speiss, H. (1990). Conical Volume-Delay Functions. *Transportation Science, 24*(2).

Spiess, H. (1990). Technical Note—Conical Volume-Delay Functions. *Transportation Science*, *24*(2), 153–158. doi:10.1287/trsc.24.2.153

Chapter 10
The Influence and Management of the Supply Chain Performance of Manufacturing SMEs in Aguascalientes

Gabriela Ortiz Delgadillo
Universidad Politécnica de Aguascalientes, Mexico

Ma. de Lourdes Yolanda Margain Fuentes
Universidad Politécnica de Aguascalientes, Mexico

Octavio Hernández Castorena
Universidad Politécnica de Aguascalientes, Mexico

Francisco Javier Ornelas Zapata
Universidad Politécnica de Aguascalientes, Mexico

ABSTRACT

The results obtained in this research confirm that currently manufacturing for SMEs by the nature of its operational activities, the handling of Supply Chain Management (SCM), is significant in the performance of this type of organizations. Because of that, the answers of the business man were analyzed with the support of polls through a field work, finding that for them, is essential that the SCM considering its complexity and impact in the supply of material resources, which is significantly influence for the manufacturer SMEs to have a bigger performance. This research, of empiric origin,, and with support of the statistic software EQS where the structural equations have been used, permit conclude that the SCM is essential for the manufacturer SMEs performance. For the investigation the sample of 288 enterprises of the state Aguascalientes has been used, obtained between August and December Of 2013, which contain from 11 to 250 workers.

1. INTRODUCTION

Nowadays, enterprises like the Small and Medium enterprises (SMES), manufacturer create partnerships and networks leaning in the Supply Chain Management (SCM), where the handling and supply of the material resources in this type of organizations is an important activity to control (Dyer, Sung y

DOI: 10.4018/978-1-4666-9779-9.ch010

Chu 1998). For that, the SCM because of its importance, relation and connection that is related with the supply, intends, with its management to be a key element to improve the performance of manufacturer enterprises (Dyer et al., 1998). In this sense, is important that business man, have special attention in the SCM in order to avoid failures that are related with the supply of material resources (Wisner, 2003). It is relevant to stand out that SCM is a group of activities that include the purchasing, manufacturing and delivering of the material resources on time (Mujuni & Zheng, 2010).

The SCM for manufacturer SMEs, imply to have information that involves the sourcing and each part involved with the supply of material resources, with the finality of having more efficiency through strategies implementation that allow the implicated in supply chain, having better participation and fulfillment of delivering of supplies (Wisner, 2003). Between factors of importance that are involved with supply is the transport service, in which through its participation, have the role of controlling the flow of material resources along the SCM (Morash & Clinton, 1997), essential activity that allows through proper logistical control avoid delivery delays and significantly improve the performance of just in time deliveries adding thus the proper management of material resources with staff in charge of the transport activity.

Considering that today for manufacturing SMEs, controlling the supply is vital for the development of enterprises, this research aims to analyze the impact of Supply Chain Management efficiency of these organizations, and in this sense, is convenient to deepen in the next question: The activity that is developed by SCM in the manufacturer SMEs permits these to obtain a better performance?

This research is formed, in the first part for the introduction and materials, in the second part the method and statistic analysis and finally, the results, conclusions and barriers.

2. MATERIALS AND METHODS

2.1 Supply Chain Management

Nowadays, SCM as operational activity required for the manufacturer SMEs, it has become an important element for organizations and in this sense, and how Beamon (1998) points, it is a structured process where material resources are transformed into finished products, before being delivered to the end customer in time and in the required quantities .

Also, Trkman, Stemberger and Jaklic, (2005), defined SCM as "a chain from row materials and ending with the sale of the finished product". Furthermore, Pienaar (2009) defines SCM as "an overview of the participation process integration of the organizations to transform raw materials into finished goods and to transport them to end users".

The concept of SCM, is often linked to operational aspects, especially the dynamics of manufacturers in order to obtain inputs safely (Wisner, 2003). Having as references that competition in international markets depends on supplies of goods, coordination among supply sources and distributors involved in the flow of material resources, this operating dynamics has become an important feature so that customer satisfaction is a crucial benchmark of success of SCM (Trkman *et al.*, 2005). Importantly, market uncertainty requires an efficient and flexible SCM to changes that occur in any dynamic business, which impacts have good results, significantly in the development of manufacturing SMEs (Langley, Coyle, Gibson, Novack & Bardi, 2008).

Also, Ganeshan and Harrison (1995) conceptualize to the SCM s a chain of services and alternative distribution in the acquisition of material resources having the objective of ensuring that material goods are delivered to customers promptly and safely supply . Also, Habr and Reichel (2005) emphasize that the SCM contains activities focused on the same network management acquiring material resources transformed into intermediate products for delivery to customers through an efficient distribution. According to Christopher (1998), the research of the SCM refers to the distribution network as a value generator in the form of goods which must be finally placed by the organizations responsible for the supply in the hands of end users without any risk of delay in deliveries.

Wisner (2003) considers that the SCM practices along with their associated benefits (improved customer service and lower cost) are closely related to existing theories of strategic supply management. Strategically to Kanjiy Wong (1999), there are two essential points in the SCM: suppliers and the actual management of the organization. In this sense, the organization provides information about what is needed from the supplier and the supplier produces goods to meet the needs of requestor. Likewise, for Guinipero and Brand (1996) as well as Cooper and Ellram (1993), a good model of SCM should consider while provider satisfaction, collaboration and satisfaction that this can provide to the organization in order not to affect consistent processes including end customer.

2.2 Performance

Currently, for organizations in general, the performance is a concept containing naturally in its meaning analysis and control of resources from an investment point of view (Harold & Darlene, 2004; Kaplan & Norton, 1993; Rajendar & Jun Ma, 2005); however, it is important to note that within the concept and application performance is not without integrating aspects such as the productivity and results to be reconciled with the approach of objectives and the proposed targets substantially in organizations whose purpose is to have better performance in organizations (Aguilera & Gonzalez, 2012) .

And it is important that companies keep in mind that they should be integrated into their strategies and performance improvement, work strongly with aspects of training, management skills and attitudes best performance management (Richard, 2002).

Importantly, within the complexity of the concept of performance, Houthoofd (2009) as well as Slater and Olson (2000), stress that the performance has a special bond with the economics that every company needs to control and manage which naturally it has a strong impact on the success and efficiency of organizations so that their participation in the markets will allow companies to have greater development and performance. And for that, it is worth mentioning that all companies within its organizational development required analyzing in detail those factors that allow them to have significant improvements to the final results of those strategies that have been implemented in their organizations have results that impact on the fulfillment of the goals that are of interest in the performance of organizations (Quinn & Rohrbaugh, 1983).

As already mentioned, the performance is based on efficiency that organizations have to achieve the goals that are designed internally and part of this efficiency is related to the performance generated in organizations (Pérez, 2008). And it is important to emphasize that efficiency is understood as the profits in businesses for the proper use of resources especially if we take special care in managing the material resources, and by the way how strategies are set to achieve the objectives implemented internally within companies and by the way how the skills and the knowledge that their assembly will allow greater business performance, especially with the SME manufacturing (Mojica, 2012).

While and as seen in the literature, there are several concepts on performance, generally for a company refers to an organization that reaches levels in terms of profitability, productivity, added value and better benefits in terms of business competitiveness (Schwarz, Kalika, Kefi & Schwarz, 2010).

However it is important to note that performance can be financial or non-financial, and secondly, the performance is related to increased productivity, improved cost control, have a greater capacity for innovation and have a reactive capacity to the demands and constant rafters that are generated in the market (Mojica, 2012) .

2.3 Management of Supply Chain and Manufacturing SME Performance

Today, for manufacturing SMEs, entrepreneurs need to focus on improving delivery systems in order to streamline the strategies as well as the handling of materials for areas such as the production process, to have the materials on time and without quality problems which leads to the commissary management to be more effective and a lot of coordination on the supply side (Cachon & Netessine, 2004), and in this sense, the SCM integrates to improve the information management, inventory control, have a good control of stock and better manage of transportation and efficient flow of material resources (Cruz 2008; Stadler, 2005), which would enable these organizations to have greater performance.

Chow and Heaver (1999) mention that the SCM consumers is formed by a group of manufacturers, suppliers, distributors, retailers and carriers, and where information relating to sources of supply, logistics management services dedicated to the provision of goods allow the company to have better performance.

SCM is comprised of various stakeholders who are directly involved in various activities related to the management and flow of material resources, so this activity by nature, is a process related to the supplies, the handling of information and financial flows, which aims to meet the final requirements that require them to offer consumers products supplies on time and thus avoid problems in delivery (Aguilera, Hernandez & Lopez, 2012; Ayers, 2001).

On the other hand, Mentzer, Witt, Keebler, Min, Nix, Smith and Zacharia (2001) defines the SCM is as a set of elements that participate directly in the flow of supply and distribution of material resources considering the source of supply to their destination in order to have a positive impact on organizations to have better performance. This approach to coordinate various activities related to the provision, complicates the way to define the GCS and it is important to note that there are different distribution systems organized to work through networking with various actors directly involved through adequate SCM in order that the manufacturing SMEs, may have better results and better business performance (Mbang, 2012).

Therefore, it is important to mention that for businesses, the SCM required to have a strategic coordination and efficiency of the functions of the business where integrated strategies in operational activity through the functions that the company has related to the SCM, allow to meet the objectives are key elements to the performance in companies, allowing them to continue their development (Deshpande, 2012).

For this, and according to what Grant, Lambert, Stock and Ellram, (2006) say referring to the integration of activities focused on meeting the needs that the end users through the contribution and intervention by supply sources, provide information that allows them to have greater control of the entire supply network and better control with the actors involved in the supply of material resources .

Therefore, and as described, it is important for SME manufacturing efficiency of the management of the supply chain through better controls and management strategies across the path of the management and supply of material resources allows companies to have better benefits and higher performance (Chun Choon, Laosirihongthong & Keong, 2011; Wisner, 2003; Shin, Collier & Wilson, 2000), so that based on the above description may pose the following hypothesis:

H1: The Management of the Supply Chain in SME Manufacturing allows higher performance.

2.4 Methods

In this research, is examined if whether the SCM is a key element for manufacturing SMEs Performance and to do this, data from INEGI (2009) which shows that reference was made in the state of Aguascalientes, It has the record of 442 SMEs in the manufacturing sector companies which have between 11 and 250 workers.

The work is empirical, exploratory and co relational in which data of 288 SMEs from the manufacturing sector were sampled to whom we applied a measuring instrument type custom survey for managers or owners of such organizations in the period between August and December 2013 transversely and a quantitative approach.

To prepare the instrument used to Manufacturing SMEs in Aguascalientes 2 blocks which are mentioned below were designed: For the I block identified as Management Supply Chain 20 items (coded as CS01 to CS20) were used, which they are measured 1 to 5 Likert scale which refer from low to high importance (Wisner, 2003). And for block II, measurement Performance Scale 12 items (coded RE01 to RE12) were used, also with the Likert scale 1-5, Quinn and Rohrbaugh (1983), which is very unfavorable to very favorable. In Figure 1 theoretical model designed for this research it is shown.

2.5 Statistical Analysis

In the present investigation, a Confirmatory Factor Analysis was performed (CFA) in order to evaluate the reliability and validity of the scales of each of the blocks (SCM and Performance). Also, a structural equation model was used (SEM) in order to check whether the structure of the model is properly designed, and for this, the maximum likelihood method in the software EQS version 6.1 was used ; for this, the reliability was evaluated considering the Cronbach α and composite reliability index (IFC) (Bagozzi & Yi, 1988).

Figure 1. Theoretical Research Model
Source: Adapted from Wisner (2003) along with Quinn and Rohrbaugh (1983).

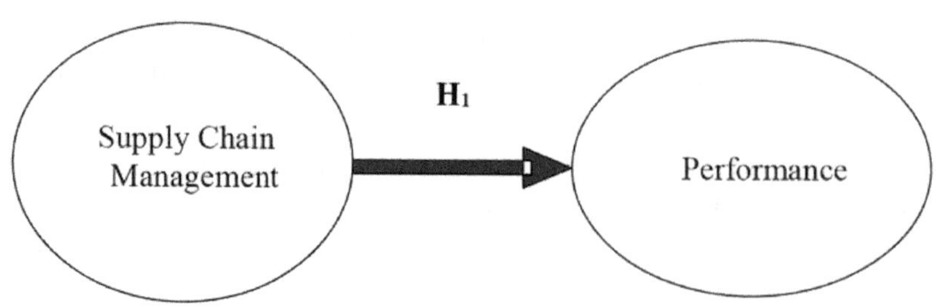

As a result of the statistical analysis, in Table 1 it shows that all values IFC 0.7 exceed the recommended level, which shows evidence of reliability in each of the items integrated in the theoretical model (Nunnally & Bernstein, 1994; Hair, Anderson, Tatham & Black, 1995), likewise, once adjusted the theoretical model in the CFA, the following results assume that the model provides a good fit is obtained ($S\text{-}BX^2 = 1187.4872$; $df = 433$; $p = 0.0000$; $NFI = 0.981$; $NNFI = 0.987$; $CFI = 0.988$; y $RMSEA = 0.078$), where it is concluded that all the items related factors in each of the main factors (SCM and Performance), are significant ($p < 0.05$) It is also important to note that the size of all the factor loadings are greater than 0.6 (Bagozzi & Yi, 1988) and the index of the extracted variance (IVE) of each pair of constructs is greater than 0.5 and Larcker recommended by Fornell (1981).

Table 2 presents the results describing discriminant validity using two tests. First, with a range of 95% confidence, none of the individual elements of the factors containing the value 1.0 (Anderson & Gerbing, 1988). Second of all, the extracted variance between each pair of constructs of the model is higher than the corresponding IVE (Fornell & Larcker, 1981) . Therefore, we can conclude that this research shows based on statistical analysis of their results enough evidence of reliability and convergent validity discriminant as well.

3. RESULTS

One SEM (Structural Equation Model) was performed to check the structure of the conceptual model and to test the hypothesis, using the blocks contained in the evaluation instrument which are described as follows: First block consisting of variables measuring the Management of Supply Chain and the second block consists of variables related to performance. Nomological validity of the model was tested by performing the Chi square test, in which the theoretical model was compared with the measurement model (Anderson & Gerbing, 1988; Hatcher, 1994).

The hypothesis in this research shows favorable results that allow to highlight results described below: Regarding the hypothesis H1, the results presented in Table 3 ($\beta = 0.448$, $p < 0.001$), indicating that the management Supply Chain have a positive impact on SME Manufacturing performance considering this results for these businesses, SCM is the key to Aguascalientes companies have increased performance element .

4. CONCLUSION

The results obtained in this investigation show that the SME manufacturing in the state of Aguascalientes, consider that SCM however complex it may be, is an essential element in the dynamics related to the supply of material resources. In this respect, entrepreneurs, have highlighted that information to improve and streamline all activities related to the SCM that will allow to have a greater impact on the performance of their organizations.

Of course, if this information is not complete and available for everyone involved in the supply of material resources, there may be problems of delay in providing that regardless of the quality problems, it is important that the management in handling allow resource materials because you must not forget that the ultimate goal is to meet in a timely manner with customer commitments. For entrepreneurs who

Table 1. Internal consistency and convergent validity of the theoretical model

Contract	T Robust	Indicator	FW>0.6 Factor Weight	Average of Factor Weight	Alpha Cronbach> to 0.7	FI > to 0.7 Feasibility Index	IVE > to 0.5 Index Variance Extracted
Supply Chain Management	1.000	CS01	0.734***	**0.706**	0.970	0.965	0.941
	22.083	CS02	0.790***				
	22.995	CS03	0.792***				
	20.909	CS04	0.814***				
	22.572	CS05	0.795***				
	16.584	CS06	0.741***				
	12.750	CS07	0.714***				
	19.346	CS08	0.768***				
	21.686	CS09	0.795***				
	21.574	CS10	0.812***				
	17.958	CS11	0.725***				
	19.111	CS13	0.758***				
	19.888	CS14	0.767***				
	21.098	CS15	0.794***				
	19.590	CS16	0.732***				
	18.060	CS17	0.764***				
	21.714	CS18	0.790***				
	14.179	CS19	0.739***				
	18.571	CS20	0.752***				
		Σ	***8.466***				
Performance	1.000	RE01	0.753***	**0.756**	0.939	0.934	0.885
	21.238	RE02	0.773***				
	20.555	RE03	0.746***				
	23.663	RE04	0.751***				
	17.010	RE05	0.696***				
	21.647	RE06	0.756***				
	17.831	RE07	0.715***				
	20.536	RE08	0.752***				
	20.536	RE09	0.768***				
	21.420	RE010	0.751***				
	14.542	RE011	0.676***				
	16.633	RE012	0.687***				
		Σ	***17.397***				

S-BX2 (df = 433) = 1187.4872; $p < 0.0000$; NFI = 0.981; NNFI = 0.987; CFI = 0.988; RMSEA = 0.078

[a]= parameters constrained to that value in the identification process *** = $p < 0.001$

Table 2. Discriminator validity of measurement of the theoretical model

	F1		F2
F1: Supply Chain Management	0.941		(0.399)2
			0.159
F2: Performance	0.399	0.032	0.885
	0.355	0.463	

The diagonal represents the extracted index variation, while above the diagonal is showed part of the variance (Correlation to the chart). Below the diagonal, the estimation of the correlation of factors with a confidence interval of 95%.

Table 3. Results of SEM Conceptual Model Management and Supply Chain Performance

Hypothesis	Structural Connection	Standardized Coefficient	t Robust Value	FIT Measure
H1: The Management of the Supply Chain in SME Manufacturing allows higher performance.	Supply Chain Management Performance	0.448***	17.469	S-BX2 = 1206.2488; p= 0.0000 NFI=0.982 NNFI=0.987 CFI=0.988 RMSEA=0.079

coordinate the operational activities of manufacturing SMEs is elementary that the GCS is effective and flexible so that supplies do not have compliance issues, it is recommended that those responsible for these organizations consider that the objectives and strategies requiring to be implemented in the organizations are in order to have a better performance so it is recommended to have more attention in each step and processes related to the flow and supply of material resources. In this sense, if entrepreneurs have the ability to have good communication, collaboration and agreements with sources of supply and are able to share information related to inventory, stock management and production needs, it will be easier to have a company with the highest performance.

Finally, it is important to note that while employers are aware of the importance of the control and management of information related to the SCM and this has a positive impact on their business performance, should further deepen the study of how to improve the efficiency of SCM course considering the nature of its complexity and material resources that manufacturing SMEs requires supplied to meet targets agreed with customers. In this regard, it is important to note that the limitations of this study are the sample obtained, the analyzed sector and the type of items that have been selected to measure the SCM and performance in such organizations. As future research, it is important to analyze the impact of other factors and sectors to study to thereby determine whether the SCM still remains key to the increased performance of manufacturing SMEs in Aguascalientes element.

5. RECOMMENDATIONS

Given the importance for the SME Manufacturing supply and enterprise development, it is elementary that employers keep in mind that the relationship we have with organizations requiring responsible as sources of supply to establish close cooperation and agreements in order to avoid delays in deliveries of supplies and thereby avoid problems within the process having organizations. For this it is important to coordinate with commissary management of demand, establish control systems for the development of suppliers through exercises process audits and establish mechanisms for information management focused on demand, requirements and flexibility in orders as well should take special care in looking after the needs of customers.

Precisely for that manufacturing SMEs do not have major problems of control in the SCM and this ensures that the management of material resources will take smoothly, it is important that the information that is generated naturally in such a complex process as it is SCM, can be controlled ethically and responsibly considering that the essence of this information is valid and important for both supply sources and companies responsible for the SCM. Having no seriousness in handling the information, it can benefit and harm the operational activity across the network involving various actors committed to the effectiveness of SCM.

It is also necessary that employers do not lose sight of that today, business management is more dynamic and this requires trained staff and committed to both the organization and with the responsibility to comply fully with the requirements of the company and certainly with the customer. Handling of materials should be properly analyzed so that within the SCM delays or problems with the dynamics of supply are avoided because a communication error may hurt significantly the provider's relationship with the requestor company as well with end customers. That is why managers should have the vision domain and prevent companies are affected in performance having an efficient and competitive SCM.

REFERENCES

Aguilera, E. L., & González, A. M. (2012). Estrategias y Rendimiento de las Micro, Pequeñas y Medianas Empresas Familiares (Aguascalientes, México). *Global Conference on Business and Finance Proceedings*, *1*(7), 877-883.

Aguilera, E. L., Hernández, C. O., & López, T. C. G. (2012). La gestión de las cadenas de suministro y los procesos de producción. *Mercados y Negocios.*, *13*(2), 43–66.

Anderson, J., & Gerbing, D. (1988). Structural equation modeling in practice: A review and recommended two-step approach. *Psychological Bulletin*, *13*(3), 411–423. doi:10.1037/0033-2909.103.3.411

Ayers, J. B. (2001). *Handbook of Supply Chain Management*. Boca Raton, FL: The St. Lucie Press/ APICS Series on Resource Management.

Bagozzi, R., & Yi, Y. (1988). On the evaluation of structural equation models. *Journal of the Academy of Marketing Science*, *16*(1), 74–94. doi:10.1007/BF02723327

Beamon, B. M. (1998). Supply chain design and analysis: Models and methods. *International Journal of Production Economics*, *55*(3), 281–294. doi:10.1016/S0925-5273(98)00079-6

Cachon, G. P., & Netessine, S. (2004). Game theory in supply chain analysis. In *Handbook of Quantitative Supply Chain Analysis: Modeling in the e-Business Era*. Kluwer. doi:10.1007/978-1-4020-7953-5_2

Chow, D., & Heaver, T. (1999). *Logistics strategies for North America* (3rd ed.). Global Logistics and Distribution Planning.

Christopher, M. (1998). *Logistics and Supply Chain Management* (2nd ed.). New York; Financial Times/ Prentice Hall.

Chun, H. Ch., Choon, T. K., Laosirihongthong, T., & Keong, L. G. (2011). Entrepreneurial SCM competence and performance of manufacturing SMEs. *International Journal of Production Research*, *49*(22), 6629–6649. doi:10.1080/00207543.2010.537384

Cooper, M. C., & Ellram, L. M. (1993). Characteristic of supply chain management and the implications for purchasing and logistics strategy. *International Journal of Logistic Management*, *4*(2), 13–24. doi:10.1108/09574099310804957

Cruz, J. M. (2008). Dynamics of supply chain networks with corporate social responsibility through integrated environmental decision-making. *European Journal of Operational Research*, *184*(3), 1005–1031. doi:10.1016/j.ejor.2006.12.012

Deshpande, A. (2012). Supply Chain Management Dimensions, Supply Chain Performance and Organizational Performance: An Integrated Framework. *International Journal of Business and Management*, *7*(8), 2–19. doi:10.5539/ijbm.v7n8p2

Dyer, J. H., Sung, C., & Chu, W. (1998). Strategic supplier Segmentation: The next "best practice" in Supply chain management. *California Management Review*, *40*(2), 57–77. doi:10.2307/41165933

Fornell, C., & Larcker, D. (1981). Evaluating structural equation models with unobservable variables and measurement error. *JMR, Journal of Marketing Research*, *18*(1), 39–50. doi:10.2307/3151312

Ganeshan, R. & Harrison, T. P. (1995). *An Introduction to Supply Chain Management*. Department of Management Sciences and Information Systems.

Grant, D., Lambert, D., Stock, J., & Ellram, L. (2006). *Fundamentals of Logistics Management* (European Edn.). Berkshire: McGraw-Hill Book Co.

Guinipero, L. C., & Brand, R. R. (1996). Purchasing's role in supply chain management. *International Journal of Logistics Management*, *5*(1), 29–38. doi:10.1108/09574099610805412

Habr, S., & Reichel, A. (2005). Identifying performance measures of small ventures – The case of the tourism industry. *Journal of Small Business Management*, *43*(3), 257–286. doi:10.1111/j.1540-627X.2005.00137.x

Hair, J. F., Anderson, R. E., Tatham, R. L., & Black, W. C. (1995). *Multivariate Data Analysis with Readings*. New York, NY: Prentice-Hall.

Harold, D. F., & Darlene, B. S. (2004). Managing for Value: Developing a Performance Measurement System Integrating Economic Value Added and the Balanced Scorecard in Strategic Planning. *The Journal of Business Strategy*, *21*(1), 1–17.

Hatcher, L. (1994). *A Step by Step Approach to Using the SAS System for Factor Analysis and Structural Equation Modeling*. Cary, NC: SAS Institute Inc.

Houthoofd, N. (2009). Business Definition and Performance Implications: Te Case of The Belgian Construction Sector. *Construction Management and Economics*, (july): 27, 639–652.

Kanji, G. K., & Wong, A. (1999). Business excellence model for supply chain management. *Total Quality Management, 10*(8), 1147–1168. doi:10.1080/0954412997127

Kaplan, R. S., & Norton, D. P. (1993). Putting balanced scorecard to work. *Harvard Business Review, 71*(5), 134–142.

Langley, C., Coyle, J., Gibson, B., Novack, R. & Bardi, E. (2008). *Managing Supply Chains: A Logistics Approach.* Canada: South-Western Cengage Learning.

Mbang, J. A. (2012). A New Introduction to Supply Chains and Supply Chain Management: Definitions and Theories Perspective. *International Business Research, 5*(1), 194–207.

Mentzer, J., Witt, W. D., Keebler, J., Min, S., Nix, N., Smith, D., & Zacharia, Z. (2001). Defining Supply Chain (SC) management. *Journal of Business Logistics, 22*(2), 1–25. doi:10.1002/j.2158-1592.2001.tb00001.x

Mojica, C.E.P. (2012). *La Influencia de la Orientación al Mercado y la Innovación en la Capacidad de Internacionalización y Desempeño de las Pymes del Estado de Aguascalientes.* Tesis Doctoral, Noviembre, 2012. 1-397.

Morash, E.A. & Clinton, S. R. (1997). The Role of Transportation Capabilities in International Supply Chain Management. *International Supply Chains, 1*(1), 5-17.

Mujuni, K. T., & Zheng, Q. (2010). Tanzanian SMEs' Perceptions towards Adoption of Supply Chain Management (SCM) Strategy. *International Journal of Business and Management, 5*(10), 42–50.

Nunnally, J. C., & Bernstein, I. H. (1994), Psychometric Theory (3rd ed.). New York: McGraw-Hill.

Pérez, V. (2008). *El Desempeño Empresarial.* Retrieved from http://www.degerencia.com/articulo/el_desempeno_empresarial

Pienaar, W. (2009). *Introduction to Business Logistics. Southern Africa.* Oxford University.

Quinn, R., & Rohrbaugh, J. (1983). A Spatial model of Efectiveness Criteria: Towards a Competing Values Aproach to Organizatinal Analysis. *Management Science, 29*(3), 363–377. doi:10.1287/mnsc.29.3.363

Rajendar & Jun Ma. (2005). Benchmarking Culture and Performance In Chinese Organizations, *Benchmarking. International Journal (Toronto, Ont.), 12*(3), 260–274.

Richard, C. (2002). Experiments with New Teaching Models and Methods. *International Public Management Review.*

Schwarz, A., Kalika, M., Kefi, H., & Schwarz, C. (2010). A Dynamic Capabilities Approach to Understanding the Impact of IT-Enabled Business Processes and IT-Business Alignment on the Strategic and Operational Performance of the Firm. *Communications of the Association for Information Systems, 26*(4), 57–84.

Shin, H., Collier, D. A., & Wilson, D. D. (2000). Supply management orientation and supplier/buyer performance. *Journal of Operations Management, 18*(3), 317–333. doi:10.1016/S0272-6963(99)00031-5

Slater, S., & Olson, E. (2000). Strategy Type and Performance: The Influence of Sales Force Management. *Strategic Management Journal*, *21*(8), 813–829. doi:10.1002/1097-0266(200008)21:8<813::AID-SMJ122>3.0.CO;2-G

Stadler, H. (2005). Supply chain management and advanced planning – basics, overview and challanges. *European Journal of Operational Research*, *163*(3), 575–588. doi:10.1016/j.ejor.2004.03.001

Trkman, P., Stemberger, M., & Jaklic, J. (2005). *Information transfer in Supply Chain (SC) management*. Available: http://2005papers.iisit.org/I46f91Trkm.pdf

Wisner, J. D. (2003). A Structural Equation Model of Supply Chain management Strategies and Firm Performance. *Journal of Business Logistics*, *24*(1), 1–26. doi:10.1002/j.2158-1592.2003.tb00030.x

Chapter 11

Comparison of Two Random Weight Generators for Multi-Objective Optimization

Victor M. Carrillo
Universidad Autónoma de Ciudad Juárez, Mexico

German Almanza
Universidad Autónoma de Ciudad Juárez, Mexico

ABSTRACT

There exist two general approaches to solve multiple objective problems. The first approach belongs to the classical mathematical methods: The weighted sum method, goal programming, or utility functions methods pertain to this approach. The output of mathematical methods is a single optimal solution. In the second approach are the heuristic methods, like the multiple objective evolutionary algorithms that offer the decision maker a set of optimal solutions usually called non- dominated or, Pareto-optimal solutions. This set is usually very large and the decision maker faces the problem of reducing the size of this set to a manageable number of solutions to analyze. In this paper the second approach is used to reduce the Pareto front using two weights generator for the non-numerical ranking preferences method and their performance is compared.

1. INTRODUCTION

Most optimization problems in the real world are multi-objective and its functions are in conflict with each other. Multi-objective optimization problems appear in economics, biology, engineering, military, etc. For example, a high performance electronic product is a high cost item, and every customer is always looking for a device with high performance, but at the lowest rates. Therefore the development of a method for finding a solution to several optimization goals has been the subject of numerous research and guides documents (Branke, Kalyanmoy, Miettinen & Slowinsky, 2008). Currently, two main approaches are known to solve multi-objective optimization problems: the first is known as classic mathematical methods. Some examples of classical mathematical methods are: the weighted sum, weighted metric,

DOI: 10.4018/978-1-4666-9779-9.ch011

rotated weighted metric, constraint method, goal programming, lexicographic method, and utility functions among others. Meanwhile the second approach involves populating a number of feasible solutions along the Pareto frontier, and the final solution is called the set of Pareto-optimal solutions, or the set of all optimal solutions of the multi-objective optimization problem. Some evolutionary methods are: Tabu-search, neural networks, particle swarm, ant colony and genetic algorithms, among others. The optimization procedure terminates once the Pareto front is obtained, but if the Pareto optimal size is very large the decision maker won't be able to select comfortably a manageable subset of the optimal solutions. The problem of obtaining a small subset of the Pareto-optimal front led to the development of methods for reducing the size of the Pareto fronts. Most of evolutionary methods obtain as a solution a set of Pareto optimal solutions based in the Pareto dominance concept. The idea of the Pareto-Front is to compare all solutions against each other, where the best fitted solutions dominate the weaker which in turn are said to be dominated (Zitzler & Thiele, 1999). The set of non-dominated solutions is known as the Pareto-Front.

Without loss of generality any multi-objective optimization problem can be presented as a minimization problem as shown in Equation 1

$$\begin{aligned} &\text{minimize} \quad f(x) = \left(f_1(x), \ldots, f_n(x)\right) \\ &s.t. \qquad g_j(x) \le 0 \quad j = 1, \ldots, \mathrm{m} \\ &\qquad\qquad x \in R^k \end{aligned} \tag{1}$$

where n is the number of objective functions $f_n(x)$, n is the number of inequality constraints $g(x) \in \mathrm{R}^m$, $x \in R^k$ is a vector of design variables, and $f(x) \in \mathrm{R}^k$ is a vector of n objective functions, $f_i : R^k \to R \;\; for \;\; i = 1, \ldots, n$.

The feasible or decision space is defined as $X = \left\{x \in R^k : g_j(x) \le 0 \;\; j = 1, \ldots, \mathrm{m}\right\}$. The objective space is defined as the image of the feasible decision space X i.e. $Z = \left\{f(x) : x \in R^k\right\}$. Having just a single solution that optimizes all objective functions simultaneously is difficult, and such a solution usually does not exist.

For example, consider the following bi-objective optimization problem Equation 2

$$\begin{aligned} &\text{Minimize} \quad f_1(x) \\ &\text{Minimize} \quad f_2(x) \\ &s.t. \qquad g_j(x) \le 0 \quad j = 1, \ldots, \mathrm{m} \\ &\qquad\qquad x \in R^3 \end{aligned} \tag{2}$$

The image of the decision space under the objective functions is shown in Figure 1. From this perspective it is hard, if not impossible to decide which is or are the optimum values of the problem. The yellow points in the purple region represent optimal solutions or non-dominated points $f(x^*) = \left(f_1(x^*), f_2(x^*)\right)$ and the yellow points in the blue region, are the corresponding arguments denoted by x^*.

Figure 1. Pareto-optimal set and Pareto-front representations (yellow dots)

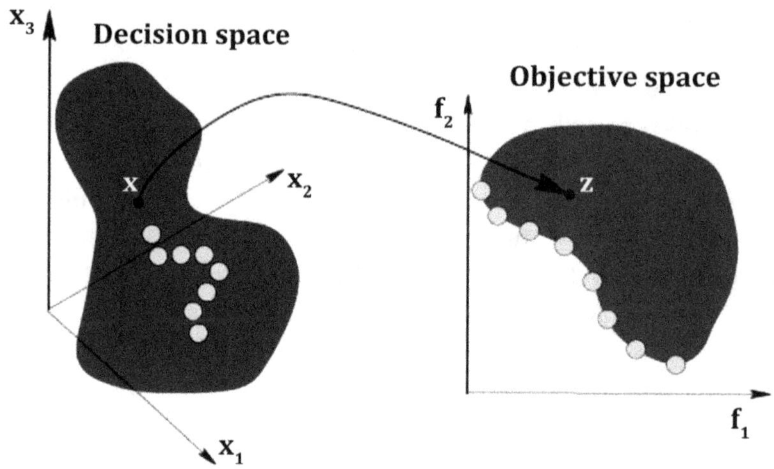

In a simple optimization problem, the notion of optimality is direct. The search is for the best (the minimum or the maximum) value of an objective function. In a multi-objective optimization problem, the notion of optimality is not as obvious as can be seen on Figure 1.

Definition: A solution x_1 dominates a solution x_2 if and only if the two following conditions are true:

- For all objectives i.e. $f_i(x_1) \leq f_i(x_2)\ \ \forall i \in \{1,2,...,n\}$
- And for at least one index $j: f_j(x_1) \leq f_j(x_2)$

The solution of a multiple objective optimization is a set of non-dominated solutions or Pareto-front. Sometimes, the Pareto front can contain a large number (in some cases, thousands) of solutions. From the decision-maker's perspective, choosing some good optimal points from all the non-dominated solutions can be prohibitive and inefficient.

Optimality Conditions for Multi-Objective Optimization Problems

The classic mathematical method to solve a MOOP applies the following

Theorem: Karush-Kuhn-Tucker (KKT) sufficient condition for Pareto-optimality

Let the objective functions of the MOOP in Equation (1) be convex and the constraint functions concave. Let be the objective and constraint functions continuously differentiable at a feasible solution x* .A sufficient condition for x* to be a Pareto-optimal is that there exist vectors $\lambda > 0$ and $\mu \geq 0$ (where $\lambda \in R^k, \mu \in R^m$) such that the following equations are true:

$$\begin{aligned} &1.\ \sum_{i=1}^{k} \lambda_i \nabla f_i(x^*) + \sum_{j=1}^{m} \mu_j \nabla g_j(x^*) = 0 \\ &2.\ \mu_i g_j(x^*) = 0 \quad \text{for all } \mathrm{j} = 1,...,\mathrm{m} \end{aligned} \tag{3}$$

A Multi-objective optimization problem with more than two objective functions is hard to solve with just the KKT optimality conditions (3). Thereby multi-objective optimization problems were solved by scalarization before the evolutionary methods were sufficiently developed. Scalarization means that objective functions of the problem are transformed into a single (scalar) function or a real valued function i.e. a transformation $F : R^n \rightarrow R$ is applied to $f(x)$ such that problem (1) is now written as Equation 4:

$$\begin{aligned} &\text{minimize} \quad F\left(f(x)\right) \\ &s.t. \quad g_j(x) \leq 0 \quad j = 1,\ldots,\text{m} \\ &\qquad x \in R^n \end{aligned} \tag{4}$$

Once the Multi-objective optimization problem has been scalarized, the single nonlinear optimization methods can be applied. One of the most used methods is the weighting technique.

The weights of the objective functions are usually chosen in proportion to its relative importance in the problem. Otherwise they are selected at random.

$$\begin{aligned} &\mathit{Minimize} \qquad F\left(x\right) = \sum_{i=1}^{k} w_i f_i(x) \\ &\qquad s.t. \quad g_j\left(x\right) \leq 0 \;\; ; j = 1, \ldots, m \\ &\qquad\qquad x \in R^n \;\; , \;\; \sum_{i=1}^{k} w_i = 1 \quad , \;\; w_i > 0 \end{aligned} \tag{5}$$

The optimal points obtained when solving the above problem with single nonlinear optimization are Pareto-optimal points due to the following

Theorem: The solution to the problem in Equation (5) is Pareto-optimal if all of the weights are positive (Mietinen, 1999)

In Figures 2 and 3 is depicted the search of Pareto points using the weighted sum. The source of all figures: "Classic methods for multi-objective optimization" by Giuseppe Narzizi.

One common characteristic of multi-objective optimization algorithms with single objective optimization problems is that Pareto-optimal points search is through updating a single solution per iteration and that mainly use a deterministic transition rule to approach the optimum solution. Every classical optimization algorithm although convergent is designed to solve a specific type of problem (Reklaitis *et al.*, 1983). The Geometric method is designed to solve only constrained polynomial-type objective functions, but cannot be applied easily to solve other types of problems. The conjugate gradient methods work well in problems having one optimal solution, but not expected to work well in problems having multiple optimal solutions. The Frank-Wolfe's successive linear programming method (Reklaitis *et al.*, 1983) works efficiently on a constrained linear function, but on nonlinear objective functions its performance depends mainly on the initial conditions. Thus, one algorithm may be best suited for one class of problems but not for those problems outside that class. This requires users to know a number of optimization algorithms to solve different optimization problems.

Figure 2. Weighted sum method

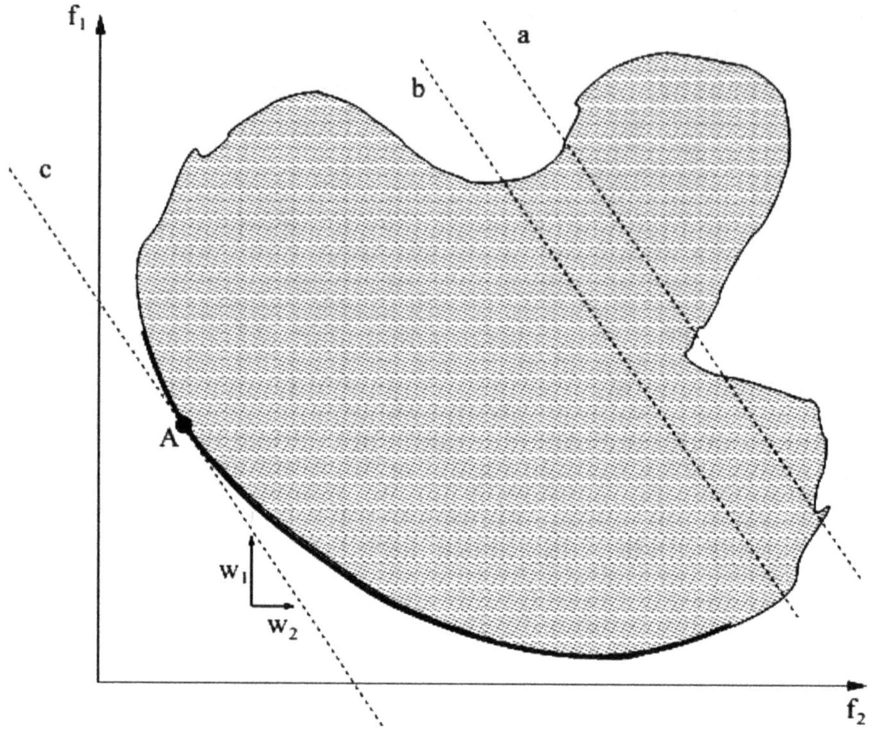

Figure 3. Disadvantage of the weighted a convex Pareto-front method for non-convexity

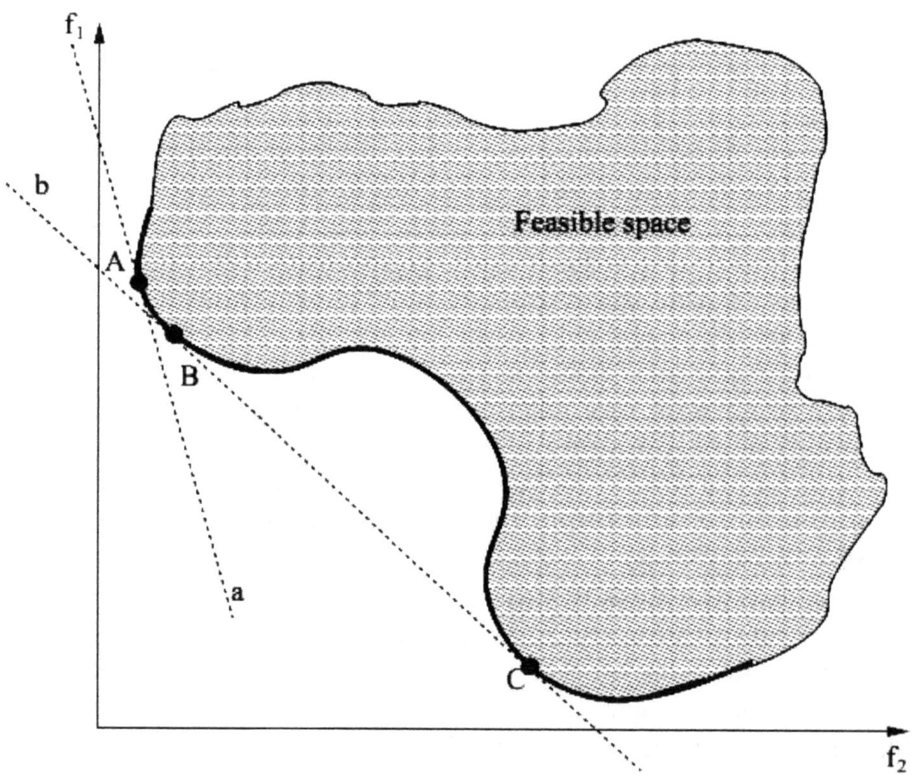

Another search approach to obtain Pareto-optimal solutions is with evolutionary algorithms (EA's). Evolutionary algorithms generate sets of Pareto-optimal points or non-dominated points.

Most classical optimization methods use the point to point approach where only one solution gets updated to a new solution by iteration so the advantages of parallel systems cannot be fully exploited.

On the other hand we have that:

Evolutionary algorithms (EA's) work with a population of solutions instead of a single solution Figure 4; EA's do not require any auxiliary information except the objective functions values ; EA's use probabilistic rules to guide their search; EA's are robust since can be applied to a wide variety of problems, and finally as early mentioned EA's can be easily used in parallel systems. One striking difference between classical optimization methods (COM) and EA's is in their convergence theory development. Whereas in COM there is plenty of convergence results that supports it. In EA's, until recently (Yu,Xiufen, & Xie, 2001) proved necessary and sufficient conditions on EA's convergence with Markov's chain theory.

The selection of solutions from the Pareto set is called post-Pareto optimality analysis, in the present work a non-uniform weight method NUWM (Carrillo, Aguirre, & Taboada, 2011) is used in the non-numerical ranking preferences method NRPM (Taboada, Coit, & Wattanapongsakorn, 2007) with the purpose to prune and obtain a smaller sub-set of preferred solutions from the large Pareto-optimal set. In section 2, the literature review about post-Pareto optimality analysis is presented, in section 3 the Non-numerical ranking preferences method is demonstrated for 3 weights and tested generating a set of random weights sample. In section 4 the Non-Uniform Weight Generator Method is demonstrated and tested and sampled as well.

In this analysis the Pareto-optimal points obtained with the evolutionary algorithm NSGA (Taboada, Coit, & Wattanapongsakorn, 2007) was used to compute the weighted function $f = w_1 f_1 + w_2 f_2 + w_3 f_3$, which afterwards was bounded using numerical thresholds to find the minimum value of such aggregated values. The weight values were previously obtained utilizing the non-uniform weights algorithm according to the following non-numerical ranking $f_1 \succ f_2 \succ \cdots \succ f_n$ i.e., $w_1 > w_2 > \cdots > w_n$.

Figure 4. Evolutionary algorithm optimal search

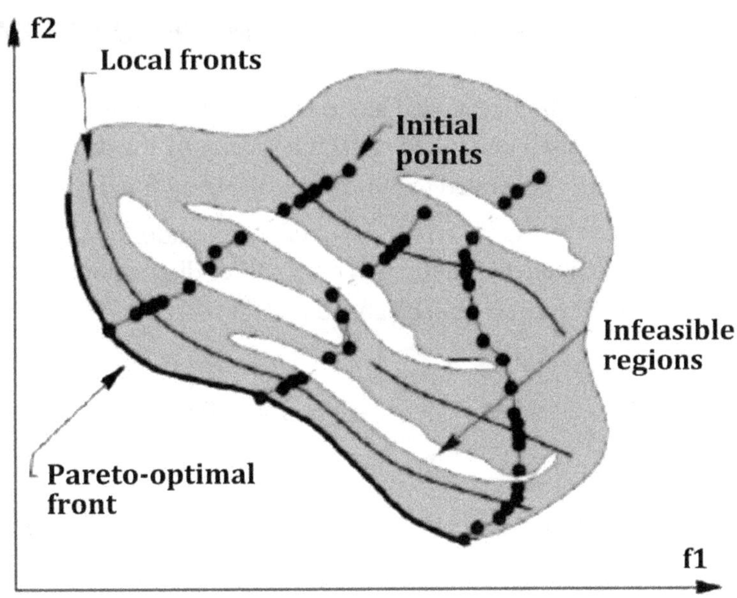

2. LITERATURE REVIEW

In 2011 Carrillo,Aguirre and Taboada proposed a method based on preferences rankings provided by the decision maker to bound an scalarized function made out of Pareto-optimal values for Pareto-optimal set reduction purposes. Carrillo & Taboada in 2012 developed an algorithm that generates randomly n non-uniformly distributed weights (NURW) to be used in the method based on preferences rankings for any MOOP problem size. Besides Carrillo & Taboada in 2012 presented the generalization of a post-Pareto optimality method known as the non-numerical ranking preferences (NNRP) method originally proposed by Taboada & Coit in 2008. In Carrillo Phd Thesis, 2013 is proposed another method to obtain weights for the scalarization function $f = w_1 f_1 + w_2 f_2 + w_3 f_3$. In 2014 Abubaker, Baharum,& Alrefai applied successfully the NRPM to validate the results of the OCBA+MOSA hybrid algorithm.OCBA, (C.H. Chen, J. Lin, E. Yaesan & S. E. Chick, 2000),(B. Ulungu, J. Teghem and P. Fortemps, 1995)

Other known pruning techniques that do not rely precisely on scalarizing are for instance: the k-means clustering methods which groups solutions with similar characteristics, for example, Taboada & Coit (2007) ; Another similar procedure was presented by Runkler,(2009) using a fuzzy c-means clustering using the k-means technique; Another approach based in clustering was proposed by Li,Liao, &Coit,(2009) where a data envelopment analysis is performed, by comparing the relative efficiency of the solutions to determine the final representative solutions for the overall problem. Aguirre & Taboada, (2011) developed an algorithm that offers two main advantages: there is no need to provide an initial number of clusters, and at each hierarchical level, the algorithm optimizes the number of clusters, and can reassign data from previous hierarchical levels in order to rearrange missed clustered data. The clustering method is the dynamically growing self-organizing tree (DGSOT) algorithm.

The first proposed weight generator method in this work is based on a non-numerical ranking of the objective functions based on the relative importance assigned by the decision maker. Next section explains all the details of the proposed algorithm and how the pruning of optimal solutions from the Pareto front is performed.

3. NON-NUMERICAL RANKING PREFERENCES METHOD (NRPM)

This is a post-Pareto optimality method proposed and developed by Taboada and Coit (2008).Initially the objective functions are scaled and then ranked non-numerically by the decision maker, and a collection of weights $w = (w_1, w_2, \ldots, w_n)$ is generated to build the scalarizing function $f = w_1 f_1 + w_2 f_2 + w_3 f_3$ required for reducing the size of the Pareto front. To obtain the random weight for a multi-objective optimization problem for three component functions is built a three dimensional p.d.f which will provide the w_i values is shown in Equation 6 (Carrillo & Taboada (2011))

$$f_W(w_1, w_2, w_3) = \begin{cases} c, & w_1 > w_2 > w_3 \\ 0, & otherwise \end{cases} \quad \text{where } w_1 + w_2 + w_3 = 1 \text{ and } 0 \le w_i \le 1 \tag{6}$$

The above p.d.f. can be reduced to a two dimensional case Equation 7 simplifying the calculations to get

$$f_W\left(w_1, w_2\right) = \begin{cases} c, & w_1 > w_2 > 1 - w_1 - w_2 \\ 0, & otherwise \end{cases} \quad (7)$$

Thus considering the following properties

$$\begin{aligned} a)&\quad 1 - w_1 \geq w_2 \\ b)&\quad w_1 \geq \frac{1}{3} \qquad , \\ c)&\quad w_2 \leq \frac{2}{3} \end{aligned}$$

derived from $w_1 + w_2 + w_3 = 1$ and $0 \leq w_i \leq 1$, its p.d.f. is Equation 8

$$f_W\left(w_1, w_2\right) = \begin{cases} 0 & , 0 \leq w_1 < \frac{1}{3} \\ 12 & , \frac{1}{2} - \frac{1}{2} w_1 \leq w_2 \leq w_1 \quad \& \frac{1}{3} \leq w_1 < \frac{1}{2} \\ 12 & , \frac{1}{2} - \frac{1}{2} w_1 \leq w_2 \leq 1 - w_1 \quad \& \frac{1}{2} \leq w_1 < 1 \end{cases} \quad (8)$$

Moreover its cumulative distribution function (c.d.f) is Equation 9:

$$F_{w_1}\left(w_1\right) = \int_{-\infty}^{w_1} f_{w_1}\left(w_1\right) dw_1 = \begin{cases} 0 & , \quad 0 \leq w_1 < \frac{1}{3} \\ 9w_1^2 - 6w_1 + 1 & , \quad \frac{1}{3} \leq w_1 < \frac{1}{2} \\ -3w_1^2 + 6w_1 - 2 & , \quad \frac{1}{2} \leq w_1 < 1 \end{cases} \quad (9)$$

Finally we are ready to calculate w_1 using the probability integral transformation Equation 10

$$w_1 = F_{w_1}^{-1}\left(u\right) = \begin{cases} 0 & , otherwise \\ \frac{\sqrt{u} + 1}{3} & , 0 \leq u < \frac{1}{4} \\ 1 - \sqrt{\frac{1 - u}{3}} & , \frac{1}{4} \leq u < 1 \end{cases} \quad (10)$$

Similarly to calculate w_2 , derive and use the inverse probability transformation of its (c.d.f) to obtain Equation11

$$w_2 = F_{w_2}^{-1}(u|w_1) = \begin{cases} 0 & , \quad 0 \le w_1 < \frac{1}{3} \\ \frac{(3w_1 - 1)u + 1 - w_1}{2} & , 0 \le u \le 1 \quad \frac{1}{3} \le w_1 < \frac{1}{2} \\ \frac{(1 - w_1)(u+1)}{2} & , 0 \le u \le 1 \quad \frac{1}{2} \le w_1 < 1 \end{cases} \tag{11}$$

To finalize the process solve for w_3 from the equation $w_1 + w_2 + w_3 = 1$. Once the weights are obtained the algorithm used to prune Pareto optimal solutions is shown below:

3.1 Pseudo Code

Randomly generate weights based on the following scheme:

1. Randomly generate an $u \in [0,1]$
2. Generate $w_1 = F_{w_1}^{-1}(u) = \begin{cases} 0 & , otherwise \\ \frac{\sqrt{u}+1}{3} & , 0 \le u < \frac{1}{4} \\ 1 - \sqrt{\frac{1-u}{3}} & , \frac{1}{4} \le u < 1 \end{cases}$
3. Then calculate $w_1 = \frac{\sqrt{u}+1}{3}$ if $0 \le u < \frac{1}{4}$ and get $w_2 = \frac{(3w_1 - 1)u + (1 - w_1)}{2}$;
4. Otherwise calculate $w_1 = 1 - \sqrt{\frac{1-u}{3}}$ if $\frac{1}{4} \le u \le 1$ and get $w_2 = \frac{(u+1)(1-w_1)}{2}$
5. Whatever path is selected above calculate, $w_3 = 1 - w_1 - w_2$
6. Repeat all steps until *n* iterations to generate sets of weights w_1, w_2 *and* w_3.

In the following section is demonstrated the second weight generator method. The pseudo code to generate the weights $w_1 > w_2 > w_3$ is resumed as:

1. $u \in (0,1)$, is randomly generated, such that

   If $0 \le u \le \frac{1}{4}$, then calculate $w_1 = \frac{\sqrt{u}+1}{3}$ and obtain $w_2 = \frac{(3w_1 - 1)u + 1 - w_1}{2}$

If $\frac{1}{4} < u \leq 1$, then calculate $w_1 = 1 - \sqrt{\frac{1-u}{3}}$ and obtain $w_2 = \frac{(u+1)(1-w_1)}{2}$

2. $w_3 = 1 - w_1 - w_2$ is calculated

Example: Ten values were generated using Matlab,with a Pc Pentium® Dual-Core CPU E5400 2.7GHz and are shown in the Table 1.

In the following section is demonstrated the second method.

4. NON-UNIFORM WEIGHT GENERATOR METHOD

The non-uniform weight generator method (NUWM) is a general procedure to obtain n increasingly ordered uniform distributed random numbers as is shown: Let be

$X_1, \ldots, X_n$ *random variables uniformly distributed* $X_i \sim U(i-1, i)$,

thus its correspondent probability density functions are:

$$f(x_i) = \begin{cases} 1 & if \ x_i \in (i-1, i) \\ 0 & otherwise \end{cases} \quad for \ i = 1, \ldots, n$$

Since $X_1, \ldots, X_n$ are independent, its joint probability density function (p.d.f.) is

$$f(x_1, x_2, \ldots, x_n) = \prod_{i=1}^{n} f(x_i) = \begin{cases} 1 & if \ x_i \in (i-1, i) \ for \ i = 1, \ldots, n \\ 0 & otherwise \end{cases}$$

Table 1. Weights generated with NRWG

	w1	w2	w3	Sum
1	0.66124773	0.28044296	0.05830931	1
2	0.396325113	0.060414146	0.543260741	1
3	0.775745164	0.207338094	0.016916741	1
4	0.851668443	0.143436115	0.004895442	1
5	0.672756541	0.27467755	0.05256591	1
6	0.715828767	0.24974959	0.034421643	1
7	0.707386756	0.255031823	0.037581421	1
8	0.549898907	0.313321451	0.136779641	1
9	0.661118452	0.280505455	0.058376093	1
10	0.471249043	0.144798596	0.383952361	1

We can see that $\int_{-\infty}^{\infty} \cdots \int_{-\infty}^{\infty} \int_{-\infty}^{\infty} f(x_1, x_2, ..., x_n) dx_1 dx_2 \cdots dx_n = 1$ which proves that $f(x_1, x_2, ..., x_n)$ is a p.d.f.

Indeed, computing

$$\int_{-\infty}^{\infty} \cdots \int_{-\infty}^{\infty} \int_{-\infty}^{\infty} f(x_1, x_2, \ldots, x_n) dx_1 dx_2 \cdots dx_n$$
$$= \int_{-\infty}^{\infty} \cdots \int_{-\infty}^{\infty} \int_{-\infty}^{\infty} \left(\prod_{i=1}^{n} f(x_i) \right) dx_1 dx_2 \cdots dx_n$$
$$= \int_{n-1}^{n} \cdots \int_{1}^{2} \int_{0}^{1} dx_1 dx_2 \cdots dx_n = 1$$

To generate each one of the X_i values such that $X_1 < X_2 < \cdots < X_n$ let us apply repeatedly the Probability Integration Theorem as follows:

To get the first weight, let's recall that the random variables $X_i \sim U(i-1, i)$ are uniformly distributed with probability density functions given by:

$$f(x_i) = \begin{cases} 1 & \text{if } x_i \in (i-1, i) \\ 0 & \text{otherwise} \end{cases} \quad \text{for } i = 1, \ldots, n$$

Thus for the first case its p.d.f. is:

$$f(x_1) = \begin{cases} 1 & \text{if } x_1 \in (0, 1) \\ 0 & \text{otherwise} \end{cases}$$

and its cumulative distribution function (c.d.f) is:

$$F(x_1) = \int_{0}^{x} ds = x_1 \quad .$$

By the Integral probability Theorem $F(x_1) = u_1$ has a uniform distribution $u \sim U(0,1)$ .Therefore for a random $x_1 = u_1$. To obtain the second value x_2 given that we already know the first value x_1 let us calculate the conditional p.d.f

$$f(x_2 \mid x_1) = \frac{f(x_1, x_2)}{f(x_1)} = \begin{cases} 1 & \text{if } x_i \in (i-1, i) \text{ for } i = 1, 2 \\ 0 & \text{otherwise} \end{cases}$$

Similarly to the previous x_1 we have to obtain its cumulative distribution function:

$$F(x_2 \mid x_1) = \int_1^{x_2} ds = x_2 - 1 = u_2 \quad where\ u_2 \in U(0,1) \quad .$$

Solving for x_2 we get $x_2 = u_2 + 1$. For the third case x_3 proceed as in the previous cases to obtain:

$$f(x_3 \mid x_1, x_2) = \frac{f(x_1, x_2, x_3)}{f(x_1)} = \begin{cases} 1 & if\ x_i \in (i-1, i)\ for\ i = 1,2,3 \\ 0 & otherwise \end{cases}$$

and its c.d.f

$$F(x_3 \mid x_1, x_2) = \int_2^{x_3} ds = x_3 - 2 = u_3 \quad where\ u_3 \in U(0,1)$$

solving for x_3 we get $x_3 = u_3 + 2$.

Since this is an iterative procedure the k-th case is:

$$f(x_k \mid x_1, x_2, \ldots x_{k-1}) = \frac{f(x_1, x_2, \ldots x_{k-1})}{f(x_1, x_2, \ldots x_k)} = \begin{cases} 1 & if\ x_i \in (i-1, i)\ for\ i = 1,2,\ldots k-1 \\ 0 & otherwise \end{cases}$$

its cumulative distribution function is:

$$F(x_k \mid x_1, x_2, \ldots x_{k-1}) = \int_2^{x_k} ds = x_k - (k-1) = u_k \quad where\ u_k \in U(0,1)$$

Solving for x_k we get the general iterative formula for each one of the x_k values: $x_k = u_k + (k-1)$

Given values to k starting with k=1, 2, 3,.., n the x_k sequence obtained is:

$$x_1 = u_1 < x_2 = u_2 + 1 < x_3 = u_3 + 2 < \cdots < x_k = u_k + (k-1) < \cdots < x_k = u_n + (n-1)$$

Adding up all the terms we get the property:

$$\sum_{k=1}^{n} x_k = \sum_{k=1}^{n} u_k + \sum_{k=1}^{n} (k-1) = \sum_{k=1}^{n} u_k + \frac{n(n-1)}{2}$$

As an immediate application of the method we can obtain a similar sequence of values inside the interval (0,1).

Let be $M = \sum_{k=1}^{n} x_k$, then we can obtain an increasing finite sequence

$$\{w_k\}_{k=1}^{n} \quad such\ that \quad 0 < w_1 < w_2 < \cdots < w_n < 1 \quad and \quad \sum_{k=1}^{n} w_k = 1$$

of values; just define $w_k = \dfrac{x_k}{M}$ $for \quad k = 1,\ldots,n$

This method can be applied to generate weights in the non-numerical ranking procedure where

The objective component functions are non-numerically ranked according to the decision maker needs, for example $f_1 \succ f_2 \succ \cdots \succ f_n \Rightarrow w_1 > w_2 > \cdots > w_n$ the first function is considered more important compared to the second all and so on, etc. The weights for the weighting function must be generated in a numerical order suggested by the "$\succ$" decision maker preference. Before starting to apply the method in a MOOP let us show some iterations of the algorithm:

Example: Ten values will be generated using the generating formula $x_k = u_k + (k-1)$ using a hand-pocket calculator random number generator.

For k=1, $u_k =, x_1 = u_1 + (1-1) = 0.943597$; next cases are shown in Table 2.

Table 2. Ten iterated x_k values

k	$u_k = rand(\ )$	$x_k = u_k + (k-1)$
1	$u_1 = 0.943597$	$x_1 = u_1 + (1-1) = 0.943597$
2	$u_2 = 0.0203$	$x_2 = u_2 + (2-1) = 1.0203$
3	$u_3 = 0.3885$	$x_3 = u_3 + (3-1) = 2.3885$
4	$u_4 = 0.3672$	$x_4 = u_4 + (4-1) = 3.3672$
5	$u_5 = 0.2330$	$x_5 = u_5 + (5-1) = 4.233$
6	$u_6 = 0.1903$	$x_6 = u_6 + (6-1) = 5.1903$
7	$u_7 = 0.5218$	$x_7 = u_7 + (7-1) = 6.5218$
8	$u_8 = 0.3266$	$x_8 = u_8 + (8-1) = 7.3266$
9	$u_9 = 01018$	$x_9 = u_9 + (9-1) = 8.1018$
10	$u_{10} = 0.0152$	$x_{10} = u_{10} + (10-1) = 9.0152$
	M	48.1083

With the values above obtained we can generate an increasing finite sequence of values

$$\left\{w_k\right\}_{k=1}^{n} \quad such\ that \quad 0 < w_1 < w_2 < \cdots < w_n < 1 \quad and \quad \sum_{k=1}^{n} w_k = 1$$

For that, let be $w_k = \dfrac{x_k}{M} \quad for \quad k = 1, \ldots, n$

The corresponding table is Table 3.

Ten values were generated using Matlab, with a Pc Pentium® Dual-Core CPU E5400 2.7GHz.

5. CONCLUSION

The non-uniform weight (NUWG) generator and the non-numerical ranking preferences (NRPM) weight generators was successfully applied in reducing the size of a Redundancy allocation (RAP) Pareto front of 75 points problem. The NUWG code is simpler than the NRPM code for its weight generator, and can be extended its use to any multi-objective optimization program. On the other hand the NRPM weight generator code method is harder to develop as number of objective functions in a multi-objective optimization problem increases, for instance for an eight multi-objective optimization functions problem. The results obtained by comparing the two Pareto-optimal pruned subsets gives the decision maker similar in importance but no equal collections of values to select simplifying his or her election task. Both methods make use of the Integral probability theorem. Is fair to mention that the NUWG method was suggested by the way the NRWG weight generator was developed.

Table 3. Ten iterated w_k for M=48.1083

k	x_k	$w_k = \dfrac{x_k}{M}$
1	0.943597	0.019614017
2	1.0203	0.021208399
3	2.3885	0.049648397
4	3.3672	0.06999208
5	4.233	0.087988975
6	5.1903	0.107887828
7	6.5218	0.135564965
8	7.3266	0.152293887
9	8.1018	0.168407531
10	9.0152	0.187393859

REFERENCES

Abubaker, Baharum, & Alrefai. (2014). Good solution for multiobjective optimization problem. *AIP Conference Proceedings*. Retrieved from http://scitation.aip.org/content/aip/proceeding/aipcp/1605?ver=pdfcov

Aguirre, O., & Taboada, H. (2011). A Clustering Method Based on Dynamic Self organizing Trees for Post-Pareto Optimaliyy Designs. In Complex Adaptive Systems. Missouri University of Science and Technology.

Branke, J., Deb, K., Miettinen, K., & Słowi'nski, R. (2008). *Multiobjective Optimization*. Interactive and Evolutionary Approaches, Springer-Verlag Berlin Heidelberg. doi:10.1007/978-3-540-88908-3

Carrillo, A., Aguirre, O., & Taboada, H. (2011). Applications and performance of the non-numerical ranking preferences method for post-Pareto optimality. *Procedia Computer Science*, *6*, 243–248. doi:10.1016/j.procs.2011.08.045

Carrillo, T., & Taboada, H. (2012). A Post-Pareto Approach for Multi-Objective Decision Making Using a Non- Uniform Weight Generator Method. *Procedia Computer Science*, *12*, 116–121. doi:10.1016/j.procs.2012.09.040

Carrillo, T., & Taboada, H. (2012). A General Iterative Procedure of the Non-Numerical Ranking Preferences Method, for Multiple Objective Decision Making. *Procedia Computer Science*, *12*, 135–139. doi:10.1016/j.procs.2012.09.043

Carrillo. (2013). *Post-Preto optimality methods for the analysis of large Pareto sets in multi-objective optimization*. PhD Thesis. Id 11353 ProQuest.

Chen, C. H., Lin, J., Ycesan, E., & Chick, S. E. (2000). Article. *Discrete Event Dynamic Systems*, *10*(3), 251–270. doi:10.1023/A:1008349927281

Chen, Y., Zou, X., & Xie, W. (2001). Conditions for the convergence of evolutionary algorithms. *Journal of Systems Architecture*, *47*(7), 601–612. doi:10.1016/S1383-7621(01)00018-2

Li, L., Liao, H., & Coit, D. W. (2009). A two-stage,apprach for multi-objective decision making with applications to system reliability optimization. *Reliability Engineering & System Safety*, *94*(10), 1585–1592. doi:10.1016/j.ress.2009.02.022

Mietinen, K. (1999). *Nonlinear Multi-objective Optimization*. Boston: Kluer Academic Publishers.

Reklaitis, G. V., Ravindran, A., & Ragsdell, K. M. (1983). *Engineering optimization methods and applications*. New York: Wiley.

Runckler, T. (2009). *Fuzzy c–Means*. Herding, IFSA-EUSFLAT 2009.

Taboada, B., Baheranwala, F., Coit, D. W., & Wattanapongsakorn, N. (2007). Practical solutions for multi-objective optimization:An application to system reliability design problems. *Reliability Engineering & System Safety*, *92*(3), 314–322. doi:10.1016/j.ress.2006.04.014

Taboada, H., & Coit, D. (2007). Data Clustering of Solutions for Multiple Objective System Reliability Optimization Problems. *Quality Technology & Quantitative Management Journal*, *4*(2), 35–54.

Taboada, H., & Coit, D. (2008). Multiple Objective Scheduling Problems: Determination of Pruned Pareto Sets. *IIE Transactions*, *40*(5), 552–564. doi:10.1080/07408170701781951

Zitzler, E., & Thiele, L. (1999). Multi-objective Evolutionary Algorithms: A Comparative Case Study and the Strength Pareto Approach. *IEEE Transactions on Evolutionary Computation*, *3*(4), 257–271. doi:10.1109/4235.797969

Chapter 12
A Hybrid Metaheuristic Algorithm for the Quay Crane Scheduling Problem

Laura Cruz-Reyes
National Mexican Institute of Technology, Mexico

Claudia Gómez
National Mexican Institute of Technology, Mexico

Arturo Lam Alvarez
National Mexican Institute of Technology, Mexico

Nelson Rangel Valdez
National Mexican Institute of Technology, Mexico

Marcela Quiroz-Castellanos
National Mexican Institute of Technology, Mexico

Guadalupe Castilla Valdez
National Mexican Institute of Technology, Mexico

Javier Gonzalez Barbosa
National Mexican Institute of Technology, Mexico

ABSTRACT

This chapter presents a novel hybrid algorithm for the quay crane scheduling problem (QCSP). QCSP consists of scheduling a sequence of unloading and loading movements for cranes assigned to a vessel, minimizing the total completion time of all the tasks. The proposed algorithm integrates two well-known metaheuristics: Greedy Randomized Adaptive Search Procedure and Ant Colony System; it also incorporates a repositioning strategy of idle cranes to reduce the interference generated by the quay cranes. The experimental results show that the proposed method is able to find quality solutions in short times. In experiments reported in the literature with crafted instances of QCSP, heuristic running time varies from seconds in small instances to hours in instances of medium size. Currently, the industry requirements are up to a maximum of approximately five minutes. The hybrid algorithm presented in this chapter allows addressing these requirements, by finding good quality solutions in significantly shorter time for larger problems, which represents an advantage in real environments.

DOI: 10.4018/978-1-4666-9779-9.ch012

INTRODUCTION

Container terminals are a critical asset to a nation's economy, infrastructure, and quality of life. They provide the vital link between domestic markets to international customers and vice-versa. Therefore, their efficiencies are critical to the success of the supply chain. Terminals face increasing demand from container shipping customers, and often operate in fierce competition with alternative terminals. It is vital for a terminal to effectively utilize its resources such as the ship berthing areas, container yard, cranes, and manpower, to achieve high productivity and customer satisfaction. The primary objective is to achieve rapid flow of containers at a minimum cost. As such, the time to load/unload a vessel has generally been the terminal's highest priority; the time spent by a vessel at berth is known as vessel turn time. Many studies have focused on improving the vessel turn time by addressing issues associated with berth scheduling, quay crane scheduling, stowage planning and sequencing, storage activities, and allocation and dispatching of yard cranes and transporters (Kaveshgar et al., 2012; Lu et al., 2012). The study of the operations within container terminals can identify different optimization problems which influence the quality of service of these terminals; metaheuristic algorithms have been widely applied in terminal operation optimization.

The problems identified in the port terminals are classified according to the components involved in them. These components are the containers where merchandise is stored, the terminal areas destined for storage of these containers, the equipment required for transportation and location within the terminal, the vessels where containers will be transported, and quay cranes responsible for loading and unloading containers from vessels.

A key part of the operations carried out in the Port Terminal is the movement of containers. This process starts when a container properly labeled, enters in the installations of the company. Later it is located in a yard, and after that into a vessel. The loading and unloading of containers that are made in the quay, where vessels, cranes and personnel required for these processes are located, are operations that require more attention and care because they have the greatest cost in a port terminal.

A port terminal has mainly five areas (see Figure 1): truck operations area, containers yard, warehousing yard, and the area for vessel operations with empty container. In Figure 1, we can identify three types of operations: (1) ground operation, (2) movements in yard and (3) operation in the quay. The processes addressed in this work are related to operations that take place in yard and quay.

The management of the operations required to service a vessel begins with the arrival of the containers to the terminal, this generates a Container Positioning Problem (CPP). The containers do not have an order of arrival and must be stored in such a way that minimizes the operations of gantry cranes. These cranes stow the containers onto trucks that carry them to the quay (Ünlüyurt & Aydın, 2012).

Once the process of accommodation of containers in the yard has been completed, they are waiting for the arrival of the corresponding vessel. Port terminals only have a very limited space to attend vessels simultaneously. When the arrival of more vessels than can be attended a Berth Allocation Problem (BAP) is generated. Solving BAP requires knowledge of certain characteristics of the vessel as well as the distribution of loading and unloading tasks of containers. This is evaluated according to the amount of available quay cranes.

Planning the stowage of a vessel is a two-step process. The first step is executed by the shipping company. In the second step, the containers by category are assigned to specific positions within the boat. This problem is known as Container Ship Stowage Problem (CSSP), and only recently Stowage Planning (Ding & Chou, 2015).

Figure 1. Container flows at a port terminal

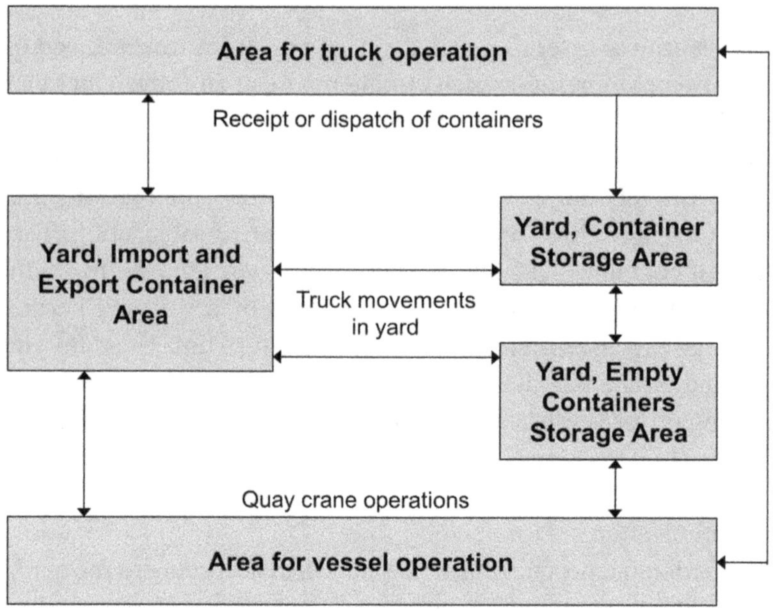

Quay cranes are critical resources in container terminals, and their operating efficiency has direct and indirect impacts on the throughput of container terminal; they are the most expensive equipment at terminals, and often represent the bottleneck on throughput. When a vessel is berthed, a certain number of quay cranes are designated to serve this vessel, i.e., to perform container unloading and loading operations. Containers to be handled are currently stacked or are to be stacked at different locations on vessel, and quay cranes move along the quay on rails to convey these containers (Lu et al., 2012).

The assignment of n cranes to arriving vessels is known as the Quay Crane Assignment Problem (QCAP). These cranes being assigned to the vessel require scheduling to meet the requirements for loading and unloading tasks (Raa et al., 2011).

Once the scheduling of loading and unloading tasks to be carried on the vessel have been established, it is necessary to obtain an optimal sequence of crane movements that minimizes the vessel's length of stay. This problem is called Quay Crane Scheduling Problem (QCSP), whereby the time required for loading and unloading the vessel is obtained (Kim & Park, 2004).

Problems described above have a NP-hard computational complexity, this characteristic makes impractical to solve them using deterministic algorithms. Several authors propose mathematical programming models to solve problems that arise in the processes of a terminal port. Because the exact solutions are generally limited to instances of small to medium scale, alternative solutions based on approximate algorithms have been proposed. From all the problems, QCSP is considered the most important in view of the high impact on the attention span of a vessel (Meisel, 2011). Considering the complexity of the problem, fast and good quality heuristic solutions are required to be implemented in a real environment. This chapter presents a method that is referred to as HACS-GRASP that improves the time required to obtain good solutions. The proposed algorithm combines a Greedy Randomized Adaptive Search Procedure (GRASP) and an Ant Colony System (ACS) with a repositioning strategy of idle cranes that reduces the interference generated by the quay cranes.

BACKGROUND

The concepts of a container terminal are very wide, so a thorough review to narrow the list was made to only the most relevant issues for this work, without affecting the understanding of the general problematic. Table 1 summarizes the concepts that are part of the everyday language of the staff of a container terminal.

The Processes of a Container Terminal

A container terminal plays an essential role in a port as an intermodal interface, in which container vessels dock on berths, load outbound (export), and unload inbound (import) containers. The main productivity of a container terminal can be measured in terms of two factors: (1) Ship operations where containers are discharged from and onto a ship; and (2) receiving and delivery operations where containers are transferred to and from outside trucks (Kim & Park, 2004).

The loading and unloading operations of containers also called containers logistics is regulated by international standards that specify how the goods should be stored in the containers and how they need to be stacked in a proper place (according to ISO 6346 standard). The main areas of container operations can be seen clearly in Figure 1. North America and Europe follow a standard called Direct Transfer System (DTS). In this standard, the same equipment is used to pick up (put down) a container from (into) the yard, to deliver it to (from) the apron, and to transfer it to (from) a quay crane. DTS requires a large operation area because of the exclusive lanes in the yard that are needed to allow vehicles to have access to the positions of the yard where the containers will be placed. DTS requires a large surface due to the lines in the yard that are needed to allow vehicles to have access to the positions of the yard where the containers will be placed (Cordeau et al., 2007).

The containers are usually built under the ISO 6346 standard. Common dimensions are 20' x 8' x 8' (called a TEU or twenty-foot equivalent unit) or 40' x 8' x 8', called an FEU. Variations are collapsible

Table 1. Basic concepts of a container terminal

Concept	Definition
Container	A box made of aluminum, steel or fiberglass with specific dimensions according to ISO 6346 standards, which is used to transport various goods according to their particular characteristics.
Yard	Area within the container terminal used for stacking containers
Quay Crane	Usually, a rail-mounted gantry crane located on a wharf for the purpose of loading and unloading containers on vessels.
Gantry crane	Track-mounted, shoreside crane utilized in the loading and unloading of breakbulk cargo, containers and heavy lift cargo in the yard of the container terminal.
Consolidated Container	Any container that has within it cargo.
Tare	Specific weight of an empty container.
Vessel	A ship or large boat designed to carry containers.
Container Terminal	A specialized facility where ocean container vessels dock to discharge and load containers, equipped with cranes and booms. Most terminals have direct rail access and container storage areas, and are served by highway carriers.
Container Logistics	All operations related to the transport, loading and unloading of containers.
Bay	Delimited space previously used for the accommodation of containers, see Figure 2.2, 2.3 and 2.4

containers, tank containers (for liquids) and "rag tops" (open-topped containers covered by a tarpaulin for cargo that sticks above the top of a closed box). In the container industry, containers are usually simply called boxes.

The distribution of the containers in the terminal is a primordial part in the operations of the container terminal. Maldistribution of the containers could affect the productivity of the ship operation (Kaveshgar et al., 2012). The containers are accommodated by choosing the appropriated bay for each container, and this is achieve by taking into account the following points: each bay must include containers of the same size and preferably of the same type.

In sequencing the containers, load planners usually attempt to minimize the handling efforts of quay cranes and yard equipment, maintain the stability of the vessel, and satisfy loading requirements that are specified by destination and size of container to be loaded onto each slot in the stowage plan.

It must be considered that the containers have a random income and even with a detailed list of the containers that will arrive or are expected to arrive at the terminal, the containers do not always arrive in a particular order or sequence or may never come. The containers can be grouped into various areas: empty containers, export containers, withheld containers and import containers.

Figure 2 includes an example of the terminal area (yard) where the containers are placed. In the example, there are six rows and seven bays with 3 levels. The bays have a maximum height of containers stacked one above the other, which is restricted by the capacity of the cranes used for stacking. There are no strict rules on the space planning process. However, the next recommendations should be considered to obtain an efficient load sequence. A widely accepted principle to have an efficient load sequence is to increase the density of the tacking (number of containers places one on top of another) and assigning all the containers to yard-bays located near the berthing position of the corresponding vessel (Kim & Park, 2004).

Handling of Containers by Vessel

Vessels just as the yard are divided into bays, rows and levels. Bay corresponds to a section with enough space to be occupied by a 20-feet container. The distribution of spaces on the vessel is detailed in Figure 3. This distribution will consist of two rows or more according to the type of vessel. The number of stacked containers in a row-bay intersection determines the level of a container within the vessel.

Figure 2. View of the yard-bays

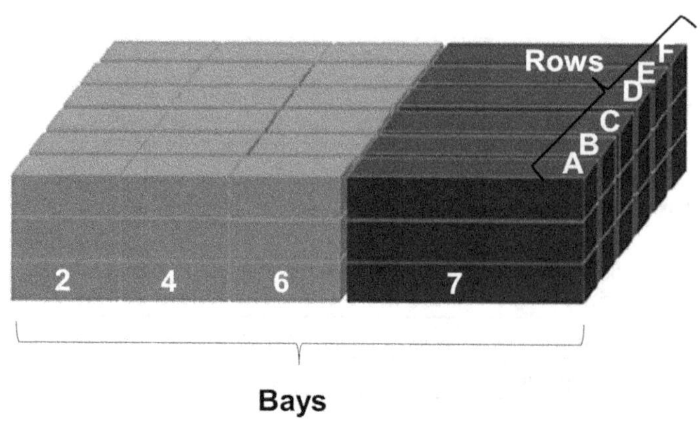

Figure 3. View of the vessel from its top

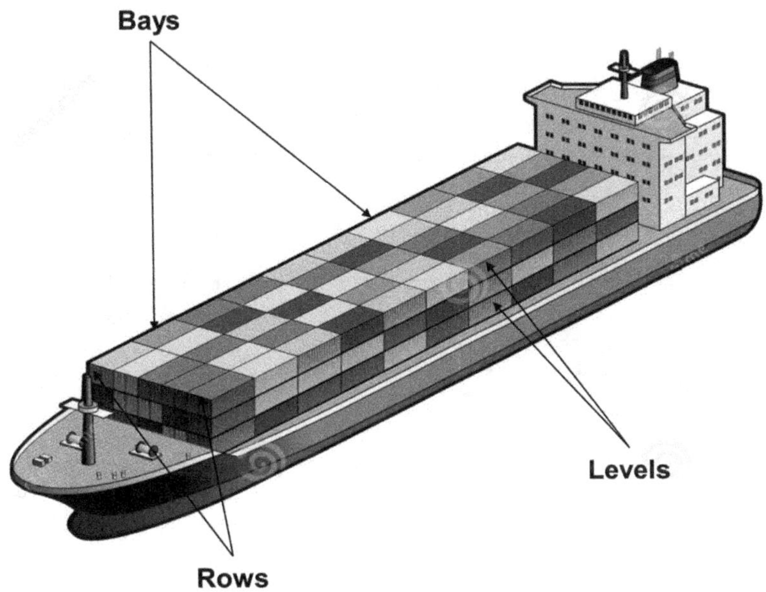

The load must be uniformly distributed respecting the stability of the vessel, whose limit varies according to the specifications provided by the vessel's personnel. The loading configurations must respect the choices of distribution of the containers and the maximum weight that can be placed on one level. It is always recommended to place heavier containers in the lower center of the vessel. The load on the vessel must be downloaded from the closest to the farthest point to the crane, reversing the process for loading.

Containers Logistic Problems

Container logistics covers a wide range of problems that have been modeled by several authors in the literature. An important part of these problems is originated during the passage of the containers through the terminal. In Figure 4 the flow of containers within the areas of a port terminal is detailed.

The container flow that occurs in a port terminal (see Figure 4), starts from the time a container arrives to it. The next step is to take the data from the container and conduct a review of it on the basis of information provided to the shipping agency to identify the destination of the container. Subsequently the yard section where it will be located temporarily the container is determined.

Containers not only arrive by truck operations area. They can be consolidated in store with various goods and introduced into the flow of containers as shown in Figure 4 representing an overall scheme flow of containers within the port terminal. This flow generates various problems described in Table 2.

General Scheme of Solution for Containers Logistic Problems

The main productivity of a container terminal can be addressed with the solution of the five optimization problems shown in Table 3. The methods used to solve the main problems of a container terminal have been raised in the literature in different ways, each of which coincides with the particular characteris-

Figure 4. Analysis of problems in the flow of containers at a port terminal

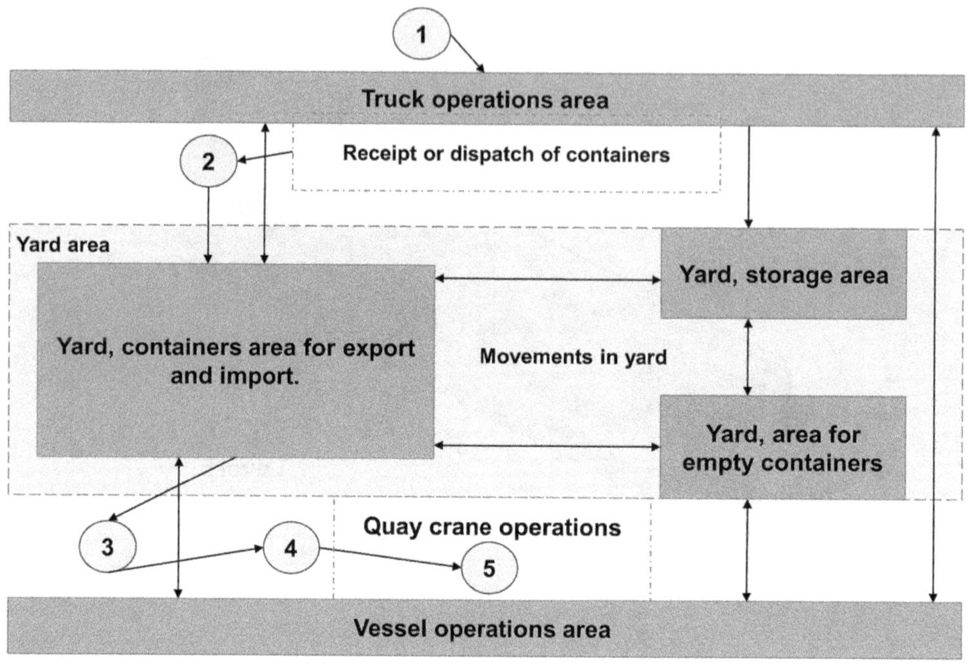

Table 2. Description of the terminal containers flow

Problem	Description	Location
1	Arrival of the containers at the terminal	Truck operation area
2	Location of the containers in the yard, according to their destination area	Terminal yard area
3	Planning containers movements from yard to vessel and its location in the vessel	Terminal yard area
4	Allocation of resources to the attention of vessels. Quay cranes and vessels location.	Vessel operation area
5	Planning sequences of cranes loading	Vessel operation area

tics of the case study or port terminal where the research was conducted. Solution models are based on evaluating only the time required to execute the solutions obtained, leaving for future work measuring the quality of the results in the test problems. Figure 5 presents a solution model that was obtained based on several interviews with the specialized staff of a port terminal, which are regarded within the industry as experts in the field.

This chapter addresses the quay crane scheduling problem (QCSP), which manages the most important equipment in port terminals. This study proposes a hybrid metaheuristic algorithm obtain a good solution of the QCSP in a short amount of time.

Table 3. Problems generated by the flow of containers and restrictions

Problem	Description	Objective	Restriction
CPP	Determination of the Container Positioning in the yard	Minimize set of bays, required for a stack of several containers.	• Container type • Cargo of the container • Container destination
CSSP	Distribution of the containers in the vessel	Generate the loading plan of the vessel	• Container destination • Container size • Container weight • Physical characteristics of the vessel
BAP	Allocation of berth position to vessels	Minimize the time schedule for the loading of a vessel	• Amount of available quays in the port
QCAP	Assignment of quay cranes to each vessel	Assign the amount of resources needed for each vessel to minimize their time loading and unloading	• Amount of available quay cranes
QCSP	Scheduling of quay cranes movements	Minimize the quay cranes movements	• Arrangement of the containers • Space between cranes

Figure 5. General scheme of solution for containers logistic problems

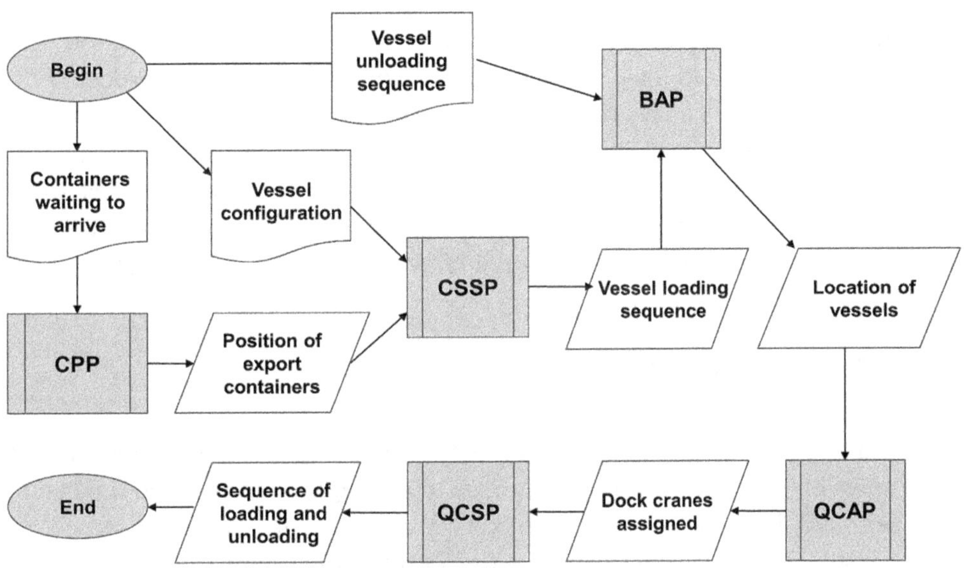

QUAY CRANE SCHEDULING PROBLEM

The planning process of the vessel operation area consists of berth planning, quay crane or work scheduling, and discharge and load sequencing. During the process of berth planning, the berthing time and the berthing position of a container ship on a wharf must be determined. A quay crane schedule specifies the service sequence of bays in a ship by each quay crane and the time schedule for the services (Tavakkoli-Moghaddam et al., 2009). The goal of QCSP is to determine the sequence of discharging and loading operations that a quay crane will perform so that the completion time of a ship operation is minimized (Kim & Park, 2004).

Input data for quay crane scheduling consists of a stowage plan of a ship, the ready time of each quay crane, and a yard map showing the storage locations of containers bound for the ship. Finally, during discharge and load sequencing, the discharge and load sequence of individual containers are determined based on a quay crane schedule (Kim & Park, 2004). QCSP receives: the initial condition of the ship, with their corresponding containers to download, and the output state of the ship containers that are loaded to be sent to other ports. Each of the above parameters includes the specific location of each container. And by way of simplification, it will be given an average loading and unloading time for all containers.

Each task constitutes a discharging or loading operation for a bay of containers and shall be assigned to a specific crane, which will have an average time to perform the task. The correct sequence of tasks allows cranes working in a part of the ship simultaneously, performing various tasks, without having to wait that other cranes complete a task to perform the following. QCSP looks for a complete schedule for the quay cranes serving a vessel, and its decisions include assigning container handling tasks to specific quay cranes, sequencing these tasks on each quay crane, and specifying each task's operation start time. This creates a problem of synchronization or schedule of m parallel machines, whose mathematical model was proposed by Kim and Park (Kim & Park, 2004; Lu et al., 2012).

Loading and unloading operations of containers follow a logical precedence relationship. Specifically, the unloading operation precedes the loading for the tasks located in the same ship bay. Also, for the tasks in the same ship bay the unloading operation must start first with tasks on the deck before preceding to the tasks in the hold (below deck). Conversely, for the loading operation of tasks in the same ship bay, the tasks in the hold must be loaded first before those tasks on the deck. For the QCSP, it is typically assumed that once a quay crane starts to load/unload containers to/from the cluster, it will continue to do so until all slots in the cluster are loaded/emptied. For security reasons, quay cranes move along shared quayside guiding rails and cannot pass one another (non-crossing constraint). Additionally, a safety distance must always be maintained between two adjacent quay cranes whenever they are handling containers or moving along the guiding rail (Kaveshgar et al., 2012; Lu et al., 2012).

PRIOR RESEARCH ON QCSP

Many studies have addressed the QCSP. Daganzo (1989) studied the problem of quay crane assignment among different holds on multiple vessels to minimize total weighted completion time and proposed a principle-based heuristic. In a related study, Peterkofsky and Daganzo (1990) proposed a branch and bound algorithm for solving the static problem with the objective of minimizing total weighted tardiness. These two studies assumed one task per ship-bay and did not take into account the interference between the quay cranes. Kim and Park (2004) formulate the QCSP as a mixed integer programming model and propose a branch and bound method and a GRASP algorithm for solving the QCSP. Moccia et al. (2006) proposed a revised model which incorporated travel times for the quay cranes that subsequently process tasks in the same bay, authors propose a branch-and-cut algorithm to solve large size instances. Sammarra et al. (2007) considered the precedence and non-simultaneity constraints between tasks and proposed a Tabu search heuristic as the solution methodology. Bierwirth and Meisel (2009) developed a heuristic called Unidirectional Scheduling (UDS) based on a tree search which takes advantage of unidirectional movement of the quay cranes and thereby reduced the feasible solution space. Lee et al. (2008) studied QCSP with non-interference constraints and they proposed a genetic algorithm to obtain near-optimal solutions. Tavakkoli-Moghaddam et al. (2008) propose a genetic algorithm (GA) to solve the mixed-integer programming (MIP) model for the quay crane scheduling and assignment problem.

The works described above did not use standard instances for evaluation. Meisel and Bierwirth (2011) proposed a unified approach for evaluating the performance of different model classes and solution procedures, resulting in a generator that allows to reproduce experiments and have a frame of reference. Authors proposed a set of benchmark instances with different values for number of quay cranes, number of tasks, task location, and safety margin between the quay cranes. These benchmark instances can be downloaded from http://prodlog.wiwi.unihalle.de/qcspgen along with a document providing the computational results for the each instance. In addition to the benchmark, they proposed a Branch and Bound scheme considered the best of state of the art for the proposed benchmark.

Chung and Choy (2012) propose a Genetic Algorithm that applies several crossover and mutation operators to reach high-quality schedules. Legato et al. (2012) presented a model that incorporates ready times and due dates for the quay cranes and crane-individual processing times and developed an extension of the UDS that incorporates new bounds and branching criteria. Expósito-Izquierdo et al. (2013) proposes a hybrid Estimation of Distribution Algorithm with local search that includes a priori knowledge about the problem to reach promising regions of the search space.

In published experiments, the execution times are of the order of seconds to hours for small and medium sized instances, respectively. Currently, the industry requirements are up to a maximum of approximately five minutes. To address these requirements, this chapter proposes a hybridization of a Greedy Randomized Adaptive Search Procedure (GRASP) with an Ant Colony System (ACS); this hybridization is called HACS-GRASP. The hybridization of the two heuristics and a waiting crane repositioning that reduces the interference generated by the quay cranes, improve the time required to obtain a good solution.

HYBRID METAHEURISTIC ALGORITHM FOR QCSP

The proposed solution model is based on the mathematical model proposed by Kim and Park (2004). The model defines a ''task'' as a discharging or loading operation for a cluster and assumes that once a quay crane starts to load (or discharge) containers into (from) a cluster of slots, it continues to do so until all the slots in the cluster become filled (empty). Therefore, the model considers handling work for a cluster to be a task. The QCSP addresses scheduling multiple quay cranes to perform discharging and loading operations in order to minimize the makespan of the operation (Kim & Park, 2004).

The proposed algorithm receives, as input data, an instance such as that in Table 4. In each input instance, the tasks are sorted in ascending order, including the time required to perform each task and the location of the task according to its bay number. The instances also include a group of restrictions, where Φ is the set of ordered pairs of tasks between which there is a precedence relationship (when i must be done before j it is said that $(i, j) \in \Phi$), Ψ is the set of pairs of tasks that cannot be performed simultaneously and K includes the quay cranes sorted in ascending order according to their position in the corresponding bay. This information is used to determine the sequence of loading that requires the shortest possible time to perform all tasks.

The instance in the Table 4 includes ten tasks. Each crane can select one task at a time. When all the tasks are selected a solution sequence is generated. There is parallelism in performing the tasks when

Table 4. Input instance for the QCSP

Task	1	2	3	4	5	6	7	8	9	10
Required time	10	5	5	3	6	10	8	4	10	7
Location	1	3	3	3	5	1	3	3	5	7
Restrictions	$\text{Ö} = \langle\{6,1\},\{7,8\},\{8,4\},\{3,2\},\{9,5\}\rangle$, $\text{Ø} = \langle\varnothing\rangle$, $\text{K} = \{1,7\}$									

each crane select a task in a given time. For example, crane k^1 can select the tasks 1, 2, 3, 4 and 5; while crane k^2 can select the other tasks 6, 7, 8, 9 and 10; generating the following sequence.

$k^1 = 1, 2, 3, 4, 5$

$k^2 = 6, 7, 8, 9, 10$

By joining the two groups the work sequence {1,2,3,4,5,6,7,8,9,10} is obtained, which, even is not feasible (because of the order of the tasks), it is one of the possible solutions that can be generated. The *n* tasks to be performed generate *n* factorial combinations of possible solutions. For the instance Table 4 there are 3,628, 800 probable combinations, of which only a few are feasible.

Kim and Park propose a Branch and Bound method to generated feasible solutions, limiting the amount of selectable task for each crane according to their current position and the restrictions on their model (Kim & Park, 2004). In the instance of Table 4 Crane k^1 can choose between the task 6, 7, 9 or 10 which meet certain restrictions of the model. However, tasks 10 breaks the constraints of the position of the crane because it is located in the same bay as the crane k^2, for that reason task 10 cannot be held. By this simple method crane k^1 will randomly select a task for the set {6, 7, 9}. Selecting this task will update its current location, which will change the tasks available to the other cranes.

The selection of the tasks for the cranes is done as follows. All cranes will select an initial task from the set of potential solutions respecting the set of constraints. Once all cranes have an initial task, the crane with the shortest time of completion is selected. A new feasible task is assigned to the selected crane and the time of that task is subtracted to the pending tasks. The process continues until all tasks have been assigned. The generated solution will indicate the operations to be performed by each crane. A feasible solution to the above example would be as follows:

$k^1 = \{6,1,7,8,4,3,2\}$

$k^2 = \{9,5,10\}$

Table 5 shows the sequence of movements generated for the above solution. The first column represents the minimum time of completion of the tasks assigned to the quay cranes. The next columns include, for each quay crane, the sequence of tasks, bays and the required time to finish the corresponding task. The time required to perform all the tasks is the sum of the minimum times generated in the solution. In the example, the solution requires 47 units of time.

Table 5. Solution for the QCSP instance of Table 4.

Minimum Time	Quay Cranes					
	k^1			k^2		
	Task	Bay	Time	Task	Bay	Time
7	6	1	10	10	7	7*
3	6	1	3*	5	5	6
3	1	1	10	5	5	3*
7	1	1	7*	5	5	9
2	1	1	0	5	5	2*
8	7	2	8*	0	7	0
4	8	2	4*			
3	4	2	3*			
5	3	2	5*			
5	2	2	5*			
TOTAL: 47	No more tasks available					

A Branch and Bound Method

Kim and Park (2004) proposed a Branch and Bound (B & B) method to obtain the optimal solution of the QCSP. In the Branch and Bound method a solution is represented as a sequence of all tasks. Once a sequence of all tasks is given, a QC schedule can be constructed by assigning tasks to QCs in such a way that the first unassigned task in the sequence is assigned to the QC with the earliest completion time for already assigned tasks. The method explores a solutions tree from the root 0, to the nodes in the next levels, where:

- The depth of the tree is the number of tasks
- A node at level *i* of the tree corresponds to a partial solution consisting of the first task to the task *i*.

Variables in the root node of the tree begin with the following values:

$$\forall k, W_k = \{\varnothing\}, S_k = (\ ), c_k = 0, l_k^c = l_k^0$$

where W_k is the set of tasks assigned to the crane *k*, S_k is the sequence of tasks assigned to the crane *k*, c_k is the completion time of the crane *k* at the current node in the tree, and l_k^c is the position of the crane *k* after the last task in S_k of the current node *c* is performed.

At each node a solution is represented as $\{S_k\}$; $1 \le k \le |K|$. The overall procedure is shown in Figure 6. A detailed explanation for each step can be found in the work of Kim and Park (2004).

The B & B method provides the optimal solution within a short computational time when the number of tasks is relatively small. However, the computational time increases rapidly when the problem size increases, which makes the B & B method difficult to use in practice. To address this problem, we

Figure 6. The overall procedure of the Branch and Bound method of Kim and Park (2004)

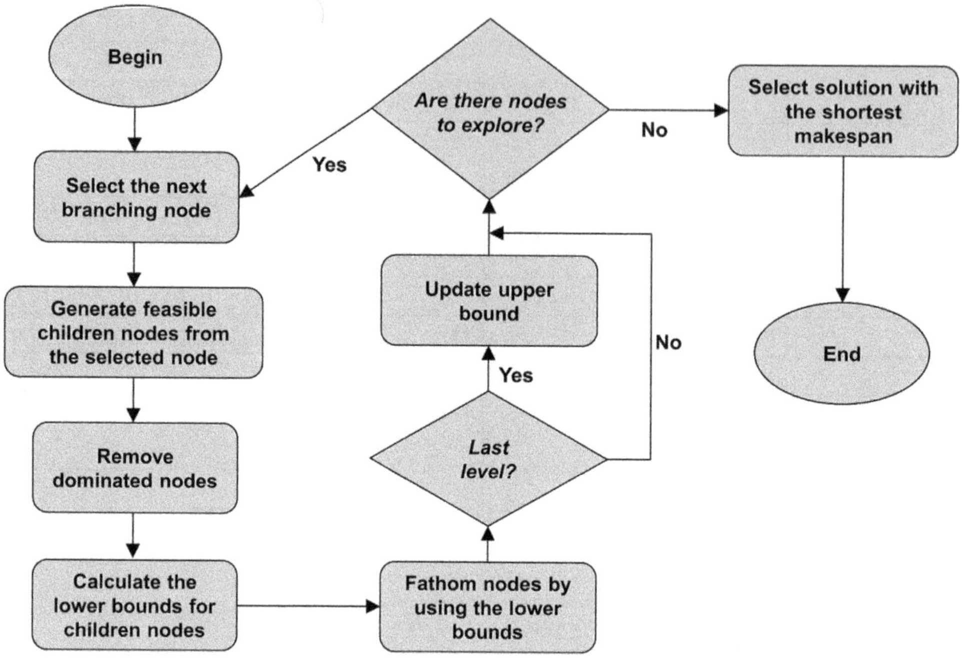

propose a hybridization of GRASP (Greedy Randomized Adaptive Search Procedure) with ACS (Ant Colony System); this hybridization is called HACS-GRASP.

A Hybrid Algorithm of ACS with GRASP

Kim and Park (2004) proposed a greedy randomized adaptive search procedure (GRASP) that generates a set of feasible tasks F for each available quay crane k^i. F is generated by obtaining the tasks accessible to k^i that do not violate any of the restrictions of precedence or simultaneity of the instance. Also, all the tasks $j \in F$ for which the following condition (Equation 1) holds are also excluded from F:

$$1 / \left| l_i^c - l_j \right| < r\max_{l \in F} \left\{ 1 / \left| l_i^c - l_l \right| \right\}, \quad (1)$$

where r is a constant between 0 and 1.

Such pruning is based on the distance, measured in number of bays, between the current position of the quay crane k^i and the location of the task j. F is formed by all those tasks whose probability is in the defined threshold constant r. Subsequently, each task $j \in F$ is associated with the following selection probability (Equation 2):

$$\left(1 / \left| l_i^c - l_j \right|\right) / \sum_{l \in F} \left(1 / \left| l_i^c - l_l \right|\right) \quad (2)$$

The selection of each task is made in a random way based on the probability distribution defined by Equation 2. Each time that a task j is selected the position of the quay crane k^i and the time when it will be available are updated. Due to security restrictions in certain periods of times any task will be assigned to a quay crane k^i; this happens when there is no work available on the location of the quay crane or when an there is an interference with other crane. A detailed explanation of this GRASP procedure can be found in the work of Kim and Park (2004).

The interference between quay cranes can avoid creating a feasible solutions. To solve this problem, reported by Meisel (2011), HACS-GRASP includes a simple heuristics that avoids interference by re-positioning the cranes to a feasible location that allow, a in a subsequent motion, to have the greatest number of feasible tasks available. This step is performed when $c_i = 0$ and $F = \{ \varnothing \}$. The selection of tasks and the repositioning procedure are made until all tasks are assigned.

Kim and Park (2004) proposed a phase of improvement based on the exchange of tasks, this phase is not implemented in HACS-GRASP because it represents a high computational cost. Instead, the solutions obtained by GRASP are improved using a traditional ant colony system (ACS). The ACS generates a first set of random solutions by means of the GRASP procedure described above. Subsequently, each ant of the colony uses the same GRASP procedure, replacing the function of randomness by the pheromone.

The HACS-GRASP procedure if described next. One solution S to QCSP is represented as a two-dimensional table, of size n x q, where n is the number of tasks to perform and q is the number of quay cranes that are available. The ant colony will consist of n ants, each ant will be a solution generated by the GRASP procedure. This size of the colony is n because in the worst case only a quay crane will perform all tasks of the instance.

```
Procedure HACS-GRASP
Output: S
P ← GRASP();
F ← update_ pheromone(P);
P ← sort(evaluate(P));
S ← best_solution(P);
while stop condition do
P ← new_colony();
P ← sort(evaluate(P));
S' ← best_solution(P);
if S' ≤ S then
S ← S';
end;
update_ pheromone(P);
end;
return S;
end HACS-GRASP.
```

The pheromone table have a size of n x q x n. This size is used to validate all possible moves of the ants. HACS standard parameters, both evaporation and exploration and development were used. The exploration is performed by randomly selecting a task in the corresponding quay crane and performing a tour in the solutions tree starting from the selected node until complete the solution.

The pheromone table is generated using the first generation of solutions generated by GRASP. This table enables each ant to select a task form the set of potential solutions, selecting the one which has a higher value in the table. It updates its state with every move that the ant performs, if the solution generated is feasible, a booster is made in the table, otherwise it continues with the evaporation of the pheromone.

The procedure iterates the maximum number of generations, *n*, and it stops when, after *n* /2 generations it comes to a deadlock in the solution.

Results from the HACS-GRASP for QCSP

The HACS-GRASP algorithm was configured with a colony of size 100 (where an ant per task is generated in each colony), an evaporation rate $T^0 = 0.00009$, and an exploration/exploitation rate of $q = 0.65$. To measure the performance of the strategy, the algorithm was run over a time limit of 360 seconds per instance, using the set of instances proposed by Meisel (2011).

Through the QCSPgen tool by Meisel (2011), each of the instances shown in Table 6 was generated.

Table 7, Table 8, Table 9, Table 10, Table 11, Table 12, and Table 13 are a summary of the results obtained for each group of instances, where it is shown that HACS-GRASP improves the required time by the state-of-the-art algorithm of Bierwirth and Meisel (2009).

Table 6. Combinations used to generate the set of instances

Set	n	b	c	*f*	Loc	d	q	s	#inst
A	[10,15,...,40]	10	200	0.5	uni	1	2	1	70
B	[45,50,...,70]	15	400	0.5	uni	1	2	1	60
C	[75,80,...,100]	20	600	0.5	[cl 1,cl 2, uni]	1	4	1	60
D	50	15	400	[0,2,0,8]	uni	1	6	1	60
E	50	15	400	0.5	uni	[0.80,0.85,...,1]	4	1	50
F	50	15	400	0.5	uni	1	[2,3,...,6]	1	50
G	50	15	400	0.5	uni	1	4	[0,1,...,4]	50

Table 7. Obtained Results for Group A

Group	HACS-GRASP		Meisel B&B	
	Result	Time	Result	Time
A1	5225	360	5164	1
A2	5117	360	5099	1
A3	5137	360	5081	1
A4	5144	360	5070	1
A5	5151	360	5068	1
A6	5181	360	5071	1
A7	5199	360	5063	1

Table 8. Obtained Results for Group B

Group	HACS-GRASP		Meisel B&B	
	Result	Time	Result	Time
B1	8361	360	7758	713
B2	8372	360	7709	1,251
B3	8633	360	7719	1,078
B4	8247	360	7711	1,317
B5	8335	360	7690	2,118
B6	8360	360	7621	2,231

Table 9. Obtained Results for Group C

Group	HACS-GRASP		Meisel B&B	
	Result	Time	Result	Time
C1	12202	360	11429	3,600+
C2	12156	360	11204	3,248
C3	12130	360	10847	3,600+
C4	11933	360	10689	3,600+
C5	11804	360	10833	3,600+
C6	11853	360	10865	3,600+

Table 10. Obtained Results for Group D

Group	HACS-GRASP		Meisel B&B	
	Result	Time	Result	Time
D1	5001	360	4750	3,240
D2	3956	360	3676	2,212
D3	3773	360	3495	2,100
D4	12806	360	12137	7
D5	12954	360	12091	508
D6	12870	360	12090	430

Table 11. Obtained Results for Group E

Group	HACS-GRASP		Meisel B&B	
	Result	Time	Result	Time
E1	8352	360	7690	3,306
E2	8371	360	7693	1,958
E3	8307	360	7697	870
E4	8332	360	7697	481
E5	8386	360	7709	41

Table 12. Obtained Results for Group F

Group	HACS-GRASP		Meisel B&B	
	Result	Time	Result	Time
F1	15316	360	15097	11
F2	10422	360	10084	199
F3	8387	360	7709	1248
F4	7624	360	7373	2892
F5	7487	360	7361	3600+

Table 13. Obtained Results for Group G

Group	HACS-GRASP		Meisel B&B	
	Result	Time	Result	Time
G1	8092	360	7584	2,440
G2	8297	360	7709	1,248
G3	10118	360	9954	3,252
G4	12549	360	12399	3,600+
G5	14275	360	14054	3,600+

Statistical Tests

Using the system SAS 2009, it was performed a Wilcoxon test. The results are illustrated in Table 14, showing, with a significance level defined by α=0.05 and a p-value=0.06, that there is no significant difference, i.e. the null hypothesis about similarity in the performance of the algorithms is not rejected.

Even though the evaluated algorithms are statistically similar in quality, the proposed algorithm uses only 8% of the time reported by the approach reported by Meisel (2011), which represents an advantage in real environments. Such a running time allows extra time for a decision maker to choose among several options, when dealing with resources to each of the attended vessels.

Finally, due to the running time is the indicator that commonly the algorithms use as evaluation strategy in the process of loading a vessel, it can be considered the factor with the greatest impact in the cost of the overall logistic process involving the containers.

Table 14. Results derived from the Wilcoxon test

Wilcoxon Two-Sample test
Statistic 1782
Normal Approximation Z 1.5541 One-Sided Pr > Z 0.0601 Two-Sided Pr > |Z| 0.1202
t Approximation One-Sided Pr > Z 0.0621 Two-Sided Pr > |Z| 0.1241

CONCLUSION

Container terminals are important nodes for global freight transportation. It is vital for the success of a terminal to effectively utilize its resources to achieve high productivity and customer satisfaction. To help in the improvement of the container terminals process, many studies have applied operations research models to optimize processes within container terminals. Among the processes within a container terminal, the Quay Crane Scheduling Problem (QCSP) is a critical problem because the efficiency in the quay cranes operations has direct and indirect impacts on the final performance of the terminal.

This chapter presented a description of the principal problems that are part of a container terminal operation, focusing on the QCSP and the methods that have been proposed for improves its solution. This study introduced a hybridization of a Greedy Randomized Adaptive Search Procedure (GRASP) with an Ant Colony System (ACS) and a waiting crane repositioning that reduces the interference generated by the quay cranes. The proposed method, referred to as HACS-GRASP, allows to address the industry time requirements by improving the time required to obtain a good solution.

ACKNOWLEDGMENT

This work has been partially supported by CONACYT Project 254498 and Prodep.

REFERENCES

Bierwirth, C., & Meisel, F. (2009). A fast heuristic for quay crane scheduling with interference constraints. *Journal of Scheduling*, *12*(4), 345–360. doi:10.1007/s10951-009-0105-0

Chung, S. H., & Choy, K. L. (2012). A modified genetic algorithm for quay crane scheduling operations. *Expert Systems with Applications*, *39*(4), 4213–4221. doi:10.1016/j.eswa.2011.09.113

Cordeau, J. F., Gaudioso, M., Laporte, G., & Moccia, L. (2007). The service allocation problem at the Gioia Tauro maritime terminal. *European Journal of Operational Research*, *176*(2), 1167–1184. doi:10.1016/j.ejor.2005.09.004

Daganzo, C. F. (1989). The crane scheduling problem. *Transportation Research Part B: Methodological*, *23*(3), 159–175. doi:10.1016/0191-2615(89)90001-5

Ding, D., & Chou, M. C. (2015). Stowage planning for container ships: A heuristic algorithm to reduce the number of shifts. *European Journal of Operational Research*, *246*(1), 242–249. doi:10.1016/j.ejor.2015.03.044

Expósito-Izquierdo, C., González-Velarde, J. L., Melián-Batista, B., & Moreno-Vega, J. M. (2013). Hybrid estimation of distribution algorithm for the quay crane scheduling problem. *Applied Soft Computing*, *13*(10), 4063–4076. doi:10.1016/j.asoc.2013.05.006

Kaveshgar, N., Huynh, N., & Rahimian, S. K. (2012). An efficient genetic algorithm for solving the quay crane scheduling problem. *Expert Systems with Applications*, *39*(18), 13108–13117. doi:10.1016/j.eswa.2012.05.091

Kim, K., & Park, Y. (2004). A crane scheduling method for port container terminals. *European Journal of Operational Research*, *156*(3), 752–768. doi:10.1016/S0377-2217(03)00133-4

Lee, D.-H., Wang, H. Q., & Miao, L. (2008). Quay crane scheduling with noninterference constraints in port container terminals. *Transportation Research Part E, Logistics and Transportation Review*, *44*(1), 124–135. doi:10.1016/j.tre.2006.08.001

Legato, P., Trunfio, R., & Meisel, F. (2012). Modeling and solving rich quay crane scheduling problems. *Computers & Operations Research*, *39*(9), 2063–2078.

Lu, Z., Han, X., Xi, L., & Erera, A. L. (2012). A heuristic for the quay crane scheduling problem based on contiguous bay crane operations. *Computers & Operations Research*, *39*(12), 2915–2928. doi:10.1016/j.cor.2012.02.013

Meisel, F., & Bierwirth, C. (2011). A unified approach for the evaluation of quay crane scheduling models and algorithms. *Computers & Operations Research*, *38*(3), 683–693. doi:10.1016/j.cor.2010.08.001

Moccia, L., Cordeau, J. F., Gaudioso, M., & Laporte, G. (2006). A branch-and-cut algorithm for the quay crane scheduling problem in a container terminal. *Naval Research Logistics*, *53*(1), 45–59. doi:10.1002/nav.20121

Peterkofsky, R. I., & Daganzo, C. F. (1990). A branch and bound solution method for the crane scheduling problem. *Transportation Research Part B: Methodological*, *24*(3), 159–172. doi:10.1016/0191-2615(90)90014-P

Raa, B., Dullaert, W., & Van Schaeren, R. (2011). An enriched model for the integrated berth allocation and quay crane assignment problem. *Expert Systems with Applications*, *38*(11), 14136–14147.

Sammarra, M., Cordeau, J. F., Laporte, G., & Monaco, M. F. (2007). A tabu search heuristic for the quay crane scheduling problem. *Journal of Scheduling*, *10*(4-5), 327–336. doi:10.1007/s10951-007-0029-5

Tavakkoli-Moghaddam, R., Makui, A., Salahi, S., Bazzazi, M., & Taheri, F. (2009). An efficient algorithm for solving a new mathematical model for a quay crane scheduling problem in container ports. *Computers & Industrial Engineering*, *56*(1), 241–248. doi:10.1016/j.cie.2008.05.011

Ünlüyurt, T., & Aydın, C. (2012). Improved rehandling strategies for the container retrieval process. *Journal of Advanced Transportation*, *46*(4), 378–393. doi:10.1002/atr.1193

Chapter 13
Temperature Modeling of a Greenhouse Environment

Sergio Steve Juárez-Gutiérrez
Universidad Politécnica de Aguascalientes, Mexico

Tania Aglaé Ramírez delReal
Universidad Politécnica de Aguascalientes, Mexico

Araceli Gárate-García
Universidad Politécnica de Aguascalientes, Mexico

Ervin Jesús Alvarez-Sánchez
Universidad Veracruzana, Mexico

ABSTRACT

This chapter presents the methodology to model the temperature inside a zenith greenhouse, without crop, using a Wireless Sensor Network (WSN) and the Recursive Least Square (RLS) Technique to estimate the modeling parameters. The greenhouse size is 6mx18m, the actuators are the motors located at the windows. The WSN collects the measurements and communicates the centralized control and the actuators located in different parts of the greenhouse, it is implemented using National Instruments devices, the graphical interface is developed in LabVIEW system design. There exist a lot of work about greenhouses in the literature; however, most of them did not use mathematical techniques to model the temperature or other environment parameter. Surprisingly, just few works use RLS techniques to estimate the parameters. The main contribution of this project is to integrate the National Instruments technology and the RLS techniques in a real experimental application.

INTRODUCTION

Several reasons motivate the continuous interest and development in protected agriculture around the world, one of them is the impact of climate change, which affects the open-field agricultural activities, as well as the dynamics and competitiveness of marketing systems (Moreno et al, 2011). For several countries, the adoption of new technologies has allowed to increase production and availability of vegetables for population, at affordable prices during different year seasons, which is possible by means of controlling environmental requirements of the crops.

DOI: 10.4018/978-1-4666-9779-9.ch013

The common parameters to fulfill the ideal growth of the crop are temperature, relative humidity, soil moisture, ventilation, light and solar radiation, carbon dioxide, oxygen and fertigation, amongst others. It is possible to control them effectively in a greenhouse; however, a mathematical model is required to implement most of the control techniques (Leal et al, 2006).

The modeling and control of weather parameters inside a greenhouse has been studied in different parts of the world, such as: Almería, España (Díaz, 2002), Mexico (Leal et al, 2006), Israel (Seginer et al., 1993) and Japan (Morimoto & Hashimoto, 2000). An emphasis must be done in the studies of the Wageningen's University (Udink Ten Cate 1983, Tap, 2000), since Holland is the principal producer of Central Europe.

This topic has been studied since the 60's with the work of (Businger, 1963), where it was suggested to use methods of energy balance to model the greenhouses behavior, however the existing models were useful in stable conditions of the climate, after that (Takakura et al, 1971) started to studied the subject using Fourier series. The work of (Udink Ten Cate, 1983) was one of the first to include a mathematical model considering the temperature inside the greenhouse, and it was improve by (Tchamitchian et al, 1992) considering the temperature changes in the greenhouse pipes.

There exist several literature about mathematical models for greenhouses based on different techniques, such as: (Gupta & Chandra, 2002, Lafont & Balmat, 2002, Ghosal & Tiwari, 2004, Luo et al, 2005, Salgado & Cunha, 2005, Abdel-Ghany & Kozai, 2006, Singh et al, 2006, Álvarez-Sánchez et al, 2014). Some works, like (Tap, 2000), consider several parameters to model the greenhouse conditions and have been the basis for other researches. Unfortunately, these models are generally obtained in an empirical form, and considering only one kind of greenhouse.

Moreover, several works in the last decade consider the automation of greenhouses, they are called intelligent, and they use neuronal networking, fuzzy control or similar techniques (see for instance, Laribi et al, 2006, Salazar et al, 2007, Ferreira & Ruano, 2008, Mohammad et al, 2010, Pengzhan & Baifen, 2010, Ma et al, 2010, Eredics & Dobrowiecki, 2011, Eredics et al, 2011). However, artificial networking has the inconvenient that the internal variables did not have necessarily a physical interpretation (Ferreira & Ruano, 2008).

Most of the automated greenhouses use wireless technologies, like the zigbee protocol (see Qiu et al, 2014 and references included), or mainly the Wi-Fi protocol (Lakshmi et al, 2007, Kang et al, 2008, Pawlowski et al, 2008, Ahonen et al, 2008, Chuanan & Yongchang, 2010, Fezari et al, 2011, Xiaoyan et al, 2013, Mad et al, 2014). Wired protocols are implemented in minor quantity, though they can be observed in works as (Du et al, 2013, and Yulong and Jiaqiang, 2011) which uses the CAN-bus protocol and the RS485 bus, respectively.

Furthermore, there are just few works with remote monitoring system that use the National Instruments® technology, or LabVIEW™ as system design software (see for example, Guofang et al, 2010; Fang & Wang, 2011, and references therein). This work integrate the National Instruments® technology and the Recursive Least Square (RLS) techniques in a real experimental application.

The proposal of this chapter is to:

1. Estimate the parameters of the interior temperature of a classic zenithal greenhouse with natural ventilation to know his behavior without crop, using the RLS method.
2. Integrate the Wireless Sensor Network (WSN) devices of National Instruments®, like the NI WSN-3202 analog input node and the NI WSN-9791 ethernet gateway, to collect the measurements and the graphical interface to manipulate them, which was developed in LabVIEW™ system design.

3. Obtain the mathematical model of the temperature changes in the greenhouse, using the RLS technique, analyzing the samples obtained by the system.

A general description of the considered greenhouse, the problem statement and some preliminaries about the RLS method are given in the following section. More details can be seen in Juárez-Gutiérrez (2015).

BACKGROUND

This section states definitions and the mathematical background, as well as the system considered throughout this work.

General Greenhouse Description

The dynamical system is a zenith greenhouse located at the Universidad Politécnica de Aguascalientes in Mexico. The greenhouse size is 6x18 m., the structure used galvanized Rectangular Hollow Section (RHS) steel pipes covered by ultraviolet treated white plastic film and anti-aphids mesh on the windows, see Figure 1.

The measurements of the temperature inside the greenhouse will be obtained using the technology for WSN of National Instruments® with devices like the NI WSN-3202 analog input node, the NI WSN-

Figure 1. Zenith greenhouse

9791 ethernet gateway, the LM35 sensors of Texas Instruments®, and the graphical interface developed in LabVIEW™ system design. The WSN collects the measurements and communicate the centralized control with the actuators and sensors located in different parts of the greenhouse.

The Least Square Method

The concept of the least squares regression is associated with Legendre, who was the first to publish the theory in 1805 and was the first who used the term least squares. However, it was Gauss who developed the method as a statistical tool incorporating least squares in a context in which a probabilistic treatment observation error was given (Solera, 2003).

Gauss indicates that "The sum of the squares of the differences between the actually observed and the computed values, multiplied by numbers that measure the degree of precision, is a minimum" (Aström, 2008).

The least-squares method is very flexible so it can be applied to solve a large type of problems. This mathematical model can be written in the form of (1).

$$y(i) = \varphi_1(i)\theta_1^0 + \varphi_2(i)\theta_2^0 + \cdots + \varphi_n(i)\theta_n^0 = \varphi^T(i)\theta^0,$$

where y(i) is the observed variable, θ_1^0, θ_2^0 … θ_n^0, are parameters of the model to be determined, and φ_1, φ_2 … φ_n are known functions that may depend on other known variables.

The vectors (2) and (3):

$$\varphi^T(i) = \left[\varphi_1(i), \varphi_2(i), \ldots, \varphi_n(i)\right],$$

$$\theta^0 = \left[\theta_1^0, \theta_2^0, \ldots, \theta_n^0\right].$$

The model is indexed by the variable *i*, which in this case denotes time. It will be assumed initially that the index set is a discrete set. The variables φ_i are called the regression variables or the regressors, and the model in (1) is also called a regression model (Aström, 2008).

The least square problem is to find values for (2) so that the obtained parameters to achieve the output of (1) is as close as possible to the output of the system, this is achieved by minimizing the cost function:

$$V(\theta, t) = \frac{1}{2}\sum_{i=1}^{t}\left(y(i) - \varphi^T(i)\theta\right)^2.$$

Due to the linearity of the parameters θ and the criterion of least squares is quadratic, the problem allows an analytical solution, having:

$$Y(t) = \left[y(1), y(2), \ldots, y(t)\right]^T,$$
$$E_r(t) = \left[\varepsilon(1), \varepsilon(2), \ldots, \varepsilon(t)\right]^T,$$
$$\Phi(t) = \begin{bmatrix} \varphi^T(1) \\ \vdots \\ \varphi^T(t) \end{bmatrix}^T,$$
$$P(t) = \left[\Phi^T(t)\Phi(t)\right]^{-1} = \left(\sum_{i=1}^{t} \varphi(i)\varphi^T(i)\right)^{-1}.$$

Where the residuals $\varepsilon\left(i\right)$ are defined by:

$$\varepsilon(i) = y(i) - \hat{y}(i) = y(i) - \varphi^T(i)\theta.$$

Then (4) can be written by:

$$V(\theta, t) = \frac{1}{2}\sum_{i=1}^{t} \varepsilon(i)^{2} = \frac{1}{2} E_r^T E_r = \frac{1}{2}\left\|E_r\right\|^2.$$

where E_r can be written as:

$$E_r = Y - \hat{Y} = Y - \Phi\theta.$$

The solution of the problem of least-squares is then given by the following Theorem:
Theorem 1 (Aström, 2008). The function (4) is minimal for parameters $\hat{\theta}$ such that:

$$\Phi^T \Phi \hat{\theta} = \Phi^T Y.$$

If the matrix $\Phi^T \Phi$ is nonsingular, the minimum is unique and given by:

$$\hat{\theta} = \left(\Phi^T \Phi\right)^{-1} \Phi^T Y.$$

Recursive Least-Square Estimation

Adaptive controllers realized the observations sequentially, in real time. Therefore, it is then desirable to do the computations recursively to save computation time, this concept is applied in the RLS estimation (Aström, 2008).

The RLS estimation is summarized in the following Theorem:

Theorem 2 (Aström, 2008). Assume that the matrix $\phi(t)$ has full rank, that is, $\phi^T(t)\phi(t)$ is non-singular, for all $t \geq t_0$. Given $\hat{\theta}(t_0)$ and $P(t)=\left(\phi^T(t_0)\phi(t_0)\right)^{-1}$, the least-squares estimate $\hat{\theta}(t)$ then satisfies the recursive equations:

$$\begin{aligned}
\hat{\theta}(t) &= \hat{\theta}(t-1)+K(t)\left(y(t)-\varphi(t)^T\hat{\theta}(t-1)\right),\\
K(t) &= P(t)\varphi(t)=P(t-1)\varphi(t)\left(\mathrm{I}+\varphi(t)^T P(t-1)\varphi(t)\right)^{-1},\\
P(t) &= P(t-1)-P(t-1)\varphi(t)\left(\mathrm{I}+\varphi(t)^T P(t-1)\varphi(t)\right)^{-1}\varphi(t)^T P(t-1),\\
&= \left(\mathrm{I}-K(t)\varphi(t)^T\right)P(t-1).
\end{aligned}$$

Returning to equation (10), which is rewritten as

$$\hat{\theta}(t)=\left(\sum_{k=1}^{t}\varphi(k)\varphi^T(k)\right)^{-1}\left(\sum_{k=1}^{t}\varphi(k)y(k)\right)=P(t)\left(\sum_{k=1}^{t}\varphi(k)y(k)\right),$$

$$P^{-1}(t)=\sum_{k=1}^{t-1}\varphi(k)\varphi^T(k)+\varphi(t)\varphi^T(t)=P^{-1}(t-1)+\varphi(t)\varphi^T(t).$$

To estimate time *t*, now is possible write as:

$$\hat{\theta}(t)=\hat{\theta}(t-1)+P(t)\varphi(t)\left(y(t)-\varphi^T(t)\hat{\theta}(t-1)\right), \tag{14}$$

$$\hat{\theta}(t)=\hat{\theta}(t-1)+K(t)\varepsilon(t).$$

where:

$$K(t)=P(t)\varphi(t). \tag{15}$$

$$\varepsilon(t)=y(t)-\varphi^T(t)\hat{\theta}(t-1).$$

The residual $\varepsilon(t)$ can be interpreted as the prediction error signal $y(t)$ a step forward based on the estimated $\hat{\theta}(t-1)$. To proceed, you need to derive a recursive equation $P(t)=P^{-1}(t-1)+\varphi(t)\varphi^T(t)$. Using this equation for $P(t)$, (15) and an algebraic property given in Lemma 2.1 of Aström, 2008, we get

$$K(t) = P(t)\varphi(t) - P(t-1)\varphi(t)\left(\mathrm{I} + \varphi^{T}(t)P(t-1)\varphi(t)\right)^{-1}.$$

Once calculated this, the exponential forgetting factor », can be added in the following Theorem:

Theorem 3 (Aström, 2008). Assume that the matrix $\phi(t)$ has full rank for $t \geq t_0$. The parameter θ, which minimizes the following equation:

$$V(\theta, t) = \frac{1}{2}\sum_{i=1}^{t} \text{»}^{t-i}\left(y(i) - \varphi^{T}(i)\theta\right)^{2},$$

is given recursively by:

$$\hat{\theta}(t) = \hat{\theta}(t-1) + K(t)\left(y(t) - \varphi^{T}(t)\hat{\theta}(t-1)\right),$$

$$K(t) = P(t)\varphi(t) = P(t-1)\varphi(t)\left(\text{»}\mathrm{I} + \varphi^{T}(t)P(t-1)\varphi(t)\right)^{-1},$$

$$P(t) = \frac{\left(\mathrm{I} - K(t)\varphi^{T}(t)\right)P(t-1)}{\text{»}}.$$

The previous theory will be applied in the following sections to estimate the mathematical model of the greenhouse temperature.

Remote Monitoring System

The architecture of a WSN operates with three principal elements: The node of transmission, the Gateway module and the interface. The interaction of these elements, necessary to implement experimentally the methods previously described, is explained in this section (See Figure 2).

The temperature measures are obtained using sensors LM35, they are wired connected to the WSN, and use a 5V power supply. The connection node NI WSN-3202 needs a 12V power supply, it has the capacity to read four analogical inputs and to transmit digital outputs, which are sending through the Wi-Fi protocol to the WSN Gateway NI 9791.

The information is received by the WSN Gateway, connected via Ethernet to a PC and it is manipulated with an interface developed in LabVIEW™ software design, which allows the storage of the date, hour, and value magnitude of each sample in a text file. The interface is shown on Figure 3, besides storing the information, the interface also allows to manipulate the actuators of the greenhouse.

The samples obtained by the installed equipment will be analyzed in the following section to obtain the mathematical model which describes the temperature greenhouse behavior.

Figure 2. Remote Monitoring System interconnections

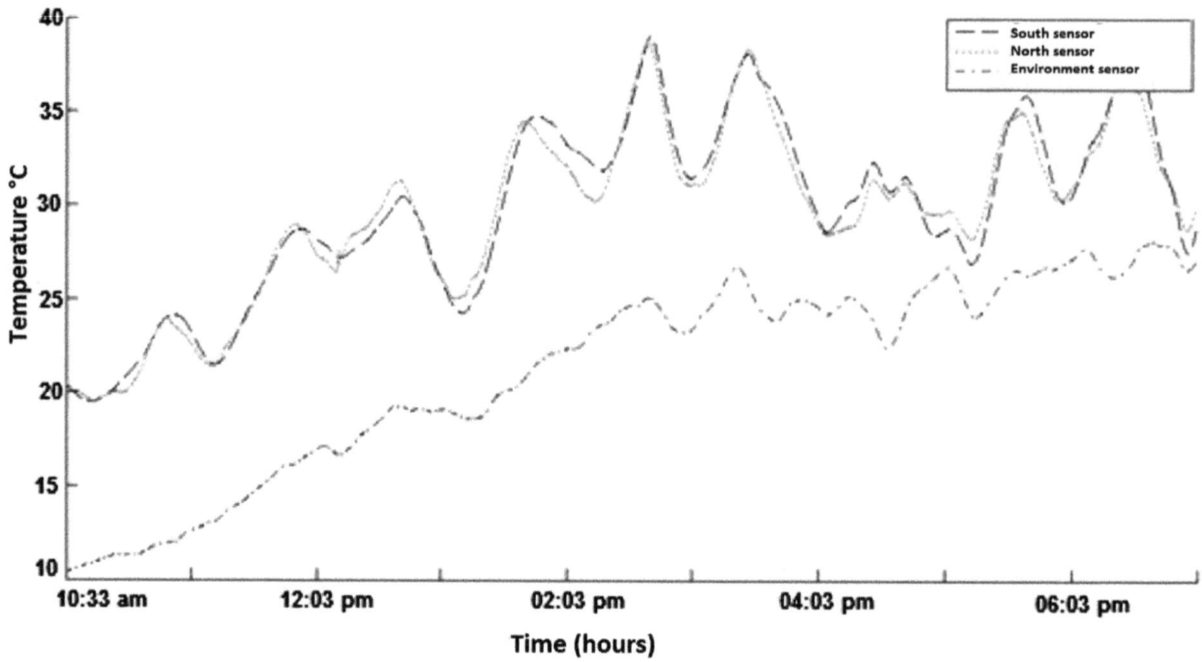

Figure 3. Remote Monitoring System interface

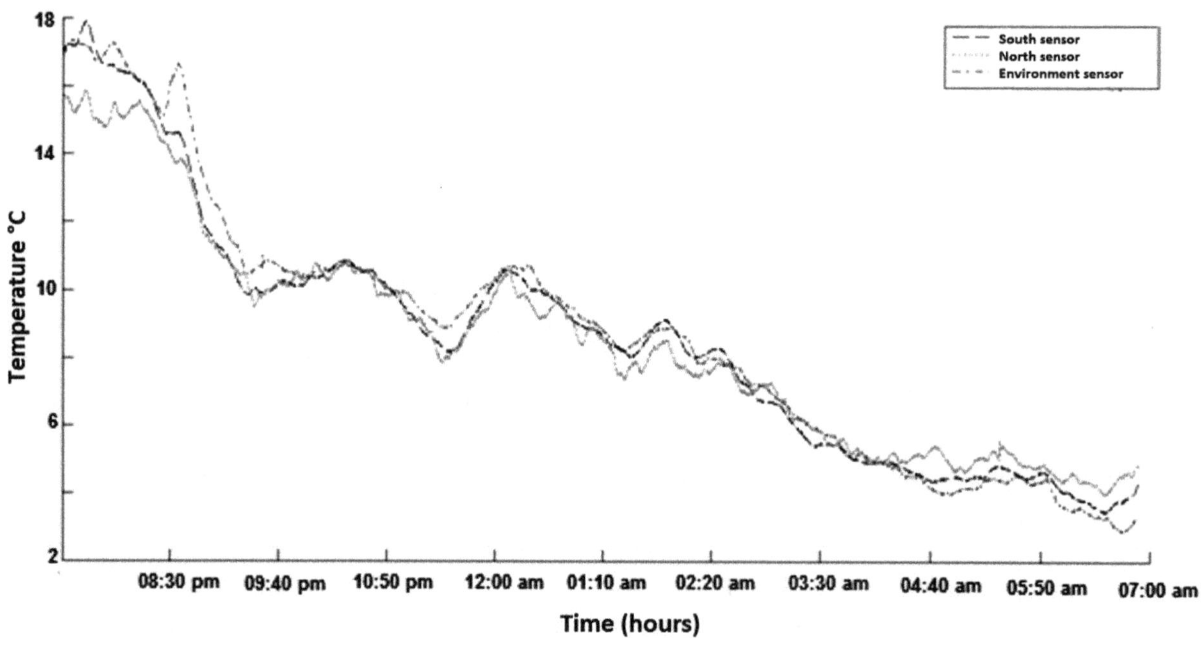

MAIN RESULTS

The results of the temperature measurements are analyzed in this section to determine the system behavior in the presence of different climatological conditions. A comparison between the simulations of the obtained transfer function and the real experiments is shown.

Diurnal and Night Response of the Greenhouse without Ventilation

The sensors of temperature in a greenhouse must be located at 2 meters as maximum height on the level of the soil, due to the distribution of temperature inside the structure, according to (Feng et al, 2013). The following figures are calculated with results of four sensors located inside the greenhouse at 1.5 meters.

The temperature response of the greenhouse with the ventilation curtains closed, in a period of almost 24 hours of measurement, with a sampling period of 1 s., appears in Figure 4. It is possible to observe the warming after the sunrise. The samples for this illustration were taken on October 10, 2014, the maximum temperature of that day was 22.6 Celsius degrees and the minimum is 13 Celsius degrees (Albanil et al, 2014), however the temperature inside the greenhouse was 28.9 degrees at 1pm and it was incremented to 49 degrees.

The greenhouse temperature was measured for five days continuously and it shows a similar behavior each day, see Figure 5.

Figure 4. Sampling with the ventilation curtains closed

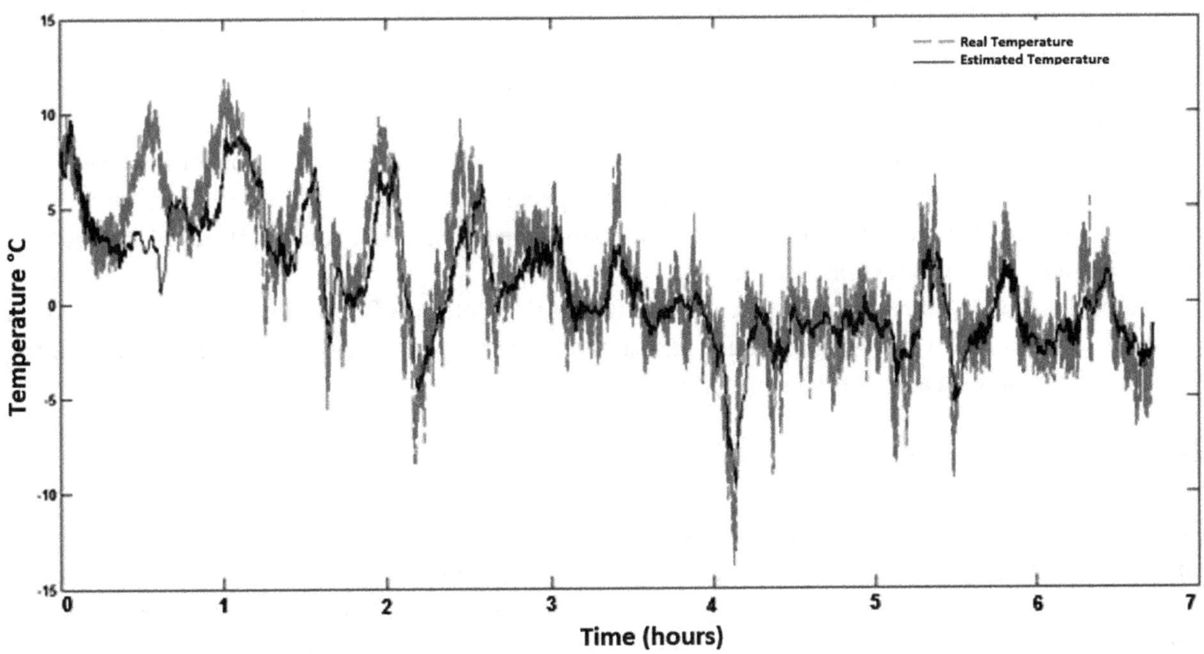

Figure 5. Five days of temperature sampling

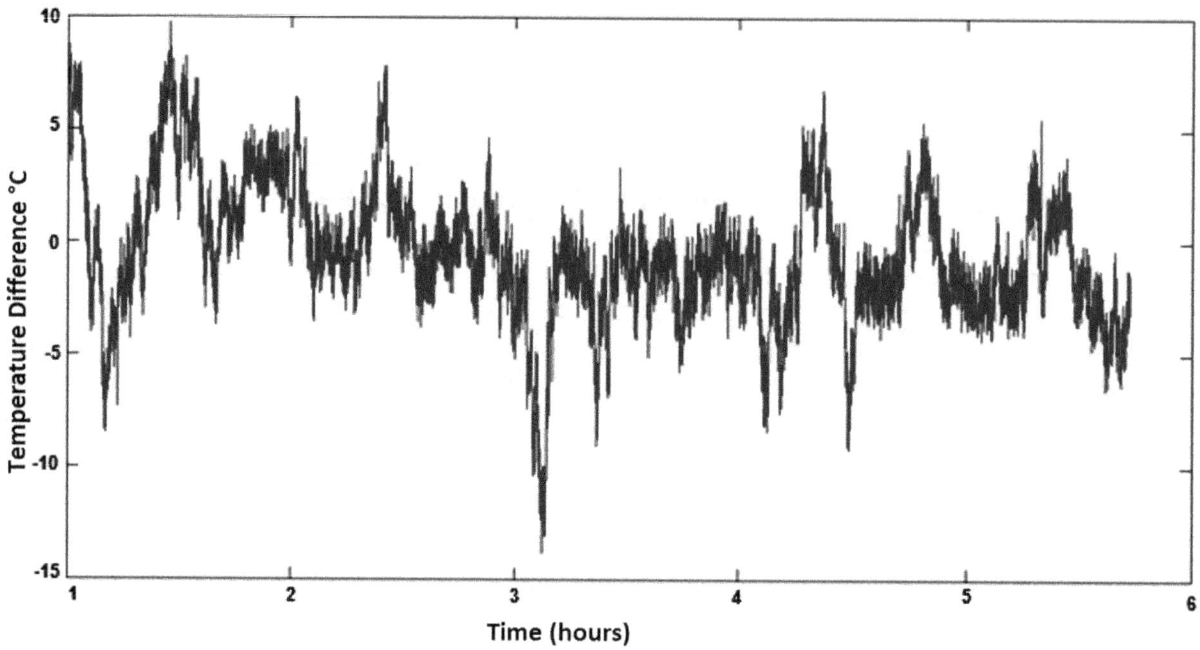

Proofs of Cooling and Warming with Natural Ventilation

The Figure 6 shows the samples gathered by the north temperature sensor located in the left side of the greenhouse with both ventilation curtains open (cooling) it is possible to observe in the temperature response that the stabilization time is 400 sec. approximately. The tests were realized on November 10, 2014 from 2:33 pm.

The next step was to test the response to the warming, the curtains of ventilation were closed. The response to the increase of the temperature (warming) appears in the Figure 7, created from the samples thought by the south sensor of the left side, the samples were taken on November 10, 2014 at 3:30 pm. The response of the north sensor was similar.

Two of the sensors were eliminated, considering that the lectures of both sensors, right and left, in the north and south part of the greenhouse had similar lectures, and one it was relocated outside the greenhouse, the new location of the sensors is shown on Figure 8.

Recursive Least-Square Estimation

The RLS estimation applied to the greenhouse samples is developed in this section. The proposed system input is shown on Figure 9. it allows to analyze the temperature greenhouse dynamics at different hours of the day. The input is modified every 300 seconds, increasing 25% the position of the ventilation curtains, hereby 1 is 25%, 2 is 50%, 3 is 75% and 4 is 100% open. The temperature is clearly affected by the environmental temperature, as we can observe in Figure 10. The samples were taken on march 23rd 2015. The time period for sampling to RLS algorithm with exponential forgetting factor is considered according to the following criteria: $5\tau/100<T<5\tau/20$.

Figure 6. Cooling response with natural ventilation, north sensor

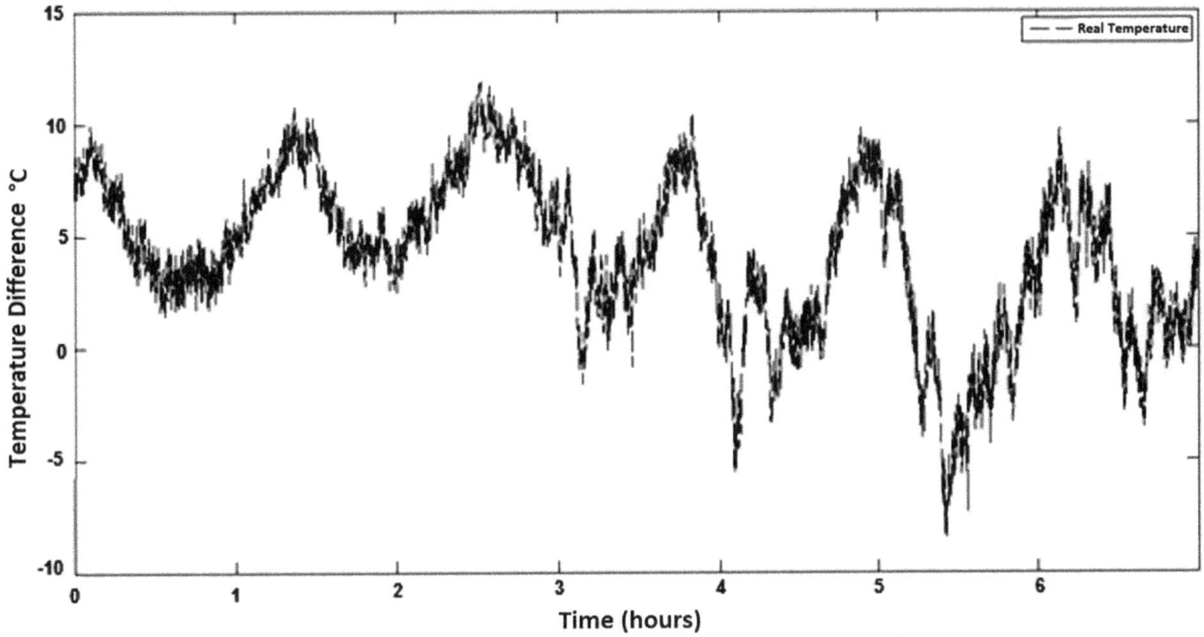

Figure 7. Warming response without ventilation, south left sensor

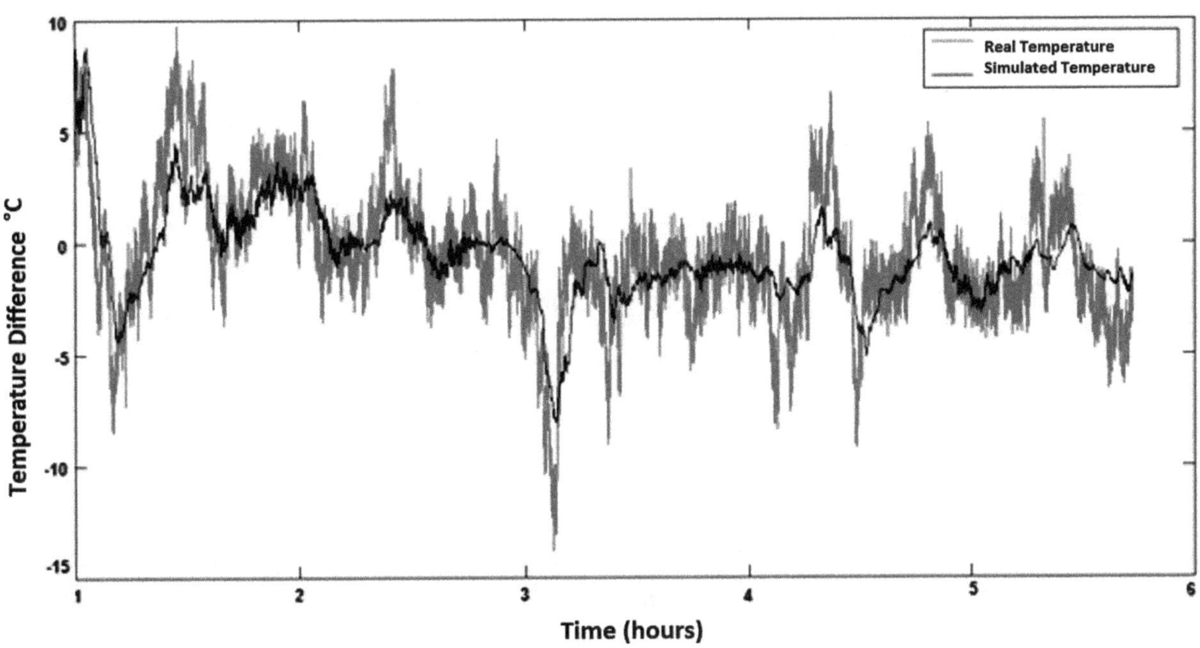

Figure 8. Location of the sensors

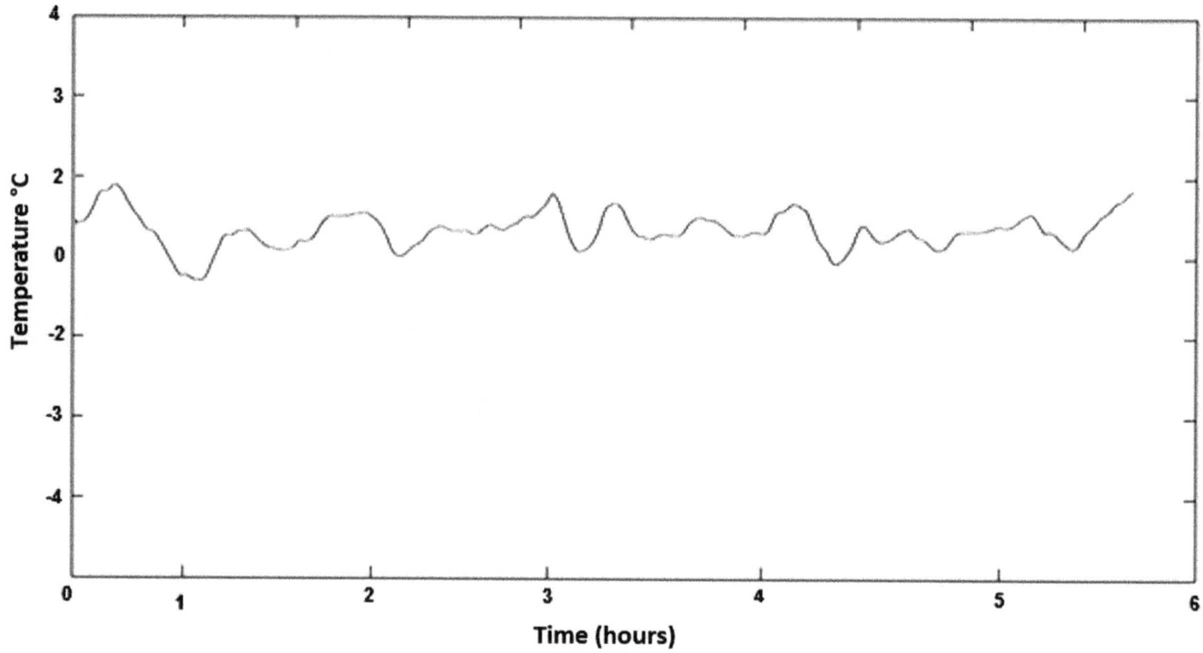

Figure 9. Input designed to observe temperature changes

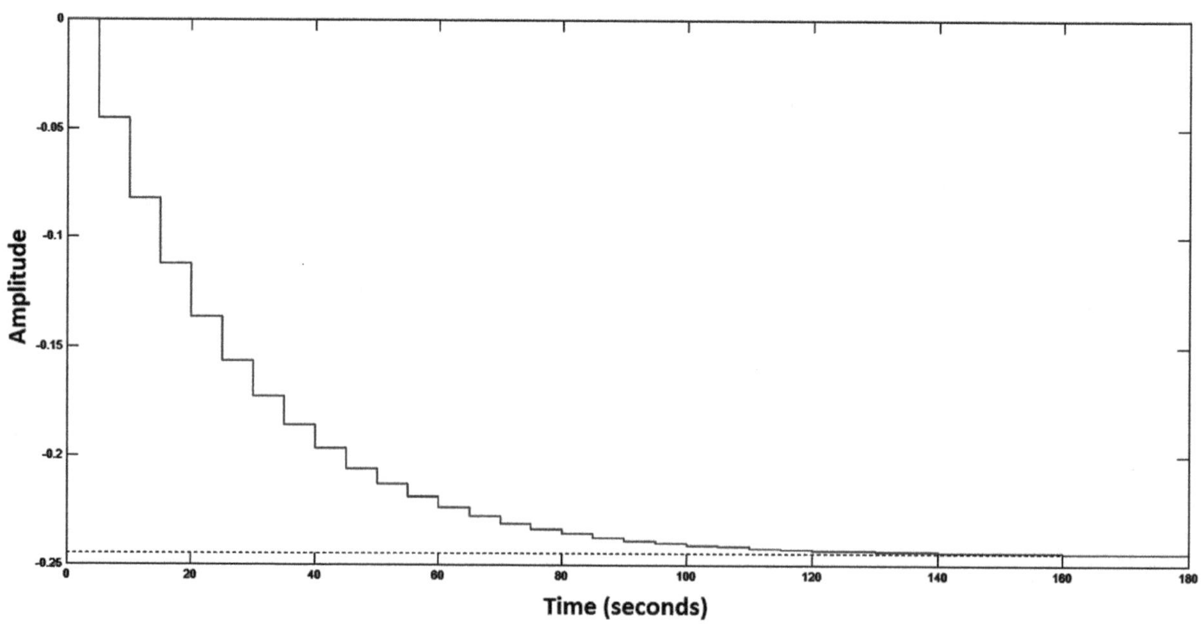

Using the same input shown in Figure 9, the night measures of the greenhouse indicate us that variation does not exist, the system does not meet affected by the changes. Both interior sensors indicate the same temperature that the environmental sensor therefore, it is possible to reject the night behavior of the greenhouse for modeling and control of temperature for natural ventilation, this can be observed in

Figure 10. Diurnal samples of the changes of temperature in the greenhouse

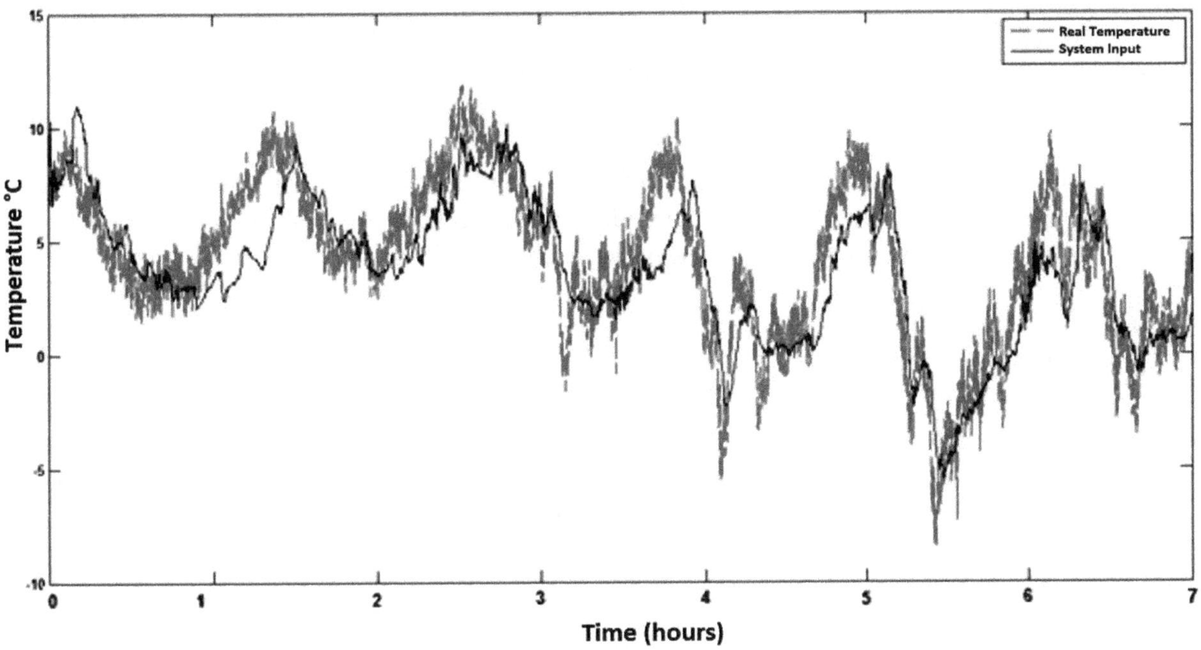

the Figure 11. The Recursive Least-Square estimation was realized with the samples obtained on March 12th 2015.

The variable to model is the greenhouse temperature then it is necessary to consider the first law of thermodynamics, which enunciates that a balance is had between the energy that is obtained and that gets

Figure 11. Night samples of the changes of temperature in the greenhouse

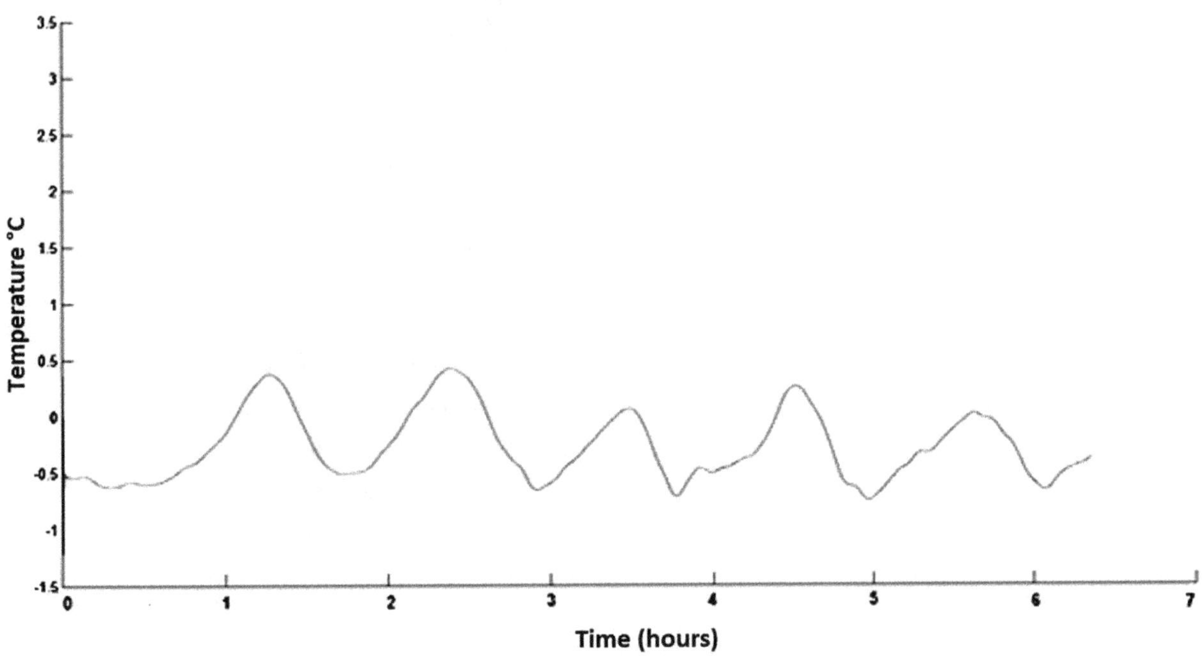

lost. The previous statement is reflected in the energetic exchanges for conduction-convection between the interior and the exterior temperature of the greenhouse, and the difference between them.

The values obtained by the subtraction between the temperature inside the greenhouse and the environmental temperature were considered to be the system's output. The Figure 12 shows the comparison between the real temperature and the RLS estimation, the system was sampled for 23 hours to get this illustration.

It is possible to observe that during the first 3.5 hours, the estimation answer is not very attached to the behavior of the real system, the error magnitude reaches 8 celsius degrees (Figure 12). Due to this behavior, the system's response was separated in two sections:

- After the zenith (Figure 13),
- Before the zenith (Figure 14).

The proofs were realized using an exponential forgetting factor algorithm for RLS estimation of 0.98.

The behavior of the RLS estimation for values after the zenith, is shown on Figure 15, and the error is shown on Figure 16, where it is possible to observe better approximations. The transfer function constructed from the parameters estimated by RLS's program, with a sampling period τ=5s is:

$$G\left(z\right) = \frac{-0.04519}{z - 0.8156}.$$

Figure 12. Real answer compared with the temperature estimated with the RLS method

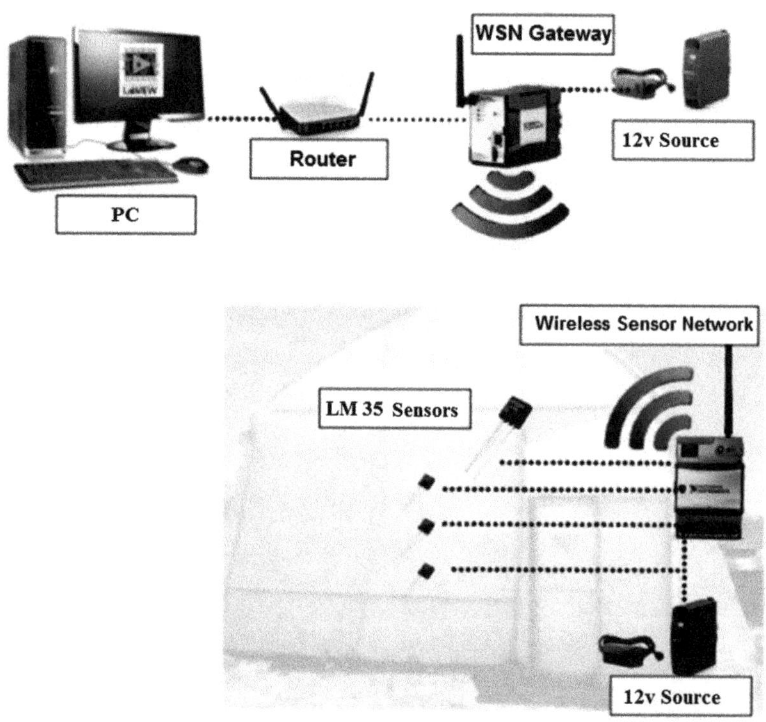

Figure 13. Temperature greenhouse response after the zenith

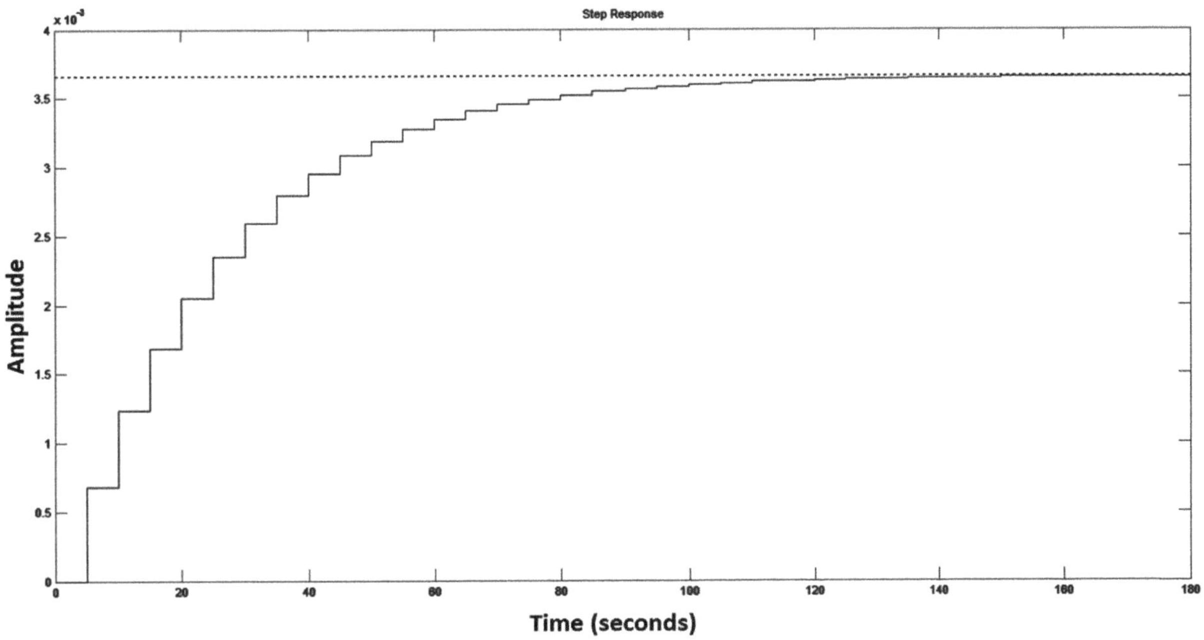

Figure 14. Temperature greenhouse response before the zenith

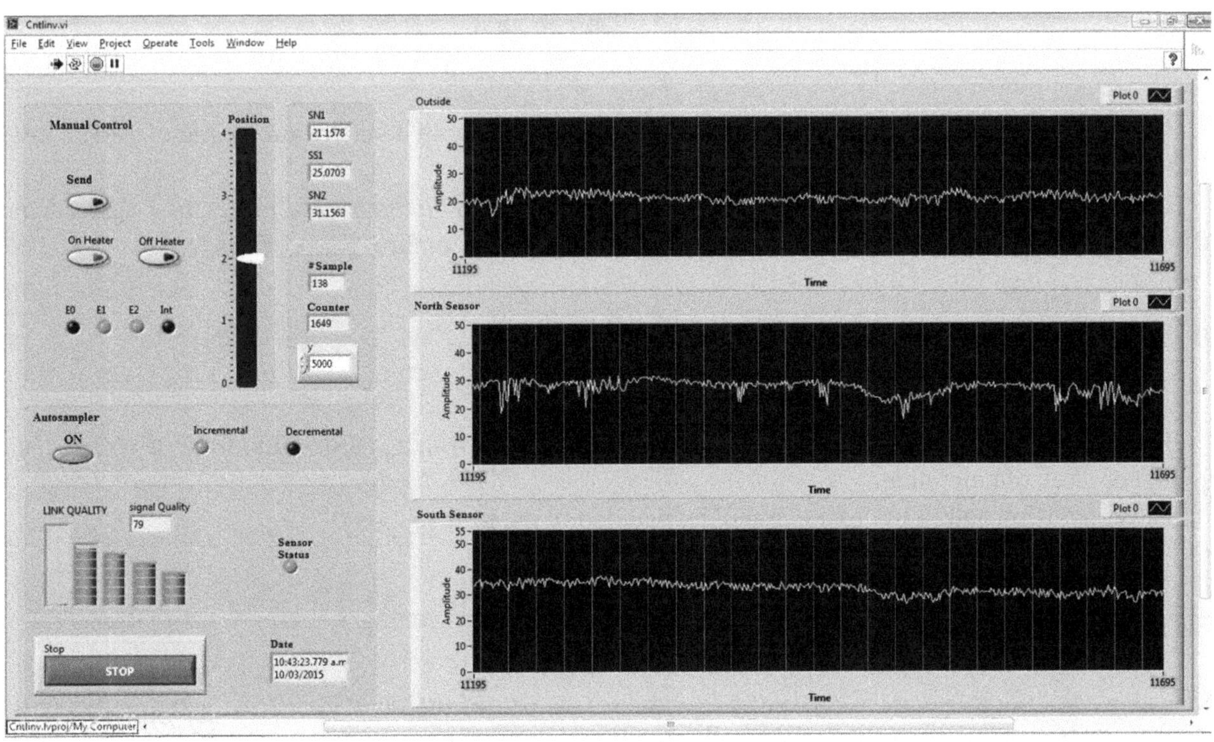

Figure 15. Real and estimated greenhouse response, after the zenith

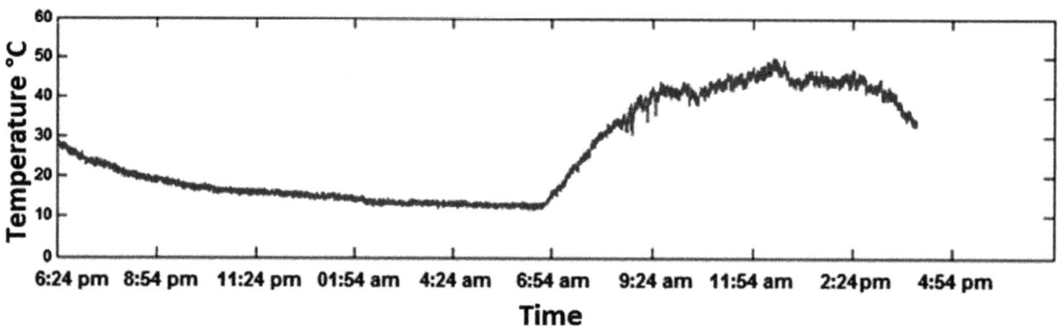

Figure 16. Average error response

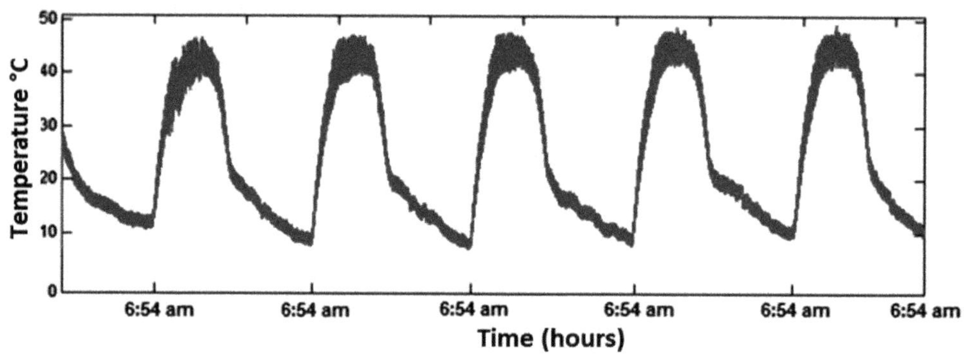

The unit step response of the system is show on Figure 17, with a stabilization time of 140 s., and 0.45 degrees of amplitude.

The RLS estimation for the values before the zenith is shown on Figure 18. The error magnitude is shown on Figure 19, and the transfer function obtained, with a sampling period τ=5s is:

$$G\left(z\right)=\frac{0.0006811}{z-0.8137}.$$

The unit step response of the system is show on Figure 20, with a stabilization time of 160 s., and 0.0035 degrees of amplitude.

The conclusion and future research will be detailed in the next sections.

FUTURE RESEARCH DIRECTIONS

Nowadays, the optimization of economic resources, it is one of the main concerns of the researchers in the area, this due to the fact that the technology to automate the greenhouses is expensive, also the electric power consumed on having been always active the sampling is high.

Figure 17. Unit step response of the transfer function proposed, after the zenith

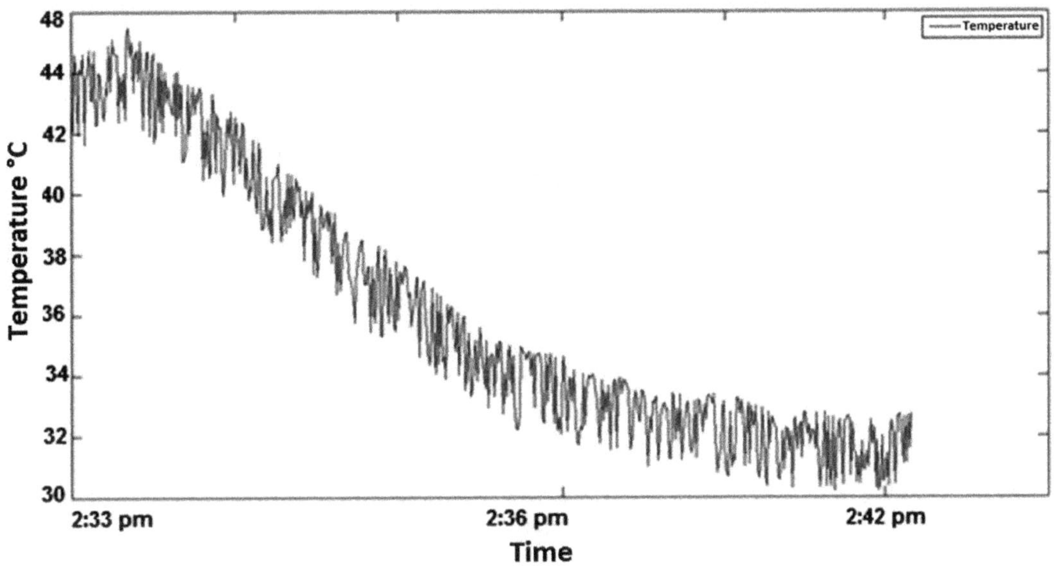

Figure 18. Real and estimated samples of the greenhouse, before the zenith

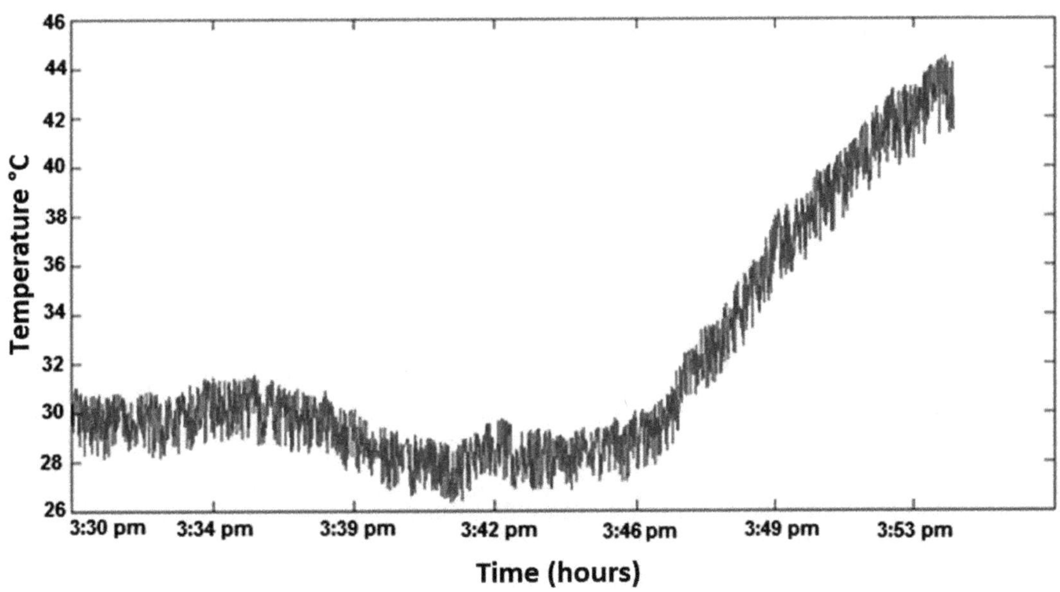

The optimization of natural resources, also it is important, for what the implementation of alternative energy is a good option to have intelligent self-sustaining greenhouses, amicable with the environment, that take advantage of the resources of every region. There exist zones where the lack of water is a problem, therefore hydroponic greenhouses offer a new solution economizing on water and land.

Particularly the presented research is just a beginning, though the natural ventilation is very common, the whole potential of an automated greenhouse cannot be exploited to control the parameters that

Figure 19. Average error response

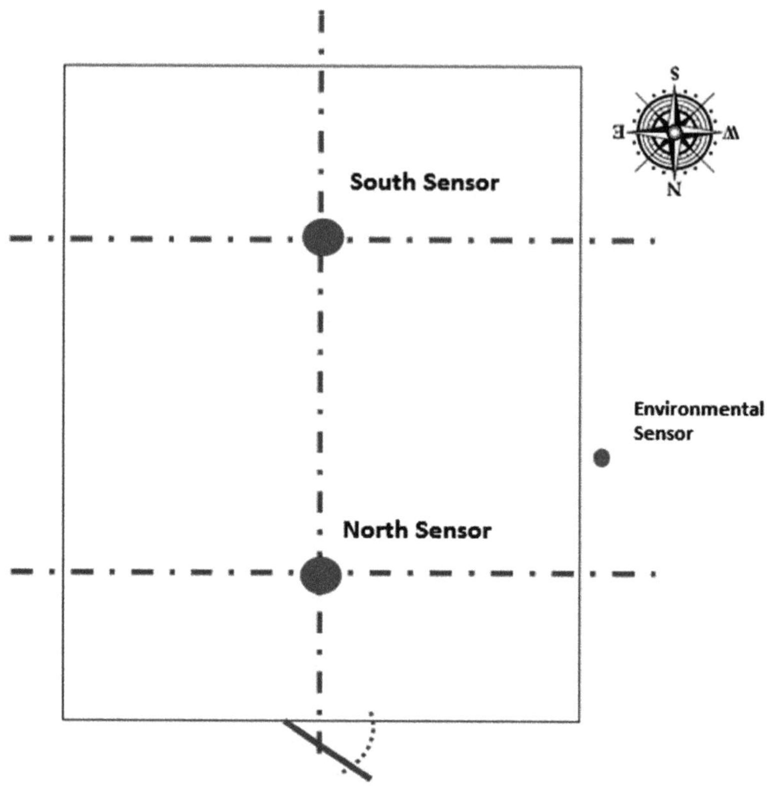

Figure 20. Unit step response of the transfer function proposed, before the zenith

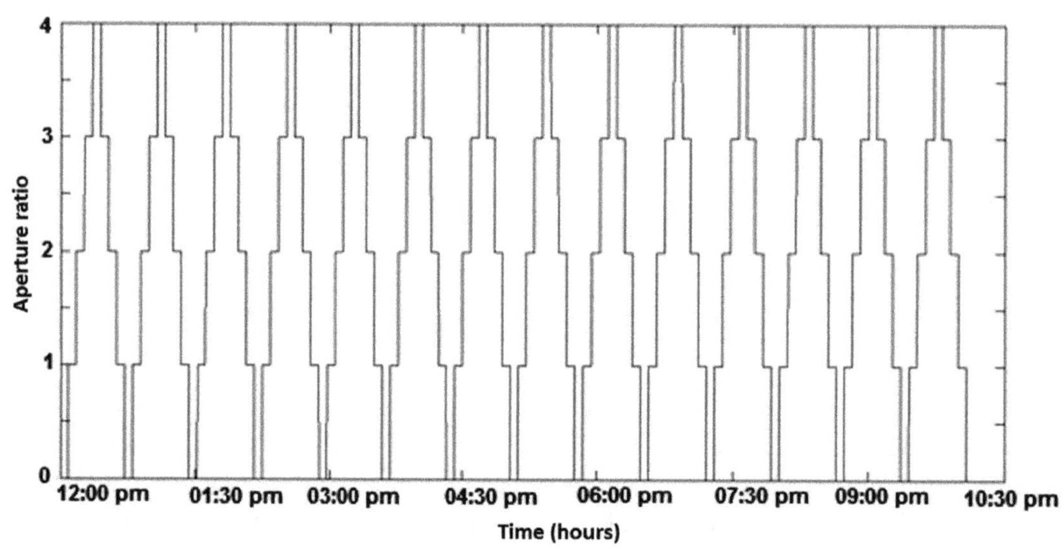

affect the crops. The implementations of actuators like ventilators and heaters to extend the scale of temperature that can be manipulated is important, as well as the incorporation of technology that allows to control another type of variables, like humidity, that is very interesting because it allows to work with the Vapor Pressure Deficit (VPD). Certainly, the comparison between a model with and without crop must be realized. That is the next direction of our research.

The RLS estimation techniques are a powerful tool for systems with dynamics that changes continuously, therefore is ideal for greenhouses, because they have different behavior in different seasons of the year. However, it will be interesting to compare this parameter identification method with some artificial intelligence technique.

CONCLUSION

The RLS estimation does not allow getting a mathematical model to characterize the behavior of the temperature greenhouse 24 hours of the day continuously due to the influence of the environment, principally, the solar radiation.

Different behavior patterns are observed before and after the zenith during the day, therefore the RLS estimation was done independently. Using natural ventilation during the night it was observed that significant variations does not exist comparing with the environment temperature, then the night behavior was discarded in this stage of the research.

The RLS technique with exponential forgetting factor allows to estimate the parameters of the transfer function of the temperature variations inside the considered zenith greenhouse. The above mentioned function is of first order.

REFERENCES

Abdel-Ghany, A., & Kozai, T. (2006). Dynamic modeling of the environment in a naturally ventilated, fogcooled greenhouse. Elsevier.

Ahonen, T., Virrankoski, R., & Elmusrati, M. (2008). Greenhouse Monitoring with Wireless Sensor Network. IEEE.

Albanil, A., Pascual, R., Quiroz, C., & Sánchez, O. (2014). *Reporte del Clima en México, Octubre, Año 4, Núm. 10*. CONAGUA.

Álvarez-Sánchez, E., Leyva-Retureta, G., Portilla-Flores, E., & López-Velázquez, A. (2014). Evaluation of thermal behavior for an asymmetric greenhouse by means of dynamic simulations. *DYNA*, *81*(188), 152–159. doi:10.15446/dyna.v81n188.41338

Aström, K. (2008). *Adaptive Control*. Addison Wesley.

Businger, J. (1963). *The glasshouse (greenhouse) climate*. North-Holland Publ. Co.

Chuanan, Y., & Yongchang, Y. (2010). Implementation of Greenhouse Monitoring System Based on RF Transceiver. *International Conference on Industrial Mechatronics and Automation*. doi:10.1109/ICINDMA.2010.5538107

Díaz, F. (2002). Modelado y Control Jerárquico de Crecimiento de Cultivos en Invernadero. *Tesis Doctoral Universidad de Almería Escuela Politécnica Superior*, 390.

Du, Y., Xue, Z., Zhu, Q., Liu, X., Feng, Y., & Zhang, S. (2013). Design and Application of Intelligent Control System for Greenhouse Environment Based on CAN bus. *Proceedings of International Conference on Modeling, Identification & Control (ICMIC)*.

Eredics, P., & Dobrowiecki, T. P. (2011). Data Cleaning for an Intelligent Greenhouse. *International Symposium on Applied Computational Intelligence and Informatics*.

Eredics, P., Gáti, K., Dobrowiecki, T. P., & Horváth, G. (2011). Neural Models for an Intelligent Greenhouse - A CMAC Global Greenhouse Model. *International Symposium on Computational Intelligence and Informatics*. doi:10.1109/CINTI.2011.6108495

Fang, J., & Wang, F. (2011). *Design of Greenhouse Remote Monitoring System based on LabVIEW*. Harbin Science and Technology Agency Foundation of Returned Overseas.

Feng, L., & Zhi, Y. (2013). *Greenhouse CFD Simulation for Searching the Sensors Optimal Placements*. Special Fund for Agro-Scientific Research in the Public Interest of China. doi:10.1109/Argo-Geoinformatics.2013.6621972

Ferreira, P. M., & Ruano, A. E. (2008). Application of Computational Intelligence Methods to Greenhouse Environmental Modelling. IEEE.

Fezari, M., Khati, A., & Boumaza, M. S. (2011). Implementation of Wireles Sensor Network for Automatic Greenhouse Monitorign. In Communications, Computing and Control Applications (CCCA).

Ghosal, M., & Tiwari, G. (2004). *Mathematical modeling for greenhouse heating by using thermal curtain and geothermal energy* (Vol. 76). Elsevier.

Guofang, L., Yubin, Q., Shengtao, L., & Junyu, X. (2010). Remote Monitoring System of Greenhouse Environment Based on LabVIEW. *International Conference on Computer Design and Applications (ICCDA 2010)*. doi:10.1109/ICCDA.2010.5541107

Gupta, M., & Chandra, P. (2002). *Effect of greenhouse design parameters on conservation of energy for greenhouse environmental control* (Vol. 27). Pergamon.

Juárez-Gutiérrez, S. S. (2015). *Modelado matemático de la variación de temperatura dentro de un invernadero clásico cenital utilizando el método de Mínimos Cuadrados Recursivos*. (Unpublished Master's thesis). Universidad Politécnica de Aguascalientes, Mexico.

Kang, B., Park, D., Cho, K., Shin, C., Cho, S., & Park, J. (2008). A Study on the Greenhouse Auto Control System based on Wireless Sensor Network. *International Conference on Security Technology*. doi:10.1109/SecTech.2008.60

Lafont, F., & Balmat, J. (2002). *Optimized fuzzy control of a greenhouse*. Academic Press.

Lakshmi, V., Arvind, A., & Bever, K. (2007). Greenhouse Asset Management Using Wireless Sensor Actor Networks. *International Conference on Mobile Ubiquitous Computing, Systems, Services and Technologies*.

Laribi, I., Homri, H., & Mhiri, R. (2006). *Modeling of a Greenhouse Temperature: Comparison Between Multimodel and Neural Approaches*. IEEE ISIE.

Leal, J., Alcorta, E., & Rodríguez, H. (2006). *Modelado del clima en invernaderos: Respuesta de la temperatura a cambios de humedad*. Ingenierías.

Luo, W., Dail, J., Wang, X., Stanghellini, C., & Bu, C. (2005). *Simulation of greenhouse management in the subtropics, Part I: Model validation and scenario study for the winter season* (vol. 90). Academic Press.

Ma, L., Ji, J., & He, Ch. (2010). The Study of Greenhouse Temperature Modeling Based on Fuzzy Neural Network. *International Conference on E-Product E-Service and E-Entertainment*. doi:10.1109/ICEEE.2010.5660748

Mad, S., Munirah, L., Kamarudin, K., Mohd, W., Muhammad, S., Zakaria, A., & Nor, M. (2014). A Real-Time Greenhouse Monitoring System for Mango with Wireless Sensor Network (WSN). *International Conference on Electronic Design (ICED)*.

Mohammad, R., Siamak, H., Hossein, M., & Maryam, G. (2010). A Hybrid Neuro-Fuzzy Approach for Greenhouse Climate Modeling. IEEE.

Moreno, A., Aguilar, J., & Luévano, A. (2011). Características de la Agricultura Protegida y su Entorno en México. Revista Mexicana de Agronegocios, 15(29), 763 - 744.

Morimoto, T., & Hashimoto, Y. (2000). An Intelligent Control for Greenhouse Automation, Oriented by the Concepts of SPA and SFA. An Application to a post-harvest proces. *Computers and Electronics in Agriculture*, 555–567.

Pawlowski, A., Guzmán, J. L., Rodríguez, F., Berenguel, M., Sánchez, J., & Dormido, S. (2008). Event-Based Control and Wireless Sensor Network for Greenhouse Diurnal Temperature Control: A Simulated Case Study. IEEE.

Pengzhan, C., & Baifen, L. (2010). Construction of Intelligent Greenhouse Control System Based on CAN bus. *International Conference on Computer Application and System Modeling*. doi:10.1109/ICCASM.2010.5622630

Qiu, W., Dong, L., Wang, F., & Yan, H. (2014). *Design of Intelligent Greenhouse Environment Monitoring System Based on ZigBee and Embedded Technology*. IEEE.

Salazar, R., López, I., & Rojano, A. (2007). A neural network model to control greenhouse environment. *Sixth Mexican International Conference on Artificial Intelligence, Special Session*. doi:10.1109/MICAI.2007.33

Salgado, P., & Cunha, J. (2005). *Greenhouse climate hierarchical fuzzy modeling*. Academic Press.

Seginer, I., Boulard, T., & Bailey, B. (1993). *Optimal Greenhouse Temperature Trajectories for a Multi-State-Variable Tomato Model. In The Computerized Greenhouse* (pp. 153–172). San Diego, CA: Academic Press.

Singh, G., Singh, P., Lubana, P., & Singh, K. (2006). *Formulation and Validation of a Mathematical model of the microclimate of greenhouse* (Vol. 31). Elsevier.

Solera, A. (2003). *El Filtro de Kalman.* Banco Central de Costa Rica, División Económica, Departamento de Investigaciones económicas.

Takakura, T., Jordan, A., & Boyd, L. (1971). Dynamic simulation of plant growth and environment in the greenhouse. *Trans. ASAE 14* (vol. 964).

Tap, F. (2000). *Economics-based optimal control of greenhouse tomato crop prod.* (PhD Thesis). Wageningen Agr.

Udink Ten Cate, A. J. (1983). *Modeling and (adaptive) control of greenhouse climates.* Wageningen UR.

Xiaoyan, Z., Xiangyang, Z., Chen, D., Zhaohui, C., Shangming, S., & Zhaohui, Z. (2013). The Design and Implementation of the Greenhouse Monitoring System Based on GSM and RF Technologies. IEEE.

Yulong, J., & Jiaqiang, Y. (2011). Design an Intelligent Environment Control System for Greenhouse Based on RS485 Bus. *Second International Conference on Digital Manufacturing & Automation.*

KEY TERMS AND DEFINITIONS

Actuator: It is something, typically a mechanism, which converts energy to motion. The most common example is a motor, but it can be a pump, switch or valve.

Greenhouse: Structure that protects crops from sudden temperature changes.

Least Squares: Mathematical parameter estimation method that uses the response of a system to operate.

Mathematical Model: Mathematical representation of a system to describe the behavior of certain variables for an indeterminate time.

Natural Ventilation: Apertures in buildings that allow inflow of outside air.

Remote Monitoring System: Provides information of some system variables remotely.

System: Components that act together and realize certain function. From the control point of view is the process, machine, that will be manipulated to complain a task, it not necessary is physical, it can be a phenomena.

Vapor Pressure Deficit: Roughly speaking, it measures the reaction of the plant to the humidity in the growing environment, and it is an important factor that allows to control diseases and can be manipulated.

Variable: It is the quantity or condition of the system that can be measured or can be manipulated to complain the task that is need.

Section 3

Applications of Innovative Logistics

Chapter 14
Multi-Objective Simulated Annealing Algorithms for General Problems

Juan Frausto Solís
Tecnológico Nacional de México, Mexico

José Carlos Soto-Monterrubio
Tecnológico Nacional de México, Mexico

Héctor Joaquín Fraire
Tecnológico Nacional de México, Mexico

Rodolfo Pazos-Rangel
Tecnológico Nacional de México, Mexico

ABSTRACT

Simulated Annealing is an analogy with the annealing of solids, which foundations come from a physical area known as statistical mechanics. This chapter presents a review of the literature on multi-objective simulated annealing (MOSA). There are several multi-objective approaches to solve optimization problems with simulated annealing such as hybridizations, implementation of strategies from different metaheuristics. Modern MOSA research includes populations and adaptive rules, and are briefly described in this chapter. We discuss different approaches in multi-objective and we revise the modern MOSA framework.

1. INTRODUCTION

Optimization has many applications in almost any area of the human activity and has two kind of research areas: single optimization and multi-objective optimization. These areas study single-objective problems or SOP (with only one objective function) and multi-objective problems or MOP (with more than one objective functions) respectively (Coello Coello, Lamont, & Van Veldhuizen, 2007). Even though single optimization is far to be a closed research area, many deterministic and stochastic strategies have been studied since the second world war. In addition, several reviews are available for the problems of this area. In the case of problems belonging to multi-objective optimization area, nature and physical principles are applied to develop algorithm for solve them. Multi-objective Evolutionary Algorithms (MOEAs) and multi-objective Simulated Annealing (MOSA) are among the most common of these algorithms (Deb, 2001). We will use the next model for distinguish single to multi-objective optimization problems:

DOI: 10.4018/978-1-4666-9779-9.ch014

$$\min/\max f_m(x_i); m = 1, \ldots, M; i = 1, \ldots, N$$

subject to

$$g_j(x) \geq 0; j = 1, \ldots, J$$

$$h_k = 0; k = 1, \ldots, K$$

$$x_i^{Lower} \leq x_i \leq x_i^{Upper}; i = 1, \ldots, N$$

In single optimization problems, M is always equal to one, while M is always greater than one for multi-objective optimization problems. The rest of this model is the same for SOP and MOP with the next features: There are $N \geq 1$ decision variables x_i. There are two set of algebraic expressions g_j and h_k known as inequality and equality constraints respectively. The number of these constraints are $J \geq 0$ and $K \geq 0$. Finally, x_i^{Lower} and x_i^{Upper} are respectively the lower and upper bound of each decision variable x_i.

Most of the practical problems belongs to the second category and the NP-Hard problems. For the last kind of problems, many multi-objective evolutionary algorithms (MOEAs) have been published since their review (Deb, 2001); the most popular MOEAs are using Pareto Archived Evolution Strategy (PAES), this use a file where dominated solutions, which are employed for new solutions (Knowles & Corne, 2000), are updated.

Current applications of MOSA can be found in the development of proportional integral derivate (PID) controller design (Hung, Shu, Ho, Hwang, & Ho, 2008) which are used in industrial process control. Other application of MOSA is in the management inventory problem and it will be applied to a large-scale supply chain at a major steel producer (Alrefaei, Diabat, Alawneh, Al-Aomar, & Faisal, 2013). MOSA is used in telecommunications networks to optimize the multicast routing problem (MRP) (Xu & Qu, 2011; Xu, Qu, & Li, 2013). Other application of MOSA is in the mobile telecommunications network's air interface with the optimization of a Code Division Multiple Access (CDMA) (Smith, Everson, Fieldsend, Murphy, & Misra, 2008). MOSA is used to solve the redundancy allocation problem (RAP) in the reliability design of parallel sub-system which have applications on networks designs (Zaretalab, Hajipour, Sharifi, & Shahriari, 2015). In (Safaei, Jardine, & Banjevic, 2008) the MOSA is used to solve a real maintenance workforce scheduling problem (MWSP) in a steel company. In this case MOSA minimizes the workforce cost and the flow time of the work request. The MOSA is used for solving an unrelated parallel machine scheduling problem in (Lin & Ying, 2015). Other scheduling application of MOSA can be found in (Frutos & Tohmé, 2015; Frutos, Olivera, & Tohmé, 2010; Loukil, Teghem, & Tuyttens, 2005).

This chapter is organized as follows. In Section 2 the multi-objective optimization is presented. In Section 3 the fundaments of Simulated Annealing algorithm and multi-objective optimization theory is presented. In Section 4, the multi-objective Simulated Annealing framework. In Section 5, the current approaches in multi-objective annealing methods that we found in the state of the art is revised. The focus of this chapter is to review the multi-objective simulated annealing algorithms in the literature to apply the most promising of them to the scheduling problem.

2. MULTI-OBJECTIVE OPTIMIZATION

In the last two decades, there have been various evolutionary algorithms EA (Evolutive Algorithms). The MOEAs (Multi-objective Evolutionary Algorithms), proposed by (Deb, 2001) are among the most popular EAs. There are also algorithms called PAE (Pareto archived evolution strategy) that maintains an archive of dominated solutions which is used for validating new solutions (Knowles & Corne, 2000).

The MOEAs typically employ penalty techniques for weighting the individual objective functions; in this way, the problem can be solved using traditional techniques for SOPs, which are problems with only one objective function. In the last two decades, the application of MOEAs has been the most popular technique (Coello Coello, Lamont, & Van Veldhuizen, 2007). However, there are reported some issues when MOEAs are applied to MOPs with certain structure (Bandyopadhyay, Saha, Maulik, & Deb, 2008):

1. The MOEAs have a very poor performance when the number of objectives grows, for instance beyond 15, 10 or even 5 goals (Bandyopadhyay, Saha, Maulik, & Deb, 2008).
2. The way to determine the effective weights used for combining several objectives remains an open question. In addition, for some combinations of weights the Pareto Optimal (PO) obtained by the algorithms could not represent practical solutions.

3. SIMULATED ANNEALING

Even though the problems mention in the last section, the MOEAs has been extensive applied over the last two decades (Coello Coello, Lamont, & Van Veldhuizen, 2007). Moreover, Simulated Annealing (SA) is a very effective way to solve very complex single optimization problems (Kirkpatrick, 1984). SA is an analogy with the annealing of solids, which foundations come from a physical area known as statistical mechanics. In this area, a set of solutions represents the particles of the solid moving from one position to another because the internal energy and temperature of the system. This energy is the Gibbs energy and the solid reaches the optimal state when this energy is minimal. The internal energy can be modified because the particles change their position and/or the temperature is changed. The situation is far to be simple because for a given temperature the particles change their position which in turn modify the Gibbs energy and the temperature. The process continues until a thermal equilibrium is reached; then, a new temperature is achieved and the process is started again. In this process, everything happens as an intelligent algorithm were applied to the set of particles. When a new macro state with lower internal energy is proposed to this algorithm, this state is accepted and the particles move to the new state. However, for a greater energy, the new macro-state is not accepted in most of the cases. In fact, there are some exceptions, and worse macro states are accepted with greater energy than the previous one. These worse states follows the famous Boltzmann distribution, and allows that the set of particles do not be trapped in a local optima. In this way, nature achieves the best macro state. It is fantastic to know that nature has a device to optimize this physical problem. A multi-objective optimization problem or MOP has the same elements that the later physical problem. The energy represents the f_m objective functions defined by equation (1). The macro states of the physical problem represents the x_i decision variables. In the following sub-sections, SA is explained with more details.

3.1 Simulated Annealing Algorithm

Simulated Annealing (SA) was designed as a single objective algorithm in the 1980s using the travel salesman problem to illustrate its application of SA (Kirkpatrick, 1984). However, in nowadays, SA is also defined for many decision variables and/or objective functions. SA starts from a given initial solution x , which can be given as a random one. Then a new y solution is obtained by an action over x that is named perturbation function over x . The perturbation function finds a new solution in the neighborhood of x and represented by $N(x)$. This perturbation function is a stochastic rule. This rule is usually as simple as a random modification of the original solution x or as complex as the application of another algorithm; for instance when x has several decision variables, the perturbation function can be the application of genetic operators over the definition of the variables in their binary representation. In most of the cases, the perturbation function is defined as the permutation of the values in x. Thus, $y \in N(x)$ and is chosen with the Boltzmann probability (Serafini, 1993).

$$p_{x,y}(T) = \min\{1; e^{\left(\frac{f(x)-f(y)}{T}\right)}\}$$

In (5), T is the temperature parameter, $f(x)$ and $f(y)$ are the values of the objective function for the old a new solution respectively. The principal components of SA are the next:

- **Initial Temperature (T_i):** This is a parameter that defines where the algorithm starts. If the initial temperature is too high, the algorithm could spend a lot of time in the first iterations. If the initial temperature is too low, the probability to be trapped in local optimum is very high.
- **Final Temperature (T_f):** This parameter indicates the stop criterion of the algorithm. When the final temperature is too small, the algorithm can be last a lot of time. If the final temperature is too high, the algorithm can finish very soon and the optimal cannot be reached.
- **Cooling Scheme:** This is not really a parameter but a criterion that allows the convergence to the global optima. The cooling scheme specifies how the temperature is decreased. If the temperature is decreased very slowly, the quality solution is increased, but a lot of execution time is required. The most common cooling scheme is the geometric function $T_{k+1} = \alpha T_k$, where $0.7 < \alpha < 1.0$.
- **Metropolis Length or Markov Chain:** This indicates the number of iterations done in each temperature T_k . During these iterations, the optimum is searched until the stochastic equilibrium is reached. The Markov chain length can be static or dynamic. In the static approach, the length is the same in each temperature. In the dynamic approach, the length depends on some parameters and functions.

- **Acceptance Criteria:** A probability function for acceptance a new solution given the previous one, also known as a transition rule. In classical SA, the Boltzmann distribution defined by equation (5) is used. This acceptance criteria uses energy differences from the current solution and the new solution.

The classic simulated annealing algorithm is shown in the Algorithm 1. The Line 1 receives the initial temperature (T_i), the final temperature (T_f), and the cooling rate (α), T_i, and T_f are used in Line 4. In Line 5, the metropolis length needs to be specified. In Line 6, a new solution S_j is generated. In line 7, if the solution S_j is better than the current solution S_i then the current solutions is accepted. In Line 8 the Boltzmann probability is used if the solution S_j is worse than S_i. The α value is used in Line 10 to decrease the temperature.

3.2 Multi-Objective Optimization

The basic concepts of multi-objective optimization are presented in this section. Without loss of generality, we will assume only minimization functions.

Definition 1. *Multi-objective optimization problem (MOP).*
Given:

1. A vector function $\vec{f}(\vec{x}) = [f_1(\vec{x}), f_2(\vec{x}), \ldots, f_k(\vec{x})]$ and
2. A feasible solution space Ω,

The MOP consists in to find a vector $\vec{x} \in \Omega$ that optimizes the vector function $\vec{f}(\vec{x})$.
Definition 2. *Pareto dominance.* A vector $\vec{x}$ dominates $\vec{x}\,'$ (denoted by $\vec{x} \prec \vec{x}\,'$):

1. If $f_i(\vec{x}) \le f_i(\vec{x}\,')$ for all i functions in $\vec{f}$, and
2. There is at least one i such that $f_i(\vec{x}) < f_i(\vec{x}\,')$.

Algorithm 1. Simulated Annealing algorithm

```
1: Initialize Simulated Annealing (T_i,T_f,α)
2: T_k = T_i
3: S_i = generateSolution()
4: While T_k ≥ T_f do
5:      While Metropolis length do
6:          S_j = N(S_i)
7:          Δ = E(S_j)-E(S_i)
8:          If e^(-Δ/Tk) < random (0, 1) then
9:          S_i = S_j
10:     T_k+1 = αT_k
```

Definition 3. *Pareto optimal.* A vector $\vec{x}^*$ is Pareto optimal if not exists a vector $\vec{x} \in \Omega$ such that $\vec{x}\,' \prec \vec{x}^*$.

Definition 4. *Pareto optimal set.* The Pareto optimal set for a MOP is defined as $P^* = \{\vec{x}^* \in \Omega\}$.

Definition 5. *Pareto front.* Given a MOP and its Pareto optimal set P^*, the Pareto front is defined as $PF^* = \{f(\vec{x}) \mid \vec{x} \in P\}$.

4. MULTI-OBJECTIVE SIMULATED ANNEALING FRAMEWORK

The first adaptation of simulated annealing to multi-objective optimization is proposed in (Serafini, 1993). One of the difficulties to adapt simulated annealing to multi-objective optimization is to find an adequate acceptance criteria. Multiple criteria acceptance has been proposed in (Serafini, 1993; Nam & Park, 2000).

The multi-objective simulated annealing algorithm or MOSA is shown in the Algorithm 2. This MOSA algorithm is very similar to the classical one with only one objective function. In fact the only difference of MOSA with SA is the Line 7 of Algorithm 2, where is used the dominance criteria to change the current solution as follows:

- If the new neighbor generated by the perturbation function dominates the current solution (i.e. the old one) then the current solution is updated.
- In the Line 9, if the neighbor solution is dominated by the current solution (the old one) then the Boltzmann distribution function is applied to accept or reject the new solution.

Notice that the Boltzmann distribution function for MOPs is a function applied for all the objectives of the problem. As in the single optimization problem, this function measure the quality solution as a criterion for decide if a new solution is or not accepted.

Algorithm 2. Multi-objective Simulated Annealing algorithm

```
1: Initialize Multi-objective Simulated Annealing (T_i,T_f,α)
2: T_k = T_i
3: S_i = generateSolution()
4: While T_k ≥ T_f do
5:      While Metropolis length do
6:           S_j = N(S_i)
7:           Δ = E(S_j)-E(S_i)
8:           If S_j ≺ S_i then
9:                S_i = S_j
10:          Else If S_i ≺ S_j then
11:     If e^(-Δ/Tk) < random (0, 1) then
12:          S_i = S_j
13:     Else
14:          S_i = S_j
15:     T_k+1 = αT_k
```

4.1 Acceptance Criteria

General acceptance criteria cannot be applied directly to multi-objective problems because acceptance criteria only supports scalar cost function. To adapt acceptance criteria to multi-objective optimization has been proposed a wide variety of acceptance criteria which made use of weighted aggregative functions as in (Serafini, 1993; Nam & Park, 2000). The most common rules are the product and the sum functions. Also other strategies can be considered in the acceptance criteria as can be seen in the next section, for example the total of dominated solutions.

5. MODERN RESEARCH ABOUT MOSA

Modern MOSA research includes diverse approaches. Between them are hybrid algorithms, new strategies in the acceptance criteria, diverse concepts to accept candidate solutions in the Pareto set, evolutive techniques, populations and adaptive rules. These approaches are briefly described below:

- The MOSA proposed in (Teghem, Tuyttens, & Ulungu, 2000) is an interactive algorithm and uses a wide diversified set of weights vector. The initial solutions are generated by a greedy problem dependent method. Uses a structure to store all the efficient solutions and includes a filter to maintain all the non-dominated solutions. The interactive part implements a dialog phase with the decision maker. The decision maker discards the unsatisfying solutions and keeps the preferred solutions. Taking into account the information of the decision maker a new set of weight vectors are generated. The decision maker can update the parameters of the MOSA to intensify the research in directions of interesting solutions. When the decision maker is completely satisfied the interactive MOSA stops.
- In (Smith, Everson, Fieldsend, Murphy, & Misra, 2008) proposed an energy measure. This function is defined on terms of the current estimate Pareto front. The function considers the difference of the number of dominated solutions between the new and the current solution. This function does not rely on weighting vectors for the objective functions. If a new solution is accepted then the dominated solutions from the estimate Pareto front are removed. This energy measure has a good convergence properties and produces a good Pareto front.
- In (Haidine & Lehnert, 2008) a Multi-case MOSA (MC-MOSA) is proposed. This algorithm subdivides the MOSA in three cases. These cases occurs when a deterioration and an improvement happens on different objectives. That means that the candidate solution is indifferent from the current solution. These cases are considered to accept a candidate solution. The first case considers the number of solutions that the candidate solution dominates in the Pareto set (dominance counter). If the dominance counter from the candidate solution is bigger than cero then the candidate solution is accepted. The second case considers the number of solutions in the Pareto set that dominates the candidate solution (dominance rank). If the dominance rank is equals to cero then the candidate solution is accepted. In the third case the candidate solution cannot be part of the Pareto set and considers the dominance rank difference between the candidate solution and the actual solution. If the dominance rank from the candidate solution is less than the dominance rank from the actual solution then the candidate solution is accepted. If the dominance rank are equals then an acceptance criteria is applied.

- In (Li & Landa-Silva, 2008) an Evolutionary MOSA (EMOSA) is proposed. This algorithm uses the ideas of MOEA/D to maintaining diverse weight vectors. Implements a two-phase strategy for maintaining the diversity of search directions. This two-phase is a combination from (Ulungu, Teghem, Fortemps, & Tuyttens, 1999) and (Czyzzak & Jaszkiewicz, 1998). In the first phase, fixed vectors are used and in the second phase an adaptive search direction is applied to examine the solution space.
- An algorithm called Archive MOSA (AMOSA) is proposed in (Bandyopadhyay, Saha, Maulik, & Deb, 2008). This algorithm uses an archive to store all the non-dominated solutions. This archive has a two limits size: the strict limit and the soft limit. All the non-dominated solutions generated are stored in the archive. When the archive reaches the soft limit a clustering is applied to reduce the number of solutions to the hard limit. The algorithm incorporates the concept of amount of dominance to distinguish between solutions which are more dominated and less dominated. Besides, the clustering is used to increase the solutions diversity in the archive. The amount of dominance is used to accept a candidate solution to be added to the archive.
- In (Hung, Shu, Ho, Hwang, & Ho, 2008) is proposed an Intelligent MOSA (IMOSA) based on orthogonal experimental designs. This MOSA uses a scoring function which considers the dominance counter and the dominance rank. A solution has a high score if it dominates many solutions. Conversely a solution has a low score if many individuals dominate it. This algorithm utilizes an elite set with high score solutions to maintain the best non-dominated solutions. IMOSA uses a three-level orthogonal array and a total experiments of 3^N for the one-factor-at-once study.
- In (Frutos, Olivera, & Tohmé, 2010) a memetic algorithm is implemented. This algorithm utilizes the MOSA algorithm as local search in NSGA-II. MOSA allows the exploration of a larger portion of the search space. The metropolis length is $M(T) = 1/\mathit{temperature}$.
- A hybridization of NSGA-II and MOSA is proposed in (Erol, 2011). In this hybridization the MOSA is used as local search in NSGA-II to escape from local optimal.
- In (Sim, Lee, & Lee, 2012) the Multi-objective Conformational Space Annealing (MOCSA) is proposed and it has the next characteristics:
  - A classification based on dominance relationship and distance between solutions in the objective space are used as fitness measure,
  - Update rules are based on the fitness, and uses a constrained local minimizer. In this approach the solutions are represented by hyper-spheres. Each solution has a hyper-sphere of radius D in the decision space. MOCSA introduces a distance parameter called D_{cut} which plays the role of temperature. This D_{cut} is initialized in a large value and is slowly decreased to small values. The diversity is controlled by distance measure between two solutions and comparing it with D_{cut}.
  - The use of a gradient-based constrained minimizer, feasible sequential quadratic programming (FSQP), for local search.
- In (Alrefaei, Diabat, Alawneh, Al-Aomar, & Faisal, 2013) an algorithm called α-MOSA is proposed. This algorithm uses the concept of α-dominance and a hill-climbing technique to escape from local minimal. The α-dominance represents a confidence interval of dominance. This algorithm constructs a Pareto set that consists of α-non-dominated solutions. The candidate solution is accepted if it α-dominates one of the solutions in the Pareto set, then the candidate solution enters to the Pareto set and the α-dominated solutions are removed from the Pareto set. Also the temperature is constant all the time to provide more chance to explore the set of feasible solutions.

- The hybrid MOSA based genetic local search (MOSAGLS) is proposed in (Xu, Qu, & Li, 2013). This algorithm uses simulated annealing strategies within the genetic evolutionary process to evolve. MOSAGLS adapts the mutation rate to guide the search directions. In this algorithm each solution is associated with a random weight vector. The weight vector and the temperature are considered in the solution acceptance and in the tuning of search directions. The genetic local search maintains a set of all non-dominated solutions. MOSAGLS uses the evaluation method of the Strength Pareto.
- A MOSA of two stages is presented in (Abubakera, Baharuma, & Alrefaei, 2014). This MOSA uses in the first stage the MOSA designed by (Ulungu, Teghem, Fortemps, & Tuyttens, 1999). In the second stage the optimal computing budget allocation (OCBA) technique is used to obtain a good solution based on the Pareto set. This algorithm considers the ranking of the decision maker preferences. The non-numerical ranking preferences (NNRP) method is used for the ranking. In NNRP the decision maker does not need to assign weight vectors for the objective functions. The decision maker only needs to rank the objective function based on their importance. The OCBA technique is used to determine the number of samples for each solution that maximizes the probability of selecting a candidate solution.
- Other hybrid is proposed in (Pham, Nguyen, & Tran, 2015). This hybrid combines the Particle Swarm Optimization with the MOSA behavior (MOPSO-SA). The MOSA is used to avoid a convergence in the local optimal solution. MOPSO-SA takes the advantages of the global search provided by the PSO and the local search from SA. This algorithm store the best particles, non-dominated particles, in an external repository and the fitness sharing of each particle is calculated based on them. The PSO part helps to approach the global optimal area. The SA helps to do the gradient search within a region. The acceptance criteria uses the root mean squared residual (RMSR) to accept a candidate solution. In the metropolis cycle, to avoid an endless loop the last position is accepted if the metropolis cycle reach the length of 100.
- In (Lin & Ying, 2015) is implemented a Multi-objective Multi-point Simulated Annealing (MOMSA). Multi-point provides a mechanism to escape from local optimal. The set of initial solutions denoted as multi-points are randomly selected from the non-dominated solutions. This multi-point solution is used to generate a neighbor. Then the differences from each objective is used in the acceptance criteria.
- In (Zaretalab, Hajipour, Sharifi, & Shahriari, 2015) the Knowledge based AMOSA (KBAMOSA) algorithm is presented. This is a modification of AMOSA. It differs from AMOSA in neighborhood structure and memory updating. In this algorithm, a memory matrix is implemented. This memory saves the best and worse changes in the solutions. This matrix sign the effective changes of the neighborhood function. To avoid converging in local solutions only the 80 percent of solutions are generated using the matrix.
- Some hybrids are proposed in (Pham, Nguyen, & Tran, 2015; Xu & Qu, 2011; Suman & Kumar, 2006; Lin & Ying, 2015; Frutos, Olivera, & Tohmé, 2010). These hybrids combines simulated annealing algorithm with other metaheuristics: NSGA-II, PSO, Tabu Search, and VNS.
- MOSA for scheduling problems are published in (Osorio, Lasso, & Ruiz, 2012; Erol, 2011; Loukil, Teghem, & Tuyttens, 2005; Lin & Ying, 2015).

The Table 1 summarizes the reviewed MOSAs algorithms. The first column represents the paper title. The second column contains the authors' names. The third column mentions the contribution from the paper.

Table 1. Summarize of multi-objective Simulated Annealing algorithms in the state of the art

Title	Author	Contribution
An interactive heuristic method for multi-objective combinatorial optimization (2000).	Jacques Teghem, Daniel Tuyttens, and Ekunda L. Ulungu	An interactive MOSA algorithm. Uses a dialog phase with the Decision Maker.
A survey of simulated annealing as a tool for single and multi-objective optimization (2006).	Balram Suman, and Prabhat Kumar	Review of WMOSA, PSA, UMOSA, PDMOSA, and SMOSA.
Dominance-Based Multi-Objective Simulated Annealing (2008).	Kevin Smith, Richard M. Everson, Jonathan E Fieldsend, Chris Murphy, and Rashmi Misra	The energy function considers the difference of the number of dominated solutions between the new and the current solution.
Multi-case multi-objective simulated annealing (mc-mosa): New approach to adapt simulated annealing to multi-objective optimization (2008).	Abdelfatteh Haidine, and Ralf Lehnert	MC-MOSA algorithm. Modifies the acceptance decision rule.
Evolutionary multi-objective simulated annealing with adaptive and competitive search direction (2008).	Hui Li, and Dario Landa-Silva	EMOSA algorithm. Incorporate a search direction and a competitive scheme.
A simulated annealing-based multi-objective optimization algorithm: AMOSA (2008).	Sanghamitra Bandyopadhyay, Sriparna Saha, Ujjwal Maulik, and Kalyanmoy Deb.	AMOSA algorithm. Use a grade of dominance and an archive to store the non-dominated solutions.
A novel intelligent multi-objective simulated annealing algorithm for designing robust PID controllers (2008).	Ming-Hao Hung, Li-Sun Shu, Shinn-Jang Ho, Shiow-Fen Hwang, and Shinn-Ying Ho	IMOSA algorithm. Implements an orthogonal experimental design.
A memetic algorithm based on a NSGA-II scheme for the flexible job-shop scheduling problem (2010).	Mariano Frutos, Ana Carolina Olivera, and Fernando Tohmé	Hybridization of NSGA-II and MOSA.
Solving multi-objective multicast routing problems by evolutionary multi-objective simulated annealing algorithms with variable neighborhoods (2011).	Ying Xu, and Rong Qu	The VEMOSA algorithm. A hybridization of EMOSA with variable neighborhoods.
Comparison of different heuristic search methods in a design space exploration (2011).	Ibrahim Erol	Hybridization of NSGA-II and Simulated Annealing.
MOCSA: multiobjective optimization by conformational space annealing (2012).	Sangjin Sim, Juyong Lee, and Jooyoung Lee	The MOCSA, an algorithm based on conformational space annealing.
Simulated annealing for multi-objective stochastic optimization (2013).	Mahmoud Alrefaei, Ali Diabat, Ameen Alawneh, Raid Al-Aomar, and Mohd Nishat Faisal.	The α -MOSA algorithm. Use a grade of dominance.
A simulated annealing based genetic local search algorithm for multi-objective multicast routing problems (2013).	Ying Xu, Rong Qum, and Renfa Li	MOSAGLS. A hybrid algorithm based on genetic local search.
Good solution for multi-objective optimization problem (2014).	Ahmad Abubakera, Adam Baharuma, and Mahmoud Alrefaei	Combines the MOSA with the optimal computing budget allocation (OCBA) technique.
A Hybrid Multi-objective PSO-SA Algorithm for the Fuzzy Rule Based Classifier Design Problem with the Order Based Semantics of Linguistic Terms (2015).	Phong Dinh Pham, Thuy Thanh Nguyen, and Thanh Xuan Tran	The MOPSO-SA algorithm. A hybrid multi-objective particle swarm optimization with simulated annealing.
A multi-point simulated annealing heuristic for solving multiple objective unrelated parallel machine scheduling problems (2015).	Shih-Wei Lin, and Kuo-Ching Ying	The MOMSA algorithm. Combines the advantages of the multi-point method and simulated annealing.
A knowledge-based archive multi-objective simulated annealing algorithm to optimize series-parallel system with choice of redundancy strategies (2015).	Arash Zaretalab, Vahid Hajipour, Mani Sharifi, and Mohammad Reza Shahriari.	The KBAMOSA algorithm. Modification of AMOSA using a memory matrix.

These algorithms covers or combines the following approaches: niche Pareto, decomposition, adaptive algorithms, hybrids, genetic algorithms, evolutive algorithms, and parameters tuning.

6. CONCLUSION

In this chapter diverse MOSA were summarized to describe the current contributions which aims to improve this algorithm. These improvements include hybridizations which made use of the MOSA as local search and complete combinations of the mechanism in other algorithms as the acceptance criteria and the temperature control. The MOSA algorithm has been applied to solve diverse real problem as mentioned in Section 1 and it is still been applied to solve modern problems.

From the author's perspective future work is still need to be research in the acceptance criteria and the generation of weight vectors or the computation of the energy, auto-tuning parameters in the metropolis cycle, and the use of modern technologies as parallel programming and multi-threads.

The chapter presents a review of Multi-objective Simulated Annealing algorithms for multi-objective problems. The chapter shows the general model formulation of MOP and the applicability of simulated annealing algorithms for these problems; the statistical mechanics approach for MOP is discussed. In addition, a general review of the main MOSA problems is presented.

ACKNOWLEDGMENT

The authors would like to acknowledge with appreciation and gratitude to CONACYT. This work has been partial supported by CONACYT Project.254498 and PRODEP.

REFERENCES

Abubakera, A., Baharuma, A., & Alrefaei, M. (2014). Good solution for multi-objective optimization problem. In Proceedings of the 21st National Symposium on Mathematical Sciences (SKSM21): Germination of Mathematical Sciences Education and Research towards Global Sustainability. AIP Publishing.

Alrefaei, M., Diabat, A., Alawneh, A., Al-Aomar, R., & Faisal, M. N. (2013). Simulated annealing for multi objective stochastic optimization. *International Journal of Science and Applied Information Technology*, *2*, 18–21.

Bandyopadhyay, S., Saha, S., Maulik, U., & Deb, K. (2008). A simulated annealing-based multiobjective optimization algorithm: AMOSA. *IEEE Transactions on Evolutionary Computation*, *12*(3), 269–283. doi:10.1109/TEVC.2007.900837

Coello Coello, C. A., Lamont, G. B., & Van Veldhuizen, D. A. (2007). *Genetic and Evolutionary Computation. Evolutionary Algorithms for Solving Multi-Objective Problems* (J. R. David Goldberg, Ed.). New York: Springer.

Czyzzak, P., & Jaszkiewicz, A. (1998). Pareto simulated annealing—a metaheuristic technique for multiple-objective combinatorial optimization. *Journal of Multi-Criteria Decision Analysis*, *7*(1), 34–47. doi:10.1002/(SICI)1099-1360(199801)7:1<34::AID-MCDA161>3.0.CO;2-6

Deb, K. (2001). *Multi-Objective Optimization using Evolutionary Algorithms*. New York: John Wiley & Sons.

Erol, I. (2011). *Comparison of different heuristic search methods in a design space exploration*. Academic Press.

Frutos, M., Olivera, A. C., & Tohmé, F. (2010). A memetic algorithm based on a NSGAII scheme for the flexible job-shop scheduling problem. *Annals of Operations Research*, *181*(1), 745–765. doi:10.1007/s10479-010-0751-9

Frutos, M., & Tohmé, F. (2015). Choice of a PISA selector in a hybrid algorithmic structure for the FJSSP. *Decision Science Letters*, *4*(2), 247–260. doi:10.5267/j.dsl.2014.11.001

Haidine, A., & Lehnert, R. (2008). Multi-case multi-objective simulated annealing (mc-mosa): New approach to adapt simulated annealing to multi-objective optimization. *International Journal of Information Technology*, *4*(3), 9.

Hung, M.-H., Shu, L.-S., Ho, S.-J., Hwang, S.-F., & Ho, S.-Y. (2008). A novel intelligent multiobjective simulated annealing algorithm for designing robust PID controllers. *Systems, Man and Cybernetics, Part A: Systems and Humans. IEEE Transactions on*, *38*(2), 319–330.

Kirkpatrick, S. (1984). Optimization by simulated annealing: Quantitative studies. *Journal of Statistical Physics*, *34*(5-6), 975–986. doi:10.1007/BF01009452

Knowles, J. D., & Corne, D. W. (2000). Approximating the nondominated front using the Pareto archived evolution strategy. *Evolutionary Computation*, *8*(2), 149–172. doi:10.1162/106365600568167 PMID:10843519

Li, H., & Landa-Silva, D. (2008). Evolutionary multi-objective simulated annealing with adaptive and competitive search direction. In *Evolutionary Computation, 2008. CEC 2008.(IEEE World Congress on Computational Intelligence). IEEE Congress on*, (pp. 3311-3318). IEEE.

Lin, S.-W., & Ying, K.-C. (2015). A multi-point simulated annealing heuristic for solving multiple objective unrelated parallel machine scheduling problems. *International Journal of Production Research*, *53*(4), 1065–1076. doi:10.1080/00207543.2014.942011

Loukil, T., Teghem, J., & Tuyttens, D. (2005). Solving multi-objective production scheduling problems using metaheuristics. *European Journal of Operational Research*, *161*(1), 42–61. doi:10.1016/j.ejor.2003.08.029

Nam, D., & Park, C. H. (2000). Multiobjective simulated annealing: A comparative study to evolutionary algorithms. *International Journal of Fuzzy Systems*, *2*(2), 87–97.

Osorio, J. C., Lasso, D. F., & Ruiz, G. A. (2012). Ob shop scheduling biobjetivo mediante enfriamiento simulado y enfoque de pareto. *Revista Ingenierias Universidad de Medellin*, *11*(21), 113–126.

Pham, P. D., Nguyen, T. T., & Tran, T. X. (2015). A Hybrid Multi-objective PSO-SA Algorithm for the Fuzzy Rule Based Classifier Design Problem with the Order Based Semantics of Linguistic Terms. *VNU Journal of Science: Computer Science and Communication Engineering, 30*(4).

Safaei, N., Jardine, A. K., & Banjevic, D. (2008). *Multi-objective simulated annealing for a maintenance workforce scheduling problem: a case study*. INTECH Open Access Publisher. doi:10.5772/5558

Serafini, P. (1993). Simulated annealing for multi objective optimization problems. In Multiple criteria decision making (pp. 283-292). Springer.

Sim, S., Lee, J., & Lee, J. (2012). *MOCSA: multiobjective optimization by conformational space annealing*. arXiv preprint arXiv:1209.0549

Smith, K., Everson, R. M., Fieldsend, J. E., Murphy, C., & Misra, R. (2008). Dominance-based multiobjective simulated annealing. *Evolutionary Computation. IEEE Transactions on, 12*(3), 232–342.

Suman, B., & Kumar, P. (2006). A survey of simulated annealing as a tool for single and multiobjective optimization. *The Journal of the Operational Research Society, 57*(10), 1143–1160. doi:10.1057/palgrave.jors.2602068

Teghem, J., Tuyttens, D., & Ulungu, E. L. (2000). An interactive heuristic method for multi-objective combinatorial optimization. *Computers & Operations Research, 27*(7), 621–634. doi:10.1016/S0305-0548(99)00109-4

Ulungu, E., Teghem, J., Fortemps, P., & Tuyttens, D. (1999). MOSA method: a tool for solving multi-objective combinatorial optimization problems. *Journal of Multicriteria Decision Analysis, 8*(4), 221.

Xu, Y., & Qu, R. (2011). Solving multi-objective multicast routing problems by evolutionary multi-objective simulated annealing algorithms with variable neighbourhoods. *The Journal of the Operational Research Society, 62*(2), 313–325. doi:10.1057/jors.2010.138

Xu, Y., Qu, R., & Li, R. (2013). A simulated annealing based genetic local search algorithm for multi-objective multicast routing problems. *Annals of Operations Research, 206*(1), 527–555. doi:10.1007/s10479-013-1322-7

Zaretalab, A., Hajipour, V., Sharifi, M., & Shahriari, M. R. (2015). A knowledge-based archive multi-objective simulated annealing algorithm to optimize series--parallel system with choice of redundancy strategies. *Computers & Industrial Engineering, 80*, 33–44. doi:10.1016/j.cie.2014.11.008

Chapter 15
Emergency Department Logistic Optimization Using Design of Experiments:
From Triage to High Quality of Service

Jose Antonio Vazquez-Ibarra
Universidad Politecnica de Aguascalientes, Mexico

Rodolfo Rafael Medina-Ramirez
Universidad Politecnica de Aguascalientes, Mexico

Irma Jimenez-Saucedo
Universidad Politecnica de Aguascalientes, Mexico

ABSTRACT

Public healthcare services face a growing demand and Emergency department is the main entrance to these services. Waiting times at Emergency departments are increasing at risky levels, causing that people die in wait rooms due to a lack of staff to serve timely every patient. Present chapter describes one research project conducted in a mexican public hospital which was in the process of adopting a triage systems in order to reach the goal of a maximum wait time in department. Design of experiments is the tool proposed to analyze waiting time factors and define the best levels to reduce the response variable value.

INTRODUCTION

Public healthcare services have around the globe an ever-growing demand along with a shrinking budget (Brandeau, Sainfort & Pierskalla, 2004).

Recently healthcare service providers are adopting operations research techniques in order to improve logistics of services reducing costs and enhancing medical facilities capacities. These research efforts are classified as patient flow improvement. Metaphors use to be useful thinking tools and conceiving patients going from one service point to another as a flow, helps to model its dynamic as a flow and build and abstract model with mathematics and logic.

DOI: 10.4018/978-1-4666-9779-9.ch015

The main research categories on patient flow improvement are the following (Medina, 2010a):

1. Operations changes to increase patient flow;
2. Lean thinking in hospitals;
3. Queueing theory application on healthcare;
4. Discrete event simulation and design of experiments;
5. Integration of Systems Dynamics and Discrete event simulation;
6. Agent Based Modeling in healthcare;
7. Analytical methods to define required beds, rooms capacity and required staff;
8. Hospital design methodology and
9. Mathematical modeling of patient flow in hospitals.

It is common to define the Emergency Department as the access gate to hospitals. The outpatient department capacity cannot provide timely service to patients so they look for an access point to medical services and they usually find it at Emergency Department.

This chapter presents one applied research in an Emergency Department (ED) designed to identify and quantify the main factors in waiting time using Design of Experiments in order to improve the quality of Mexican public healthcare services.

BACKGROUND

Healthcare access is a constitutional guarantee for all citizens in Mexico; most of healthcare services are delivered by public institutions. Work regulation demands employers to pay for healthcare service for his employees and their families. Healthcare delivery is the responsibility of both, the federal and the state governments. Federal health care delivery is achieved through IMSS (Mexican Institute of Social Security) or ISSSTE (Institute of Social Security and Services for State [government] Workers), for most of Mexican employees in private and public offices. Some government employees (oil industry, army and navy) are covered by PEMEX or SEDENA healthcare delivery subsystems (see Figure 1). A set of Mexican workers had no access to any of the previous agencies (mostly small business owners,

Figure 1. Mexican social security providers

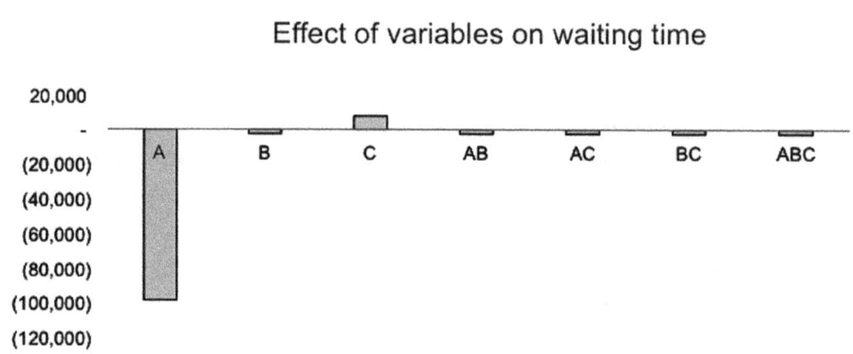

freelancers) so *Seguro Popular* (SPSS, Popular healthcare insurance) was created to help those persons and its families to have access to health care services, which are delivered through SSA (state governments' health ministry, SSA is the acronym to agency's name in Spanish).

Mexican Health System is aimed to reach universal coverage. Every single Mexican must have access to health services. In ten years the percentage of people with social security grew more than 50% as is shown in Figure 2.

Due to this tendency on public healthcare services demand, it is mandatory to understand the dynamic of that system and design viable solutions to the challenge of provide quality services to general public.

Mathematical modeler has to deal with a trade-off between model attributes and resources available to develop the model. Model attributes are fidelity, resolution and scale. Fidelity defines how closely the model matches or behaves like the real system it is representing. Resolution is the degree of detail with which the real world is simulated. A high resolution model includes a more detailed description of the system. Scale is the size of the overall scenario or even the simulation represents; the system subject of the research study defines the scale of the simulation model, the larger the system, the greatest the effort to collect information on scenario components and components details (Sokolowski & Banks, 2010).

Modeling health care services usually demands high fidelity models in order to assess medical units performance or medical techniques effectiveness without threatening patient's safety. Resolution and scale are chosen on decision level involved, research budget available and data availability.

Figure 2. Mexican population social security

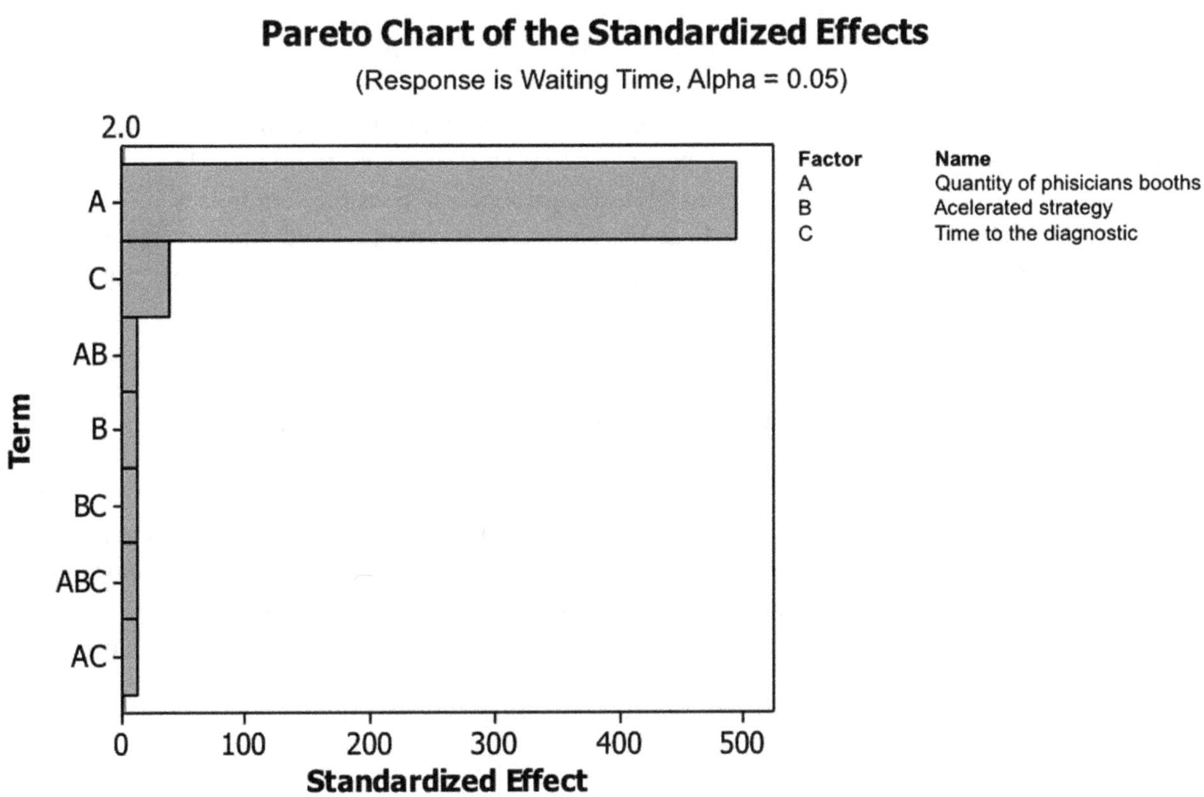

Variables Which Affects Waiting Time

Quality is measured through a set of quantitative variables and their values are compared against performance goals. In Spain, Ministry of Health and Social Policies have conducted research to define performance measures (Ministerio de Sanidad y Política Social, 2010); in Mexico SSA has published a set of performance goals in ED which includes: waiting time, percentaje of satisfied patients, among others (Secretaria de Salud, 2002). Waiting Time at ED is a main research objective due to its relevance in patient health and safety (Ramirez, 1980).

The influencing variables in waiting time can be classified in five analysis groups (Llorente, 2000):

1. Acuity level variables (in a general sense a classification system divide medical urgencies in actual urgencies (AU) and perceived urgencies (PU) and such a system may be a triage);
2. Physicians' variables (example: time to diagnosis);
3. Patient related variables (example: age, gender, social security);
4. Resources demand related variables (ancillary medical services, blood tests, x rays studies);
5. Exogenous waiting time (considering ED as reference);

This chapter describes one research project in which logistics is applied to a public healthcare service and includes DOE and patient flow concepts to model ED; model is aimed to identify main factors of waiting time in order to define the optimal ED configuration which minimize wait time.

EMERGENCY DEPARTMENT WAITING TIME: DOE ANALYSIS

Healthcare services providers usually do not have skills to mathematical modeling and optimization of services. They propose changes to services based on experience and intuition only. They need expert help to conduct deeper analysis of services and build confidence on possible solutions to it.

DOE is a statistical tool easy to use, so healthcare manager can adopt it and propose enhancements to medical services with higher level of confidence on the possible efficacy and efficiency of proposal.

In Mexican hospitals is possible to find some experiences on statistical tools use to analyze and face services challenges.

Mexican Experiences on Statistical Analysis of Healthcare Services

A public hospital at Mexico's north border (Baja California) developed a discrete-event simulation model of obstetrics and gynecology department. Model includes from check-in process up to recovery room discharge. An scenario evaluation was conducted to identify the best solution strategy and hospital achieved a 76% reduction of wait time (Medina, 2010b).

Medina propose an analysis and improvement model of patient flow in hospitals using lean manufacturing, basic statistical tools and even discrete event simulation with DOE through eight steps (Medina, 2012).

DESIGN OF EXPERIMENT

A Design of Experiment (DOE) was conducted using three factors, as described Figure 3:

Factor one of DOE has three levels and the other two has two levels each. Levels of each factor are described next:

Factor 1

Quantity of Physician's Booths (A): Defines the number of physical physician's booths at ED. It is different from physician's booths available (BA) since availability of booths is a dynamic variable affected by the occurrence of system events through time.

This factor has three levels considering the number of booths, going from the basic capacity (one booth) up to the maximum capacity (three booths). Factor values are 1, 2 and 3.

Factor 2

Accelerated Strategy (B): Defines if physicians are working at an accelerated pace in booths based on number of patients waiting in ED.

This factor has two levels stating if physicians are using an accelerated strategy, described in Table 1, or not. Factor values are 1 if accelerated strategy is being used and -1 otherwise.

Figure 3. DOE developed to optimize ED

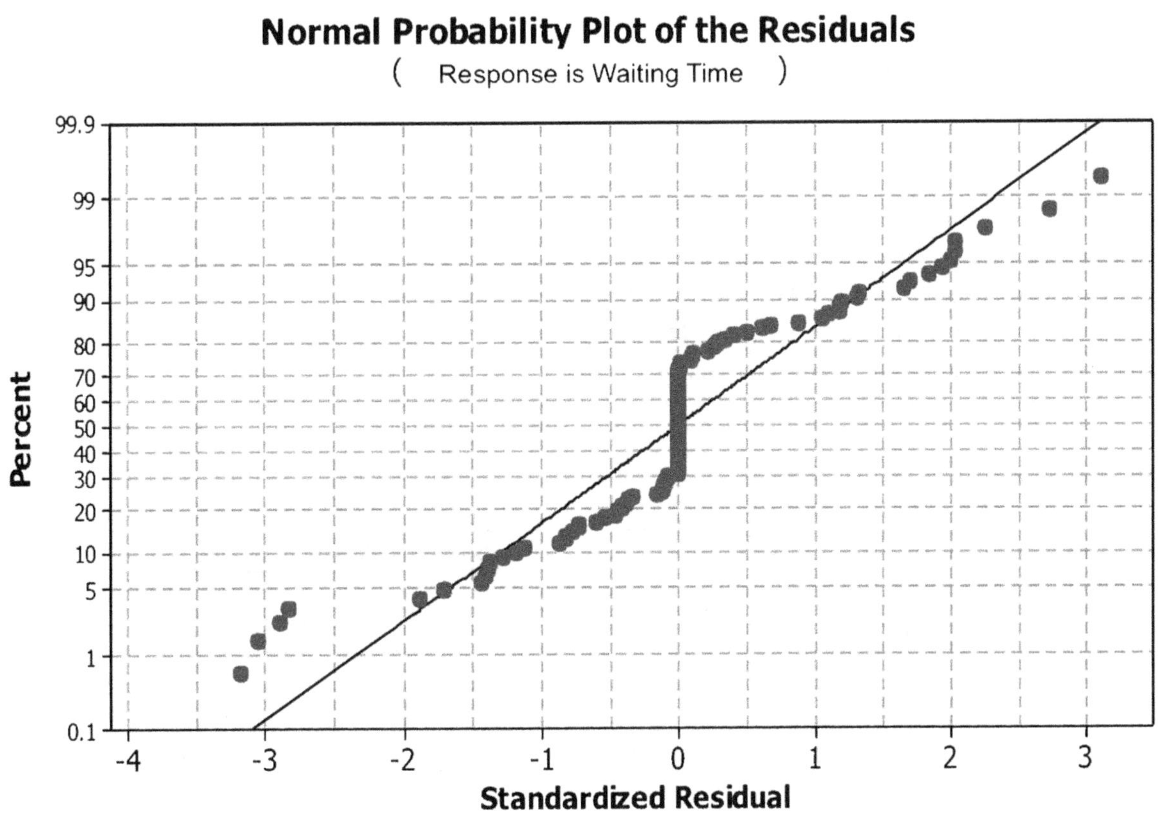

Table 1. Diagnosis strategies used in ED (Vazquez-Ibarra, 2014)

Size of Patients Queue	0 – 5	6 – 10	11 – 15	16+
Acceleration Factor	0%	15%	30%	50%
% of Time to Diagnosis	100%	85%	70%	50%

Factor 3

Time to Diagnosis (C): Defines the time invested by physicians on diagnosing patient's health.

This factor has two levels considering a fixed lapse of time to diagnosis (physicians will spend the same time on diagnosing without influence of patient acuteness nor illness complexity) or a variable diagnosis time (patient acuteness or illness complexity define length of required diagnosis time). Factor values are 1 for variable diagnosis time and -1 otherwise.

Healthcare Provision under Possible Scenarios

Combining three factors and seven levels, twelve different scenarios are possible. This scenarios are described in Table 2:

DOE model equation is Equation (1).

$$WT = \propto + \beta_A A + \beta_B B + \beta_C C + \beta_{AB} AB + \beta_{AC} AC + \beta_{BC} BC + \beta_{ABC} ABC \quad (1)$$

Equation 1: DOE Equation

Where:

WT = wait time (minutes)
α = rate of change constant of factors
β_A = Quantity of Physician' Booths factor coefficient
A = Quantity of Physician's Booths factor level
β_B = Accelerated Strategy factor coefficient

Table 2. Scenarios of healthcare provision

	Accelerated Strategy (B)	Quantity of Physicians' Booths		
		1	2	3
Time to Diagnosis (C)	w/strategy	fixed	fixed	fixed
		variable	variable	variable
	wo/strategy	fixed	fixed	fixed
		variable	variable	variable

B = Accelerated Strategy factor level
β_C = Time to Diagnosis factor coefficient
C = Time to Diagnosis factor level
β_{AB} = Quantity of Physician's Booths and Accelerated Strategy factors interaction coefficient
β_{AC} = Quantity of Physician's Booths and Time to Diagnosis factors interaction coefficient
β_{BC} = Accelerated Strategy and Time to Diagnosis factors interaction coefficient
β_{ABC} = Quantity of Physician's Booths, Accelerated Strategy and Time to Diagnosis factors interaction factor coefficient

A three factors DOE were created using Minitab® based on Table 2 scenarios, DOE which is described in Table 3:

Standard order column defines the way the different scenarios were generated based on the levels of each factor, so first row correspond to one physician booth, no accelerated strategy and a fixed time to diagnosis, this scenario has 27 as id number. Every scenario has then runs so 120 experiments were conducted.

Run order column is a random order generated by Minitab® and defines the order in which every experiment should be run.

Using ED simulation model developed in (Vázquez, 2014) each scenario were run ten times, the model was run on ProModel® and was adapted accordingly with each scenario features, following run order column. Table 4 comprises results of wait time variable:

Scenarios Analysis

Table 3. ED Optimization DOE (first twelve experiments)

Standard Order	Run Order	Quantity of Physicians Booth	Accelerated Strategy	Time to Diagnosis
27	1	1	wo/strategy	fixed
94	2	3	w/strategy	variable
113	3	2	w/strategy	fixed
91	4	2	wo/strategy	fixed
117	5	3	w/strategy	fixed
43	6	2	wo/strategy	fixed
32	7	2	wo/strategy	variable
87	8	1	wo/strategy	fixed
55	9	2	wo/strategy	fixed
57	10	3	w/strategy	fixed
79	11	2	wo/strategy	fixed
25	12	1	w/strategy	fixed

Considering Wait Time scenarios are stochastics experiments, a test for residual normality were led so experiments results can be considered random and independent. Figure 4 shows residuals fitting stan-

Table 4. ED Optimization experiments results (wait time in minutes)

Standard Order	Run Order	Quantity of Physicians Booth	Accelerated Strategy	Time to Diagnosis	Wait Time
45	1	1	wo/strategy	variable	107286
32	2	3	w/strategy	variable	90
41	3	1	wo/strategy	fixed	98219
47	4	1	w/strategy	variable	107286
20	5	3	w/strategy	fixed	29
35	6	1	w/strategy	fixed	98219
51	7	1	w/strategy	fixed	95858
72	8	3	w/strategy	variable	93
19	9	1	w/strategy	fixed	94760
14	10	3	wo/strategy	variable	9431

dardized line what can be taken as a proof of statistic independence of the experiments with enough confidence.

In order to assess relevance of factors or a combination of them, a Pareto analysis was conducted finding out that Quantity of Physicians Booth has the bigger influence on wait time, followed by Time to Diagnosis. The other factors and combinations have a similar and less significant influence on wait time. Figure 5 shows Pareto chart of wait time factors.

Figure 4. Test for residual normality of wait time

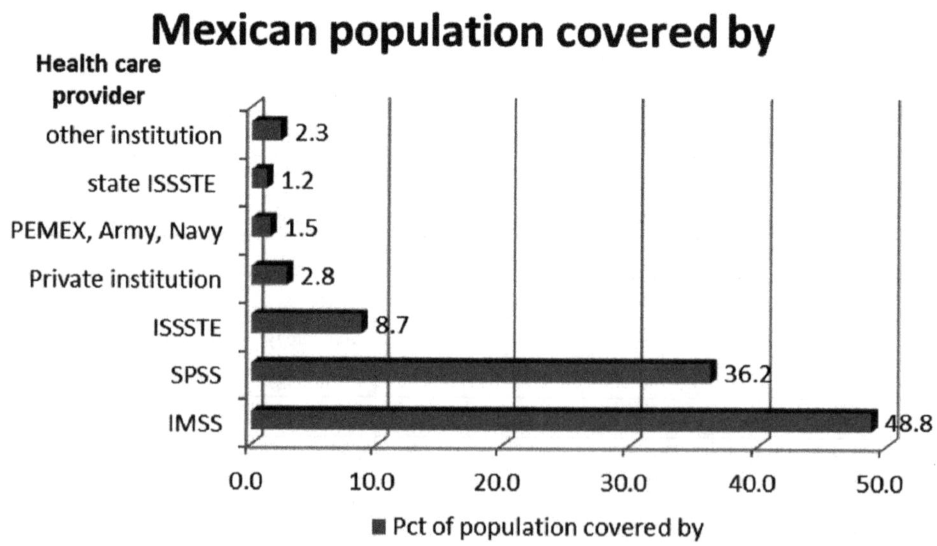

Total is greater than 100% due to those who are covered by more than one institution

To determine the direction of influence of relevant factors detected through Pareto chart, another was built to measure the size of variable effect on wait time. Figure 6 shows magnitude and direction of each factor effect on wait time, factor A: quantity of physicians booths has a negative effect, this means that if the quantity of physicians booths is increased, wait time will be reduced significantly. Time to diagnosis factor (factor C) has a positive effect; if time to diagnosis is variable, wait time will increase slightly.

One effect chart deepens insight of each factor effect on wait time variable (figure 7).Quantity of physicians' booths is a line with a significant slope, showing that as long as such quantity increases, wait time reduces in a significant amount.

The following factor in effect magnitude is accelerated strategy. If ED uses a strategy to accelerate time to diagnosis based on size of patients queue, wait time will be reduced, otherwise, the variable of interest shows a strong increasing tendency.

Time to diagnosis is the factor with the minor impact on wait time. Slope of time to diagnosis line if small so, seems like switching from a fixed amount of time to diagnosis to a variable one based on patient's illness issues, will have a reduced effect on the variable of interest, and the effect will be an increase of it for the variable time to diagnosis.

Figure 5. Pareto chart of wait time factors

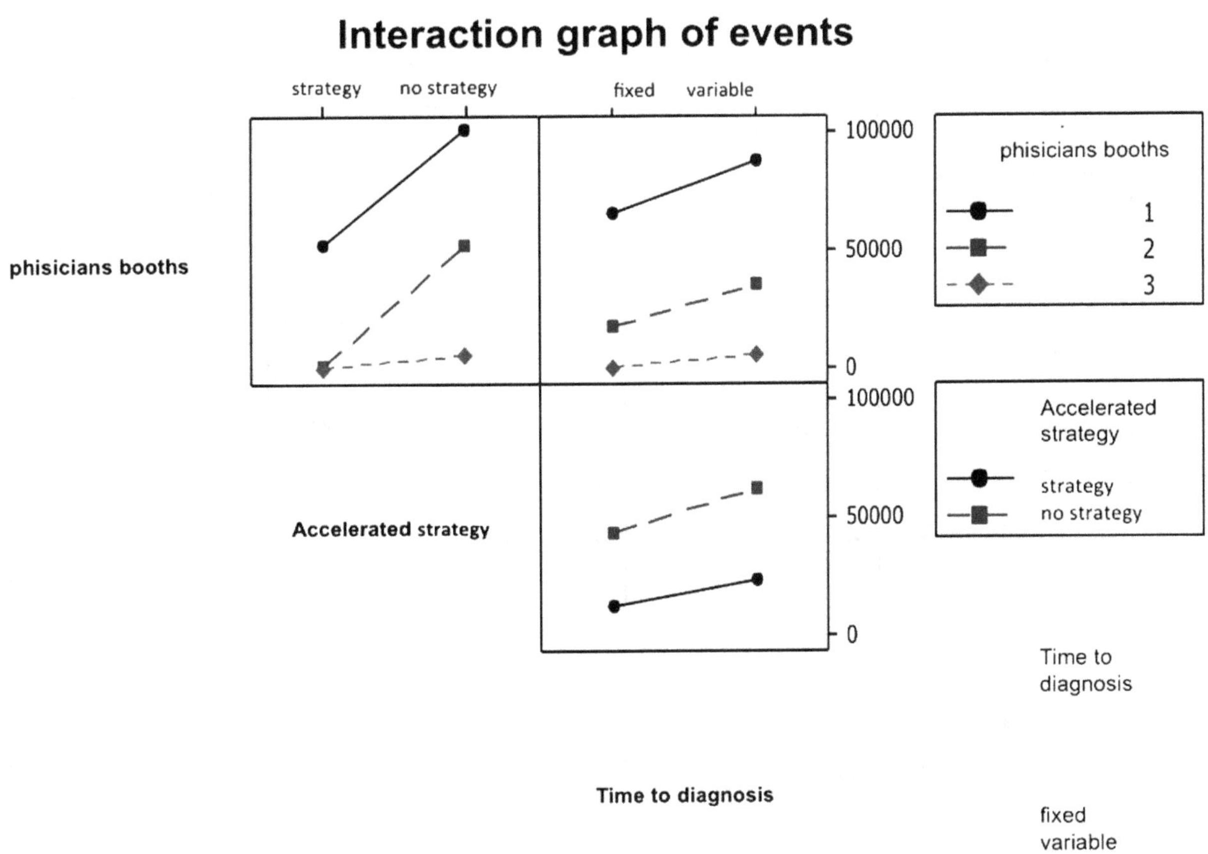

Figure 6. Size and direction of factors effect on wait time

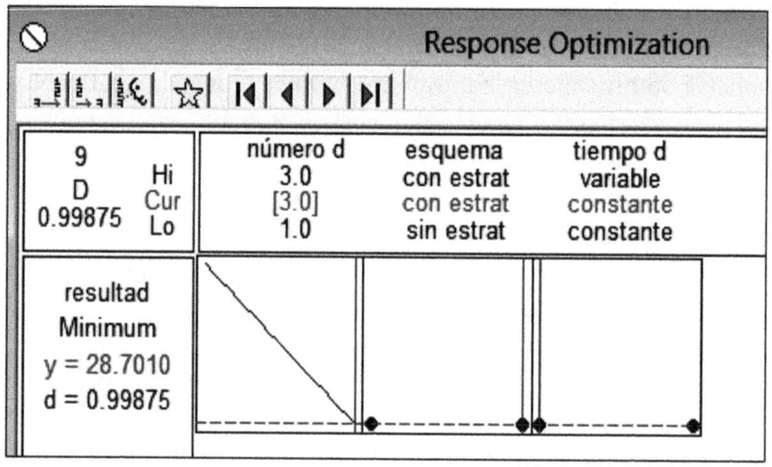

Figure 7. Effect chart of factors effect on wait time

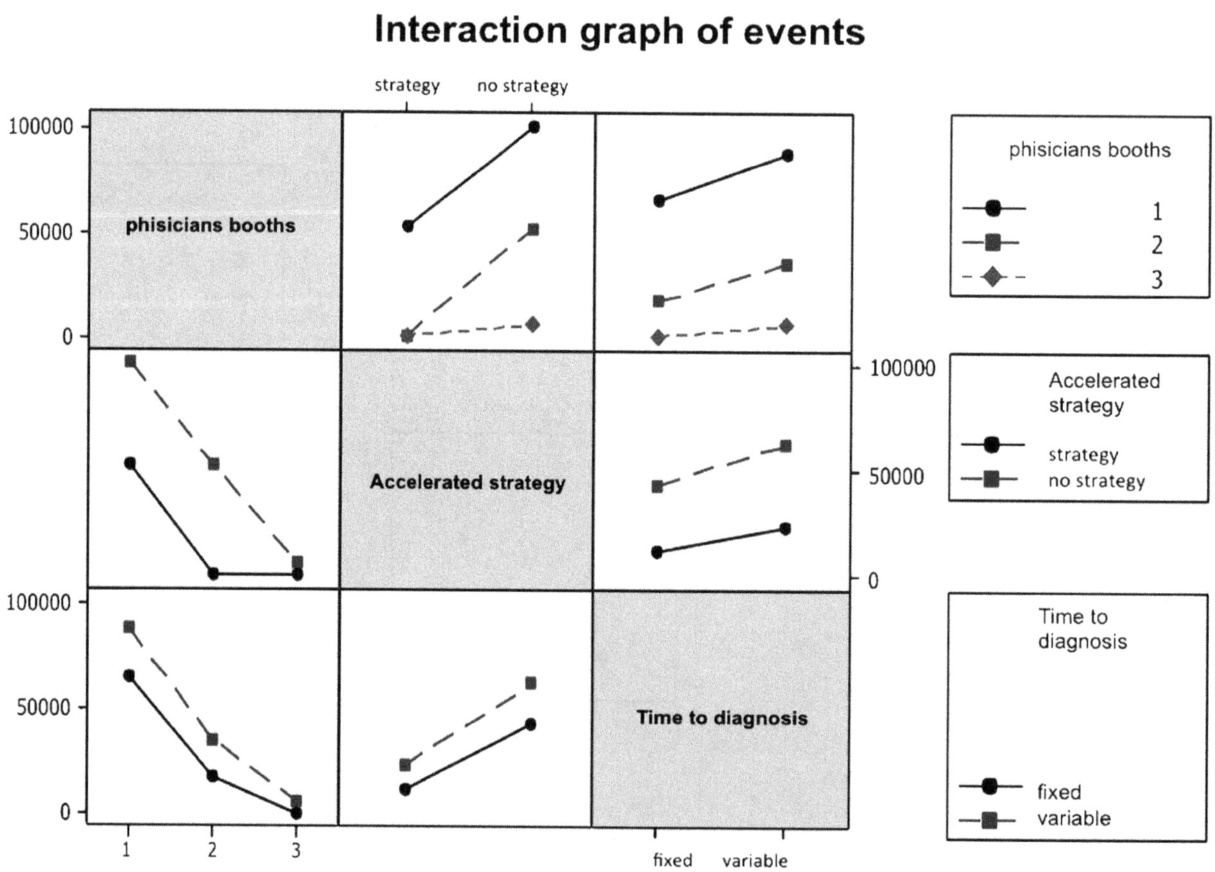

When using design of experiments it is easy to identify which combination of factors has an impact on wait time and its magnitude. Figure 8 shows factors combinations and its effect on wait time.

To identity interaction between factors their lines must intersect otherwise no interaction may be identified, the impact of this combination on the variable of interest is null.

Figure 6 shows two significant combinations. The first is the quantity of phisician's booths and accelerated strategy. If an accelerated strategy is being used and physician's booths goes from one to two, wait time is reduced in a significant amount of time, but adding a third booth has no impact on wait time (row 3 column 1 of figure 8).

If no accelerated strategy is used, every step increasing quantity of physician's booths has a constant impact on reducing wait time (row 3 column 1 of figure 8).

One conclusion is an accelerated strategy is required to reduce wait time if ED has two or three physician's booths. Figure 9 shows quantity of physician's booths and accelerated strategy intersecting at three booths and accelerated strategy applied.

The following questions arise: using an accelerated strategy wait time will not change even if a fourth physician's booth is added to ED? If an accelerated strategy is not used and the quantity of physician's booths is increased far beyond three the wait time will be significantly reduced? These questions may be easily answered running additional experiments using four or more physician's booths.

Figure 8. Effects of Combinations of factors

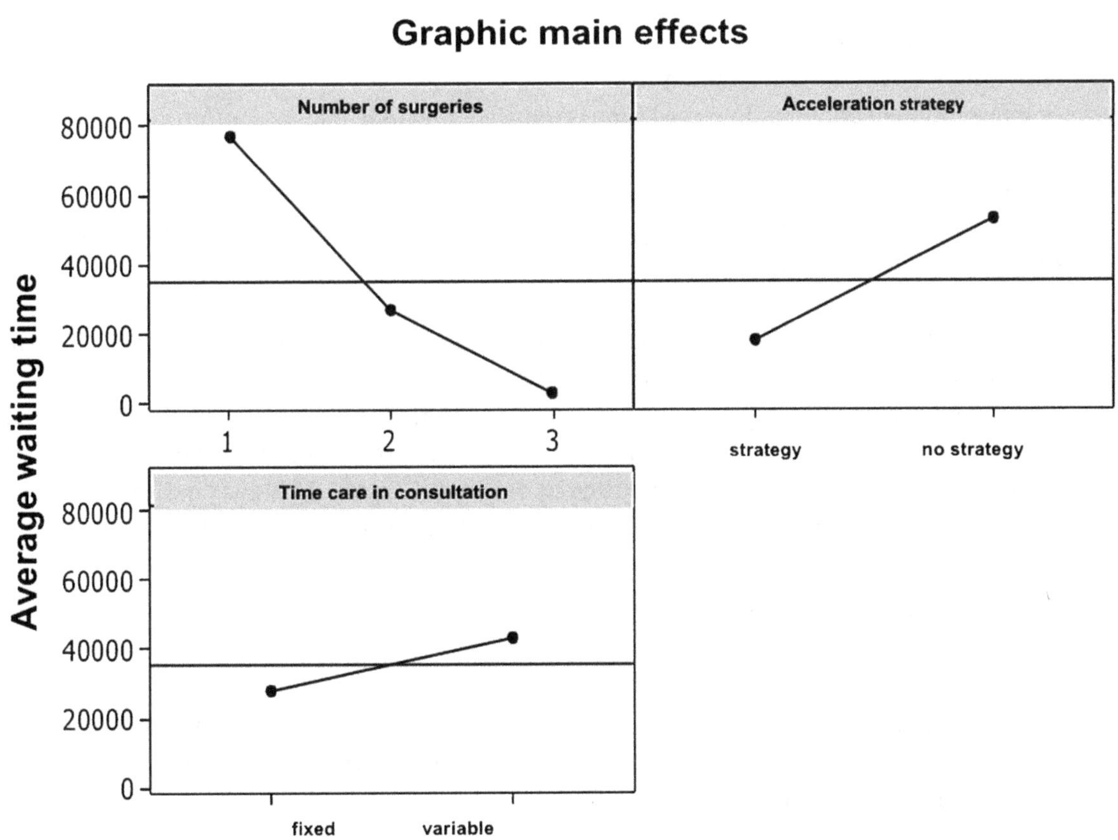

Figure 9. Interaction of factors

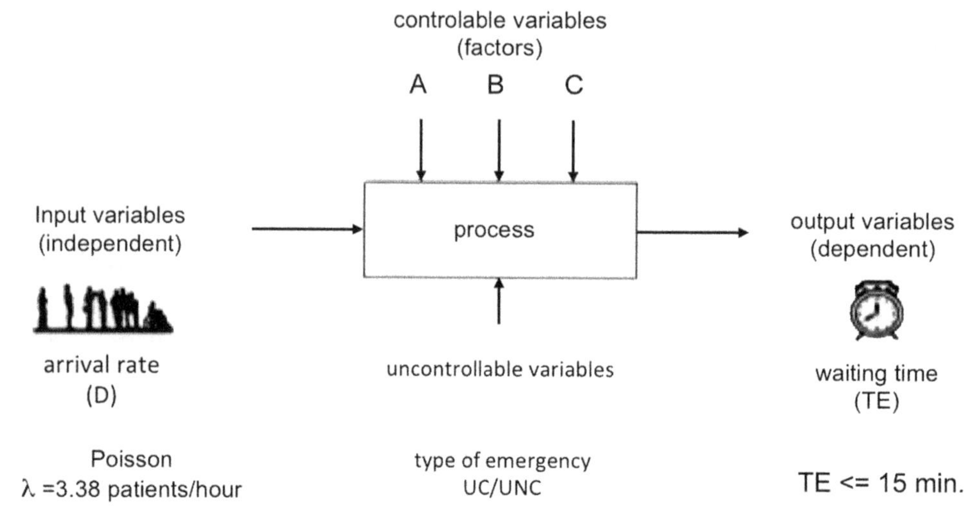

The quantity of physician's booths and time to diagnosis is a combination with significant interaction. Looking at row 1 column 3 of figure 6seems evident that a fixed time to diagnosis has the same effect on wait time as a variable one, but when quantity of physician's booths goes to three bot factors have an intersection.

Accelerated strategy and time to diagnosis combination has no interaction and neither quantity of physician's booths with accelerated strategy and time to diagnosis combination has.

In every intersection of combinations wait time is reduced, meaning that a solution designed to reduce wait time must consider this variables.

Design of experiments helps to find optimal values of factors to optimize the variable of interest, in this case, it helps to define the optimal values of quantity of physician's booths, accelerated strategy and time to diagnosis in order to minimize wait time at ED.

Cube graph is the tool to find out which are the limits of wait time. Quantity of physician's booths must be adjusted to two levels (minimum, one, and maximum, three) to build cube corners. Figure 10 shows wait time under ED configurations defined by factors.

Under current resources of ED the minimum wait time is 29 minutes which far above the 15 minutes goal government is demanding. To achieve the minimum wait time ED must have three physician's booths, use an accelerated strategy and a fixed time to diagnosis. Figure 11 shows best response times.

For this particular ED optimal coefficients of wait time(2) defined by design of experiments are:

$$WT = 150370 - 49222.6A + 1321.15B + 6494.06C - 1321.15AB - 1276.4AC + 1316.85BC - 1316.85ABC \quad (2)$$

Having a correlation coefficient R of 99.97%. (2) is the model of wait time in ED considering A physician's booths; B accelerated strategy (used or not) and C time to diagnosis (fixed or variable); AB, AC, BC and ABC the interactions between previous factors.

Figure 10. Box graph of factors

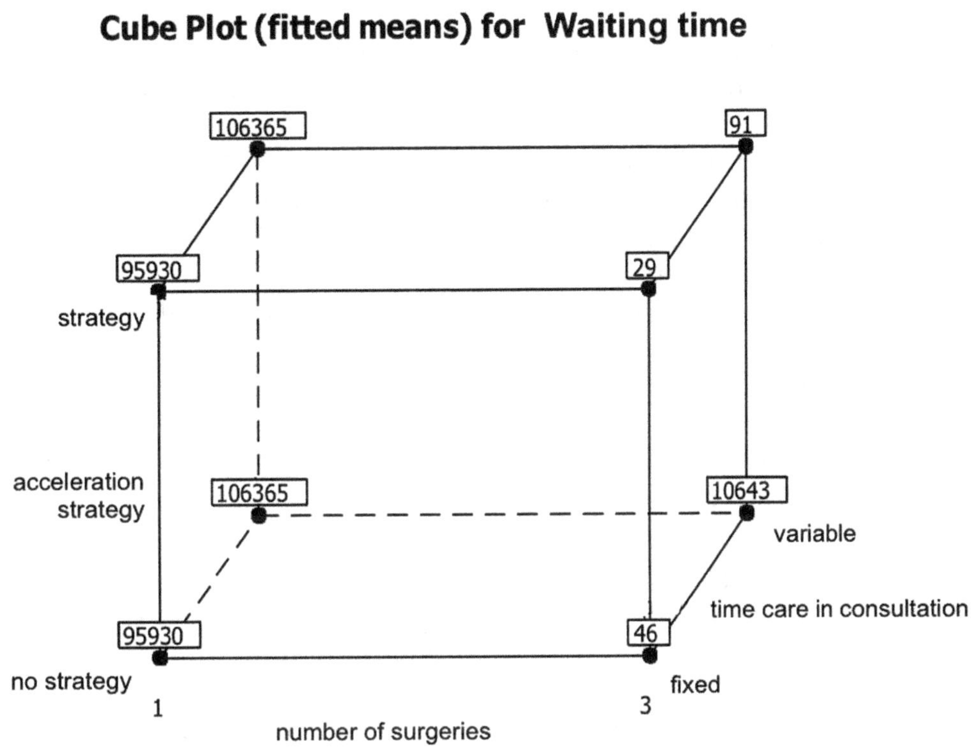

Figure 11. Best response times

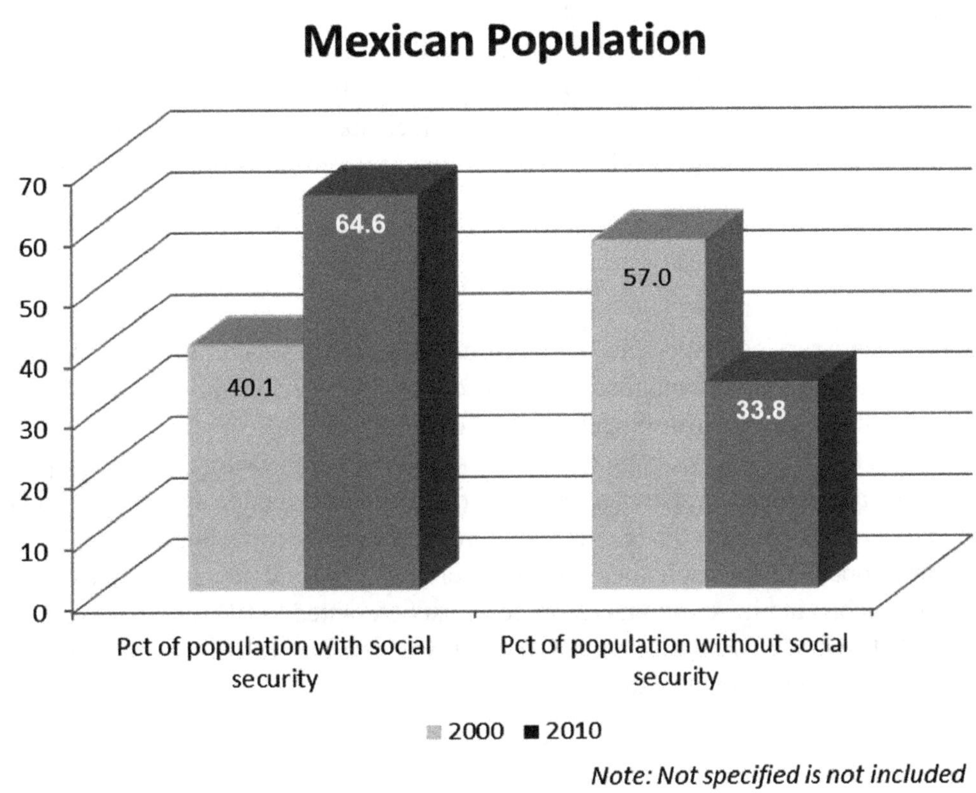

FUTURE RESEARCH DIRECTIONS

Public hospitals are facing challenges on providing quality services to a growing population. These hospitals must find a way to invest wisely their limited resources and they need analysis and modeling tools, designed to be used by their staff that usually lacks of technical training in the use of such tools. Developing automated integrated environments of analysis and modeling is a main field of research.

Some issues at public healthcare services demand attention from researches, one of them is renege rate. In queueing theory, renege is when someone in queue waiting for service quits from system, leaving without being served. Which is the threshold of people waiting for health services? Reneges quit from service or look for alternative ways to get demanded service? What percentage of people reneging from outpatient department goes to ED?

Another relevant issue is Acuteness. Even though among medical community elective procedures turned into emergency ones are a well-known phenomenon, no data is collected on this subject to characterize it, measure it and understand it.

No suggestion on this matter is available since no work has been done, but a comprehensive project which links medical records, Outpatient Department, ED and management of health services will lead to a reliable network which recognizes patients on wait lists demanding ED services related to his ongoing treatment.

Around the world as life expectancy increases, and birth rate lowers near to zero, population is integrated mainly by old people, transforming the demand of healthcare services focusing it on geriatrics services. This kind of demand use to be processed through ED.

Upper age society groups require more specialized health services in a higher frequency and for longer periods. This is true if no preventive health care programs are designed and developed previous to society transformation. A not-preventive health care approach will demand from health system a costly equipment program almost for sure not feasible to develop.

Mexican public healthcare system may be benefited with a standardized DRG system which helps to follow patients through public health subsystems and to standardize medical procedures fares.

To conclude this section, medical schools must consider update their majors including statistical modeling and operations research to develop scientific management skills in their students.

CONCLUSION

Managers of healthcare services experience stress from performance standards above managed facility capacities. Modeling their systems and measuring the capacity of their facilities will help them to find out is goal is unfeasible under their resources availability.

One challenge in healthcare services modeling is understanding key factors on service performance. Managers of healthcare services must implement comprehensive information systems to gather information from operative to tactical and strategic levels in order to enhance modeling efforts.

The model developed in this chapter demonstrates performance level demanded to ED is unfeasible, and ED is far from achieving it. Additional resources must be granted to ED if management is committed with that goal. The model is a useful tool to find out which resources will be the most effective and which has the best cost/benefit relationship.

If you want to improve something you must measure it. Measuring healthcare services capacity, timeliness, reliability and other quality measures is an unsolved challenge, but projects like this one show one way to advance in that direction.

REFERENCES

Brandeau, M. L., Sainfort, F., & Pierskalla, W. P. (2004). *Operations Research and Health Care: A Handbook of Methods and Applications* (M. L. Brandeau, F. Sainfort, & W. P. Pierskalla, Eds.). Boston, MA: Kluwer Academic Publisher.

Fabila, G. (1998). *Diseño y análisis de experimentos industriales*. México: Universidad Iberoamericana.

Llorente, S. (2000). *Factores que influyen en la demora del enfermo en un servicio de urgencias hospitalarias*. Emergencias.

Medina, S. (2010a). *Reducir tiempos de espera de pacientes en el departamento de emergencias de un hospital utilizando simulación.* (Unpublished doctoral dissertation). Mexicali, Baja California, México.

Medina, S. (2010b). *Reducir tiempos de espera de pacientes en el departamento de emergencias de un hospital utilizando simulación*. Revista de la facultad de Ingeniería Industrial.

Medina, S. (2012). *Esquema genérico para el análisis y mejora del flujo de pacientes*. Mexicali, México: UABC.

Medina-Ramirez, R. (2013). *Main Factors of Surgical Deferment in Elective Surgery: A System Dynamics Analysis.* (Unpublished doctoral dissertation). University of Guadalajara, Jalisco, México.

Ministerio de Sanidad y Política Social. (2010). *Unidad de urgencias hospitalaria: estándares y recomendaciones*. España, Ministerio de Sanidad y Política Social.

Ramírez, M. (1980). *Reorganización de los servicios de urgencias del hospital universitario Miguel Hidalgo*. Durango, México: Instituto Tecnológico de Durango.

Secretaría de Salud. (2002). *Lineamientos operativos del sistema de monitoreo para la gestión de la calidad*. México: Secretaría de Salud.

Sokolowski, J. A., & Banks, C. M. (2010). *Modeling and Simulation Fundamentals. Theoretical Underpinnings and Practical Domains*. Hoboken, NJ: Wiley. doi:10.1002/9780470590621

Vazquez-Ibarra, A. (2014). *Modelo computacional del servicio de urgencias de un hospital público y análisis de las variables relacionadas con el factor humano y con los tiempos de atención en consulta.* (Unpublished doctoral dissertation). University of Guadalajara, Jalisco, México.

KEY TERMS AND DEFINITIONS

Actual Urgency (AU): A patient condition in which patient health or even patient life is under risk of suffering permanent damages.

Design of Experiments (DOE): A technique to optimize a system based on the impact of levels of the main factors with influence on response variable.

Diagnosis-Related Group (DRG): A technique to classify diagnoses with similar healthcare services requirements in other to optimize resources (mainly financial resources). Different countries have a working DRG system. No DRG system is adopted in Mexico.

Emergency Department (ED): The more endangered department in the whole hospital and the one with the bigger demand. This department offers healthcare services 24 hours a day every day of the year.

Perceived Urgency (PU): A patient condition which causes pain or worries a patient leading to ED but neither patient's life nor health is in great risk.

Triage: Means to classify. It is a technique to define priorities on healthcare service delivery when demand is above service capacity; it is intended to assure critical viable cases go first; cases are classified into acuity levels. Different countries have different triage levels. Mexico doesn't have a unified triage system.

Chapter 16
Implementing a Container Ship Stowage Problem for Humanitarian Aid in Palestine Based on Cultural Algorithms

Alberto Ochoa-Zezzatti
Juarez City University, Mexico

Kyrk Haltaufoerhyde
Cardiff Technological University, UK

Julio Arreola
Juarez City University, Mexico

Vinicius Scarandangotti
Sassari University, Italy

ABSTRACT

Bioinspired algorithms are a generic term used to refer to the solution of computational problems, based on the planning and implementation based on existing models in the evolutionary process-related nature. Most evolutionary algorithms proposed paradigms that occur in the biology of living things and concepts of natural selection, mutation and reproduction. However, other paradigms that can be taken in the creation of evolutionary algorithms also exist such as the forces of nature, which have been many algorithms based on water, gas and wind reactions. Many of the environments involving unstructured problems in this case a problem of accommodation of containers of humanitarian aid to a company with limited resources, which can be considered from the perspective of cultural paradigms, because the cultural paradigms offer a wide range categorized models that ignore the possible solutions to the problem-a common situation in real life. The purpose of this research is to apply evolutionary computation properties of cultural technology; in this case, to corroborate through data mining analysis of how low the support of various companies use technology for their own benefit to propose a solution to a given problem, in our case carry different types of goods deemed humanitarian aid . The mentioned above, to carry out an adaptation from the standpoint of the modeling societies. An environment for conducting tests for this type of analysis in our case a model arrangement of containers was developed in order to enable learning about not very conventional characteristics of a cultural technology. This environment was named Allaliyah in Palestinian culture means "Together we can achieve everything."

DOI: 10.4018/978-1-4666-9779-9.ch016

1. INTRODUCTION

Most of computer problems and especially those related to intelligent optimization are located in the real world, are characterized by not having a definitive (final) solution [1]. Cultural Algorithms use culture as a vehicle for storing relevant information so accessible to members of the population in our case a company artificially for many generations, were developed to model the evolution of the cultural component in time and to demonstrate how This learns and acquires such knowledge [10]. A cultural algorithm can described with the following pseudocode. Initially, a population of individuals representing the solution space is represented as a set of solutions within the search space, which is generated at random to create the first generation of the company. In our example, the solution space contains a list of attributes that can be used in the classification procedure. The space of beliefs is empty. For each generation, the Cultural Algorithm may imply a population of individuals using the "frame" Vote-inherite-Promote (VIP). During the voting phase of this process, members of the population are evaluated to identify their contribution to the belief space, using the function of acceptance. These beliefs allowed contribute most of the solution of the problem and are selected or placed on the ballot to contribute to the current space of beliefs. The belief space is modified when combined with inherited beliefs are beliefs that have been added by the current generation, this is done using a process of reasoning that allows updating the belief space (see Figure 1).

```
Begin
t=0;
Initialize POP(t); /* Initialization of population */
Initialize BLF(t); /* Initialization of believing space */
Repeat
Evaluate POP(t);
Vote (BLF (t), Accept (POP(t))));
Adjust (BLF (t));
```

Figure 1. Pseudo code of Cultural Algorithms

```
Begin
t=0;
Initialize POP(t); /* Initialization of population */
Initialize BLF(t); /* Initialization of believing space */
Repeat
Evaluate POP(t);
Vote (BLF (t), Accept (POP(t))));
Adjust (BLF (t));
Evolve(POP(t), Influence(BLF(t)));
t = t +1;
Select POP(t) from POP(t-1);
Until (Term condition is reached)
End
```

Evolve(POP(t), Influence(BLF(t)));
t = t +1;
Select POP(t) from POP(t-1);
Until (Term condition is reached)
End

Then the updated belief space is used to influence the evolution of the population in our case the artificial society. Belief space is used to influence during the time in which the population of combinations of variables is modified. During the last phase of a new population that is played using a basic set of evolutionary operators. This new population may be evaluated and the cycle goes on. The VIP cycle ends when a stop condition is introduced and is attained. This condition is usually achieved when only a small or no change is detected in the population through several times or when certain knowledge have emerged in the belief space, as shown in Figure 2. According to [9], the brain human could be an instrument of cognitive metaphorical metaphors that could "assume" the knowledge gained in previous form from past experiences and to complete the knowledge needed to solve real world problems. Research on the subject has worked gradually and differently from the models offered by biology [6]. Cultural Algorithms, for example, are based on the assumption of obtaining the best learning which extends, as is the case in an evolutionary algorithm (as in genetic algorithms) [5] adding to this an element of evolutionary pressure called "belief space" - a cultural pressure mechanism. What has often been suggested that cultural evolution allows companies to better development or to be adapted to their environment, in our case to improve the accommodation of humanitarian aid -A support from 27 companies in Europe for the Palestinian people - Cultural Algorithms based on ranges that exceed biological evolution only on the basis of genetic inheritance [10]. Another particular research topic: Artificial Societies, according to [3, 4], consists in simulating social theories or models expressed in the form of computer programs. The main contribution of this research concerns the fact that the cultural paradigm has its own properties and cultural aspects. Second, specifies that some computational problems could benefit from a cultural method of resolution, and this can lead to automation of processes through the use of intelligent agents [5].

Several authors have written about the confrontation between individuals and society, and how the latter influences their behavior (changes in society over time) [5]. Gessler, makes an approach to artificial societies with regard to cultural issues, proposing some key concepts such as in connection with artificial cultures. These are: Time -This support allows the researcher to simulate the behavior of soci-

Figure 2. Conceptual diagram of cultural algorithms

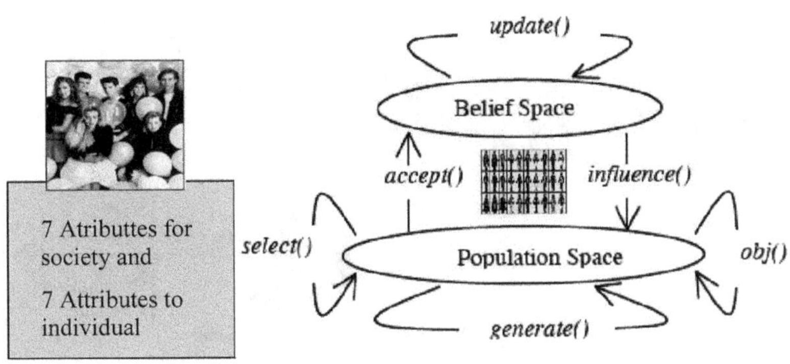

ety towards the future, the past to justify paradigm shifts, or in various scales in time according to the interest of the developer. Space-this may represent a two-dimensional array. Agents and actors interact, what will the behavioral aspects (like that can interact with the real world) and cognitive (the way you think to solve a particular problem). Some of these actions will have an external aspect, while others will be internal and in a cognitive sense. Device-through them characterized as elements, agents can interact and exchange information. Given these concepts, we can build a representation of what could be described as an environment for building a society based on artificial agents and thereby allow the proposed arrangement proposed as shown in [2] to function properly.

2. ARTIFICIAL DESIGNING AN ENVIRONMENT TO ACCOMMODATE CONTAINERS ON A CARGO SHIP

In this research, we focus on a practical problem adapted from the literature related to the arrangement of containers on a cargo ship, characterized by "negotiation towards a suitable accommodation in an adverse environment (time and distance traveled, which for us it is a set of constraints) "to a set of societies characterized by finding a safe path (a place to go with the most perishable food and humanitarian aid) for a cargo restrictions defined space within a hostile and specified with the shortest delivery time and populated by adverse occur at different times of travel and to different adverse situations including a worst situation. Goods transport is actually just a matter of moving some articles from one point to another but it is not always that simple and can quickly become complicated because an article is often produced of many different parts which hardly ever are made in the same place or not even in the same country. The reason for this is that the industrial companies think a lot about how they can save money and therefore find the countries where the material, labor and production respectively are cheapest. The production of one article can therefore involve a lot of transportation even before the distribution of the complete product to the final destination has started. This type of optimization problem can be represented by a two dimensional array, called "dimension", which is our search space, as shown in Figure 3, wherein each container represents a value of help and each position where the containers should be

Figure 3. Representation of the problem to solve the arrangement of containers on the aid ship

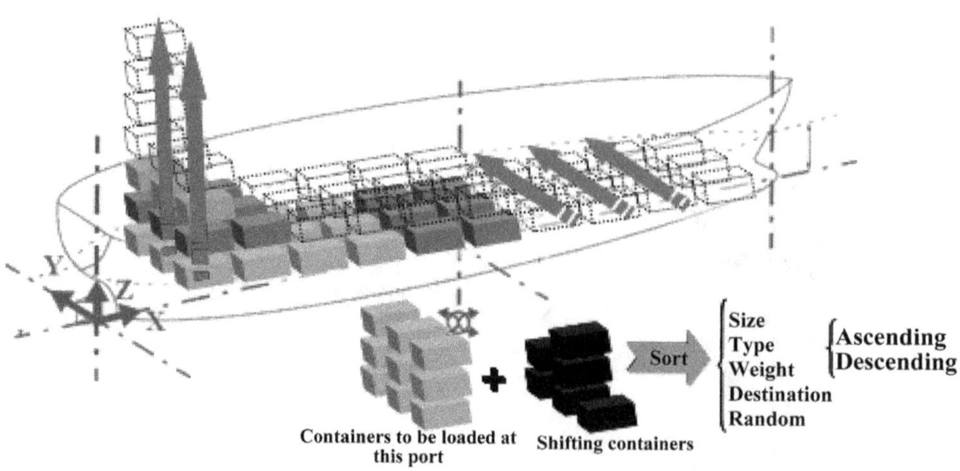

covered in a limited time are located, these situations Penalty (unknown to the companies represented by the cultural algorithms) respectively, and the numbers in dimension representing cost penalty experienced for each of the areas where the load must be located. The aim of the Cultural Algorithm is finding the ideal accommodation with the fewest penalties while the spaces are arranged where there may be "adequate load situations", characterized by sanctions not consider keeping a standard balance in the cargo ship bound for Palestine to bring humanitarian aid.

The solution to this problem is given by a sequence of generations of agents, designated as "community." Agents can only know the spaces adjacent to them to establish themselves there is an optimum load value and not receive a penalty, as each assessment taking better decisions for each artificial society known only distances and finite spaces. The group of spaces that achieve through each group of agents called "quadrant accommodation". From the point of view of the agents, this optimization problem is quite complex because we do not know the location or distance to arrive at the meta-or whether any resources regarding energy cost and burden to reach hers and you cannot see the search space beyond its quadrant. Also you do not have any previous heuristics to try to improve optimization. For a better understanding of the use of cultural algorithms, these are used to solve an optimization problem with uncertainty; we now introduce some basic concepts and representations of artificial culture related to this problem.

These representations are located between levels of abstraction (The unknown part of the agent), and the domain of the problem (Dimension) and agents. The union of these levels of abstraction is the protocol of artificial-atmosphere model that performs dynamic linking agents, the problem domain and its cultural representation. In the paradigm shift of cultural algorithm, advances in the belief space (beliefspace) are made by the best paradigm (BestParadigm) on what is set to zero, representing the fact that culture increases the amount of benefit associated with such spaces, giving an incentive for behavior associated with the best paradigm (BestParadigm).

2.1 Agents

Agents are actors, with which you can use every space in the dimension of what Freud referred to as the "Principle of satisfaction", according to this, the agent will be able to select the areas with the lowest cost of experimentation.

2.2 Paradigm

The paradigm is the personal representative of agents to the belief space (beliefspace) or his personal interpretation of cultural references. According to Gessler, this represents for each group of agents (artificial societies) cognition and its particular vision of cultural interpretation of the search space. The paradigm may be the best solution to a problem, which is denoted as the best proposed paradigm (BestParadigm).

2.3 Space Beliefs (Beliefspace)

The space of beliefs is the collective representation of the real world. In other words, this is the world as it is interpreted by a culture of community being represented, where agents interact to find your way even with moral values -quedar in last position removes them from further participation as a Model Grand Prix- ..

2.4 Dimension

The dimension is the real world, which can never be fully known by the group of agents. This contains the cost and experimentation in which the agents are able to live to make a proper optimization and this can be improved.

2.5 Exploration

The agents belonging to a community search within the dimension for the most appropriate place to be reached (Meta). The solution obtained by the agents of which are the goal in the shortest number of steps could be considered the "model" of the community, or the best paradigm (BestParadigm). According to Geertz, this model or ideology is a "diagram of the psychological and social processes". The culture then could try to direct the behavior of the new generation of agents through this best solution. The best solution to the optimization problem will be given by the sequence of movements by the agent that lets you find all the optimal number of steps and the lower cost of penalties. Each agent community is led by an objective function that lets you select the blanks with the fewest penalties. If more than one space quadrants of the race possess a minimum penalty and this is identical, the agent will select one freely, in our case, is represented by a random selection. It can be seen that the principle of satisfaction does not affect the overall strategy for solving the problem at the collective level (culture). Instead, it connects the agent with an autonomous entity. The culture controls behavior to be adopted as a model, creating a strategy for action-a common ideology for the group of agents-in terms of a given problem domain. The group of agents selects the cell with the minimum penalty, as indicated by the belief space (Beliefspace) adding to this the cultural value, such as:

beliefspace (x) = beliefspace (x) + dimension (x)

Where x is a set of dimension spaces

In this research the functions represent the group of agents and interaction with their culture to be selected according to the problem to solve, in our case the best accommodation and the load is distributed in an appropriate and balanced manner. Therefore, we cannot try to establish a mathematical model of how the cultural process occurs in the real world. The adoption of a random function as it has tried to show and explain how to include this in the process, a system of multiple interactions between the agent and culture. We try to analyze other mathematical representations in our future work.

3. CULTURAL ALGORITHM SIMULATOR

To test and validate the theoretical concepts previously presented, Cultural Algorithms simulator was developed to simulate our ideal arrangement of cargo ship (Allaliyah). Initially we intended only create a capable of performing the analysis and the respective experimental environment. When cultural algorithms are used becomes more difficult to understand the peculiarities of each solution. Each time the system has a precise answer, the solution obtained can hardly be duplicated exactly. This property of evolutionary algorithms in general and cultural algorithms, in particular, has been little explored or discussed in

Figure 5. Evaluating an issue optimization problem related with Ship Stowage using Allaliyah

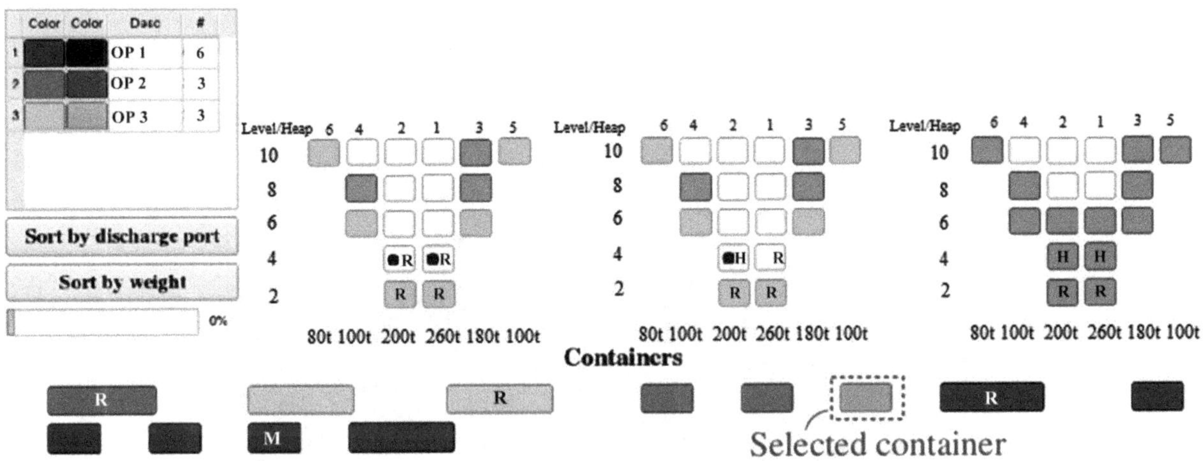

the literature. The creation of systems with an individuality or "Soul" is our contribution in the area. To this end, 27 companies that can send different supplements such as those characterized by procurement rules Ferengi [7], which were characterized based on their behavior through seven basic attributes (agility, ability to fight, intelligence, strength is selected, endurance, speed and emotional control), allowing well describe both society and the individual. Allaliyah development was based on our desire to share an intuitive understanding about the treatment of an optimization problem for a new class of systems, with individuals able to possess unexpected creativity, a typical feature of the living entities. Allaliyah shown in Figure 5, the user has the possibility to choose the point and purpose of accommodating from the start, coupled with the places where one can receive a penalty for a bad arrangement made, and the amount of the penalty associated with each space dimension in which societies live (communities of agents). Our prototype was developed using JBuilder X platform.

4. COMPLEMENTARY METHODOLOGY

Data mining is the search of global patterns and related what existing between the data bases of huge data, but these data are hidden between them within the immense amount of information [5]. These relationships represent the knowledge of the value of each object and that are in the database to be analyzed. This information is not necessarily a true copy of the information stored in the database. Rather, it is the information that can be deduced from it. One of the main problems of data mining is that the number of possible relationships extracted is exponential [5]. Therefore, there are a variety of heuristic learning machine have been proposed for knowledge discovery in databases [5]. One of the most popular methods for representing the results of data mining is the use of decision trees [8]. A decision tree provides a method for recognizing a particular case of a concept. Which is a strategy of "divide and rule" for the acquisition of the concept (instance). Decision trees have been useful for a variety of case studies in science and engineering, in our case we use data mining to characterize individuals in society (community of agents) and to understand how you can get the best paradigm.

Figure 4. Allaliyah, environment developed to show the ideal accommodations for our cargo ship container color contemplating the type of supplement that is sent

5. EXPERIMENTS

In this section, developed experiments using Allaliyah described. We hope to contribute in the sense of making evident the importance of creating a new methodology to test and analyze the results. This was not a trivial task, given the diversity of behaviors performed by previous solutions Allaliyah because it resembles more a descriptive anthropology that a simple test software. In the first experiment, the results of 27 communities with 50 agents were compared, and secondly with 500 agents to 27 communities each one. The points associated with the beginning and the lens is shown in Figure 5. The optimal number of steps from the beginning to the target is 57.

One of the most interesting features observed in this experiment is the diversity of established cultural patterns for each community. For solutions with the same number of steps the result provided by the "beliefspace" is entirely different. Structured scenarios associated agents cannot be reproduced in general, because they belong to a particular instant in time and space. They represent a unique and accurate model resolution and the innovative form of adaptive behavior that solves a problem of calculation for complex change existing relationships. Configurations can be generated metaphorically and related community knowledge and behavior with respect to an optimization problem (to make alliances, to recommend possible solutions in time and change according to the need to contextualize the problem) or a tradition of accommodation with which emerge from experience and with which they begin a dynamic process. Comparison of the 50 agents of the first community on the community of 500 agents, the latter performs best in terms of the average number of steps from the start to the finish (54.82 versus 57.96) and a smaller standard deviation (3.27 compared to 3.99). Also had a higher average paradigm shifts (5.85 vs. 4.25), indicating that even a minor negotiation regarding the intervening periods, which explored the less interesting parts of the space within the dimension search, which could improve their search for better results. In the second experiment, the same scenarios for the experiment considered one except after obtaining a solution from a community of 50 agents, five placed near the target areas and starting a new community of 500 agents. The new community was informed earlier cultural settings, but must take into account the new scenario. The comparison between the two solutions is not immediate, from the standpoint of trying to solve different problems. In this experiment, was initially surprising to see how the community of 500 agents using the solution offered by 50 agents, provided that these solutions were near optimum, instead of finding new fully complete solutions. These results highlight the conservation of a global action strategy that is regulated by the agents, as proposed in [11]. This can be compared metaphorically with the concept of culture mentioned in the introduction.

6. CONCLUSION

Using Data Mining improves the understanding of change for the better substantial paradigm that ranks communities agents appropriately in terms of their relationship attribute approach; this allows us to understand that there is the concept of "situational leadership" based on determining the function accepted by the other to a particular solution proposed communities for each community. Cultural algorithms provide a powerful alternative to optimization problems. For that reason is that it provides a comprehensive overview of the cultural phenomenon. This technology leads to the possibility of the generation of experimental knowledge, created by the community of agents for a given application domain. In much the extent of this knowledge is cognitively community of agents is a topic for future work. The answer may be similar to the hard work involved in communication between two different cultures. The specification of each artificial environment would be a contradiction to digital systems, characterized by a wide reproduction, and lacks the concepts of original and copy. In contrast to traditional digital systems, characterized by a wide reproduction, and determined from concepts associated with the original or a copy, as well as the implicit technological culture to the unique individuality of each system, which is totally different today industrial and standards. A new artificial intelligence could be in charge of these systems is still far in the horizon, in the same way that we still lack methods to understand the original and peculiar aspects of each company. In future work will attempt to analyze the discussion of priority values for each type of supplement and what could be the reason for being prioritized over others.

REFERENCES

Alpha, M. (n.d.). *Proyecto de Wikipedia relacionado con Star Trek*. Retrieved from http://www.memory-alpha.org/en/wiki/Main

Desmond, A., & Moore, J. (1995). *Darwin-la vida de un evolucionista atormentado*. Sao Paulo, Brazil: Generacion Editorial.

Geertz, C. (1989). *La interpretacion de las culturas*. RJ, Brazil: Editorial Guanabara.

Gessler, N. (1999). *Artificial Culture-experimemts in Syntethic Anthropology*. Retrieved from http://www.soc.surrey.ac.uk/JASSS/1/1/4.html

Hue et al. (2006). Towards Robust Indexing for Ranked Queries. In *VLDB' 06*.

Ochoa et al. (2007). Implementing Data Mining to improve a Game Board based on Cultural Algorithms. In *Proceedings of HAIS'2007*.

Pinker, S. (1999). *Como funciona nuestra mente?* Sao Paulo, Brazil: Compania de Letras.

Reynolds, R. G. (1998). An Introduction to Cultural Algorithms. *Cultural Algorithms Repository*. Retrieved from http://www.cs.wayne.edu/~jcc/car.html

Reynolds, R. G. (2010). Networks Do Matter: The Socially Motivated Design of a 3D Race Controller Using Cultural Algorithms. *International Journal of Swarm Intelligence Research*, *1*(1), 17–41.

Suaremi, T. (2003). *Understand social groups. In NDAM'2003*, Reykiavik, Iceland.

Zevzevic, I. (2005). Similar cultural relationships in Montenegro. In JASSS'2005.

Chapter 17
A Hybrid Intelligent Risk Identification Model for Configuration Management in Aerospace Systems

Jose Nava
Pinnacle Aerospace, Inc., Mexico

Alejandro Osorio
Pinnacle Aerospace, Inc., Mexico

ABSTRACT

This chapter proposes a multi-dimensional patterns recognition model for Configuration Management's Risk Identification in Aerospace Safety Critical Systems. This work has been designed for Aerospace software systems where companies require full compliance with the Aerospace Standard DO-178b. The solution focuses on Risk Identification for the Configuration Management Process Area. An Anomaly Detection Solution has been designed through the modeling of statistics and artificial intelligence algorithms, following CRISP-DM model standard for data mining solutions. A dimensional architecture was designed to model the problem through three dependent and interconnected dimensions. The first dimension, Behavioral Biometrics, which this model has extended to Human Behavioral Patterns. The second dimension is Infrastructure, which represents all physical specialized equipment, environments, networking, and its configurations. The third dimension is space-time, which in this model represents a time dimension against all geographical information project related (code, files, among others).

INTRODUCTION

Software products are deployed across a wide range of global industries, but when it comes to Aerospace safety critical systems; security and criticality are strictly rigorous. In these kinds of systems, a software failure or malfunction could result in loss of life and significant property damage.

Aviation systems development has been evolving for almost a hundred years now. One of the more notorious development evolving areas is the instruments and displays that were electro-mechanical in the past and now are composed by electronic parts, which make them more reliable, secure and lighter. The

DOI: 10.4018/978-1-4666-9779-9.ch017

biggest priority of this evolution has always been to avoid aircraft accidents and providing security in the air. The rapid increase in the use of software in airborne systems and equipment used on aircraft and engines in the early 1980s resulted in a need for industry-accepted guidance for satisfying airworthiness requirements. DO-178, "Software Considerations in Airborne Systems and Equipment Certification", was written to satisfy this need.

Developing reliable software products becomes more and more challenging for Safety Critical Industries due to specific regulations, short production cycles or high complexity design.

One of the more challenging aspects when working with Aerospace Software is maintaining a good Configuration Management control. A well planned strategy for this area can result in the fault tolerant system we have designed since the beginning.

Configuration Management has always been a complex, dynamic and constantly changing area, which not only interacts with the project team members and the IT department of the service provider, but also with those entities from the client side. This configuration makes it more complex because each company has its own Configuration Management practices, tools, and processes. Other important risk factor is that no matter the technical level of the involved teams, each employee works differently when it comes to practices like code design, SVN activities like commits, check in/out, updates, merges, syncs, cleanups, or other activities like estimation, activities tracking culture, among others.

The Federal Aviation Administration and the Joint Aviation Authorities have defined a list of Failures Condition Categories within DO-178b, as well as a list of Software Level Definitions based upon the contribution of software to potential failure conditions.

The Software level implies that the level of effort required showing compliance with certification requirements varies with the failure condition category (Radio Technical Commission for Aeronautics, 1992). These software level definitions are described below:

- **Level A (Catastrophic):** Software whose anomalous behavior would cause or contribute to a failure of system function resulting in a catastrophic failure condition for the aircraft, and multiple deaths. An example of this anomalous behavior can be the Turbine Engine Monitoring Software, which malfunction can cause a fatal accident.
- **Level B (Hazardous/Severe-Major):** Software whose anomalous behavior would cause or contribute to a failure of system function resulting in a hazardous/severe-major failure condition for the aircraft, and some deaths or severe injuries. An example of this anomalous behavior can be the Landing Gear Control Software, which malfunction can cause a big accident.
- **Level C (Major):** Software whose anomalous behavior would cause or contribute to a failure of system function resulting in a major failure condition for the aircraft, and severe injuries. An example of this anomalous behavior can be the FMS/NAV/COM System, which malfunction can cause damages and accidents.
- **Level D (Minor):** Software whose anomalous behavior would cause or contribute to a failure of system function resulting in a minor failure condition for the aircraft, and minor injuries. An example of this anomalous behavior can be the A/C monitoring system software, which malfunction can cause a failure condition.
- **Level E (No Effect):** Software whose anomalous behavior would cause or contribute to a failure of system function with no effect on aircraft operational capability or pilot workload. An example of this anomalous behavior can be the Entertainment audio system software, which malfunction can cause just a minimal failure condition.

We will describe the software level definitions that we are covering in this work, and the reasons to cover these levels.

The general objective of designing this model is to avoid Configuration Management issues during Aerospace software development and verification projects through artificial intelligence and statistical techniques like anomaly detection and pattern recognition in exploratory data analysis.

The specific objectives of this model are:

- Based on the most threatening issues in Configuration Management, to identify user behaviors through the project life cycle.
- To have a real-time intelligent pattern recognition algorithms tool based on these behaviors to detect anomalies.
- To have a real-time risks notification and diagnosis tool.
- Based on the previous intelligent pattern recognition algorithms, to have valuable statistical information, which cannot be detected or accessed through traditional mathematical methods.
- To use this statistical information to improve real-time decision making, and have informational tools to build performance improvement plans in real-time without having an impact in the project.

BACKGROUND

Over time, Aerospace companies have always put special attention at the moment of creating the Configuration Management strategy for each project at the initial phases. In this sense, special controls, monitoring strategies, resources and processes are put in place to overcome the big risks that the development and testing teams are phasing in a daily basis. These controls are not enough because there are several factors that can trigger a failure in a system; some examples are listed below:

- A failure in a build
- Bad version control
- Poor requirements traceability, bad hooks strategy
- Design complexity
- Disaster recovery issues
- Change control confusions
- Confusing environments' configuration
- Bad release management or deployment management
- High estimations' deviations
- Poor activities tracking
- High overtime rates

The Software Configuration Management Process, in cooperation with other processes during the software life cycle process assists in satisfying general objectives to:

1. Provide a defined and controlled configuration of the software throughout the software life cycle.
2. Provide the ability to consistently replicate the executable Object Code for Software Manufacture or to regenerate it in case of a need for investigation or modification.

3. Provide control of process inputs and outputs during the software life cycle that ensure consistency and repeatability of process activities.
4. Provide a known point of review, assessing status, and change control by control of configuration items and the establishment of baselines.
5. Provide controls that ensure problems receive attention and changes recorded, approved, and implemented.
6. Provide evidence of approval of the software by control of the outputs of the software life cycle processes.
7. Aid the assessment of the software product compliance with requirements.
8. Ensure that secure physical archiving, recovery and control are maintained for the configuration items.

The daily execution of each of the hundreds of configuration management processes and good practices each engineer has to perform for these complex systems can lead to a minimal to moderate deviation in the execution's exactness. For this reason, we can say that the avionics process could be perfect, and the way a team of several engineers execute within a project could not be exactly the same in detail. In aerospace systems, a minimal deviation is a risk and a possible failure. For creating this model, we did an extensive research in Artificial Intelligence techniques to find a way to use patterns recognition to derive with behavioral biometrics mathematical solutions which can perform as a live system focused in each micro-technical movement of the engineers within a project, to detect these deviations and avoid issues within the configuration management process.

Below we provide detailed background information in Safety Critical Systems, Configuration Management in DO-178, Artificial Intelligence, Data Mining and Anomaly Detection techniques.

AEROSPACE SAFETY CRITICAL SYSTEMS

Safety Critical Systems have been evolving exponentially during the last 50 years. Technology advances are helping to have better processes and better controls over these kinds of Systems.

The ultimate objective of any system is the performance of any function. This function is frequently called the mission. The term often used to describe the overall capability of a system to accomplish its mission is system effectiveness. For consumer products, system effectiveness is related to customer satisfaction, which is related to the overall concept of quality. Quality of products and services is defined and evaluated by the customers and users of the product. System effectiveness is defined as the probability that the system can successfully meet an operational demand within a given time when operating under specified conditions. (Salvendy, 2001).

Safety-critical systems are introduced into a system to prevent, or stop the development of an accident. Safety-critical functions carried out by these systems may range from simple physical barriers to very complex systems including a lot of software. The safety-critical functions can be of technical nature, but the interaction with administrative controls and human actions may need to be included to fully understand how the performance is affected. Several international standards have been introduced related to design and evaluation of safety-critical functions. Among these standards are the IEC61508 standard, and its application specific versions/related standards, such as IEC 61511 for the process industry, IEC 62061 for machinery (ISO 13849), ISO 26262 for automotives, EN 50126/EN 50129/ IEC 50128 for railway (IEC versions are IEC 62278, IEC 62425, and IEC 62269), and IEC 61513 for nuclear.

Some of the more important concepts we should understand to have an accurate idea of what Safety Critical Systems mean for us are: Safety is freedom from accidents or losses. Safety is not reliability.

Reliability is one of the major attributes determining system effectiveness. It is generally defined as the probability that a system will perform its intended function satisfactorily, for its intended life, under specified operative conditions (Salvendy, 2001).

Safety is not security, is protection or defense against attacks, interference, or espionage. Accident is a loss of some kind, such as injury, death, or equipment damage. Risk is a combination of the likelihood of an accident and its severity:

$$risk = p(a) * s(a)$$

Hazard is a set of conditions and/or events that leads to an accident. A failure is the nonperformance or inability of a system or component to perform its intended function for a specified time under specified environmental conditions. Failures are events. e.g.. a component failure. An error is a systematic fault, which is a design error.

Errors are states or conditions, e.g. a software bug. A fault is either a failure or an error.

Safety must be considered in the context of the system, not the component or the software. It is less expensive and far more effective to build in safety early than try to tack it on later. The Hazard Analysis ties together hazards, faults, and safety measures.

Some of the most common types of Hazards are described below:

- **Actions**
  - inappropriate system actions taken
  - appropriate system actions not taken
- **Timing**
  - too soon
  - too late
- **Sequence**
  - skipping actions
  - actions out of order
- **Amount**
  - too much
  - too little

System safety is a planned, disciplined and systematic approach to preventing or reducing accidents throughout the life cycle of a system.

As an example of a Safety Critical System we can mention a common instrument system which is a configurable color display that provides backup information for attitude, altitude, airspeed, heading, and navigation all plains have installed in the cockpit. This instrumentation can have the capability of interfacing with several types of sensors and protocols that can be installed in different airplane models.

CONFIGURATION MANAGEMENT IN DO-178

Having a good Configuration Management control is critical in any Industry; a company's quality level can be measured by its configuration management process and control. This is a very complex area because it involves different departments as Development, Testing, IT, and Quality. Maintaining a perfect log database of a picture of every second of the life of the project in terms of configurations it is always very complex and challenging.

The objectives for Software Configuration Management are independent of software level. However, two categories of software life cycle data may exist based on the Software Configuration Management controls applied to the data.

The Software Configuration Management process includes the activities of configuration identification, change control, baseline establishment, and archiving of the software product, including the related software life cycle data. The following guidelines define the objectives for each Software Configuration Management process activity. The Software Configuration Management process does not stop when the software product is accepted by the certification authority, but continues throughout the service life of the airborne system or equipment.

The objective of the Configuration Identification activity is to label unambiguously each configuration item (and its successive versions) so that a basis is established for the control and reference of configuration items. Guidance includes:

1. Configuration identification should be established for the software life cycle data.
2. Configuration identification should be established for each configuration item, for each separately controlled component of a configuration item, and for combinations of configuration items that comprise a software product.
3. Configuration items should be configuration-identified prior to the implementation of change control and traceability data recording.
4. A configuration item should be configuration-identified before being used by other software life cycle processes, referenced by other software life cycle data, or used for software manufacture or software loading.

Baselines and Traceability

It is important to state that at the beginning of a project, we need to establish baselines when planning. The objective of baseline establishment is to define a basis for further software life cycle process activity and allow reference to. Control of configurations and traceability in a project is complex. Guidance includes:

1. Baselines should be established for configuration items used for certification credit. (Intermediate baselines may be established to aid in controlling software life cycle process activities).
2. A software product baseline should be established for software product and define in the software configuration index.
3. Baselines should be established in controlled software libraries (physical, electronic, or other) to ensure integrity. Once a baseline is established, it should be protected from change.
4. Change control activities should be followed to develop a derivative baseline from an established baseline.

5. A baseline should be traceable to the baseline from which was derived, if certifications credit is sought for software life cycle process activities or data associated with the development of the previous baseline.
6. A configuration item should be traceable to the configuration item from which it was derived, if certification credit is sought for software life cycle process activities or data associated with the development of the previous configuration item.
7. A baseline of configuration item should be traceable either to the output it identifies or to the process with which it is associated.

Problem Reporting, Tracking and Corrective Action

The objective of problem reporting, tracking and corrective action is to record process non-compliance with software plans and standards, to record deficiencies of outputs of software life cycle processes, to record anomalous behavior of software products, and to ensure resolution of these problems.

Guidance includes:

1. A problem report should be prepared that describes the process non-compliance with plans, output deficiency, or software anomalous behavior, and the corrective action taken.
2. Problem detection should provide for configuration identification of affected configuration items or definition of affected process activities, status reporting of problem reports, and approval and closure of problem reports.
3. Problem reports that require corrective action of the software product or outputs of software life cycle process should invoke the change control activity.

Change Control

The objective of the change control activity is to provide control for recording, evaluation, resolution and approval of changes throughout the software life cycle.

Guidance includes:

1. Change control should preserve the integrity of the configuration items and baselines by providing protection against their change.
2. Change control should ensure that any change to a configuration item requires a change to its configuration identification.
3. Changes to baselines and to configuration items under change control should be recorded, approved, and tracked. Problem reporting is related to change control, since resolution of a reported problem may result in changes to configuration items or baselines.
4. Software changes should be traced to their origin and the software life cycle processes repeated from the point at which the change affects their inputs. For example, an error discovered at hardware/software integration, that is shown to result from an incorrect design, should result in design correction and repetition of the associated integral process activities.
5. Throughout the change activity, software life cycle data affected by the change should be updated and records should be maintained for the change control activity.

ARTIFICIAL INTELLIGENCE

The term Artificial Intelligence was conceived in 1956 used by researches in universities within different scientific areas. Artificial Intelligence has acquired several definitions along the years. One of the most known is "A field of study that seeks to explain and emulate intelligent behavior in terms of computational processes" (Schalkoff R., 1990). We defined Artificial Intelligence as the science that tries to emulate evolution, and within evolution, natural selection and cognitive modeling.

Artificial Intelligence has several branches, the most known are shown below:

- **Logical AI:** What a program knows about the world in general the facts of the specific situation in which it must act, and its goals are all represented by sentences of some mathematical logical language. The program decides what to do by inferring that certain actions are appropriate for achieving its goals (McCarthy, 2007).
- **Search:** AI programs often examine large numbers of possibilities, e.g. moves in a chess game or inferences by a theorem proving program. Discoveries are continually made about how to do this more efficiently in various domains.
- **Pattern Recognition:** When a program makes observations of some kind, it is often programmed to compare what it sees with a pattern. For example, a vision program may try to match a pattern of eyes and a nose in a scene in order to recognize a face. More complex patterns, e.g. in a natural language text, in a chess position, or in the history of some event are also studied. These more complex patterns require quite different methods than do the simple patterns that have been studied the most.
- **Representation:** Facts about the world have to be represented in some way. Usually languages of mathematical logic are used.
- **Inference:** From some facts, others can be inferred. Mathematical logical deduction is adequate for some purposes, but new methods of non-monotonic inference have been added to logic since the 1970s. The simplest kind of non-monotonic reasoning is default reasoning in which a conclusion is to be inferred by default, but the conclusion can be withdrawn if there is evidence to the contrary. For example, when we hear of a bird, we can infer that it can fly, but this conclusion can be reversed when we hear that it is a penguin. It is the possibility that a conclusion may have to be withdrawn that constitutes the non-monotonic character of the reasoning. Ordinary logical reasoning is monotonic in that the set of conclusions that can be drawn from a set of premises is a monotonic increasing function of the premises. Circumscription is another form of non-monotonic reasoning.
- **Common Sense Knowledge and Reasoning:** This is the area in which AI is farthest from human-level, in spite of the fact that it has been an active research area since the 1950s. While there has been considerable progress, e.g. in developing systems of non-monotonic reasoning and theories of action, yet more new ideas are needed.
- **Learning from Experience:** Programs do that. The approaches to AI based on connectionism and neural nets specialize in that. There is also learning of laws expressed in logic. The field of Machine Learning is concerned with the question of how to construct computer programs that automatically improve with experience (Mitchell, 1997). Programs can only learn what facts or behaviors their formalisms can represent, and unfortunately learning systems are almost all based on very limited abilities to represent information.

- **Planning:** Planning programs start with general facts about the world (especially facts about the effects of actions), facts about the particular situation and a statement of a goal. From these, they generate a strategy for achieving the goal. In the most common cases, the strategy is just a sequence of actions.
- **Epistemology:** This is a study of the kinds of knowledge that are required for solving problems in the world.
- **Ontology:** Ontology is the study of the kinds of things that exist. In AI, the programs and sentences deal with various kinds of objects, and we study what these kinds are and what their basic properties are. Emphasis on ontology begins in the 1990s.
- **Heuristics:** A heuristic is a way of trying to discover something or an idea embedded in a program. The term is used variously in AI. Heuristic functions are used in some approaches to search to measure how far a node in a search tree seems to be from a goal. Heuristic predicates that compare two nodes in a search tree to see if one is better than the other, i.e. constitutes an advance toward the goal, may be more useful.
- **Genetic Programming:** Genetic programming is a technique for getting programs to solve a task by mating random Lisp programs and selecting fittest in millions of generations.

We will explain which Artificial Intelligence techniques we have chosen to design this model and help solving this problem, from the most mentioned nowadays like Anomaly Detection, to using a clustering-based technique called K-means for Pattern Recognition and descriptive statistics techniques.

Data Mining

Data mining is a complex process that involves the interaction of the application of human knowledge and skills, and technology. This must be supported by clearly defined processes and procedures.

The main purpose of data mining is the extraction of hidden and useful knowledge from large volumes of raw data. However data mining is not simply about technology, it brings together different disciplines like software engineering, computer science, business intelligence, human-computer interaction and analysis techniques. Phases of these disciplines must be combined for data mining project outcomes. CRISP-DM methodology defines its processes hierarchically at four levels of abstraction, from general to specific, allowing a project to be structured modularly, being more maintainable, scalable and the most important, to reduce complexity. At the first level we have the phases, the second level consists of generic tasks, the third level is the specialized task level, and the fourth level is the process instance.

CRISP-DM describes the life cycle of a data mining project consisting of six phases: business understanding, data understanding, data preparation, modeling, evaluation and deployment. The sequence of the phases is not rigid, moving back and forth between phases is always required and allowing a data mining project to have a flexible development during its life cycle.

The most important and core in CRISP-DM methodology is that defines its processes hierarchically, consisting of a set of tasks at four levels of abstraction, from general to specific, explained below:

- Using levels of abstraction, at the top level we have the phases, the data mining process is organized into a number of phases, where each phase consists of several second-level generic tasks.
- At the second level we have the generic tasks, which must be general enough to cover all possible data mining situations and must be as complete and stable as possible. Complete means covering

all possible applications and the whole process of data mining. Stable means this model and process must be valid for any kind of problems, and for old and new modeling techniques.

- The third level is the specialized task level, which describe how actions in the generic tasks should be carried out in specific situations. Ideally, the phases and tasks must be described as discrete steps and must be performed in a specific order. In practice, many of the tasks can be performed in a different order and it will often be necessary to repeatedly backtrack to previous tasks and repeat certain actions.
- The fourth level is the process instance, a record of the actions, decisions and results of an actual data mining engagement. A process instance is organized according to the tasks defined at the higher levels, but represents what actually happened in a particular engagement, rather than what happens in general.

According to CRISP-DM, the life cycle of a data mining project contains the phases of the project and the respective tasks and relationships between these tasks. At this description level, it is not possible to identify all relationships (see figure 1).

Essentially, relationships could exist between any data mining tasks depending on the goals, the background and interest of the user and most importantly on the data.

A data mining project involves more than modeling. The modeling phase is just one phase in the implementation process of a data mining project. Steps of critical importance proceed and follow model building and have a significant effect on the success of the project (Chorianopoulos & Tsiptsis, 2009).

The CRISP-DM process model contributes to data mining as a process, which is reflected in its origins (McCue, 2007).

The life cycle of a data mining project consists of six phases. Figure 2 shows the phases of a data mining process. The sequence of the phases is not rigid. Moving back and forth between different phases is always required. It depends on the outcome of each phase which phase or which particular task of a phase, has to be performed next. The arrows indicate the most important and frequent dependencies between phases.

The outer circle in Figure 2 symbolizes the cyclical nature of data mining itself. Data mining is not over once a solution is deployed. The lessons learned during the process and from the deployed solution

Figure 1. Four level breakdown of the CRISP-DM methodology

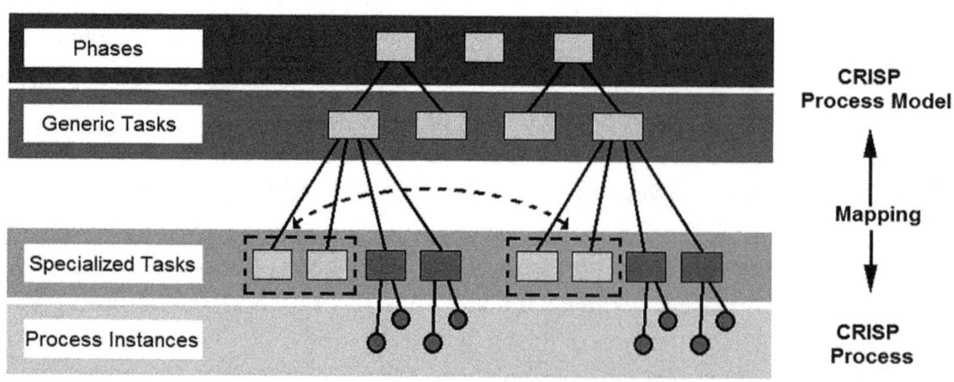

Figure 2. Phases of the CRISP-DM reference model

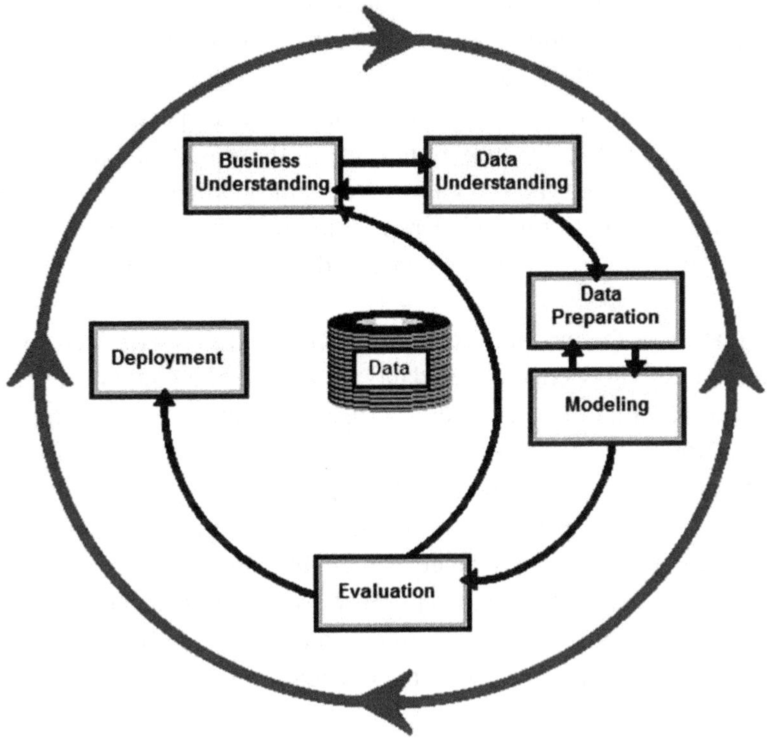

can trigger new, often more focused business questions. Subsequent data mining processes will benefit from the experiences of previous ones.

The Data Mining Phases

- **Business Understanding:** This initial phase focuses on understanding the project objectives and requirements from a business perspective, then transforming this knowledge into a data mining problem definition and a preliminary plan designed to achieve the objectives. This phase also includes assessing the current situation, establishing data mining goals, and developing a project plan (Delen & Olson, 2008).
- **Data Understanding:** The data understanding phase starts with an initial data collection and proceeds with activities in order to get familiar with the data, to identify data quality problems, to discover first insights into the data or to detect interesting subsets to form hypotheses for hidden information.
- **Data Preparation:** The data preparation phase covers all activities to construct the final dataset (data that will be fed into the modeling tool(s)) from the initial raw data. Data preparation tasks are likely to be performed multiple times and not in any prescribed order. Tasks include table, record and attribute selection as well as transformation and cleaning of data for modeling tools.
- **Modeling:** In this phase, various modeling techniques are selected and applied and their parameters are calibrated to optimal values. Typically, there are several techniques for the same data

mining problem type. Some techniques have specific requirements on the form of data. Therefore, stepping back to the data preparation phase is often necessary.

- **Evaluation:** Model results should be evaluated in the context of business objectives established in the first phase (Delen & Olson, 2008). At this stage in the project you have built a model (or models) that appears to have high quality from a data analysis perspective. Before proceeding to final deployment of the model, it is important to more thoroughly evaluate the model and review the steps executed to construct the model to be certain it properly achieves the business objectives. A key objective is to determine if there is some important business issue that has not been sufficiently considered. At the end of this phase, a decision on the use of the data mining results should be reached.
- **Deployment:** Creation of the model is generally not the end of the project. Even if the purpose of the model is to increase knowledge of the data, the knowledge gained will need to be organized and presented in a way that the customer can use it. It often involves applying "live" models within an organization's decision making processes, for example in real-time personalization of Web pages or repeated scoring of marketing databases. However, depending on the requirements, the deployment phase can be as simple as generating a report or as complex as implementing a repeatable data mining process across the enterprise. In many cases it is the customer, not the data analyst, who carries out the deployment steps. However, even if the analyst will not carry out the deployment effort it is important for the customer to understand up front what actions need to be carried out in order to actually make use of the created models.

Anomaly Detection

Anomaly Detection is an artificial intelligence area which focuses in detecting patterns in data that are not complying with the expected behavior. These anomalies are also known as outliers in the data. These kinds of solutions are widely used in critical industry applications like intrusion detection in security, fraud detection, stock market, insurance, fault detection in critical electronic systems, among others.

Anomaly detection is about finding what we don't know to look for. We are searching for anomalies, but we don't know what their characteristics or properties will be present in the information. When we know the data in detail, we can use machine learning techniques, like classification, we can just write specific rules to find the anomalies. Classification is a form of supervised learning where in the data; we have examples of each kind of thing we are looking for. We can use a learning algorithm for these examples to build a model that can use features of new data to classify them into categories that represent each kind of data of interest. When we have examples of normal and some number of abnormal situations, classifiers can help us to mark new situations as normal or abnormal. Even when we know about some kinds of anomalies characteristics, it is good to keep an eye to monitor new kinds that we don't know about. That is where anomaly detection is applied, so we use the unsupervised-learning approach of anomaly detection when you don't know exactly what you are looking for. Outliers (anomalies) are those events that don't match what you already know. Together, anomaly detection and classification make for a useful pair when it comes to finding a solution to real-world problems. We will use the anomaly-detection model to spot outliers.

So the first step in anomaly detection would be using our model to discover what is normal. This also requires human insight both to structure the model mathematically as well as to interpret what aspect of a pattern is of interest and whether or not it represents a reasonable view of a normal situation.

We then must experiment to determine at what sensitivity you want your model to flag data as anomalous. In one of the algorithms used for this problem, this is possible by calibrating a variable. If it is set too sensitive, random noise will get flagged, and with huge amounts of data, it will be essentially impossible to find anything useful beyond all the noise. Even if you've adjusted the sensitivity to a coarser resolution such that your model is automatically flagging actual outliers, you still have a choice to make about the level of detection that is useful to you.

For an anomaly detection solution, we need to identify the cost of false-positives and false-negatives, in order to know how we will be calibrating our solution. To exemplify a case, we can say that if a false-positive is too expensive, we will try to calibrate the algorithms to get a low quantity of anomalies.

For this problem, we will need to define a strategy to calculate these costs, and design the architecture based on this. There could be a need of a very high performance infrastructure in case the cost of false-positives is low for the solution, this means that the algorithms will end up with a results of a high volume of anomalies.

SOLUTION ARCHITECTURE

An Anomaly Detection Solution (within the solutions spectrum of Data Mining) has been designed through the modeling of statistics and artificial intelligence techniques, following CRISP-DM model standard for data mining solutions.

We describe how we have implemented CRISP-DM model and its activities below, for each phase of the methodology.

- **Business Understanding:** For this initial phase, we have analyzed and understood the project objectives, which are described in the introduction of the chapter.
- **Data Understanding:** For data understanding and preparation we used the same approach, looking at the initial data collection and analyzing the data. Also we have built a dimensional architecture.
- **Data Preparation:** During this phase, the modeling of a properties vector took place, looking for properties adding value to help to have rich data in the vector.
- **Modeling:** During the Modeling Phase, the selection of algorithms and creation of the model design were done. Iterations for the algorithms selected were run and it was necessary to iterate the design of the model. In this phase, various modeling techniques are selected and applied and their parameters are calibrated to optimal values. Typically, there are several techniques for the same data mining problem type. Some techniques have specific requirements on the form of data. Therefore, stepping back to the data preparation phase is often necessary.
- **Evaluation:** In this Phase, we were stressing the algorithms with real data, to iterate over results, and to see how real the data was behaving, and how efficient the algorithms were.
- **Deployment:** As we mentioned before, the creation of the model is generally not the end of the project, so we modeled this system for the results of data to be organized in a way that the information is easily understandable and valuable.

A dimensional architecture was designed to model the problem through three dependent and interconnected dimensions. The first dimension, Behavioral Biometrics, is a scientific area defined by the Human-Computer-Interaction (HCI), which this model has extended to Human Behavioral Patterns. The second dimension is Infrastructure, which represents all physical specialized equipment, environments, networking. The third dimension is space-time, which in this model represents a time dimension against all geographical information project related (code, file, systems and environments and its physical and logical locations).

This dimensional architecture will allow us to identify, separate and relate the different problem components, to understand and abstract each entity separately and be able to design and develop the problem in an organized way (see figure 3).

The dimensional entities are explained below:

- Behavioral Biometrics will represent Human Behavioral Patterns, where we will identify the individuals activities as team members of each project:
    a. Configurations
    b. Development Activities
    c. Testing Activities
    d. Technical Activities
    e. Others...
- Infrastructure is representing the physical entity of the problem:
    a. Equipment
    b. Environments
    c. Network Architecture
    d. Others...
- Space-Time is representing the dimension with which we will be able to contrast the other dimensions:
    a. Geographical information
    b. Time

Figure 3. Dimensional Architecture

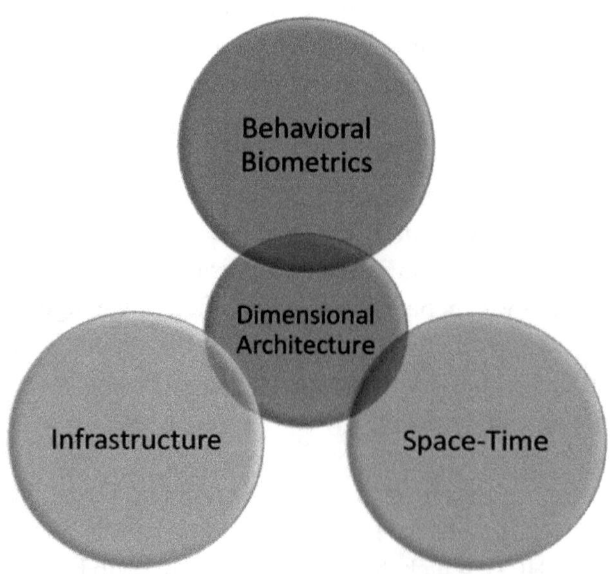

It is very important to emphasize that the problem would not have a possible solution if one the dimensions was not existing, we exemplify this below with two possible scenarios:

1. Having only Infrastructure and Space-Time dimensions; in this scenario we will not be able to identify who is performing activities, neither to know what this individual is doing (behavior).
2. Having only the Behavioral Biometrics and Infrastructure dimensions; in this scenario we will not be able to contrast these dimensions with time, so we will not have the "when", and the information will not be complete.

Now we can see that each dimension is critical for the problem, and that all possible data we can enter to each dimension will be important for the model.

For the Model Design and Architecture, a General Solution Framework was designed with two architectural stacks. The first stack is the Algorithms Stack, which will contain the Business Logic of the model containing a Statistics module, a Classifier module and a User Profiling module (see figure 4).

The second stack is the Utilities Stack, which will perform as an ETL standard process, containing a Preprocessing module in charge of the readers and parsers of the information and a Configurations module (see figure 5).

Figure 4. Algorithms Stack, The Business Logic

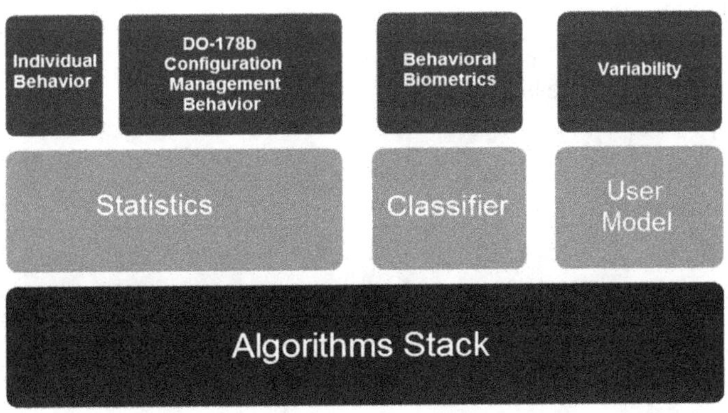

Figure 5. Utilities Stack

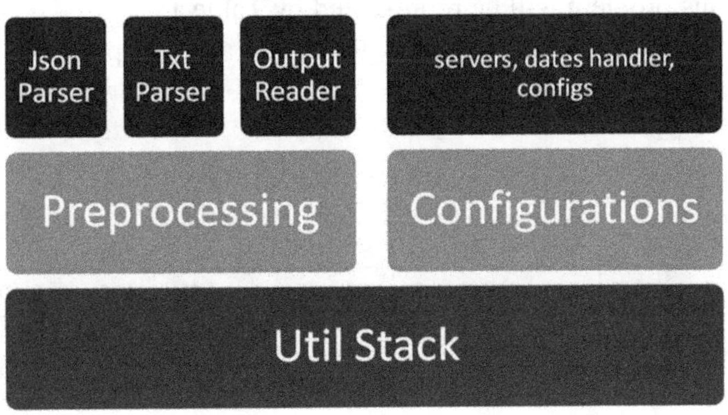

To explain the algorithms we have used in our architectural design, we will develop the methods in detail. In the first step of the model, a filter has been designed implementing a partitioning-based clustering algorithm, ***k-means,*** which will attempt to find k non-overlapping clusters to recognize patterns and identify types of behaviors of each individual. Each cluster is represented by a mean point in the data, called centroid. This algorithm will iterate and group the data based on behavioral patterns until its convergence, this means that the algorithm has grouped the information.

In order to mathematically model the information to be processed, a properties vector has been designed containing Configuration Management behavioral data for each active team member of the project.

Attributes for the Properties Vector:

A1. Commit Size.
A2. Failed Changes Frequency.
A3. Unauthorized configurations Frequency.
A4. Incidents caused by releases.
A5. Number of Bugs.
A6. Code integration issues.
A7. Unit testing´s code coverage.
A8. Release document correctness.
A9. Overtime factor.
A10. Standards Compliance.
A11. Code design accuracy.
A12. Requirements traceability process execution.
A13. Tracking accuracy.
A14. Trunk destabilization.
A15. SVN File-Sharing problems.
A16. Big average file size.
A17. Unexecuted code.

A properties vector with this design would be formed as follows:

$$v = \left\{3,4,2,5,0,2,5,3,1,0,1,1,2,3,3,2,0\right\}$$

Then data matrix of this problem will be represented by Table 1:

Based on DO-178b Configuration Management guidelines and best practices, specific data from the properties vector are labeled (Table 1).

Table 1. Data matrix example

	A1	A2	A3	A4	A5	A6	A7	A8	A9	A10	A11	A12	A13	A14	A15	A16	A17
Data 1	4	2	1	...	...	...	...	...	...	...	...	...	...	...	...	...	...
Data 2	...	...	...	...	...	...	...	...	...	...	...	...	...	...	...	...	...

The information is analyzed separately for Aerospace Software Development teams, and Aerospace Verification or Testing teams; this is because the Configuration Management activities within both groups can vary significantly and cause noise in the system. The objective of this separation is to avoid false-positives in the system results.

The clustering process of the algorithm is as follows (see figure 6):

Our K-means implementation for this problem is detailed in the procedure shown below:

1. The Data Matrix

To explain the methodology we have designed with a visual example for representation purposes, we have chosen three of the seventeen properties of the vector model, shown in the following list (this way, a three dimensional figure can visually represent the individuals in our system):

Figure 6. K-means Algorithm Process

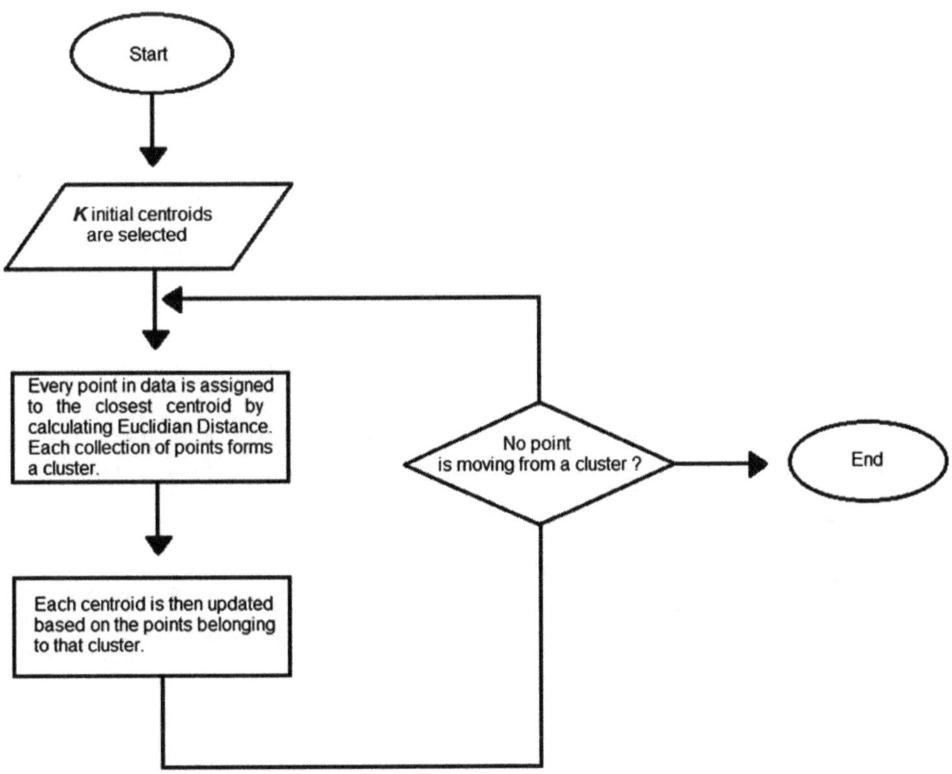

Table 2. Data matrix

	Tester 1	Tester 2	Tester 3	Tester 4	Tester 5	Tester 6	Tester 7	Tester 8	Tester 9	Tester 10
A1	4	3	2	4	4	3	2	4	2	4
A2	2	0	1	1	2	0	1	1	5	2
A3	1	3	2	2	1	3	2	2	0	1

1. Commit Size.
2. Failed Changes.
3. Number of unauthorized configurations.

The data Matrix of ten individuals or team members will be represented by the following table (see table 2):

2. Centroids Selection for Iteration 0:

For the centroids quantity selection, we will determine this number based on what we are looking for at this moment, which is patterns or types of behavior to determine anomalous data for each tester. We defined 4 centroids as the first set of centroids (the first four testers). We will denote the coordinates of the centroids by:

$C_1(4,2,1)$
$C_2(3,0,3)$
$C_3(2,1,2)$
$C_4(4,1,2)$

3. Objects-Centroids Distance for Iteration 0:

For this step of the algorithm, Euclidean Distance formula is used:

$$dij = \sqrt{\sum_{k=1}^{n} (x_{ki} - x_{kj})^2}$$

In this case, we have *x,y,z* coordinates to be used in the formula to obtain the Distance Matrix for iteration 0, calculating the distance between each centroid to each object (testers data):

$$D^0 \left[\begin{array}{ccccc|l} 0 & 3 & 2.45 & 1.41 & 0 & C1(4,2,1) group1 \\ 3 & 0 & 1.73 & 1.73 & 3 & C2(3,0,3) group2 \\ 2.45 & 1.73 & 0 & 2 & 2.45 & C3(2,1,1) group3 \\ 1.41 & 1.73 & 2 & 0 & 1.41 & C4(4,1,2) group4 \end{array}\right]$$

$$\begin{array}{ccccc} t1 & t2 & t3 & t4 & t5 \end{array}$$

$$\left[\begin{array}{ccccc|l} 4 & 3 & 2 & 4 & 4 & x \\ 2 & 0 & 1 & 1 & 2 & y \\ 1 & 3 & 2 & 2 & 1 & z \end{array}\right]$$

4. Objects Clustering for Iteration 0.

Now we have to assign each object to a group based on the minimum distance. The elements of the Group matrix below has a value of 1 if and only if the object has been assigned to that group or cluster.

$$G^0 \begin{bmatrix} 1 & 0 & 0 & 0 & 1 \\ 0 & 1 & 0 & 0 & 0 \\ 0 & 0 & 1 & 0 & 0 \\ 0 & 0 & 0 & 1 & 0 \end{bmatrix} \begin{matrix} group1 \\ group2 \\ group3 \\ group4 \end{matrix}$$
$$t1 \quad t2 \quad t3 \quad t4 \quad t5$$

5. Iteration 1, Determine centroids.

Now we know the members of each group or cluster, we calculate the new centroid of each cluster based on the new memberships, where the centroid of each group is the average coordinate among the members. In groups with one member, the centroid is the only object.

$$C1 = \frac{4+4}{2}, \frac{2+2}{2}, \frac{1+1}{2} = \frac{8}{2}, \frac{4}{2}, \frac{2}{2} = (4,2,1)$$

$$C2 = (3,0,3)$$
$$C3 = (2,1,2)$$
$$C4 = (4,1,2)$$

6. Objects-Centroids Distance for Iteration 1:

We will repeat step 3 with the new centroids coordinates we just obtained.

7. Objects Clustering for Iteration 1.

We will repeat step 4 with the new centroids coordinates we just obtained.

8. When the group $Gn = Gn - 1$, the algorithm has converged, and we have a result, no objects are moving, and we have our clusters defined.

9.

$$G^0 \begin{bmatrix} 1 & 0 & 0 & 0 & 1 \\ 0 & 1 & 0 & 0 & 0 \\ 0 & 0 & 1 & 0 & 0 \\ 0 & 0 & 0 & 1 & 0 \end{bmatrix} \begin{matrix} group1 \\ group2 \\ group3 \\ group4 \end{matrix}$$
$$t1 \quad t2 \quad t3 \quad t4 \quad t5$$

The following figure (figure 7) illustrates visually a converged k-means for this problem, but with 91 objects we have added, in order to have an understandable representation of our 4 clusters.

The following step is the cluster selection performing an exploratory analysis, in order to discriminate the rest of the data.

An analysis to try to identify a threshold and categorize a good behavior and a bad behavior with all seventeen attributes of data together will be performed to the different components of the properties vector, this is, k-means clusters results will be analyzed to find which cluster is maximizing or minimizing behavioral values, this is, which cluster represents a good or a bad behavior. This part of the design can be categorized within the Non-Supervised Learning techniques. The cluster with more anomalous characteristics will be selected and will be entering to the next statistical filter which will utilize the exploratory data represented in the figure below (see figure 8).

Figure 7. K-means Results Represented in a 3 Dimensional Plot

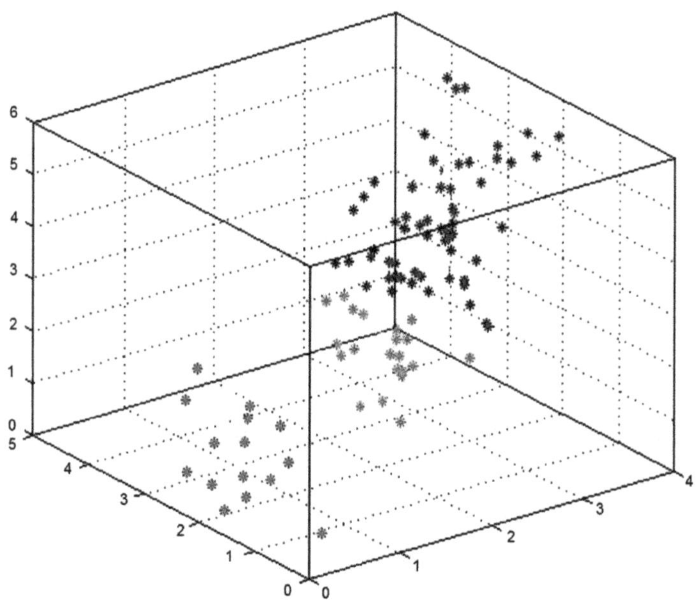

Figure 8. Exploratory Data

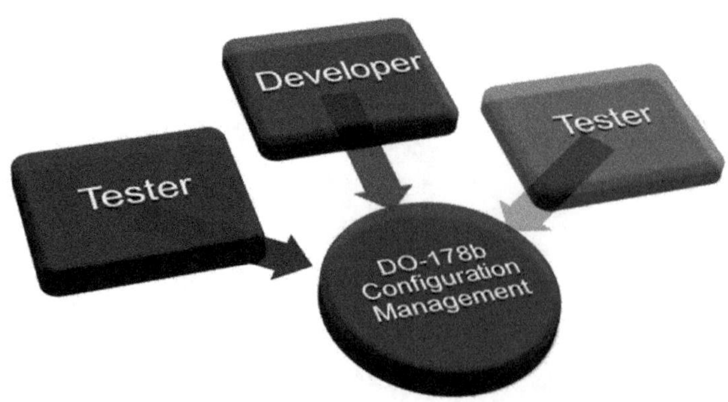

As we can see in the above image, exploratory data will come from different entities or individuals; testers and developers working with DO-178 standard in its b revision.

For the next step of this solution, the anomaly detection, an exploratory analysis is performed by using an anomaly detection technique of the descriptive statistics area. In this area, a box plot or boxplot is a convenient way of graphically depicting groups of numerical data through their quartiles. Box plots display variation in samples of a statistical population without making any assumptions of the underlying statistical distribution. The spacings between the different parts of the box indicate the degree of dispersion (spread) and skewness in the data, and show outliers. In addition to the points themselves, they allow one to visually estimate various L-estimators, notably the interquartile range, range, mid-range, and trimean. Boxplots can be drawn either horizontally or vertically (McGill, et al., 1978). John W. Tukey, an American mathematician, introduced this type of visual data display in 1969.

A graphical representation of a boxplot and a probability density function, emphasizing the quartiles, is shown in the figure below (see figure 9).

Figure 9. Boxplot and a probability density function of a Normal N(0,1σ2) Population

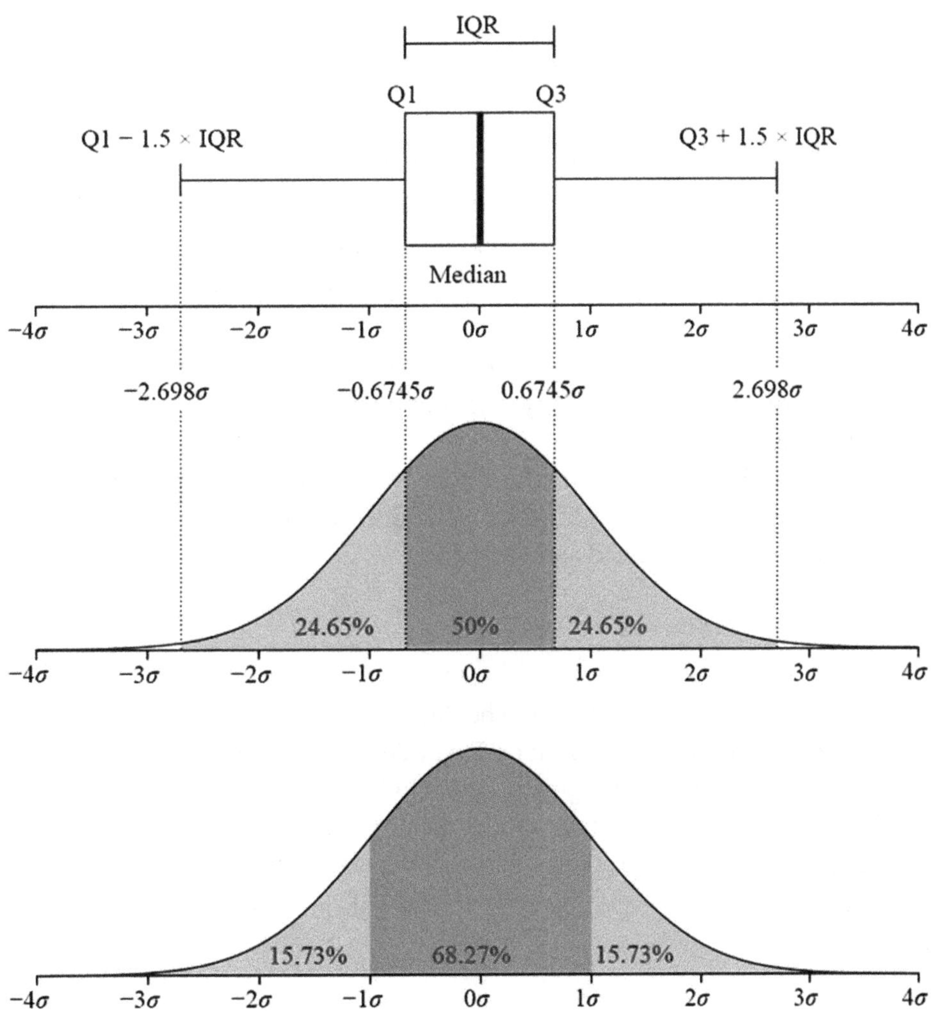

The outliers' detection will be performed by an exploratory data analysis through Tukey Criteria, a statistical method constructing a lower and a higher thresholds to flag data as anomalous, where the lower and upper Tukey limits (lower and higher thresholds) correspond to a distance of 4.5 σ (standard deviations) from the sample mean. Any observations falling beyond these limits are considered as anomalous.

To explain how the probability distribution is formed by the problem data, we have modeled the information with vectors identified by two dimensions: *i* and *j*, where *i* represents the individual ID (i.e. 1714), and *j* represents the day ID (i.e. 01.25.2015).

Vectors of different users and dates with the form $V_{ij}\left\{x, x1, x2...\right\}$, would be represented as follows:

$$V_{1714,05.19.2015} = \left\{3,4,2,5,0,2,1,4,2,5,4,2,0,2,1,4,2\right\}$$
$$V_{2302,05.19.2015} = \left\{1,4,2,5,4,2,0,4,1,1,0,2,3,5,4,2,0\right\}$$
$$V_{1995,05.20.2015} = \left\{0,2,1,4,2,5,4,2,2,5,1,4,1,4,2,5,4\right\}$$
$$V_{0988,05.20.2015} = \left\{3,3,2,0,3,2,5,4,2,2,5,1,4,3,2,0,3\right\}$$
$$V_{1554,05.21.2015} = \left\{1,4,2,2,5,0,2,1,4,2,5,4,2,4,2,2,5\right\}$$
$$V_{1645,05.21.2015} = \left\{3,0,2,4,5,2,2,5,0,2,1,4,2,2,4,5,2\right\}$$

This data will be arranged in an ascending order from the smallest to the largest observation. The ordered data is broken into four quarters, the boundary of each quarter defined by Q_1, Q_2, and Q_3, called the 1st quartile, 2nd quartile, and 3rd quartile respectively. The difference $\mid Q_3 - Q_1 \mid$ is called the inter-quartile range. The Tukey upper and lower thresholds for anomalies respectively are: $ltl = Q_1 - 3 \mid Q_3 - Q_1 \mid$ and $Q_3 + 3 \mid Q_3 - Q_1 \mid \cdot$ Observations falling beyond these limits are called serious anomalies and any observation, X_i, $i = 1,2,...,n$ such that $Q_3 + 1.5 \mid Q_3 - Q_1 \mid \leq X_i \leq Q_3 + 3.0 \mid Q_3 - Q_1 \mid$ will be called a possible anomaly (Wang Ch. et al. 2011).

To illustrate one of the 17 distributions types we have worked with, the following distribution is representing the Commit Size Attribute Frequency from our data matrix in a DO-178 Project, in a period of time of 2 weeks (see figure 10).

With this frequency distribution we can see that we are working with an Exponential Distribution (see figure 11). The Tukey criteria formula for this the type of distribution ideal for this project (exponential probability distribution) would be defined as follows:

$$Q3 + 1.5IQR = \ln(4) / \lambda + 1.5\ln(3) / \lambda$$

The results for the Tukey Criteria Analysis will be thrown for each Vector´s Attribute Frequency Distribution which we will refer to as "Rules". After a Tukey Analysis has been performed, we can observe the results in a bar chart represented in the image below to identify anomalies per rule (see figure 12):

Tukey method allows the user flexibility in setting the threshold limits (see table 3). This way we can calibrate the strictness level of the algorithms. This will depend on the problem, the model efficiency, and the needs of the user. The upper and lower limits for Tukey can be set to $Q_3 + k \mid Q_3 - Q_1 \mid$ and $Q_1 - k \mid Q_3 - Q_1 \mid$.

Figure 10. Frequency Distribution of Attribute 1 in our Data Matrix (2 weeks)

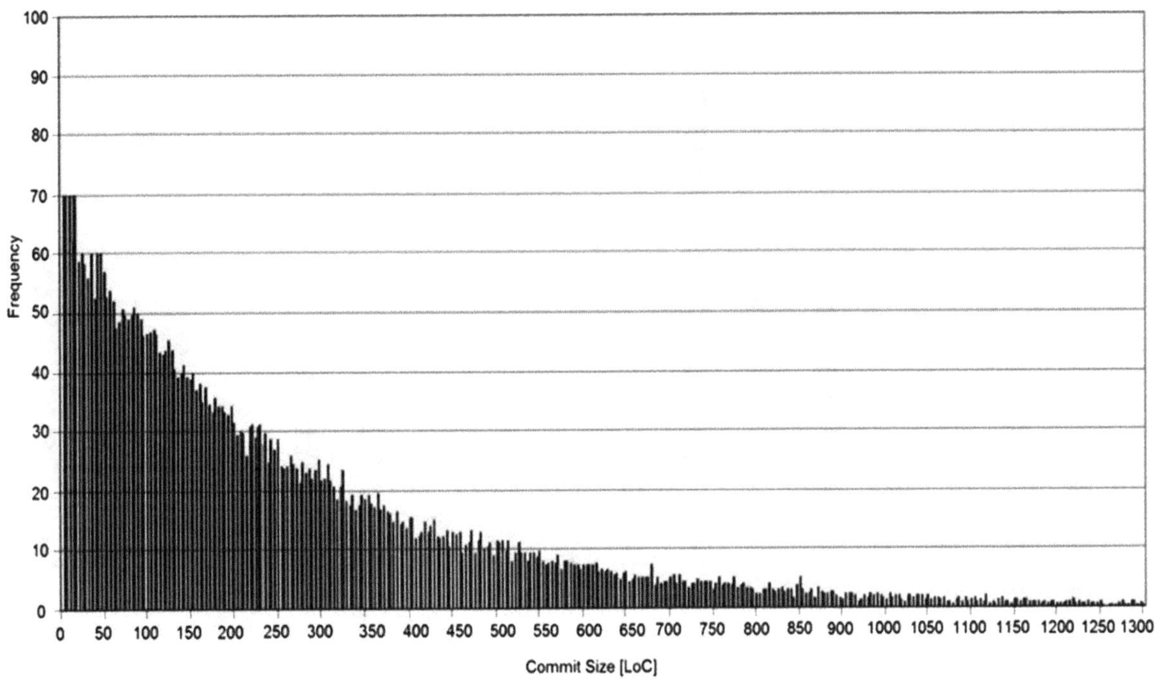

Figure 11. Exponential Probability Distribution Function

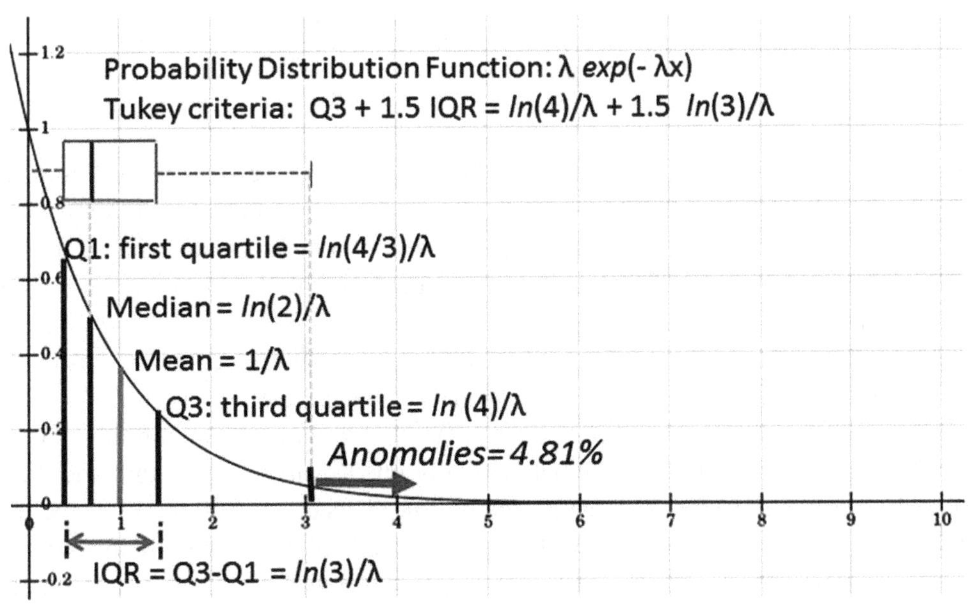

Figure 12. Anomalous Users per Attribute Rule

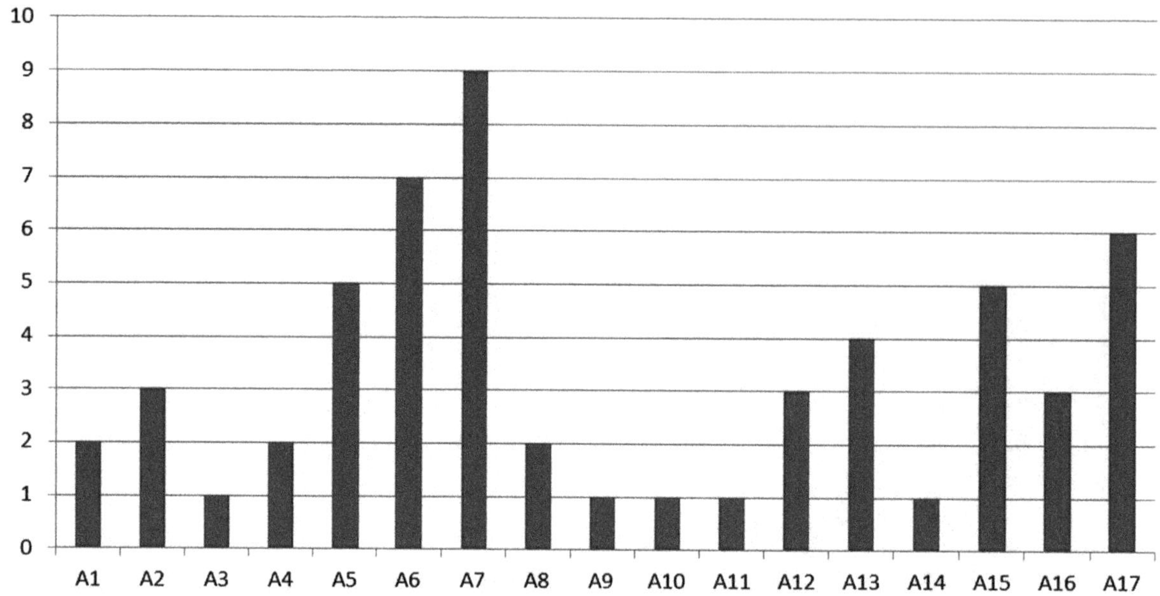

Table 3. Tukey Threshold Calibration

Tukey Criteria	Threshold	Anomalous Population
$Tk_1 = 1.5$	$X = 3.034213 / \lambda$	4.81125%
$Tk_2 = 0.834044$	$X = 2.302585 / \lambda$	10.0%
$Tk_3 = 0.464974$	$X = 1.897120 / \lambda$	15.0%
$Tk_4 = 0.203114$	$X = 1.609438 / \lambda$	20.0%

It is important to mention that at the beginning of each project, there is a phenomenon we call calibration period for the algorithms that take place during the first two weeks (initiation period). During this period, the system's uncertainty is high because there is not enough data to have a good level of efficiency in the algorithms results. This will be decreasing in parallel with increasing historical data after this period (see figure 13).

A Remediation Process for noise reduction purposes has been designed; this process is a quarantine that takes place after each algorithm's cycle has delivered results of anomalous data. This Remediation process takes the outliers individuals and start a pre-defined period of time quarantine of deep detailed supervision, where the system compares the behavioral biometrics in deeper detail, and the results for each individual are compared against a longer period of time. This process allows the system to look for false-positives or noise in the system, this way the system is able to eliminate possible noise.

The pre-defined period of time for the remediation process will depend on the combination of two constants of the system, the false-positives cost and the true-negatives cost for the needs of the project.

The results of this Remediation Process can represent two scenarios: The first one is when the outlier(s) was detected as noise in the system, then this individual is extracted of the remediation process. The

Figure 13. Calibration Period Representation

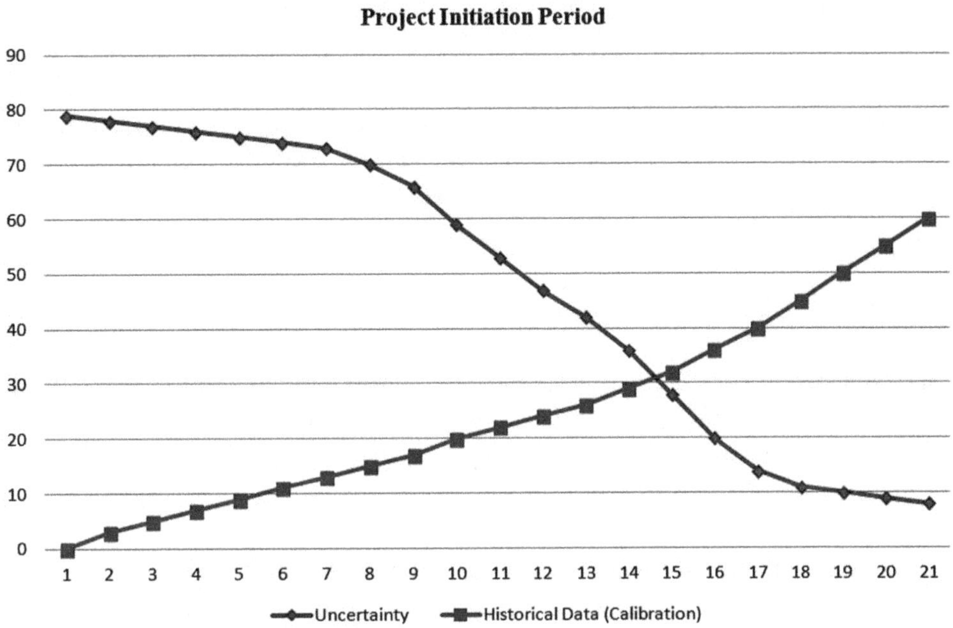

second scenario is when the remediation process result is a true-positive, then the individual is a real outlier, and specific actions are to be taken.

A real-time notifications system has been designed, which will notify when an individual enters or exits Remediation, this will trigger an alert to the manager, so a deep analysis of the case can be performed with the information provided by the system. The outliers' notification will report a graphical behavioral diagnosis to demonstrate the results (see figure 14).

FUTURE RESEARCH DIRECTIONS

The design of solutions for risk identification within the spectrum of complex systems will be always involving different areas of the science and different approaches.

Figure 14. Solution Framework

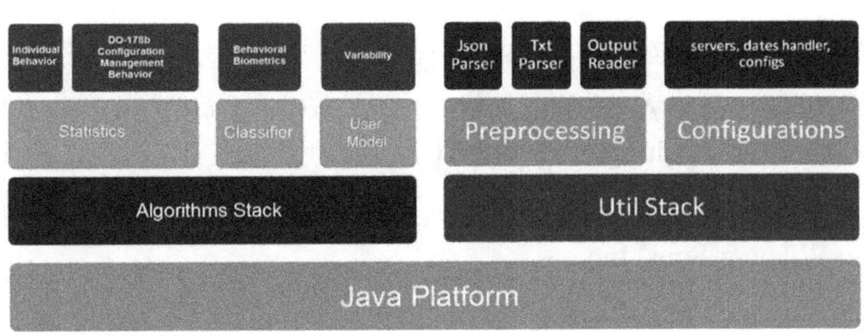

After working with this project, we were identifying possible future work for scope augmentation on this problem. We can add some other techniques like Principal Component Analysis if we increase the number of factors and variables to identify risks company wise, instead of identifying risks at a project level. We also identify the needs of a prediction model in the future, to be able to exploit a big quantity of projects historical information with forecasting techniques like Time Series, Linear Regression, Support Vector Machines or Neural Networks. A prediction model would let a company to be prepared in advanced, and create strategies to improve its projects efficiency.

CONCLUSION

Configuration Management provides all the important control elements to the process: configuration items (CIs), environments' configurations, project libraries, version control, records and other configurations to identify parts and software that should be perfectly working in compliance with the approved versions, but this does not have an impact in the real-time detection of process and human errors, or requirements' specifications during the execution.

Configuration Management risks and issues have not been decreasing over time or with technology improvements, on the contrary, its complexity is growing with time. This model represents a significant help in risks identification for software development and testing projects in the aerospace industry. The behavioral diagnosis when reporting outliers are a valuable visual tool not only to avoid impacts in projects, but also to help to create real-time performance improvement plans for engineers, and to optimize and continuously improve the Configuration Management strategy, with valuable statistical data obtained by this model.

REFERENCES

Chorianopoulos, A., & Tsiptsis, K. (2009). *Data Mining Techniques in CRM: Inside Customer Segmentation*. Wiley.

Delen, D., & Olson, D. (2008). *Advanced Data Mining Techniques*. Springer. doi:10.1002/9780470172339

McCarthy, J. (2007). *What Is Artificial Intelligence?* Stanford, CA: Stanford University.

McCue, C. (2007). *Data Mining and Predictive Analysis: Intelligence Gathering and Crime Analysis*. Butterworth-Heinemann.

McGill, R., Tukey, J. W., & Larsen, W. A. (1978). Variations of Box Plots. *The American Statistician*.

Mitchell, T. M. (1997). *Machine Learning*. McGraw-Hill.

Radio Technical Commission for Aeronautics. (1992). *Software Considerations in Airborne Systems and Equipment Certification*. Washington, DC: RTCA, Inc.

Salvendy, G. (2001). *Handbook of Industrial Engineering: Technology and Operations Management*. New York, NY: John Wiley & Sons, Inc.

Schalkoff, R. J. (1990). *Artificial Intelligence: An Engineering Approach*. McGraw-Hill College.

Wang, Ch., Viswanathan, K., Choudur, L., Talwar, V., Satterfield, W., & Schwan, K. (2011). Statistical Techniques for Online Anomaly Detection in Data Centers. *12th IFIP/IEEE International Symposium on Integrated Network Management (IM 2011)*. doi:10.1109/INM.2011.5990537

KEY TERMS AND DEFINITIONS

Anomaly Detection: A Data Mining technique in charge of the identification of events, items and observations not expected as a pattern.

Biometrics: In Computer Science, this term refers to the metrics related to human characteristics.

Clustering: An Unsupervised Learning technique of Data Mining used for identification and grouping of objects with similar characteristics.

Configuration Management: An area of the Software Industry related to configuration control activities.

Data Mining: An area of Artificial Intelligence related to the Knowledge Discovery through exploiting big quantities of data.

Pattern Recognition: A science that focuses in finding regularities in data.

Safety Critical Systems: Systems where a failure can result in big damages like loss of life, environment or property damage.

Chapter 18
Hybrid Multi-Annealing Simulated Annealing Applied to Vehicle Routing Problem:
A Case of Study

Ernesto Liñán Garcia
Universidad Autónoma de Coahuila, Mexico

Carlos Gonzalez-Flores
Universidad Autónoma de Coahuila, México

Linda Crystal Cruz-Villegas
Universidad Autónoma de Coahuila, Mexico

ABSTRACT

In this chapter, a hybrid meta-heuristic algorithm is proposed, which is based on simulated annealing in order to solve the Vehicle Routing Problem (VRP) with stochastic demands. The proposed algorithm has several annealing phases, which are named Simulated Multi-annealing (SMA). This algorithm is combined with the Clarke and Wright saving algorithm (CW). This last algorithm was placed inside the Metropolis Cycle of Simulated Multi-annealing algorithm. The hybrid algorithm is applied to obtain optimal routes for a cleaning distribution company. The initial solution of VRP is generated by the CW Algorithm, and it was improved by the multi-annealing phases. The solutions into the metropolis cycle were generated by a solution perturbation approach, each solution was compared with a previous one in order to obtain an optimal one.

1. INTRODUCTION

The Planning of optimal routes for the distribution of goods to several business´s customers can generate important savings for companies. In order to improve the delivery services of goods to customers with different demands, distribution centers must arrange optimal vehicle routes to ensure lowest transportation and distribution costs. The Vehicle Routing Problem (VRP) is an important task in many private and public corporations, because several benefits can be obtained if it is solved.

DOI: 10.4018/978-1-4666-9779-9.ch018

This problem was introduced by Dantzig and is applied to design optimal routes, with the goal of servicing a number of customers with a fleet of vehicles subject to the following conditions and constraints: each customer is served by only one vehicle, the demands of all customer must be met, and the capacity of the vehicles may not be violated (Dantzig, 1954).

The objective of this problem is to determine the optimal route for each vehicle in order to serve multiple customers with known demands. There is a fleet of identical vehicles, which supply goods to a set of customers. In this problem, each vehicle of the fleet is assumed to have a common start point, a home base, which is called depot, so, each vehicle must start and finish at the depot. All customers are served by exactly one vehicle, such that the overall transportation cost of the routes are minimized (Dantzig, 1959). The cost of traveling between each pair of customers and between the depot and each customer is given.

There are different variants to the VRP, which are formulated based on the nature of the transported goods, the quality of service required, and the characteristics of the customers and the vehicles. Some of these variants are multi-depot VRP (MDVRP) (vidal, 2013), periodic VRP (PVRP), stochastic VRP (SVRP) (Flatberg, 2005), VRP with indows (VRPTW) among others.

The Vehicle Routing Problem and its variants are classified as NP-Hard Problems (Chiang, 1996; Braysy, 2004; Nagy, 2007). This problem has become one of the most widely studied problems in combinatorial optimization. There are many factors considered and many possibilities of permutation and combinations in order to find an optimal solution of a particular instance of this problem, VRP becomes more complex as constraints and the number of customers increases.

There are several computational algorithms, which are clustered in exact, heuristic and meta-heuristic methods. Those algorithms have been applied to solve VRP, but it is best solved by heuristics and meta-heuristics (Braysy, 2005; Laporte, 2014). Some exact algorithms have been proposed for solving the Vehicle Routing Problem (Laporte, 1987; Laporte, 1992). The advantage of using exact algorithms is that they always obtain optimal solutions to a certain size of the problem, but they become time inefficient for large instances. Due to this reason, several meta-heuristic methods have been designed to obtain sub-optimal solution of VRP variants, e.g., Ant Colony Optimization (Bell, 2010), Simulated Annealing (Hassan, 1993; Czech, 2002), Genetic Algorithm (Thangiah, 1995), (Homberger, 2005) and Tabu Search (Garcia, 1994; Ta, 2013) among others. The disadvantage of meta-heuristics is that they do not guarantee optimal solutions, but can generate solutions very close to the optimal in a reasonable processing time.

This chapter is organized as follows: section II, the Capacity Vehicle Routing Problem is formally described; the classical Simulated Annealing algorithm is detailed in section III, in section IV, the simulated multi-annealing algorithm is described. In section V, the Saving´s algorithm is described, in section VI, the proposed algorithm is explained in detail. The experimentation and results are shown in section VII; and finally, in section VIII the conclusions are discussed.

2. THE FORMAL DEFINITION OF THE VEHICLE ROUTING PROBLEM

The Capacity VRP is formally defined as follows. Let $G\left(V, A\right)$ be an undirected graph where V denotes a set of $n+1$ vertices. A denotes an arc set, which is defined as $A = \left\{\left(v_i, v_j\right) : v_i, v_j \in V, i \neq j\right\}$. V defines the set $V = \left\{0, 1, 2, \ldots, n\right\}$ where n represents the number of customers. The vertex number $\left\{0\right\}$

represents a depot while the remaining $\{1,\ldots,n\}$ corresponds to n customers. Each (v_i, v_j) of set A is associated to nonnegative cost $C_{ij}\,(C_{ij} \geq 0)$. Each customer i of the set $\{i,\ldots n\}$ has a nonnegative demand q_i, which are the units supplied from depot 0 to customer i. The demand of depot q_0 is equal to zero $(q_0 = 0)$. A set $m = \{1,2,\ldots,m\}$ of identical vehicles of capacity Q must be used to supply goods to the n customers. The m vehicles must start and finish at depot. A route is defined as the least cost simple cycle of graph G passing through depot 0 and such that the total demand of the vertices visited does not exceed the vehicle capacity. The objective of this problem is to minimize the total travelled distance, time or cost by the m vehicles, subject to: (1) each route starts and ends at the depot, (2) each customer is visited exactly once by exactly one vehicle and (3) the total demand of each route does not exceed Q. The CVRP is formulated as follows:

$$\min\left(\sum_{i\in V}\sum_{j\in V, j\neq i} C_{ij}X_{ij}\right) \quad 1$$

where $i,j \in V, V = \{0,1,2,\ldots,n\}, X_{ij} \in \{0,1\}$ subject to

$$\sum X_{0j} = m \quad 1.1$$

$$\sum X_{i0} = m \quad 1.2$$

$$\sum_{j=1} X_{ij} = 1, (i = 0,1,\ldots,n) \quad 1.3$$

$$\sum_{i=1} X_{ij} = 1, (j = 0,1,\ldots,n) \quad 1.4$$

$$m \geq 1 \quad 1.5$$

The objective function (1) is the total cost of the solution. Constraints 1.1 and 1.2 indicate that m is the number of vehicles used in the solution and that all vehicles that leave the depot should be returned. Constraints 1.3 and 1.4 ensure that every client is an intermediate node of a route. Constraints 1.5 indicates that the number of vehicles must be equal or greater than 1.

3. THE CLASSICAL SIMULATED ANNEALING

The Simulated Annealing algorithm (SA) is applied to solve complex problems, which have a large solution space to obtain results close to the global optimum value in a short period of time. The SA concept was independently presented by Kirkpatrick (Kirkpatrick, 1983) and Cerny (Cerny, 1985). This algorithm is based on the solids annealing physical process and Boltzmann distribution, it is a Monte Carlo method generalization, which was developed by Nicholas Metropolis (Metropolis, 1953), whom introduced a simple algorithm called Metropolis Algorithm (MA) to emulate the evolution of a solid in a heat bath until thermal equilibrium, at a given temperature, is reached.

The classical Simulated Annealing emulates the equilibrium state of a collection of atoms at any given temperature. Begins with an initial solution at a very high temperature, which is slowly decreased, allowing to the MA to generate configurations of the solution until the final temperature is reached, simulating the gradual metal cooling for crystallization (Kirkpatrick, 1983).

This algorithm starts at a high temperature, and then this parameter is decreased until a final temperature is reached. The final temperature is typically very close to zero $\left(temperature \cong 0\right)$ (Kirkpatrick, 1983; Cerny, 1985). The temperature is decreased by a cooling function from the initial to the final temperature.

There are cooling functions that have been used in the simulated annealing algorithm (Aarts,1989), (Ingber,1993), (Kjaerul,1991), (Laarhoven,1987), the most common cooling function is defined by $T_{t+1} = \alpha T_t$, where T_t is the value temperature at t time, T_{t+1} is the value temperature at $t+1$ time, and α is the decrement factor of temperature. This cooling function decreases the temperature value by an α factor, which does a range of $0.70 < \alpha < 1.0$. A gradual cooling is applied when α is very close to 1 $\left(\alpha \cong 1\right)$, and a fast cooling is applied when α is very close to $0.70\left(\alpha \cong 0.70\right)$.

A has two cycles; the first one is named Temperature Cycle, which the temperature value is decreased by a cooling function. The second cycle is named Metropolis Cycle (MC), and it is applied to generate, accept or reject solutions in order to optimize a problem.

The algorithm number 1 shows the SA algorithm, which is described in three sections. The first one is the setting section (lines 1-7), where all initial parameters are calculated; the second section correspond to the Temperature Cycle (lines 9-29), which contain the Metropolis Cycle (lines 11-27). MA is an internal procedure of the Temperature Cycle; and the algorithm results are shown in the last section (line 31).

Several parameters are determined at setting section. An initial solution $\left(S_{current}\right)$ is calculated by random way (line 1) and its associated energy value $\left(E_{current}\right)$ as well (line 2). The best solution $\left(S_{best}\right)$ and best energy $\left(E_{best}\right)$ are variables assigned to store the best values of all algorithms execution (lines 3 and 4); they are setting at the beginning with their respective current values. An important algorithm parameter is the temperature, which determines the initial and stop criteria. Initial temperature $\left(T\right)$ is tuned in this section (line 5), as well the final temperature $\left(T_{final}\right)$ (line 6). Finally, the cooling factor has the task of decreasing the temperature, each time that Metropolis Cycle is completed (line 28), in order to reach the final one. Therefore, α is set before the Temperature Cycle begins (line 7).

The Second SA section correspond to the Temperature Cycle, also called external cycle, which contains the Metropolis Cycle and the temperature decreasing function; once MC is completed, the function sets a new temperature value (line 28) and the Temperature Cycle is executed until the final temperature is reached.

The Metropolis Cycle starts with the temperature one. The first step is to make a li0ttle change to $S_{current}$ and assign the perturbed solution to the new solution $\left(S_{new}\right)$ (line 1). Also, the new solution energy $\left(E_{new}\right)$ is calculated (line 12). The difference between E_{new} and $E_{current}$ (line 13), the S_{new} is accepted, if the difference is minor than zero (at the minimization case) (line 14), the $S_{current}$ and $E_{current}$ are replaced by S_{new} and E_{new} values respectively (line 15). This means that the new solution is better

Algorithm 1. Pseudo code of classical SA

```
// setting section
1: Scurrent = Initial Solution
2: E(Scurrent) = Energy of Initial Solution
3: Sbest = Scurrent
4: E(Sbest) = E(Scurrent)
5: T = Initial Temperature value
6: Tfinal = Final Temperature value
7: α= Alpha value
8: // Second section
9: While (T >Tfinal) do // Temperature Cycle
10: While (stop criteria) // Metropolis Cycle
11: Snew= perturbation (Scurrent)
12: Enew = Energy(Snew)
13: Obtain difference between Enew and Ecurrent
14: If (difference <= 0) then
15: Accept Snew
16: If E(Snew) < E(Sbest) then
17: Sbest = Scurrent
18: Ebest = Ecurrent
19: End if
20: End if
21: Else
22: Boltzmann probability = exp(-difference/T)
23: If (Boltzmann probability > random(0,1) then
24: Accept Snew
25: end if
26: end if
27: end while
28: Decrease T by a cooling function
29: end while
30: // Third section
31: Show The best solution (Sbest) of CVRP
```

than the current one, and a new difference is calculated, but now between $S_{current}$ and S_{best}. If this difference is minor than cero, the current solution is better than S_{best} and its value is replaced by the $S_{current}$ value (lines 17). It also happens with the energy values, where $E_{current}$ is assigned to E_{best} (line 18). The MC starts again if the stop criteria is not reached yet. On the other hand, when the current solution is better than the new one, the Boltzmann probability is calculated (line 22) and if this value is greater than a random number between zero and one (line 23), the new solution (a bad solution) is accepted (line 24) and replaces the $S_{current}$ value and E_{new} replaces the $E_{current}$ value; if it is not better, the thermal equilibrium (stop criteria) is evaluated and the Metropolis Cycle is repeated until the thermal equilibrium state is reached. Finally, the best solution obtained by the Simulated Annealing algorithm is stored at S_{best} as well as its evaluation at E_{best} (line 31).

4. THE SIMULATED MULTIANNEALING ALGORITHM

The Simulated Multi-annealing Algorithm is based on the classical Simulated Annealing. This algorithm is divided into sub-phases and each phase has a different cooling factor α. The α value is set depending on the quality of the solutions obtained in Metropolis Cycle. The generated solutions into this cycle can be accepted or rejected. A relationship between accepted and generated solutions is represented by ρ. When the temperature is very high, ρ is very close to 1 $(\rho \approx 1)$. It means than there are accepted solutions when the temperature is very high. When the temperature is very low, ρ is very close to 0 $(\rho \approx 0)$. It means than there are very few accepted solutions when the temperature is very low. The factor cooling causes that the Simulated Multi-annealing algorithm generates more or fewer solutions within the Cycle Metropolis.

In the algorithm 2, the first eight lines of pseudocode is the initialization section. In this section, the initial solution is created and the initial and final temperature are established. The initial cooling factor is set. The number of accepted and generated solutions are set with zero value.

Lines 9-33 section is the algorithm simulated multiannealing. In this section, the metropolis cycle is executed and the temperature is decreased until it is smaller than the final temperature. In the cycle, the number of accepted solutions is increased when a solution is better than the previous one (see line 16) or when a bad solution is accepted (see line 26). The number of generated solutions is increased too when a solutions is created (see line 12). After the metropolis cycle is finished, the quality of solutions quality (ρ) is calculated. This variable is calculated by the follow formula:

Algorithm 2. Pseudo code SMA

```
1: Scurrent = Initial Solution
2: E(Scurrent) = Energy of Initial Solution
3: Sbest = Scurrent, E(Sbest) = E(Scurrent)
4: T = Initial Temperature value
5: Tfinal = Final Temperature value
6: α= Initial Alpha value
7: Saccepted = 0
8: Sgenerated = 0
9: While (T > Tfinal) do // Temperature Cycle
10: While (stop criteria) // Metropolis Cycle
11: Snew = perturbation (Scurrent)
12: Sgenerated = Sgenerated + 1
13: Obtain difference between Snew and Scurrent
14: If (difference <= 0) then
15: Accept Snew
16: Saccepted = Sacepted + 1
17: If E(Scurrent) < E(Sbest) then
18: Sbest = Scurrent
19: Ebest = Ecurrent
20: End if
21: End if
22: else
23: Boltzmann probability = exp(-difference/T)
24: If (Boltzmann probability > random(0,1) then
25: Accept Snew
26: Saccepted = Sacepted + 1
27: end if
28: end if
29: end while
30: Quality = Saccepted/Sgenerated
31: α= change_α(Quality)
32: Decrease T by a cooling function
33: end while
34: Show the Sbest
```

$\rho = numberS_{accepted} \,/\, numberS_{generated}$ (see line 30). This value of quality determines the cooling factor (see line 31). The temperature value is decreased by applied the equation $T_{t+1} = \alpha T_t$ (see line 32). The temperature cycle is executed again. After this cycle is finished, the better solution find is shown (see line 34).

5. THE CLARKE AND WRIGHT SAVINGSALGORITHM

The Clarke and Wright savings algorithm is one of the most widely used heuristic for the VRP (Clarke, 1964). This heuristic is applied to solve problems for which the number of vehicles is not fixed. This algorithm calculates all the savings S_{ij} between customers i and j. In order to maximize the saving, two different routes $(0,\ldots,i,0)$ and $(0,\ldots,j,0)$ can be combined to form a new one. In the new route, the arcs $(i,0)$ and $(0,j)$ will not be used and will be added the arc (i,j). Therefore, the route $(0,\ldots,i,j,\ldots,0)$ is created, and a distance saving is generated. In this algorithm, it is assuming that C_{i0} is the cost of travelling from the depot to customer i and C_{ij} is the cost of travelling from customer i to j.

In the algorithm 3, a pseudocode of Saving´s algorithm is described. The customer coordinates are read from a file. The customer demands are read too. The distance between pairs of customers is calculated.

The cost matrix is used in order to calculate the savings by the equation $S_{ij} = C_{i0} + C_{0j} - C_{ij}$. The values of savings are sorted. There is a cycle, which is done while there are savings without being process. Into this cycle, the savings are processed in order to add customers to routes or merge two routes. Customer i and j are from savings matrix. If customer i or j have been processed, the other one can be added to customer processed route.

6. THE HYBRID SIMULATED MULTIANNEALING ALGORITHM

In order to solve the Capacity Vehicle Routing Problem, a new hybridized algorithm has been designed and implemented. This new approach combines the Simulated Multi-annealing with the Clarke and Wright

Algorithm 3. Pseudo code of Saving Algorithm

```
Read customer coordinates
Read customer demands
Calculate of distances between pair customer
Calculate the savings matrix
Sort the savings matrix
While (there are savings without to process) do
customers i and j of the maximum savings not processed
if customer i or j has not been processed then
Obtain Ri or Rj (routes of i or j respectively)
Modify Route Rj or Ri
else // merge two routes
Merge two routes
end if
End while
```

savings algorithm Saving´s algorithm. The last algorithm is applied to generate new CVRP solutions with customer stochastic demands and it is applied to generate both initial and new solutions into Metropolis Cycle. The Simulated Multi-annealing is applied to reduce the temperature value and to accept or reject solutions. If a new CVRP solution is better than previous one, the new solution is accepted; on the other hand, the new solution can be accepted or rejected by applying the Boltzmann probability.

The pseudocode of SMA algorithm is shown in algorithm 4. The initial and final temperatures ($T_{initial}$ and T_{final}) is set (see line 1). The cooling factor alpha $\left(\alpha\right)$ is set. The $S_{accepted}$ and $S_{generated}$ are set with zero (see lines 4 and 5). The demands of all customers are generated by a stochastic way (see line 6). A feasible initial solution $\left(S_{initial}\right)$ of CVRP is created (see line 7), it can be created by applying Saving´s algorithm or a random way. The CVRP initial solution´s cost $\left(C_{S_{initial}}\right)$ is calculated (see line 8). The CVRP current solution $\left(S_{current}\right)$ is set to initial solution (see line 9). The best CVRP solution is set to initial solution (see line 10). The variable T is set to initial temperature $\left(S_{initial}\right)$ (see line 11). The maximal length of Metropolis Cycle is setting (see line 12).

In the simulated annealing section, the two cycles of SA are described. The first one is named temperature cycle, which contains the metropolis cycle and it reduces the temperature values. While the variable T is greater than final temperature, this cycle is executed. The Metropolis Cycle length is started to 1, this length is increased until the maximal length is reached. The metropolis is executed while the length is less than the maximal length of MC. Into MC, stochastic demand of a customer is generated in random way, this change of demand is used for generating a new CVRP solution $\left(S_{new}\right)$ by applied alternatively the Clarke-Wright or random way. The cost's difference between S_{new} and $S_{current}$ is calculated. In order to solve a minimization problem, new CVRP solution is accepted if the difference is less or equal than zero. If the new CVRP solution is better than the best CVRP solution then the best solution is replaced by S_{new}.

If the difference between S_{new} and $S_{current}$ is greater than zero, Boltzmann probability is calculated by $e^{-Difference/T}$. If this probability is greater than a random number within range of [0,1] then the new CVRP solution is accepted else this solution is rejected.

7. EXPERIMENTATION AND RESULTS

The proposed algorithm was implemented using powerbuilder software tool version 10.5. The parameters of Simulated Multiannealing were established in an experimental way. The initial temperature is set at 1000, and final temperature is set at 1. The initial alpha factor is set to 0.85. In this section, the results obtained are shown and analyzed. The algorithm is tested with a case study which considered 219 customer. The Global Positioning System of depot is at $Longitude = -101.431133$ and $Latitude = 26.8459551$. The demands of each customer is defined with the range from 6 to 12 kilograms. The distances between customers i and customer j is calculated by the haversine formula, which use the longitude and latitude of customer i and j. In the tables 1, 2 and 3, the information of GPS of each customer is shown. The maximum capacity of each vehicle is 500 kilograms.

Algorithm 4. Pseudocode of Hybrid SMA.

```
// Setting section
T = Initial Temperature value
Tfinal = Final Temperature value
α= Initial Alpha value
Saccepted = 0 // accepted solutions
Sgenerated = 0 // generated solutions
Generate stochastic demands of all customer
Create a factible initial solution of CVRP
Calculate the cost of this initial solution of CVRP
Scurrent is initialized with the initial solution of CVRP
Sbest is initialized with initial solution of CVRP
T is equal to Tinitial
Lcm_max is initialized
// simulated multi-annealig section
While (T > Tfinal) do // Temperature Cycle
lcm = 1
While (lcm < Lcm_max) // Metropolis Cycle
Generate stochastic demand of a customer in random way
Create new solution (Snew)of CVRP using Saving´s Algorithm
Sgenerated = Sgenerated + 1
Obtain difference of costs between Snew and Scurrent
If (difference <= 0) then
The new solution of CVRP is accepted
Saccepted = Sacepted + 1
If the new solution is best than the best solution then
Sbest is equal to Snew
End if
Else
Boltzmann probability = exp(-difference/T)
If (Boltzmann probability > random(0,1) then
The new solution of CVRP is accepted
else
The new solution of CVRP is rejected
End if
End if
lcm = lcm + 1
End while
Quality = Saccepted/Sgenerated
α= change_α(Quality)
Decrease T by cooling function T = alpha * T
End while
Show Sbest, of CVRP
```

The best solution obtained is shown in the table 4. This solutions includes three routes, the first one is formed by 75 customers; the second one is formed by 69 customers, and the third one is formed by 75 customers. The load of the vehicle used in the first route is 495 kg and a maximum distance of 37.12 kilometers. The load of the vehicle used in the second one is 470 kg and a maximum distance of 34.16 kilometers. The load of the vehicle used in the third route is 482 kg and a maximum distance of 37.29 kilometers. As shown in Table 4, the first vehicle load is 99%, the second vehicle load is 94% and the third vehicle load is 96.4%.

Table 1. Customers (1-100) with longitude and latitude

Customer	Longitude	Latitude	Customer	Longitude	Latitude
1 (depot)	-101.431133	26.8459551	51	-101.433312	26.8540881
2	-101.441354	26.8761448	52	-101.444961	26.8679301
3	-101.406684	26.9227788	53	-101.441008	26.869237
4	-101.393587	26.9236169	54	-101.44334	26.8662781
5	-101.46056	26.9136101	55	-101.401284	26.8827813
6	-101.458598	26.9292265	56	-101.449981	26.874013
7	-101.406632	26.8880786	57	-101.423032	26.8504414
8	-101.429139	26.8865003	58	-101.421718	26.8500822
9	-101.438945	26.8703361	59	-101.453917	26.909953
10	-101.388748	26.9144672	60	-101.431728	26.8570942
11	-101.424937	26.8557525	61	-101.43471	26.8732909
12	-101.395946	26.9114632	62	-101.434859	26.8733083
13	-101.468238	26.9216308	63	-101.457061	26.940927
14	-101.449113	26.9082926	64	-101.456807	26.9330759
15	-101.397575	26.9205593	65	-101.455186	26.9337134
16	-101.462443	26.9243755	66	-101.440307	26.9382087
17	-101.463553	26.9217919	67	-101.45635	26.9340571
18	-101.456115	26.9129111	68	-101.457972	26.9283791
19	-101.437348	26.8865439	69	-101.445173	26.9521005
20	-101.434241	26.8933776	70	-101.442127	26.8904313
21	-101.433986	26.8848209	71	-101.42857	26.9408072
22	-101.439616	26.856057	72	-101.452853	26.9357561
23	-101.434465	26.8845506	73	-101.447046	26.9338757
24	-101.429166	26.8751246	74	-101.457135	26.9427365
25	-101.423584	26.8494029	75	-101.45949	26.9351128
26	-101.45993	26.9086861	76	-101.442635	26.9316923
27	-101.446059	26.9110203	77	-101.445463	26.9528958
28	-101.44731	26.9130041	78	-101.427725	26.8910056
29	-101.457436	26.9130336	79	-101.465254	26.92604
30	-101.452823	26.9185007	80	-101.448995	26.9466289
31	-101.451627	26.9186094	81	-101.428249	26.8855428
32	-101.457556	26.9178256	82	-101.446412	26.9303994
33	-101.460633	26.9128171	83	-101.452751	26.9557125
34	-101.460521	26.9064272	84	-101.436348	26.8902628
35	-101.449345	26.9155882	85	-101.438311	26.895475
36	-101.456874	26.9160404	86	-101.446579	26.89741
37	-101.434837	26.8624841	87	-101.458309	26.9162513
38	-101.467461	26.9194904	88	-101.446844	26.92048
39	-101.458068	26.9351667	89	-101.425578	26.945972

Table 1. Continued

Customer	Longitude	Latitude	Customer	Longitude	Latitude
40	-101.452566	26.9071148	90	-101.444558	26.9172491
41	-101.4208	26.8460105	91	-101.441576	26.9123443
42	-101.444831	26.8643434	92	-101.431059	26.8930375
43	-101.443943	26.8689931	93	-101.43853	26.8921943
44	-101.439067	26.8758621	94	-101.39418	26.9130163
45	-101.438757	26.8680052	95	-101.408937	26.9302578
46	-101.423352	26.8550707	96	-101.43622	26.862535
47	-101.43948	26.8691828	97	-101.426992	26.8612024
48	-101.42799	26.8684106	98	-101.405709	26.9305803
49	-101.439655	26.8744048	99	-101.402432	26.9302097
50	-101.425111	26.8561342	100	-101.404388	26.9241521

Table 2. Customers (101-200) with longitude and latitude

Customer	Longitude	Latitude	Customer	Longitude	Latitude
101	-101.404523	26.9269437	151	-101.444032	26.8619432
102	-101.405325	26.9189069	152	-101.451728	26.9119227
103	-101.411806	26.9126909	153	-101.454789	26.9078502
104	-101.401145	26.9129182	154	-101.428924	26.8866334
105	-101.400776	26.887789	155	-101.409965	26.9078484
106	-101.394548	26.8904267	156	-101.409977	26.9140974
107	-101.398262	26.8945241	157	-101.471862	26.917137
108	-101.396932	26.9097958	158	-101.405178	26.9132131
109	-101.402604	26.8798304	159	-101.427427	26.936367
110	-101.400145	26.9193789	160	-101.3978	26.9093929
111	-101.405958	26.8833816	161	-101.386782	26.921788
112	-101.40353	26.8804975	162	-101.467415	26.9207223
113	-101.396772	26.8848122	163	-101.399466	26.913265
114	-101.414026	26.8883405	164	-101.448256	26.8571477
115	-101.431567	26.8713195	165	-101.400228	26.88016
116	-101.402246	26.8848913	166	-101.460295	26.9233648
117	-101.413587	26.8920675	167	-101.408223	26.8930397
118	-101.396352	26.9169028	168	-101.403458	26.9297976
119	-101.402063	26.9301321	169	-101.462724	26.9218226
120	-101.388883	26.9301257	170	-101.438272	26.8565039
121	-101.407209	26.9175663	171	-101.40667	26.9120815
122	-101.397736	26.9209875	172	-101.399671	26.921594

Table 2. Continued

Customer	Longitude	Latitude	Customer	Longitude	Latitude
123	-101.398212	26.9164991	173	-101.389291	26.9145538
124	-101.407218	26.9118771	174	-101.455883	26.9067616
125	-101.402075	26.9025937	175	-101.445396	26.9272612
126	-101.385762	26.9323868	176	-101.453215	26.8568236
127	-101.393456	26.9250528	177	-101.423486	26.8506265
128	-101.386313	26.9254822	178	-101.433769	26.8536957
129	-101.385266	26.9194571	179	-101.443883	26.8723686
130	-101.388441	26.9195131	180	-101.40439	26.8947569
131	-101.434986	26.8542908	181	-101.435567	26.898206
132	-101.396575	26.901448	182	-101.446313	26.9455226
133	-101.393642	26.9098534	183	-101.444258	26.9091526
134	-101.389863	26.9134439	184	-101.432238	26.8719954
135	-101.401819	26.906914	185	-101.414062	26.8929433
136	-101.402221	26.9047994	186	-101.432234	26.8713441
137	-101.465028	26.9205353	187	-101.448032	26.9100541
138	-101.455721	26.9344976	188	-101.45499	26.9382037
139	-101.473711	26.9152803	189	-101.421061	26.938806
140	-101.405856	26.9086157	190	-101.46662	26.9247102
141	-101.400282	26.8868128	191	-101.400691	26.9203358
142	-101.441559	26.8971161	192	-101.448708	26.8595077
143	-101.459412	26.9358256	193	-101.449852	26.9140578
144	-101.387861	26.9307319	194	-101.464984	26.9211217
145	-101.448612	26.8591986	195	-101.456588	26.9100722
146	-101.426872	26.8881534	196	-101.459903	26.9363794
147	-101.443485	26.8755093	197	-101.409945	26.9113123
148	-101.464245	26.9197896	198	-101.409236	26.9325247
149	-101.440622	26.8564495	199	-101.434669	26.8852357
150	-101.397493	26.872945	200	-101.453542	26.9070466

8. CONCLUSION

The new proposed approach to solve Capacity Vehicle Routing Problem with stochastic demands generates better high quality solutions for a case study, which is for a cleaning distribution company. The hybrid simulated multiannealing algorithm was applied to this particular problem. The best solution obtained for this algorithm is very high quality, which has three routes with minimum total distance of 108.57 kilometers. In order to generate high quality solution of capacity Vehicle Routing Problem, Simulated Multiannealing and Saving´s Algorithm, combination works better than they by apart. As a future work,

Table 3. Customers (201-220) with longitude and latitude

Customer	Longitude	Latitude	Customer	Longitude	Latitude
201	-101.431808	26.8757763	211	-101.450112	26.8582051
202	-101.450552	26.916397	212	-101.459985	26.9339927
203	-101.403892	26.8844734	213	-101.44152	26.9329257
204	-101.397986	26.8731893	214	-101.452161	26.8559838
205	-101.385468	26.9314352	215	-101.428514	26.9408989
206	-101.40602	26.9087923	216	-101.463962	26.9219956
207	-101.406485	26.9084023	217	-101.463653	26.9216301
208	-101.433468	26.8601216	218	-101.40801	26.8919069
209	-101.433532	26.8597562	219	-101.402379	26.8902887
210	-101.431948	26.857127	220	-101.456955	26.9432887

Table 4. The best solution obtained

Route Number	Route of Customers	Total Demand	Total Distance
1	Depot - 153 - 195 - 59 - 152 - 14 - 187 - 27 - 90 - 28 - 193 - 202 - 35 - 88 - 175 - 76 - 213 - 82 - 73 - 31 - 30 - 18 - 29 - 36 - 87 - 32 - 166 - 169 - 216 - 17 - 217 - 194 - 137 - 148 - 38 - 162 - 13 - 190 - 79 - 16 - 68 - 6 - 65 - 138- 67 - 64 - 212 -75 - 196 - 143 - 39 - 63 - 74 - 220 - 188 - 72 - 182 - 80 - 83 - 77 - 69 - 66 - 89 -71 - 215 - 159 - 189 - 95 - 198 - 98 - 168 - 99 - 119 - 101 - 100 - 3 - Depot	495.00	37.12
2	Depot -131 -170 - 22 - 149 - 214 - 176 - 164 - 211 - 145 - 192 - 151 - 54 - 42 - 52 - 56 - 43 - 179 - 53 - 47 - 45 - 186 - 115 - 184 - 62 - 61 - 201 - 24 - 146 - 81 - 154 - 8 - 78 - 92 - 20 - 181 - 85 - 142 - 86 - 91 - 183 - 40 - 200 - 174 - 26 - 33 - 5 - 157 - 139 - 34 - 19 - 84 - 93 - 70 - 23 - 199 - 21 - 9 - 49 - 147 - 2 - 44 - 37 - 96 - 208 - 209 - 60 - 210 - 51 - 178 - Depot	470.00	34.16
3	Depot - 41 - 177 - 57 - 58 - 25 - 46 - 11 - 50 - 97 - 150 - 204 - 111 - 203 - 112 - 165 - 109 - 113 - 55 - 116 - 141 - 219 - 105 - 106 - 107 - 132 - 125 - 136 - 135 - 160 - 108 - 133 - 134 - 173 - 94 - 12 - 163 - 104 - 123 - 118 - 130 - 161 - 129 - 10 - 128 - 127 - 4 - 120 - 144 - 205 - 126 - 122 - 15 - 172 - 110 - 191 - 102 - 121 - 103 - 156 - 158 - 124 - 171 - 140 - 206 - 207 - 155 - 197 - 180 - 167 - 218 - 7 - 185 - 117 - 114 - 48 - Depot	482.00	37.29

this new approach can be modified in order to generate high quality solution of CVRP. This algorithm can be hybridized with other heuristics or metaheuristics.

REFERENCES

Aarts, E., & Korst, J. (1989). *Simulated annealing and Boltzmann machines: A stochastic approach to combinatorial optimization and neural computing*. John Wiley and Sons.

Bell, J. E., & McMullen, P. R. (2004). Ant colony optimization techniques for the vehicle routing problem. *Advanced Engineering Informatics*, *18*(1), 41–48. doi:10.1016/j.aei.2004.07.001

Bräysy, O., Dullaert, W., & Gendreau, M. (2004). Evolutionary algorithms for the vehicle routing problem with time windows. *Journal of Heuristics*, *10*(6), 587–611. doi:10.1007/s10732-005-5431-6

Bräysy, O., & Gendreau, M. (2005). Vehicle routing problem with time windows, part i: Route construction and local search algorithms. *Transportation Science*, *39*(1), 104–118. doi:10.1287/trsc.1030.0056

Cerny, V. (1985). Thermodynamical approach to the traveling salesman problem: An eficient simulation algorithm. *Journal of Optimization Theory and Applications*, *45*(1), 41–51. doi:10.1007/BF00940812

Chiang, W. C., & Russell, R. (1996). Simulated annealing metaheuristics for the vehicle routing problem with time windows. *Annals of Operations Research*, *63*(1), 3–27. doi:10.1007/BF02601637

Clarke, G., & Wright, J. W. (1964). Scheduling of Vehicles from a Central Depot to a Number of Delivery Points. *Operations Research*, *12*(4), 568–581. doi:10.1287/opre.12.4.568

Czech, Z. J., & Czarnas, P. (2002). Parallel simulated annealing for the vehicle routing problem with time windows. In *10th Euromicro Workshop on Parallel, Distributed and Network-based Processing*, Canary Islands-Spain. doi:10.1109/EMPDP.2002.994313

Dantzig, G., Fulkerson, R., & Johnson, S. (1954). Solution of a large-scale traveling-salesman problems. *Operations Research*, *2*, 393–410.

Dantzig, G., & Ramser, J. (1959). The truck dispatching problem. *Management Science*, *6*(1), 80–91. doi:10.1287/mnsc.6.1.80

Flatberg, T. (2005). *Dynamic and Stochastic Aspects in Vehicle Routing: A Literature Survey. SINTEF rapport*. SINTEF ICT.

Garcia, B. L., Potvin, J. Y., & Rousseau, J. M. (1994). A parallel implementation of the tabu search heuristic for vehicle routing problems with time window constraints. *Computers & Operations Research*, *21*(9), 1025–1033. doi:10.1016/0305-0548(94)90073-6

Homberger, J. H. G., & Gehring, H. (2005). A two-phase hybrid metaheuristic for the vehicle routing problem with time windows. *European Journal of Operational Research*, *162*(1), 220–238. doi:10.1016/j.ejor.2004.01.027

Ingber, L. (1993). Simulated annealing: Practice versus theory. *J. Mathl. Comput. Modelling*, *18*(11), 29–57. doi:10.1016/0895-7177(93)90204-C

Kao, Y., Chen, M. H., & Huang, Y. T. (2012). A Hybrid Algorithm Based on ACO and PSO for Capacitated Vehicle Routing Problems. *Mathematical Problems in Engineering*, *2012*, •••.

Kirkpatrick, S., Gelatt, C., & Vecchi, M. (1983). Optimization by simulated annealing. *Science*, *4598*(4598), 671–680. doi:10.1126/science.220.4598.671 PMID:17813860

Kjaerul, U. (1991). Optimal decomposition of probabilistic networks by simulated annealing. *Statistics and Computing*, *2*(1), 7–17. doi:10.1007/BF01890544

Laarhoven, V., & Aarts, E. H. L. (1987). *Simulated annealing: Theory and applications*. Kluwer Academic Publishers. doi:10.1007/978-94-015-7744-1

Laporte, G. (1992). The vehicle routing problem: An overview of exact and approximate algorithms. *European Journal of Operational Research*, *59*(3), 345–358. doi:10.1016/0377-2217(92)90192-C

Laporte, G., & Nobert, Y. (1987). Exact algorithms for the vehicle routing problem. *Surveys in Combinatorial Optimization*, *31*, 147–184. doi:10.1016/S0304-0208(08)73235-3

Laporte, G., Ropke, S., & Vidal, T. (2014). Heuristics for the Vehicle Routing Problem, in Vehicle Routing: Problems, Methods, and Applications. Society for Industrial and Applied Mathematics.

Mendoza, J.E., & Villegas, J. G. (n.d.). A multi-space sampling heuristic for the vehicle routing problem with stochastic demands. *Optimization Letters*, 1-14.

Metropolis, N., Rosenbluth, A., Rosenbluth, M., Teller, A., & Teller, E. (1953). Equation of state calculations by fast computing machines. *The Journal of Chemical Physics*, *21*(6), 1087–1092. doi:10.1063/1.1699114

Nagy, G., & Salhi, S. (2007). Location-routing: Issues, models and methods. *European Journal of Operational Research*, *177*(2), 649–672. doi:10.1016/j.ejor.2006.04.004

Osman, I. H. (1993). Metastrategy simulated annealing and tabu search algorithms for the vehicle routing problem. *Annals of Operations Research*, *41*(1-4), 421–451. doi:10.1007/BF02023004

Pichpibul, T., & Kawtummachai, R. (2012). An improved Clarke and Wright savings algorithm for the capacitated vehicle routing problem. *ScienceAsia*, *38*(3), 307–318. doi:10.2306/scienceasia1513-1874.2012.38.307

Ta, D., Dellaert, N., Woensel, T. V., & Kok, T. D. (2013). Vehicle routing problem with stochastic travel times including soft time windows and service costs. *Computers & Operations Research*, *40*(1), 214–224. doi:10.1016/j.cor.2012.06.008

Thangiah, S. R. (1995). *Vehicle routing with time windows using genetic algorithms*. Academic Press.

Vidal, T., Crainic, T. G., Gendreau, M., & Prins, C. (2013). A hybrid genetic algorithm with adaptive diversity management for a large class of vehicle routing problems with time windows. *Computers & Operations Research*, *40*(1), 475–489. doi:10.1016/j.cor.2012.07.018

Chapter 19
Maritime Logistics

Aylin Caliskan
Yasar University, Turkey

Yucel Ozturkoglu
Yasar University, Turkey

ABSTRACT

In the literature, mostly, under the heading of maritime logistics, classical sea transportation issues have been studied. In actual fact, with broader scope, maritime logistics is a different field under the logistics theme and sea transportation is just one of the functions that must be carried effectively and in an integrated manner with other logistics functions. This chapter focuses on the maritime logistics functions which were selected by the authors in accordance with the gap in the literature, namely transportation and traffic management, inventory management, demand forecasting, material handling, and packaging and labelling. In addition to the maritime logistics functions; port logistics, green maritime logistics and decision problems in maritime logistics topics are handled deeply with the aim of completing the skeleton of the maritime logistics concept by which an understanding of the clear definition of maritime logistics can be achieved. The practitioners can gain more integrated and holistic viewpoint about maritime logistics with the help of this chapter.

INTRODUCTION

The needs and expectations of a rapidly growing world population can only be met by transporting the products. Due to the geographical extension and dispersion of production places, global transportation modes getting more attraction in the last decade. UNCTAD (2014) stated that almost 85% of international trade in products is carried at sea by shipping industry. Thus, the shape of the world economy is highly dependent on maritime industry.

Ports are the milestones of that industry in all over the world. Ports measure the volume of containers they handle in twenty-foot equivalent units (TEU), the standard size of a container. In 2014, approximately 650 million TEU were handled by all ports world-wide (UNCTAD, 2014). Based on Clarkson (2014), global containerized trade is projected to grow by 5.6% in year 2014. Lam and Yap (2011) told that containerization has fragmented the shipping operations and network integration.

DOI: 10.4018/978-1-4666-9779-9.ch019

In the maritime industry, there exist a range of actors which offer various activities and stand for different functions. Suppliers (shipyards, educational institutes, bunker, provision), ship owners and ship management companies, intermediaries (ship agents, ship brokers, freight forwarders), customers (charterers, shippers), port authorities, customs, stevedores, class and survey institutions, insurance companies, and banks are the actors in maritime transportation system (Deveci, 2013, p.24). These actors as represent different activities and stand for different functions, must act in an integrated and coordinated manner to offer efficient, effective, economic, sustainable and environmental friendly maritime activities.

The reality is that the deep meaning of maritime activities can be defined by logistics activities. Logistics is defined as the process of planning, implementing and managing the movement of products and information from one point to another. Logistics is a highly broad concept which involves many businesses and management related functions and also it is divided into many activities/functions which can be listed as follow: Transportation, inventory management, order processing, warehousing, customer services, handling, packaging, labelling, information management, returned product management, salvage scrap disposal, and site/location selection (Lambert, Stock, & Ellram, 1998; Ballou, 1999; Wood, Barone, Murphy, & Wardlow, 2002; Bowersox, Closs, & Cooper, 2010). Correspondingly maritime industry is also highly broad concept in which all maritime related fields such as ship owning, chartering, agents, brokering, forwarding, stevedoring, supply management, port management, maritime business administration, etc. are involved. The broad nature and scope of logistics and maritime industry make it more difficult to specify the borders of maritime logistics. In this chapter it is tried to draw the framework of maritime logistics concept as far as possible with the help of topics namely; transportation and traffic management, inventory management, demand forecasting, material handling, packaging and labelling, port logistics, green maritime logistics and decision problems in maritime logistics

BACKGROUND

Maritime logistics is formed by a combination of transshipment functions and maritime services at ports. But it is not easy to define 'maritime logistics', surprisingly to prove this difficulty, any clear definition, scope and role of maritime logistics have not been appeared in the academic literature yet (Lee, Nam, & Song, 2012, p.9). Hult et al. (2007) defined the value of maritime logistics as facilitating the higher efficiency and effectiveness of a maritime logistics service. According to Panayides (2006), "the integrated demand for maritime transport brings on a maritime logistics concept". Lee and his friends (2012, p.11) define maritime logistics as follow:

Maritime logistics is referred to as the process of planning, implementing and managing the movement of goods and information involved in the ocean carriage.

The main issue in maritime logistics lies down in the concept of integration which has to be occurred in physical level (intermodal or multimodal), economic/strategic level (vertical integration, management structure) or organizational level (relationship based, people and process integration between organizations) (Tuna & Arabelen, 2013).

There are three important actors in maritime logistics system: shipping, port/terminal operating, and freight forwarding (Table 1). The main function of shipping is to move cargoes of shippers from one port to another. By adding logistics activities to the main function like collecting the cargoes in one point, informing the delivery position, helping customers who want special services, preparing bill of lading,

Table 1. Main Function and Supportive Activities of Maritime Logistics (Lee, et al. 2012)

	Shipping	Port/Terminal operating	Freight forwarding
Main function	Moving cargoes between ports	• Shipping reception • Loading/unloading cargoes • Stevedoring • Connecting to inland transportation	Booking vessels and preparing for requisite documents for ocean carriage and trade on behalf of shippers
Supportive logistics activities	• Documentation relating sea trade • Container tracking and information flow • -Providing intermodal service	• Warehousing • Offering distribution center • Testing • Assembly • Repairing • Inland connections	• Inventory management • Packaging • Warehousing

container tracking, performing intermodal services and information flow (Heaver, Meersman, & Van de Voorde 2000), the shipping service gets more close to the value added shipping logistics concept.

The main port/terminal operations are; loading ships, discharging ships and preparation of cargoes for final destination. In modern logistics system ports also contain value added services like warehousing, packaging, inland connections, repairing and assembly which make ports a cluster of organizations to fully become a link in supply chain (Bichou & Gray, 2004; Carbone & De Martino, 2003).

In general, freight forwarding can be considered as a dimension or a function of logistics outsourcing which deals specifically with the transportation of freight. Lambert et al. (1998) defined freight forwarders as companies that serve both to shippers and carriers by organizing and coordinating the transportation of goods. So in the international trade process, sometimes freight forwarders act on behalf of shippers by providing the main activities like reserving a vessel, preparing bill of lading and other documents and with the supportive logistics activities like inventory management, packaging and warehousing (Table 1). On the other hand, Yang (2012) claimed that ocean freight forwarders must respond promptly to clients' complaints and cargo claims, and must provide customized logistics services to compete in this dynamic marketplace. Lately, customers are willing to pay annual fee and to make contract with the freight forwarders networks (IFLN, WCA, Pacific Shipping, Combined Logistics Network, Asian Groupage Service, Planet Logistics Network and Mac-Nels).

The maritime logistics definitions discussed previously indicate activities for which the maritime industry actors might be responsible. It can be said that maritime logistics is not a concept that just shows the performed logistics activities like warehousing in addition to the main activities like shipping. It is a concept with broader philosophy which contains integration, coordination, value added customer services, lower costs, higher flexibility, reduced response time and higher quality. The list is quite comprehensive; it is worthwhile to discuss important maritime logistics functions and their relationship with other functions in the firm's supply chain.

MARITIME LOGISTICS FUNCTIONS

Transportation and Traffic Management

Transportation is the operational main activity in the maritime logistics system. Because of its fundamental importance, all decision makers have directed significant attention toward reducing cost of transportation. Economies of scale benefit and reduction in unit cost can be achieved by carrying high volume of cargo in one voyage. Maritime transportation can be classified into three types: industrial, tramp and liner shipping. Under this topic we will handle these three maritime transportation types in general.

Industrial Shipping: Industrial shipping refers to the case where the cargo owner also owns the ship and aims to minimize shipping costs (Ronen, 1983).

Tramp Shipping: Tramp ships are similar to taxis as they try to catch up the available cargo that cargo shipped generally belongs to one owner. Tramp shipping is a kind of shipping which in the long run does not have a fixed itinerary, and which carries mainly bulk cargoes over relatively long distances and from one or more ports to one or more ports by any vessel with a tonnage of 4,000 dwt or above (Metaxas, 1972, p.271). The freight rate in tramp shipping fluctuates seasonally according to the ship size, charter durations and market conditions. The major difference between tramp and liner shipping is the unscheduled, open market mode of operation (OECD, 2002, p.20). The cargoes that carried in tramp shipping are generally in the form of semi-product or raw material that will be used in industry to produce final products. There are three main categories of bulk cargo carried in tramp shipping (Stopford, 2009, p.64):

- The five major bulks: Iron ore, coal, grain, bauxite/ alumina and phosphate rock
- Minor bulks: Agri-bulks, fertilizers, metals, minerals, steel and forest products
- Liquid bulk: Crude oil, petroleum products, and liquid chemicals cargoes

Liner Shipping: Liner shipping offers a special service for the shippers who prefer a constant rate rather than a fluctuating market rate. Liner shipping; specializes in the transport of small cargo parcels, which do not fill the hold of a ship, on regular services. The main classes of cargoes in liner shipping are loose cargo, containerized cargo, palletized cargo, pre-slung cargo, liquid cargo, refrigerated cargo, and heavy cargoes (Stopford, 2009, p.65).

As seen in the Table 2 while maritime transportation is just focusing on the individual functions relating to sea transportation, maritime logistics focuses on the effective and efficient cargo and information flow in the scope of logistics system. Maritime logistics does not only interested in sea transportation activities like shipping, loading and unloading but also in value added logistics activities like warehousing, packaging, repairing, stuffing, storage, etc. Maritime transport operators are accordingly encouraged

Table 2. Maritime logistics versus maritime transportation, (Lee, et al., 2012, p.12)

	Maritime Transportation	**Maritime Logistics**
Concept	The process of carrying and handling cargoes across the ocean.	The process of planning, implementing and managing the movement of goods and information involved in ocean carriage.
Focusing point	Maritime transportation emphasizes individual functions relating to sea transportation. Each function pursues its own aims or competitiveness.	Maritime logistics is concerned with not only individual functions relating to sea transportation, but also an effective logistics flow as a systematic entity of the logistics integration system.
Managerial function	Sea transportation activities: contracting, shipping, sea voyage, moving cargo, and loading/unloading.	Sea transportation activities: Contracting, shipping, sea voyage, moving cargo, and loading/ unloading. Additional logistics services: Stripping/stuffing, storage, warehousing, offering a distribution center, quality control, testing, assembly, packaging, repacking, repairing, inland connection, and re-use

to keep pace with other logistics functions as a central member of a global logistics integration system (O'Leary-Kelly & Flores 2002; Panayides 2006; Panayides & Song 2008).

Traffic management and routing is a part of transportation sub-function of logistics. In maritime transportation routing depends on the type of shipping. Routes in liner shipping are predetermined, the loading and unloading ports are fixed and time schedules are certainly established. In tramp shipping there is not any predetermined thing related with routing. Route, ports and the time are determined with the order of cargo carriage. Kendall and Buckley (2001) also consider the industrial shipping in which there are fixed ports but flexible sequences and schedules.

Inventory Management

Inventory management concerns with trying to reduce the stock level and relating costs as low as possible. Song and Lee (2012) suggested that maritime managers who obtain significant customer information can constantly update on market demands. Later, they can forecast the market situation and that gives them a positive effect to coordinate their inventory management systems. Inventory planning has a critical role in customer service level. Service level can be measured as an order time, lead time, case fill rate, line fill rate, order fulfillment rate or any mix combination of these.

In the old times, both shipping companies and ports offered just basic logistics functions to their customers. Now days, the competition is getting larger each day, the new logistics services have been offered by shipping companies to gain a competitive advantage. Inventory management is a kindly new service for the ports to manage customer inventory and fulfill their orders. In maritime transportation most often occur in vertically integrated companies, where the same company controls all inventories as well as the fleet of ships transporting the products (Stalhane, Andersson, Christiansen, & Fagerholt, 2014).

Many firms prefer to decentralize the inventory management in charge of the flow synchronization between overseas assembling plants and first tier suppliers in the port (Carbone & De Martino, 2003). They suggested that procurement and pre-assembly stages are becoming an important significance and shape the inventory management of ports. With using some information technologies ports should provide some inventory control activities such as; palletize racked-unracked option, lot and serial number control, cycle count review process and vendor managed inventory - wherever products are in transit from the port or moving to their final destination.

Vendor managed inventory (VIM) is a stock management method that a supplier of products is responsible for optimizing the inventory held by a distributor. VIM agreements in which the supplier takes control of their tanks inventory levels, ensuring that adequate service levels are maintained. In maritime logistics, a fleet of ships transport multiple bulk goods from a set of production ports to a set of different customers, have their facilities (Giesen, Munoz, Silva & Leva, 2007).

On the hand, inventory management also can be defined as controlling all contract affairs relating to rental, equipping and usage contracts for land areas, buildings, floating installations, as well as the clearing of contaminated sites and the removal of explosive ordinance on rented land areas which is located in the ports.

Demand Forecasting

The forecast is a specific definition of what will be sold, when, and where (Bowersox, Closs, & Cooper, 2010, p.260). Forecasting remains as a critical capability for planning. It is important for maritime sector to plan the future to minimize the failure risks because volatile characteristic of the sector. Good planning starts with the appropriate demand forecasting. Sustainability in logistics activities depends on efficient, effective and accurate demand forecasting. Like all other businesses, in maritime sector, demand forecasting acts as a decision support system. But the demand for shipping services isn't a self-sustained system. It should be considered as a part of general demand for other modes of transportation.

With the capital intensive characteristics of maritime sector, demand forecasting is the basis of capacity calculation and investment decisions. Critical questions; which ship, which size, when to buy, are tried to be answered by shipping companies with the help of demand forecasting. Spot-chartering ships, time-chartering ships, sale and purchase, budget and strategic planning, make-or-buy, outsourcing are important decisions which are required to be forecasted. While searching the answers of these questions the shipping companies and port authorities need past information and data sets.

In shipping sector there available different actors with different forecasting needs. Some of them are listed below:

- **Shipping Companies:** Depending on freight rates, new building and second hand prices shipping companies try to figure out the sale and purchase of ships decisions (Stopford, 2009, p.705). How second hand prices will develop, when to buy or sell ships are the basic questions of shipping companies that lead them to forecast demand before deciding.
- **Shipbuilders:** Shipbuilders concern about the decisions like future demand for new ships, prices, currencies, subsidies and competition from other ship builders (Stopford, 2009, p.705).
- **Port Authorities:** Intense competition in port industry pushes port authorities to invest on advanced cargo handling facilities and to expand the port area. As a result decisions about port development depend crucially on traffic forecasting to find out the volume of cargo in the future (Stopford, 2009, p.706).

The expanded list of Stopford excluded here because the authors want to focus on the logistics functions of maritime sector. Of course government, international organizations and banks also need forecasting but they are not related with demand forecasting, and mostly they want to predict the future to help setting policies, regulations and so on.

Forecasting the demand is getting more difficult in maritime sector due to the following reasons (Stopford, 2009, p.707);

- Subtle attribute of shipping markets does not allow decision makers easily place the statistical data on models. Sudden changes affect the sector unexpectedly.
- Because information sharing is not popular in maritime sector, the beneficial data is limited and arrives too late to be useful to the shipping company.
- Some information or data in sector are too volatile, unstable to use in a formal forecasting model. An experienced businessman may have better chance of grasping what is really happening.
- Little reliable data is available so decision makers generally rely on their own intuitive voice.

Demand Forecasting Time Horizon

Depending upon the plan's intended use, forecasts may be required on a daily, weekly, monthly, quarterly, semiannual, or annual basis (Bowersox et al., 2010). The usefulness of the forecast is gradually declining when the time period is shortened. The time horizon of the forecast in shipping can be classified as follows:

- **Very Short-Term Forecasts:** Very short-term forecasts are carried out every day, every week, or even every hour for the purpose of stock control, preparing job tasks and work assignments and workloads in logistics. Very short-term forecasts, or called momentary forecasts (Stopford, 2009, p.709), are more accurate and realistic in contrast to the other forecasts containing different time horizons. This type forecast mostly used by charterers, shipbrokers and traders who have to decide whether to fix a ship or cargo (Stopford, 2009, p.709).
- **Short-Term Forecasts:** Short-term forecasting aims to cover 3-6 months period and is performed to determine the supply times and order size.
- **Medium-Term Forecasts:** Medium-term forecasting is both used in seasonal fluctuation planning which contains the time horizon between 6 months and 5 years and in investment decisions which contains 5-10 years periods. While tramp and liner shipping companies and also port authorities focus on investment decisions like expanding the fleet, expanding the port area, and buying new equipment by using medium term forecasts, intermediaries in shipping sector like brokers, traders and agents focus on business planning.
- **Long-Term Forecasts:** Long-term forecasts are more helpful in setting up strategies for 20-25 years rather than developing logistics plans to meet the demand in the short operational run.

Methods Used in Demand Forecasting

Demand forecasting requires the selection of appropriate mathematical or statistical techniques to generate periodic forecasts (Bowersox et al. 2010). In demand forecasting both qualitative and quantitative methods are used (Figure 1). Qualitative methods are mostly used in shipping demand forecasting but to use both qualitative and quantitative methods would be more appropriate for more accurate results.

Maritime sector is depending on demand. In national and international markets, the sales contract between buyer and customer and the need for transportation of goods by sea depend on this contract give rise the demand for maritime sector actors like port, shipping companies, and intermediaries. In this meaning the demand for maritime sector is a kind of derived demand. That's why it is important to forecast demand in maritime sector.

The data mainly used in cargo demand forecasting for ports are; population, GDP, GNP, Consumer Price Index, capacity use rates, raw material, semi material and material carrying volume, import and export volumes and targets, regional consumption of main goods (JICA, 1998). Quantitative forecasting methods must be supported by qualitative methods. In this context, it is needed to be taken of expert opinion, evaluate the strategic agreements made in the region (like horizontal and vertical integration/merger), development in logistics sector (like logistics villages), development in regulations (like privatization of railways), global and regional crisis, the development in transportation infrastructures and investments and to foresee the possible effects to cargo demands as qualitative data resources (Esmer, 2014, p.3).

Figure 1. Methods used in demand forecasting
Source: Adapted from; Bowersox, et al. (2010, pp.265-268); Stevenson (2011, pp.80-89) and Esmer (2014, p.3)

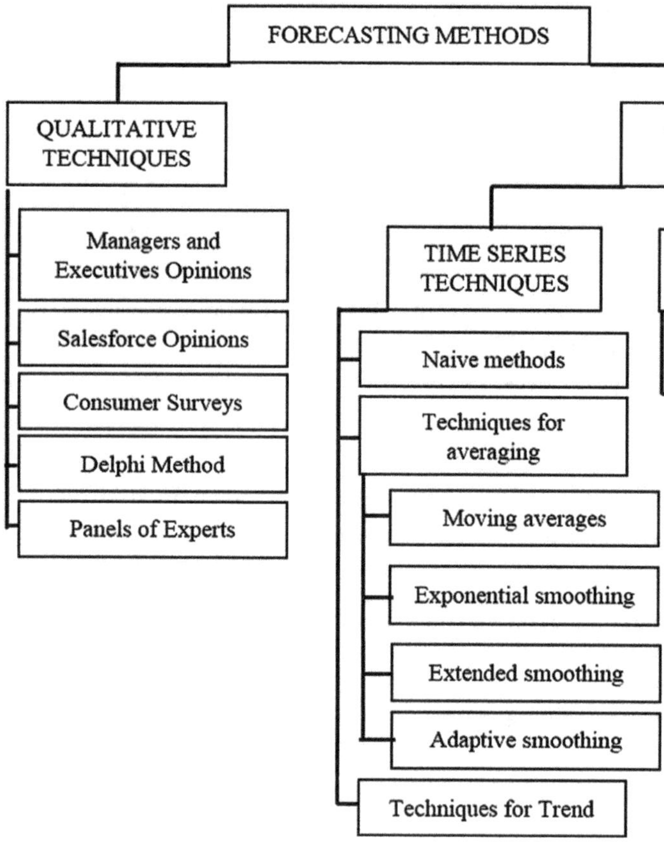

The basic principles in demand forecasting for container terminals are (Esmer, 2014, p.7):

- Local and transit cargo forecasting must be done separately.
- The data of cargo hinterland must be used.
- The forecasting must be cover a time horizon.
- The local hinterland of the port must not be used in transit cargo forecasting.

Shipping is a derived demand of the international trade, thus the type of commodities or the number of containers that a country imports and exports will vary from time to time. Therefore it is essential to frequently update the forecast taking the trends of country's external trade into consideration. The firm usually considers GDP growth rates, previous year actual loading/discharging of containers and calculate the export volume for the following year. The usefulness of the forecast is gradually declining when the time period is shortened. This is because the shipping line cannot take any precautionary measures to minimize any repercussion arising out of the latest container position. Shipping lines usually consider seasonal variations at planning stage. They carefully analyses the country's external trade pattern when preparing the annual budget.

Cargo Handling

Cargo handling involves marshalling services (receive, store, assembly and sort the product in for delivery to a ship's berth) and stevedoring services (load and unload of products from ships). Unfortunately, this function is not easy as in the former sentence because the cargoes are differed based on three groups;

- In size, type, shape, dimension, volume, height, weight, gravity and rate
- Type of cargo handling equipment, number of workers needed and number of handling machines needed
- Type of vessels, weather conditions, handling time, size of the vessels, layout of port.

In broad sense, there are two generic cargo groups; the general and the bulk cargoes. Based on that groups, also cargo handling equipment are classified as; general cargo equipment and bulk cargo equipment. Lifting and moving the cargo in a batches or discrete units is called the general cargo. Dry-bulk cargo is the backbone of international sea trade. These bulk cargoes are handled at specialized terminals with special handling systems. Conveyors, pneumatic systems are some kinds of unloading systems for dry bulks.

Each type of cargo requires specialized equipment and berthing facilities (passenger berths, oil, coal, ore, grain, timber, roll-on/roll-off, containers, chemical and gas) (European Commission, 2013). *Tyre cranes* are commonly used in large ports for loading & unloading and hoisting bulk cargo. They can also take the task of renovating, dismantling and installing large port facilities. *Portal cranes* play vital role in bulk ports that would handle loads with heavy weight and high height, lifting, laying down and luffing often during the operating process (Liu et. al, 2015). *Hammerhead crane* was firstly designed to use in shipyards to support the battleship construction. The crane has an ability to lift up to 350 tons. So it is very useful for installing large pieces of ships. Unlike most cranes, *Deck cranes* placed on both ships and ports. They are used for handling operations or ship unloading and retrieval where no shore unloading facilities are available. *Forklift* is a small industrial vehicle which can be inserted under loads to lift and move them. In maritime logistics, it should be designed to stand up to the hard use and every day wear and tear of a marine environment.

Effective port handling minimizes transportation costs per unit carried, made possible by the steady increase in size of container ships. If shipping adds the functions of a material handling into its activities, generally in ports, it implies perfect coordination of calls by its various mother and feeder ships. By taking over handling operations the shipping line is no longer dependent on a handling company that it does not control and can schedule its ships through a port that is totally dedicated to its own operations (Musso, Ferrari, & Benacchio, 1999; Haralambides, Benacchio, & Cariou, 2002).

Rao and Raghavan (1996) argued that safe movement of hazardous commodities by the sea mode is not a local problem but an international problem which impacts on every maritime nation. The world's petroleum and dangerous chemicals are most being moved by sea transportation with large tankers. Ports are environments often overloaded with hazardous materials, both in bulk and containerized. In view of these facts, public authorities are beginning to feel concerned about how safe harbors are, not only with regard to land operations but also to the possibility of ship collisions, loading-unloading accidents and especially hazardous material handling (Ronza et al. 2006). The International Maritime Dangerous Goods (IMDG) is a code that recommendation on the transport of dangerous goods and model regulation in the ports.

Packaging and Labelling

One of the functions of maritime logistics is the packaging of goods for shipments. To manage the movement of cargoes, which requires container tracking via information systems, graduate to goods tracking, or to performing direct operations (labelling, repackaging, bringing to standard) on products when they pass through the warehouse stage, becoming even more extensively involved in the maritime logistics chain (Antoine, 2008). Three purposes are served by packaging: identifying the product (automated recognition of the product through bar coding), protecting it (protect the product and to keep the product from damaging surrounding items), and aiding in handling (Wood, Barone, Murphy, & Wardlow, 2002, p.249). The choice of packaging materials is influenced by concerns for environmental protection. Containers that can be recycled, or are made of recycled materials, are enjoying increased demand.

The main functions of packaging are (David & Stewart, 2010, p.338; Lambert, et al., 1998, p.331);

- Protection of cargoes to allow them arriving undamaged to destination and keeping cargoes from damaging other items. Three purposes are hidden behind the desire of protection of goods; (1) Protecting the goods from mechanical damage (breakage, crushes, nicks and dents), (2) Protecting the goods from water damage, (3) Protecting the goods from theft and pilferage.
- Facilitating the handling issue for stevedores and the shipping lines.
- Improving customer service strategy by reflecting the quality image of the exporter and by providing aiding (e.g. quickly unpack) for the importer.
- Identifying the products (automated recognition of the product through bar coding) (Wood, et al. 2002, p.249).

In the seagoing trade container as a packaging vehicle is mostly used. The greatest number of containers are the "standard" 20-foot and 40-foot units (6.1 and 12.2 meters, respectively), called boxes; these containers are 8 feet wide, 8 feet tall (2.44 meters), and fully enclosed in steel (David & Stewart, 2010, p.322). Specialized containers are used for some kind of cargoes like liquids and refrigerated cargoes. To use for other cargo types like grain, sugar, etc., containers are required to be cleaned. Polypropylene liners are used for UV, thermal and chemical protection. Some cargo types like cacao and coffee need a level of air circulation that's why lining is a must in packaging of those. Some containers can be used for dedicated cargoes, hence the preparation costs and time can be reduced while repositioning costs are increased (Rodrigue & Notteboom, 2015, p.220).

Container types are not limited with those; here the point is to choose the right container to ship the cargo. Container types are listed below (David & Stewart, 2010, pp.322-326; CMA CGM, 2013):

- **General Purpose Containers:** This is the mostly used container type in seagoing trade. It is suitable for the carriage of dry cargoes, including boxes, pallets, cartons, drums, etc.
- **Reefer (Refrigerated) Containers:** This thermal container is designed to hold cargo at a constant temperature. It needs a power source to cool or heat the air inside and must be plugged in during the journey. It is suitable for the carriage of perishable goods.
- **Open-Top Containers:** These containers are used for cargoes that are too large to fit through the door of a regular container and therefore must be loaded from the top. Open-top containers have no roof but have movable or removable tarpaulin.

- **Flat-Rack Containers:** With no side walls but end walls, these containers are used for heavy, bulky, over height, over width items like pleasure boats, trucks, and military vehicles.
- **High Cube Containers:** These containers are designed 9.5 feet (3 meters) tall and therefore can hold more cargo. Shippers are able to load more Euro pallets than they would in a standard ISO container.
- **Tank Containers:** These containers are dedicated to carry hazardous or non-hazardous liquids (foodstuff).
- **Hanger Containers:** This container type is relatively new, which is designed to hold garments on hanger (i.e. clothes). It seems to fulfill the need for a more convenient way to ship hanging clothes, which may be damaged when they are shipped flat in boxes or may be difficult to fold.

In the case of different types of goods in the same container, the heavier ones should always be placed on the bottom to lower the center of gravity of the load. Great care should be taken to make sure that the center of gravity is somewhat in the center of the container (David & Stewart, 2010, p.343). To keep the cargo from moving inside the container hooks, straps or wood braces are used, therefore the cargoes and the container are secured. These blocking materials are called dunnage.

Through the supply chain to gain time and ease it is important to unitize the cargoes. More efficient loading and unloading processes need goods assembled in a single larger unit by placing the goods on pallets. Unitized package also includes one more layer of protection for the goods. Shrink-wrapped protects the cargo that placed on pallets better from water infiltration and from condensation in the container. It also facilitates handling and protects the goods once they have arrived at the customer's warehouse. Unitizing prevents the primary package from being crushed and therefore allows the goods to be immediately sellable. Also it is costly to repackage the goods into a new box, therefore it is generally much cheaper to properly unitize them (David & Stewart, 2010, p.340).

Although special container sizes are created to carry as much cargo as possible, a substantial proportion of cargo is still shipped as break-bulk. Break-bulk cargo should be packaged differently than containerized cargo. Wooden boxes made especially for break bulk called crates, bags, metallic drums, polymer drums and fiber drums can be used for packaging break-bulk cargoes.

Handling and care instructions must be clearly positioned on the package. To protect the cargo from poor handling, it is necessary to use many of the international pictorials for cargo handling as are relevant. To protect break-bulk or LCL (less container load) cargo from being lost or shipped to the wrong consignee, it should be clearly marked with consignee's name as well as the shipment number. It is needed to write the information on several of the sides of the load. For the reason of decreasing the probability of theft of pilferage, the marking should not include detailed information about the goods in the shipment. Many companies use codes rather than brand markings to identify their shipments and change the code on a regular basis to avoid security problems. At all incoterms conditions, the responsibility of packaging belongs to the exporter.

PORT LOGISTICS

The current global trend towards greater market concentration in shipping and port operations has been intensified, as has competition between container ports (Song & Lee, 2009). A port is composed of several terminals, or the terminals are components of a port. Seaports are the gate for foreign trade for the economies since they are the key point for the activities in maritime transportation. Moreover, they have become logistics based with many types of transportation are combined. These developments in functions of seaports lead to growing of ports` operating as an industry. The costs of shipping depend also on the efficiency of the ports of call. Traditionally, ports are considered to act as an interface between ships and shore by providing shelter and berthing space, temporary storage and the provision of superstructure and infrastructure for cargo operation and movement within port (Tongzon & Heng, 2005).

The main activities of ports are as follow:

- **Pilotage:** Is a service provided by a pilot with local knowledge and skills which enable him to conduct the navigation and maneuvering of the vessel in and approaching the harbor (European Commission, 2013, p.9).
- **Towage:** Is a service provided by tug boats which move larger ships that either should not or cannot power themselves (European Commission, 2013, p.10).
- **Cargo-Handling:** Involves the movement of cargo in and around a port. This includes marshalling services (the receipt, storage, assembly and sorting of cargo in preparation for delivery to a ship's berth) and stevedoring services (the loading of cargo onto and discharging cargo from ships) (European Commission, 2013, p.9).

Panayides and Song (2009) defined seaport terminal supply chain integration (TESCI) as the extent to which the terminal establishes systems and processes and undertakes functions relevant to becoming an integral part of the supply chain as opposed to being an isolated node that provides basic ship-shore operations. Because the ports are intersection points in the supply chain, port logistics is a two-way process (Figure 2). One logistics process is for the outgoing cargoes (exports) which move from the hinterland to destination and other logistics process is for the incoming cargoes (imports) which come from destination to the port hinterland. In this manner the inbound and outbound logistics activities change in both processes.

Robinson (2002) indicated that ports are a part of value driven chain and they can ass value to the goods while passing through them. In the literature while inbound logistics involves production scheduling, procurement and handling activities, outbound logistics involves demand forecasting, order management, packaging, labeling, documentation flow, customer service activities. But in port logistics both inbound and outbound logistics processes are *sui generis* (Figure 2). In the case of outgoing cargo, inbound logistics composes of hinterland operations like truck and train operation, transshipment by transtainers (if the cargoes are container) and yard planning, and outbound logistics involves activities like berth and yard planning, crane assignment, and loading. In the case of incoming cargo, inbound logistics involves activities like berth assignment, yard planning, crane assignment, discharging, discharging sequencing, and stowage planning while outbound logistics composes of truck and train operations, transshipment of cargoes, yard planning and sequencing of trucks to be loaded. Also as an operation part of logistics, port logistics involves yard operations which involves export stocking, packaging, stuffing and redistribution within the yard in the case of outgoing cargoes, import stocking, packaging, redistribution within the yard and unsnuffing in the case of incoming cargoes.

Other ancillary (or general) services provided in many ports include bunkering, chandelling, ship repairing, container maintenance, marine appraisals, insurance claims inspections, banking, etc. (European Commission, 2013, p.9).

Figure 2. Logistics Operations in a Container Port

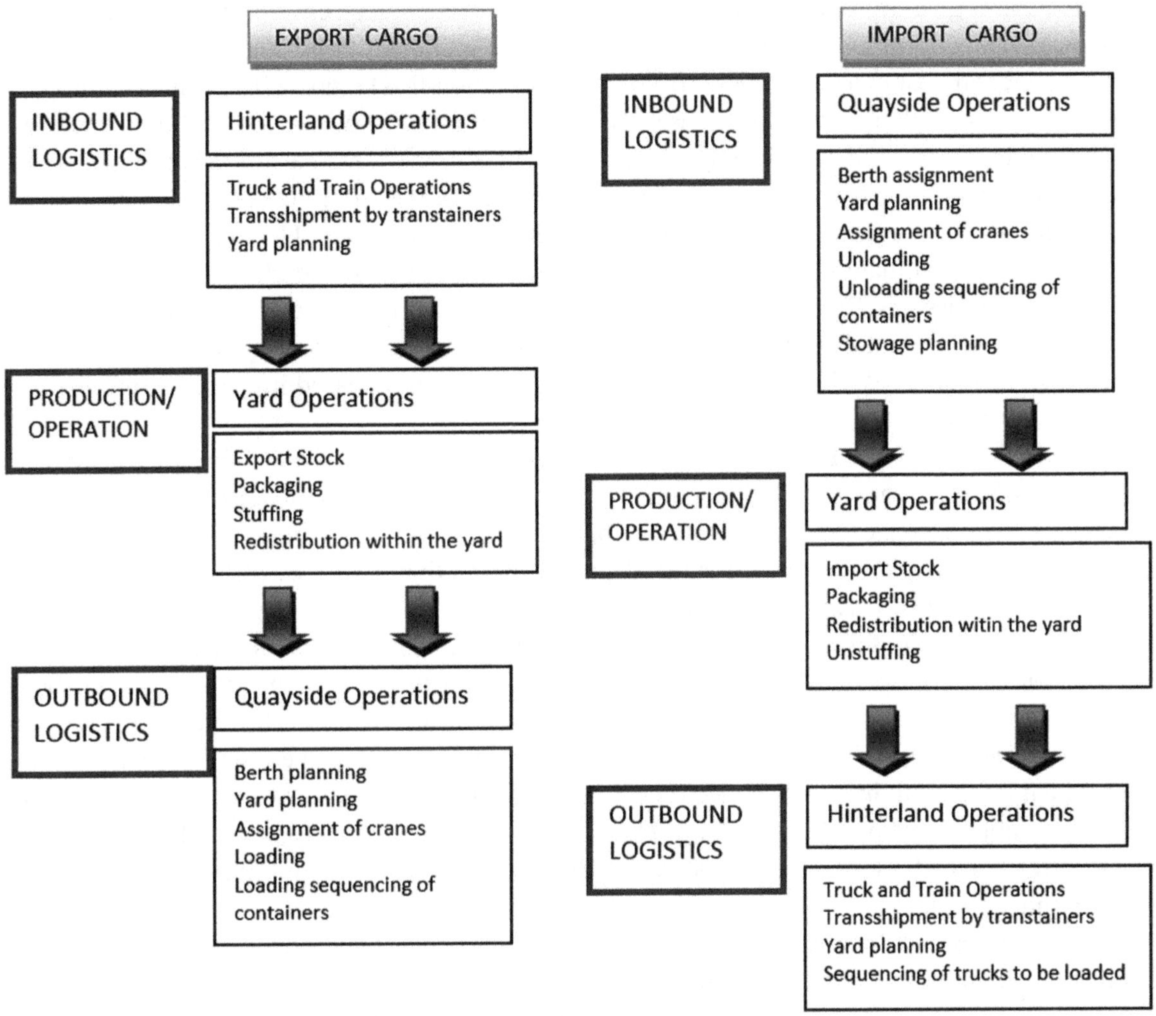

The maritime logistics activities' security also represents a critical variable in terms of competitiveness in the ports. In order to obtain secure ports, customs are willing to locate the container flows, through reducing of import/export documents and to permit the transportation of products to and from ports under bond or in a sealed container. Banomyong (2005) suggested that, the security issue is directly related to the performance measurement of port logistics that all security conditions must be met and guaranteed in order for products to move unhindered within port.

GREEN MARITIME LOGISTICS

With the present climate change in mind, the environmental aspect is very important to address when operating in any modality in the transportation industry. Green maritime logistics is an attractive area whose importance increase recently. Psaraftis and Kontovas, (2009) stated that containerships are the largest maritime CO_2 emitters. The International Maritime Organization (IMO) is an agency that maintains a comprehensive regulatory framework for shipping which include safety, environmental concerns, legal matters, technical cooperation, maritime security and the efficiency of ships. The main responsibility is to prevent and control of marine pollution from ships. Based on IMO report (2012), sea transportation accounted for 4.8% of global greenhouse gas (GHG) emissions with a total of 2031 million tons CO_2 in 2012. For making successful eco-friendly maritime logistics, important decisions should take that the decisions going from being based upon economical parameters alone, to taking the environmental aspect into consideration basing the decision on parameters related to emissions. The main questions that try to find answers are; estimation of emissions, impact of emissions on world climate and technological means to reduce emissions. To reduce maritime GHG emissions, it is important trying to optimize with respect to both environmental and traditional criteria. Psaraftis and Kontovas, (2010) analyzed that there are three ways to reduce GHG emissions; technical measures, market based instruments measures and logistics based options.

Technical measures are defined as use of more efficient options such as; eco-friendly fuels, energy saving engines, vehicle capacity, engine load, efficient propulsions, ship hulls. Emissions trading and carbon levy schemes are the categories of market based instruments. Reducing speed, optimizing route, planning fleet, changing the number of ships, degree of utilization, specific engine load are the some example of logistics measurements. Lindstad (2012) stated that to increase the fleet size and mix, the emission per freight will minimize around 30%. Endresen et al., (2007) calculated CO_2 emission based on bunker consumption and emission factor. According to their formulation one ton of marine bunker produces 3.17 tons of carbon monoxide (CO_2), 0.02 tons of sulfur oxides (SO_2) and 0.057-0.087 tons of nitrogen oxides (NO). Also, heavy metals, dioxins, formaldehyde, ozone are the other particulate matters which affect both human health and environment. Environmental Protection Agency (EPA) is expected to double severe pollution by 2020. This effect is worth studying, especially for container vessels, which represented 4% of all maritime vessels but generated 20% of emissions from international shipping (Buhaug et al., 2009).

Slow steaming is another alternative to reduce both carbon emissions and fuel costs. Cariou (2010) defined slow steaming as a term that used if the average speed of a voyage is less than the commercial speed. The term comes in three degrees: slow steaming, extra slow steaming, and super slow steaming. Reduced speed will entail additional in-transit inventory costs for the shipper, in this time may be extra vessels may be needed to fulfill the contractual obligations limiting the total obtainable gains.

Marines have large impact on the surrounding environment, including pollution from overseas vessels and land based equipment (forklifts, trucks, cranes). They are also major sources of noise and light pollution, severe traffic congestion, and solid and hazardous waste generation in surrounding communities. Tzannatos (2010) suggested that although port emissions are not significantly contributing to the overall picture of ship generated emissions, it is important to note that the impact of ship exhaust pollutants has a direct effect on the human population and built environment of many urbanized ports. All the transportation vehicles and material handling equipment should contribute to diesel emissions at ports. Bailey and Solomon (2004) argued that if emissions control on ships, shore-side power for ships while at dock and sufficient power must exist, then the pollution of ports will significantly reduce. Lattila et al. (2013) indicated the importance of dry ports in the reduction of CO_2 emission related to the seaport hinterland. Hoen et al. (2014) supported with quantitative analysis the use of inland waterways instead of short sea shipping in order to reduce CO_2 emissions. Diesel oxidation catalyst is a kindly new device which oxidizes pollutants in the exhaust stream and can be packaged with mufflers (EPA, 2015). When the device installed the equipment, the emissions will reduce around 20% (EPA, 2015).

The most important point is that limiting emissions in one part of the intermodal chain may increase emissions in another transportation mode.

DECISION PROBLEMS IN MARITIME LOGISTICS

In today's competitive marketplace manager got all kinds of tough decisions to make. As mentioned previous parts of this chapter, maritime logistics is also very different from decisions and planning perspectives. There is a larger variety in decision structures and operating environments in maritime logistics planning. Demand, volume, ship speed, fuel consumption, equipment breakdowns, rates, costs, vessels size, due dates, weather condition, mechanical problems are the some important parameters which are associated with uncertainty. Like the other decisions problems, maritime transportation problems can be classified based on planning time horizon into strategic, tactical and operational problems.

The strategic decisions are long period decisions that related with long term contracts. The most important strategic decisions are; the target market share, port choices, port design, ship design, select the trade lanes what a fleet should be composed of regarding size and mix. The studies about strategic planning in maritime logistics back to 50's (Barnhart & Laporte, 2007). Dantzig and Fulkerson (1954) first developed a model about fuel oil tankers design with different size of six ships. Then the other researchers studied various perspectives of the long decision problems about maritime industry. Some important researches; Bellmore (1968), Jansson and Shneerson (1982), Garrod and Miklius (1985), Talley et al. (1986), Jaikumar and Solomon (1987),Pope and Talley (1988), Ariel (1991), Cho and Perakis (1996), McLellan (1997), Gillman (1999), Cullinane and Khanna (1999),Xinlian et al. (2000), Crary et al. (2002), Kleywegt (2003), Christiansen et al. (2004), Christiansen et al. (2007), Zaucha, J. (2014), Sahebi et al. (2014) and Angeloudis et al. (2015) were concerned with strategic planning decisions for logistics firms which focus on maritime logistics.

The tactical planning decisions in maritime transportation is faced with medium term contract. The main problem is routing and scheduling the actors which are related to maritime. The first research review had done by Ronen (1983) in ship routing and scheduling problems. After a decade later Ronen (1993) and then Christiansen et al. (2004) made overall reviews which both were focused on ship routing and scheduling problems in maritime industry. Vis and Koster (2003) studied decision problems based on container terminal. Crainic and Kim (2007) analyzed the logistics operations of the land side of port terminals. Brown et al. (1987), Fisher and Rosenwein (1989), Perakis and Bremer (1992), Desrosiers et al. (1995), Fagerholt and Christiansen (2000), Christiansen and Fagerholt (2002), Jetlund and Karimi (2004), Persson and Göthe (2005),Hwang (2005), Bronmo et al. (2006), Kang et al. (2012), Norstad et al. (2013), Moon et al. (2014), Zhen (2014) and Sahebi et al. (2014) are some of the important researchers studied in tactical planning decisions problems in maritime logistics.

During the day, port managers should take several important operational decisions. The maritime environment is dynamic and fill by full of uncertainty. The operational decisions should be given very fast and they have very short term impact. The most important problems faced by planners are the speed selection, ship loading and unloading, environmental routing problems. The planners are responsible for a given fleet segment for each ship. Appelgren (1969), Martin et al. (1988), Lo et al. (1991), Brown et al. (1997), Bausch et al. (1998), McCord et al. (1999), Wilson and Roach (2000), Cho and Perakis (2001),

Kang and Kim (2002), Sigurdet al. (2005), Crainic and Kim (2007), Dohn et al. (2011) Rasmussen et al. (2012), Fagerholt and Ronen (2013), Coccola et al. (2014), Vilhelmsen et al. (2014)can be found in a review of literature focused on operational decisions problems in maritime logistics.

One important decision for maritime logistics is the select the best partner (freight forwarders, agent, shippers etc.) in order to exchange resources or execute different acts in order to seize common benefits. Trust, service quality, payment guarantee are the most cited criteria that rise throughout this selection problem.

Pearson (1980) studied that port itinerary, arrival &sailing date, transit item, port proximity, slot available and regularity are ranked the most important criteria for UK shippers. Brooks (1984) identified shipper criteria based on Canadian shipper choice on the basis of rates of service, reputation, transit time and directness of sailings. Collison (1984) determined that average time in transit, schedule reliability and ability of service are as important factors for both inbound and outbound ports. Suthiwartnarueput (1988) evaluated the Thailand maritime industry and found that the cost of service, punctuality, transit times, frequency and directness of sailing, past loss and damage experience are important service attributes from customer side. Tengku (1995) almost found similar criteria; freight rate, cargo care, knowledgeability, punctuality, transit time and service frequency. La Londe and Cooper (1989) identified several factors as important customer service factors for selecting and evaluating carriers, namely equipment, consistency, speed of transit time, administration, flexibility, and responsiveness. Lu (2004) determined that logistics services in Taiwanese distribution centers and identified seven logistics service dimensions: value-added services, support services, distribution services, information and transportation services, cargo-related services, consolidation services, and storage services. In generally, logistics service reliability capability, value added service capability, flexibility, information service capability factors are among the most frequently cited ones. Talley and Ng (2013) analyzed the similarity between port choice, shipping line choice and maritime transport chain choice determinants by carriers, ports and shippers. Zavadskas et al. (2015) studied multi-criteria selection of a deep-water port in the Eastern Baltic Sea which contains decisions like accessibility, storage capacity, construction cost, competitiveness, etc.

CONCLUSION AND FUTURE RESEARCH DIRECTIONS

In this chapter firstly the maritime logistics functions including transportation and traffic management, demand forecasting, inventory management, handling, packaging and labelling explained briefly. Later, ports as logistics nodes in the supply chain were handled by answering the question of what logistics processes do occur in a single port with which activities. Because achieving environmental friendliness is one of the main purpose of both logistics and sea transportation, green maritime logistics topic was studied involving the ways to reduce the GHG emissions. Also the uncertain and fluctuating environment of maritime logistics make it compulsory for the authors to handle strategic, tactical and operational decision problems. Especially the port logistics and inventory management topics in maritime logistics have not been studied in the literature. Also because the other topics are handled separately formerly, this chapter serves as holistic directory which handled all the topics under the subject of maritime logistics. Shortly, it offers wholly new findings that extend previous research.

To conclude, this chapter comprehensively reviews the maritime logistics and its functions. The authors believe that without maritime transportation, the global economy and trade would be dead in the water, and without maritime logistics, maritime transport would be grossly inefficient. It should be realized that maritime logistics functions must be planned and executed in coordination with each other and with other functions in the firm's supply chain.

In the time to come, other logistics functions can be handled with related to maritime sector, like information management and site/location selection which were excluded in this chapter due to the limited space. Also the integration level can be measured with appropriate scales and methods among different actors and different activities in maritime logistics since it is the most important matter in the rise of the 'maritime logistics' concept. In addition to these, with the help of performance indicators the efficiency and effectiveness of maritime logistics can be measured intended for both business and meeting the customer needs.

REFERENCES

Angeloudis, P., Greco, L., & Bell, M. G. (2015). Strategic Maritime Container Transport Design in Oligopolistic Markets. *Transportation Research Procedia*, *9*, 269–282. doi:10.1016/j.trpro.2015.07.015

Antoine, F. (2008). *Empirical evidence for integration and disintegration of maritime shipping, port and logistics activities*. OECD/ITF Joint Transport Research Centre Discussion Paper, No. 2009-1.

Appelgren, L. H. (1969). A column generation algorithm for a ship scheduling problem. *Transportation Science*, *3*(1), 53–68.

Ariel, A. (1991). The effect of inventory holding costs on the optimal payload of bulk carriers. *Maritime Policy & Management*, *18*(3), 217–224. doi:10.1080/03088839100000024

Bailey, D., & Solomon, G. (2004). Pollution prevention at ports: Clearing the air. *Environmental Impact Assessment Review*, *24*(7-8), 749–774. doi:10.1016/j.eiar.2004.06.005

Banomyong, R. (2005). The impact of port and trade security initiatives on maritime supply-chain management. *Maritime Policy & Management*, *32*(1), 3–13. doi:10.1080/0308883042000326102

Barnhart, C., & Laporte, G. (Eds.). (2007). Handbook in OR & MS (vol. 14). Elsevier B.V.

Bausch, D. O., Brown, G. G., & Ronen, D. (1998). Scheduling short-term marine transport of bulk products. *Maritime Policy & Management*, *25*(4), 335–348. doi:10.1080/03088839800000057

Bellmore, M., Bennington, G., & Lubore, S. (1968). A maximum utility solution to a vehicle constrained tanker scheduling problem. *Naval Research Logistics Quarterly*, *15*(3), 403–411. doi:10.1002/nav.3800150305

Bichou, K., & Gray, R. (2004). A logistics and supply chain management approach to port performance measurement. *Maritime Policy & Management*, *31*(1), 47–67. doi:10.1080/0308883032000174454

Bowersox, D., Closs, D., & Cooper, M. (2010). *Supply Chain Logistics Management* (3rd ed.). New York, NY: McGraw-Hill Companies.

Brønmo, G., Christiansen, M., & Nygreen, B. (2006). Ship routing and scheduling with flexible cargo sizes. *Journal of the Operation Research Society*, doi:.10.1057/palgrave.jors.2602263

Brooks, M. R. (1984). An alternative theoretical approach to the evaluation of liner shipping - Part I: Situational factors. *Maritime Policy & Management*, *11*(1), 35–43. doi:10.1080/03088838400000032

Brown, G. G., Cormican, K. J., Lawphongpanich, S., & Widdis, D. B. (1997). Optimizing submarine berthing with a persistence incentive. *Naval Research Logistics*, *44*(4), 301–318. doi:10.1002/(SICI)1520-6750(199706)44:4<301::AID-NAV2>3.0.CO;2-A

Brown, G. G., Graves, G. W., & Ronen, D. (1987). Scheduling ocean transportation of crude oil. *Management Science*, *33*(3), 335–346. doi:10.1287/mnsc.33.3.335

Buhaug, Ø., Corbett, J., Endresen, O., Eyring, V., Faber, J., Hanayama, S., & Yoshida, K. et al. (2009). *Second IMO greenhouse gas study*. London: International Maritime Organization.

Carbone, V., & De Martino, M. (2003). The Changing Role of Ports in the Supply Chain Management: An Empirical Analysis. *Maritime Policy & Management*, *30*(4), 305–320. doi:10.1080/0308883032000145618

Cariou, P. (2011). Is slow steaming a sustainable means of reducing CO2 emissions from container shipping? *Transportation Research Part D, Transport and Environment*, *16*(3), 260–264. doi:10.1016/j.trd.2010.12.005

Cho, S.-C., & Perakis, A. N. (1996). Optimal liner fleet routing strategies. *Maritime Policy & Management*, *23*(3), 249–259. doi:10.1080/03088839600000087

Cho, S.-C., & Perakis, A. N. (2001). An improved formulation for bulk cargo ship scheduling with a single loading port. *Maritime Policy & Management*, *28*(4), 339–345. doi:10.1080/03088830010002755

Christiansen, M., & Fagerholt, K. (2002). Robust ship scheduling with multiple time windows. *Naval Research Logistics*, *49*(6), 611–625. doi:10.1002/nav.10033

Christiansen, M., Fagerholt, K., Nygreen, B., & Ronen, D. (2007). Maritime Transportation. In C. Barnhaft & G. Laporte (Eds.), *Transportation, Handbooks in Operations Research and Management Science* (pp. 189–284). Amsterdam: North Holland.

Christiansen, M., Fagerholt, K., & Ronen, D. (2004). Ship routing and scheduling: Status and perspectives. *Transportation Science*, *38*(1), 1–18. doi:10.1287/trsc.1030.0036

Clarkson Research Services (2014). *Container Intelligence Monthly*. Author.

Coccola, M. E., Dondo, R., & Mendez, C. A. (2014). A milp-based column generation strategy for managing large-scale maritime distribution problems. *Computers & Chemical Engineering*. doi:10.1016/j.compchemeng.2014.04.008

Collison, F. M. (1984). North to Alaska: Marketing in the Pacific Northwest- Central Alaska liner trade. *Maritime Policy & Management*, *11*(2), 99–112. doi:10.1080/03088838400000003

Crainic, T. G., & Kim, K. H. (2007). Intermodal transportation. In C. Barnhart & G. Laporte (Eds.), *Transportation, Handbooks in Operations Research and Management Science* (pp. 467–537). Amsterdam: North-Holland.

Crary, M., Nozick, L. K., & Whitaker, L. R. (2002). Sizing the US destroyer fleet. *European Journal of Operational Research*, *136*(3), 680–695. doi:10.1016/S0377-2217(01)00031-5

Cullinane, K., & Khanna, M. (1999). Economies of scale in large container ships. *Journal of Transport Economics and Policy*, *33*(2), 185–208.

Dantzig, G. B., & Fulkerson, D. R. (1954). Minimizing the number of tankers to meet a fixed schedule. *Naval Research Logistics Quarterly*, *1*(3), 217–222. doi:10.1002/nav.3800010309

David, P., & Stewart, R. (2010). *International Logistics: the Management of International Trade Operations* (3rd ed.). Atomic Dog Publishing.

Desrosiers, J., Dumas, Y., Solomon, M. M., & Soumis, F. (1995). Time constrained routing and scheduling. In Network Routing, Handbooks in Operations Research and Management Science (pp. 35-139). North-Holland.

Deveci, D. A. (2013). Deniz ulaştırması: İşletmeler ve işlevler (Maritime transportation: Businesses and functions). In A. Cerit, A. Deveci, & S. Esmer (Eds.), *Denizcilik İşletmeleri Yönetimi (Maritime Business Management)* (pp. 23–34). İstanbul: Beta Basım A.Ş.

Dohn, A., Rasmussen, M. S., & Larsen, J. (2011). The vehicle routing problem with time windows and temporal dependencies. *Networks*, *58*(4), 273–289. doi:10.1002/net.20472

Endresen, Ø., Sørgård, E., Behrens, H. L., Brett, P. O., & Isaksen, I. S. A. (2007). A historical reconstruction of ships fuel consumption and emissions. *Journal of Geophysical Research*, *112*(D12), D12301. doi:10.1029/2006JD007630

EPA. (2015). *United States Environmental Protection Agency*. Retrieved from http://www.epa.gov/cleandiesel/technologies/retrofits.htm

Esmer, S. (2014). Talep tahminleri ışığında türk limanlarına yönelik kapasite analizi (Capacity analysis for Turkish ports with the light of demand forecasts). In *Proceedings of 8*. Istanbul, Turkey: Kıyı Mühendisliği Sempozyumu.

European Commission. (2013). Commission Staff Working Document, Impact Assessment, 2. Brussels: Author.

Fagerholt, K., & Christiansen, M. (2000). A combined ship scheduling and allocation problem. *The Journal of the Operational Research Society*, *51*(7), 834–842. doi:10.1057/palgrave.jors.2600973

Fagerholt, K., & Ronen, D. (2013). Bulk ship routing and scheduling: Solving practical problems may provide better results. *Maritime Policy & Management*, *40*(1), 48–64. doi:10.1080/03088839.2012.744481

Fisher, M. L., & Rosenwein, M. B. (1989). An interactive optimization system for bulk-cargo ship scheduling. *Naval Research Logistics*, *36*(1), 27–42. doi:10.1002/1520-6750(198902)36:1<27::AID-NAV3220360103>3.0.CO;2-0

Garrod, P., & Miklius, M. (1985). The optimal ship size: A comment. *Journal of Transport Economics and Policy*, *19*(1), 83–91.

Giesen, R., Munoz, J. C., Silva, M., & Leva, M. (2007). *Multi-item inventory routing problem for ship distribution of liquid oil bulk products*. Paper presented at TRISTAN VI, Phuket Island, Thailand. http://transp-or2.epfl.ch/tristan/FullPapers/099Giesen.pdf

Gillman, S. (1999). The size economies and network efficiencies of large containerships. *International Journal of Maritime Economics*, *1*(1), 39–59.

Haiwei, L., Weijian, M., Ning, Z., & Yufei, F. (2015). Modeling and simulating the operation of the harbor portal crane. *Journal of Coastal Research*, *73*(1), 89–94.

Haralambides, E. H., Benacchio, M., & Cariou, P. (2002). Costs, benefits and pricing of dedicated container terminals. *International Journal of Maritime Economics*, *4*(1), 21–34. doi:10.1057/palgrave.ijme.9100031

Heaver, T., Meersman, H., & Van de Voorde, E. (2000). Do mergers and alliances influence European shipping and port competition? *Maritime Policy & Management*, *27*(4), 363–373. doi:10.1080/030888300416559

Hoen, K. M. R., Tan, T., Fransoo, J. C., & Houtum, G. J. (2014). Effect of carbon emission regulations on transport mode selection under stochastic demand. *Flex Serv Manuf J.*, *26*(1-2), 170–195. doi:10.1007/s10696-012-9151-6

Hult, G. T., Ketchen, D. J., & Arrfelt, M. (2007). Strategic supply chain management: Improving performance through a culture of competitiveness and knowledge development. *Strategic Management Journal*, *28*(10), 1035–1052. doi:10.1002/smj.627

Hwang, S. J. (2005). Inventory constrained maritime routing and scheduling for multi-commodity liquid bulk. (Doctoral dissertation). Atlanta, GA: Georgia Institute of Technology. Available at https://smartech.gatech.edu/bitstream/handle/1853/6928/Hwang_Seungjune_200505_phd.pdf

IMO. (2012). *International Maritime Organization Annual Reports*. London: IMO.

Jaikumar, R., & Solomon, M. M. (1987). The tug fleet size problem for barge line operations: A polynomial algorithm. *Transportation Science*, *21*(4), 264–272. doi:10.1287/trsc.21.4.264

Jansson, J. O., & Shneerson, D. (1982). The optimal ship size. *Journal of Transport Economics and Policy*, *16*(3), 217–238.

Jetlund, A. S., & Karimi, I. A. (2004). Improving the logistics of multi-compartment chemical tankers. *Computers & Chemical Engineering*, *28*(8), 1267–1283. doi:10.1016/j.compchemeng.2003.08.009

JICA. (1998). *Container Terminal Planning. Port and Harbour Engineering II*. Japan: Training Course.

Kang, J. G., & Kim, Y. D. (2002). Stowage planning in maritime container transportation. *The Journal of the Operational Research Society*, *53*, 415–426.

Kang, K. W., Zhang, C., Guo, L. Y., & Ma, T. (2012). Research on ship routing and deployment mode for a bulk. In *Annual Conference Proceedings of the International Conference of Management Science Engineer*. doi:10.1109/ICMSE.2012.6414421

Kendall, L. C., & Buckley, J. J. (2001). *The business of shipping* (7th ed.). Cornell Maritime Press.

Kleywegt, A. (2003). *Contract planning models for ocean carriers*. Working paper, Georgia Institute of Technology.

La Londe, B. J., & Cooper, M. C. (1989). *Partnerships in providing customer service: a third party perspective*. Oak Brook, IL: Council of Logistics Management.

Lam, J. S. L., & Yap, W. Y. (2011). Container port competition and complementarity in supply chain systems: Evidence from the Pearl River Delta. *Maritime Economics & Logistics*, *13*(2), 102–120. doi:10.1057/mel.2011.5

Lambert, D. M., Stock, J. R., & Ellram, L. M. (1998). *Fundamentals of Logistics Management*. Singapore: McGraw Hill/Irwin.

Lättilä, L., Henttu, V., & Hilmola, O.-P. (2013). Hinterland operations of sea ports do matter: Dry port usage effects on transportation costs and CO2 emissions. *Transportation Research Part E, Logistics and Transportation Review*, *55*, 23–42. doi:10.1016/j.tre.2013.03.007

Lee, E.-S., Nam, H.-S., & Song, D.-W. (2012). Defining maritime logistics and its value in Maritime Logistics. In D.-W. Song & P. M. Panayides (Eds.), *Maritime logistics: a complete guide to effective shipping and port management*. Kogan Page.

Lindstad, H., Asbjørnslett, B. E., & Pedersen, J. T. (2012). Green maritime logistics and sustainability. *Maritime Logistics: Contemporary Issues*, 227-261.

Liu, H., Mi, W., Zhao, N., & Feng, Y. (2015). Modeling and Simulating the Operation of the Harbor Portal Crane. *Journal of Coastal Research*, *73*, 89–94. doi:10.2112/SI73-016.1

Lo, H. K., McCord, M. R., & Wall, C. K. (1991). Value of ocean current information for strategic routing. *European Journal of Operational Research*, *55*(2), 124–135. doi:10.1016/0377-2217(91)90218-K

Lu, C. S. (2004). An evaluation of logistics services' requirements of international distribution centers in Taiwan. *Transportation Journal*, *43*(4), 53–66.

Martin, G. L. Jr, Randhawa, S. U., & McDowell, E. D. (1988). Computerized container-ship loading: A methodology and evaluation. *Computers & Industrial Engineering*, *14*(4), 429–440. doi:10.1016/0360-8352(88)90045-9

McCord, M. R., Lee, Y.-K., & Lo, H. K. (1999). Ship routing through altimetry-derived ocean currents. *Transportation Science*, *33*(1), 49–67. doi:10.1287/trsc.33.1.49

McLellan, R. G. (1997). Bigger vessels: How big is too big? *Maritime Policy & Management*, *24*(2), 193–211. doi:10.1080/03088839700000070

Metaxas, B. N. (1972). The future of tramp shipping industry. *Journal of Transport Economics and Policy*, *6*(3), 271–280.

Moon, I. K., Qiu, Z. B., & Wang, J. H. (2014). A combined tramp ship routing, fleet deployment, and network design problem. *Maritime Policy & Management*. doi:10.1080/03088839.2013.865847

Musso, E., Ferrari, C., & Benacchio, M. (1999). On the Global Optimum Size of Port Terminals. *International Journal of Transport Economics*, *26*(3), 415–437.

Norstad, I., Fagerholt, K., Hvattum, L. M., Arnulf, H. S., & Bjrkli, A. (2013). *Maritime fleet deployment with voyage separation requirements*. Flexible Services and Manufacturing Journal; doi:10.1007/s10696-013-9174-7

O'Leary-Kelly, S. W., & Flores, B. E. (2002). The integration of manufacturing and marketing/sales decisions: Impact on organizational performance. *Journal of Operations Management*, *20*(3), 221–240. doi:10.1016/S0272-6963(02)00005-0

OECD (2002). *Regularity Issues in International Maritime Transport*. DSTI/DOT (2001)3.

Panayides, P. M. (2006). Maritime logistics and global supply chains: Towards a research agenda. *Maritime Economics and Logistics*, *8*(1), 3–18. doi:10.1057/palgrave.mel.9100147

Panayides, P. M., & Song, D.-W. (2008). Evaluating the integration of seaport container terminals in supply chains. *International Journal of Physical Distribution & Logistics Management*, *38*(7), 562–584. doi:10.1108/09600030810900969

Panayides, P.M. & Song, D.-W. (2009). Port integration in global supply chains: measures and implications for maritime logistics. *International Journal of Logistics Research and Applications: A Leading Journal of Supply Chain Management, 12*(2), 133-145.

Pearson, R. (1980). *Containerline performance and service quality*. Liverpool, UK: Marine Transport Centre, University of Liverpool.

Perakis, A. N., & Bremer, W. M. (1992). An operational tanker scheduling optimization system, background, current practice and model formulation. *Maritime Policy & Management*, *19*(3), 177–187. doi:10.1080/751248659

Persson, J. A., & Göthe-Lundgren, M. (2005). Shipment planning at oil refineries using column generation and valid inequalities. *European Journal of Operational Research*, *163*(3), 631–652. doi:10.1016/j.ejor.2004.02.008

Pope, J. A., & Talley, W. K. (1988). Inventory costs and optimal ship size. *Logistics and Transportation Review*, *24*(2), 107–120.

Psaraftis, H. N., & Kontovas, C. A. (2009). CO2 emissions statistics for the world commercial fleet. *WMU Journal of Maritime Affairs*, *8*(1), 1–25. doi:10.1007/BF03195150

Psaraftis, H. N., & Kontovas, C. A. (2010). Balancing the economic and environmental performance of maritime transportation. *Transportation Research Part D, Transport and Environment*, *15*(8), 458–462. doi:10.1016/j.trd.2010.05.001

Rao, P. G., & Raghavan, K. V. (1996). Hazard and risk potential of chemical handling at ports. *Journal of Loss Prevention in the Process Industries*, *9*(3), 199–204. doi:10.1016/0950-4230(96)00017-4

Rasmussen, M. S., Justesen, T., Dohn, A., & Larsen, J. (2012). The home care crew scheduling problem: Preference-based visit clustering and temporal dependencies. *European Journal of Operational Research*, *219*(3), 598–610. doi:10.1016/j.ejor.2011.10.048

Robinson, R. (2002). Ports as elements in value-driven chain systems: The new paradigm. *Maritime Policy & Management*, *25*(1), 21–40. doi:10.1080/03088839800000043

Rodrigue, J.P. & Notteboom, T (2015). Looking inside the box: evidence from the containerization of commodities and the cold chain. *Maritime Policy & Management: The flagship journal of international shipping and port research, 42*(3), 207-227.

Ronen, D. (1983). Cargo ships routing and scheduling: Survey of models and problems. *European Journal of Operational Research*, *12*(2), 119–126. doi:10.1016/0377-2217(83)90215-1

Ronen, D. (1993). Ship scheduling: The last decade. *European Journal of Operational Research*, *71*(3), 325–333. doi:10.1016/0377-2217(93)90343-L

Ronza, A., Carol, S., Espejo, V., Vílchez, J. A., & Arnaldos, J. (2006). A quantitative risk analysis approach to port hydrocarbon logistics. *Journal of Hazardous Materials*, *128*(1), 10–24. doi:10.1016/j.jhazmat.2005.07.032 PMID:16125840

Sahebi, H., Nickel, S., & Ashayeri, J. (2014). Strategic and tactical mathematical programming models within the crude oil supply chain context—A review. *Computers & Chemical Engineering*, *68*, 56–77. doi:10.1016/j.compchemeng.2014.05.008

Sigurd, M. M., Ulstein, N. L., Nygreen, B., & Ryan, D. M. (2005). Ship scheduling with recurring visits and visit separation requirements, In Column Generation, (pp. 225-245). Springer USA.

Song, D.W. & Lee, E.S. (2012). Competitive networks, knowledge, acquisition and maritime logistics value. *International Journal of logistics Research and Applications: A Leading Journal of Supply Chain Management, 15*(1), 15-35.

Song, D.W. & Lee, P. (2009). Maritime logistics in the global supply chain. *International Journal of Logistics Research and Applications: A Leading Journal of Supply Chain Management, 12*(2), 83-84.

Stalhane, M., Andersson, H., Christiansen, M., & Fagerholt, K. (2014). Vendor managed inventory in tramp shipping. *Omega*, *47*, 60–72. doi:10.1016/j.omega.2014.03.004

Stevenson, W. J. (2011). *Operations Management* (11th ed.). McGraw-Hill/Irwin.

Stopford, M. (2009). *Maritime Economics* (3rd ed.). Routledge. doi:10.4324/9780203891742

Suthiwartnarueput, K. (1988). *The exploration of sea transport efficiency: with a concentration on the case of Thailand*. (Doctoral Dissertation). Department of Maritime Studies and International Transport University of Wales College of Cardiff, UK.

Talley, W. K., Agarwal, V. B., & Breakfield, J. W. (1986). Economics of density of ocean tanker ships. *Journal of Transport Economics and Policy*, *20*(1), 91–99.

Talley, W. K., & Ng, M. (2013). Maritime transport chain choice by carriers, ports and shippers. *International Journal of Production Economics*, *142*(2), 311–316. doi:10.1016/j.ijpe.2012.11.013

Tengku, J. T. M. S. (1995). *Marketing of freight liner shipping services with reference to the far East-Europe trade: a Malaysian perspective*. (Doctoral Dissertation). Department of Maritime Studies and International Transport, University of Wales College of Cardiff, UK.

Tongzon, J., & Heng, W. (2005). Port privatization, efficiency and competitiveness: Some empirical evidence from container ports. *Transportation Research Part A, Policy and Practice*, *39*(5), 405–424. doi:10.1016/j.tra.2005.02.001

Tuna, O., & Arabelen, G. (2013). Deniz Ulaştırma Lojistiği (Maritime Logistics). In A. Cerit, A. Deveci, & S. Esmer (Eds.), *Denizcilik İşletmeleri Yönetimi (Maritime Business Management)* (pp. 533–561). İstanbul: Beta Basım A.Ş.

Tzannatos, E. (2010). Ship emissions and their externalities for the port of Piraeus-Greece. *Atmospheric Environment*, *44*(3), 400–407. doi:10.1016/j.atmosenv.2009.10.024

UNCTAD. (2014). Review of Maritime Transport. *United Nations Conference on Trade and Development*, New York.

Vilhelmsen, C., Lusby, R., & Larsen, J. (2014). Tramp ship routing and scheduling with integrated bunker optimization. *EURO Journal on Transportation and Logistics*, *3*(2), 143–175. doi:10.1007/s13676-013-0039-8

Vis, I. F. A., & De Koster, R. (2003). Transshipment of containers at a container terminal: An overview. *European Journal of Operational Research*, *147*(1), 1–16. doi:10.1016/S0377-2217(02)00293-X

Wilson, I. D., & Roach, P. A. (2000). Container stowage planning: A methodology for generating computerized solutions. *The Journal of the Operational Research Society*, *51*(11), 1248–1255. doi:10.1057/palgrave.jors.2601022

Wood, D. F., Barone, A. P., Murphy, P. R., & Wardlow, D. L. (2002). *International Logistics* (2nd ed.). AMACOM.

Xinlian, X., Tangfei, W., & Daisong, C. (2000). A dynamic model and algorithm for fleet planning. *Maritime Policy & Management*, *27*(1), 53–63. doi:10.1080/030888300286680

Yang, C. C. (2012). Assessing the moderating effect of innovation capability on the relationship between logistics service capability and firm performance for ocean freight forwarders. *International Journal of Logistics: Research and Applications*, *15*(1), 53–69. doi:10.1080/13675567.2012.669469

Zaucha, J. (2014). Sea basin maritime spatial planning: A case study of the Baltic Sea region and Poland. *Marine Policy*, *50*, 34–45. doi:10.1016/j.marpol.2014.05.003

Zavadskas, E. K., Turskis, Z., & Bagocius, V. (2015). Multi-criteria selection of a deep-water port in the Eastern Baltic Sea. *Applied Soft Computing*, *26*, 180–192. doi:10.1016/j.asoc.2014.09.019

Zhen, L. (2014). Container yard template planning under uncertain maritime market. *Transportation Research Part E, Logistics and Transportation Review*, *69*, 199–217. doi:10.1016/j.tre.2014.06.011

Chapter 20
A Mobile Application for Helping Urban Public Transport and Its Logistics

Juan J. G. Mancha
National Mexican Institute of Technology, Mexico

Javier Gonzalez Barbosa
National Mexican Institute of Technology, Mexico

Mayra S. H. Guerrero
National Mexican Institute of Technology, Mexico

Claudia Gómez
National Mexican Institute of Technology, Mexico

Ana Gpe. Velez Chong
National Mexican Institute of Technology, Mexico

Laura Cruz-Reyes
National Mexican Institute of Technology, Mexico

Gilberto Rivera
Autonomous University of Sinaloa, Mexico

ABSTRACT

Urban growth in developed countries has made highly difficult the logistics of public transport in many cities. This issue is a consequence of the increase in the number of public transport routes, which has caused that citizens do not know important information on such routes. Especially, data about stops, terminals, timetables, paths and which are the easily reachable places for each single route. On the other hand, smartphones have become very popular in the last lustra. This kind of device has high-end services such as cameras, high-tech sensors and Global-Positioning-System (GPS) navigation, to mention only a few. Thus, since mobile phones are useful and practically ubiquitous, they should be applied to collective-transport logistics, giving the edge to citizens in an economic manner. Ergo, to solve the aforementioned problem, we propose here an approach based on using technology connected to mobile phones, the GPS and the Internet.

DOI: 10.4018/978-1-4666-9779-9.ch020

INTRODUCTION

Concerns about the environmental impact of urban transport are growing along with population density and urban congestion. Thankfully, awareness of the need for sustainable urban development is also on the rise and the coordination of traffic and logistics is receiving greater attention. Yet the need is urgent for more efficient and effective transport systems that not only address costs but also fully tackle environmental issues such as noise, air pollution, vibration and visual intrusion.

Nowadays, the logistics of the transport services has become a very complicated matter in some cities. This issue is a priority governmental concern for industrialized countries, because

1. It is a relevant source of (air and noise) pollution;
2. It appears to be a decisive criterion for tourists if they consider the return to a specific city (cf. Albalate & Ben, 2010; Khadaroo & Seetenah, 2008); and
3. It affects the daily life of people, and then local governments are constantly argued by dwellers to improve public transport services.

People continue to come together in urban areas for a better quality of life, a trend that will certainly continue. Currently, about half of the world's population lives in urban areas, a figure estimated to increase to over 60% by 2030. Because the importance of this problem is manifested by the large amounts of subsidy that are invested in urban public transport systems (cf. Mackett & Edwards, 1998; Serebrisky et al., 2009; Simpsom, 1994). Besides, the so-called green organisations have an increasing concern in improving the environment in many cities. One alternative is to promote the use of public transport, but people often perceive several disadvantages of it, such as the uncertainty of when the transport will arrive, travelling time and the lack of comfort (cf. Beirao & Sarsfield, 2007).

To mitigate some of those inconveniences, we propose in this work the implementation of a public-transport geolocation system. This system can contribute to successful urban development, which entails several social, environmental, political and economic benefits (cf. Gwilliam, 2003; Alcantara-Vasconcellos, 2014).

In this chapter, we introduce a mobile application that provides helpful information on public transport routes. This app enables citizens to plan better their transportation activities. So, they do not waste too much time waiting for the bus or asking for stopping points. This application:

1. Presents the bus stops in a map with a brief description;
2. Approximates how far away the bus is from the nearest stopping point;
3. Roughly estimates the travelling time; and
4. Shows the nearest places of interest on the map.

This app monitors buses in real time, and this feature helps to identify unexpected delays. In this way, this app may be found useful, not only by citizens, but also by operators and transport managers.

BACKGROUND

Mobile Devices

Currently, mobile devices as notebooks, tablets or smartphones are able to provide wireless internet using Wi-Fi or 3G (or 4G) broadband. This feature allows users to be online and connected to their e-mail and social networks.

There is a wide variety of apps for mobile devices, almost as large as the number of personal interests of each user. For this reason, mobile phones are used as lamps, music players, video games, movie players, cameras, recorders, calculators, alarm clocks, radio, and so on.

Moreover, when these devices use the internet, the number of possible activities increases even further. Here we can mention the following: reservations (hotels, concerts, transportation), shopping, information retrieval, payments and alerts (bank, home automation, climatological).

In addition, when social networks are involved, large communities are made, and these communities help in the spreading of information, causing some social and politic changes, in several scenarios such as commerce, recommendation (for example articles, business and food), and politics (citizens' criticism against their local government).

Finally, it is important to highlight that built-in sensors into mobile devices (as cameras, microphones, gyroscopes, GPSs and accelerometers) have opened up the possibility to develop a large range of applications in the smartphones on different topics, as health care (Consolvo et al., 2008), social networks (Miluzzo et al., 2008), security (Ni et al., 2009), environment (Honicky et al., 2008) and transportation (Thiagarajan et al., 2009).

Each smartphone-software seller offers an application store in which their users can download or buy applications to increase the usefulness of their mobile phones. In other words, they use and mix the smartphone sensors' functionalities in such a way that users can have a better experience.

As a consequence, smartphones are very addictive, especially for young people. Pearson & Hussain (2015) found that 13% of participants in their study were actually addicted to their phones, with the average user spending 3.6 hours a day on their device. Most participants said their smartphone use caused distraction from many aspects of their lives, including employment, hobbies and studies. Now, people are so familiar with mobile devices that the idea of using them to get information about collective transport is completely acceptable and, even, necessary.

As matter of fact, mobile phones can obtain not only geographical coordinates, but also the altitude in which the device is. This technology has led an increase in the number of geolocation apps (e.g. Balakrishnan et al., 2009; Husted & Myers, 2010).

Social Networks

During the last two decades, we have witnessed a huge increase in the usage of social networks and the sharing of different kind of files (e.g. YouTube). Many communities have been created (e.g. Facebook, Google+ and Whatsapp) with this objective in mind, and the people have joined by means of such platforms to help themselves in a social, cultural and personal manner.

The previous ones are the most internationally popular; however, there are others applications that are focused in geolocation, which are used in big cities around the world. So, in the last years, some useful tools have been developed, including the following:

1. **Waze (2015):** It offers real-time information about traffic, but just in some places. It works better if the interaction between users and the application is while they are driving, but this trait could be very dangerous.
2. **Google Maps:** In some countries, it gives the transport timetabling. However, it does not send notifications to the user and only works for the most important cities.
3. **FourSquare (2015):** It is a geolocation network application, in which places of interest are located depending on user's position.
4. **Google Latitude (2015):** It is another similar application, which allows you to see the geographical position of your friends in Google Maps.

There have been developed other social networks that help people (or companies) in the urban transport logistics. Here we can mention Google Now (2015), which gives you a possible route to avoid traffic jams, and it also has the capacity of giving you the schedules of public transport. Citymapper (2015) generates walking routes, and gives you public transportation schedules or even the approximate taxi fare.

Other interesting projects are:

1. The MIT VTrack project (Thiagarajan et al., 2009), which is a system for travel estimation using WiFi sensors to estimate a user's trajectory and travel time along the route. Vtrack uses a hidden Markov model-based map and an interpolation method for estimating the travel time.
2. Mobile Millennium project (Hunter et al., 2011), which uses machine learning to infer traffic conditions for large metropolitan areas from crowdsourced data, and so it processes traffic information obtained from the GPS in mobile phones.

In the Table 1, some applications are described. These software tools are provided by governments or transportation companies to their users in order to plan their trips.

Table 1. Applications to plan trips

App's Name	Characteristics	Links
PTV mobile apps	• Real-time tram information • Journey planner • Interactive map • Fully online	http://ptv.vic.gov.au/getting-around/mobile-apps/
TriMet App Center	• Real-time information • Provides stops • Provide stations	http://trimet.org/apps/
Go metro Los Angeles	• Live map • Vehicle location • Realtime arrival • Trip planner	http://www.metro.net/mobile/metro-mobile-app/app-info/
Transit App	• Real-time predictions • Plan trips • Real-time position • View schedules	http://transitapp.com/
Moovit	• Shows nearby stops • Check arrival times • Plan your trips • Live reports • Over 600 cities	http://moovitapp.com/

continued on following page

Table 1. Continued

App's Name	Characteristics	Links
AT Public Transport app	• Journey planner • Real time board • Bus stop locator • Map	https://at.govt.nz/about-us/
Arrive Sydney	• Real-time location • Real-time departures times • Can announce next stop	http://www.transportnsw.info/en/travelling-with-us/keep-updated/apps.page?

PROJECT DESCRIPTION

This is geolocalization system, developed for devices with android operating system, formed by a pair of mobile applications and a web server, it intends to locate the urban public transport in the metropolitan area of southern Tamaulipas, Mexico. The bus mobile application sends its location by GPS to a web server, the bus location data is shown in the user mobile application, showing all the active buses in a map, and also the bus stops that are near the user. The user can ask for notifications to the server, so it alerts him when the public transport is near an asked stop.

MAIN FOCUS OF THE CHAPTER

The implementation of an urban transport location system helps its users to manage their time because it would not be necessary to wait for a long time at the bus stops; this is because the application alerts the user about the bus coming near. In the same way, the application will let know the urban transport routes, also the bus stops are going to be pointed in a map with a brief description that will help the user. This is very useful for tourists visiting the zone.

In addition, the users will be able to see all the buses running through the route in real time, this would help the user to estimate the time of the travel. In the same way, it would help tourists that are visiting the place and need to take the bus. The system may help to coordinate the urban transport's logistic, giving it the buses' location and helping it to offer a better quality service.

The social benefit of the application is huge, because the users can identify where the routes and bus stops are, also estimate the bus distance, manage better their time and decrease the waiting time. In addition, the owner company can keep track of where their buses are. At last, the system is developed using free software, this will give easiness to keep developing it, and also it will decrease its costs because the tools used are free.

APPLICATION DEVELOPMENT

This system is a geolocation project, in other words, about locating geographically an object, in the scenario a mobile device (smartphone) and it is divided in two modules, two mobile applications and a web server. First of all, it is needed to install the application in the device, after that, it is placed in a bus and started the application, which is going to send its location using the GPS every 5 seconds, for this it is necessary to have turned on the GPS, otherwise the application will ask to turn it on. These data is sent to the server, which stores and updates the data as the latitude and longitude of the device.

Another mobile application shows the map in which are shown some of the bus stops that the routes have, this can be given through the programming code or by a database filled by an administrator. With the location sent by the bus it also shows the buses running through the route in a Google Map and its movement.

Notifications can also be received if the user fills a brief formulary in the application, in which you can ask for a notification for the next 3 buses that get near a requested bus stop of a specified route.

TRANSITION DIAGRAM

A state transition diagram represents the life cycle of a class' instances. Describe the states, transitions that link it and the events that provoke the transitions. (Debrauwer, 2005). Transition diagram emphasize the dependent behavior of the system's time and it is very useful in real time systems.

The transition diagram for the passenger's mobile application can be seen in Figure 1.

Figure 1. Transition diagram of the passenger's mobile application

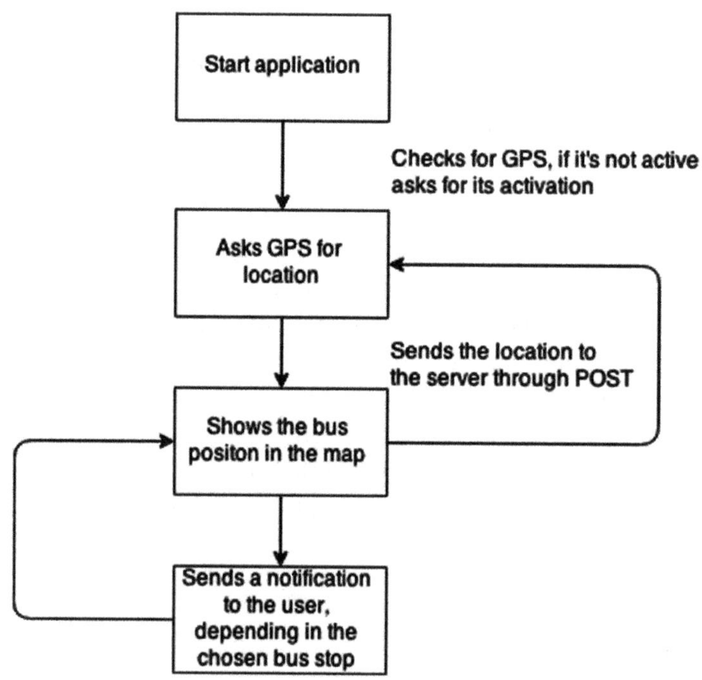

USE CASE DIAGRAM

Use cases are a technique to specify system behavior. They are a sequence of interactions between the system and someone that use one of its services. Each use case gives one or more scenarios where it indicates how the system should behave with the user or with other system to do a specific objective.

Use case diagrams allow us to see the different actors that interact with our application, the relations between them and the actions that can do each one. These kinds of diagrams are easier to understand. They represent systems' functional requisites and are used for a iterative and incremental development. Use case diagrams have three elements:

- **Actors:** They are system users. An actor can be a person, a set of persons, a hardware system or a software system. The actors represent a role, which can be performed by someone who needs to interchange information with the system.
- **Use Case:** A use case describes a specific way of using part of a system's functionality. The use cases set describe the whole systems' functionality.
- **Actors and Use Cases Communication:** Each actor does a specific number of use cases. That's why we say it exists a communication between them. Figure 2 shows how the mobile application's actors interact with the system.

In the next Tables 2, 3, 4, 5, 6 and 7, parameters related diagrams use cases are described for the mobile application actor.

Figure 2. Use case diagram of the mobile application

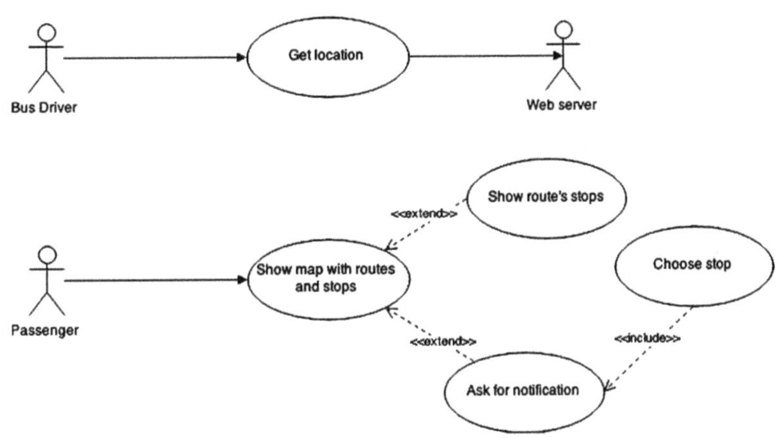

Table 2. Mobile application actors

Actor	Description
Bus Driver	He is the one who drives the bus, and uses the app that determines its location and sends it to a server.
Passenger	He is the application final user; he sees the map and asks for notifications.
Data server	This one receives bus location and passenger's notification requests, also it processes the locations so it can send the notifications.

Table 3. Use case 01_Getting location

ID:	UC – 01
Name:	Getting location (Sending data).
Description: Bus driver's application will get its location and send it to the data server so it can be show in passenger's application.	
Actors: 1. Bus driver.	
Preconditions: 1. User must have application installed on its android mobile device. 2. Mobile application must be open.	
Normal Stream: 1. User will have to click start button on application.	

Table 4. Use case 02_Show map with routes and stops

ID:	UC – 02
Name:	Show map with routes and stops.
Description: Allows the passenger to see bus position in a map, also to see the different routes and bus stops.	
Actors: 1. Passenger.	
Preconditions: 1. User must have application installed on its android mobile device. 2. Application must be open.	
Normal Stream: 1. Passenger will look at the map showing the bus, routes and stops.	

Table 5. Use case 03_ Show bus stops

ID:	UC – 03
Name:	Show bus stops
Description: Allows the actor to see the map with the bus stops that the route has.	
Actors: 1. Passenger	
Preconditions: 1. User must be using the application.	
Normal Stream: 1. User will have to start application. 2. User will choose the route. 3. User will see the bus stops.	

Table 6. Use case 04_ Get notification

ID:	UC – 04
Name:	Get notification
Description: Allows asking for a notification that will be sent when a bus is coming near the bus stop chosen.	
Actors: 1. Passenger	
Preconditions: 1. User must be using the application.	
Normal Stream: 1. Passenger will ask for the notification. 2. Passenger will answer the form with useful information to register its notification correctly. 3. Application will send form data to the server.	
Alternative Stream: 1. Without internet connection, try later.	

Table 7. Use case 05_ Choose bus stop

ID:	UC – 05
Name:	Choose bus stop
Description: Passenger can choose a bus stop to get a notification where the bus comes near it.	
Actors: 1. Passenger	
Preconditions: User must be using the application.	
Normal Stream: 1 Passenger will choose the stop from a list or the map.	

Next are described use cases that were developed in the application and were illustrated at Figure 3.

CLASS DIAGRAM

A class diagram describes the type of objects that are in a system and the different kind of static relations that exist between them.

To indicate that an association exists it will be used a solid line connecting to classes or a single line to itself. To assign a name to a relationship it is important to identify the relations between the classes and the name should be placed somewhere near the middle of the association line.

In the class diagram of Figure 3 and 4 are shown aggregation and generalization associations. It must be taken into consideration that between two classes there may be more than one relation. It is shown the bus driver application's class diagram on Figure 3 and passenger application's class diagram on Figure 4.

Figure 3. Class diagram of the bus driver's mobile application

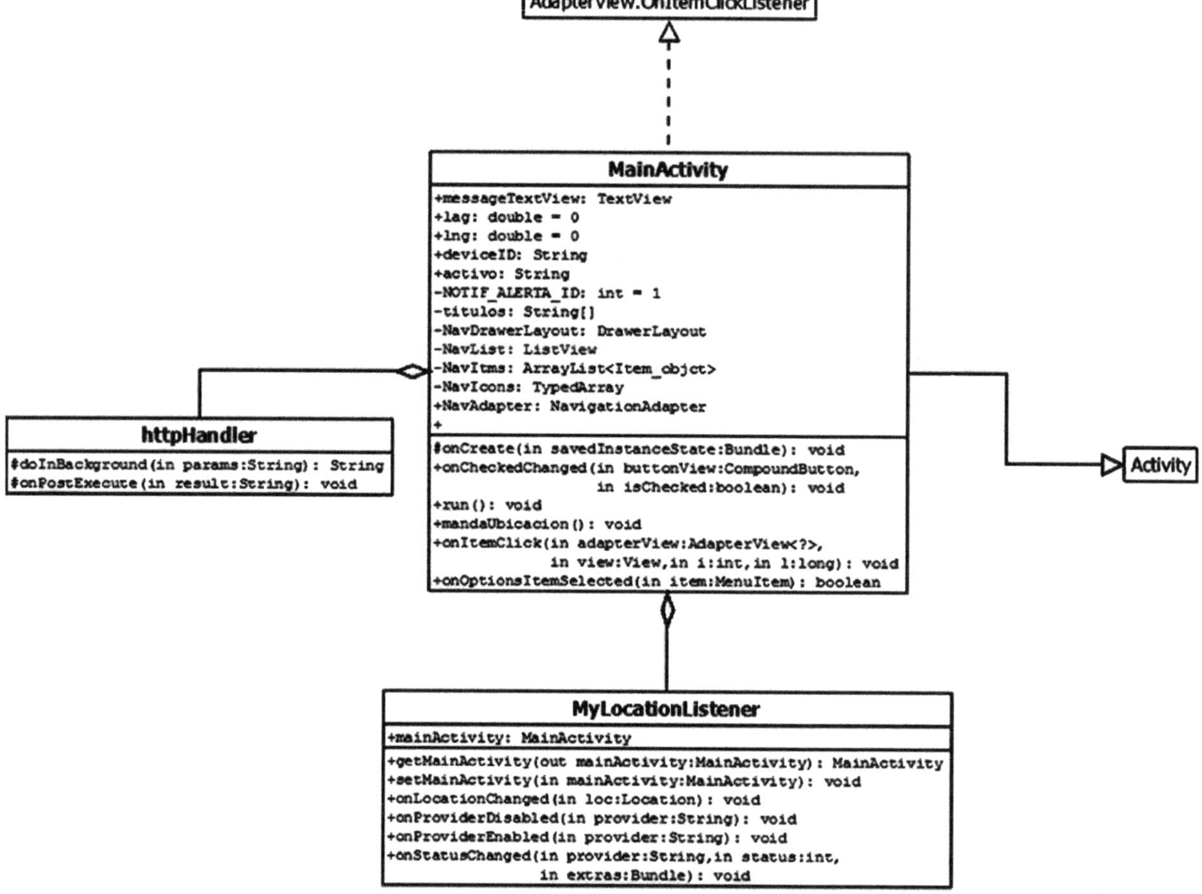

DATABASE

MySQL is used as the Database Management System (DBMS). The server has a total of six tables that represents the buses, stops, notifications, and routes as shown in Figure 5.

Buses table is named Autobus, and it stores an id, the latitude, longitude, current stop, last stop, route, an active flag, the distance to its current stop and its status (before, after or on) respectively to current stop.

Ruts is the name of the routes table, it just has a numeric id, a name for the route and a string to represent the city where it belongs to. Bus stops table has four different attributes, a numeric id, the route's name, the latitude and longitude of the route. The name of this table is Parades.

It also exists a table to model the existing relation between stops an routes: a route has many stops and a stop can be in many routes. The table's name is ruta_parada. The notification table (named notification), is the one used when a notification is requested. It stores the requester's GCM id, the route, bus stop, where it is going (north, south, east, west) and how many notifications have been sent.

There also exists a table to store to which users a notification have been sent with a specified bus, so no two or more notifications will be send to the same user and same bus.

Figure 4. Class diagram of the passenger's mobile application

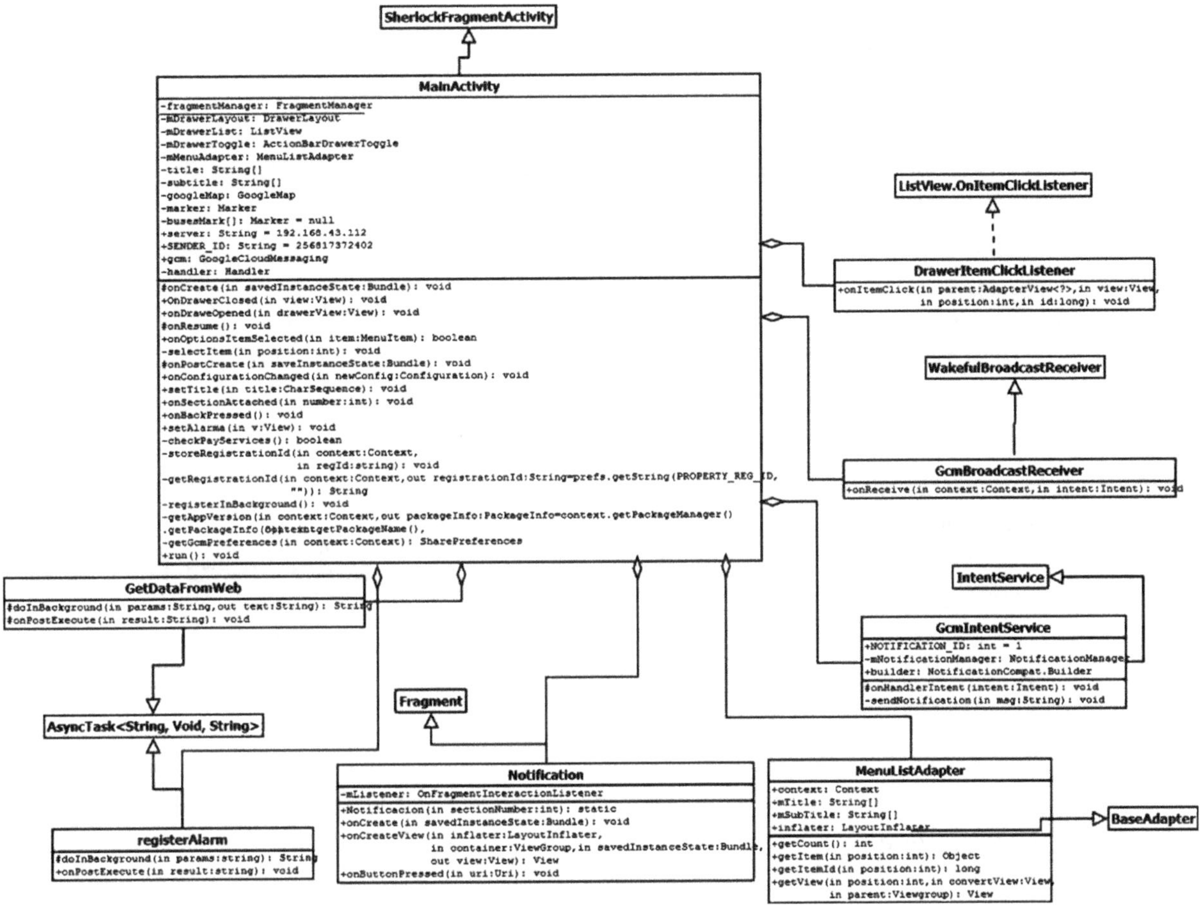

TECHNOLOGIES

Android Studio (2015) was the IDE used to develop the mobile application, it is the official Android IDE. You can write and compile your code from here, it provides some code templates to help build some common app features, also it has a layout editor where you can drag and drop different components to modify the layout.

Eclipse is another IDE to develop android applications, but it doesn't have the same compatibility with latest android versions. A web server is a program that is always waiting for connection request using HTTP protocol (Mora, 2001), those requests are made mostly by browsers or mobile devices. In this work an Apache Web Server is being used.

Apache was used to run a web server to serve different PHP scripts that connected to the database, modifying or returning data to the mobile devices. Google Cloud Messaging (GCM), it is a service that allows you to send data from your server to the Android devices where you have installed the application. As seen in Figure 6 a third-party app server (our web server) connects to GCM servers, and GCM servers connect to client app.

Figure 5. Database schema

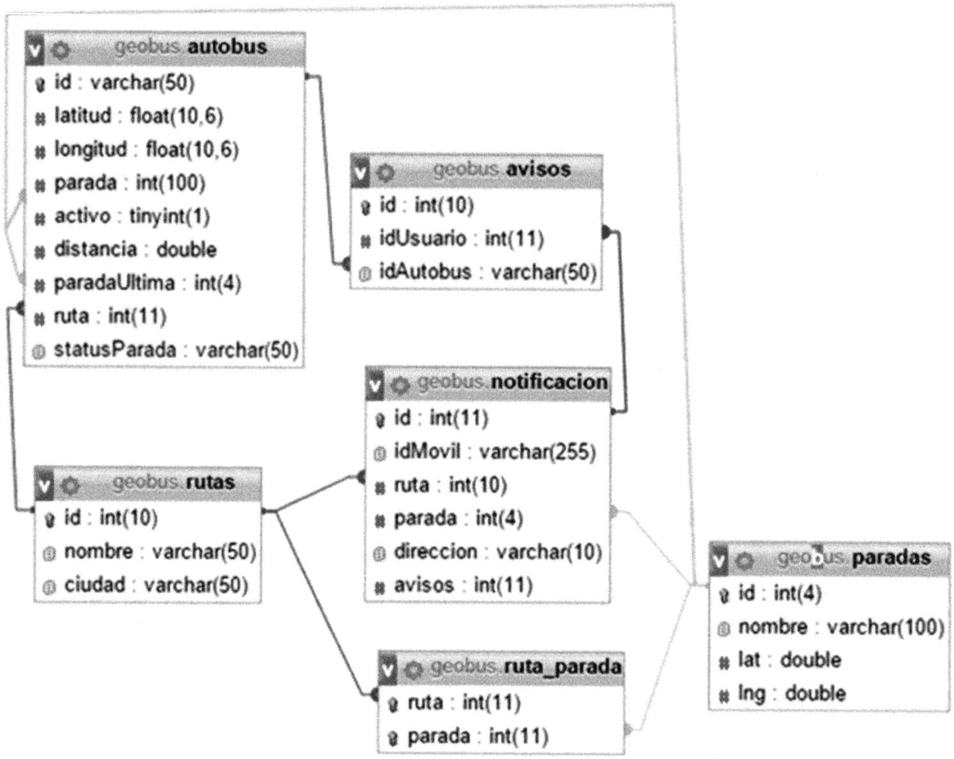

Figure 6. GCM Architecture

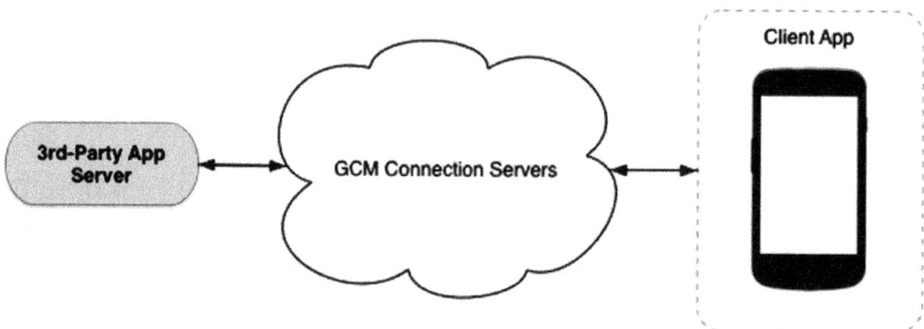

PHP Mail, this is a php function that sends a mail to a specified email address. Instead of this we used GCM that provided a fastest way to send notifications. Google Maps, it is a service that shows the Earth's map. It allows you to embed it into a mobile application and shows directions and markers.

OpenStreetMap, it is an alternative to show maps. It is fed by the community, it uses open data and it can be used freely as long as you give credit. Cron is a time-based job scheduler, it was used to schedule bash scripts that ran PHP scripts to update bus related data, also to check if sending a GCM notification was needed.

ALGORITHMS

The mobile applications and web server have different algorithms; the most important ones are those which establish the communication between the client applications and the server, the one that sends the GCM messages and the one that calculates where is the bus located.

Bus driver's application main task is to send its location to the web server which stores the location in the database, Table 8 shows the algorithm that is used to send the latitude and longitude.

Algorithm for the passenger's app is shown in Table 9, this application shows the buses on a map and let the user sign up for a notification

Webserver stores different PHP files, two of them are running every five seconds so required data gets updated. The buses location is updated with algorithm shown in Table 10, as for the notifications sent to the users algorithm can be seen in Table 11.

IMPLEMENTATION

The application has a web data server and two mobile applications. As shown in Figure 7 the web server communicates with the two android applications through the internet.

Table 8. Algorithm for sending location in bus driver's application

```
Activity implements Location Listener.
Every time location is changed
  Store latitude and longitude
  Set url where the request is going to be sent
  Store in an array data which is going to be sent
  Send a POST request to URL
```

Table 9. Main algorithm for passenger's app

```
Set a google map for the layout
Get bus stops for default or selected route
Show bus stops on map
Every 5 seconds
Retrieve buses from selected route
Display buses on map
If a notification request form button is clicked and form is filled
Send notification request to web server
```

Table 10. Algorithm for busstop.php

```
Establish database connection
Get all active buses from database
If last stop is not known
  Look for nearest stop from all stops in the route.
Else
  Look for nearest stop from just stops nearer last one
If current stop is different from last one
  Set stop status to "Before stop"
Else
  If distance to current stop is greater to last known distance
    Set stop status to "After stop"
If distance is less than 15 meters
  Set stop status to "On stop"
```

Table 11. Algorithm for avisos.php

```
Establish database connection
Get all notification requests
For each notification request
  Get all buses that are four stops away at most from requested stop
  If a notification has not been sent for that bus
    Send notification through GCM
    Increment number of notifications sent for this request.
    If three notifications have been already sent
      Delete notification request from database
```

Figure 7. System's communication description

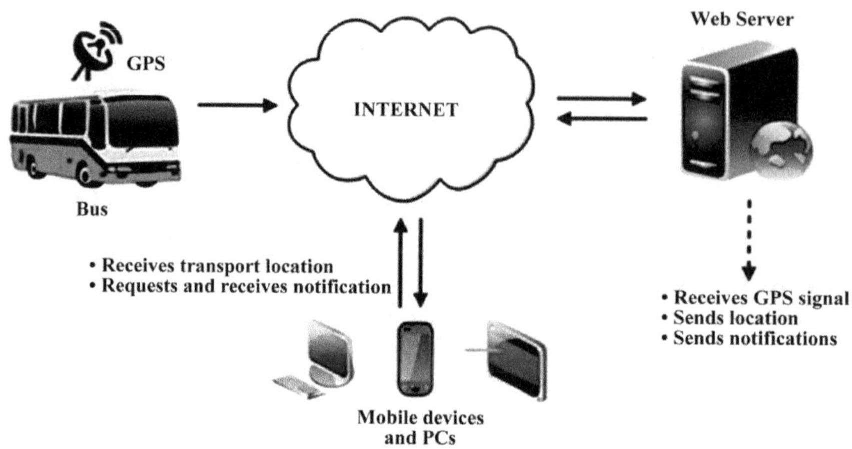

Web Server

The data web server is running an Apache HTTP Server with a GNU/Linux operating system (Ubuntu 14.04). It stores different PHP files that establish the database communication, and the ones that wait for android applications to communicate.

When the apache web server start running it start listening for the bus driver's application for the coordinates that are sent through JSON, this location is received with the PHP file written for that and stored in the database.

It also serves the buses' location to passenger's android application, so it can show them in a map. They are sent using JSON format. There is also a cron job that executes every 5 seconds a PHP script that checks if a user need to be notified about a bus coming near the requested stop.

Bus Driver's Application

The bus driver's application is written in java for android, and it is targeted to run on Android Kit Kat (version 4.4). This application gets the GPS location using Location APIs from Google Play Services, and sends it asynchronously using android's AsyncTask class to the http web server through a POST request. The location is sent every time it changes using onLocationChanged method (implemented on the Location API's interface LocationListener).

Passenger's Application

Passenger's applications is also written in java for android and targeted to run on Android Kit Kat (version 4.4). It has two different screens. The app's main screen is a google map fragment showing bus stops and active buses running in a specified route. Both data are gathered through JSON format after making a POST request to the server and displayed on the map, active buses are reloaded asynchronously every 5 seconds so the user can keep track of them in real time.

The second screen is a fragment where the user can request an alert for a specified bus stop. The user needs to specify the route, bus stop, and direction the bus is running, and he will be notified with an android push notification before the bus gets to the requested stop.

TREE DIRECTORY

Passenger's APP Tree Directory

The passenger's app shows a map with real time information of the bus location, and it allows asking for a notification when the bus is coming near a bus stop. The file tree directory is shown in Figure 8 and the files are described below.

- **GcmBroadcastReceiver:** It creates and manages a wakelock. It passes all the work to an intent.
- **GcmIntentService:** It manages the GCM message and frees the wakelock.
- **MainActivity:** Application main class, it shows the map, establishes communication with the server to obtain bus coordinates, and it register new notifications.
- **MenuListAdapter:** This class is used to show correctly sidebar in the layout.
- **Notification:** Fragment used so the user can register a new notification petition.

Bus Driver's APP Tree Directory

The bus driver's app sends the device's location to the web server so it can store it in a database. The app only has two states: on and off. Its tree directory is shown in Figure 9, and its most important files are described below it.

- **LocationServiceErrorMessages:** This class returns an error message depending in an error code.
- **LocationUtils:** It returns the latitude and longitude in a string given a Location object.
- **MainActivity:** It sends asynchronously the latitude and longitude to a server each time they are updated.

Web Server's Tree Directory

The Figure 10 shows the files stored and served by the apache with php web server, and the descriptions below the figure.

- **avisos.php:** File that runs every 5 seconds to send the notifications.
- **Busstop.php:** File that runs to determine which is the nearest bus stop to each bus.
- **Dbconfig.php:** File that contains database configuration as username, password and database name, so a connection can be established.
- **Markersjson.php:** File that generates and sends a json string that contains all the active buses to the mobile devices that request it.

Figure 8. Passenger's app tree directory

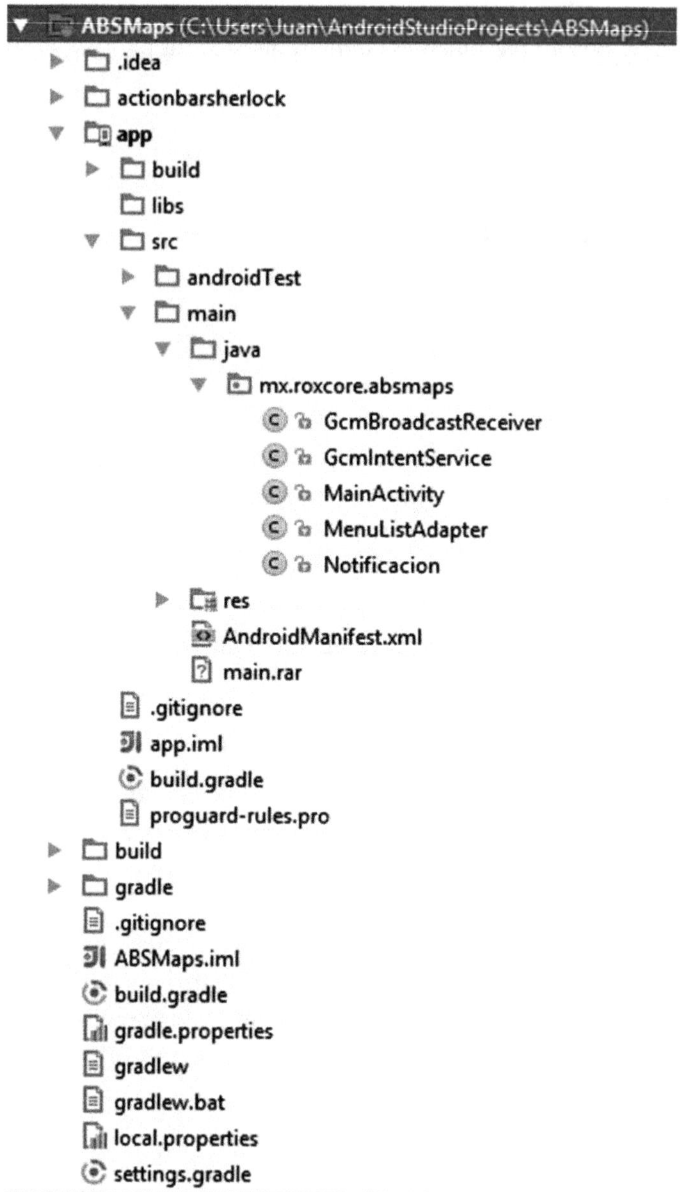

- **Notificación.php:** File that gets the data sent by the mobile devices throught a POST request to register a user asking for a notification.
- **Paradas.php:** This file returns all the bus stops in json format.

Figure 9. Bus driver's app tree directory

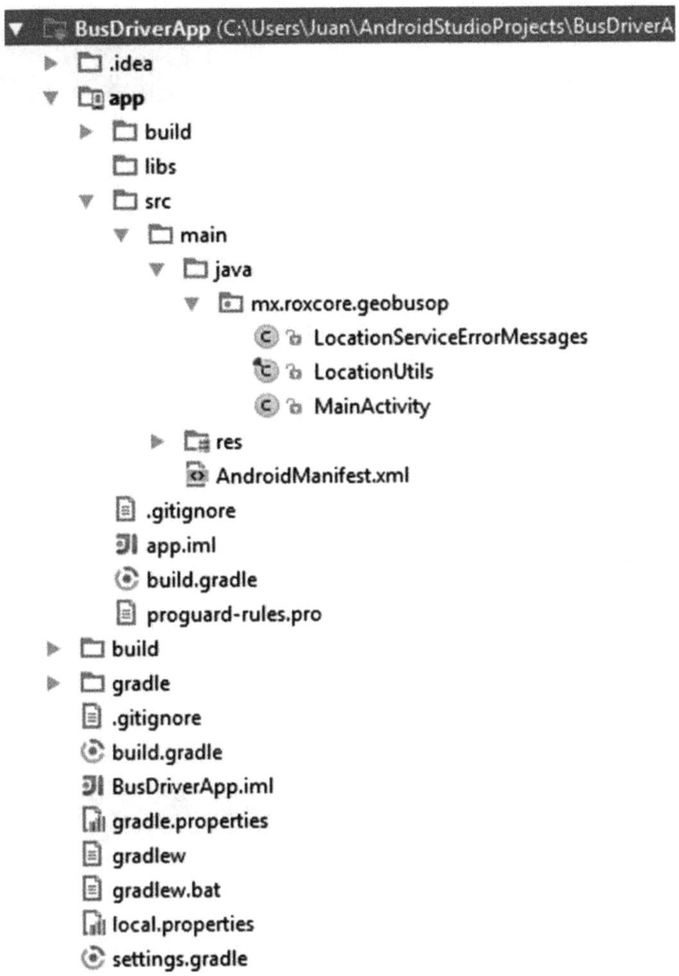

APPLICATION

The passenger's application shows a Google Map with markers as different stops or public known places, and also the buses that are running through the route at that moment, as shown in Figure 11.

To start using the passenger's application, you must install it first. When the user opens it for the first time it automatically gets an GCM ID from Google Servers that will be used if a notification is requested, then a map is shown with all the bus stops and buses currently active.

In addition, it allows asking for a notification, it is just needed to choose a route, the direction and a stop. The server will send a GCM notification (push notification) to the mobile device once the bus is near the asked stop.

While being on main screen, the user can select the option "notifications" from the slide bar so a new fragment will be open. On this screen the user must choose a route from the available ones in the select field, a direction (north or south) in the radio button fields, and a bus stop from the select field, when

Figure 10. Web server's app tree directory

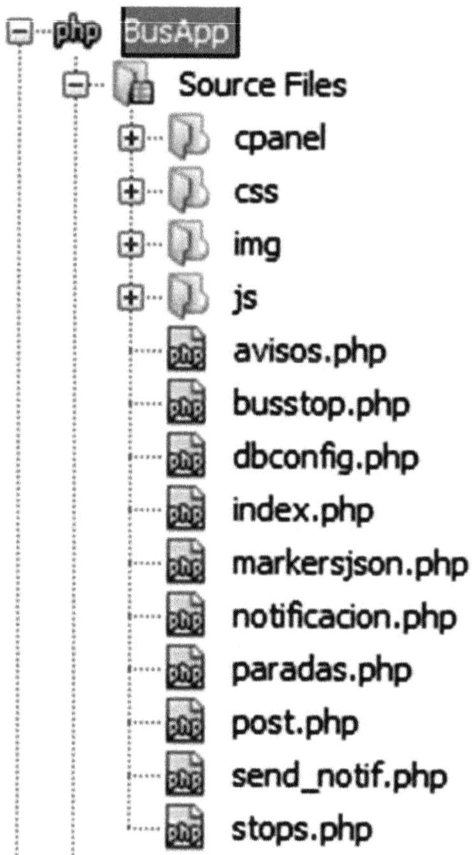

Figure 11. App's main screen

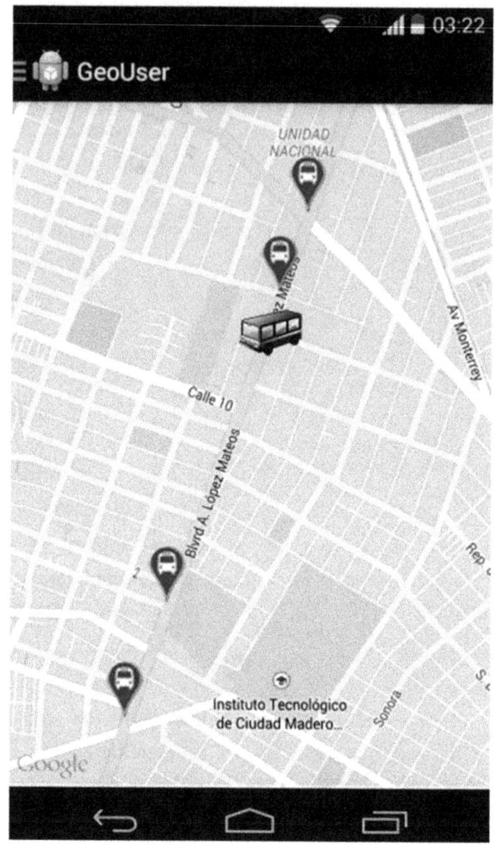

all the fields are filled the user can request the notification pushing the Request notification button. This can be seen on Figure 12 and Figure 13.

The bus needs to run its own application, once it starts, it sends the location to the server so it can be shown in the passenger's application, bus driver's application can be seen in Figure 14.

SOLUTIONS AND RECOMMENDATIONS

The goals for urban transport, in this respect, are to promote the use of cleaner cars and cleaner fuels. The need is also to reduce the number of fatalities and incidents. Yet another challenge is that the amount of traffic in urban areas has been increasing inexorably during last decades. The task of people involved in urban traffic management is to best allocate the scarce resources of road and kerbside space to potentially competing transport modes, within a network that has finite capacity. A more accessible public transport system has to be prioritized in traffic management.

Functional tests made to the system showed that it worked correctly, showing the map, routes, bus stops, and alerting when notifications where requested. Connectivity was tested running the bus driver's application in two different mobile devices and it updated its data correctly in the server's database, also

Figure 12. Notification request screen

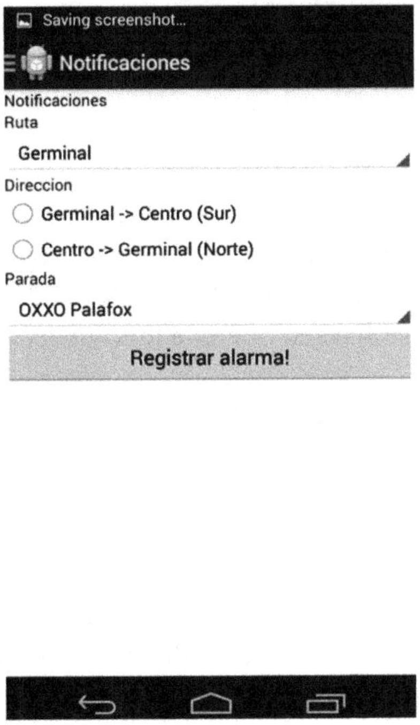

Figure 13. Choosing a bus stop

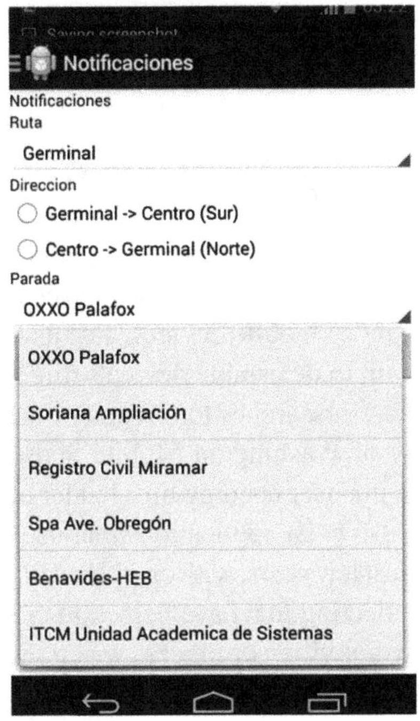

Figure 14. Bus driver's application

it showed the two respectively bus icons on the passenger's application. Notifications on the passenger's application were tested requesting multiple notifications for one user on different bus stops. The notifications arrived when expected and on time.

FUTURE RESEARCH DIRECTIONS

Nowadays many people, including children and adults, have a smartphone with them and they use it in many different activities. While the sensors keep getting smaller and cheaper, they will be embedded in more and more mobile devices. Just as the new smartphones that have accelerometers and gyroscopes and can help to determine device's direction and orientation.

There are researches to integrate other sensor on smartphones, as barometers, thermometers, Intel/University of Washington Mobile Sensing Platform (MSP), and humidity sensors, that could help to determine if a user is going up a ladder or hiking, also to know the room's temperature where the device is. There also exist applications that use earphones to get users' blood pressure.

In the last few years, with or without knowledge millions of users have collaborated in audio and pictures sample collecting that have helped to create corpus that later are used applying data mining techniques to identify activities or places. When the applications get smarter and get to predict the activities that the users are doing, it will be possible to establish a user's behavior pattern and advices could be given.

In our future work is planned to show routes schedules, help bus companies alerting them which buses are running over the speed limit, doing stops where is not allowed, doing detours, etc. It is also intended to work in the algorithm to predict when the buses are approaching to each bus stop, so the notifications can be handled in a more efficient way.

CONCLUSION

In this chapter, we introduce a mobile application that provides helpful information on public transport routes. This app monitors buses in real time, and this feature helps to identify unexpected delays. In this way, this app may be found useful, not only by citizens, but also by operators and transport managers.

We work in the mobile applications and web server have different algorithms; the most important ones are those which establish the communication between the client applications and the server, the one that sends the GCM messages and the one that calculates where is the bus located. Bus driver's application main task is to send its location to the web server which stores the location in the database.

The functional tests made to the system showed that it worked correctly, showing the map, routes, bus stops, and alerting when notifications where requested. Connectivity was tested running the bus driver's application in two different mobile devices and it updated its data correctly in the server's database, also it showed the two respectively bus icons on the passenger's application. Notifications on the passenger's application were tested requesting multiple notifications for one user on different bus stops. The notifications arrived when expected and on time.

ACKNOWLEDGMENT

This work has been partial supported by CONACYT Project 254498 and Prodep.

REFERENCES

Albalate, D., & Bel, G. (2010). Tourism and urban public transport: Holding demand pressure under supply constraints. *International Journal of Tourism Management*, *31*(3), 425–433. doi:10.1016/j.tourman.2009.04.011

Alcantara-Vasconcellos, E. (2014). *Urban Transport Environment and Equity: The Case for Developing Countries, Brussels BE-1080 Belgium*. Routledge.

Android Developers. (2015). Retrieved from http://developer.android.com/sdk/index.html

Balakrishnan, M., Mohomed, I., & Ramasubramanian, V. (2009). Where's that phone?: geolocating IP addresses on 3G networks. In *Proceedings of the 9th ACM SIGCOMM conference on Internet measurement conference* (pp. 294–300). doi:10.1145/1644893.1644928

Beirao, G., & Sarsfield, J. A. (2007). Understanding attitudes towards public transport and private car: A qualitative study. *International Journal of Transport Policy*, *24*(6), 478–489. doi:10.1016/j.tranpol.2007.04.009

Citymapper. (2015). *Citymapper, making cities usable*. Retrieved from https://citymapper.com

Consolvo, S., McDonald, D. W., Toscos, T., Chen, M. Y., Froehlich, J., Harrison, B., & Landay, J. A. et al. (2008). *Activity sensing in the wild: a field trial of ubifit garden*. In *Proceedings of the SIGCHI Conference on Human Factors in Computing Systems* (pp. 1797–1806). doi:10.1145/1357054.1357335

Debrauwer, L., & Heyde, F. v. (2005). *UML 2: iniciación, ejemplos y ejercicios corregidos*. Cornellà de Llobregat, Barcelona: Ed.ENI.

FourSquare. (2015). *Introducing the all-new Foursquare, which learns what you like and leads you to places you'll love*. Retrieved from https://foursquare.com/

Google Latitude. (2015). *Ayuda de Google Maps para móviles*. Retrieved from https://www.google.com/latitude?hl=es

Google Now. (2015). *Google Now. Información precisa en el momento adecuado*. Retrieved from https://www.google.com/landing/now/

Gwillian, K (2003). Urban transport in developing countries. *Transport Reviews: A Transnational Transdisciplinary Journal, 23*(2), 197–216.

Honicky, R., Brewer, E. A., Paulos, E., & White, R. (2008). N-smarts: networked suite of mobile atmospheric real-time sensors. In *Proceedings of the second ACM SIGCOMM workshop on Networked systems for developing regions* (pp. 25–30). doi:10.1145/1397705.1397713

Hunter, T., Moldovan, T., Zaharia, M., Merzgui, S., Ma, J., Franklin, M. J., & Bayen, A. M. et al. (2011). Scaling the Mobile Millennium system in the cloud. In *Proceedings of the 2nd ACM Symposium on Cloud Computing* (pp. 28). doi:10.1145/2038916.2038944

Husted, N., & Myers, S. (2010). Mobile location tracking in metro areas: Manets and others. In *Proceedings of the 17th ACM conference on Computer and communications security* (pp. 85–96).

Khandaroo, J., & Seetena, B. (2008). The role of transport infrastructure in international tourism development: A gravity model approach. *International Journal Tourism Management*, *29*(5), 831–840. doi:10.1016/j.tourman.2007.09.005

Macket, R. L., & Edwards, M. (1998). The impact of new urban public transport systems: Will the expectations be met? *International Journal Transportation Research Part A: Policy and Practice*, *32*(4), 231–245.

Miluzzo, E., Lane, N. D., Fodor, K., Peterson, R., Lu, H., Musolesi, M., & Campbell, A. T. et al. (2008). Sensing meets mobile social networks: the design, implementation and evaluation of the CenceMe application. In *Proceedings of the 6th ACM conference on Embedded network sensor systems*. doi:10.1145/1460412.1460445

Mora, S. (2001). Qué es una aplicación web. In *Programación en Internet: Clientes web* (p. 9). Alicante: Club Universitario.

Ni, X., Yang, Z., Bai, X., Champion, A. C., & Xuan, D. (2009). DiffUser: Differentiated user access control on smartphones. In *IEEE 6th International Conference on Mobile Adhoc and Sensor Systems*. IEEE Press.

Pearson, C., & Hussain, Z. (2015). Smartphone Use, Addiction, Narcissism, and Personality: A Mixed Methods Investigation. *International Journal of Cyber Behavior, Psychology and Learning*, *5*(1), 17–32. doi:10.4018/ijcbpl.2015010102

Serebrisky, T., Gómez-Lobo, A., Estupiñán, N., & Nuñoz-Raskin, R., (2009). Affordability and Subsidies in Public Urban Transport: What Do We Mean, What Can Be Done?, *Transport Reviews: A Transnational Transdisciplinary Journal, 29*(6), 715–739.

Thiagarajan, A., Ravindranath, L., LaCurts, K., Madden, S., Balakrishnan, H., Toledo, S., & Eriksson, J. (2009). VTrack: accurate, energy-aware road traffic delay estimation using mobile phones. In *Proceedings of the 7th ACM Conference on Embedded Networked Sensor Systems* (pp. 85–98). doi:10.1145/1644038.1644048

Waze. (2015). *Get the best route, every day, with real–time help from other drivers*. Retrieved from https://www.waze.com/

Section 4

Optimization in Logistics

Chapter 21
A Review of the Main Options of Tools for Optimizing Operations (in Companies, Manufacturing, and Supply Chains)

Arturo Cordova Rangel
Polytechnic University of Aguascalientes, Mexico

Irma Jimenez Saucedo
Polytechnic University of Aguascalientes, Mexico

ABSTRACT

Investigated methods for optimizing manufacturing operations are diverse as the theories and hypotheses that have developed over time. There have been different types of studies, using different methods, theories and hypotheses. All aimed at determining what the best methods would be more or less impact on the further optimization of processes and operations in manufacturing companies. This paper concludes that at least there are 8 methods that are most used to optimize operations and in this study are shown which ones are used most frequently and which are used moderately and which never used by the respondents. And depending on the situations to optimize, you can identify and / or develop different algorithms for optimization of operations (in companies, manufacturing and supply chains). And some examples of applications are shown.

1I. INTRODUCTION

With the internationalization of markets and open borders has forced companies' public or private order to seek and apply various methods that allow optimizing operations, that, should be more efficient to deal and l challenge global competitiveness.

DOI: 10.4018/978-1-4666-9779-9.ch021

Here are some options for the definition of operations research, which was first used during World War II to the conduct of military operations are defined. Later those ideas were adapted to improve productivity and efficiency in the civil sector. At present, this discipline is a necessary tool for decision making (Taha, 2004):

Lazat (1996) defines operations research as a group of methods and techniques to solve operational problems of systems.

According to Wittenberg (2000), operations research is a method to find the optimum ratios that best operate a system, given a specific objective.

And Moya (2003) defines operations research as a teaching unit. The development and application of quantitative techniques (Scientific Procedures) for solving problems and making decisions that managers face both public organizations and private organizations.

Meanwhile, Hillier and Lieberman (2010) state that the Operations Research is the application of scientific method by interdisciplinary teams to problems involving the control and management of organized systems (interaction between man and machine); in order to identify the best solutions to help the purposes of the system (or company) as a whole, framed in decision-making processes.

Preliminary Review of the Literature

In studies Forgionne (1983), it is mentioned that the methods used for optimization of operations are (see table 1):

As shown in the table above, this study is eight the most used methods for optimizing operations and shows, which are the most and least used.

And reviewing some examples of its application are the following:

A model study and mathematical simulation in mechanical drying of parchment coffee Parra, A.; Roa, G. and Oliveros, E., (2008), was based on the programming language Microsoft Visual Basic 6.0, based on models Thompson of the University of the State of Michigan.

Diaz and Perez (2012) on the work of optimizing inventory levels in supply chain optimization studies inventories joint supplier-buyer, compared to traditional non-collaborative policies, analyzing the advantages of this approach in the overall costs of inventories in the supply chain.

Table 1. Use of methodologies of management science and operations research frequency of use (% of responses)

	Never	Moderate	Frequent
1. Statistics	1.6	38.7	59.7
2. Computer Simulation	12.9	53.2	33.9
3. PERT/CPM	25.8	53.2	21.0
4. Linear programming	25.8	59.7	14.5
5. Queuing theory	40.3	50.0	9.7
6. Nonlinear Programming	53.2	38.7	8.1
7. Dynamic Programming	61.3	33.9	4.8
8. Game Theory	69.4	27.4	3.2

Source: Forgionne (1983).

The purpose of this research is to describe and compare traditional and most common methods used by public or private company s to find the optimization of resources used or used.

2. BACKGROUND

Following a literature review of more investigations by various methods related to optimization of operations authors, whose main purpose is to improve efficiency in production processes, operations of companies in the industrial, commercial and service sector is detailed.

During the literature review, optimization methods are, as in the following case, related to optimization of quality combined with pre study of technical feasibility economic:

1. **Choice of method for optimization of Quality and Technical Feasibility Studies pre – Economic**

The work, Taguchi method to optimize functional quality desserts for the elderly and Technical Feasibility Study pre - Economic, is to assess the quality control on 4 points: 1 Physics, 2 Chemistry, 3 and 4 Microbiological Sensory, aimed at assess the quality of the products. And for economic evaluation are used the following indicators IRR (internal rate of return on investment) and NPV (Net Present Value or present), Chuaqui, P., Wittig, E. and M. Villarroel (2004).

To optimize operations, you can also use various types of algorithms, as described in the following examples.

2. **Options methods using algorithms for optimization:**

As described in the optimization study applied to the hydrothermal coordination Argentine Electricity Market, where different techniques are used, such as dynamic programming, and two methods are proposed: 1 branch and bound type considering Lagrangian relaxation for dual dimension, 2 And one based on Lagrangian duality. The advantages of these methods is that the dual problem is separable at the expense of differentiability thereof, and also as general duality there jump to get a good solution (near-optimal) should be used primal recovery techniques as Lagrangian increased, Rubiales, A., Mayorano, F. and Lotito (2007).

In the study of Moitre and Garcia (2005), a model for Dynamic Optimization of Bilevel Hydroelectric pumping a Competitive Pool, which consists of a dynamic bilevel technique, where the first level is to maximize the benefits of used Central, where the various decision-bles are the powers of generation and pumping, while the second level is the problem of programming daily dispatch system variables to minimize operating costs, subject to the operational and technological restrictions the system where the algorithm used for the numerical simulations is also based on dynamic programming. Where under certain assumptions the bilevel optimization problem the optimal scheduling of the operations of a pumping station in a competitive pool, is redefined as a single level and different algorithms are used to solve it. Initially solution to the problem through dynamic programming, exhaustive search and found the same results in terms of volumes pumped and generated at the dam for a day. Then applied to the problem of duality that result make it a non-differentiable problem is solved by the method Haces. This way you can get a more efficient algorithm avoiding high memory requirements are the two algorithms initially proposed.

Figure 1. Network with "n" origins, "m" destinations and a port connection
Source: (Ametller, 2007)

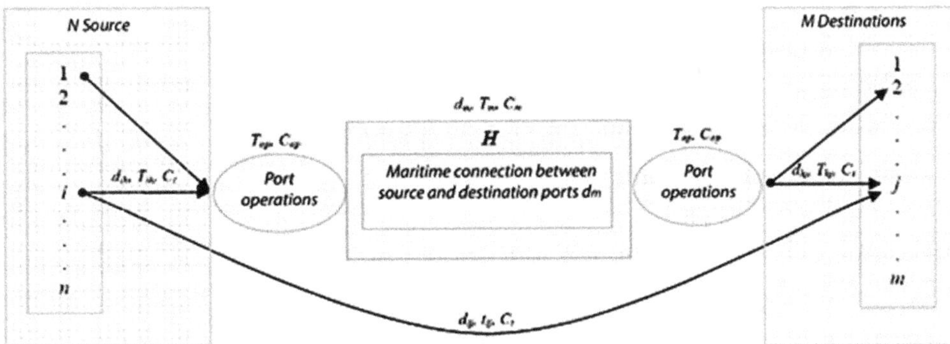

3. Options used in supply chain methods.

Optimization options operations are methods used in supply chains:

As in the work done by Ametller, (2007), says the study that is evaluated on a quantitative and qualitative distribution of goods by the short shipping distance (SSS), the countries of the Union Euro-pea. SSS emerges as an alternative and complementary to te-rrestre enabling contribute to the sustainable and economic development of the EU transport. In the optimization model freight, the problem consists of a set of pairs origin / destination assigned to the same pair of ports on which to choose for each pair is addressed, that route that minimizes the total cost of transporting goods. As seen in Figure 1, is the set of sources $i \in N = \{1,2,3 \ldots, n\}$ and the set of destinations $j \in M = \{1,2,3 \ldots, m\}$. Let *pij* platform or trailer to be transported between origin *i* and destination *j*, either directly or through the SSS. This approach does not consider the case to stop or peddling in origin or destination, see Figure 1.

Considering the factors shown in the Figure 1, to address the problem (Ametller, 2007) proposes the following expression in the form of algorithm:

Close:

$$\sum_{i=1}^{n}\sum_{j=1}^{m}\left(C_t\left(d_{ij}\right) + \int_{6a}\left(d_{ij},\alpha\right)\right)W_{ij}X_{ij} +$$

$$+\sum_{i=1}^{n}(C_t\left(d_{ih}\right) + \int_{6a}\left(d_{ih},\alpha\right))W_{ih} +$$

$$+\sum_{j=1}^{m}(C_t\left(d_{hj}\right) + \int_{6a}\left(d_{hj},\alpha\right))W_{hj} +$$

$$+\sum_{i=1}^{n}(C_{op}\left(W_{ih}\right) + \int_{6b}\left(d_{ih},\alpha\right)) +$$

$$+\sum_{j=1}^{m}(C_{op}\left(W_{hj}\right)+\int_{6b}\left(d_{hj},\alpha\right))+$$

$$+C_{m}\left(S,d_{m},\alpha,r,\delta\right)W_{h}+\int_{6c}\left(W_{ih},W_{hj},\alpha\right)$$

where, left to right:

- Costs associated with land transport raw mode

$$\sum_{i=1}^{n}\sum_{j=1}^{m}\left(C_{t}\left(d_{ij}\right)+\int_{6a}\left(d_{ij},\alpha\right)\right)W_{ij}X_{ij}:$$

Cost of the origins carries the position origin

$$\sum_{i=1}^{n}(C_{t}\left(d_{ih}\right)+\int_{6a}\left(d_{ih},\alpha\right))W_{ih}:$$

- Cost of the transports from the port bound for destinations
- Cost of port operations at source
- Cost of port operations at destination

$$\sum_{j=1}^{m}(C_{op}\left(W_{hj}\right)+\int_{6b}\left(W_{hj},\alpha\right)):$$

$$\sum_{i=1}^{n}(C_{op}\left(W_{ih}\right)+\int_{6b}\left(W_{ih},\alpha\right)):$$

- Cost of segment seagoing

$$C_{m}\left(S,d_{m},\alpha,r,\delta\right)W_{h}:$$

- Inventory cost of maritime leg

$$\int_{6c}\left(W_{ih},W_{hj},\alpha\right):$$

The algorithm designed to analyze routes offers very encouraging results for SSS in relation to the different scenarios presented in terms of modal distribution of goods (Ametller, 2007).

Another model of optimization of operations is that posed by Dominguez (2006) in the simultaneous optimization work for continuous improvement and cost reduction processes, expresses a function that describes the combination of individual goals. A first method is the multiplicative model optimization and a second additive is applied in two case studies in the industry in order to highlight the processes of continuous improvement and cost reduction. The design involves factors of interest in order to find the combination that corresponds to the best all the features of the overall product quality, design process known as multi-response optimization. Two methods exist for optimizing multiple response variables, which are used for the study of experimental strategy: first, that posed by Kros and Mastrangelo (2004), $and_j = \beta_{j0} + X^t \beta_j + X^t + \varepsilon B_j X_j$

Where:

$X^t = (x_{1,} \ldots, X_{k)}$ k factors

β_{j0} = constant

$_j = \beta\, (\beta_{j1, \ldots,}\, \beta_{jk)}$ a vector of parameters

$B_j = (\beta_{j11} \cdots {}_{J1K}\, \beta,\, \beta_{JK1} \cdots {}_{jkk}\, \beta)$ symmetric matrix of second order parameters

$_j\varepsilon \sim N\,(0, \sigma^2{}_{j)}$

j = 1, ..., r

The second presenting by Goel and Kumar (2002), which is to determine the combination of factors that produce the global optimum, i.e., that all responses give their best value:

Optimize Y_1

$$\text{Subject to} \quad \begin{array}{ccc} Y_2 & = & M_1 \\ & \vdots & \\ Y_r & = & M_{r-1} \end{array}$$

$X \in R$, R: regression experimental

Where Mj $(j = 1, \ldots, r - 1)$ represent important considerations, objectives or restrictions for answers. This theory provides a natural way to treat problems in which the print font is the absence of well-defined criteria.

In the work done by Diaz and Perez (2012) Optimization of inventory levels in a supply chain, develop an example using contributions and Liao Pan (1989), Kelle and Miller (1998) and Kelle et al, (2003):

The classic model for inventory control s EOQ (Economic Order Quantity) Schroeder (2005), is one such example and their behavior is shown in Figure 2.

Considering that the company X, builds on this model, as the following:

D = Demand product units per year

C = cost per unit of product (purchase or production)

S = Cost per order (or setup cost)

H = cost of keeping inventory unit per year

Figure 2. Behavior of EOQ inventory model
Source: (Diaz and Perez, 2012)

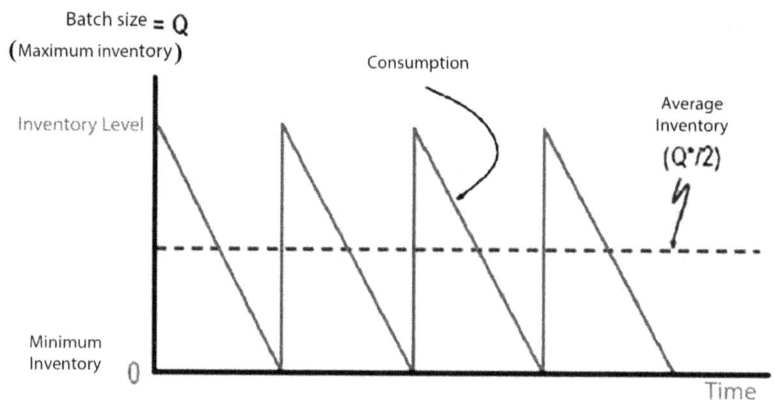

Therefore, the batch quantity or to apply optimum Q * is known from the expression:

$$Q^* = \sqrt{\frac{2DS}{H}}$$

and the minimum total cost per year:

$$CTA = \frac{SD}{Q^*} + \frac{HQ^*}{2} + DC$$

Now, assuming the company Y, has a production capacity of P units per year to meet the annual demand D above. In this case, the behavior of inventory (POQ -Period Order Quantity- model) is showed in Figure 3.

Figure 3. Behavior inventory model POQ
Source: (Diaz and Perez, 2012)

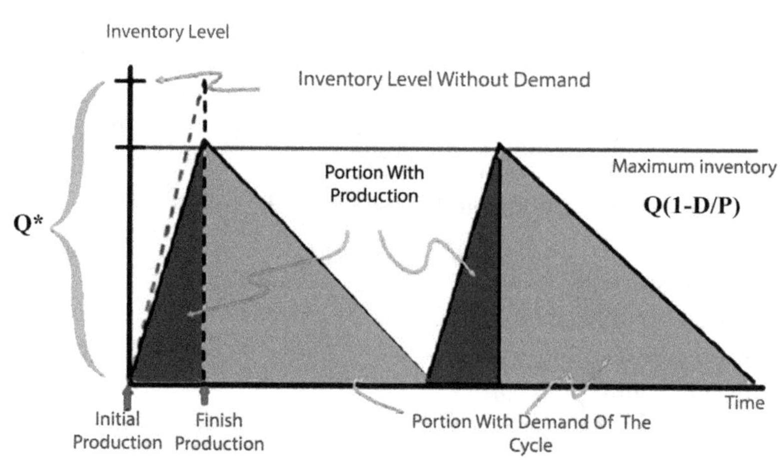

For this case, the amount of the batch to produce optimum Q * is represented by the expression:

$$Q^* = \frac{2DS}{H\left(1-\frac{D}{P}\right)}$$

Whereas a supply chain is composed of the supplier relationship (firm *Y)* and buyer (company *X)*. Due to the difference in costs and the optimal policy of each of her economic batch size for the supplier (Q_s *) and the buyer (Q_c *), if they work each by hand, this does not they will be the same. But if the policy adopted by both companies is collaborative, then it is possible to calculate the cost and optimal batch d set that minimizes the total cost entrée both.

The expression of total yearly set for the supplier and the buyer is:

$$CTA = CD + \frac{D}{Q}S_C + \frac{1}{2}H_C Q + \frac{D}{Q}S_S + \frac{1}{2}H_S Q\left(1-\frac{D}{P}\right)$$

and used to set the optimal batch Q_J *:

$$Q_J^* = \frac{2D(S_C + S_S}{H_C + H_S\left(1-\frac{D}{P}\right)}$$

Collaborative activity can lower total annual cost that non-collaborative.

The proposal Diaz and Perez (2012), supporting contributions Bread and Liao (1989), Kelle and Miller (1998) and Kelle et al, (2003). In a just-in-time environment, it requires the buyer to receive small and frequent shipments, resulting in lower inventory holding costs for this. However, this requirement may increase the costs of the supplier, then it is advisable a collaborative relationship.

Shipping size that minimizes the annual cost to the system set can be expressed in terms of m and n as shown in the following expression:

$$q_j(m,n) = \sqrt{X_j(m,n)\,y_j(m,n)}$$

Finally, the overall cost is a function of m and n shown in the expression:

$$JCT_j(m,n) = 2\sqrt{X_j(m,n)\,y_j(m,n)}$$

By identifying the value of the derivative of the X_j *(m, n)* and $_j$ *(m, n)* function in relation amine equating to 0, can be derived expressions for the optimal values of *m* and *n* Process to find. The optimal values n *, m * q_J.

3. METHODOLOGY

This research is descriptive qualitatively, the main purpose of study is to make a detailed comparison of work done by other researchers have tried to identify the best method according to the environment or circumstances to establish the path that optimizes system operations, a business, process, etc.

The aim of this chapter is to show some optimization methods of operations that have achieved better results. Therefore, this work can provide support for researchers in the field of optimization, as it shows a compilation of work done for other researchers to compare relevant aspects of the proposals submitted by their authors.

A comparison of elements or previously analyzed aspects related to the optimization methods of operation in public or private companies work according to Onset:

- *Method* (s) used for in different cases.
- *Study / Theory* (target function: Overview Method (s))
- **Contributions:** Contribution of (the) method (s) used (s).
- **Limitations:** Conditions for the development or application of the method used.
- **Author:** Researchers who conducted the study
- *Year* of the study.

4. DATA ANALYSIS

D for analysis and data, a comparison of the elements indicated in the previous point is made:

1. **Author / years:** Forgionne / 1983.
   a. **Methods:** Statistics, Computer Simulation, PERT / CPM, linear programming, queuing theory, Nonlinear Programming, Dynamic Programming and Game Theory
   b. **Stu gave / Theory (Objective Function):** Surveys to analyze the frequency of use of methodologies Management Science and Operations Research, most used in companies.
   c. **Contributions:** In this study, the percentages of use of the methods are shown:
      i. Never
      ii. Moderately
      iii. Frequently used by businesses.
   d. **Limitations:** The study is from 1983, so it would be wise to do a current study to analyze possible changes from 1983 to date.
2. **Author (year):** Parra et al. (2008).
   a. **Methods:** Mathematical simulation, using computer simulation.
   b. **Stu gave / Theory (Objective Function):** Mathematical modeling and simulation in the mechanical drying of parchment coffee.
   c. **Contributions:** Using *Microsoft Visual Basic 6.0 Programming.*
   d. **Limitations:** The study other options of mathematical simulation are described.
3. **Author (year):** Diaz and Perez (2012).
   a. **Methods:** Optimizations inventory levels, using: computer simulation.

   b. **Stu gave / Theory (Objective Function):** Inventory optimization study. Applying Operations Control.
   c. **Contributions:** Using atomization techniques Inventories sets Supplier - Buyer, compared with traditional non-collaborative policies. Analyzing the advantages of this approach in the overall cost of inventories in the supply chain.
   d. **Limitations:** The study other options to optimize inventory levels are described.
4. **Author (year):** Chuaqui et al. (2004).
   a. **Methods:** Taguchi to optimize quality, using Statistics.
   b. **Stu gave / Theory (Objective Function):** Technical feasibility study pre - Economic. Applying Process Control.
   c. **Contributions:** L assess the quality since you 4 points Control:
      i. Physical,
      ii. Chemicals,
      iii. Microbiological, and
      iv. Sensory. Using Statistics.
   d. **Limitations:** The study other options to optimize the quality are described.
5. **Author (year):** Rubiales, Mayorano and Lotito.
   a. **Methods:** optimization algorithms, using: Dynamic Programming.
   b. **Stu gave / Theory (Objective Function):** Optimization applied to the hydrothermal coordination Argentine Electricity Market. Applying Administration of Operations.
   c. **Contributions:** The use of dynamic programming using the following methods:
      i. The type branch and bound considering Lagrangian relaxation for dual quota and
      ii. One based on Lagrangian duality.

The advantages of these methods is that the dual problem is separable at the expense of differentiability thereof, and also as general duality there jump to get a good solution (near-optimal) should be used primal recovery techniques as Lagrangian increased, (2007).

   d. **Limitations:** The study other optimization options are not described.
6. **Author (year):** Moitre and Garcia (2005).
   a. **Methods:** Dynamic Optimization.
   b. **Stu gave / Theory (Objective Function):** Binevel dynamic optimization of Hydroelectric pumping a Competitive Pool. Applying Management Operations.
   c. **Contributions:** ECHNICAL bilevel dynamics, where the first is to maximize the benefits of the plant and where the decision variables are the power generation and pumping, while the second level is the problem of programming daily dispatch system for T minimizing variable operating costs, subject to the restrictions of the system, where the algorithm used for the numerical simulations is also based on dynamic programming.
   d. **Limitations:** The study other optimization options are not described.
7. **Author (year):** Ametller (2007).
   a. **Methods:** Quantitative and Qualitative Evaluation.
   b. **Stu gave / Theory (Objective Function):** Quantitative and qualitative evolution of the distribution of goods by short sea shipping (SSS). Applying, administration and Supply Chain n.

   c. **Contributions:** SSS emerges as an alternative and complementary to land that allows contributing to sustainable and economic development of the EU transport. . In the optimization model freight, the problem consists of a set of pairs origin / destination assigned to the same pair of ports on which to choose for each pair is addressed, that route that minimizes the total cost of transportation goods.
   d. **Limitations:** The study other optimization options are not described.
8. **Author (year):** Dominguez (2006).
   a. **Methods:** Simultaneous Optimization
   b. **Stu gave / Theory (Objective Function)** Simultaneous optimization for continuous improvement and **cost reduction processes.**
   c. **Contributions:** Function describing the combination of individual goals. A first method is the multiplicative model optimization and a second additive is applied in two case studies in the industry in order to highlight the processes of continuous improvement and cost reduction.
   d. **Limitations:** The study other options simultaneous optimization are described.
9. **Author (year):** Schroeder (2005).
   a. **Methods:** The classic model for inventory control.
   b. **Stu gave / Theory (Objective Function):** Inventory Optimization, using the model EOQ (Economic Order Quantity).
   c. **Contributions: Collaborative activity** can lower total annual cost that non-collaborative.
   d. **Limitations:** The study other options inventory optimization are described.

5. RESULTS

Of the studies analyzed result is obtained at least, eight are the most used for the atomization of operations, which are next in order of those most commonly used models:

Methods:

1. Statistics
2. Computer simulation
3. PERT / CPM
4. Linear Programming
5. Queueing Theory
6. Nonlinear programming
7. Dynamic Programming
8. Game theory.

And these methods are used for different purposes:

1. Operations optimizations.

   a. Maximize
   b. Minimize

They are used in:

1. Business
2. Manufacturing
3. And in supply chains
4. Other applications

6. CONCLUSION

A t least there are eight methods that are most used to optimize operations and in this study is what are those used most often. And depending on the situations to optimize, you can identify and / or develop different algorithms for optimization of operations (in companies, manufacturing and supply chains) and some examples of applications are shown:

1. Mathematical modeling and simulation in the mechanical drying of parchment, using mathematical simulation.
2. Optimizing inventory levels, using computer simulation, applying Operations Control, using computer simulation.
3. Technical feasibility study Pre - Economic Process Control applied through the Taguchi method to optimize the quality, applying Statistics.
4. Optimization of hydrothermal Coordination Argentine Electricity Marketing, Operations Management using through Dynamic Programming.
5. Dynamic optimization Binevel of Hydroelectric pumping a Competitive Pool using Operations Management by Dynamic Programming.
6. Quantitative and qualitative evolution of the distribution of goods by short sea shipping (SSS), management applied in the Supply Chain, by Mathematical Algorithms to minimize the total cost of freight transport.
7. Simultaneous optimization for continuous improvement and cost reduction processes through optimization models and multiplicative additive.
8. Inventory Optimization, using the EOQ model (Economic Order Quantity), comparing the results achieved by applying collaborative activity and not collaborative. Proving that collaborative activity, produce a lower total annual cost that non-collaborative.

REFERENCES

Ametller, X. (2007). *Optimization of freight transport by short sea shipping*. (Thesis). Polytechnic University of Catalonia. Recovered from http://upcommons.upc.edu/pfc/handle/2099.1/4942

Chuaqui, P., & Wittig Villarroel, E. M. (2004). Taguchi method for optimizing quality of functional desserts intended for elderly and technical feasibility study. *Economic Chilean Journal of Nutrition, 31*(2), 118-127. Retrieved on March 11, 2015, from http://www.scielo.cl/scielo.php?script=sci_arttext&pid=S0717-75182004000200006&lng=es&tlng=es.10.4067/S0717-75182004000200006

Diaz, J. A., & Perez, D. (2012). Optimization of inventory levels in a supply chain. *Industrial Engineering, 33*(2), 126-132. Available at: http://scielo.sld.cu/scielo.php?script=sci_arttext&pid=S1815-59362012000200004&lng=es&nrm=iso

Dictionary of Accounting. (2002). Mexico: Hara.

Dominguez, J. (2006). Simultaneous optimization for continuous improvement and cost reduction processes. *Engineering and Science-ing. Cienc., 2*(4), 145-162. Recovered from http://publicaciones.eafit.edu.co/index.php/ingciencia/article/view/473/440

Forgionne, G. A. (1983). Corporate Activities Management Science. *Interfaces, 13*(3). doi:10.1287/inte.13.3.20

Hillier & Lieberman. (2010). *Introduction to operations research*. McGraw Hill.

Kelle, P., Al-Khateebb, F., & Miller, P. A. (2003). Partnership and joint negotiation optimal support by ordering / setup policies for JIT. *International Journal of Production Economics, 81-82,* 431-441.

Kelle, P., & Miller, P. A. (1998). Quantitative support for buyer-supplier in JIT Purchasing negotiation. *International Journal of Purchasing and Materials Management, 34*(2), 25-30.

Kros, J., & Mastrangelo, M. (2004). Comparing Multi-response Design Methods With Mixed Responses. *Quality and Reliability Engineering International, 20*(5), 527–539. doi:10.1002/qre.664

Kumar, P., & Goe, P. (2002). Product Quality Using Fuzzy Optimization September Concepts: A Case Study. *Quality Engineering, 15*(1), 1–8. doi:10.1081/QEN-120006706

Moitre, D., Sauchelli, V., & Garcia, G., (2005). Bilevel Dynamic Optimization of Central Hydroelectric pumping in a Competitive Pool - Part I: Model and Algorithm *Magazine. IEEE Latin America, 3*(2), 62 - 67.

Moya, M. J. (2003). Operations research. Editorial Universidad Estatal a Distancia.

Pan, A. C., & Liao, C. (1989). An inventory model under just-in-time purchasing agreements. *Production and Inventory Management, 30*(1), 49-52. doi: 10.1108/01443579810192907

Parra, A., Roa, G., & Oliveros, E. (2008). SECAFÉ Part I: mathematical modeling and simulation in the mechanical drying of parchment coffee. *Revista Brasileira de Engenharia Agrícola e Ambiental, 12*(4), 415-427. Retrieved March 12, 2015, from http://www.scielo.br/scielo.php?script=sci_arttext&pid=S1415-43662008000400013&lng=en&tlng=es.10.1590/S1415-43662008000400013

Rubiales, A., Mayorano, F., & Lotito, P. (2007). Optimization applied to the hydrothermal coordination of the Argentine electricity market. *Computational Mechanics, 26*, 1906–1920.

Schroeder, R. G. (2005). *Operations Management*. Boca Raton, FL: McGraw-Hill.

Stoner, J. (1992). *Administration*. Mexico: Prentice Hall.

Taha, H. A. (2004). *Operations research*. Editorial Pearson Education.

Wittenberg, J. P. (2000). *Methods and models of operations research* (Vol. 1). Editorial LIMUSA.

KEY TERMS AND DEFINITIONS

Cost: Dispensing or payment in cash, other assets, in equities or services that increase their value in inventories Accounting Dictionary (2002).

CPM (Critical Path Method): An administrative process of planning, programming, implementation and control of each and every one of the component project activities (Stoner, 1992).

Economic Batch: The amount should buy or produce periodically to optimize acquisition costs (Stoner, 1992).

Just in Time: An industrial philosophy is to manage inventories, involving products manufactured strictly necessary, at the right time and in the amounts due. Anything that involves eliminating waste in the production process (Hillier and Lieberman, 2010).

PERT (Evasion Technique and Review of Programs): A network analysis system of time-event, in which the various events of a program or project with e l time planned for each (Stoner, 1992) are related.

Chapter 22
The Application of Hanoi Towers Game in Logistics Management

José Alberto Hernández Aguilar
Autonomous University of Morelos State, Mexico

Santiago Yip Ortuño
Autonomous University of Morelos State, Mexico

Augusto Renato Pérez Mayo
Autonomous University of Morelos State, Mexico

Alberto Ochoa-Zezzatti
Autonomous University of Juarez City, Mexico

Julio César Ponce Gallegos
Autonomous University of Aguascalientes, Mexico

ABSTRACT

In this chapter we discussed the application of Towers of Hanoi in logistics management applications, for this purpose. Firstly, we discussed how pile problems applications impact in logistics, later we discussed the Hanoi Towers application in Logistics Management, its mathematical model and common solutions applying different paradigms (iterative, recursive, and heuristics), we present an in depth analysis in genetic algorithms, later we present the analysis of a genetic algorithm applied to solve basic Hanoi Tower game (three pegs three discs) implemented in C language, and later we discuss its generalization from three to four discs. Finally, we discussed preliminary results and present our conclusions. Main contribution of this chapter is demonstrate game theory, specifically Towers of Hanoi, can be applied to solve logistics problems, and an approach for the generalization of basic Hanoi Tower form, three discs to four discs, by means of genetic algorithms implemented in C language.

1. INTRODUCTION

Lots of situations in real life present a configuration of items stacked up to a pile that requires to be transferred in another pile with another order of items, this type of problems can be identified in management logistics also. According Hempfel and Schmiedel (2006) the question that involves an optimization problem is: How to do it in the best way using only available resources and devices?

DOI: 10.4018/978-1-4666-9779-9.ch022

1.1. Pile Problems Applications in Logistics Management

Some examples are mentioned in (Hempel and Schmiedel, 2006): imagine a port where there are several containers and is necessary to load or to unload loading from the containerships, or suppose large concrete parts for a modern bridge or construction of buildings, or the shunt of trains in a marshaling yard, or patient plays. A container pre-marshaling problem is reported in (Shan-Huen & Tsan-Hwan, 2012) which purpose is finding a set of container movements to reach a final container arrangement by satisfying certain restrictions. Lu & Dillon (1995) discuss a parallelism implementation of Towers of Hanoi on which is permitted concurrent moves of disks in a single step. These authors analyzed three versions of parallelism considering two factors: the order of moves of disks and the resource utilization of pegs. They present a common technique for getting the non-recursive implementation by considering minimum steps. All of these problems can be modeled by applying basic principles of the Towers of Hanoi.

1.2. Mathematical Modeling

Puzzles are connected with graphs, it means can be modeled by using graphs, vertices of the graph represent configuration and graph edges define possible moves of the puzzle (Cull, Merril and Van, 2013). According (Hempel & Schmiedel, 2006) to describe pile problems more precisely is highly recommended using graph theory. For these authors:

For a given directed graph G = (V, A) and its vertices v ϵ V can be used the following notations: $N_{G+}(v)$ is the set of all successors of v, $N_{G-}(V)$ is the set of all predecessors of v. $d_{G+}(v)$ represents numbers of outgoing arcs of v and $d_{G-}(v)$ numbers of incoming arcs of v respectively. For $G_1 = (V_1, A_1)$ and $G_2 = (V_2, A_2)$ the union of G_1 and G2 is defined as G_1 U G_2:= (V_1 U V_2, A_1 U A_2). A cycle in a directed graph G = (V, A) is a closed path in G. The directed graph G = (V, A) is called a pile, if and only if G is finite and does not contain a loop or a cycle. The vertices of G represent the elements of the pile; the arcs of G represent the order in which the elements are piled up. A step is the move of an element of data structure to another place. This place can be in different pile or in an auxiliary place if it exists (Hempel & Schmiedel, 2006).

Figure 1 and Figure 2, show an example of typical pile problems, meanwhile graphs representing the starting *Gs* and final *Gf* are shown below (Hempfel and Schmiedel, 2006):

Note how arcs in Figure 3 and Figure 4 correspond to inverted graphs.

Figure 1. Starting situation

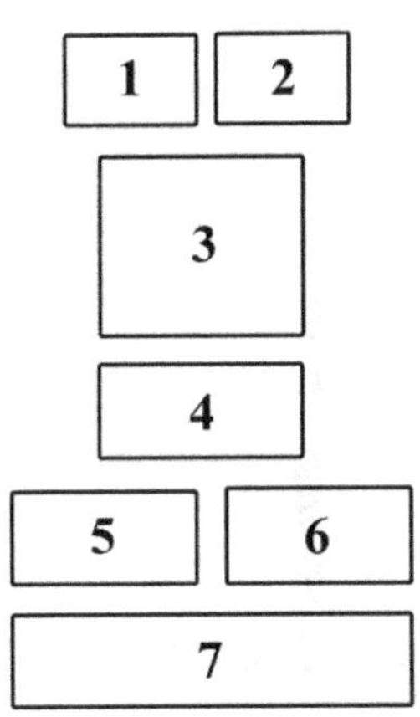

Figure 2. Required final situation

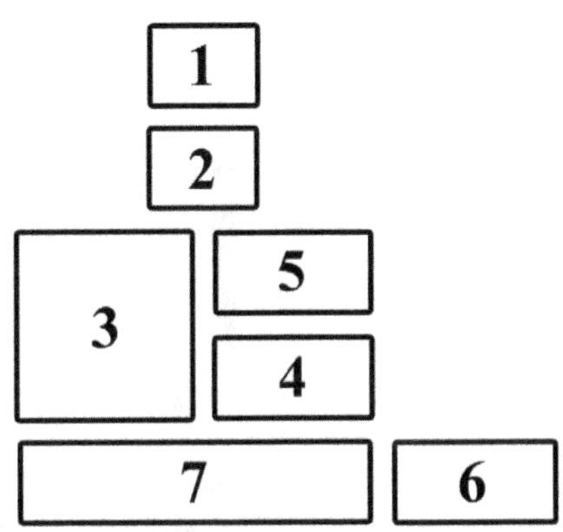

Figure 3. Starting pile Gs

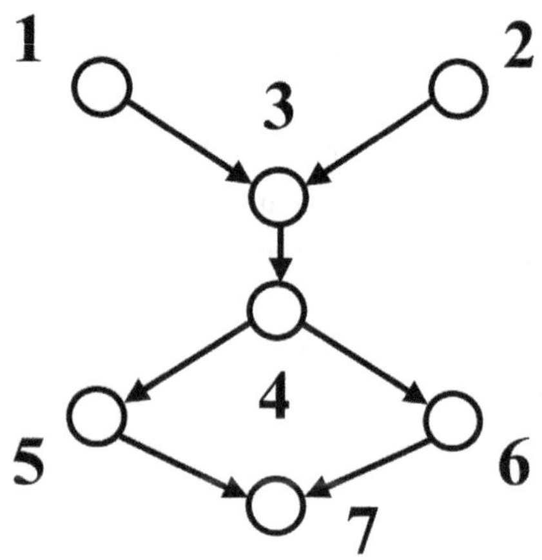

Figure 4. Final pile Gf

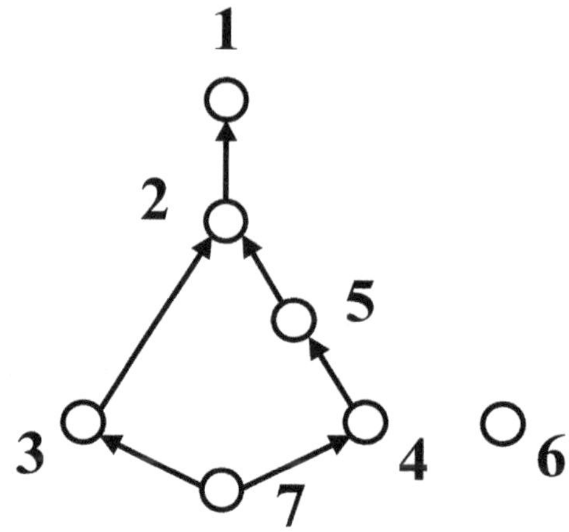

Towers of Hanoi

Also called the Towers of Brahma or Towers of Lucas; is a mathematical game, puzzle or toy that is made of three pegs (poles) and a number of disks of different sizes (in its basic form consisting of three disks), which can be moved into any peg (Hofstadter, 1996). The game starts with the disks placed in ascending order of size in one peg, the smallest at the top, forming a cone or pyramid (see Figure 5).

The purpose of the game is moving the entire pile of disks from the origin peg to the destiny peg, by considering next simple rules (Scheinerman, 2000):

Step 1: Sole one disk can be moved each turn.
Step 2: A move consists of taking the upper disk from one of the stacks and placing it on top of another stack.
Step 3: Is not allowed a disk be placed on top of a smaller disk.

According Mishra and Vishnoi (2015) with three single disks, the game can be solved in seven steps. The minimum moves required to solve a Tower of Hanoi puzzle is $2^n - 1$, where n represents number of disks, as can be seen, this is an exponential number that increases when n increases.

Figure 5. Towers of Hanoi Basic game (three pegs - three disks) source (Hofstadter, 1996)

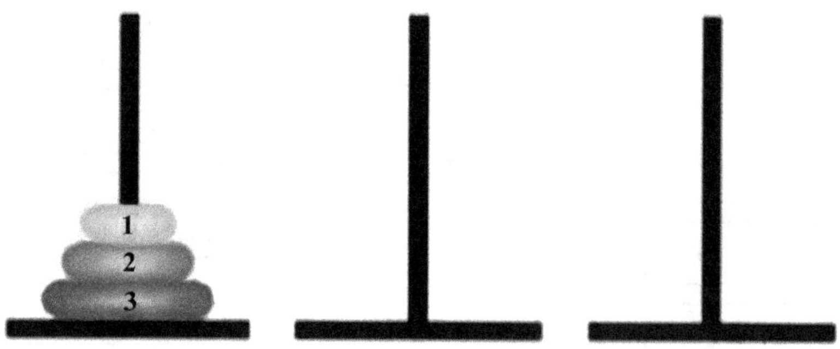

2. COMMON SOLUTIONS FOR TOWERS OF HANOI

There are several solutions for the towers of Hanoi puzzle, below we present some of them, specifically we discuss iterative and recursive solutions.

2.1. Iterative Solution

A simple approach for solving the puzzle of towers of Hanoi consists of alternating moves between the smallest and a non-smallest piece (disk). According Buneman and Levy cited in (Cull, Merrill & Van, 2013):

If n is ODD, the smallest disk should be moved to its target tower, but when n is EVEN, disk should be moved to the non target tower. The simplest way to handle this is to arrange the towers in circular order 0,2,1 when n is ODD and in circular order 0,1,2 when n is EVEN. Then moving disk 1 clockwise will always move it in the right direction.

Next is shown the complete Buneman and Levi (1980) algorithm, this assumes the towers are assigned to numbers 0,1 and 2.

```
ITERATIVE ALGORITHM
Move the smallest disk one tower clockwise
WHILE a disk (other than the smallest)
        Can be moved DO
        Move that disk
        Move the smallest disk one tower clockwise
ENDWHILE
```

In this sense, according Troshkin (2014): "if you started with three pieces, you would move the smallest piece to the opposite end, then continue in the left direction after that. When the turn is to move the non-smallest piece, there is only one legal move".

By using the algorithm of Buneman and Levi the game will be completed in the fewest number of moves.

Described in terms of rules, the shortest solution can be obtained following one of the next two procedures, which depend on the number of disks (Bornstein & Svane, 2011):

For an even number of disks:

- Make a valid move between pegs A and B
- Make a valid move between pegs A and C
- Make a valid move between pegs B and C
- Repeat until game is complete.

For an odd number of disks:

- Make a valid move between pegs A and C
- Make a valid move between pegs A and B

- Make a validmove between pegs C and B
- Repeat until complete.

For each case (even or odd) a total of 2^n-1 moves are made, where *"n"* represents the number of disks (Troshkin, 2014).

2.2. Recursive Solution

According (Havur, Haspalamutgil, Palaz, Erdem, & Patoglu, 2013) recursive is a procedure call that requires high programming skills, is a process or function in the definition and description that directly or indirectly call itself as a method. It is usually a large complicated problem into a layer with the original problem, which is modeled like smaller problems to solve. Next is shown the recursive algorithm for Towers of Hanoi that produces minimal move solution (Cull, Merrill & Van, 2013).

```
RECURSIVE ALGORITHM
PROCEDURE HANOI(i,j,n)
IF n >0
        THEN
                k = neither i nor j
                HANOI(i,k,n-1)
                Move the top disk from tower i to tower j
                HANOI(k,j,n-1)
```

In this way Hanoi towers function code, can be written to achieve its layers of recursive. Language code in C++ is as shown next (Li, 2014):

```
void move(char one, char anoth)
{
cout<<one<<"--->"<<another<<endl;
}
void hannuota(int n,char no1,char no2,char no3
{
        if(n==1)
        {
                move(no1,no3);
        }
        else
        {
                hannuota(n-1,no1,no3,no2);
                move(no1,no3);
                hannuota(n-1,no2,no1,no3);
        }
}
```

2.3. Heuristics

Genetic Algorithms

An algorithm is a sequence of steps or finite computational process whose purpose is to execute a specific action. But what is a genetic algorithm? According (Hernández Aguilar, Yip-Otuño, Vakhania, Ponce Gallegos, & Ochoa Ortiz, 2015) "the genetic algorithm is a search technique based on Charles Darwin's theory of evolution, this technique is based on the mechanisms of selection that uses the nature, according to which individuals more suitable (stronger and/or smarter) in a population are those that survive and reproduce, the more easily adapted to the changes that occur in their environment".

A genetic algorithm is defined by (Hernández Aguilar et al., 2015) as "a mathematical function that takes as an input each individual in the population and returns as output which of them should generate offspring for the next generation" (see Figure 6).

According to (Yip Ortuño, Hernández Aguilar & Zavala Díaz, 2015) Genetic algorithms can be applied to solve Towers of Hanoi and is required at least four methods for their operation:

1. **Mutation:** The stage in charge of mutation of individuals' genes.
2. **Crossover:** Performs the breeding of selected individuals.
3. **Fitness Function:** The most important function for these algorithms because on this step can be found the solution for the problem.
4. **Selection:** It is responsible for selecting the fittest individuals.

Figure 6. Flowchart of a continuous GA (Yip Ortuño et al., 2015)

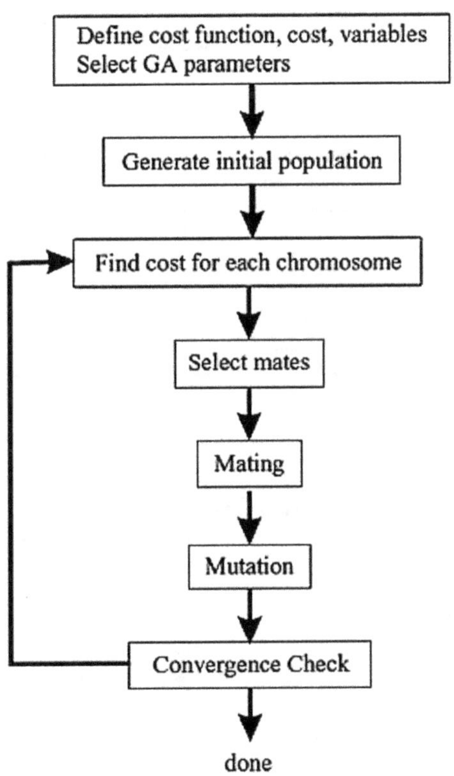

3. METHODOLOGY

We will analyze a sequential implementation solution for the Towers of Hanoi (Three pegs - three disks) in C languaje, and then, based on this analysis, we will propose a generalization for this solution (Three pegs - four disks), we will present our proposals and we will conduct experiments to verify validity of solutions.

3.1. Genetic Algorithm Implementation

Next is presented the implementation of a genetic algorithm designed to resolve the Towers of Hanoi puzzle, this implementation is based on works of (Jones, 2010).

This problem involves three pegs and three unique disks, and the goal of moving the disks from one peg to another (see Figure 7). The constraint is that a disk can be moved to a peg only if the peg is empty, or if the disc currently on the peg is larger than the peg to be moved (Jones, 2010).

If pegs are numbered from zero to two, is possible to define a space of solutions as a sequence of move, where each move encodes the source peg and the destination peg.

The chromosome for the GA can be modeled as a sequence of moves with numbers 0 to 5, which provides a simpler format to evolve.

Figure 7. Illustration of moves provided in the solution sequence (Jones, 2010: 205)

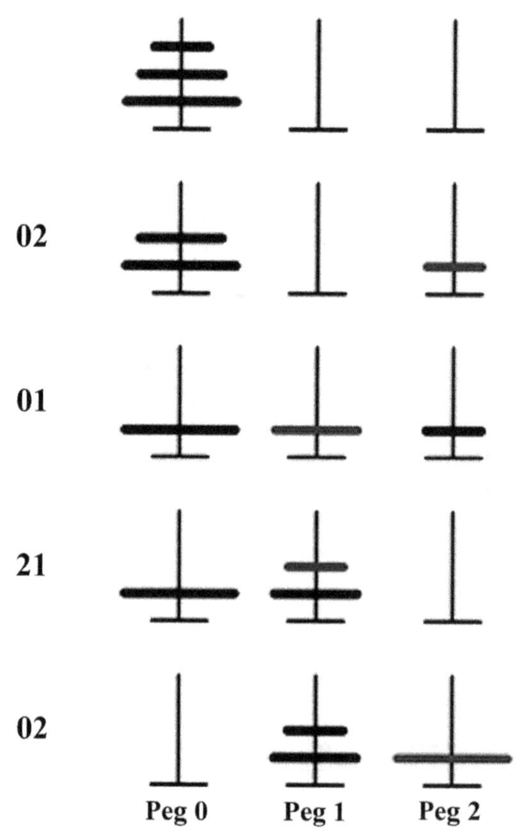

The simulation of game is provided by the pegs array, with the number of disks concurrently in the peg defined by variable count. The solution is represented by solution_t that contains the sequence of moves (plan), the number of active moves (op_count), and the fitness evaluation.

3.2. Generalization of the Solution of Towers of Hanoi from Three to Four Disks

The genetic algorithm proposed by Tim Jones (Jones, 2010) was designed to solve the puzzle of the towers of Hanoi in its single version (three disks - three pegs) only. The solution is stored in a structure, so it was necessary to modify it to use more disks. The proposed C language structure by Jones (2010) is:

```
typedef struct {
char peg[3];
int count;
} peg_t;
```

Where “peg” contains the constant value of 3, which indicates the maximum number of disks that can contain each peg or rod.

To increase this number of disks, we define a variable “NUM_DISC”

```
#DEFINE NUM_DISC N
```

In addition, the structure peg_t was modified, as shown below:

```
typedef struct {
peg NUM_DISC[char];
int count;
}peg_t;
```

Where variable N sets number of disks. On the other hand, is necessary to change the values where is establish the number of disks as general variable for entire program. Among these changes is included the compute_fitness() function, which is shown below:

```
Double compute fitness (int cur_pop, int member,  int trace){
int i, from, to, disc= 3;
int illegal_moves= 0;
int move= 0;
double fitness;
/ * Initialize the pegs * /
for(i= 0; i < 3; i ++){
        pegs[0] .peg[i] =disc--;
        pegs[1] .peg[i] = 0;
        pegs[2] .peg[i] = 0;
}
pegs[0] .count = 3;
```

```
pegs[1] .count = 0;
pegs[2] .count = 0;
... }
```

In this function, is set the variable “disc” with value 3. This variable is equal to the variable “NUM_DISC”. During the execution of this function, it initializes the peg[0] or rod, which is located on the far left with the number of disks that now contains the variable “NUM_DISC” with a specified number. The code is hereby amended as follows:

```
double compute fitness (int cur_pop, int member,  int trace){
int i, from, to, disc=NUM_DISC;
int illegal_moves= 0;
int move= 0;
double fitness;
for(i= 0; i < NUM_DISC; i ++){
        pegs[0] .peg[i] =disc--;
        pegs[1] .peg[i] = 0;
        pegs[2] .peg[i] = 0;
}
pegs[ 0] .count = NUM_DISC;
pegs[ 1] .count = 0;
pegs[ 2] .count = 0;
.
.
.
}
```

As seen in the previous code, the variable “disc” is substituted by the constant “NUM_DISC” and subsequently replaced where the constant “3” appears.

On the other hand, in the testing process, it is clear that the “plan” generated by the algorithm does not finish the game in its entirety; that is to say, the sequence generated does not move the disks defined by “NUM_DISC”. For the foregoing reasons, in the case of four disks was necessary to expand the number of operations. This variable is called “NUM_OPERATIONS”. Jones define that for the case of three disks, the maximum number of operations is 30 movements; for the case of four disks, we proposed to raise this number to 50, since that with this amount of movements, the algorithm yields a plan long enough to resolve the game with four disks.

#DEFINE NUM_OPERATIONS 30 //(Jones, 2010).

#DEFINE NUM_OPERATIONS 50 //our proposal for 4 disks.

4. EXPERIMENT

Sequential programs were run ten times, firstly for 3 disks and 30 number of operations and then for 4 disks and 50 number of operations. Two measures were obtained: the starting time and the total execution time for the complete program. The difference between these two measures is reported for each run in next table. Experiment was executed in a Lufac Workstation with 32 Gbytes in RAM, gcc 4.4.7 and Ubuntu 14.10 Operating System.

5. RESULTS AND DISCUSSION

As Table 1 shows is possible to modify program to manage 4 disks, as can be observed is necessary to modify variable number of operations (Num-op) from 30 (3 disks) to 50 (4 disks), execution time increased from 2.198 to 14.633 seconds.

CONCLUSION

Is possible to apply Tower of Hanoi principles to solve Logistic Management Problems in several areas like loading or unloading, transport and construction or play patience. Is possible to generalize basic Tower of Hanoi game (three pegs three disks) to (threes pegs four disks) using genetic algorithms and C language. This research is not in any form complete to the contrary, future work includes generalization of Tower of Hanoi game (three pegs four disks to three pegs *n* disks) using genetic algorithm. We would like to outline genetic algorithm is a time consuming process and future work includes analysis and design of a generic solution (three pegs *n* disks) and its parallel implementation (Lu & Dillon, 1995) on GPUs based on our previous works.

Table 1. Execution time comparison between three disks and four disks

Run	3 disks t Sequential	4 disks t Sequential
Num_op	30	50
1	1.86	27.49
2	6.86	0.56
3	3.5	10.9
4	1.53	6.02
5	0.72	18.51
6	1.72	11.2
7	2.29	19.78
8	1.3	32.88
9	0.54	17.71
10	1.66	1.28
Mean (seconds)	2.198	14.633

ACKNOWLEDGMENT

We want to thank to Research and Development Department of UAEM for internal financial support for the execution of this research project and the consolidation of CA-128 "Operation Research and Informatics" research group.

REFERENCES

Bornstein, N., & Svane, S. (2011). *Heuristics in Generic Planning using PDDL.* (Doctoral dissertation). Technical University of Denmark, Lyngby, Denmark.

Buneman, P., & Levy, L. (1980). The towers of Hanoi problem. *Information Processing Letters, 10*(4), 243–244. doi:10.1016/0020-0190(80)90150-7

Cull, P., Merrill, L., & Van, T. (2013). A Tale of Two Puzzles: Towers of Hanoi and Spin-Out. *Information and Media Technologies, 8*(3), 695–709.

Havur, G., Haspalamutgil, K., Palaz, C., Erdem, E., & Patoglu, V. (2013). A case study on the Tower of Hanoi challenge: Representation, reasoning and execution. In *Robotics and Automation (ICRA), 2013 IEEE International Conference on* (pp. 4552-4559). IEEE.

Hempel, L., & Schmiedel, R. (2006). Pile problems and their Mathematical Treatment. In S. T. Morlock (Ed.), Perspectives in Operations Research (pp. 257-276). Deutscher Universitäts-Verlag. doi:10.1007/978-3-8350-9064-4_15

Hernández Aguilar, J., Yip-Otuño, S., Vakhania, N., Ponce Gallegos, J., & Ochoa Ortiz, A. (2015). *Genetic Algorithm Implementation on GPUs to solve Hanoi Towers Game*. Retrieved October 03, 2015, from http://www.micai.org/2013/all-proceedings-2013/Open access/HIS/3.doc

Hofstadter, D. R. (1996). *Metamagical themas: Questing for the essence of mind and pattern*. Basic books.

Jones, M. T. (2010). *Artificial Intelligence A System Approach.* Laxmi Publications, Ltd.

Li, H. (2014). Algorithm Efficiency Research on Hanoi Problem Symmetric Solution. In *International Conference on Mechatronics, Control and Electronic Engineering (MCE 2014)* (pp. 759-762). Atlantis Press. doi:10.2991/mce-14.2014.100

Lu, X., & Dillon, T. (1995). Parallelism for multipeg towers of Hanoi. *Mathematical and Computer Modelling, 21*(3), 3–17. doi:10.1016/0895-7177(94)00210-F

Mishra, B. K., & Vishnoi, S. (n.d.). Towers of Hanoi - an iterative solution for parallel computation. *International Journal of Research Science & Management.* Retrieved November 12, 2015, from http://www.ijrsm.com/issues%20pdf%20file/Archive-2015/July-2015/1.pdf

Scheinerman, E. R. (2000). *Mathematics: A Discrete Introduction.* Brooks.

Shan-Huen, H., & Tsan-Hwan, L. (2012). Heuristic algorithms for container pre-marshalling problems. *Computers & Industrial Engineering, 62*(1), 13–20. doi:10.1016/j.cie.2011.08.010

Troshkin, M. (2014). Doomsday comes: A Nonrecursive Analysis of the Recursive Towers of Hanoi Problem. *Focus (San Francisco, Calif.)*, *95*(2), 10–14.

Yip Ortuño, S., Hernández Aguilar, J. A., & Zavala-Díaz, J. (2015). Implementación de un algoritmo genético para resolver el juego de las torres de Hanói. In Aplicaciones Modernas de Optimización: la experiencia entre cuerpos académicos. Morelos, México: UAEM y Artificio Editores.

Chapter 23
Determining Maximum Load of Passengers and Goods to an Aerotaxi in Southwestern Chihuahua

Alberto Ochoa-Zezzatti
Juarez City University, Mexico

Alfonso Uribe
Juarez City University, Mexico

Eder Fuentes
Juarez City University, Mexico

ABSTRACT

Travel in Chihuahua is a very complicated situation to realize because the distances between land routes is very far because the poor quality of road. The importance of this research is to understand from a Multivariable optimization associated with the path of a group of airplanes associated with an aero taxi company and determine the optimal flight route involve speed, storage and travel resources for determining the cost benefit have partnered with a travel plan, which has as principal basis the orography airstrip restriction, although this problem has been studied on several occasions by the literature failed to establish by supporting ubiquitous computing for interacting with the various values associated with the achievement of the group of airplanes -Cessna 208 Caravan- and their cost-benefit of each issue of the company and comparing their individual trips for the rest of group. There are several factors that can influence in the achievement of a group of Cessna 208 Caravan group for our research we propose to use Bat Algorithm, which has proven to be efficient for the convergence of several issues (artificial bats) when they have such restrictions and obstacles should use this energy to avoid in our case, a gain resource as food which in our case is represented as the use of travel passengers and goods optimally for the duration of a long trip with the uncertainty of not knowing when you have a resupply or reduce this quantity.

DOI: 10.4018/978-1-4666-9779-9.ch023

1. INTRODUCTION

The Cessna 208 Caravan, also known as Cargo master, is a regional jet / turboprop short-range utility manufactured in the United States by the company Cessna. The standard version has 10 places (9 passengers and a pilot), although a subsequent design according to new regulations of the Federal Aviation Administration (FAA) can carry up to 14 passengers. The aircraft is also widely used to make connections in freight services, so that goods arrive at smaller airports are transported to major hubs for distribution as in Figure 1.

The concept of the Cessna 208 appeared in early 1980, the first prototype flew on 8 December 1982. After two years of testing and review, in October 1984, the FAA certified the model for flight. Since then, the Caravan has experienced many evolutions. Hand international logistics company FedEx; Cessna produced first the Cargo master, which was followed by an improved and extended version called Super Cargo master and other passengers called Grand Caravan. Practitioners will be free fall boarding a Cessna 208 in the Dutch island of Texel. Currently Cessna 208B offers different configurations to meet the varied market demand. The core 208 can be supplemented with different types of landing gear and can operate in a variety of terrains. Some adaptations include skis, larger tires for unprepared runways or floats with wheels in the case of the Caravan Amphibian. In cabin seats can be placed or leave room for cargo in various configurations. The standard configuration of airline consists of 4 rows of seats 1-2

Figure 1. A Cessna 208 used to flight with passengers and goods in Southwestern Chihuahua

after two seats in the cockpit. This variant is capable of carrying up to 13 passengers; albeit only lead to 4 longer an operation rentable 1. The cabin can also be configured to a low density of passengers, in combination or alone as a freighter. Some versions include an additional compartment at the bottom to increase the capacity or luggage. In the cockpit, the 208 has standard analog gauges with a modern digital avionics equipped with autopilot and GPS, modern radio and transponder. Cessna currently offers two different packages avionics manufacturers, one of Garmin and other Bendix / King, a subsidiary of Honeywell. Routing problems vehicle (Vehicle Routing Problem - VRP) are actually a broad range of variants and customizations problems. From those that is simplest to some that remain today research as in Barbucha (2013). They generally were trying to figure out the routes of a transportation fleet to service a customer, nowadays including aerial transportation. This type of problem belongs to the combinatorial optimization problems. In the scientific literature, Dantzig and Ramser were the first authors in 1959, when they studied the actual application in the distribution of gasoline for fuel stations. The objective function depends on the type and characteristics of the problem. The most common is to try: minimize the total cost of ownership, minimize total transportation time, minimize the total distance traveled, minimize waiting time, maximize profit, maximize customer service, and minimize the use of vehicles, balance of the resource utilization.

Another considerations to this problem:

- *Design and specification of a hub for a network of high competitiveness flights using a model based on artificial intelligence*

Implement an optimal flight routes connecting southern Chihuahua, based on artificial intelligence. This is carried out by the problems in areas governed peoples of Chihuahua, aviando the need for a distribution of runways and determine the center or central distribution aeropista this through flight routes in the aircraft. In communities in southern Chihuahua the problem of air transport between the communities arises even if required to travel to the capital for an emergency either disease so the Aerotaxi basically would Cessna that have flight paths is implemented already established and the optimal connection between each airstrip to avoid the delay and make travel direct to the destination, there are plans of the regional central axis Airport at Hidalgo del Parral in Chihuahua (see location in Figure 2)..

Connections carried out between the communities of Madera, Batopilas, Temoris, Balleza Morelos and in order to implement a central axis located in the town of Parral Hidalgo is showed in Figure 3.

The algorithm also called Dijkstra shortest path algorithm is an algorithm for determining the shortest route given source vertex to the other vertices in a weighted graph in each edge. This will serve to determine the shortest flight route, this to make optimal fuel from the aircraft and minimize flight time and make it effective for passengers as the proposed in Figure 4.

The distribution of flight paths can be also just using graph theory called graph coloring that would be the area to be covered by each Cessna 208 in each community. This would be envisaged as a likely option because aside from telling us the shortest route tells us the area that can be covered (see figure 5).

The aircraft used by these communities would be a Cessna 208 with a capacity between 8-10 passengers, which is one of the best performing within fuel efficiency and speed for transportation (see figure 6).

The Cessna 208 is a short-range regional aircraft turboprop that meets the expectations routes to cover the landing strips of wood, Batopilas, Temoris, Balleza Morelos and all these with connection to the airport in Hidalgo del Parral, estimated to last approximately 45 minutes (see figure 7).

Figure 2. Airport at Hidalgo del Parral located in Cyan color

Figure 3. Aeropistas' communities and corresponding Hub

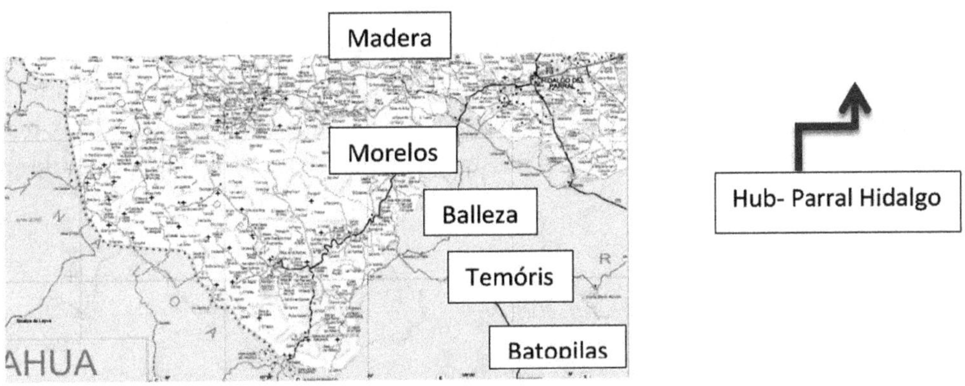

Figure 4. Distribution routes in a graph (Dijkstra algorithm)

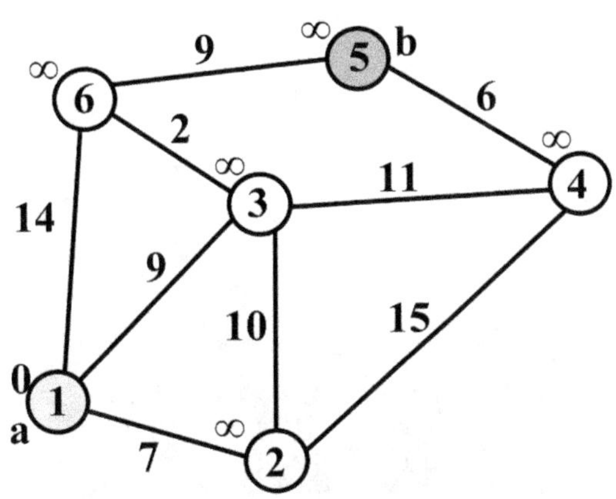

Figure 5. Graph coloring used to determine adjacency in a graph

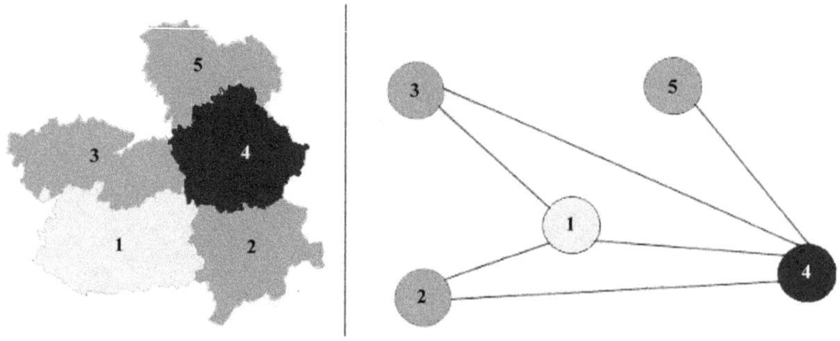

The planes that travel the state of Chihuahua in time and date will not cause any problems regarding the accumulation of air traffic in that area so it is contemplating that. It is determining the departures and arrivals of each Cessna aircraft through logistics (see figure 8).

- *Specifying the Relationship Terrestrial Optimal Distance with Respect to an Airstrip in Southwestern Chihuahua to Reduce Marginality*

Many of the villages in southwestern Chihuahua state have no access to medical services in the area and some of them if they are taking more than 1 hour to reach the destination by the same geography of the state, which can be a time vital. With the latest increase of airports in Chihuahua near the villages gives great opportunity to travel to major cities much faster, either by distance or by the same geography, by reason of being a direct flight. What needs to be solved, and it is the purpose of this research is to generate efficient ways that might bring as many passengers who need services not available and with the

Figure 6. Design and simulation of a Cessna 208

Figure 7. Views fuselage dimensions associated with Cessna 208

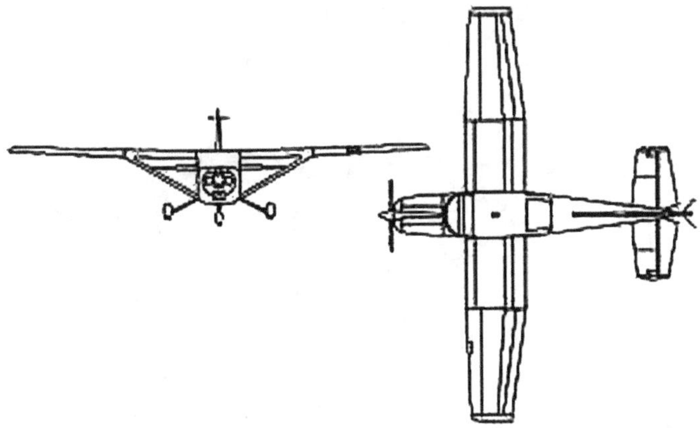

Figure 8. Flightradar application in real time

shortest time made, to the nearest airports for immediate departure. More specifically, the study will be made in the areas of the municipalities of Belisario Dominguez, Carichi, Guachochi, Nonoava (Figure 9).

Creating Points

To create the best possible routes by land will have to make three important points: Create zones that correspond to each airport, people set containing at least 50 people in it and eliminate people who have

Figure 9. Implement the study regions

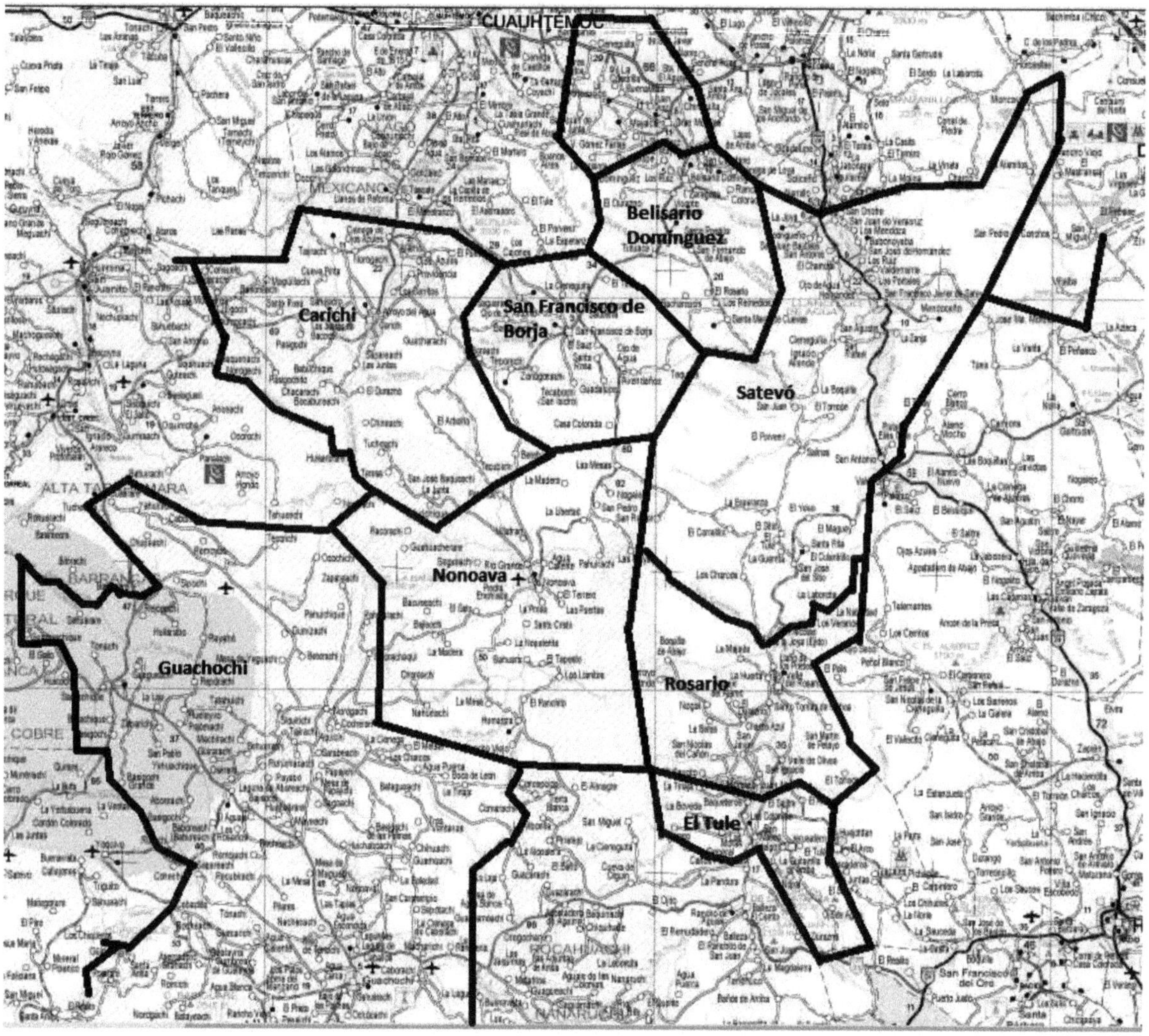

one road (road is used as input and output at a time). In establishing these points, Dijkstra's algorithm which is to establish the shortest and most efficient to review each point worth weight in each path path was implemented.

To establish such zones, according to the map, each airstrip will be assigned an area around 35km to cover the best amount of people and avoid areas "unconnected" with the airstrip. When establishing zones, you can make a better analysis if necessary orders to cross from one area to another and have to use the nearest airstrip last town to visit are taken. If it contains a request outside the established areas, the passenger will have to approach a village within the area or need to go directly to the nearest city for its location and more feasible to go directly to the city (see figure 10).

For the creation of the important points to implement the algorithm, you have to remove a lot of people to reduce the cost of using gasoline and be as profitable time, especially if it is an emergency. Whereby a minimum limit of 50 persons per village, based on a survey earlier in the INEGI statistics for inhabitant per location will be established. By combining the results of the towns within the zones and the results of elimination by residents, significant points shall be established for the algorithm.

Determine Charges

The charges are the distances that require conveys travel from the airstrip, to the point where there is demand and return to the airstrip of origin. What you have to take into account a factor: the rise, which will cause the vehicle height change and help (with decreasing elevation) or give more load (with increasing elevation) (see figure 11).

To simplify the determination of the load with altitude by this is considering the altitude start and final altitude and each way between two points is considered. A simple formulation was used to take the load:

$$c = r_{horizontal} * \left(1 + \frac{h_f - h_i}{r_{horizontal}}\right)$$

Figure 10. Zones for each Aeropista in our research

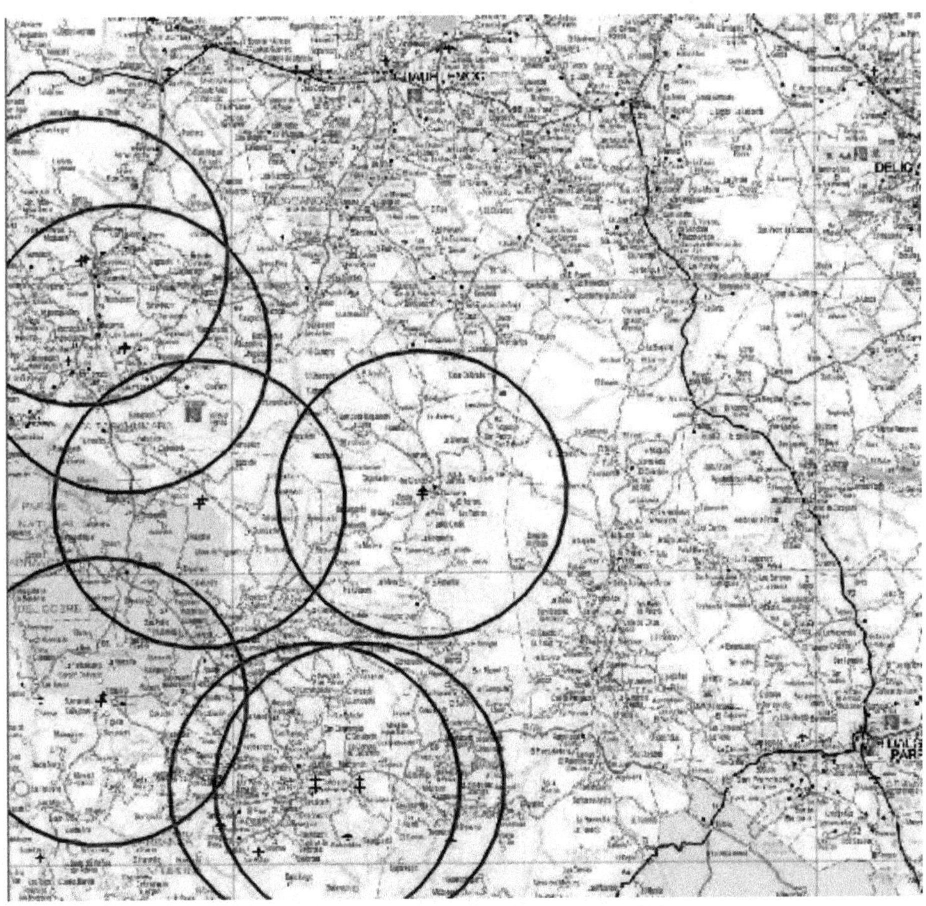

Figure 11. Topography of Chihuahua

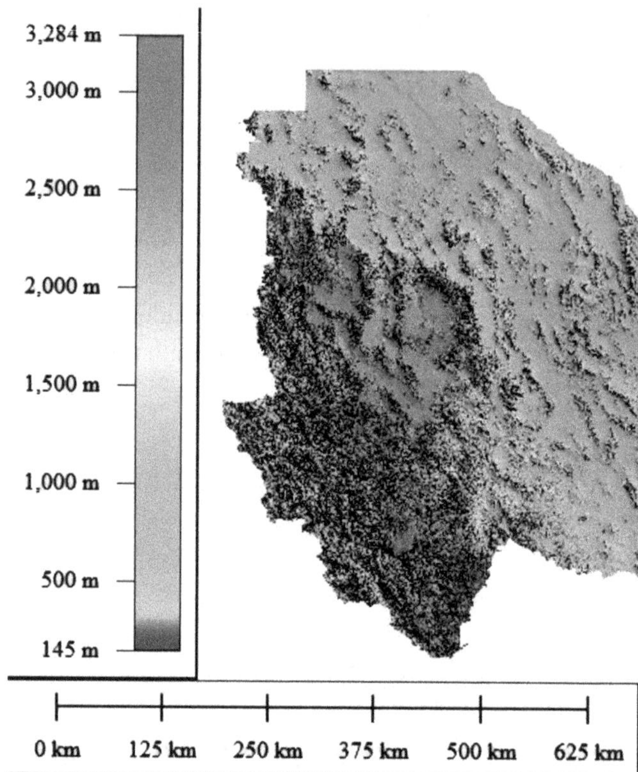

where $r_{horizontal}$ is the horizontal path, h_f is the fnal altitude and h_i is the altitude at first time. This formulation is shown in Figure 12. If the final altitude is increased if the load increases and the load decreases is smaller.

2. PROJECT DEVELOPMENT

To do this research project was developed by dividing into three sections which are modules of application development, implementation of the server and the intelligent module associated with Bat Algorithm and Data Mining. Android is the operating system that is growing in to Streak 5 from Dell, for this reason we select this mobile dispositive along with other manufacturers are propelling the Latin

Figure 12. Graphically representation associated with Charge

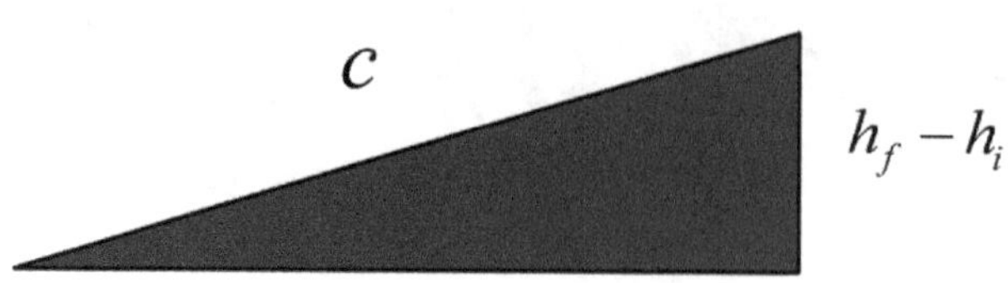

American landing on Android with inexpensive equipment, and on the other hand, some complain about the fragmentation of the platform due to the different versions. Android is free software, so any developer can download the SDK (development kit) that contains your API (Alejandro, 2011). This research tries to improve group travel related with Recreational Vehicles in Chihuahua where 7500 people conforms the Caravan Range Community.

Components of the Application

Bat Algorithm

Bats are fascinating animals and their advanced capabilities of echolocation have attracted attention of researchers from different fields, we propose Artificial bat algorithms as in Figure 13. Echolocation works as a type of sonar: bats, mainly micro-bats, emit a loud and short pulse of sound, wait it hits into an object and, after a fraction of time, the echo returns back to their ears (Griffin, Webster, & Michael, 1960). Thus, bats can compute how far they are from an object (Metzner, 1991). In addition, this amazing orientation mechanism makes bats being able to distinguish the difference between an obstacle and a prey, allowing them to hunt even in complete darkness (Schnitzler & Kalko, 2001).

Based on the behavior of the bats, Yang (2011) has developed a new and interesting meta-heuristic optimization technique called Bat Algorithm. Such technique has been developed to behave as a band of bats tracking prey/foods using their capability of echolocation. In order to model this algorithm, Yang (2011) has idealized some rules, as follows:

Figure 13. Representation of Artificial bats proposed in Bat Algorithm featured obstacles and food (prey)

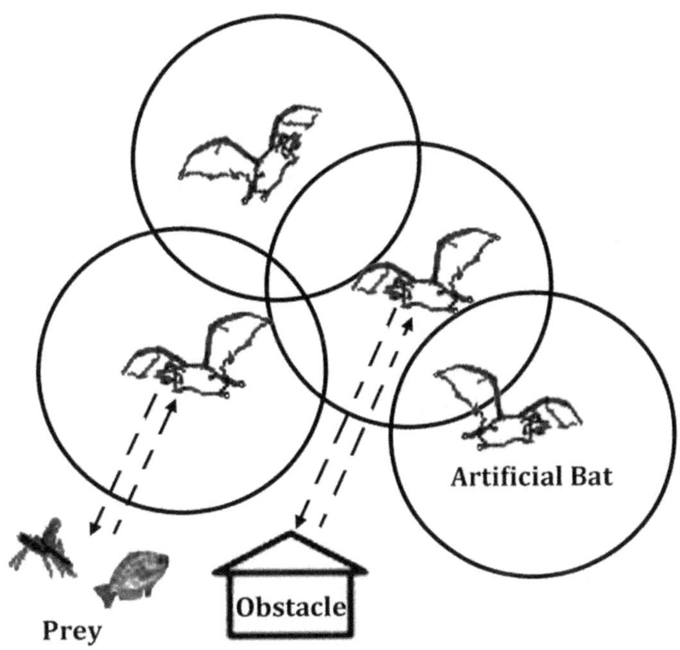

1. All bats use echolocation to sense distance, and they also "know" the difference between food/prey and background barriers in some magical way;
2. A bat bi *fly* randomly with velocity vi *at* position xi *with* a fixed frequency $f$min, *va*rying wavelength λ and l*o*udness $A0$ to se*ar*ch for prey. They can automatically adjust the wavelength (or frequency) of their emitted pulses and adjust the rate of pulse emission $r \in [0, 1]$, *d*epending on the proximity of their target;
3. Although the loudness can vary in many ways, Yang (2011) assumes that the loudness varies from a large (positive) $A0$ to a m*in*imum constant value $Amin$.

Alg*orithm 1* presents the Bat Algorithm (adapted from Yang, 2011).

Firstly, the initial position xi, velocit*y* vi and freque*ncy* fi are initiali*zed* for each bat bi. For each time *step t*, being T the maxi*mum* number *o*f iterations, the movement of the virtual bats is given by updating their velocity and position using Equations 1, 2 and 3, as follows:

$$fi = f\text{min} + (f\text{min} - fmax), \tag{1}$$

$$vji\ (t) = vji\ (t-1) + [\hat{\ }xj - xji\ (t-1)]i, \tag{2}$$

$$xji\ (t) = xji\ (t-1) + vji\ (t), \tag{3}$$

where β *d*enot*es a* r*andom*ly gener*ated* num*b*er within the *in*terval [0, 1]. Recall that $xji\ (t)$ denotes the value of decision variable j *for bat i a*t time step t. The result of fi (Equation *1*) is used *t*o control the *pa*ce and range of *the* movement of the bats. The variable $\hat{\ }xj$ represents the current global best location (solution) f*or d*ecision variable j, which is achieved comparing all the solutions provided by *the m* bats. In order to improve the variability of the possible so*lu*tions, Yang (2011) has proposed to employ random walks. Primarily, one solution is selected among the current best solutions, and then the random

Algorithm 1. Bat Algorithm

Objective function (x), $x = (x1, ..., xn)$.
Initialize the bat population xi and vi, $i = 1, 2, ...,$.
Define pulse frequency fi at xi, $\forall i = 1, 2, \ldots, m$.
Initialize pulse rates ri and the loudness Ai, $i = 1, 2, \ldots, m$.
1. While $t < T$
2. For each bat bi, do
3. Generate new solutions through Equations (1), (2) and (3).
4. If rand > ri, then
5. Select a solution among the best solutions.
6. Generate a local solution around the best solution.
7. If rand < Ai and (xi) < f(^x), then
8. Accept the new solutions.
9. Increase ri and reduce Ai.
10. Rank the bats and find the current best ^x.

walk is applied in order to generate a new solution for each bat that accepts the condition in Line 5 of Algorithm 1:

$xnew = xold + \epsilon(t)$, (4)

in which $A(t)$ stands for the average *loudness of all the* bats *at* time t, and $\epsilon \in [-1, 1]$ attempts to the direction and strength of the random w*a*lk. For *each iteration* of the algorithm, the loudness Ai and the emission pulse rate ri are updated, as follows:

$A(t + 1) = \alpha Ai(t)$ (5)

and

$r(t + 1) = ri(0)[1 - exp(-\gamma t)]$, (6)

where α and γ *are* ad-hoc *constants*. At the *first* step of *the* algori*thm, the emiss*ion rate $r(0)$ *and* the *l*oud-ness $Ai(0)$ are often randomly chosen. Generally, $A(0) \in [1, 2]$ and $ri(0) \in [0, 1]$ [3]. Bat Algori*thm is* very different from PSO Algorithm becaus*e* spec*i*fy better ca*sual* mino*r* variations when is affected by exogenous events (Yang, 2011). In addition we compare the results of another novel research as Wolf Search Algorithm (Glass, Vallipuram, & Portmann, 2010) and ideas from a proposal of Okapi Algorithm which try to understand the mimetically concept in addition of velocity and support at the rest of its herd.

3. IMPLEMENTATION OF AN INTELLIGENT APPLICATION

When designing an interface for mobile devices has to take into account that the space is very small screen, plus there are many resolutions and screen sizes so it is necessary to design an interface that suits most devices. This module explains how to work with different layout provided by the Android API. The programming interface is through XML. Obtaining the geographical position of a device can be made by different suppliers; the most commonly used in this project through GPS and using access points (Wi-Fi) nearby, and perform the same action but differ in some as accuracy, speed and resource consumption. Data Server Communication is the module most important because it allows communication with the server, allowing you to send the GPS position obtained by receiving the processed image and map of our location, thus showing the outcome of your application that is the indicator of insecurity. To communicate to a server requires a HTTP client which can send parameters and to establish a connection using TCP / IP, client for HTTP, can access to any server or service as this is able to get response from the server and interpreted by a stream of data. The Android SDK has two classes with which we can achieve this, HttpClient and HttpPost. With the class HttpClient is done to connect to a remote server, it needs HttpPost class will have the URI or URL of the remote server. This method receives a URL as a parameter and using classes HttpPost HttpClient and the result is obtained and received from the server, in this specific case is only text, which can be JSON or XML format. Here, the server responds with a JSON object which would give the indicator is then used to create the map, as is shown in the Figure 14.

For the construction of the polygon that indicates the rate of incidents in a certain radius of the current position is not possible to create it using the GPS coordinates that yields, as these are specified in "degrees" and requires the unit to convert to meters. For this you need to know how an arc equals the terrestrial sphere, which depends on the place on earth where it is located and the address where you are, the simplest case is to measure an arc in Equator, considering that the earth is 3670 km radius, the perimeter of serious Equator radio 2, which would be equal to 40.024 miles. With this you can get a relationship that would be as follows. If 360 degrees is 40.024 miles then a degree is 111,000.18 miles, this relationship can add and subtract yards to the position, as shown in the figure 15, this process related

Figure 14. Intelligent Tool recommend a group travel associated with limited resources and optimize energy (oil), time and comfort

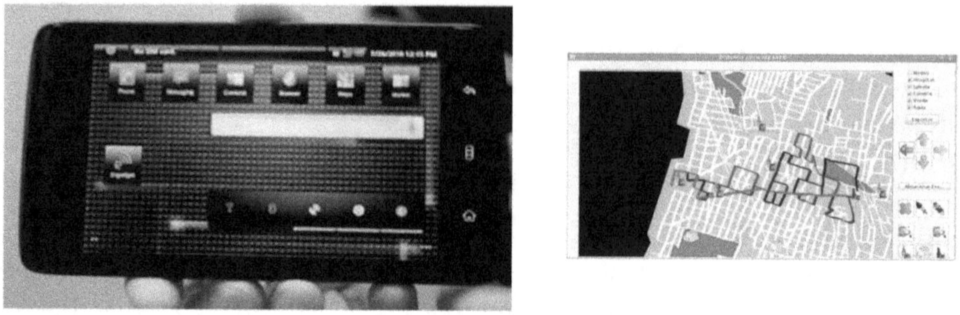

a specific point of interest in our case the National Park of Basaseachi with the most biggest waterfall in Chihuahua State, the information is display on the mobile device recommend inclusively clothes more adequate with the weather by the date of the travel.

For the preparation of graphics, we propose use a class supported with Bat Algorithm supported which facilitates to manipulation of data to express visually using different types of graphs, as in the figure 4 in where each iteration proposed is related with a generation. To the 6th Iteration a specific

Figure 15. Acquisition of polygon map with the position of the Android application

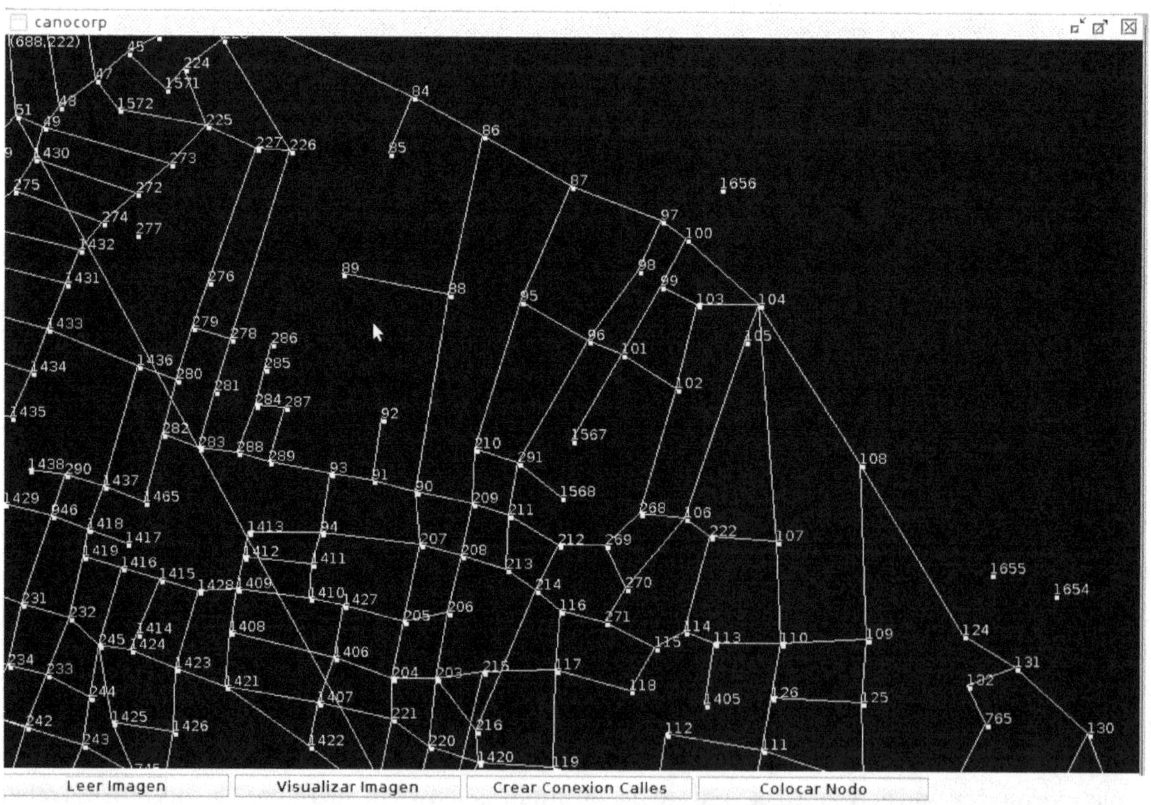

behavior is described to the most possible places where is possible [ad]vance the most of vehicles in the Caravan Range Community.

To implement the application is installed in operating system devices with Android 2.2 or higher, which tests the system in different areas of the three different parks and natural reserve in Chihuahua based on the previously research related with Cultural Algorithms on Urban Transport Cruz, Ochoa, et al., 2010), by answering a questionnaire of seven questions to all users related with the Caravan Range Community have elapsed since installing the application, the questions are to raise awareness of the performance, functionality and usability of the system, the demonstrate use of this application is shown in figure 5. To understand in an adequate way the functionality of this Intelligent Tool, we proposed evaluate our hybrid approach and compare with only data mining analysis and random select activities to protect in the city, we analyze this information based on the unit named "époques" used in Bat Algorithm

Figure 16. Statistics Graphics related with the solution proposed by Bat Algorithm

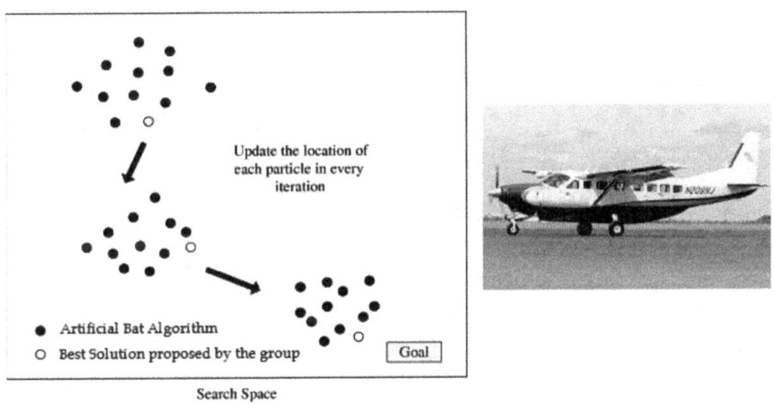

Figure 17. Hybrid Intelligent Application based on Bat Algorithm and Data Mining

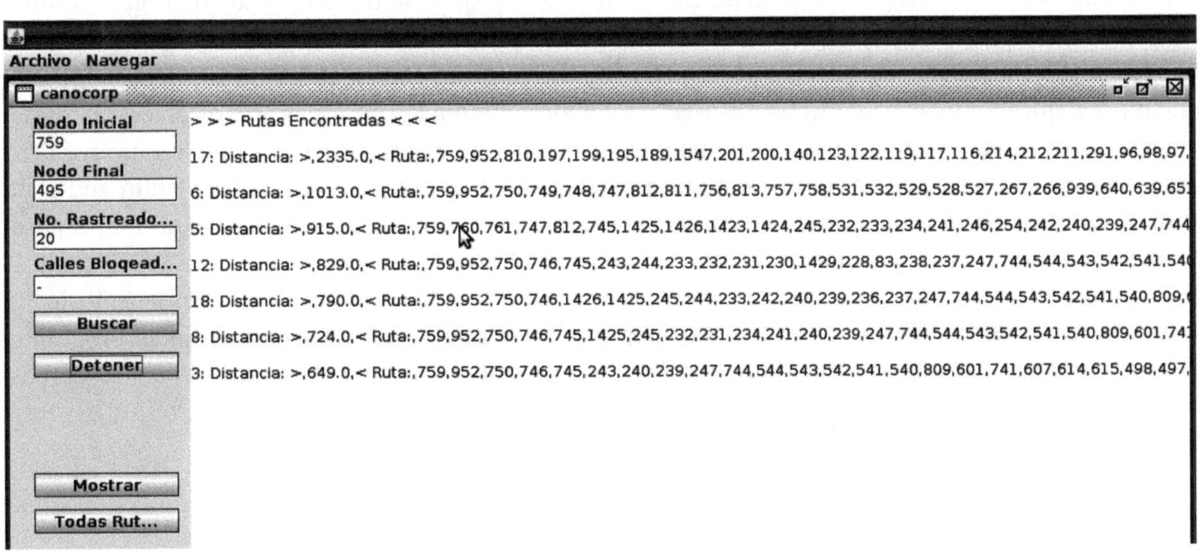

Figure 18. Solutions proposed to best trips associated with the request to travels: (circle) hybrid approach; (light triangle) only using data mining analysis and (dark triangle) using randomly actions to improve the time and quantity of goods

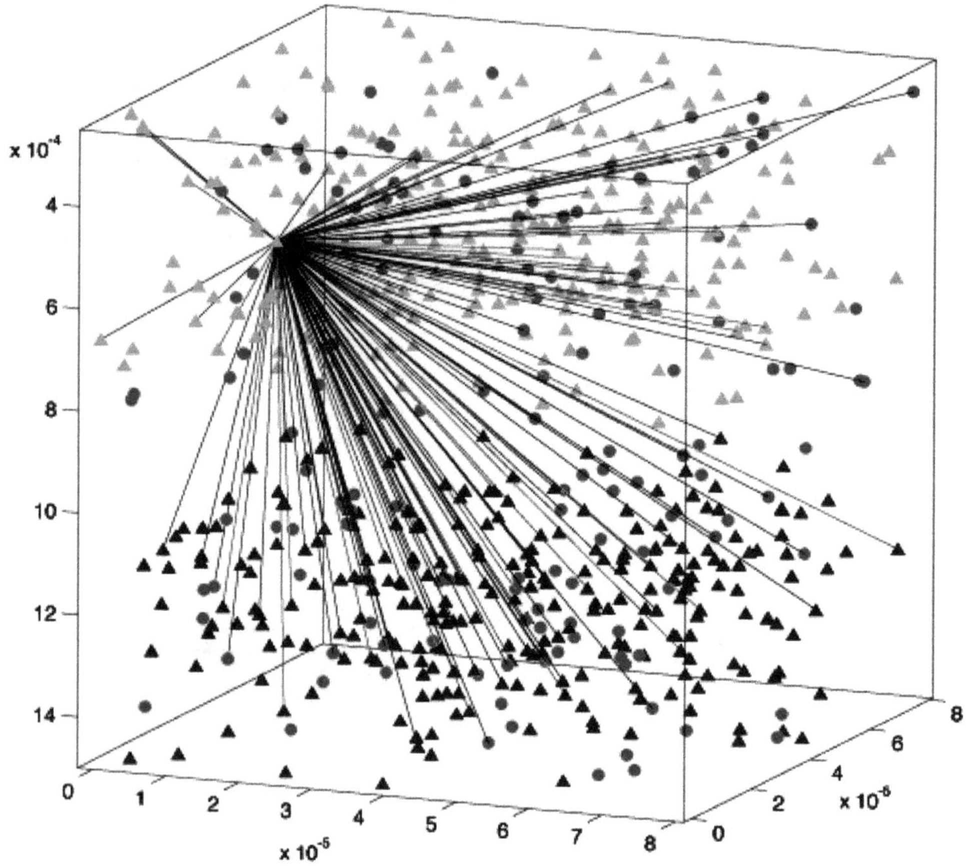

Algorithms, which is a variable time to determine if exist a change in the proposed solution according at different situation of different routes with better use of restricted resources.

We consider different scenarios to analyze during different time, as is possible see in the Figure 6, and apply a questionnaire to a sample of users to decide search a specific number of places to travel, when the users receive information of another past travels (Data Mining Analysis) try to improve their space of solution but when we send solutions amalgam our proposal with a Bat Algorithm and Data Mining was possible determine solutions to improve the resources of the group, the use of our proposal solution improve in 91.72% against a randomly action and 14.56% against only use Data Mining analysis the possibilities of recommend arrive in a specific time to use less energy, these messages permits in the future decrease the possibility of deplete the supply of food and spend time rerouted in an orography terrain with more obstacles and uncertainty of the weather conditions and the use of limited resources.

4. CONCLUSION

With the use of this innovative application combine Bat Algorithm and Data Mining based on a mobile dispositive is possible determine the places where is possible improve a group travel in Chihuahua's natural parks by an alert sent to a mobile device with GPS, providing statistical information through a Web server that returns the level of insecurity in the area consulted (Glass, Vallipuram; & Portmann, 2010) a specific factor decisive in this part of Mexico. The future research will be to improve the visual representation of group travel to a social networking to this we proposed an Intelligent Dyoram –An Intelligent display in 3D used on Social Networking- with real on time information of each one of their integrants and establishment a travel including save money and time and enjoy with the rest of Community. The most important contribution is prevent spend time and money in this kind of travels because drive in a unsafe place at night on a wrong time and suffer an assault, our future research is adequate the information to actualize from the central server of security of the State government, to the users, considering that the number of people travel at this parks during this last year was of 57000 people, because the Ecotourism is very high as recreational activity, this innovative application is possible to use in another natural parks in Wales, Sardinia or West Australia –This society has 87000 people living in Caravan Range Communities and realize an accumulate of 97500 travels during one year- with similar conditions of travel by their respective Caravan Range Communities, this Intelligent Tool will be used by different kind of people whom requires travel together. In addition this application will be used as Recommender System when travel to another cities or places in different societies (Souffriau, Maervoet, Vansteenwegen, Vanden Berghe, & Van Oudheusden, 2009) and explain different scenarios according time, limited resources and location. Another field topic will be benefited with more adequate organization is Logistics of product or service as in Ochoa, Garcí, & Yañez, (2011) which describes the use of Cultural Algorithms to improve a Logistics networking associated with the deliveries of a bottle product.

ACKNOWLEDGMENT

The authors were supported with funds from Social Research Center in Juarez City University and used data from a Caravan Range Community in Chihuahua whom permits compare the simulation with real travels realized by them. The Dell mobile device was bought with funds from a Promep project supported by SEP.

REFERENCES

Alejandro, A. R. (2011). "Estudio del desarrollo de aplicaciones RA para Android," Trabajo de fin de Car*rera. Catalunya: España.*

Barbucha. (2013). Experimental Study of the Population Parameters Settings in Cooperative Multi-agent System Solving Instances of the VRP. T. Computational Collective Intelligence, 9, 1-28.

Cruz, L., & Ochoa, A. *et al.. (2010). A Cultural Algorithm for th*e Urban Public Transportation. HAIS, 2010, 135–142.

Glass, S., Vallipuram, M., & Portmann, M. (2010). *The* In*sec*urity of Time-of-Arrival Distance-Ranging in IEEE 802.11 Wireless Networks. ICDS Workshops, 2010, 227–233.

Griffin, D. R., Webster, F. A., & Mi*chael, C. R. (*1*960).* The echolocation of flying insects by bats. Animal Behaviour, 8(34), 141–154. doi:10.1016/0003-3472(60)90022-1

*Metzner, W. (*1991). Echolocation behaviour in bats. Science Progress Edinburgh, 75(298), 453–465.

Ochoa, A., Garc*í, Y., & Yañez, J. (2011).* L*o*gistics Optimization Service Improved with Artificial Intelligence. Soft Computing for Intelligent Control and Mobile Robotics, 2*011, 57–65.*

Schnitzler, H.-U., & Kalko, E. K. V. (2001, J*uly).* Echolocation by insect-eating bats. Bioscience, 51(7), 557–569. doi:10.1641/0006-3568(2001)051*[0557:EBIE*B]2.0.CO;2

Souffriau, W., Maervoet, J., Vansteenwegen, P., Vanden Berghe, G., & Van Oudheusden, D. (2009). A Mobile Tourist Decision Support System for Small Footprint Devices. IWANN, 2009(1), 1248–1255.

Yang, X.-S. (2011). Bat a*lgori*th*m for* multi-objective optimisation. International Journal of Bio-inspired Computation, 3*(5), 267–274. doi:10.1504/IJBIC.2011.042259*

Chapter 24
Use of GVRP as a Model of Two Specific Real World Problems and Its Bioinspired Solution

Jorge Rodas
Universidad Autónoma de Ciudad Juárez, Mexico

Alberto Ochoa-Zezzatti
Universidad Autónoma de Ciudad Juárez, Mexico

Daniel Azpeitia
Universidad Autónoma de Ciudad Juárez, Mexico

Raymundo Camarena
Universidad Autónoma de Ciudad Juárez, Mexico

Tania Olivier
Universidad Autónoma de Ciudad Juárez, Mexico

ABSTRACT

The aim of this chapter is about the inclusion of real world scenarios, viewed as a Generalized Vehicle Routing Problem (GVRP) model problem, and treated by bio inspired algorithms in order to find optimum routing of product delivery. GVRP is the generalization of the classical Vehicle Routing Problem (VRP) that is well known NP-hard as generalized combinatorial optimization problem with a number of real world applications and a variety of different versions. Due to its complexity, large instances of VRP are hard to solve using exact methods. Thus a solution by a soft computing technique is desired. From a methodological standpoint, the chapter includes four bio inspired algorithms, ant colony optimization and firefly. From an application standpoint, several factors of the generalized vehicle routing are considered from a real world scenario.

INTRODUCTION

This chapter is interested in biologically inspired computing, a branch of natural computing (de Castro 2006) where algorithms inspired from nature are developed to solve highly complex problems, in particular problems that cannot be addressed in a satisfactory way with traditional approaches. Under this paradigm, algorithmic models of processes observed in nature are developed and implemented on computers to explore solution spaces.

In the scientific literature, there is a growing interest in the exchange of ideas between natural systems and computational systems. Bio-inspired algorithms are representative of this trend and have been

DOI: 10.4018/978-1-4666-9779-9.ch024

applied to a large variety of problems, including optimization problems. Among them, Vehicle Routing Problems (VRP) have generated a lot of attention, because they are inherently complex and hold a central place in real-world distribution activities.

Dantzig and Ramser (1959) proposed VRP for the first time with the objective to design optimum routing of a fleet of gasoline delivery trucks between a depot and a large number of service stations. VRP is a generic name given to a whole class of problems in which a set of routes for a fleet of vehicles based at one or several depots must be determined for a number of geographically dispersed customers (Laporte et al., 1989, 2002; Calvete et al., 2004; Baldacci et al., 2010). A depot, a set of geographically dispersed customers with demands, and a set of vehicles define it with capacity. In addition, customers must be visited exactly once and the total customer demand of a route must not exceed the vehicle capacity (Clarke & Wright, 1964; Laporte et al., 2002; Novoa et al., 2006). Its objective is to find a collection of K simple ways each corresponding to a vehicle route with minimum cost, distance or time.

To achieve this objective, high numbered of research has been conducted on VRP to design and develop different models and optimization methods. VRP models are differed by inclusion or exclusion of different restrictions. The commonly used restrictions on capacity of vehicle, total time, windows time, precedence relations between pairs of destinations, the number of depots (Laporte, 1992; Madsen et al., 1995; Paolo & Daniele, 2002; Chenghua & Xiaofeng, 2011) and others.

It is common to find the well-known travelling salesman problem (TSP), a canonical combinatorial optimization problem, widely studied in the scientific literature mentioned above. Given a graph with a number of vertices and a cost associated with each arc, the TSP looks for a least cost tour that visits each vertex exactly once (where the cost of a tour is the sum of its arc costs) (Lawler et al., 1985). In vehicle routing problems, side constraints are introduced to limit the number of vertices that can be visited with a single tour, thus leading to solutions that cover all vertices with different routes that start from and end at a particular vertex, called the depot. These problems are used to model various real-world applications. In the manufacturing sector, for example, they model transportation activities within the supply chain, that is, the movement of goods from suppliers to factories and then to customers. In the service sector, they include activities such as mail delivery, garbage collection, and public transportation, in another. (Potvin, 2009).

The chapter is organized as follows. Section 2 introduces a brief overview of the general Vehicle Routing Problem (VRP). Then, Sections 3 to 6 introduce the selected four bio-inspired algorithms, namely Ant Colony Optimization (ACO) and Firefly (FF, in this order, and describe their application to two real world logistic problems treated as VRPs. Section 7 results obtained from the 4 algorithms application to real world problems are discussed. Concluding remarks follow in Section 8.

VEHICLE ROUTING

The aim of vehicle routing is to determine optimal collection or delivery routes for a fleet of vehicles in a transportation network (Laporte, 1992, 1997; Toth & Vigo, 2001). These NP-hard problems generalize the classical Traveling Salesman Problem (TSP) by requiring both an assignment of vertices to vehicles and a sequencing of these vertices within each vehicle route to obtain a solution. A large number of variants are reported in the literature, depending on the objective function to be optimized and the types of constraints to be satisfied. These variants are reviewed in the following.

Figure 1. Transport model with nodes and arcs

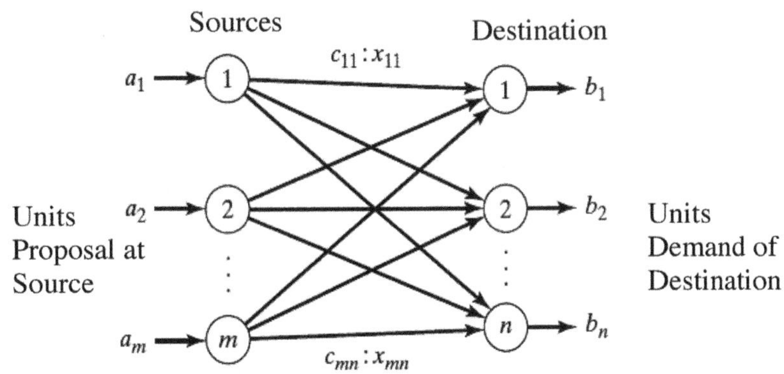

TRANSPORT MODEL AND ITS VARIANTS

The transport model is a special class of linear programming that has to do with transport an item from their sources (factories) to their destinations (depot). The objective is to determine the transportation program that minimizes the total cost of transport and while meeting the limits of supply and demand. The model assumes that the transportation cost is proportional to the number of units shipped in certain route.

The general problem is represented in the network of Figure 1. There are m sources and n destinations, each source and each destination represented by a node. The arcs represent the routes between sources and destinations. The arc (i, j) connecting the source to the destination i j conducts two kinds of information: c_{ij} transportation costs per unit, and x_{ij} amount transported. The amount proposal at source i is a_i and the amount of demand in the destination j is b_j. The objective of the model is to determine the unknowns x_{ij} which minimize the total cost of transport, and at the same time meet the demand and supply constraints (Taha, 2004).

TRAVELLING SALESMAN PROBLEM

In the Traveling Salesman Problem (TSP) provides one vehicle you should visit all customers in a single route and minimum cost. There is usually a depot (and if he had not distinguished from customers), no demand associated with customers and there are no time restrictions. The problem can be formulated as:

$$\min \sum_{k \in K} \sum_{(i,j) \in E} c_{ij}^k x_{ij}^k \tag{1.1}$$

$$\text{s.a.} \sum_{j \in \Delta+(i)} x_{ij} = 1$$

$$\forall i \in V \tag{1.2}$$

$$\forall j \in V$$

$$\sum_{j \in \Delta - j} x_{ij} = 1$$

$$\sum_{\substack{i \in S, j \in \Delta + (i)/S \\ X_{ij} \in \{0,1\}}} x_{ij} \geq 1 \tag{1.3}$$

$$\forall S \subset V$$

$$\forall (i, j) \in E \tag{1.4}$$

This formulation was proposed by Dantzig, Fulkerson and Johnson (Dantzig, G., Fulkerson, D., Johnson, 1954).The binary variables x_{ij} indicate if the arc (i, j) is used in the solution. The objective function (1.1) states that the total cost of solution is the sum of the costs of the arcs used. Restrictions 1.2 and 1.3 indicate that the route must arrive and leave each node exactly once. Finally, restrictions 1.4 delete restrictions are called sub-tours and indicate that every subset of nodes S must be left at least once.

VEHICLE ROUTING PROBLEM

VRP is an important transportation problem that seeks an optimal solution for constructing delivery routes given a depot, a fleet of vehicles and a some number of geographically dispersed customers, each having a demand that must be fulfilled. The problem also incorporates characteristics such as travel times and/or distances as well as side constraints such as a maximum vehicle load. This problem is important due to both its widespread application and its complexity in solving (Spliet & Desaulniers, 2015).

The VRP is a problem that can be formulated mathematically from a directed graph G (V, E). The problem considers a set of vertices V = $\{_{v0, v1, ..., vn}\}$ where v_0 represents the warehouse or store, while v_i (i = 1, ..., n) represents each of the n customers (or nodes) to be visited. The set of edges is given by: A = $\{ (v_i; v_j) / vi, v_i, v_j \in V; i \neq j \}$. Meanwhile C = $\{c_{ij}\}$ is a matrix of distances or negative costs between each pair of vertices v_i and v_j including deposit and customers. When $c_{ij} = c_{ji}$ for each $(v_i, v_j) \in A$ the problem is said to be symmetrical and then replace it as a set of vertices A by the set E = $\{(v_i, v_j) / v_i, v_j \in V; i < j\}$. The vector d represents the demands of customers and where d_i is the amount of goods required by the client v_i. The number of vehicles is given by m and assume that they are all identical, with a Q per vehicle capacity. Each vehicle i was assigned to a route $_{Ri}$. Without loss of generality, and for simplicity, the i_{cj} cost needed to go v_i to v_j is the time tij, considered also equal to the distance (constant speed and unit for vehicles) (Antonio & Santana, 2011).

Sometimes a fixed cost associated with each vehicle. The goal is then directed to minimize the weighted sum of fixed costs and travel costs. A maximum distance or time constraint can also be considered, in addition to the capacity constraint, for each vehicle route (Paolo & Daniele, 2002).

Summarizing, the VRP is a method that aims to optimize the resources needed for the delivery of any kind of products to the consumer or customer, or the provision of services. This method combines multiple factors such as:

- **Depot:** The place where production is centred, where the distribution of the product begins or where it starts and stops the service.

- **Customers:** They are the input and output distribution and the establishment of product demand.
- **Fleet:** Refers to vehicles that are available, capacity and features.

Planning is an important factor to generate the optimal route. The Figure 2 shows an example of routing.

CAPABILITY VEHICLE ROUTING PROBLEM

The capability vehicle routing problem (CVRP) (Toth, 2002) is one of the basic vehicle routing problem, which is that every customer must be sent a certain amount of goods, starting already known from a single tank . To distribute it has a certain number of vehicles with some ability to transport product, which depart and arrive at a single tank. Each vehicle must visit exactly to a single client, each client must also be visited by one vehicle and the sum of the demands of customers visited by a vehicle must not exceed

Figure 2. Describes vehicle routes that visit every city exactly once and that obeys user-specified vehicle volume and distance limits. This is a Screen Shot from http://coral.ie.lehigh.edu/~larry/software/vrp-solver/

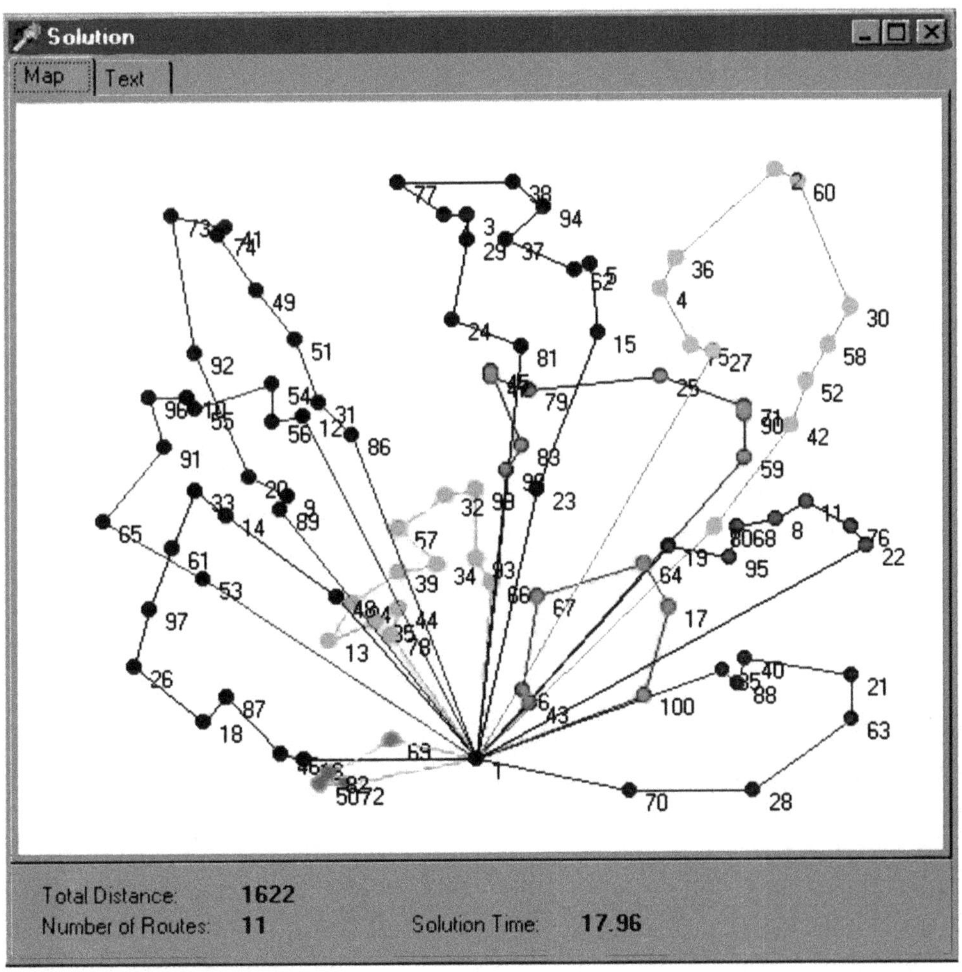

the capacity. The aim of this type of problem is usually to minimize the total cost, where such cost can be seen as a function of time or the total distance traveled, or to minimize the number of vehicles used.

VEHICLE ROUTING PROBLEM TIME WINDOWS

In this variant of the problem, in addition to capacity, each customer $i \in V \setminus \{0\}$ has an associated time window $[e_i, l_i]$ establishing a service schedule allowed for a vehicle arrives at the time and service delay if. If (i, j) is an arc of the solution and t_i and t_j are the hours of arrival in the i and j customers time windows mean that you necessarily must be fulfilled and $t_j \leq l_i$ and $t_j \leq l_j$. On the other hand, if $t_i < e_i$, then the vehicle must wait until the customer "open" and necessarily $t_j = e_i + s_i + t_{ij}$.

Using nodes 0 and n + 1 to represent the deposit and the set K to represent the vehicle (not the types of vehicles as in the previous section), the problem is formulated for a fleet of vehicles possibly heterogeneous, as follows (Cordeau, Desaulniers, Desrosiers, Solomon, & Soumis, 1999)

$$\min \sum_{k \in K} \sum_{(i,j) \in E} c_{ij}^k x_{ij}^k \tag{1.5}$$

$$\text{s.a.} \sum_{k \in K} \sum_{j \in \Delta-(i)} x_{ij}^k = 1 \; \forall i \in V \setminus \{0, n+1\} \tag{1.6}$$

$$\sum_{j \in \Delta+(0)} x_{0j}^k = 1 \; \forall k \in K$$

$$\sum_{j \in \Delta+(i)} x_{ij}^k - \sum_{j \in \Delta-(i)} x_{ji}^k = 0 \; \forall k \in K, i \in V \setminus \{0, n+1\} \tag{1.7}$$

$$\sum_{i \in V \setminus \{0,n+1\}} di \sum_{j \in \Delta+(i)} x_{ji}^k \leq q^k \; \forall k \in K \tag{1.8}$$

$$y_j^k - y_i^k \geq s_i + t_{ij}^k - M\left(1 - x_{ij}^k\right) \; \forall i, j \in V \setminus \{0, n+1\}, k \in K \tag{1.9}$$

$$e_i \leq y_i^k \leq l_i \; \forall i \in V \setminus \{0, n+1\}, k \in K \tag{1.10}$$

$$x_{ij}^k \in \{0,1\} \; \forall (i,j) \in E, k \in K \tag{1.11}$$

$$y_i^k \geq 0 \; \forall i \in V \setminus \{0, n+1\}, k \in K$$

The variables x^k_{ij} indicate if the arc (i, j) is traversed by the vehicle k. The variables y^k_i indicate the arrival time customer i when visited by the vehicle k (if the client is not visited by said vehicle the variable value has no meaning). The objective (1.5) function is the total cost of the routes.

The restriction 1.6 indicates that all customers should be visited. Restrictions 1.7 and 1.8 determine that each carrier $k \in K$ travels a path from 0 to n + 1. The capacity of each vehicle is imposed on 1.9. Where M is a constant large enough, the restriction ensures that if a 1.10 vehicle k travels from i to j, j cannot come before $y_i + s_i + t^k_{ij}$, and also act as restrictions deleted by sub-tours. Finally, the limits of the time windows are tax 1.11.

A BRIEF VIEW OF OTHER VEHICLE ROUTING PROBLEMS

Without being exhaustive, it is necessary to discuss the following VRP cases:

- **Time-Dependent:** The travel times are not fixed but rather depend both on the distance between two vertices and the time of the day (e.g., it takes longer to get from one location to another during rush hours). This case is thus aimed at more closely modelling situations observed in the real world.
- **Pick-Up and Delivery Problem (PDP):** In this problem, each customer request corresponds to a pair of vertices. That is, for a given pair of vertices i+ and i−, the demand must be picked up at vertex i+ and delivered at vertex i−. Clearly, both vertices must be in the same vehicle route (pairing constraint) and vertex i+ must be visited before i− (precedence constraint). When time windows are associated with the vertices, the PDPTW is obtained. When people are transported, as in transportation-on- demand services, trip duration constraints must also be taken into account. The latter problem is known as the dial-a-ride problem (DARP).
- **Multiple-Depot Problems:** Where each vehicle can start (end) its route from (to) a different depot.
- **Mixed Fleet and Size Problems:** Where the fleet size must be determined based on different types of vehicles with different characteristics (e.g., capacity).
- **Periodic Vehicle Routing Problems:** Where routes are determined over an horizon that spans a number of periods and where each vertex must be visited at a given frequency within this horizon.
- **Location-Routing Problems:** Where strategic decisions about the location of different facilities (e.g., depots, warehouses) are taken concurrently with the determination of vehicle routes.
- **Inventory Routing** where each vertex has an inventory and delivery routes are determined to replenish the inventory so as to avoid any inventory shortage.
- **Stochastic Vehicle Routing Problems:** Where some data are not known with certainty. For example, the demand might not be known with certainty at a given vertex, but the probability distribution is known and can be used to minimize the expected solution value.
- **Dynamic Vehicle Routing Problems:** Where new information is obtained as the vehicle routes are executed. For example, new service requests occur and must be integrated into the current routes or the travel times suddenly change due to unforeseen events.

There are also a number of vehicle routing problems where arcs must be visited instead of vertices, but these problems are not addressed here (see Dror, 2000; Eiselt et al., 1995). Arc routing problems are useful to model applications where the customers are located along the streets of the transportation network, as in postman delivery routes.

GVRP ON REAL WORLD SCENARIOS

The inclusion of real world scenarios on VRP models resulted in many variants of VRP that evolved from the general VRP model (GVRP). For instance, the most commonly studied variant is VRPTW (Madsen et al., 1995; Cordeau et al., 2001, 2002, 2007; Kenyon & Morton, 2003; Kallehauge, 2008). It is the same problem as VRP but with additional restriction on time called *window time*, which associated a defined time interval wherein the customer, has to be supplied (Balakrishnan, 1993; Taillard et al., 1997; Calvete et al., 2004; Stefan & Natalia, 2009; Xiangyong et al., 2010).

Consideration of time interval at the depot is similar to the case, which is mainly characterised by using vehicles of the same capacity based at a central depot that will be optimally routed to supply customers with known demands knows as CVRP (capacitated vehicle routing problem) with Pick-up and Deliveries and time Windows (CVRPPDTW). There are several variants in the time restriction as: Multiple Depot VRP (MDVRP) (Blanton & Wainwright, 1993; Mingozzi, 2003), Multiple Depot VRP with Time Windows (MDVRPTW), Split Delivery VRP with Time Windows (SDVRPTW) and Periodic VRP with Time Windows (PVRPTW) (Dror and Trudeau, 1989, 1990; Dror et al., 1994)...

In addition to the above variants, more additional side constraints are also added in the problem structures that lead to have more variants of VRP. Some of them are multiple depots (Bertsimas and Van Ryzin, 1990), a paired-vehicle recurs strategy (Ak & Erera, 2007), route duration limit (Christopher, 2010; Erera et al., 2010), single vehicle and limited capacity and priority on customers' visit (Bertsimas, 1991, 1992; Smith et al., 2010). Another strategy to account for the demand uncertainties is to develop a waiting strategy for vehicles (Ghiani et al., 2008) and uncertainty demand with adverse weather conditions (Sumalee et al., 2011).

However, the classical VRP model usually does not capture an important aspect of real world transportation and distribution-logistic problems. This is due to the fact that several of the problem parameters such as demand, time, distance, and others are stochastic by their nature (do not have the certainty when they are present); but they are often oversimplified and treated as deterministic (Ak & Erera, 2007).

BIO-INSPIRED ALGORITHM AND THEIR APPLICATION IN VEHICLE ROUTING PROBLEM

Natural Computing arises as a possible substitute to classical computing in the pursuit of new paradigms that can provide an effective solution to the restrictions of conventional models. Currently, Natural Computing covers a set of models whose common characteristic is the simulation of how nature works, that is, Natural Computing studies how the various laws of nature produce changes in certain systems (from habitats to joint molecules, to living organisms) that can be interpreted as calculation processes on its elements (Pérez J. 2003).

Bio-inspired Computing is a branch of Natural Computing, where bio-inspired algorithms are developed to solve highly complex problems, particularly dealing with problems that cannot be approached adequately by traditional means. Under this paradigm, algorithmic models of processes observed in nature are developed and implemented in computers to explore a wider array of solutions. These new ways of finding solutions to different kinds of problems, remains in strong growth thanks to the increasing interest in exchanging ideas among researchers of Natural Computing and Traditional Computing. Bio-inspired algorithms are representative of these trends and have been applied to a broad range of problems that simulate, to a large degree, the way in which nature deals with optimization problems. Mainly it comes to emulate the natural evolution of species and the response of social systems to outside challenges. Bio-inspired algorithms group a set of heuristics that models natural processes, like the evolution of species and social behavior among insects or animals, to solve complex problems of exploration. There is an extensive field of application such as logistics, design of escape routes and military tactics. As well as structure design such as aircraft wings, communication antennas, cranes, pharmaceutical research, chemical and material compounds, financial markets forecasts, robotics, pattern recognition and data mining, just to name the most common uses and applications as it continues to grow due to the excellent results provided (C.E. Borges 2011).

Different ways to approach VRP problems exists; currently meta-heuristic techniques are high-level strategies to create processes that are less likely to get stuck on local solutions and perform robust searches, these methodologies are integrated in multiple ways, local improvement procedures are subject of constant study due to the excellent results they provide. Within these techniques there is a group of algorithms that are inspired by nature, such as the Ant Colony Optimization Algorithm, bio-inspired algorithms, and artificial neural networks, among others. This chapter approaches the ant colony and firefly algorithms, which they will be explained deeply on their respective segments within this same chapter.

FOOD DELIVERY

The vehicle routing problem (VRP) involves finding a set of routes, starting and ending at a depot, that together cover a set of customers. The objective is to minimize the total distance traveled. We consider the vehicle routing problem with time windows (VRPTW), which is a generalization of the VRP where the service at any customer starts within a given time interval, called a time windows.

The travel begin in a pizzeria in the Ciudad Juárez, Chihuahua, México. The map analyzed is seen in Figure 3.

The points shown in Figure 3 represents the street crossings, this information is very important in the searching of the optimal route.

The firefly algorithm have to search optimal delivery route within 30 minutes after the client called.

DATA TREATMENTS

Firefly algorithm need to read the information of the map. The code was implemented with a matrix with 131 columns and 131 rows.

The point matrix [1,1] is the starting point of the pizzeria.

Figure 4 is an example of a matrix and a network nodes used in the algorithm.

Intersection matrix with zero mean that these points don't have communication.

Figure 3. Representation of model graph on City

Figure 4.

	1	2	3	4
1	0	34	0	0
2	34	0	40	28
3	0	40	0	0
4	0	48	0	0

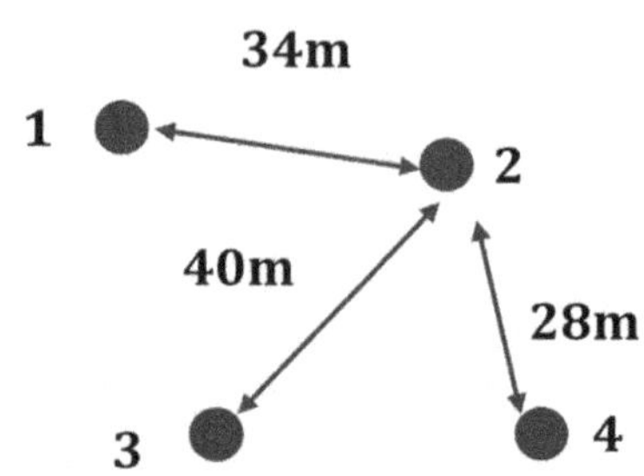

ANT COLONY OPTIMIZATION ALGORITHM

In the Ant Colony Optimization when an ant begin the search for new sources of food ants look around in a random manner, if they find a source then they come back to the nest, laying down pheromone on the ground. The amount of this substance is related to the quantity of the food and its distance from the nest. The next ants will thus search less randomly. Paths where the pheromone is higher attract more ants and for that will continue adding higher amounts of pheromone.

In the ant system we suppose that we have a m number of ants, where each ant is assigned to a starting location on a vertex, then each ant constructs a tour by their movements to a new vertex applying the probabilistic nearest neighbor transition rule. And then the pheromone is updated in each optimal route found. This procedure is repeated until a stopping criterion is met, for example a fixed number of iterations.

The simple pseudo-code is shown below:

```
1. Assign each ant to a starting vertex;
2. Put some initial amount of pheromone on every edge;
3. For T iterations do:
        3.1 For each ant do: constructs a tour;
        3.2 Update the amount of pheromone on each edge;
4. Output the best tour found.
```

When the ant moves to a next vertex the probabilistic transition rule is:

$$p_{ij}^{k} = \frac{\left[\tau_{ij}\right]^{\alpha}\left[\eta_{ij}\right]^{\beta}}{\sum_{l \in N^{k}}\left[\tau_{ilk}\right]^{\alpha}\left[\eta_{il}\right]^{\beta}} \quad \text{if} \quad j \in N^{k}. \tag{1.12}$$

In step 3.2 the pheromone is update after ant has constructed optimal tour with the equation 1.13, the parameter (1- ρ) is the pheromone evaporation to avoid accumulation of pheromone on the edges and favors the exploration of new ways. The equation $\Delta\tau_{ij}^{k}$ is the pheromone deposited by ant k on edge (i,j) (Potvin 2009):

$$\tau_{ij} \leftarrow \left(1-\rho\right)\tau_{ij} + \sum_{k=1}^{m} \Delta\tau_{ij}^{k} \tag{1.13}$$

On the Ant Colony System (ACS), when an ant must decide where to go next, the edge leading to the closest vertex is considered first and with high probability. The other edges will only be considered according to equation (1.12) if this edge is not selected. In the so-called global update rule, pheromone is deposited or removed only on the edges of the best tour, reducing the complexity of this update. There is also a local update rule which favors the exploring of new paths by reducing the amount of pheromone on each edge an ant crosses while creating a new tour (Potvin 2009).

ANT-BASED SYSTEMS FOR VEHICLE ROUTING

The first work was done on a VRP with maximum distance constraints. In this work the ant is forced to return to the depot, and start a new tour, whenever the next vertex is infeasible due to capacity or distance constraints. Further improvements are proposed: to account for location, a savings term is added to the transition rule. The probability to go to a vertex j is higher when the savings is larger.

In another work by the same authors, the visibility term is replaced by a parameterized savings term. In both papers, a local search heuristic based on 2-opt arc exchanges is applied to every route in every solution constructed by the ants for further improvement.

A further development is aimed to reduce the computational effort by decomposing the problem into smaller subproblems made of subsets of geographically close routes and by solving each subproblem separately.

Ant-based systems have also been applied to the Vehicle Routing Problems with Time Window (VRPTW). The proposed multiple ant colony system (MACS) defines two colonies: ACS-VEI that minimizes the number of vehicles and ACS-TIME that minimizes the total travel time. ACS-VEI searches for a solution that maximizes the number of visited vertices with a fixed number of vehicles. Then ACS-TIME constructs solutions with that number of vehicles, while trying to minimize the total travel time. Also, ACS-VEI is restarted with one fewer vehicle. This is repeated until it cannot be reduced anymore.

In [1], the nearest neighbor rule is replaced with by a randomized variant of Solomon's I1 sequential insertion heuristic [2]. This approach is used to solve the VRPBTW, where the backhaul and line haul vertices can be mixed. At each iteration of the construction process, the transition rule selects an unvisited vertex i and inserts it at a feasible place between two consecutive vertices j and j' in the current route. Thus, the attractiveness of vertex i is higher when it is located far from the depot, and when it introduces a small detour and a small service delay in a route. The transition formula thus accounts for both the attractiveness of vertex i and the amount of pheromone on the edges (Potvin 2009).

FIREFLY ALGORITHM

In the logistics area we have an opportunity to develop tools that help optimize decision making for the management of shipping routes, which will help reduce costs, delivery times and delivery distances. Therefore, when analyzing algorithms nature-inspired, we decided to study the behavior of fireflies and apply it in the logistics area with an algorithm considering the delivery time windows for customers. In this module you can observe the advantages of using firefly algorithm and the analysis of the results of tests made with this algorithm.

Firefly algorithm (FA) is a nature-inspired algorithm based of the behavior on the fireflies in their environment, the agents follow a simple rules to survive in the nature.

The flashing light of fireflies is produced by a process or bioluminescence. Their purpose of flashing are attract mating partners, attract potential prey and protective warming mechanism.

After analysis of behavior of the fireflies the researchers developed the firefly algorithm to solve a problems in the different industry areas, especially for the NP-hard problems such as the travelling salesman problem (Ivanell, Sørensen, & Henningson, 2007).

Firefly algorithm is developed based in a specific characteristic of the firefly behavior, their luminescence.

Firefly algorithm have specific rules:

1. All fireflies are unisex so that one firefly will be attracted to other fireflies regardless of their sex.
2. Attractiveness is proportional to their brightness, thus for any two flashing fireflies, the less brighter one will move towards the brighter one. The attractiveness is proportional to the rightness and they both decreases as their distance increases. If there is no brighter one than a particular firefly, it will move randomly.
3. The brightness of a firefly is affected or determined by the landscape of the objective function.

Based on these three rules, the basic steps of the firefly algorithm can be summarized as the pseudo code:

```
Objective function f(x), x = (x1, ..., xd)T
Generate initial population of fireflies xi (i = 1, 2, ..., n)
Light intensity Ii at xi is determined by f(xi)
Define light absorption coefficient
while (t <MaxGeneration)
for i = 1: n all n fireflies
        for j = 1: i all n fireflies
               if (Ij > Ii), Move firefly i towards j in d-dimension; end if
               Attractiveness varies with distance r via exp[-r]
               Evaluate new solutions and update light intensity
        end for j
end for i
Rank the fireflies and find the current best
end while
Postprocess results and visualization
```

FIREFLY ALGORITHM APPLIED TO REAL WORD PROBLEM

The following pseudo code represent the firefly algorithm implemented (Box 1).

RESULTS

The tests were run using the firefly algorithm and ant colony algorithm using a population of 500 agents, with the following results (see figure 5).

We can see in the table with the algorithm fireflies are faster optimal objectives with the algorithm of ant colony, for example: The point 108 algorithm fireflies optimal target was found in the search 115 500 fireflies and the algorithm of ant colony point 108 was found to search 345 of 500 ants sent.

Box 1. Pseudo code

1. Declare *firefly_algorithm* function which is involving all algorithm code.

**Into *firefly_algorithm* function:

2. Declare the following variables:
 a. *destination=4* - Final node.
 b. *fireflies=500* - Amount of fireflies.
 c. *actual_firefly=1* - Variable to save the actual position.
 d. *Iterations=200* - Amount of iterations.
 e. *node=0* - number of the new node.
 f. *node_distance=0* - distance of the new node.
 g. *stop* - flag to stop iterations.
 h. *search_matrix=113* - Amount of columns and rows of the matrix.
 i. *last_position=1* - Variable to save the last position selected.
 j. *map* - [matrix]
 k. *answer=0* - Possible next node.
 l. *answer_distance=0* - Distance of possible next node.
 m. *route=1* - Variable to save all route.
 n. *vector_routes=zeros(fireflies,search_matrix)* - Vector to save each firefly route (filled with zeros)
 o. *vector_routes_distance=zeros(fireflies,1)* - Vector to save distance of each firefly route
 p. new_route
 q. new_routes_distance
 r. *optimal_distance=0* - The best distance
 s. *optimal_route=0* - The best node
3. Declare *attractionfirefly(node_distance)* function, it receive a parameter about the actual node distance, and then will be equalize to *[firefly_selected_distance,firefly_selected]*, where:
 a. *Firefly_selected* variable is assigned a random value generating a 1-by-1 column vector of uniformly distributed random integers from the interval: [2 - length of *node_distance* variable].
 b. *Firefly_selected_distance* variable is equalized to the node distance value of the current firefly.
4. Declare *select_optimal_distance()* who receive 5 parameters (*vector_routes*, vector_routes_distance, *fireflies*, *search_matriz*, *destination*). Where:
 a. A *For* loop does *n* iterations as *fireflies* variable.
 i. If the current firefly route is minor of the best distance or the best distance is equal zero, and destination is equal to current firefly route or current firefly route is equal zero then:
 1. The best distance will be equal to current firefly distance and
 2. The best node will be equal to current firefly route.

5. First *For* loop used to repeat all instructions in each firefly (it depends of the *fireflies* variable)
 a. Second *For* loop searches in each iteration:
 i. If stop flag is zero:
 1. A For loop searches in current row to find next node
 • If this node is mayor of zero then current iteration is assigned to *node* variable, and its distance is saved in *node_distance*.
 2. The *attractionfirefly()* function is invocated (it receive *node_distance* variable as parameter) and it will sent the possible next node and its distance as parameters.
 3. Then distance value of the new node will be equalize to the distance of possible next node.
 4. The new node value will be equal to the possible next node value.
 5. The new node will be save in the route variable.
 6. The current node value will be saved in the firefly route vector too.
 7. Now is necessary to check if the current node is equal to final node, if it is true, the stop flag will be 1 and the search will be completed.

Figure 5. Comparison of both algorithms

	Firefly Algorithm	Ant System Algorithm
Point	108	108
Route	1 2 5 97 82 9 45 46 56 107 57 61 62 108	1 2 5 97 82 9 45 46 56 107 64 65 66 108
Distance	696.3	699.3
The first agent to find the optimal objective	115	346
Point	14	14
Route	1 2 3 99 13 100 14	1 2 5 97 10 12 99 13 100 14
Distance	291.4	533.9
The first agent to find the optimal objective	4	271
Point	35	35
Route	1 2 3 6 7 35	1 2 3 6 7 35
Distance	550.4	550.40
The first agent to find the optimal objective	2	21
Point	71	71
Route	1 2 5 97 82 9 45 47 104 73 72 71	1 2 5 97 82 9 45 46 56 107 57 61 60 70 71
Distance	591.4	854.0
The first agent to find the optimal objective	5	388
Point	7	7
Route	1 2 3 6 7	1 2 3 6 7
Distance	299.4	299.4
The first agent to find the optimal objective	12	31

Figure 6. Valid solutions found Firefly algorithm vs. Ant algorithm

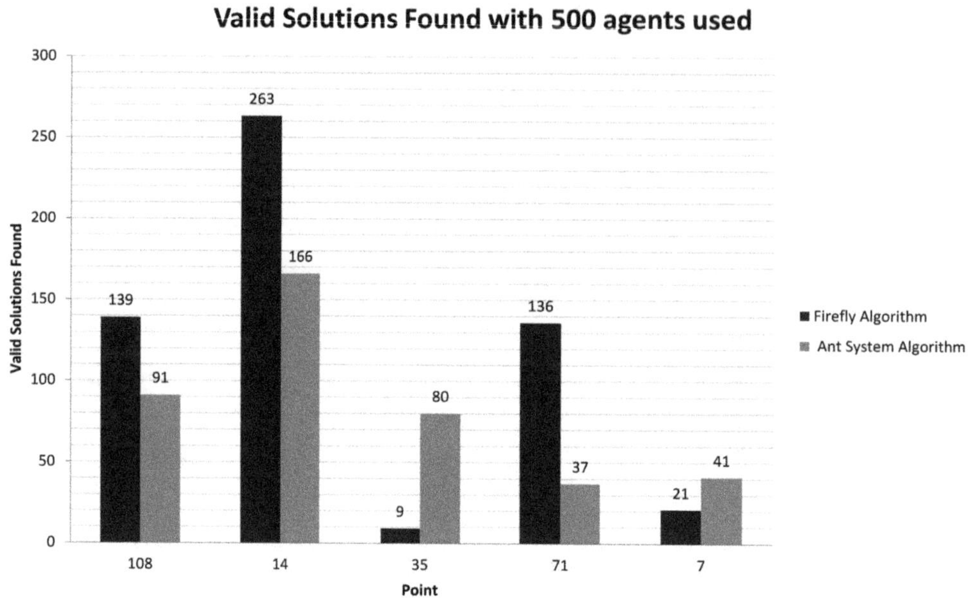

Firefly algorithm and ant colony algorithm was tested with 5000 agents where their function was looking for a specific point in the map, but only some agents found the objective, the following graphic shows the number of agents that found the objective point (see figure 6).

CONCLUSION

We test two different algorithms based in the behavior of fireflies and ant colony where we analyze its similarities and differences with particle swarm optimization. Then we implemented and compared these algorithms. Our results of the simulation of VRP to find the global optimal of various test functions suggest that the firefly algorithm is more efficient than ant colony algorithm.

This implies that firefly algorithm is potentially more powerful in solving NP-hard problems where we are looking for a point in the map.

REFERENCES

Ak, A., & Erera, A. L. (2007). A paired-vehicle recurs strategy for the vehicle routing problem with stochastic demands. *Transportation Science*, *41*(2), 222–237. doi:10.1287/trsc.1060.0180

Balakrishnan, N. (1993). Simple heuristics for the vehicle routing problem with soft time windows. *The Journal of the Operational Research Society*, *44*(3), 279–287. doi:10.1057/jors.1993.53

Baldacci, R., Vigo, D., & Toth, P. (2010). Exact solution of the capacitated vehicle routing problem. In *Wiley Encyclopaedia of Operations Research and Management Science*. John Wiley & Sons, Inc.

Bertsimas, D. (1991). A vehicle routing problem with stochastic demand. *Journal of Operations Research, 40*(3), 554–585.

Bertsimas, D. (1992). A vehicle routing problem with stochastic demand. *Operations Research, 40*(3), 574–585. doi:10.1287/opre.40.3.574

Bertsimas, D., & Van Ryzin, G. (1990). A stochastic and dynamic vehicle routing problem in the Euclidian plane. *Operations Research Society of America, 39*(4), 600–615.

Bertsimas, D., & Van Ryzin, G. (1993). Stochastic and dynamic vehicle routing with general demand and inter-arrival time distributions. *Advances in Applied Probability, 25*(4), 947–978. doi:10.2307/1427801

Blanton, J. L., & Wainwright, R. L. (1993). Multiple vehicle routing with time and capacity constraints using genetic algorithms. In *Proceedings of the Fifth International Conference on Generic Algorithms (ICGA-93)*, (pp. 452–459).

Borges, C. E. (2011). Algoritmos Bioinspirados. C.E. Borges, J.L. Montaña. Departamento de Matemáticas, Estadística y Computación Universidad de Cantabria. Retrieved from paginaspersonales.deusto.es/cruz.borges/Papers/10APIAXXI.pdf

Calvete, H. I., Carmen, G., Mara, J. O., & Beln, S. V. (2004). Vehicle routing problems with soft time windows: An optimization based approach. *Journal of Monógrafas del Seminario Matemático García de Galdeano, 31*, 295–304.

Chenghua, S., & Xiaofeng, Z. (2011). Research on model fitting capacity of vehicle routing problem. *International Journal of Advancements in Computing Technology, 3*(11), 185–193. doi:10.4156/ijact.vol3.issue11.24

Clarke, G., & Wright, J. (1964). Scheduling of vehicles from a central depot to a number of delivery points. *Operations Research, 12*(4), 568–581. doi:10.1287/opre.12.4.568

Cordeau, J. F., Gendreau, M., Laporte, G., Potvin, J. Y., & Semet, F. (2002). A guide to vehicle routing heuristics. *The Journal of the Operational Research Society, 53*(5), 512–522. doi:10.1057/palgrave.jors.2601319

Cordeau, J. F., Laporte, G., Martin, W. P. S., & Daniele, V. (2007). Vehicle Routing. In Handbooks in Operations Research and Management Science. Elsevier B.V.

Cordeau, J. F., Laporte, G., & Mercier, A. (2001). A unified tabu search heuristic for vehicle routing problem with time windows. *The Journal of the Operational Research Society, 53*(8), 928–936. doi:10.1057/palgrave.jors.2601163

de Castro, L. N. (2006). *Fundamentals of natural computing: Basic concepts, algorithms, and applications*. Boca Raton, FL: CRC Press.

Dror, M. (1991). Modeling vehicle routing with uncertain demands as a stochastic program: Properties of the corresponding solution. *European Journal of Operational Research, 68*, 432–441.

Dror, M. (1993). Modeling vehicle routing with uncertain demands as a stochastic program: Properties of the corresponding solution. *European Journal of Operational Research*, *64*(3), 432–441. doi:10.1016/0377-2217(93)90132-7

Dror, M. (Ed.). (2000). *Arc routing: Theory, Solutions and Approximations*. Berlin: Springer. doi:10.1007/978-1-4615-4495-1

Dror, M., Laporte, G., & Francois, V. L. (1993). Vehicle routing with stochastic demands and restricted failures. *Methods and Models of Operations Research*, *33*, 271–283.

Dror, M., Laporte, G., & Trudeau, P. (1994). Vehicle routing with split deliveries. *Discrete Applied Mathematics*, *50*(3), 229–254. doi:10.1016/0166-218X(92)00172-I

Dror, M., & Trudeau, P. (1986). Stochastic vehicle routing with modified savings algorithm. *Journal of Operational Research*, *23*(2), 228–235. doi:10.1016/0377-2217(86)90242-0

Dror, M., & Trudeau, P. (1989). Savings by split delivery routing. *Transportation Science*, *23*(2), 141–145. doi:10.1287/trsc.23.2.141

Dror, M., & Trudeau, P. (1989). Savings by split delivery routing. *Transportation Science*, *23*(2), 141–145. doi:10.1287/trsc.23.2.141

Eiselt, H. A., Gendreau, M., & Laporte, G. (1995). Arc routing problems, Part 1: The Chinese Postman problem. *Operations Research*, *43*(2), 231–242. doi:10.1287/opre.43.2.231

Eiselt, H. A., Gendreau, M., & Laporte, G. (1995). Arc routing problems, Part 2: The rural Postman problem. *Operations Research*, *43*(3), 399–414. doi:10.1287/opre.43.3.399

Ghiani, G., Laporte, G., Manni, E., & Musmanno, R. (2008). Waiting strategies for the dynamic and stochastic travelling salesman problem. *International Journal of Operations Research*, *5*(4), 233–241.

Kallehauge, B. (2008). Formulations and exact algorithms for the vehicle routing problem with time windows. *Computers & Operations Research*, *12*(7), 2307–2330. doi:10.1016/j.cor.2006.11.006

Kenyon, S. A., & Morton, P. D. (2003). Stochastic vehicle routing with random travel times. *Journal of Transportation Science*, *37*(1), 69–82. doi:10.1287/trsc.37.1.69.12820

Laporte, G. (1992). The vehicle routing problem: An overview of exact and approximate algorithms. *European Journal of Operational Research*, *59*(3), 345–358. doi:10.1016/0377-2217(92)90192-C

Laporte, G. (1997). Vehicle routing. In M. Dell'Amico, F. Maffioli, & S. Martello (Eds.), *Annotated bibliographies in combinatorial optimization*. Chichester, UK: Wiley.

Laporte, G., Louveaux, F., & Hamme, V. L. (2002). An integer l-shaped algorithm for the capacitated vehicle routing problem with stochastic demands. *Operations Research*, *50*(3), 415–423. doi:10.1287/opre.50.3.415.7751

Laporte, G., Louveaux, F., & Mercure, H. (1989). Models and exact solutions for a class of stochastic location-routing problems. *European Journal of Operational Research*, *39*(1), 71–78. doi:10.1016/0377-2217(89)90354-8

Lawler, E. L., Lenstra, J. K., Rinnooy Kan, A. H. G., & Schmoys, D. B. (Eds.). (1985). *The traveling salesman problem*. Chichester, UK: Wiley.

Madsen, O. G., Ravn, F., & Rygaard, M. (1995). The vehicle routing problem with time windows part ii: Genetic search. *Journal of Computing*, *8*, 165–172.

Mingozzi, A. (2003). *An exact algorithm for period and multi- depot vehicle routing problems. Technical report*. Bologna, Italy: Department of Mathematics, University of Bologna.

Novoa, C., Berger, R., Linderoth, J., & Storer, R. (2006). *A set-partitioning-based model for the stochastic vehicle routing problem. Technical report*. Texas State University and Lehigh University.

Novoa, C., & Storer, R. (2009). An approximate dynamic programming approach for the vehicle routing problem with stochastic demands. *European Journal of Operational Research*, *196*(2), 509–515. doi:10.1016/j.ejor.2008.03.023

Paolo, T., & Daniele, V. (Eds.). (2002). *The vehicle routing problem. SIAM Monographs on Discrete Mathematics and Applications*. Philadelphia, PA: Society for Industrial and Applied Mathematics.

Pérez Jiménez. (2003). *Maquinas Moleculares Basadas en ADN*. M. de J. Pérez Jiménez F. Sancho Caparrini.

Potvin, J. Y. (2009). A Review of Bio-inspired Algorithms for Vehicle Routing. Springer-Verlag Berlin Heidelberg. doi:10.1007/978-3-540-85152-3_1

Smith, S. L., Pavone, M., Bullos, F., & Frazzoli, E. (2010). Dynamic vehicle routing with priority classes of stochastic demands. *IAM Journal on Control and Optimization*, *48*(5), 3224–3245. doi:10.1137/090749347

Stefan, B., & Natalia, K. (2009). *An overview on vehicle scheduling models*. Decision Support & Operations Research Lab, University of Paderborn. Available at: https://wiwi.uni-paderborn.de/leadmin/lehrstuehle/de-partment-3/wiwi-dep-3-ls-5/Forschung/Publikationen/buntekliewer anoverviewonvehicle-schedulingmodels2006.pdf

Stewart, W. Jr, & Golden, B. (1983). Stochastic vehicle routing: A comprehensive approach. *European Journal of Operational Research*, *14*(4), 371–385. doi:10.1016/0377-2217(83)90237-0

Sumalee, A., Uchida, K., & Lam, H. K. W. (2011). Stochastic multi-modal transport network under demand uncertainties and adverse weather condition. *Transportation Research Part C, Emerging Technologies*, *11*(2), 338–350. doi:10.1016/j.trc.2010.05.018

Taillard, E., Badezu, P., Gendreau, M., & Guertin, F. (1997). A tabu search heuristic for the vehicle routing problem with soft-time window. *Journal of Transportation Science*, *31*(2), 170–186. doi:10.1287/trsc.31.2.170

Toth, P., & Vigo, D. (Eds.). (2001). *The vehicle routing problem. SIAM Monographs on Discrete Mathematics and Applications*. Philadelphia, PA: Society for Industrial and Applied Mathematics.

Xiangyong, L., Tian, P., & Leung, S. C. H. (2010). Vehicle routing problems with time windows and stochastic travel and service times: Models and algorithm. *International Journal of Production Economics*, *125*(1), 137–145. doi:10.1016/j.ijpe.2010.01.013

Chapter 25
Strategic Designing and Optimization of Mixed Flow Impeller Blades for Maritime Applications

Apurba Kumar Roy
Birla Institute of Technology, India

Supriyo Roy
Birla Institute of Technology, India

Kaushik Kumar
Birla Institute of Technology, India

ABSTRACT

Mixed flow impellers are extensively used in turbomachines either to convert mechanical energy to fluid energy or to convert fluid energy to mechanical energy. According to the geometry of flow passage, turbo machines can be classified as radial, axial and mixed flow. Mixed flow turbomachines are widely used for engineering applications like cooling water duties, water intake impellers for maritime applications, flood water draining, irrigation and other application fields. The design of mixed flow impellers of high specific speed is a direct extension of the well-established methods of the designing of radial flow impellers but the introduction of near diagonal flow layout at a still larger specific speed stimulated the incorporation of axial impeller design techniques in mixed flow impeller technology. Here, an attempt has been made to design a mixed flow turbo machine blade from the basic principle of turbomachinery and fluid mechanics. On the basis of stress analysis, the blade positioning in the meridional annulus was selected and validated using artificial neural network.

1. INTRODUCTION

Over the past decades, the mean line analysis method based on the conventional empirical loss correlations (assuming that the flow characteristics averaged over the cross-section of flow passage could represent the three-dimensional flow phenomena through the passage of turbomachinery) has been widely employed

DOI: 10.4018/978-1-4666-9779-9.ch025

to determine the overall geometric design variables for each component of turbomachinery and to predict the on and off-design performance characteristics. In recent past, mixed flow impellers are extensively used in turbomachines either to convert mechanical energy to fluid energy or to convert fluid energy to mechanical energy as torque. Mixed flow impellers are unification of axial and radial characteristics. As per geometrical shape analysis, turbo machines can be classified as radial, axial and mixed flow. In axial flow machines, flow is mainly parallel to the axis of rotation and is used where low pressure head and high volumetric discharge are primary requirements. On the contrary, centrifugal turbomachines, comprise a very wide range for generation of pressure head and the flow through the impeller is largely normal to the axis of rotation. Mixed flow turbo machine are the intermediate types which are in between the axial and centrifugal. Since the flow through impeller has both radial and axial components; the mixed flow impeller blades have very complex in its geometry. But owing to its geometrical complexities, mixed flow turbomachines are extensively used for cooling water duties and specifically in water intake impellers for maritime applications. Apart from these, mixed flow turbomachines are also widely used in flood water draining, irrigation, urban water supply, cooling water system for various power plants, fire fighting systems and many other engineering and management fields.

Design of mixed flow impeller blades are quite complex as a large number of variables play a significant role on the overall performance of blades. The design of mixed flow impellers of high specific speed is a direct extension of the well-established empirical methods of the designing of radial flow impellers. Extension of similar methods serves for the design of mixed flow impellers, but introduction of near diagonal flow layout at a still larger specific speed stimulated the incorporation of axial impeller design techniques in mixed flow impeller technology.

Usual industrial design methods for mixed flow impellers are basically based on the use of empirical correlations and design constants. Industrial design method often obscures or ignores the actual happening within the mixed flow turbo machine flow passage and consequently a poor guide when the question of new design and development of such turbo machine comes into the picture. In this present work, an attempt has been made to design a mixed flow turbo machine blade from the basic principle of turbomachinery and fluid mechanics. As compared to radial flow and axial flow turbomachines, mixed flow type turbomachines got less importance due to its complexity. The flow through the mixed flow type turbomachines is very complex due to its highly complicated blade and fluid flow passage geometry. A schematic of a mixed flow turbomachine impeller is shown in Figure 32 in Appendix1. The present work relates to designing of impeller blade primarily. Entire research can be divided into four basic sections: namely Design, Model Development, FEM Analysis and validation / optimization of proposed design via Artificial Neural Network in ANSYS.

2. NOTATIONS AND ASSUMPTIONS

To design a mixed flow turbo machine blade from the basic principle of turbomachinery and fluid mechanics, the following notations are used (Srivastava et al. 2014):

Ω = specific speed, rad. / sec., Q = discharge of the fluid, m^3/sec.
g = acceleration due to gravity, m/s^2; H = Pressure head developed, m; N_s = specific speed, rpm.
γ = specific weight of the fluid, kg/m^3; D_{m2} = mean blade diameter at the outlet.
u_2 = circumferential velocity at the impeller exit; D_{2o} = outlet blade diameter at tip.

D_{2i} = outlet blade diameter at hub; r_{1o} = inlet blade diameter at tip.
r_{1i} = inlet blade diameter at hub; r_{2o} = outlet blade diameter at tip.
r_{2i} = outlet blade diameter at hub; α_1 = fluid inlet angle; α_2 = fluid outlet angle.
θ_1 = inlet camber angle; θ_2 = outlet camber angle; i = incidence angle; δ= deviation angle.
β_1 = inlet blade angle (α_1-i); β_2 = outlet blade angle (α_2-δ); C = chord length; S = blade spacing.
λ= blade stagger angle; α_m = vector mean flow angle = $\tan^{-1}$ (tan α_1 + tan α_2)/2.
$C_{\theta 2}$ = tangential component of absolute velocity at outlet in m/s.
η = hydraulic efficiency; ω = Angular velocity, rad./sec.; C_{m1} = Inlet meridional velocity.
C_{m2} = Outlet meridional velocity; Z = no. of blades; θ_1 = camber angle; S = pitch of blades.
C_1 = actual chord; m = a constant whose value is due to Horlock (1958).

3. LITERATURE SURVEY

Though lot of academicians work on turbomachining, but as compared to radial and axial flow turbo machine impellers, literature on mixed flow turbo machine impeller is very scanty. Preliminary design of a mixed flow impeller was developed by Wislicensus in the year 1965. Myles (1965) improved the design of mixed flow impeller designed by Wislicensus (1965). A formula to calculate the slip velocity for a mixed flow impeller was developed by Busemann (1928). For calculation of meridional streamline, a design methodology was developed by Senoo and Nakase (1972). A similar design methodology developed by Senoo and Nakase (1972) was adapted by Inoue et al. (1980) for design of mixed flow impeller. A modified procedure to design a mixed flow impeller was given by Stepanoff (1957). Step wise design methodology for a mixed flow impeller design was suggested by Neumann (1991). Gahlot and Nyiri (1993) adapted a similar procedure but with few modifications in the blade design.

A multi-objective optimization technique of designing a mixed flow impeller with the aid of inverse three-dimensional design approach was presented by Takayama and Watanabe (2009). To find a procedure for optimizing the design of mixed-flow impellers, Kim and Kim (2011) adopted certain optimization tools. In a recent study, Bing et al. (2013) experimentally studied the effects of geometry of flow passage along meridional plane on overall characteristics of mixed-flow impeller. Using FEM technique, Sham Sunder (1987) put forward a stress analysis method to determine the distribution of stresses in various types of impellers. Ramamurti and Balasubramanian (1987) studied steady state stress analysis of centrifugal impellers using triangular plate element. Impeller stresses were calculated by him who was in good agreement with the experimental results obtained using a strain gauge technique. Using FEM technique Jonker and Van Essen (1997), Samir Lemeš and Nermina Zaimović- Uzunović (2002) studied the stress distribution pattern on various types of impellers. Using experimental methods, Bhope and Padole (2004) and Hao et al. (2011) studied the Von Mises stress distribution on various types of impeller blades. Mehta and Patel (2013) and Srivastava et al. (2014) studied the stress analysis and natural frequencies of various types of impellers having highly complex blades.

Sham Sunder (1987) presented a three-dimensional method of stress analysis using finite element techniques for determining the stress distribution in centrifugal impellers. Many researchers (Jonker and Van Essen (1997), Samir Lemeš and Nermina Zaimović-Uzunović (2002), Bhope and Padole (2004), Arewar and Bhope (2013)) contributed on the stress analysis for highly complex blades of various turbomachines. Samanta (2008) used ANN technique to validate the result of CNC machining result. Karnik (2008) did the delamination analysis in high speed drilling is performed by developing an artificial neural network (ANN) model with spindle speed, feed rate and point angle as the affecting parameters. Tosun (2002) estimated the tool life using Artificial Neural Network (ANN) with Back Propagation (BP) algorithm where as Somashekhar (2010) showed that a well-trained neural network model was effective in estimating the MRR and also used the same to improve the MRR using optimized machining parameters. Panda (2005) used an artificial feed forward neural network based on the Levenberg-Marquardt back propagation technique of appropriate architecture of the logistic sigmoid activation function to predict the material removal rate.

4. DESIGNING OF IMPELLER BLADE

The present work relates to designing of impeller blade primarily (Srivastava et al. 2014). Entire research can be divided into four basic sections: namely Design, Model Development, Finite Element analysis; required in order to pick the best option out of the designs made and finally validation / optimization of proposed design via Artificial Neural Network in ANSYS.

4.1 Design of Mixed Flow Impeller Blade

Here, a mixed flow impeller blade has been designed to be used for incompressible fluid with no change of temperature of the fluid medium. The dimensional specific speed has chosen to be 1.998 rad / sec where rotational speed of the impeller has been taken as 1000 rpm as suggested by Stepanoff (1957). Free vortex theory is adopted for designing the impeller blades.

4.2 Model Development Procedure

Development of mixed flow impeller blade involves the following steps in a sequential manner:

1. Selection of diameters at inlet and outlet of the impeller.
2. Analysis of meridional flow resulting in surface of revolution.
3. Using free vortex theory calculation of blade angles from hub to tip.
4. Selection of cascades to deliver required flow angles and blade loading.
5. Geometrical transformation of each surface of revolution into a plane with flow angles.
6. Reverse transformation of cascade geometry and flow data to the back to back intersections on the surfaces of revolution.
7. Recalculation of meridional axisymmetric flow for convergence check.
8. Span wise superimposition of the various blade sections one over another to satisfy blade stagger angle on the conical surface of the impeller. Here standard NACA 10C4 circular arc profile has been used for the blade profile.
9. 3D model Development.

4.3 Design Procedure of the Mixed flow Turbomachinery Impeller Blade

Before proceeding to design a mixed flow impeller blade, we need to calculate the specific speed of the turbomachinery, so as to choose type of impeller to be selected for its best performance. In this present work, following specification of turbomachinery has been selected to suite the maritime applications. Discharge, Q=125 l/sec, total head developed, H = 5m, and rotational speed of the turbomachine N = 1000 rpm.

The specific speed (Ω) is calculated using the following relation (Eqns. 1-2):

$$\Omega = \frac{\omega * \sqrt{Q}}{\left(gH\right)^{\frac{3}{4}}} \tag{1}$$

$$N_s = \frac{N * \sqrt{Q}}{H^{\frac{3}{4}}} \tag{2}$$

From equations (1 and 2, it is found that the specific speed of the turbomachine (Ω) = 1.998 rad. / sec. or N_s = 105.74 rpm (see table 1).

The following is the range of specific speed (Ω) for various turbomachinery as suggested by Csanady (1964).

Ω = 1.25 to 2.27 for mixed flow impeller.
Ω = 2.27 to 5.33 for axial flow impeller.

Considering range of specific speed as appearing in table 1 and values obtained using equation 1 and 2, it can be concluded that the '*turbomachinery impeller should be of mixed flow type'*.

4.4 Preliminary Estimate of Design Parameters

Test code (I.E.C -1976) specifies a minimum impeller size (Impeller tip diameter) of 200mm where as the Hydraulic Institute Standards (1961) recommends a minimum impeller tip diameter at outlet of 300mm. Initially the power, head and flow were calculated for pumps of different specific speeds running at different rpm. The calculations were done for a range of inlet diameter of 200mm to 300mm. The flow rate and power of the impeller depends on the type, speed and size of the turbomachinery to be studied. These were estimated a range of specific speeds and inlet diameters, varying from 200 mm

Table 1. Variation of specific speed (N_s) rpm, and (Ω) rad. /sec. for three different speeds

N (rpm)	750	1000	1250	1500
Ns(rpm)	79.30	105.74	132.17	158.61
Ω(rad/s)	1.498	1.998	2.497	2.996

to 300 mm and speed varying from 750 to 1500 rpm. The design constants are function of specific speed and are produced in metric unit written in equation (3.3).

$$Q = \left[\frac{N_S}{N} * H^{\frac{3}{4}} \right]^2 \tag{3}$$

The pressure head is given in equation (4)

$$H = \frac{u_2^2}{2gK_2^2} \tag{4}$$

where, K_u = empirical constant in equation (4)
Equation (4) can be written as,

$$H = \frac{1}{2gK_2^2} * \left[\frac{\pi D_2 N}{60} \right]^2 \tag{5}$$

$$= \frac{1}{2gK_u^2} * \left[\frac{\pi D_1 N}{\left(\frac{D_1}{D_2} \right) XN} \right]^2$$

$$= \frac{1}{2gK_u^2} * \left[\frac{\pi D_1 N}{(e) XN} \right]^2$$

$$= \frac{1.397 * 10^{-4} * N^2 * D_1^2}{e^2 K_u^2}$$

$$\therefore H = \frac{1.397 * 10^{-4} * N^2 * D_1^2}{e^2 K_u^2} \tag{6}$$

The variation K_u and e = (D_1 / D_2) with specific speed are adapted from Stepanoff (1957) and are reproduced in graphs as shown Figure 33 (Appendix 1).

For a given specific speed (N_s) and inlet diameter, 'H' can be obtained from equation (6) and discharge from equation (3) as a function of speed.

The power required is obtained from the following relation (Eqn. 7):

$$P = \frac{\gamma QH}{75} hp \tag{7}$$

$$= \frac{\gamma QH}{75} * 0.7355 kW$$

$$P = \frac{\gamma QH}{102} kW \tag{8}$$

The computation of flow rate (Q), head (H) and power (P) is carried out for a range of speeds N=750 to 1500 rpm and for impeller diameter 200mm, 250mm and 300mm and is shown in Figure 34, Figure 35, Figure 36, Figure 37, Figure 38, Figure 39, Figure 40, Figure 41 and Figure 42 given in Appendix 2.

4.5 Design of Mixed Flow Impeller Blade

For a specific speed (N_s) of 105.74 rpm, consulting (Q vs. Ns) and (H vs. Ns) plots [Figure 37] the inlet tip diameter, D1 = 250mm is selected. Referring the (e vs. Ns) plot, the value of 'e'= (D1/D2) is found to be 0.775, where e = diameter ratio [Figure 33].

Hence D_2 = 324mm (outer diameter) which is also referred as D_{2o}. Again, from (P vs. Ns) plot Figure A6, Motor power reqd. P=15 kW (ideal); Expected efficiency, η = 76%; Actual motor power = 19.74 (kW) ≈ 20 kW

So the adopted values are,

Specific speed, Ns = 105.74 rpm; D_1 = 250 mm; $D_2 = D_{2o}$= 324 mm; P = 20 kW.

From Stepanoff (1957)

$$\frac{D_1}{D_{m2}} = 0.9$$

where

$\therefore D_{m2}$ = 278 mm

Again,

$$D_{m2} = \sqrt{\frac{D_{2o}^2 + D_{2i}^2}{2}} \tag{9}$$

or $(278)^2 \times 2 = (324)^2 + (D_{2i})^2$

$\therefore D_{2i}$ = 224 mm

Now the layout of the blade is drawn by choosing a suitable semi-cone angle of 30° as suggested by Stepanoff (1957). The blade height "l" is measured from the layout drawing (Figure 1) and found to be =58 mm. This is also known as span of the blade. Then mean radius r_{m1} and r_{m2} is calculated from the equations (10 & 11.

$$r_{m1} = \sqrt{\frac{r_{1o}^2 + r_{1i}^2}{2}} \tag{10}$$

$\therefore r_{m1} = 102$ mm

$$r_{m2} = \sqrt{\frac{r_{2o}^2 + r_{2i}^2}{2}} \tag{11}$$

$\therefore r_{m2} = 139$ mm

Figure 1. Layout of impeller and blade (in meridional plane)

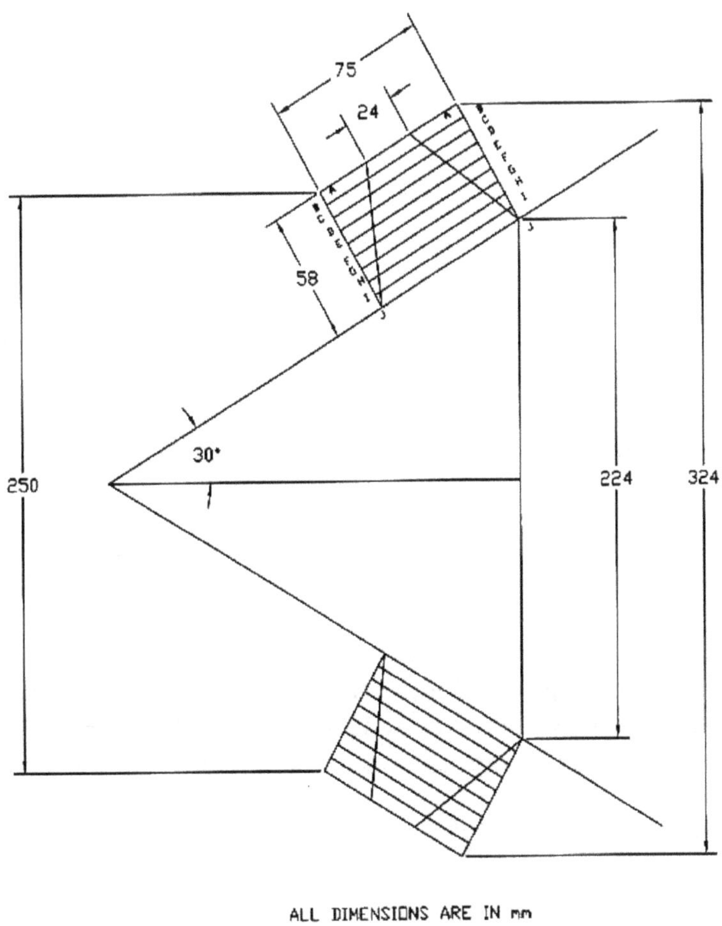

Now

$$\frac{C_{m1}}{C_{m2}} = \frac{r_{m1}}{r_{m2}} = 1.363$$

Myles (1965) suggested that C_{m1}/ C_{m2} should be 1.2 for stream line designs. Therefore the value of C_{m1}/C_{m2} is required to be modified to get the stipulated value. Here a trial and error method has been adapted to modified the values of r_{m1} and r_{m2} as to get $C_{m1}/ C_{m2} = 1.2$. The revised value of C_{m1} and C_{m2} was found to be 109 mm and 132 mm respectively.

The present design is based on Free Vortex distribution at inlet and outlet of the impeller. The meridional velocity C_m is taken to be along the inlet and outlet of the impeller from hub to tip and equal to the meridional velocity at mean radius. The meridional velocity C_m at inlet and outlet are calculated have been calculated using equations (12 - 14).

$$C_m = \frac{Q}{A} \tag{12}$$

$$C_{m1} = \frac{Q}{2\pi r_{m1} l} \tag{13}$$

$$= \frac{0.125}{2\pi * 0.109 * 0.058} = 3.15\, m/s$$

$$C_{m2} = \frac{Q}{2\pi r_{m2} l} \tag{14}$$

$$= \frac{0.125}{2\pi * 0.132 * 0.058} = 2.6\, m/s$$

This gives, $\frac{C_{m1}}{C_{m2}} = 1.21$

So the rectangular profile is changed to trapezoidal one and the meridional cross-section of the impeller has been divided into ten equal sections, marked A-A, B-B, C-C,D-D, E-E, F-F, G-G, H-H, I-I and J-J, which are shown in the Figure 1. Figure 2 shows the definition sketch for cascade flow parameters.

Figure 2. Definition sketch for cascade flow parameters

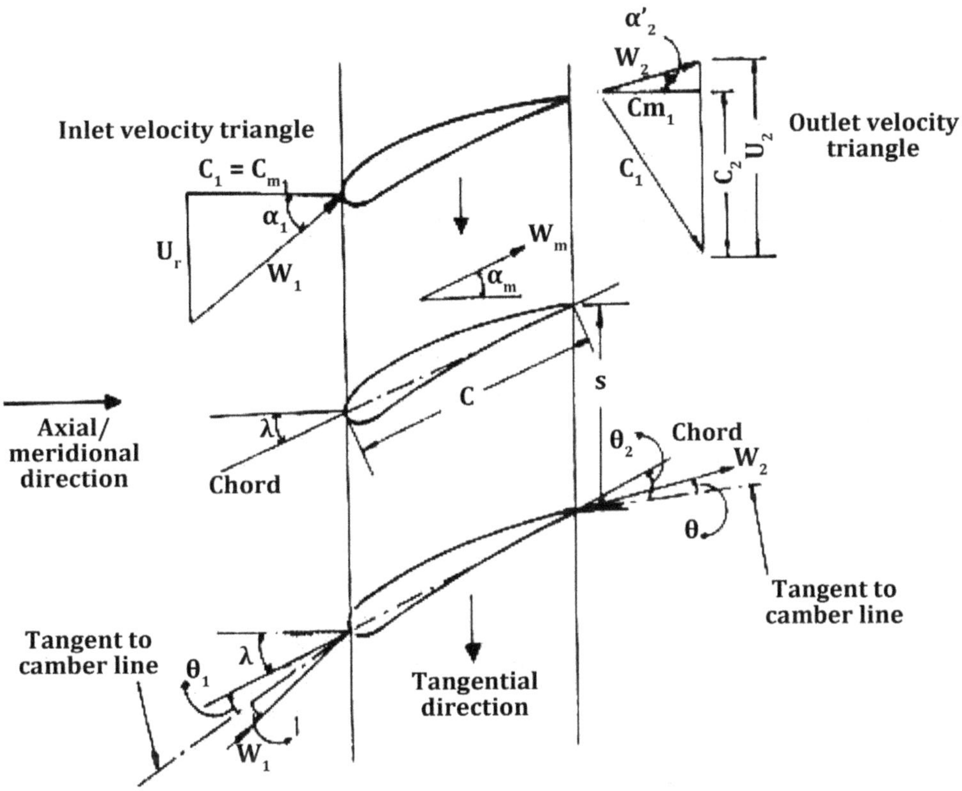

4.5.1 Sample Calculations for Blade Section at A-A

Inlet circumferential velocity,

$$u_1 = \frac{\pi D_{1A} N}{60} \tag{15}$$

$$= \frac{\pi * 0.274 * 1000}{60} = 14.35\, m/s$$

$$\tan \alpha_1 = \frac{u_1}{C_{m1}} = \frac{u_1}{C_1} = \frac{14.35}{3.15} \text{ (Since } C_{m1} = C1 = 3.15 \text{ m/s)}$$

$$\therefore \alpha_1 = \tan^{-1} 4.55 = 77.62^0$$

Here D_{1A} and D_{2A} are obtained by measuring them from the Figure 2.

Again, as $H = \frac{\eta u_2 C_{\theta 2}}{g}$ (16)

where,

H = 5m
$C_{\theta 2}$ = tangential component of absolute velocity at outlet in m/s.
g = 9.81 m/s^2
η = 76%

or,

$$5 = \frac{0.76 * 15.6 * C_{\theta 2}}{9.81}$$

$\therefore C_{\theta 2} = 4.14$ m/s

$$\therefore \tan\alpha_2 = \frac{u_2 - C_{\theta 2}}{C_{m2}} = \frac{15.6 - 4.14}{2.6} = 4.41$$

or

$$\alpha_2 = \tan^{-1} 4.41 = 77.22^0$$

Similarly, velocity triangles at both inlet and outlet can fully be described for all the other sections viz. B-B, C-C, D-D, E-E, F-F, G-G, H-H, I-I and J-J . Figure 3 and Figure 4 shows the inlet and out velocity triangles. In Figures (3-4), W_1 represents the relative velocity of the flow.

Figure 3. Inlet Velocity Triangle

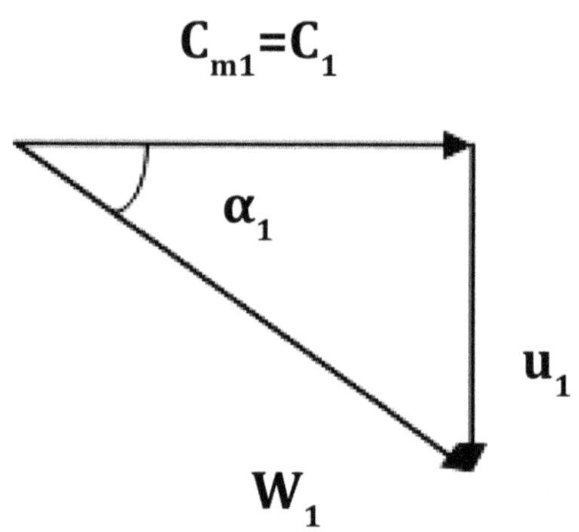

Figure 4. Outlet Velocity Triangle

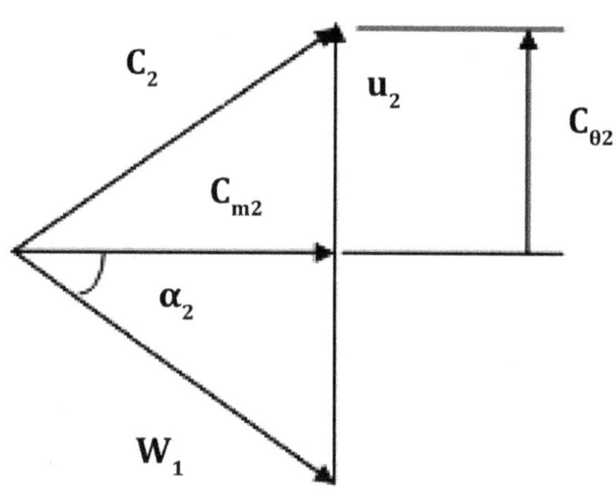

4.5.2 Slip Velocity Consideration

In the design it is assumed that the flow is irrigational inside the impeller. But the outlet velocity from the impeller will have on angular velocity due to rotation of the impeller. Every particle in the impeller channel will have relative angular velocity. The result is relative eddy in each passage, opposite in direction to the rotation of the impeller. The outer edge of the eddy will have a velocity V_S that produces the changed circumferential velocity $C_{\theta 2}$. Here, V_S is called the slip velocity. In case of mixed flow impeller the relative eddy can be resolved into two components, one parallel and the other normal to the meridional surface. The sin ψ component (where 'ψ' is the semi-cone angle of the impeller) is responsible for generating the slip velocity. According to Stodola (1945), for a given impeller tip speed and flow rate, the eddy can be reduced or eliminated by increasing the number of blades, i.e., by decreasing the spacing of blades. Based on the above recommendation 8 blades has been chosen and slip velocity consideration is calculated based on Stodola (1945) slip model as given in equation (17). Stodala slip velocity,

$$V_s = \frac{\pi u_2 \operatorname{Cos} \alpha_2 \operatorname{Sin} \psi}{Z} \tag{17}$$

$$= \frac{\pi * 15.6 * \operatorname{Cos} 77.22^0 * \operatorname{Sin} 30^0}{8}$$

so,

$$V_s = 0.6775 m/s$$

As

$$\tan \alpha_2' = \tan \alpha_2 - \frac{V_s}{C_{m2}} \tag{18}$$

$$= 4.41 - \frac{0.6775}{2.6} = 4.15$$

$$\therefore \alpha_2' = \tan^{-1} 4.15 = 76.45^0$$

Here α_2' is the modified absolute flow angle at outlet. Using the same procedure, the blade outlet angles for every section of the blade have been modified.

Now again,
Mean modified blade angle,

$$\alpha_m = \tan^{-1}\left[\frac{\tan\alpha_1 + \tan\alpha_2}{2}\right] = 77.06^0 \tag{19}$$

and pitch S is

$$S = \frac{\pi D_{mA}}{Z} \tag{20}$$

where Z= 8

$$D_{mA} = \frac{D_{1A} + D_{2A}}{2} \tag{21}$$

so,

$$D_{mA} = \frac{0.274 + 0.298}{2} = 286mm$$

Here D_{1A} and D_{2A} are measured from the Figure 2.

$$S_{A-A} = \frac{\pi * 286}{8} = 112.31mm$$

The axial chord, at section A-A, C' = 24 mm [measured from the Figure 1]. Based on that, the blade angles and stagger angles are calculated using the following iteration method:
For 1st approximation, Actual chord:

$$C_1 = \frac{C'}{\text{Cos}\,\lambda} = \frac{C'}{\text{Cos}\,\alpha_m} = \frac{24}{\text{Cos}\,77.06^0} = 107.18mm$$

[Where λ = stagger angle = α_m (assumed for the first iteration)]

∴Camber angle $\theta_1 = \beta_1 - \beta_2 = \alpha_1 - \alpha_2' + \delta^* = 77.62° - 76.45° + 0 = 1.17°$

Now deviation

$$\delta_1^* = m\theta_1\sqrt{\frac{S}{C_1}} \text{ [Carter's correlation]} \tag{22}$$

The variation of 'm' vs. 'λ' is shown in Figure 43 (Appendix 3).

$$\delta_1^* = m\theta_1\sqrt{\frac{S}{C_1}} = 0.47*1.17*\sqrt{\frac{112.31}{107.18}} = 0.563^0$$

∴ modified stagger angle

$$\lambda_1 = \alpha_2^{'} + \frac{\theta_1}{2} - \delta_1^* = 76.45^0 + \frac{1.17^0}{2} - 0.563^0 = 76.472^0$$

For 2nd iteration
Actual chord,

$$C_2 = \frac{C^{'}}{\text{Cos}\,\lambda_1} = \frac{24}{\text{Cos}\,76.472^0} = 102.6mm$$

∴Camber angle $\theta_1 = \beta_1 - \beta_2 = \alpha_1 - \alpha_2^{'} + \delta^* = 77.62° - 76.45° + 0.563° = 1.733°$

Now deviation

$$\delta_2^* = m\theta_2\sqrt{\frac{S}{C_2}} \text{ [Carter's correlation]} \tag{23}$$

$$\therefore \delta_2^* = 0.46*1.733*\sqrt{\frac{112.31}{102.6}} = 0.834^0$$

∴ the modified stagger angle

$$\lambda_2 = \alpha_2^{'} + \frac{\theta_2}{2} - \delta_2^* = 76.45^0 + \frac{1.733^0}{2} - 0.834^0 = 76.48^0$$

The iterations were carried out till the difference between the old stagger angle (λ) and the modified stagger angle was 0.01.Likewise, the values of α_1, α_2, α_2', θ, δ^*, S, C, λ are calculated for each sections from B-B to J-J and checked that $\alpha_1 > \alpha_2 > \alpha_2$'.

The blade sections were also verified to satisfy the Leiblein blade loading criteria to avoid separation on the suction side of the cascade aerofoil. Leiblein blade loading diffuser factor D_L is given as:

$$D_L = \left[\left(1 - \frac{\cos \alpha_1}{\cos \alpha_2^{'}} \right) + \frac{S}{C} * \frac{\cos \alpha_1}{2} \left(\tan \alpha_1 - \tan \alpha_2 \right) \right] > 0.6 \tag{24}$$

The blade angles at inlet and outlet for all 10 sections are tabulated in table 2. Table 3 shows the span wise variation blade section of chord lengths, stagger and camber angles.

Table 2. Blade angles at inlet and outlet for different sections

Position	D_1(mm)	D_2(mm)	α_1(mm)	$\alpha_{2'}$(mm)	α_m (mm)	S(mm)
A-A(tip)	274	298	77.62	76.45	77.06	112.31
B-B	260	290	76.97	75.74	76.39	108
C-C	244	280	76.15	74.77	75.49	102.89
D-D	230	272	75.34	73.84	74.62	98.57
E-E	218	264	74.57	72.82	73.73	94.64
F-F	204	256	73.57	71.6	72.65	90.32
G-G	190	248	72.43	70.23	71.39	86
H-H	176	240	71.14	68.58	69.95	81.68
I-I	162	232	69.62	66.55	68.18	77.36
J-J(hub)	150	224	68.14	64.09	66.32	73.43

Table 3. Span wise variation blade section of chord lengths, stagger and camber angle

Position	Actual Chord, C (mm)	S/C	λ(°)	θ(°)	δ(°)	Incidence i (°)	$\beta_2=\alpha_2'-\delta$ (°)
A-A (tip)	102.6	1.095	76.48	1.733	0.834	0	75.616
B-B	122.55	0.881	75.86	1.751	0.748	0	74.992
C-C	134.86	0.743	74.98	1.911	0.749	0	74.021
D-D	153.5	0.642	74.15	2.234	0.806	0	73.034
E-E	161.77	0.585	73.19	2.335	0.795	0	72.025
F-F	172.25	0.524	72.11	2.788	0.878	0	70.722
G-G	176.27	0.488	70.84	3.061	0.919	0	69.311
H-H	180.98	0.451	69.34	3.489	0.984	0	67.596
I-I	183	0.423	67.52	4.167	1.111	0	65.439
J-J(hub)	180.23	0.407	65.43	5.4	1.361	0	62.729

Figure 5. Spanwise variation of blade angles from hub to tip

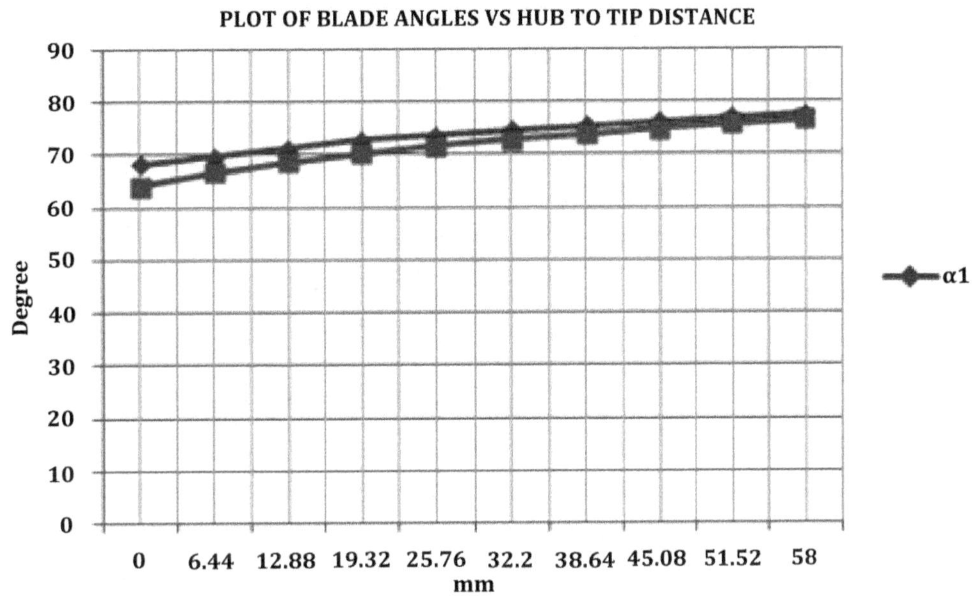

Figure 5 shows the span wise variation of blade angles from hub to tip. The variation of mean blade angles from hub to tip is shown in Figure 6. The spanwise variation of camber angle, S/C, Actual Chord Length is shown in Figures (7-9). From the above Figures, it can be observed that the span wise variation of those quantities are quite smooth, which is very much desired for a good design.

Figure 6. Variation of mean blade angles from hub to tip

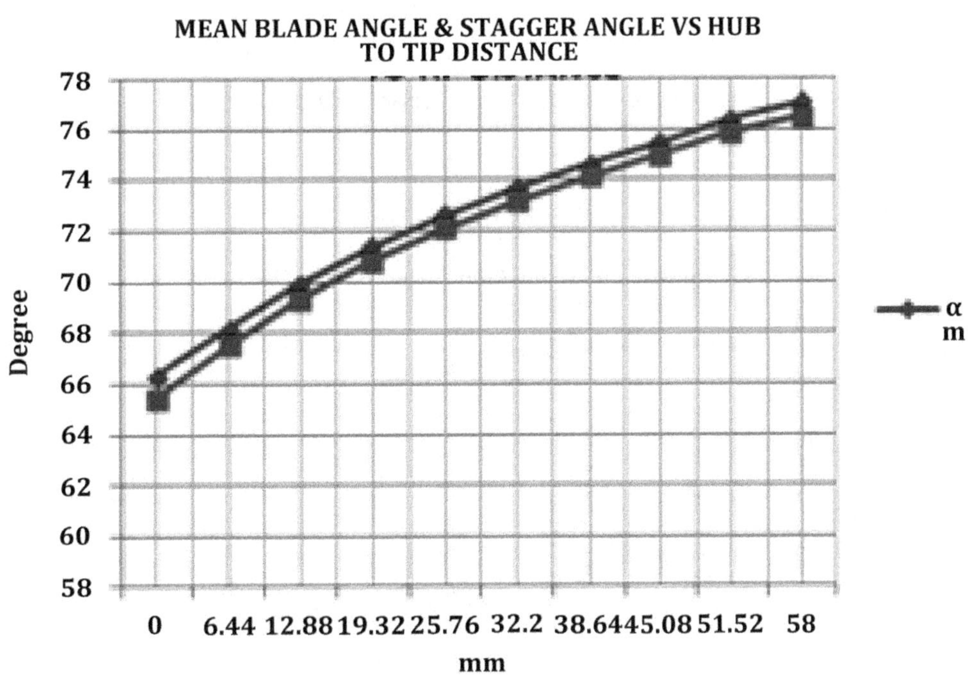

Figure 7. Variation of camber angle from hub to tip of the blade

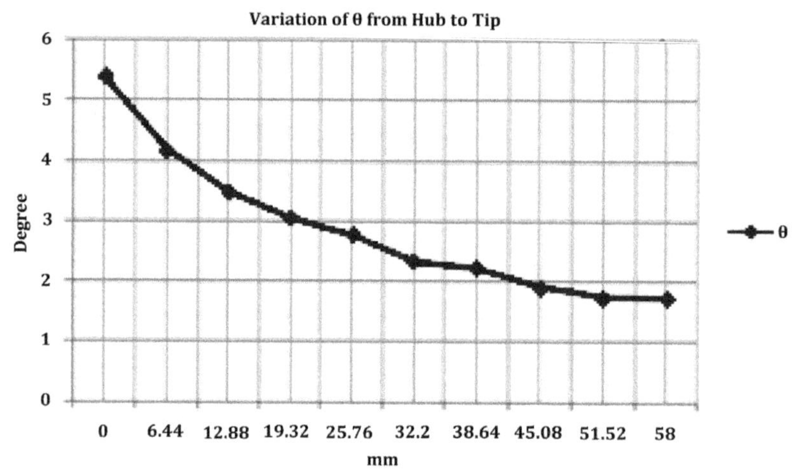

Figure 8. Variation of S/C from hub to tip of the blade

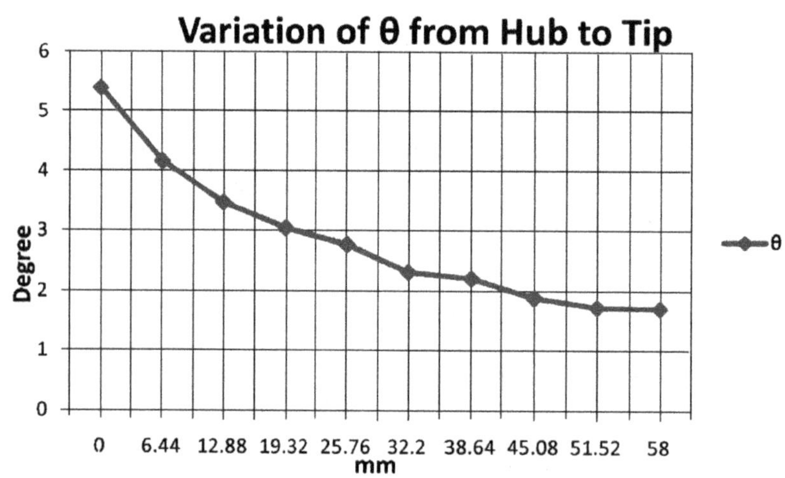

Figure 9. Variation of Actual Chord Length from Hub to Tip

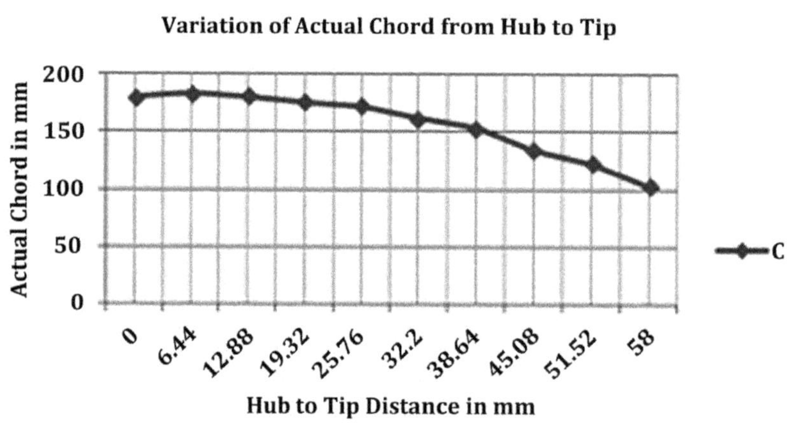

4.5.3 Transformation of Straight Blade Cascade to Cambered Cascade

In this present work, 10C4 blade profile is used for the impeller blade. Table 4 shows the uncambered 10C4 data profile and the corresponding blade thickness, as percentage of chord length. The uncambered 10C4 profile is also shown graphically in Figure 10 for convenience.

From table 4, upper and lower surface coordinates of the blade can be obtained only when the actual chord length 'C' of a 10C4 aerofoil is already known.

Here, a circular arc profile is considered as camber profile for the blade as shown in Figure 11. To determine the upper surface and lower surface point coordinates of the aerofoil after giving a specified

Table 4. 10C4 Blade Profile Data

Distance from Leading Edge	Thickness	Distance from Leading Edge	Thickness	Distance from Leading Edge	Thickness
0	0	20.0	9.66	90.0	3.20
1.25	3.30	30.0	10.0	95.0	2.12
2.5	4.54	40.0	9.78	100.0	0
5.0	6.16	50.0	9.14		
7.5	7.24	60.0	8.01		
10.0	8.04	70.0	6.74		
15.0	9.10	80.0	5.08		

Leading edge radius = 1.2% of chord length and Trailing edge radius = 0.6% of chord length.

Figure 10. 10C4 blade profile without camber and stagger

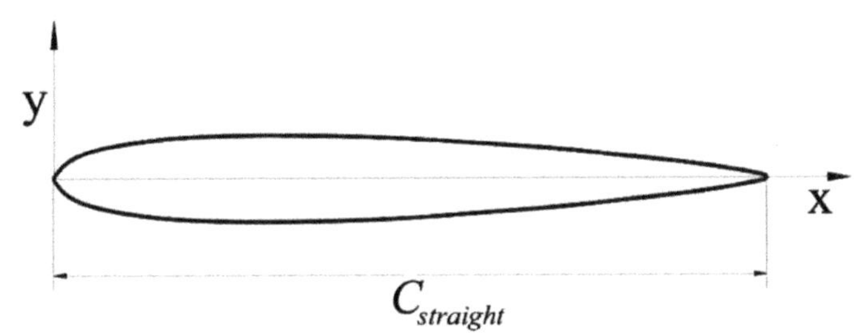

Figure 11. 10C4 blade profile with specified camber

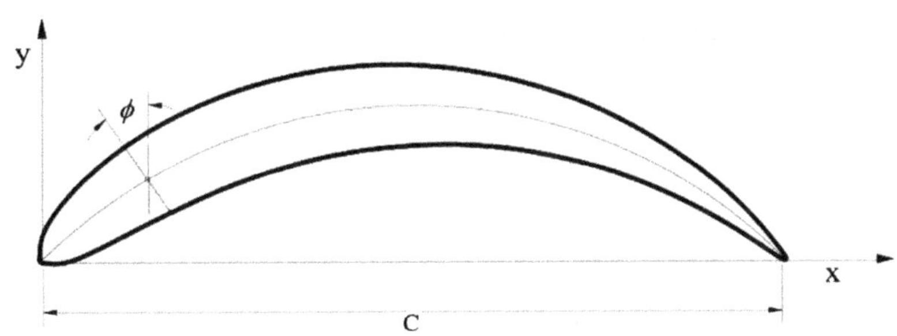

Figure 12. Geometrical description of a circular camber line

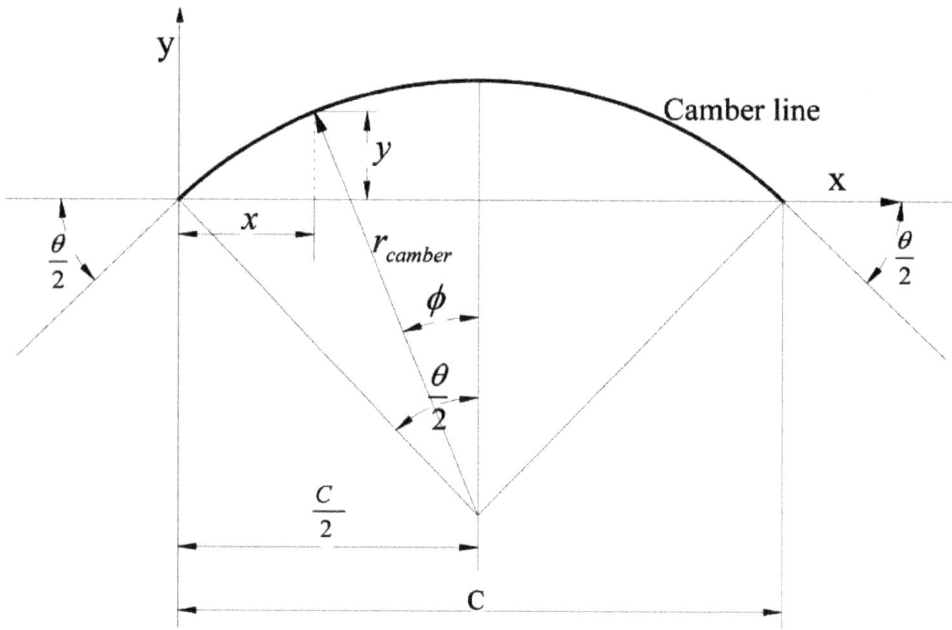

camber, the coordinates of points on camber line should be determined first. Then the required coordinates can be found out with the help of predetermined straight aerofoil coordinates data and camber line coordinates. Geometrical description of a circular camber line is shown in Figure 12.

Here, (x_c, y_c) is the Cartesian coordinate of any point on the centre line of the cambered aerofoil.

From Figure 12, where, θ = camber angle,

Now, for circular arc camber (Figure 12),

$$\theta_1 = \theta_2 = \frac{\theta}{2} \tag{25}$$

Again,

$$\text{Sin}\frac{\theta}{2} = \frac{\left(\frac{C}{2}\right)}{R} \tag{26}$$

where, R= radius of curvature of the camber line

$$\text{or, } R = \frac{C}{2\,\text{Sin}\frac{\theta}{2}} \tag{27}$$

Now at any distance 'x' from the leading edge,

$$\operatorname{Sin}\alpha = \left[\frac{\frac{C}{2} - x_c}{R}\right] \tag{28}$$

or

$$\alpha = \operatorname{Sin}^{-1}\left[\frac{\frac{C}{2} - x_c}{R}\right] \tag{29}$$

After getting coordinates of points on camber line, coordinates of points on upper and lower surface of the aerofoil can be determined with the help of the following geometric relations:

$$x_u = x_c - \left(\frac{t}{2}\right)\sin\alpha \tag{30}$$

$$y_u = y_c + \left(\frac{t}{2}\right)\cos\alpha \tag{31}$$

$$x_l = x_c + \left(\frac{t}{2}\right)\sin\alpha \tag{32}$$

$$y_l = y_c - \left(\frac{t}{2}\right)\cos\alpha \tag{33}$$

where t = thickness of the aerofoil and (x_u, y_u) or, (x_l, y_l) is the Cartesian coordinate of any point on the upper surface or lower surface of the aerofoil with a certain camber respectively.

5. 3D MODELING OF IMPELLER BLADE

Once the blade camber, blade angles, blade stagger angles are known from the calculations, a 3D model of the mixed flow turbomachinery blade has been prepared using CATIA V5 software. The salient steps involved in modeling of the blade are shown below:

1. Creating the various airfoil sections using UDF (Figure 13)
2. Extruding various sections and stacking them to desired stagger angle (Figure 14)
3. Surface profile was rotated to maintain desired stagger angle (Figure 15)
4. Connecting the trailing points of airfoil with a spline curve (Figure 16)
5. Joining all the surfaces to form a single 3-D surface of the blade (Figure 17)

Figure 13. Airfoil section using UDF

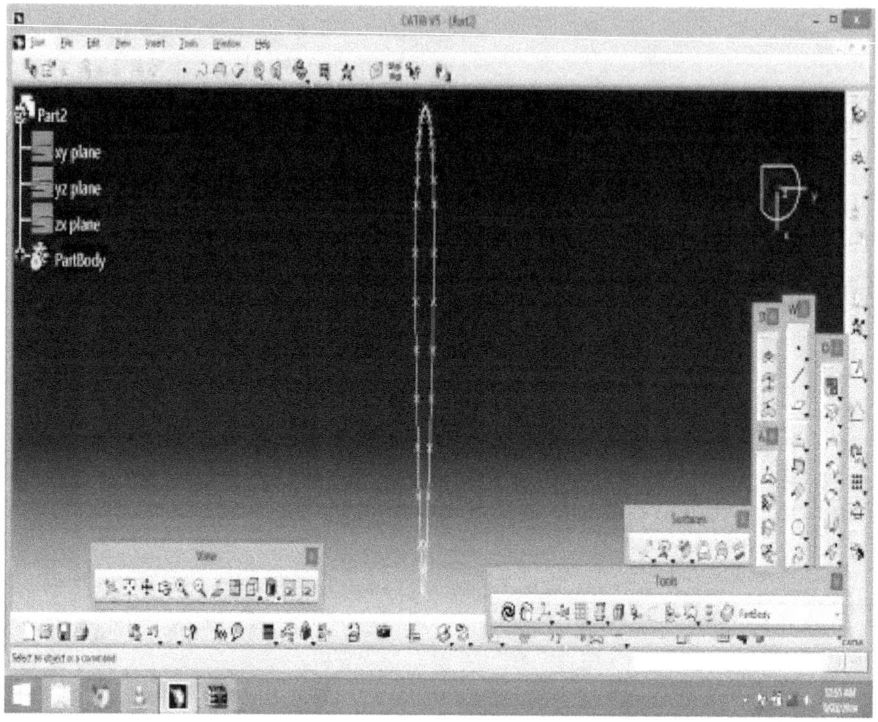

Figure 14. Picture of extruded and stacked

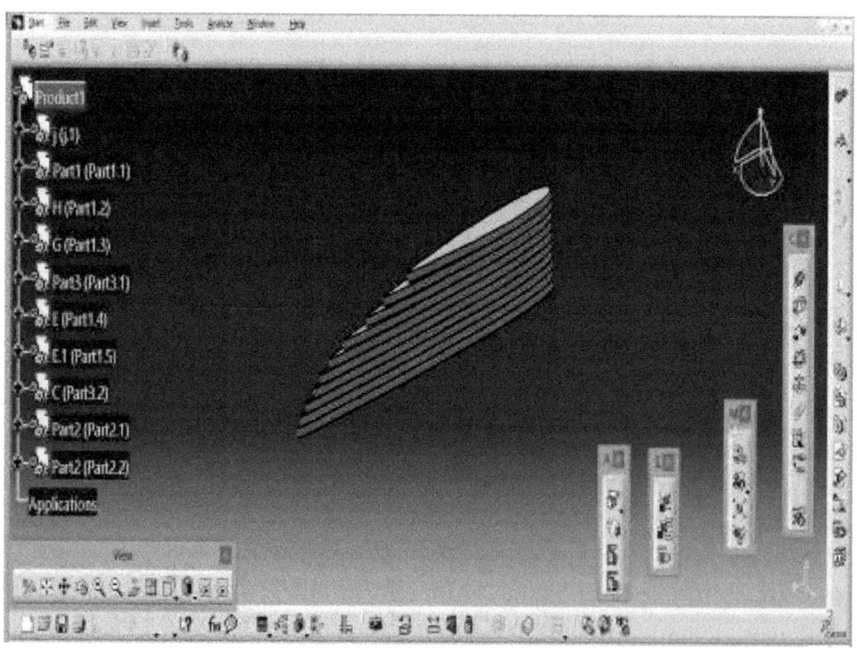

Figure 15. Surface profile rotated

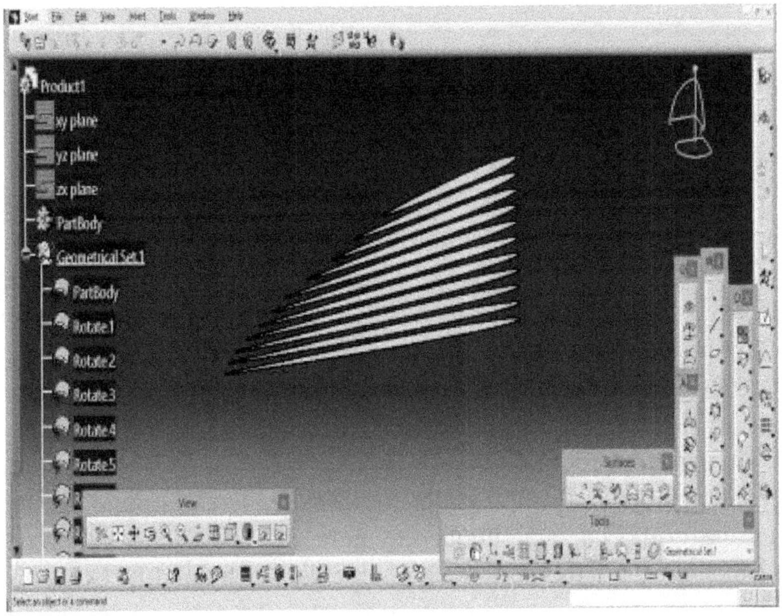

Figure 16. Trailing points of airfoil

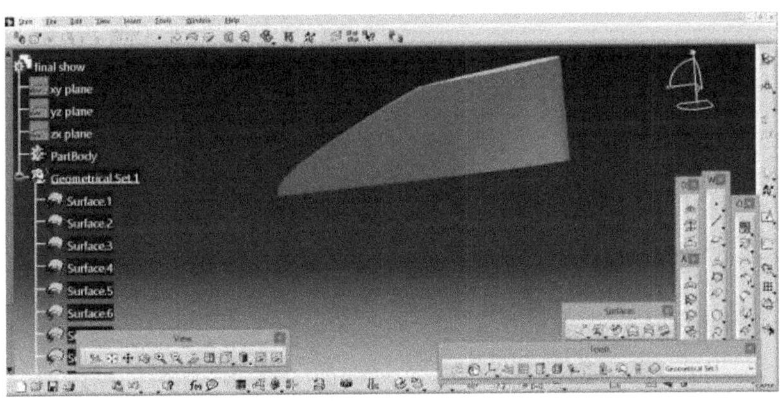

Figure 17. 3D surface of blade profile

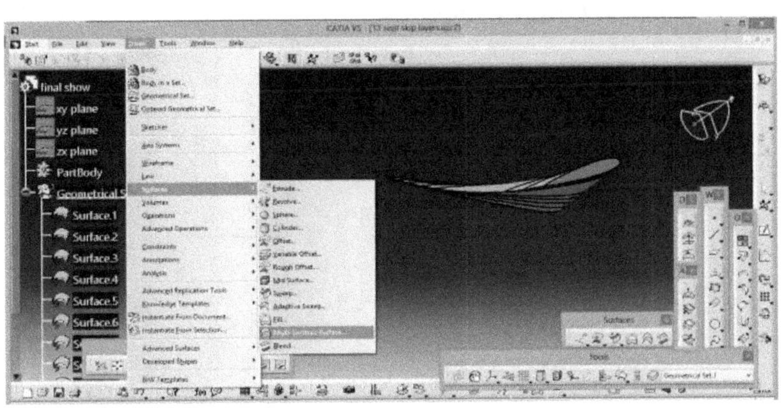

6. FINITE ELEMENT METHOD ANALYSIS

The Finite Element analysis of a mixed flow pump impeller blade is required in order to pick the best option out of the designs made in the earlier sections. The fundamental designs were developed based on the principles of fluid mechanics and turbomachinery and the design and development was carried out. The blade positioning on the hub can be under three categories, namely inlet inclined (case-I), trapezoidal (case-II) and outlet inclined (case-III). It has been observed that the situation in case III is quite unstable and hence was not considered in the present work. The pictorial representation of the two cases can be visualized in Figure 18 and Figure 19 respectively and Figure 20, Figure 21 and Figure 22.

The top view of the different blade sections for inlet inclined and trapezoidal blade positioning in the meridional annulus are shown in Figure 20 and Figure 21 respectively. The 3D models of the inlet inclined blade positioning (case-I) and trapezoidal blade positioning (case-II) in the meridional annulus are also shown in Figure 22 and Figure 23 respectively.

Since the blades are extremely complicated in geometry and as well as the complex loading characteristics, the finite element size selection is done with an equivalent plate having rectangular cross section acting like a cantilever. The data obtained by numerical method is validated with ANSYS and the element size at which both the methods synchronize, is determined. It is assumed that each blade of the mixed flow pump impeller acts as a cantilever fixed to the hub and due to torque applied on the impeller; a surface force acts on the blade.

Torque applied on the plate:

$$T = w * l * b * \frac{l}{2} \tag{34}$$

Also, torque applied:

$$T = (P.(60).1000)/(2.\pi.N) \tag{35}$$

where, P = 20 Kw, N = 1000 rev. /min., n = number of blades in the impeller = 8, L = span of each plate= 58 mm, b = mean width of each plate= 162 mm, t = thickness of each plate = 17 mm. So from the above equation,

$$w = \frac{2*T}{L^2*b*n} = \frac{2*(191000)}{58^2*(162)*8} = 0.088N/mm^2 \tag{36}$$

Bending stress is given by the relation:

$$\sigma_b = \frac{M}{I} \tag{37}$$

Figure 18. Different positions of mixed flow pump impeller blades in the meridional annulus (Case-I: Forward)

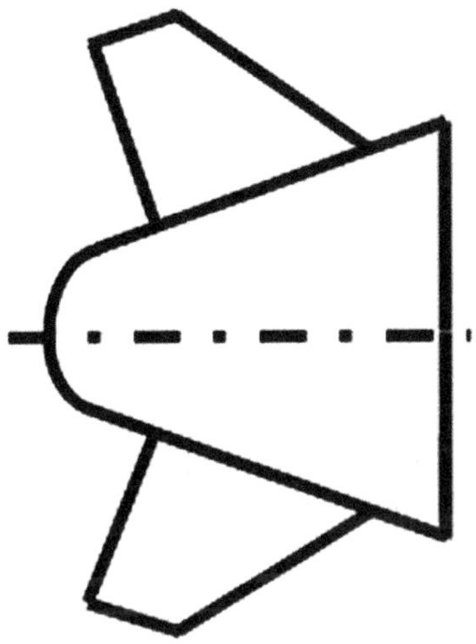

Figure 19. Different positions of mixed flow pump impeller blades in the meridional annulus (Case-II: Trapezoidal)

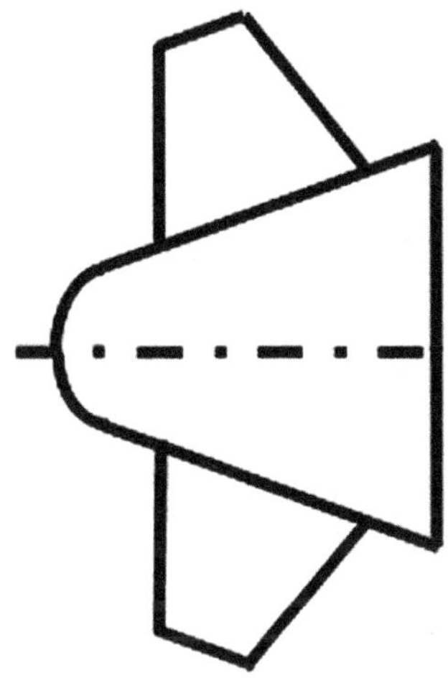

Figure 20. Top view of different blade sections (case-I)

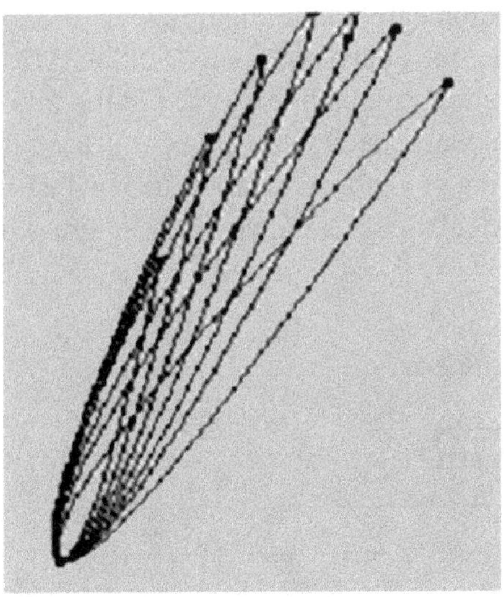

Figure 21. Top view of different blade sections (case-II)

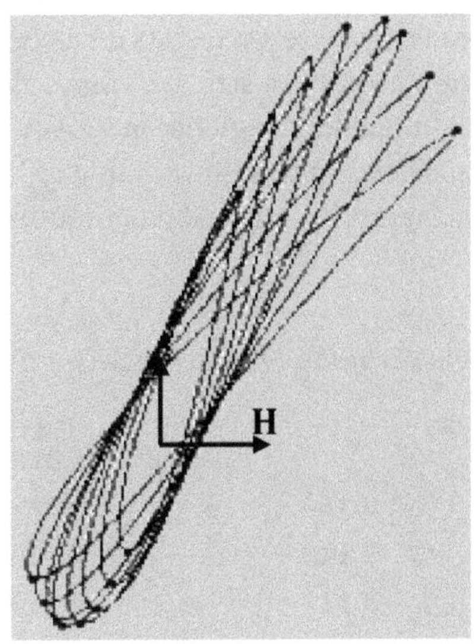

Figure 22. 3D model of the impeller blade (case-I)

Figure 23. 3D model of the impeller blade (case-II)

where, M = bending moment = w . b N-m, I = moment of inertia = (b . t^3) /12 m^4, y = varying distance from neural axis, m and shear stress can be written as:

$$\tau = \frac{6 * w * b * l}{b * t^3} * \frac{(t^2)}{(4 - y^2)} \tag{37}$$

So, Von Mises stress is given by the relation:

$$\sigma_b = \sqrt{\sigma^2 + 3\tau^2} \tag{39}$$

So, the calculated value of the Von Misses stress at the root of the plate was calculated as 3.07299x 10^6 N/m^2. While calculating the Von Misses stress the following material properties were taken into consideration.

Material: Bronze; Young Modulus: 1.1 x 10^5 MPa; Poisson Ratio: 0.341; Density: 8.86 x 10^{-9} kg/mm^3; Thermal Expansion: 1.78x 10 $^{-5}$/°K; Yield Strength: 5.2 x 10^8 N/m^2

To finalize the mesh size in ANSYS convergence tests were conducted. The von Mises stress of the cantilever beam was calculated with seven independent mesh sizes, viz. 4.15mm, 4.2mm, 4.25mm, 4.3mm, 4.35mm, 4.40mm and 4.45mm. Results are shown graphically in the Figure 24

From the above seven results it can be clearly observed that the analysis done by ANSYS 11.0 software with global mesh size 4.25 mm validate with the theoretical calculations, hence in the following analysis of impeller the global mesh size of 4.25mm is taken. Here, tetrahedral element is used for the present analysis. Span wise variation of Von Mises stress for the plate for an element size of 4.25 mm is shown in Figure 25. Using the optimum element size, numerical analysis for Von Misses stress on the

Figure 24. Variation of max.von Mises Stress for different mesh size

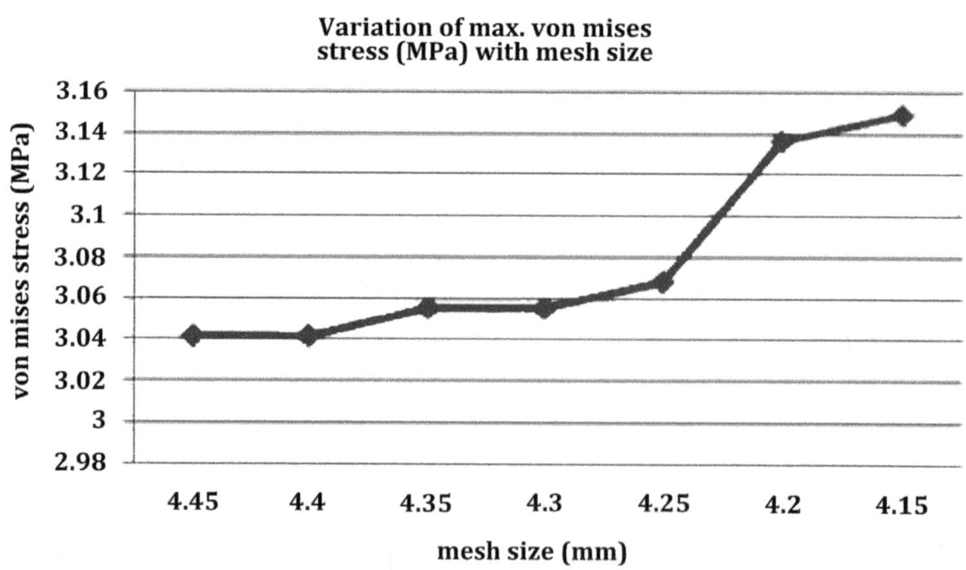

Figure 25. Span-wise variation of Von Misses stress for plate

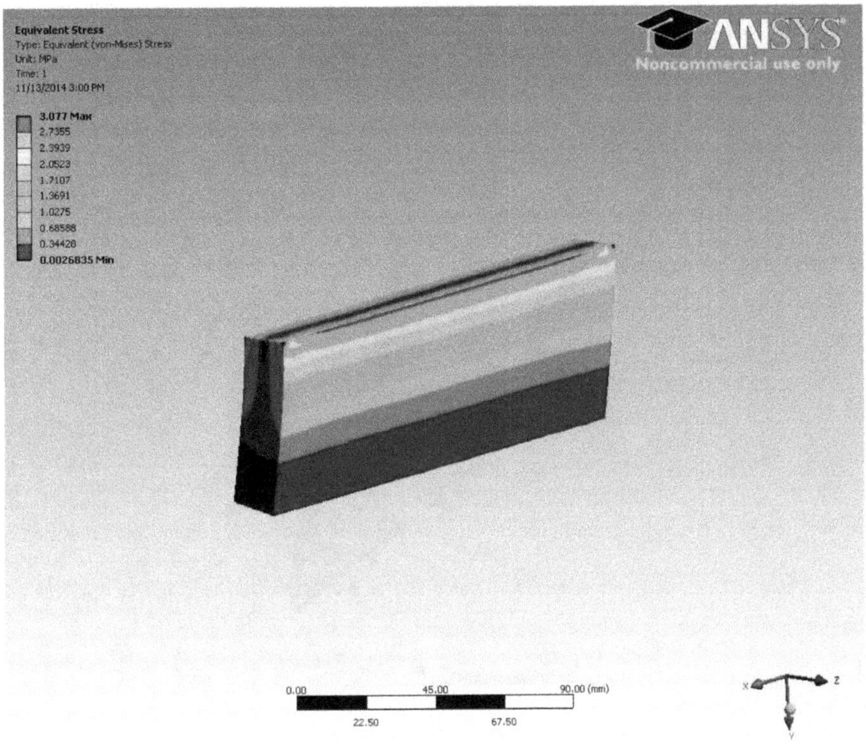

pump impeller blades has been carried. The span wise variation of Von Misses stress for case-I and II are shown in Figures 26 and Figure 27 respectively.

From the Figure 26 and Figure 27 it is observed that the Von Misses stress at the root of the pump blade for inlet inclined blade position is 12.131 MPa, whereas for trapezoidal blade positioning in the meridional annulus it is 24.322 MPa. This indicates that the pump blade with inlet inclined position is more stable than the same with trapezoidal position.

7. VALIDATION OF BLADE POSITION USING ARTIFICIAL NEURAL NETWORK (ANN)

In both the cases of mixed flow pump impeller NACA10C4 aero foil cross-section blade is treated as cantilever beam. And in these two cases Von mises stress was analyzed by ANSYS 11.0 under UDL loading. In order to ascertain the validity of the results obtained by FEM analysis, Artificial neural network (Figure 28-29) was performed by cutting the plane of the blade along and across in von mises analysis. The lower profile of the aero foil cross section of the blade was fixed and varying thickness was measured through it. Number of nodes taken from the meshed blade was 25 for von mises. Von mises value was taken as an input variable in artificial neural network and the thickness as target value, each of the input variables were used to train Neural Network with feed forward technique done with RAPID MINER 3.5.

Figure 26. Span-wise variation of Von Misses stress (case-I)

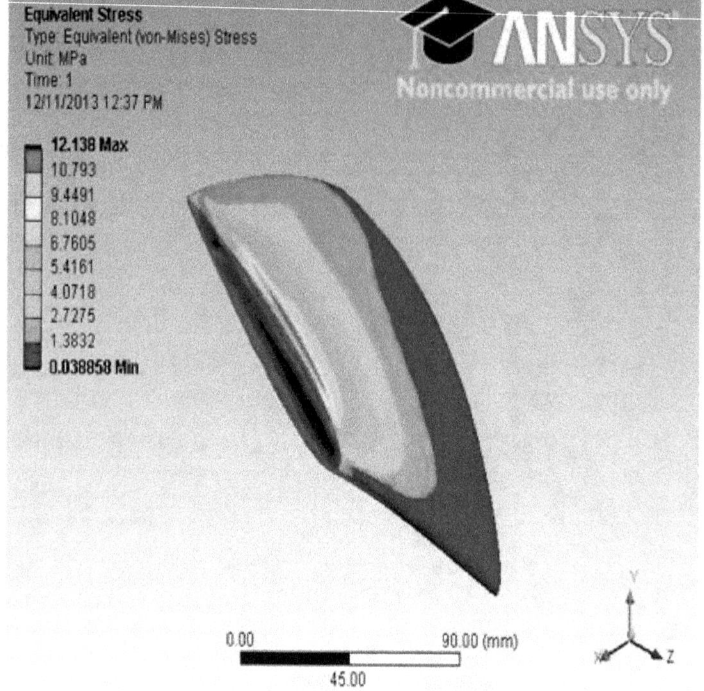

Figure 27. Span-wise variation of Von Misses stress (case-II)

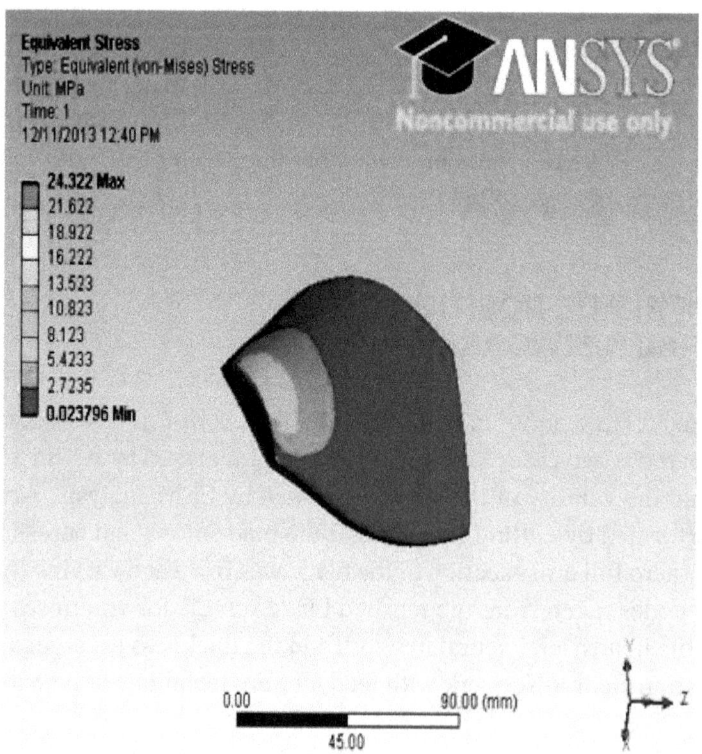

Fully trained neural network predicted the new varied thickness along and across the profile, with the help of these predicted values Mean Square Error is calculated below and tabulated in Tables 5 and 6.

$$MSE = \frac{1}{n}\sum_{i=1}^{n}\left(\hat{Y}i - Yi\right)^2 \tag{40}$$

where, $\hat{Y}$ is a vector of n predictions, and Y is the vector of the true values.

Predicted thickness value and true value is compared against both the cases (I & II) results are shown below in Figure 30 and Figure 31

As shown in Figures 30 and 31 it can be clearly seen that, for minimum Von-misses stress, the critical thickness of for case-I is 2.5mm, whereas for case-II, it is 6.7mm. This estimated result indicates that for trapezoidal blade thickness needs much improvement than that for the inlet inclined blade as depicted in Table 5 and Table 6. These data confirms the suitability of inlet inclined blade thickness over trapezoidal blade thickness with reference to the respective von-misses stress across the span.

Figure 28. Model applied in Rapid miner 5.3

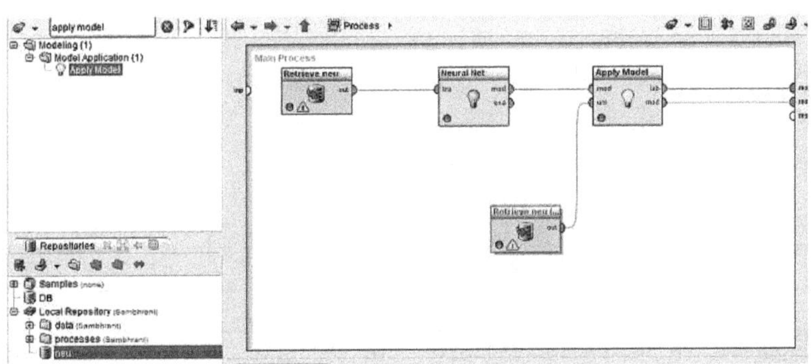

Figure 29. Neural network diagram with 1 input and 2 hidden layer

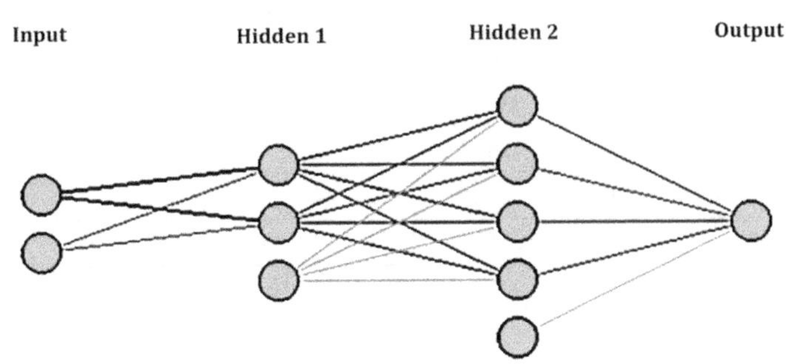

Table 5. Estimation of mean square error of thickness for inlet inclined blade

Node NO.	Von-Mises Stress (MPa)	Thickness (mm)	Predicted Thickness (mm)	Mean Square Error %	Node NO.	Von-Mises Stress (MPa)	Thickness (mm)	Predicted Thickness (mm)	Mean Square Error %
1	12.138	0	10.6873	277.389	14	9.9481	10.506	10.6551	6.86335
2	5.1323	1.395	4.50384	169.261	15	9.8773	11.057	10.6518	20.9233
3	3.3975	1.72	3.03054	143.993	16	10.484	11.507	10.6722	23.5135
4	2.0427	2.765	2.9432	93.4562	17	10.485	10.506	10.6722	4.86007
5	1.9386	3.53	2.94121	62.4388	18	10.066	11.994	10.6599	36.4132
6	2.5211	4.797	2.95718	33.3122	19	10.189	12.233	10.6643	54.2451
7	5.3412	6.463	5.00675	8.96395	20	10.272	12.651	10.6668	81.9798
8	5.9452	6.781	6.77455	8.69095	21	10.173	12.76	10.6637	102.528
9	6.8917	7.7	9.11683	2.81809	22	9.3354	12.547	10.6136	111.36
10	8.0137	7.18	10.2792	3.9576	23	8.5445	12.1755	10.4799	113.61
11	8.8443	7.052	10.5467	15.9824	24	6.6713	8.847	8.69232	23.4275
12	9.9153	7.875	10.6536	8.4396	25	3.3419	0	3.02235	147.827
13	9.5582	9.572	10.6325	0.00976					

Table 6. Estimation of mean square error of thickness for Regular trapezoidal blade

Node NO.	Von-Mises Stress (MPa)	Thickness (mm)	Predicted Thickness (mm)	Mean Square Error %	Node NO.	Von-Mises Stress (MPa)	Thickness (mm)	Predicted Thickness (mm)	Mean Square Error %
1	23.257	0	-1.632	977.300	14	7.7642	9.6257	6.773	144.250
2	7.2213	1.337	6.779	817.262	15	6.9358	9.7825	6.781	136.815
3	3.213	2.791	6.797	659.452	16	6.1315	9.8856	6.787	132.034
4	2.0573	3.546	6.799	584.179	17	6.1358	9.9811	6.787	127.681
5	1.9057	4.277	6.799	515.644	18	5.7113	9.9153	6.789	130.672
6	2.182	5.103	6.799	443.347	19	5.0342	9.8342	6.792	134.407
7	2.7823	5.827	6.798	384.466	20	4.5621	9.6526	6.794	142.960
8	3.3872	6.5149	6.797	332.406	21	4.5743	9.09609	6.794	170.815
9	4.2571	7.132	6.795	288.925	22	3.3498	8.1737	6.797	222.439
10	4.0175	7.5667	6.795	260.125	23	2.5792	7.9438	6.798	236.366
11	6.3007	8.1083	6.786	226.358	24	2.0482	4.6218	6.799	484.801
12	7.4087	8.5309	6.777	201.640	25	4.2348	0	6.795	977.300
13	8.7812	9.3956	6.759	155.516					

Figure 30. Predicted thickness variation (Case-I)

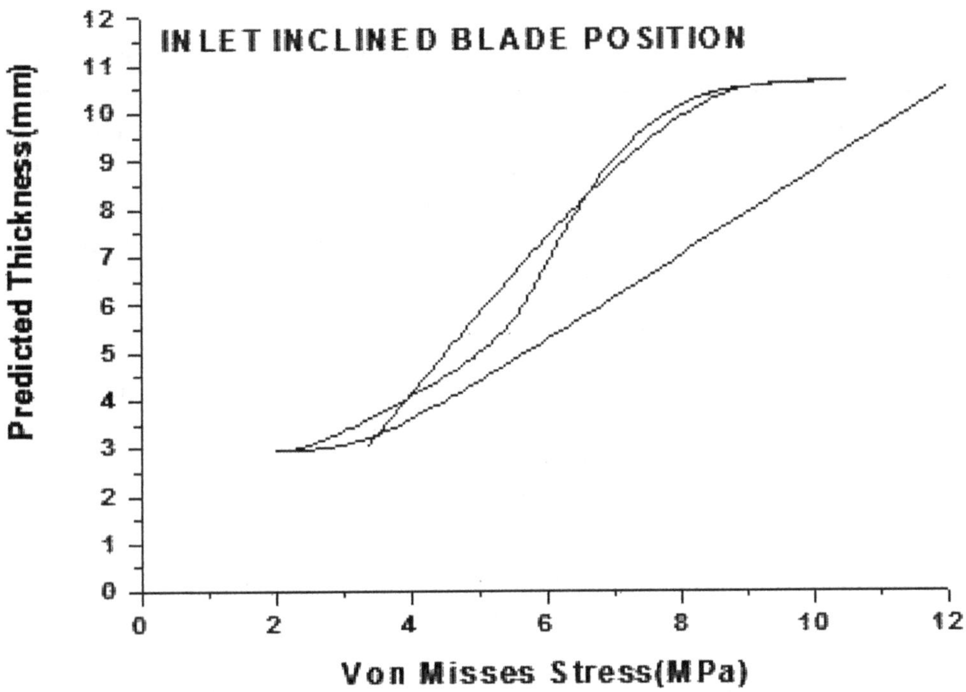

Figure 31. Predicted thickness variation (Case-II)

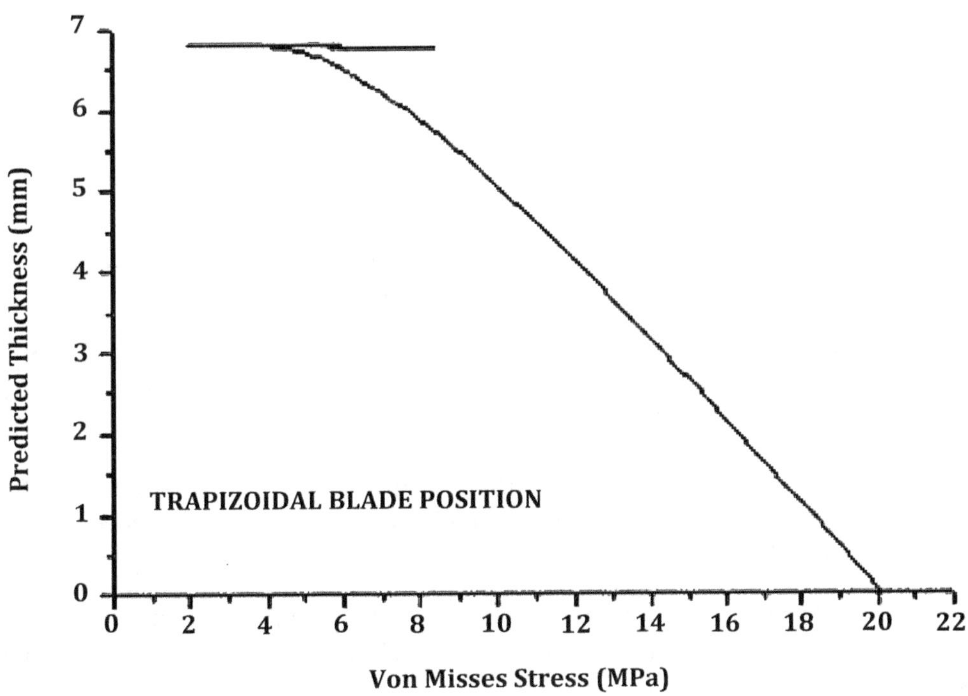

8. CONCLUSION

Here, in this research paper, a design of mixed flow turbo machine is discussed by taking into consideration of the basic principle of turbomachinery and fluid mechanics. The developed model is then optimized and the best design would be proposed having a specific material, blade orientation and thickness in support of the objectives for usage in maritime applications. The effectiveness of the design of the blade profile and the positioning depends on the stresses developed at the root of the blades. From the design analysis it is observed that the Von Misses stress at the root of the pump blade for inlet inclined blade position was much lower than that for trapezoidal blade positioning in the meridional annulus which indicates that the inlet inclined blade position is a better option than the other one for the application. The observation was supported by using the artificial neural network analysis where the estimated mean square error was less in case of inlet inclined blade position than that for trapezoidal blade positioning in the meridional annulus. This also indicated that the improvement required in the inlet inclined blade position would be less than that for trapezoidal blade positioning.

A number of research work, though, are available on the flow analysis through mixed flow pumps, but research on design of mixed flow pump impeller blades using stress analysis and validation of the same via heuristics network are very less. This is considered as one of very little research work where a practical design and performance analysis procedure of a mixed-flow pump, in which the conceptual approach to turbomachinery design using the mean line analysis is followed by the detailed design and analysis, has been presented. Design and predictive procedure, including cavitations, employed in this study may serve as a reliable tool for the detailed design optimization and assist in understanding operational characteristics of general purpose hydraulic and compressible flow turbomachinery. This work can be further developed / modified in the line of validation by other sophisticated heuristics technique. Hybridization of one or more heuristics may validate this model in an '*even more accurate direction*' which may explore even better area of application in engineering and management … research works in this direction is highly envisaged.

REFERENCES

Arewar, A. P., & Bhope, D. V. (2013). Stress analysis of axial flow fan impeller. *International Journal of Engineering Research and Applications*, *3*(1), 2086–2090.

Bhope, D. V., & Padole, P. M. (2004). Experimental and theoretical analysis of stresses, noise and flow in centrifugal fan impeller. *Mechanism and Machine Theory*, *39*(12), 1257–1271. doi:10.1016/j.mechmachtheory.2004.05.015

Bing, H. C., Shuliang, T. L., & Zhu, B. (2013). Effects of meridional flow passage shape on hydraulic performance of mixed-flow pump impellers. *Chinese Journal of Mechanical Engineering*, *26*(3), 469–475. doi:10.3901/CJME.2013.03.469

Busemann, A. (1928). The pump head ratio of radial centrifugal pumps with logarithmic spiral blades. [In German]. *Zeitschrift für Angewandte Mathematik und Mechanik*, *8*(5), 372–384. doi:10.1002/zamm.19280080504

Casandy, G. T. (1964). *Theory of Turbomachines*. McGraw Hill Book Company.

Gahlot, V. K., & Nyiri, A. (1993). *Impeller Pumps, Theory and Design*. Bhopal: M.A.C.T.

Horlock, J. H. (1958). *Axial Flow Compressor-Fluid mechanics and thermodynamics*. London: Butterworth Scientific Publications.

Inoue, M., Ikui, T., Kamada, Y., & Tashidaro, M. (1980). A quasi three dimensional design of diagonal flow impellers by use of cascade data. *IAHR, 10th Symposium*.

Jonker, J. B., & Van Essen, T. G. (1997). A finite element perturbation method for computing fluid-induced forces on a centrifugal impeller, rotating and whirling in a volute casing. *International Journal for Numerical Methods in Engineering*, *40*(2), 269–294. doi:10.1002/(SICI)1097-0207(19970130)40:2<269::AID-NME63>3.0.CO;2-G

Karnik, S. R. (2008). *Delimitation analysis in high speed drilling of carbon fiber reinforced plastics (CFRP) using artificial neural network model*. Elsevier Ltd.

Kim, J.-H., & Kim, K.-Y. (2011). Optimization of Vane Diffuser in a Mixed-Flow Pump for High Efficiency Design. *International Journal of Fluid Machinery and Systems*, *4*(1).

Lemeš & Zaimović-Uzunović. (2002). Mode shapes of centrifugal pump impeller. *6th International Research / Expert Conference on Trends in the Development of Machinery and Associated Technology, TMT 2002*. B&H.

Mehta, M. P., & Patel, P. M. (2013). Performance analysis and optimization of mixed flow pump. *International Journal of Emerging Trends in Engineering and Development*, *1*(3), 647–661.

Myles, D. J. (1965). *A design method for mixed flow fans and pumps*. Report No. 117. National Engineering Laboratory.

Neumann, B. (1991). *The interaction Between Geometry and Performance of a centrifugal Pump*. London: Mechanical Engineering Publications.

Panda, D. K., & Bhoi, R. K. (2005). Artificial Neural Network Prediction of Material Removal Rate in Electro-Discharge Machining. *Materials and Manufacturing Processes*, *20*(4), 645–672. doi:10.1081/AMP-200055033

Ramamurti, V., & Balasubramanian, P. (1987). Steady state stress analysis of centrifugal fan impellers. *Computers & Structures*, *25*(1), 129–135. doi:10.1016/0045-7949(87)90224-0

Samanta, B. (2008). Surface roughness prediction in machining using Soft computing. *International Journal of Computer Integrated Manufacturing*, 2008.

Senoo, Y. & Nakase, Y., (1972). *An analysis of flow for Power*. Paper No. 71-GT-2.

Sham Sunder, K. (1987). *Finite Element Analysis of Centrifugal Impellers*. (Ph.D. Thesis). School of Mechanical Engineering, Cranfield Institute of Technology, Cranfield.

Somashekhar, K. P., Ramachandran, N., & Mathew, J. (2010). Optimization of Material Removal Rate in Micro-EDM Using Artificial Neural Network and Genetic Algorithms. *Materials and Manufacturing Processes*, *25*(6), 467–475. doi:10.1080/10426910903365760

Srivastava, S., Roy, A. K., & Kumar, K. (2014). Design of a mixed flow pump impeller and its validation using FEM analysis. *Procedia Technology*, *14*, 181–187. doi:10.1016/j.protcy.2014.08.024

Stepanoff, A. J. (1957). *Centrifugal and Axial flow pumps* (2nd ed.). New York: John Wiley and Sons.

Stodola, A. (1945). *Steam and gas turbine*. Academic Press.

Takayama, Y., & Hiroyoshi, W. (2009). *Multi-Objective Design Optimization of a Mixed flow impeller*. ASME, Fluids Engineering Division.

Tosun, N. (2002). *A study of tool life in hot machining using artificial neural networks and regression analysis method*. Elsevier Science B.V.

Wislicensus, G. F. (1965). The Design of mixed flow pumps. *Proceedings of the symposium held at the N.E.L.*

APPENDIX 1

Figure 32. Schematic of a mixed flow impeller

Figure 33. Impeller Constants as function of Specific Speed (Stepanoff, 1957)

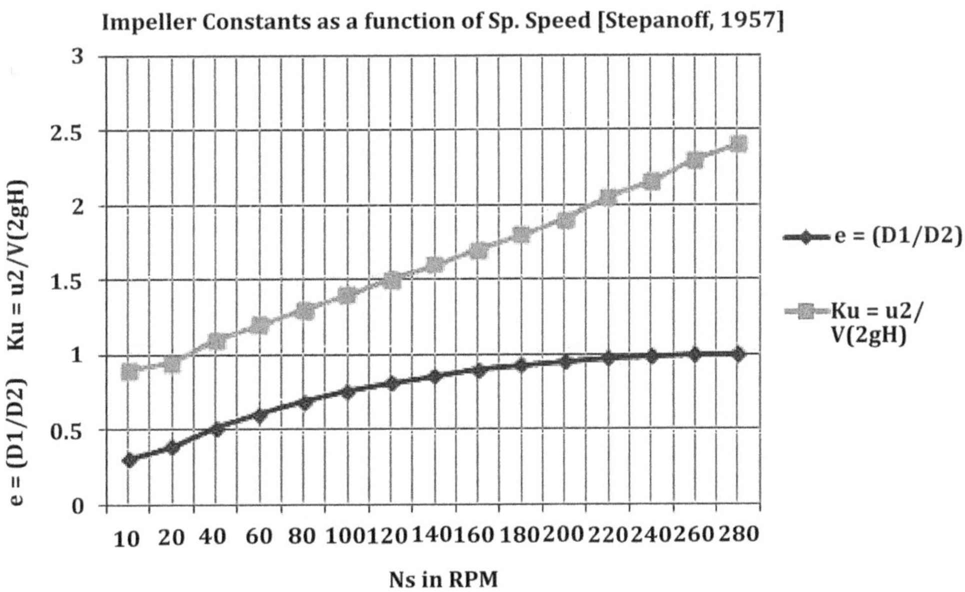

APPENDIX 2

Figure 34. Discharge vs. Sp. Speed plot for D1=200 mm

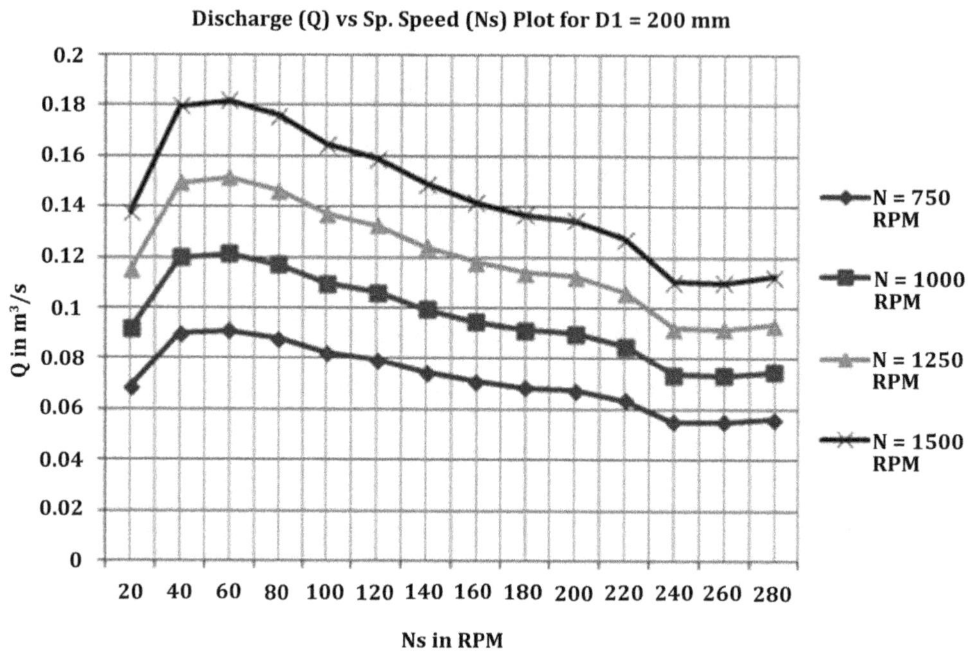

Figure 35. Head vs. Sp. Speed plot for D1=200 mm

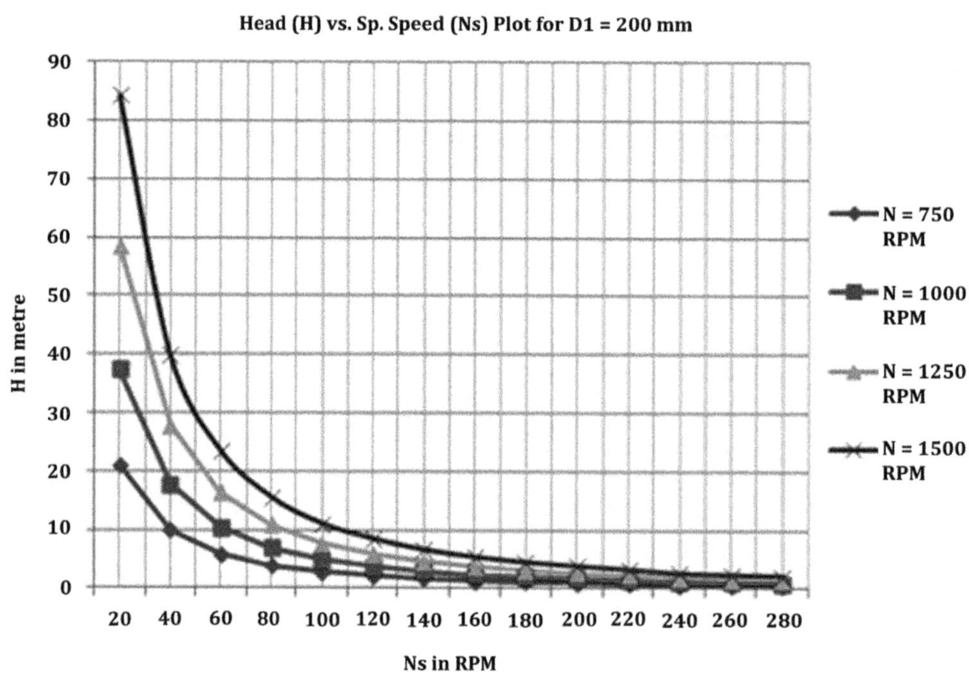

Figure 36. Power vs. Sp. Speed plot for D1=200 mm

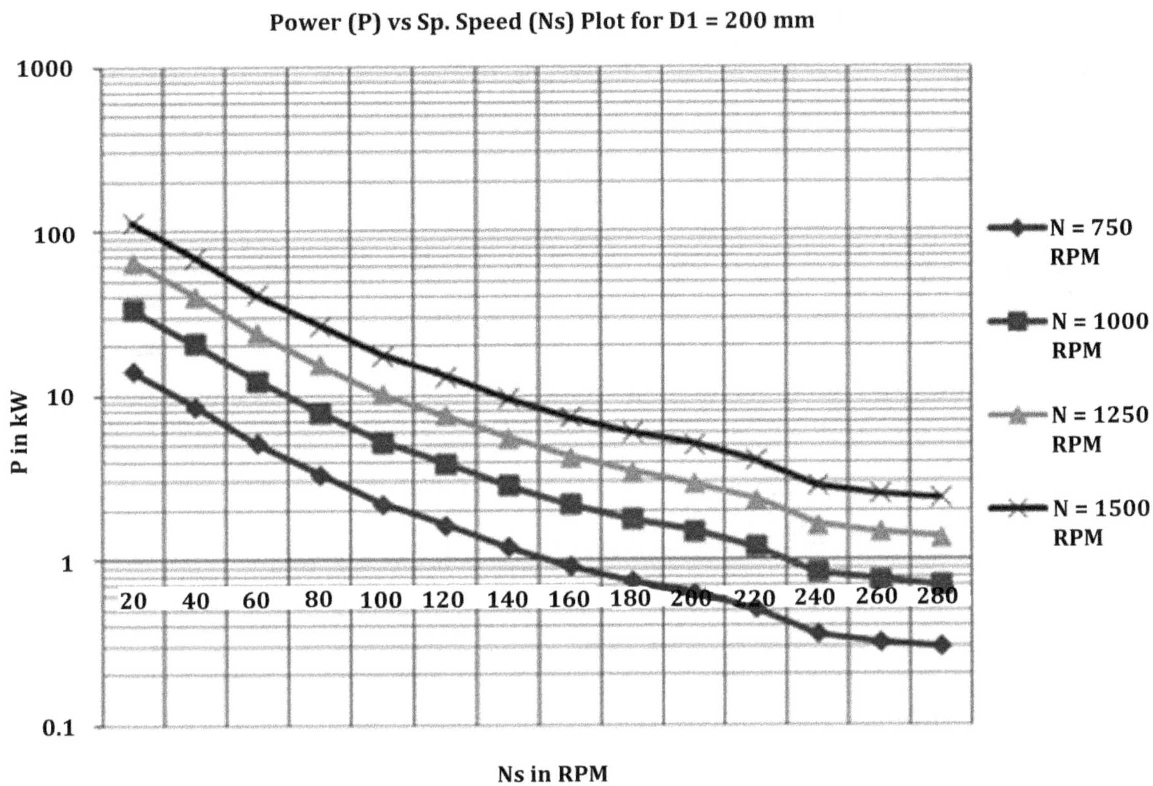

Figure 37. Discharge vs. Sp. Speed plot for D1=250 mm

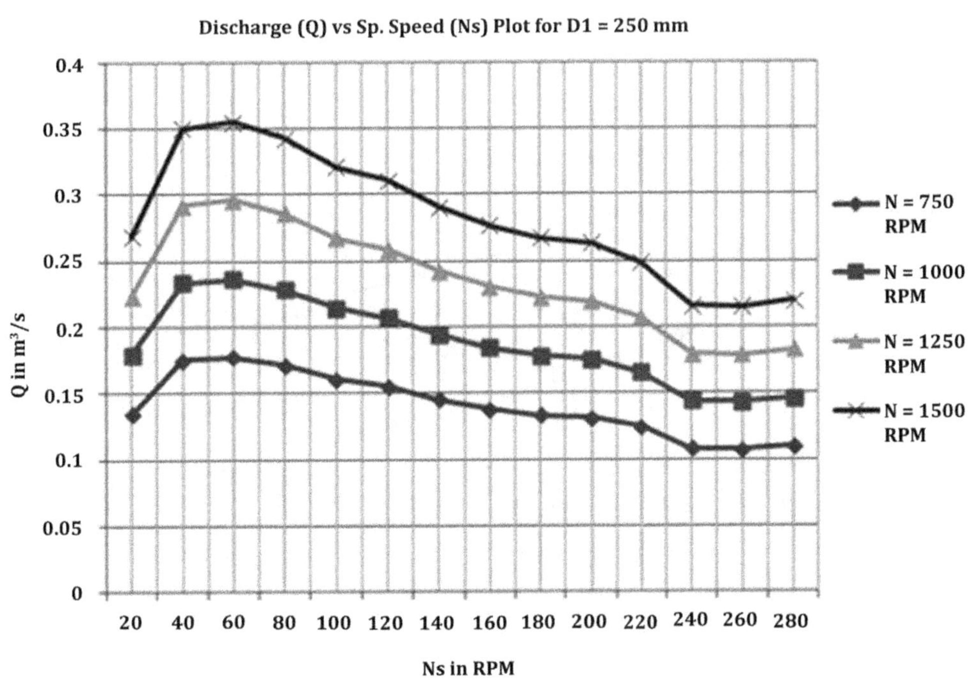

Figure 38. Head vs. Sp. Speed plot for D1=250 mm

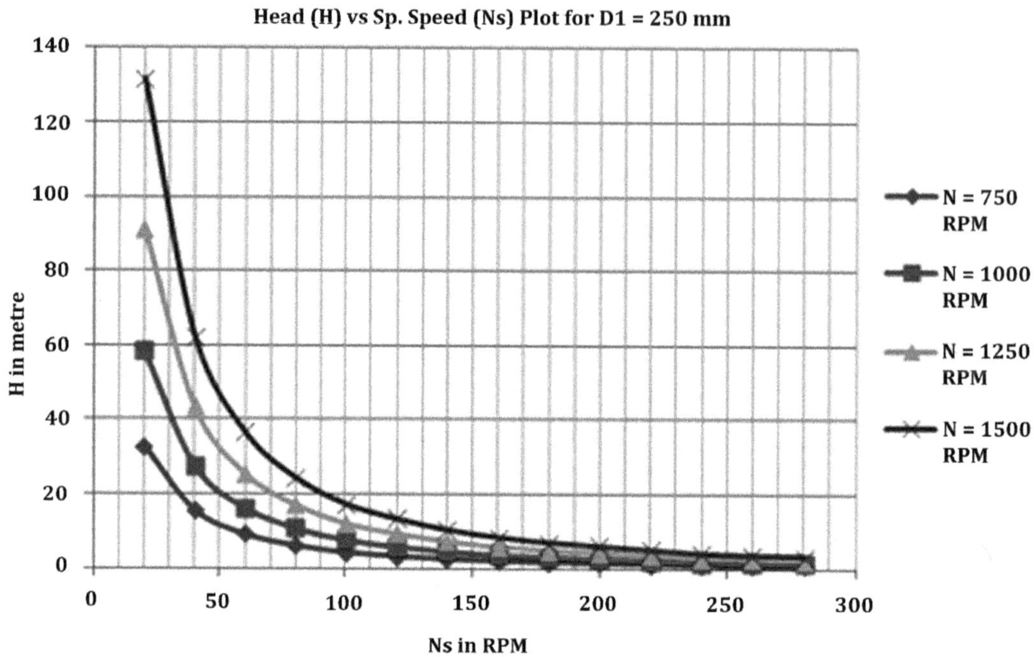

Figure 39. Power vs. Sp. Speed plot for D1=250 mm

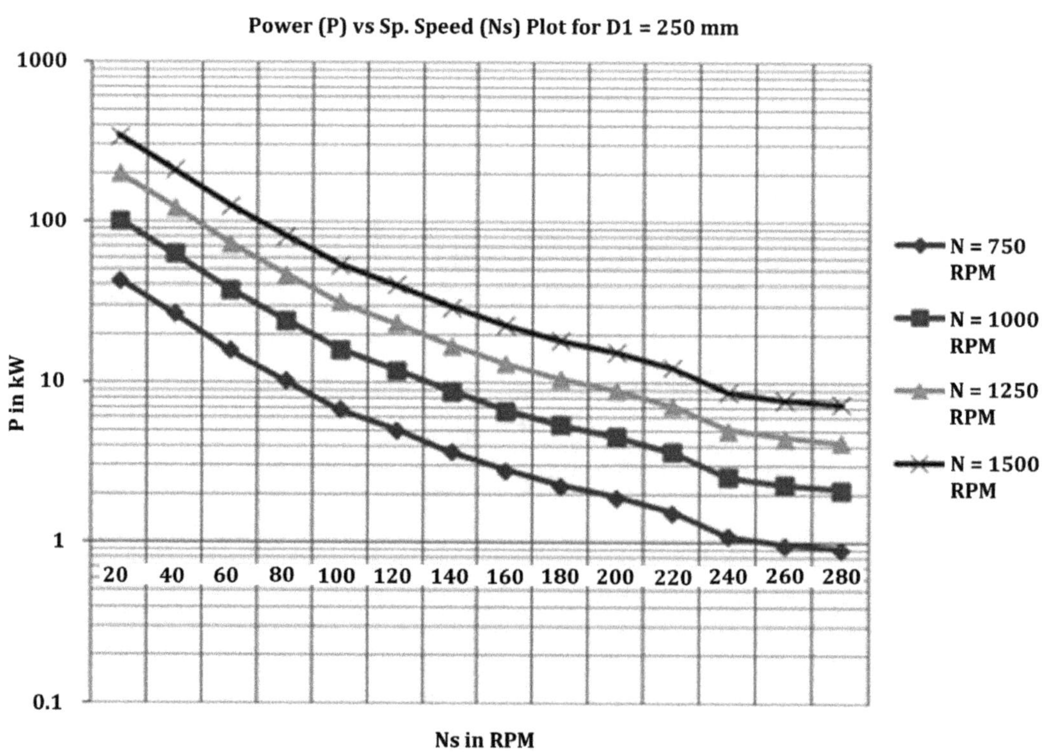

Figure 40. Discharge vs. Sp. Speed plot for D1=300 mm

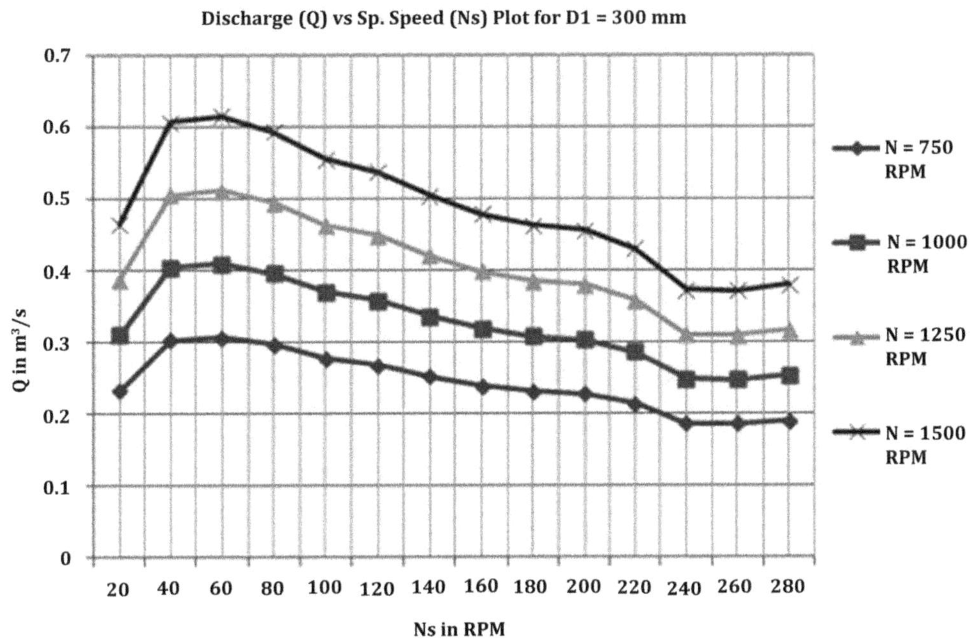

Figure 41. Head vs. Sp. Speed plot for D1=300 mm

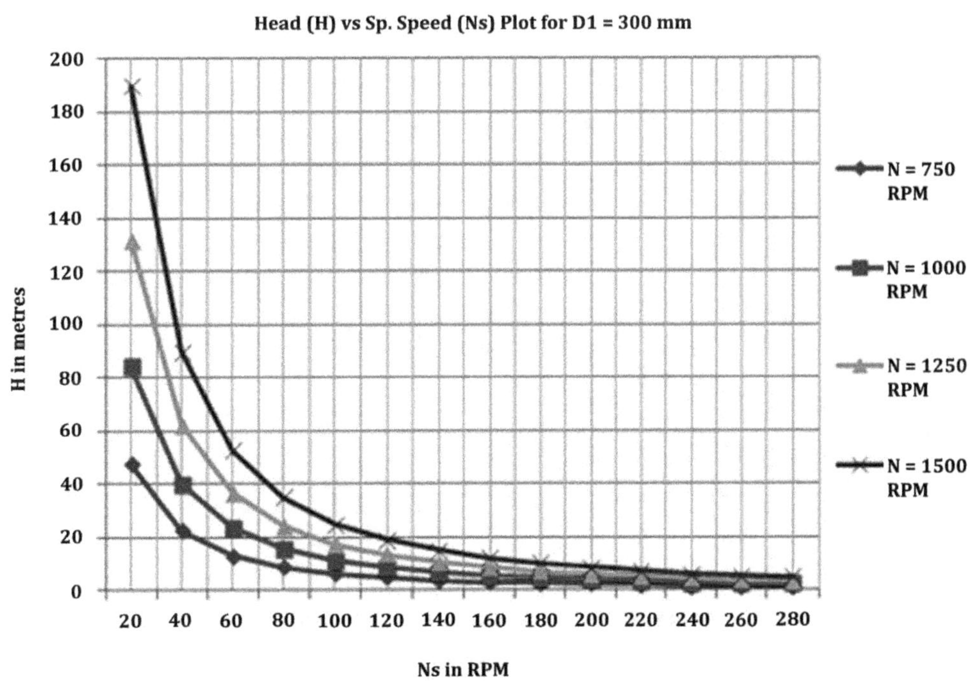

Figure 42. Power vs. Sp. Speed plot for D1=300 mm

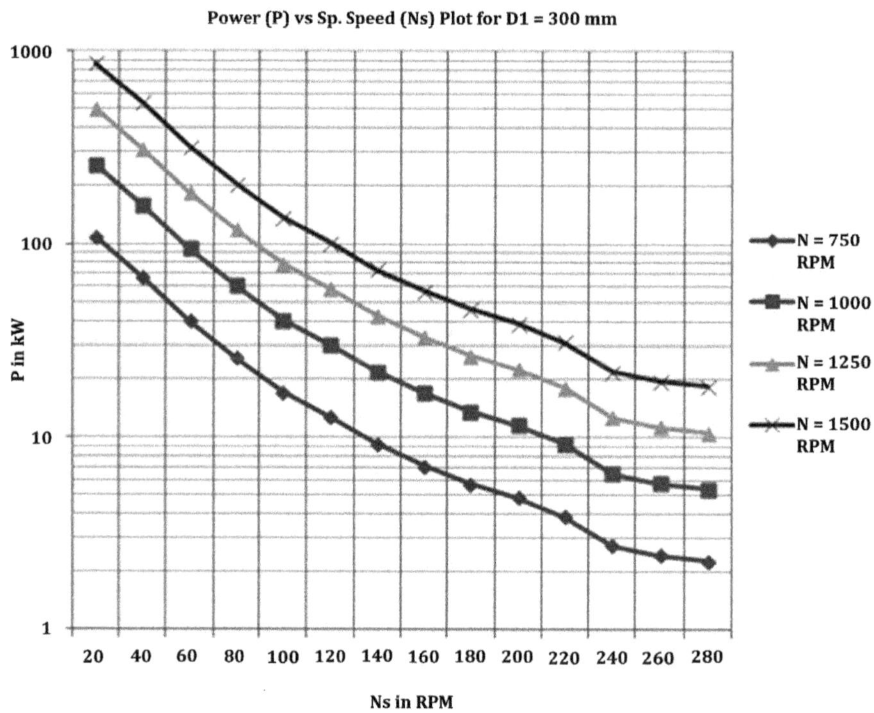

APPENDIX 3

Figure 43. Variation of 'm' with stagger angle 'λ' (Horlock (1958))

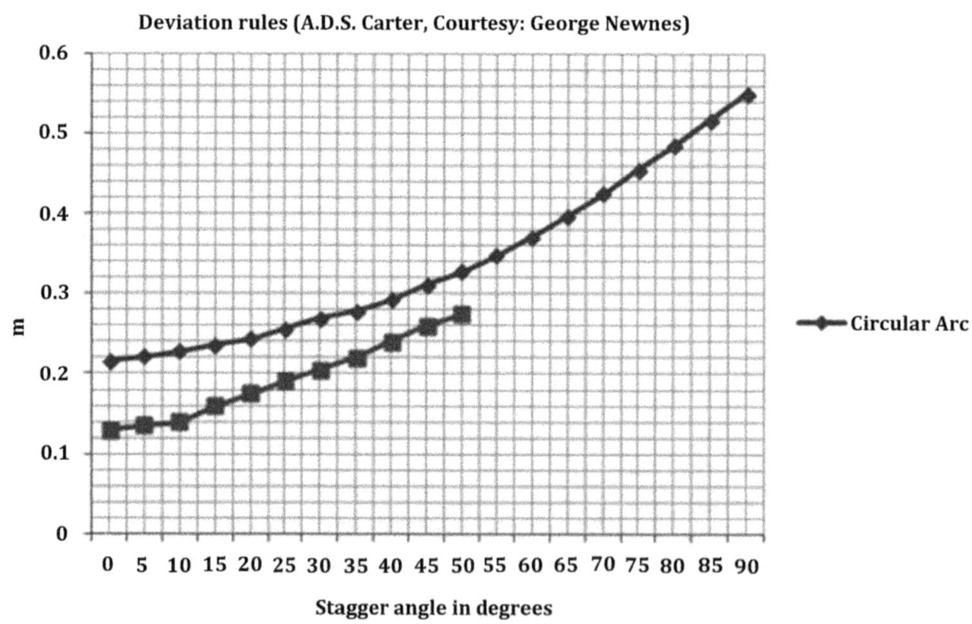

Chapter 26
An Estimation of Distribution Algorithm-Based Approach for the Order Batching Problem:
An Experimental Study

Ricardo Pérez-Rodríguez
Center for Research in Mathematics, Mexico

Arturo Hernández-Aguirre
Center for Research in Mathematics, Mexico

ABSTRACT

In the supply chain and the planning and control of warehouse processes, the order picking is an aspect critical. Combining customer orders into picking orders to minimize the picking time is known such order batching. Extensive evolutionary algorithms haven been proposed to build better batches for the order picking. The authors think that any algorithm should preserve batches that appear frequently in all members of the population in order to keep track and inherit these characteristics exhibited by the parents to the next generation. However, the traditional evolutionary operators used in current research sometimes lose the characteristics mentioned. In order to describe the characteristics exhibited by the parents as a distribution of the solution space, the authors build a probability model. An acceptable performance using the model proposed is shown against different evolutionary algorithms known in the literature in a series of extensive numerical experiments.

INTRODUCTION

Order picking systems involving operators are common in warehouse environments. The order picking arises when different articles are stored in unit loads such as pallets, and the customers require a few quantities of different articles (de Koster et al. 2007). One of the manual order picking systems is picker-to-parts (Wäscher, 2004), where pickers normally walk through the warehouse and collect articles. Batches are used when the number of arriving orders is too large for processing each customer order

DOI: 10.4018/978-1-4666-9779-9.ch026

separately; therefore batching is the process of grouping a set of customer orders into picking orders. It means that different customer orders can be simultaneously released for picking. In this chapter the authors analyze manual picking systems such as picker-to-parts where the activity of transformation of customer orders into batches exists.

Order pickers are guided by pick lists. A pick list comprises a set of order lines, each one identifying a particular article, the quantity of the article requested and the respective storage location. The order lines are already sorted into the sequence according to which the order picker is meant to collect the items; therefore the pick list is useful for making tours through the warehouse.

The order batching problem features some similarities with the capacitated vehicle routing problem. Both are combinatorial issues, however differs with respect to the customer order integrity condition, i.e. all items of a customer order must be picked on the same tour. Thus, traditional solutions approaches to the capacitated vehicle routing problem cannot be applied directly to the order batching problem (Bozer, 2008).

Different evolutionary algorithms such as Genetic Algorithms (GA), Memetic Algorithms (MA), and Tabu Search (TS) have studied and solved the order batching. An example can be found in Öncan (2013) where the algorithm proposed preserves batches or sequences of batches that appear frequently in all members of the population in order to keep track and inherit these characteristics exhibited by the parents to the next generation. However, the traditional evolutionary operators used in current research sometimes lose the characteristics mentioned.

The authors of this research propose an alternative approach consists of using Estimation of Distribution Algorithms (EDA), introduced by Mühlenbein and Paaß (1996). EDA is a relatively new paradigm in the field of evolutionary computation. Compared with other evolutionary algorithms, the EDA reproduces new population implicitly instead of using traditional evolutionary operators. In the EDA, a probability model of the most promising area is built by statistical information based on the search experience to describe the characteristics exhibited by the parents and then the probability model is used for sampling to generate new individuals. The EDA makes use of the probability model to describe the distribution of the solution space. The updating process reflects the evolutionary trend of the population. To the best of knowledge of the authors, this kind of algorithm has not been used to tackle the order batching problem in order-picking warehouses.

LITERATURE REVIEW

A discussion about the most current research on the order batching process is outlined below.

A practical batching problem where greeting cards are retrieved from a warehouse was analyzed by Kamin (1998). In that environment, the pickers use automated guided vehicles on a fixed course collecting the items according to given customer orders. The Kamin's research focuses on the minimization of average turnover times where the orders arrive throughout the study horizon.

The optimal number of customer orders that should be assigned to a batch such that the average turnover time is minimized is focused on Chew et al. (1999). They employ a queuing network with two queues. In the first queue, customer orders arrive according to a Poisson process and batches are generated by means of the FCFS rule (First Come First Serve). If a particular number of customer orders are in the first queue, those orders are assigned to a batch and move onto the second queue. Those orders are released according to the availability of pickers.

A study of the average turnover time of a random customer order for a two-block layout is carried out by Le-Duc et al. (2007). A corresponding model for all customer orders arriving during a particular time interval are assigned to batches in a two-block layout is presented by van Nieuwenhuyse and de Koster (2009). Yu and de Koster (2009) explain an order picking area that is divided into several zones of identical size. The articles of each batch are picked up sequentially by zones. For this picking process, the researchers give an estimation of the average turnover times and observe that an optimal batch size exists.

An online order batching problem in a walk-and-pick warehouse in which the completion times of all (dynamically arriving) customer orders (or the makespan) are to be minimized is described by Henn (2010). In order to deal with the online situation, the author also shows modifications of solution approaches for offline order batching.

A GA for the order batching problem considering traversal and return routing policies is introduced by Öncan (2013). The proposed GA is tested on randomly generated instances and compared with the well-known savings algorithm. The proposed GA yields promising solutions in acceptable computation times according to the author's extensive computational experiments.

In all this current research the authors employ discrete vectors where the number of elements equals the total number of orders to pick up and where each element contains an integer value that represents the batch to be assigned is the common representation of the solution such main characteristic. The researchers explained in the introduction section above that these algorithms try to preserve batches or sequences of batches that appear frequently in all members of the population in order to keep track and inherit these characteristics exhibited by the parents to the next generation. However, the traditional evolutionary operators used in current research sometimes lose the characteristics mentioned.

BACKGROUND

The order batching problem and an optimization model are explained below based on Henn (2010).

Let

v_travel: travel velocity, i.e. number of length units the picker can cover in the warehouse per time unit.
v_pick: pick velocity, i.e. number of items the picker can search and pick per unit time.
n: be the number of customer orders known,
m: the number of batches to be processed,
a_i: the arrival time of order i for all $i \in \{1,\ldots,n\}$,
k_i: the number of articles of order i for all $i \in \{1,\ldots,n\}$,
K: the maximal number of articles that can be included in any batch (device capacity),
S_j: the start time of the batch j for all $j \in \{1,\ldots,m\}$,
$x_{ji} = \{1$ if order i is assigned to batch j, 0 otherwise$\}$,
$\mathbf{x}_j = (x_{j1},\ldots,x_{jn})^T$ is a vector where some customer orders are associated to the batch j, for all $j \in \{1,\ldots,m\}$,

An optimization model can be formulated as follows:

The total length of the picking tour for a particular batch involving the S-shape routing method is determined by d as a function for determining the length of tour given a batch. The problem can be modeled as follows:

$$\min \max_{j\in\{1,\dots,m\}} \{S_j + \frac{d(\mathbf{x}_j)}{v_{travel}} + \frac{\sum_{i=1}^{n} k_i x_{ji}}{v_{pick}}\} \quad (1)$$

$$\sum_{j=1}^{m} x_{ji} = 1, \forall i \in \{1,\dots,n\} \quad (2)$$

$$\sum_{i=1}^{n} k_i x_{ji} \leq K, \forall j \in \{1,\dots,m\} \quad (3)$$

$$S_j \geq \max_{i\in\{1,\dots,n\}} \{a_i \cdot x_{ji}\}, \forall j \in \{1,\dots,m\} \quad (4)$$

$$S_j \geq S_{j-1} + \frac{d(\mathbf{x}_{j-1})}{v_{travel}} + \frac{\sum_{i=1}^{n} k_i x_{j-1i}}{v_{pick}}, \ \forall j \in \{2,\dots,m\} \quad (5)$$

$$S_j \geq 0, \forall j \in \{1,\dots,m\} \quad (6)$$

$$x_{ji} \in \{0,1\} \quad (7)$$

In the objective function (1) the expression after S_j represents the service time of a batch. This sum is composed of the time the picker needs to travel and the time he/she needs to pick the items. Equations (2) ensure the assignment of each order to exactly one batch. Inequalities (3) guarantee that the capacity of the picking device is not violated. The conditions (4) indicate that a batch is started after all customer orders assigned to that batch are known. From (5) follows that a batch is started after the previous one is completed. Finally (6) and (7) indicate that the start times are non-negative and the variables are binary.

The picker starts at a depot, (s)he walks through the warehouse and picks up articles from different storage locations. Afterwards, the picker returns to the depot and hands over the picked articles. In this research the route considered by the picker for finding and collecting articles is the S-Shape route. It has been widely used in industry. Figure 1 shows the S-Shape route for a set of articles to be picked. The black rectangles represent pick locations, i.e., the corresponding locations where articles have to be picked.

Figure 1. S-Shape route
Source: the authors

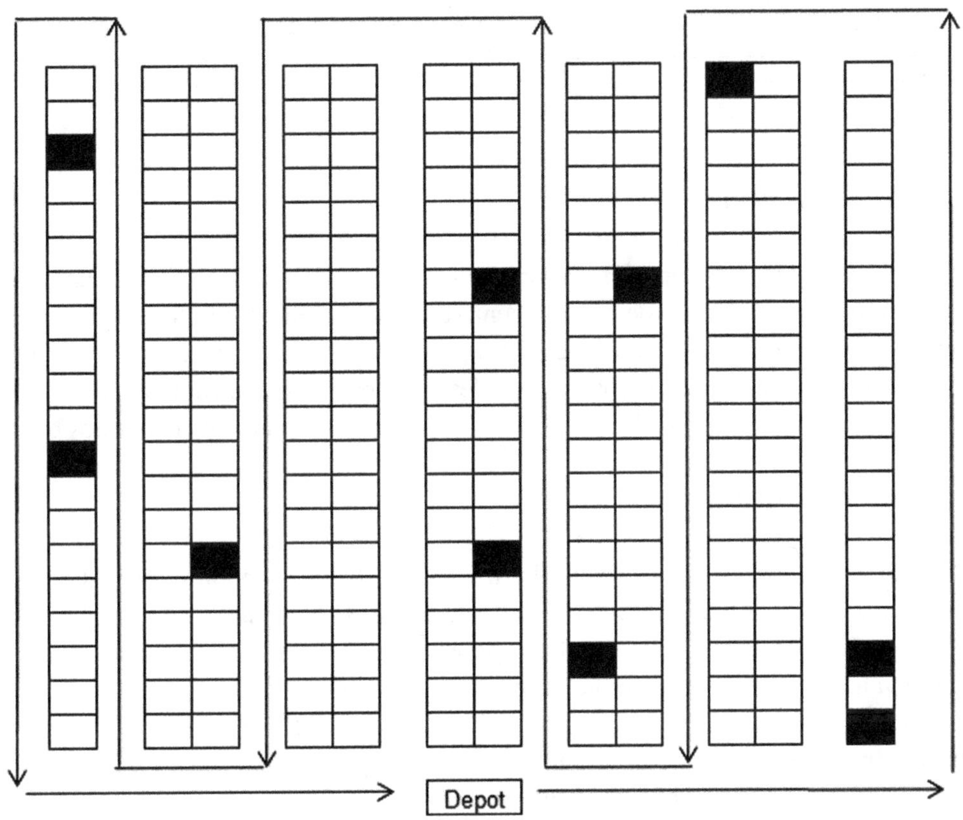

EDA FOR THE ORDER-BATCHING PROBLEM

The EDA are a set of algorithms that belong to the field of Evolutionary Computation. The EDA have been successfully used to solve complex combinatorial optimization problems such as order batching. The main characteristic of the EDA is the use of techniques inspired by the natural evolution of the species. The specie is represented by individuals. In the world of computation, the individuals represent a solution for the problem to be solved.

Basically, the EDA produces offspring implicitly by using a probability model according to the knowledge of the obtained solution space during the search process instead of using crossover and mutation operators of GA. At the beginning of each generation, a subset of the population is determined and the obtained best solution is updated. Then, the probability model is updated based on the subset mentioned and the offspring is generated by sampling the probability model. The EDA repeats the above process until a stopping condition is met.

Based on this general framework, several EDA approaches have been developed in the last years such as Larrañaga and Lozano (2002) and Pelikan et al. (2002). Many works in the literature confirm the good performance of EDA in the solution of problems from diverse fields.

Figure 2. Solution vector
Source: the authors

	Customer Orders									
	1	**2**	**3**	**4**	**5**	**6**	**7**	**8**	**9**	**10**
Batch	3	1	1	1	2	2	2	3	3	2

Solution Representation

A solution to the order batching problem previously mentioned is expressed by the assignment of customer orders to batches, i.e., an order assignment vector represents a solution where the number of elements equals the total number of orders to pick up, where each element contains an integer value. This representation is shown in Figure 2 with 10 orders. Observe that the first batch consists of orders {2; 3; 4}; the second batch includes orders {5; 6; 7; 10} and the third batch incorporates order {1; 8; 9}.

Probability Model Proposed

A probability matrix is designed as the probability model, i.e., batch probability matrix P. The probability that the batch *j* be used for the *i* customer order where $j = \{1,\ldots,m\}$ and $i = \{1,\ldots,n\}$ is represented by the element p_{ji} of the probability matrix. The opportunity of a customer order on a certain batch is indicated by the value of p_{ji}. To compute the number of times that batch *j* appears for the customer *i* in the selected population is necessary to build the probability matrix P. The result obtained previously is divided by the number of individuals considered in the selected population. Via sampling according to the probability matrix new promising individuals may be generated. For every position *i*, batch *j* is selected with the probability p_{ji}. If the batch *j* has already filled according to the device capacity, it means the assignment of the batch *j* has been finished. Then, the whole column $p_{j1}, p_{j2}, \ldots, p_{ji}$ of probability matrix will be set as zero. This updating mechanism considers the previous assignments.

The characteristics exhibited by the parents as a distribution of the solution space is described by the probability model. Therefore, the algorithm proposed does not need to hold members of the population to keep track and inherit the characteristics mentioned in each generation. The characteristics persist for all members of the population in the evolutionary progress by the probability model.

The major procedure of the EDA is listed as follows.

EDA Pseudo-Code

Step 1: Set the generation index g = 0. Initialize an initial population S(0) of size M.
Step 2: Select a subset D from S(g) of size N, where N≤M.
Step 3: Establish the probabilistic model P which describes the distribution characteristics of D.
Step 4: Generate a set K of size M of new individuals by sampling P.
Step 5: A binary tournament is done to select the best individuals from K ∪ S(g) and assign them to the next generation S(g+1).
Step 6: Let g = g+1. If g<GN, where GN is the maximum number of generations return to step 2. Otherwise, output the best solution in S(g).

The EDA tries to determine the relationship or interaction among variables of the problem as primary objective to describe the distribution of the solution space. The traditional evolutionary operators such as crossover and mutation are substitute by the probability model which is built with information on relationships and interactions among variables. The probability model on step 3 plays an important role in the EDA. Therefore, it should be well designed according to the characteristics of the considered problem. Step 4 makes another main difference between the EDA and other evolutionary algorithms. Sampling offspring from the probability matrix instead of using classical operators is the critical feature.

RESULTS AND COMPARISON

The researchers established a workload that contains different customer orders and differing types of articles required in a workday, replicating the order-picking process conditions to evaluate and find the minimum total completion time. The experiments were based on a single-block warehouse with two cross aisles. It is assumed that there is one in the front and one in the back of the picking area. This layout was used by Henn (2009). In this layout, the picking area consists of 900 articles and storage locations. The storage locations are arranged into 10 aisles. The authors further assume that an order picker walks through 10 storage locations in 30 seconds and they need 10 seconds to search and collect an article from a storage location. For the capacity of the picking device K, the researchers assume three different values, namely 45, 75 and 100 articles. For the total number of orders n, 30, 60 and 90 were considered. The orders should arrive within a planning period of eight hours. The inter arrival times, i.e., the time between the arrival of order i and order $i + 1$, are exponentially distributed with a parameter λ called arrival rate. Let $X(t)$ be the number of incoming orders in the time interval $[0,t]$. In the case of exponentially distributed inter arrival times $E[X(t)] = \lambda \bullet t$ holds. In the numerical experiments λ was chosen in a way that the expectation $E[X(t)]$ is equal to n for $t = 8[h]$. In summary, the authors use the following values for λ: for $n = 30$: $\lambda = 0.0625$, for $n = 60$: $\lambda = 0.125$, for $n = 90$: $\lambda = 0.1875$.

A fixed time window batching is defined for all the experiments, i.e. all customer orders arriving during a particular time interval are assigned to batches. In this research, the fixed time interval is each hour. Therefore, the process of building batches is executed eight times in a workday.

The flow for each experiment is detailed below.

Experimental Flow

```
1: Define a capacity of the picking device K.
2: Define a number of orders n.
3: Set λ.
4: For each order
        Set an arrival time
        Set the corresponding articles to pick up according to K
5: For each hour (the fixed time interval)
        Identify the number of unprocessed orders
        Build batches with the unprocessed orders
6: For each minute
```

```
If there is processed batches. Release them and collect statistics
If there are unprocessed batches. Allocate one by one to the picker
```

Three evolutionary algorithms have been proposed as benchmark for comparison with the EDA scheme. A GA based on Whitley et al. (1990) where the chromosome is represented by integer values, i.e. each element shows a specific batch for using in the corresponding unprocessed order. It is the same representation for all the algorithms. A selection rule for the best chromosomes is defined through the tournament selection, i.e. two parents are compared according to their fitness value (the minimum total completion time). The "edge recombination operator" is used as a cross operator and a mutation operator changes batches among different positions in the chromosome. The process is executed until the best offspring is reached in only ten generations. The best solution is considered for each experiment.

A TS based on Henn (2010) works with the operator ¨swap move¨, meaning the interchanging of two customer orders from different batches. A tabu list is built on ten neighbors. A MA based on Moscato (1989) works with the same operator for the local search phase.

The tests are executed in a Dell® Vostro® 3500 computer, Intel® CoreTM i3 processor, 2.6 GHZ, 4 GB of RAM, Windows® 7 for 64 bits to run each algorithm.

To account for the stochastic nature of the order-picking warehouse, the authors ran 50 trials for all algorithms. Table 1 details the minimum total completion time obtained for each case among all the trials.

CONCLUSION AND FUTURE WORK

Based on the experimental results shown, the researchers conclude that the EDA can be an efficient algorithm to handle different order-picking conditions such as capacity and customer orders. When the amount of orders increase the EDA has an acceptable performance according to the results detailed above. The probability model is useful in order to preserve batches or sequences of batches that appear frequently in all members of the population. The EDA is able to describe the characteristics exhibited

Table 1. Comparison of results

Capacity	Orders	Algorithms			
		GA	MA	TS	EDA
45	30	659	659	659	665
	60	840	784	834	**784**
	90	1926	1738	1818	**1716**
75	30	694	694	694	694
	60	1422	1329	1391	**1328**
	90	2294	2109	2231	**2096**
100	30	711	711	712	712
	60	1316	1214	1259	**1214**
	90	2011	1892	1981	**1891**

by the parents in each generation. Finally, this research contributes by using an EDA as an optimization method for the order-picking process if the S-shape routing method is implemented in a warehouse. Future research work will use higher probabilistic models in order to model higher interactions or relationships between variables of the order-picking performance. In addition, different routing methods will be used in order to analyze the performance of the EDA in complex environments.

ACKNOWLEDGMENT

The authors would like to express their gratitude to Elizabeth O'Shaughnessy for reviewing the manuscript.

REFERENCES

Bozer, Y., & Kile, J. (2008). Order Batching in Walk-and-Pick Order Picking Systems. *International Journal of Production Research*, *46*(7), 1887–1909. doi:10.1080/00207540600920850

Chew, E., & Tang, L. (1999). Travel time analysis for general item location assignment in a rectangular warehouse. *European Journal of Operational Research*, *112*(3), 582–597. doi:10.1016/S0377-2217(97)00416-5

de Koster, M., Le-Duc, T., & Roodbergen, K. (2007). Design and Control of Warehouse Order Picking: A Literature Review. *European Journal of Operational Research*, *182*(2), 481–501. doi:10.1016/j.ejor.2006.07.009

Henn, S. (2009). Metaheuristics for the order batching problem in manual order picking systems. *Ottovon-Guericke-University Magdeburg. Faculty of Economics and Management. Verband der Hochschullehrer für Betriebswirtschaft e.*, *3*(1), 82–105.

Henn, S. (2010). Algorithms for on-line order batching in an order-picking warehouse. In *Proceedings of the 3rd International Conference on Information Systems, Logistics and Supply Chain ILS 2010*. Casablanca: Business Process Consulting.

Kamin, N. (1998). *On-line optimization of order picking in an automated warehouse*. (Ph.D. Dissertation). Aachen: Shaker Verlag.

Larrañaga, P., & Lozano, J. A. (2002). *Estimation of distribution algorithms: a new tool for evolutionary computing*. Dordrecht: Kluwer. doi:10.1007/978-1-4615-1539-5

Le-Duc, T., & de Koster, R. (2007). Travel time estimation and order batching in a 2-block warehouse. *European Journal of Operational Research*, *176*(1), 374–388. doi:10.1016/j.ejor.2005.03.052

Moscato, P. (1989). *On Evolution, Search, Optimization, Genetic Algorithms and Martial Arts: Towards Memetic Algorithms*. Caltech Concurrent Computation Program (report 826).

Mühlenbein, H., & Paaß, G. (1996). From recombination of genes to the estimation of distributions: I. binary parameters. In H. Voigt, W. Ebeling, I. Rechenberg, & H. Schwefel (Eds.), *Parallel Problem Solving from Nature PPSN IV* (pp. 178–187). Berlin: Springer. doi:10.1007/3-540-61723-X_982

Öncan, T. (2013). A Genetic Algorithm for the order batching problem in low-level picker-to-part warehouse systems. In *Proceedings of the International MultiConference of Engineers and Computer Scientists IMECS 2013 1*. Hong Kong.

Pelikan, M., Goldberg, D. E., & Lobo, F. G. (2002). A survey of optimization by building and using probabilistic models. *Computational Optimization and Applications*, *21*(1), 5–20. doi:10.1023/A:1013500812258

van Nieuwenhuyse, I., & de Koster, R. (2009). Evaluating order throughput time in 2-block warehouses with time window batching. *International Journal of Production Economics*, *121*(2), 654–664. doi:10.1016/j.ijpe.2009.01.013

Wäscher, G. (2004). Order Picking: A survey of planning problems and methods. In H. Dyckhoff, R. Lackes, & J. Reese (Eds.), *Supply Chain Management and Reverse Logistics* (pp. 323–347). Berlin: Springer. doi:10.1007/978-3-540-24815-6_15

Whitley, D., Starweather, T., & Shaner, D. (1990). The traveling salesman and sequence scheduling: quality solutions using genetic edge recombination. In L. Davis (Ed.), *Handbook of genetic algorithms* (pp. 350–372). New York.

Yu, M., & de Koster, R. (2009). The impact of order batching and picking area zoning on order picking system performance. *European Journal of Operational Research*, *198*(2), 480–490. doi:10.1016/j.ejor.2008.09.011

KEY TERMS AND DEFINITIONS

Estimation of Distribution Algorithm: A group of evolutionary algorithms based on probabilistic models to solve combinatorial optimization problems.

Online Optimization: The method for finding the best decision for online batching.

Order Batching: It consists in grouping different orders in only one batch.

Order Picking: It collects different articles in a warehouse environment.

Pickers: Operators in a warehouse.

Probability Matrix: The layout of the probability model.

Probability Model: The representation of the solution to describe the distribution of the solution space.

Warehouse Management: It improves the performance of the warehouse.

Chapter 27
Determining the Relationship between Time of Construction and Cranes Organization by Applying the Bees Algorithm for Reducing the Time of Construction of Houses

Alberto Murillo
Universidad Autónoma De Nayarit, Mexico

Perla Aquilar
Universidad Autónoma De Nayarit, Mexico

Daniel Azpeitia
Juarez City University, Mexico

Yolanda Camacho
Universidad Autónoma De Nayarit, Mexico

Alberto Ochoa-Zezzatti
Juarez City University, Mexico

ABSTRACT

You can get a wide range of strong cranes able to lift and move heavy weights with constant acceleration, that result in an optimization in the construction time. However, it is possible that factors such as lack of cranes, incorrect cranes for work, lack of coordination, lack of communication, which result in a material is two transportation or more times occur, that once a job is done when another crane at the same time could be doing the following process. This is known as deferred construction time.

INTRODUCTION

Everybody continues in its state of rest or uniform rectilinear unless it is forced to change its state by forces impressed upon him move mentioned at Newton's First Law, or Law of Inertia. The change of movement is proportional to the printed driving force and occurs along the line along which that force line is printed (Gabay, 1974).

DOI: 10.4018/978-1-4666-9779-9.ch027

Force = mass * acceleration.

(Second Law of Newton or law of force)

With every action it always happens with an equal and opposite: it means that the mutual actions of two bodies are always equal and oppositely directed mentioned at Newton's third law or principle of action and reaction as in Figure 1.

HONEY BEE APIS MELLIFERA

Putting workers, drones and queen together in a box (hive) get a bee colony. These social insects, they live very close and yet have a division of labor created by them. The Queen is considered the most important bee in the colony. She has the function of laying eggs and propagate the species. (García, 2007). They secrete special pheromones that occur in their mandibular glands and other glands, which are social

Figure 1. Problem representation where was applied Bee Algorithm to improve the solution

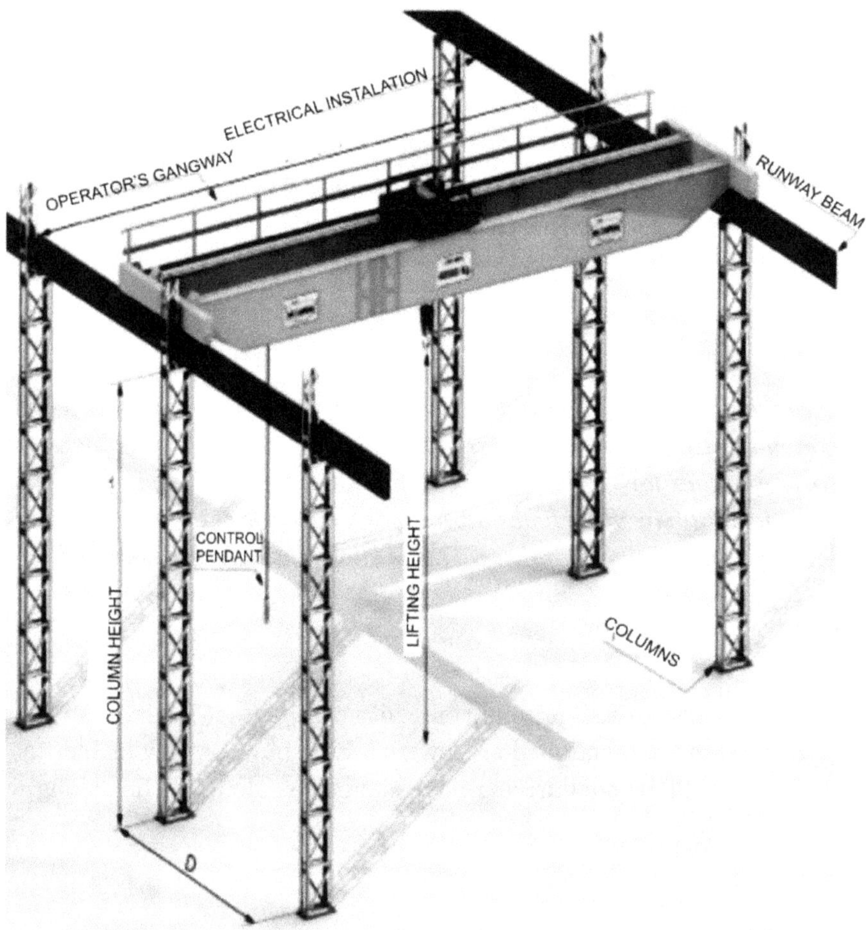

paste colony because consolidate the unity and cohesion of their family and that control certain aspects of the physiology and behavior of workers.

In addition, they live eight months in tropical areas and almost two years in temperate areas, usually there is only one queen in a hive but during times of replacement queen sometimes there are one or more daughters (virgin queens), Cries and queens in the cells real.

The shape of the queen differs from other bees, as your body is longer, her wings seem shorter in relation to body size, its tools and brushes devoid of legs look longer. It has a curved and ready sting only used in fighting another queen (Bellmunt, 1998). No wax glands or basket (corbiculae) in the third leg to transport pollen. Your abdomen (the third of its body) is golden and sometimes darker, no rings of different colors. Their movements are slow and lively and is able to put about 1,500 eggs a day.

The queen lays two types of eggs: fertilized eggs (that will produce a female - another queen or a worker according to the power that will be given during the larval stage) and eggs unfertilized (be born a drone = type of development pathogenesis name). During the first week of his adult life, the virgin queen leaves the nuptial flight and gets sperm from a few (a dozen or more) drones to keep in your sperm inside her body for all her life. The drone, the adult male is unique and its mission is the fertilization of the Virgin Queen, dies after such activity. He lives less than a month and does not work within the hive. There are several hundred in a hive, only when conditions (temperature, climate, resources of nectar and pollen) are good (Bellmunt, 1998).

The workers are numerous and play countless roles in the hive (his work). They are females but not development of their ovaries. They live only 3-6 weeks in peak periods (spring and summer), a little more in other seasons. In areas with long winters they live 2-4 months. The number of workers in a beehive may vary from 10,000 (winter) to 60,000 (during the increased flow of nectar).

The workers are the smallest inhabitants of the hive and form the majority of the population. The functions of a worker, as its name suggests is to perform all tasks in the hives. These works vary with age: when young dealing with internal honeycombs.

FUNCTIONS OF WORKERS

Clean cells the first task is to clean the combs of the brood chamber, removing the scales and other debris found inside the cells of the honeycomb. Their bodies also produce heat for breeding. In the other hand Nurse after a few days, the working works as a nurse, feeding breeding. Within the body of the mother, hypo pharyngeal glands, preparing food for the larvae. The nurse is giving special food, real queen larva jelly. This food is different to the larvae of workers that in the last three days these are a mixture of honey diluted with water and pollen (Bellmunt, 1998)..

Also, cerera manufactured and wax honeycombs build or repair as needed. After nine days of age, the body begins to produce a working laminitis wax. These are produced with wax glands that are in your abdomen, bottom. Bees use their legs and mouth to form and mold wax cells or opercular nymphs (pupae). There are three types of cells - most are 5mm and are used for worker brood or honey. There are other larger cells (6.5mm diameter), the cells to drones. The third type of cell are special cells for the queen (the queen cell). Worker cells and drone are horizontal (the proper honeycomb) and the queen cells (which are in forms of glasses when empty) to raise a new queen, are vertical in orientation.

Like the other, winemaking receives nectar forage (foraging or gatherer) to the hive. They carry the nectar to the cells of the brood chamber where they become mature honey - usually within a few days.

They also store pollen in cells next to the nursery. And defender monitors their entrances to the hives income for anyone disturbs the colonies. To defend the defenders (the guards) bite and do not let robber bees and other animals that try to steal honey.

Jobs workers are responsible for caring for the queen and for the distribution of social pheromone in hives. When the weather is very hot they are ventilators driving the cooler outside air into the hives, while inside other bees come to expel hot air using their wings.

Forage (other foraging or collecting names) - when he reaches the age of 2-3 weeks, the workers begin to leave the nest to forage for nectar, pollen, water and propolis (4 things a colony needs outside their hive) and lead to hive these materials that the colony needs to live. During the first flights they learn the location of their hive to return only to their own beehive.

LIFE CYCLE - METAMORPHOSIS

The bee is an insect of complete metamorphosis, with a life cycle consisting of four stages: egg (3 days), larva (variation between breeds, pupa (or nymph) in capped condition and adulthood. Bees have their own form of communication. They exchange messages like: There is much nectar close; "The queen is laying eggs well"; "Danger, no enemies" and others. They use smell, touch and dance in their communication (Sammataro & Avitabile, 2005).

Bees have two antennae coming out of his head. These have small holes that serve as nose and may feel the smells of flowers, their enemies and bees of the colony, and special scents of pheromones. Some important examples are:

1. **The Smell of the Queen:** The queens have a special smell. When this young and gets well, the smell is very strong. All workers of the colony know that the queen there and need not raise a new queen. The workers recognize their queen.
2. **Alarm Smell:** If a person or animal hive mind, defending the sly. The sting that left stuck in the skin of the enemy has an odor (smell alarm = ethyl, iso-pentyl) that attracts other bees to sting him too.
3. **Orientation Smell:** When bees swarm, go to a new nest. The first workers entering the new gap (cavity as a tree) wings begin to move very quickly and simultaneously, open the gland Nasanof his abdomen (upper in recent rings). Air movement doing with wings spread the scent around the new nest.
4. **Scent of Flowers:** Forage take nectar and pollen from flowers. Each flower has a smell, color and different design. By the smell of flowers the bees dress, suck the nectar and collect pollen.
5. **Hives Smell:** In the mixture of smells of flowers and each colony queen bee has a particular smell different from the other colonies odor. That's why the defenders of the colony recognize and leave the robber entering the hive to steal honey.
6. **Danzas:** Forage return to the hive carrying nectar and / or pollen. If a forage is a place that has much nectar or pollen, communicated to others through dance above the honeycomb. These dances (forms a circle nearby resources or figure 8 when the flowers are further) indicate the place to the sun and away from flowers. Bees use these dances to indicate water whips and sites for a new nest cavities when there swarming.

ANATOMY BEE

The body of the bee is divided into three parts like insects: head, thorax and abdomen. Your body has a hard chitinous external skeleton covered with dense hair. They have 3 pairs of legs, one pair of antennae and two pairs of membranous rise. (See figure 2 - From The Beekeeper Book Manual for Sammataro And Avitabile 2005 Letemendia Publishing House Buenos Aires, Arg..)

The head has a hexagonal shape in Queens, triangular in labor and travel within the drone; It contains the sensory organs: 2 compound eyes, one on each side of the head and 3 simple eyes (ocelli), located at the top of the head; a pair of antenna segments and very flexible elbow joint as a human and mouthparts. The palate is adapted to function lick and suck (siphon) and consists of the proboscis [sometimes called tongue or gloss] with two pairs of maxilla and lip each with a pair of palpo, and two mandibular (which opens side) and over one labrum. When the bee is at rest, the whole complex is retracted mouth under the head and thorax.

The central part, the thorax, is regarded as the locomotive center, as it is provided with strong and short muscles, ensuring the movement of the wings and the rapid movement through his legs. The thorax consists of three segments, from front to back, prothorax, mesothorax and metatórax, naming each pair of legs that sit on it and holding in his two rear wings.

The six legs have brushes to collect pollen; forelegs called "palms" are cleaning especially for antennas. The legs in half at its bottom has a tip or rectum for transfer ram wax leaves the jaws of the mouth to prepare wax honeycombs with building cover or mature cells with nymphs or honey. The hind legs, the third pair, have the widened tibia and hair forming a small basket (or basket), called corbiculae in depositing and transport pollen from flower to the hive.

The abdomen has 9 segments (rings) retractable six of which are visible (the initial segment is part of the center of the body); the drone 7 are visible and have the shape of a barrel. Each ring has two parts,

Figure 2. Representation of anatomy of a bee

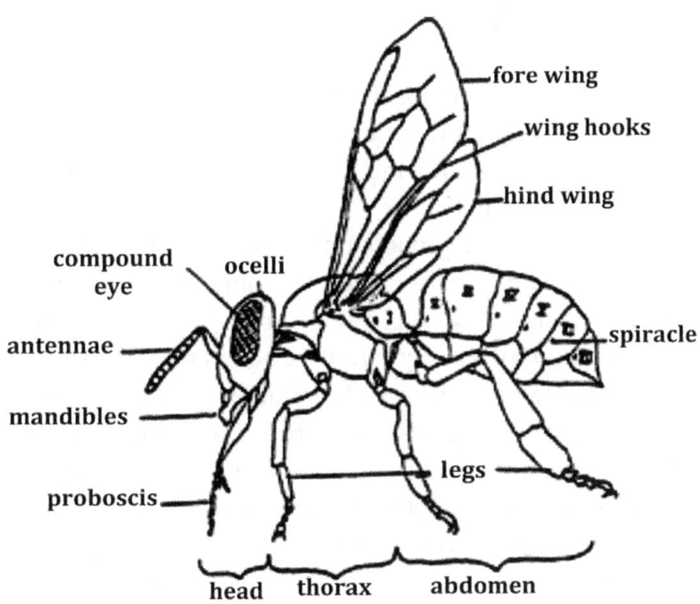

one longer than the other with larger dorsal part of the ventral part. The segments are joined together by highly flexible thin membranes that allow lengthen or contract it, what is observed in breathing or expansion when bees have much nectar within its Melaria bag.

At the bottom of the workers they have eight Cerera glands and at its upper end are located Nasanof glands (above) the sting (below). Within the abdomen is the biggest digestive system of the bee (García, 2007).

Differences between the queen, workers and drones, queen differs both drone bees as to have a more elongated abdomen, due to the development of the ovaries and the stomach fails to be covered by the wings. The drone's abdomen is thick, heavier and larger than the queen or the workers. The eyes do not come together above the head and the queen and the drones are smaller compared to the eyes of the workers. Queen legs are longer than those of a worker bee and the third pair of legs lacks corbiculae; the drone also lacks these special legs. Internal systems have several organs of the bee working in concert to perform a specific function. Most are in the abdomen. Another anatomy systems:

1. Food system.
2. Respiratory system.
3. Circulatory system.
4. Nervous system.
5. Reproductive system.

About glandular system, pheromones are produced in glands with open ducts away from the body. Examples are the mandibular scent of Nasanof, dorsal abdominal glands. Another hypopharyngeal glands have products (royal jelly), salivary (enzymes for digestion of food and processing wax), Cerera (wax). There also glands that produce hormones (endocrine glands): The brain youth 1. and 2. Hormones Hormones are signals to the brain that the epidermis cells during metamorphosis skeleton together with ecdysone prothorax gland and its product with the message for the next stage.

GENERAL CHARACTERISTICS

The tower crane is a machine used for lifting loads by a hook suspended from a cable, and transport, in a radius of several meters, at all levels and in all directions. A metal tower, a rotating horizontal arm, and yaw motors, distribution or lifting and moving load also providing a drive motor when the crane is disposed on rails essentially constitutes it.

The tower crane can be embedded in the ground, immobilized without wheels or movable on straight or curved paths. Qualified personnel must perform the assembly. Also maintenance and conservation is carried out in accordance with the norms issued by the manufacturer (see Figure 4).

In an optimization problem, the goal is usually to find the best of several alternatives. In some problems, this can be done within a reasonable time visiting all possible solutions of the search space (eg search a database) or by any technique that reduces the problem (eg, minimization of continuous functions). However, there are problems not simplify cables with a search space too large to explore it completely with comprehensive techniques. In these cases it makes sense to apply metaheuristics techniques. Briefly we can define metaheuristics as high-level strategies for exploring search spaces. The particular strategy used depends on the philosophy of the technique itself. Some of them can be considered "smart" versions

Figure 3. Parts of a tower crane

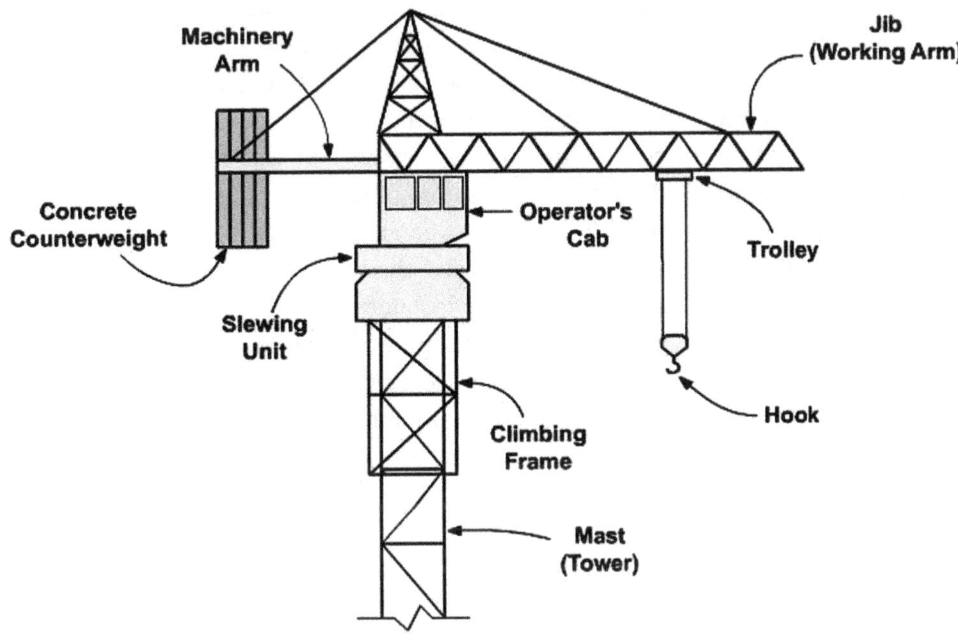

Figure 4. Konecranes Goliath gantry crane will supply the world's largest, to a shipyard in Brazil. With a lifting capacity of 2,000 tons and a width between channels of 210 meters, this crane will be large enough to place a football stadium between their legs!
Source: Konecranes USA (on Facebook) https://gruasytransportes.wordpress.com/

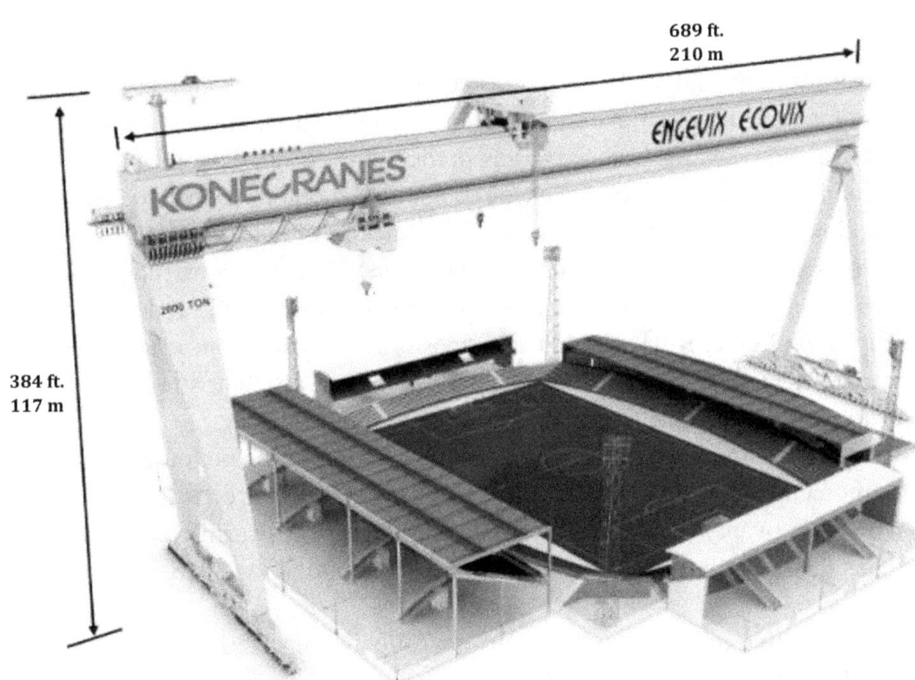

of local search algorithms (simulated annealing). Others, however, try to learn the correlation between the decision variables to identify areas of high quality in the search space (evolutionary computation).

Algorithms Swarm intelligence (SI) are metaheuristics techniques artificial intelligence based on pre Sentes studying collective behavior in natural systems, usually decentralized and autorganizativo character. Social behavior that defines the movements of the decision variables in the search space and oriented towards'optimas solutions. Swarm Intelligence expression was introduced by Gerardo Beni, Suzanne Hackwood and Jing Wang in 1989, in the context of mobile robotic systems.

Individuals must maintain a minimum distance between them the rest of the cluster throughout the process. Moreover, each individual should avoid excessive distance with the rest of the group, so it tends to be attracted to other individuals, thus avoiding being isolated and resulting in different neighborhoods depending on the attraction. As shown in Figure 3, the radius of influence of each individual is composed of an initial zone of repulsion (zor - to avoid collisions), an orientation zone (zoo - the individual directs their movement) and an outer area Attraction (zoa - the individual avoid being isolated). Depending on the configuration of each individual radio area, different behaviors and ways generate global neighborhoods.

Within the collective intelligence algorithms, it goes on to describe briefly the Particle Swarm Optimization and Ant Colony Optimization, which are of great importance for the applicability that are showing and the number of publications that have been generating in recent years. As another example, reserves a small space based algorithms pollination processes unique in its design.

IF TWENTY WORKSHOP

The construction is located in the San Luis street number 18 city of Tepic, from its inception was intended for construction. The developer offers design, planning and construction of the works. The construction has a swarm of four cranes to build a house 15 meters long by 6 meters wide and it takes 15 days to set up the house roof. In tune details such as doors, windows, paintings, etc. One month is full to the delivery of the full house.

One of the main causes of the delay of construction time in twenty Workshop construction is the lack of cranes, as three cranes are cement mixers, driven by a gear system, and a backhoe that practically does all the work, download building material stores in the honeycomb, digs, despalma, remove debris and leveling the ground. The builder does not have the necessary cranes to form a swarm, rather it is a queen bee that does almost all the work, workers with 3 cranes limited to stir.

The current hypothesis is that there is a relationship between the time of construction and organization of the cranes. If you manage to coordinate the movements of each crane, it will reduce construction time. Another relationship is the Force. If the number of cranes is correct and your organization is perfect, it continues to mutate cranes increasing the number of pulleys to increase their strength, the same that can be controlled so as to contribute to the reduction of construction time.

Deferral build time is the time from starting the work, after completing negotiations with suppliers of materials, until the moment when the work is finished, considering deferred if the period is longer than 15 days time. (“Workshop twenty”). Cranes are a representation of a swarm created with a purpose. It notes that a crane has the following properties: strength and organization. Bees algorithm is a method of harvesting.

In nature many insects live in colonies and swarms groups which facilitates them to perform their collective activities. However each member of the group has limited ability to perform tasks colony exhibits an emerging behavior arises from the interaction of individuals. (Arnoely Araujo)

METHODOLOGY FOR APPLICATION OF THE ALGORITHM OF BEES

An important step in implementing the algorithm of bees is to determine each of the cranes involved and their environment and the interaction between them. "Each hive has its own morality. Simply one of the workers to realize that you can get more resources with less effort for the hive becomes war. "

1. Generate a set (population) solutions (individuals) the problem.
2. Evaluate each solution in the objective function.
3. Select the best solutions of the population according to the value of the objective function.
4. Generate new solutions from the best, using variation operators.
5. Evaluate new solutions.
6. Choose the solutions that come to the next iteration (generation).

RECEIPT OF MATERIALS

Cranes and conveyor systems consist mainly of pulleys and gears which work together to form a very strong crane capable of traveling long distances until their energy runs are set. A crane carries Cement Sand the second crane transports The transports Record third crane, the crane transports water fourth, and fifth Fuel crane transports.

These cranes highways passing through the width of the crane. When they reach the construction area are directed each in its own download area, there will be waiting for the bridge cranes, who is in charge of holding, lifting and moving the load. While the crane are shifting the burden to the dock, crane and conveyor returns unburdened his journey to continue transporting and replaces another conveyor crane standing in the discharge position to be just below the crane.

Then, when receiving the bridge crane charging, rises and moves to a discharge zone, then back to repeat the process. While the crane back for more cargo, another tower crane takes charge and distributed among the spin materials to mix and transport cranes.

Honeycomb live within at least 5 other bridge cranes are responsible for downloading and ordering the materials needed for construction, have a movement in all directions, this allows them to step aside if they clog the stage tower cranes for sharing materials for tops, often happens that the load is just below the bridge crane beam and if this is not downloading anything can change its movement to expose the charge and the tower crane can take it without problems, everything depends on a Good communication between the operators of cranes.

CONSTRUCTION

A crane backhoe is responsible for leveling the ground and creates the necessary digging to lay the foundation. Then a tower crane is responsible for placing the castles in the correct position and guide crane hoses mixers crane to place the concrete and create a foundation with the dalas or pillars, and then continue to give rise to construction.

Finally honeycomb tower cranes loaded with the necessary materials for the production of concrete to cranes with a rotating top, its design allows mesclar the load during transport and prevent contents from spilling, these are transferred to a mixer which is responsible being pumped to build the roof.

ALGORITHM IMPLEMENTATION BEES TWENTY CASE WORKSHOP

1. Generate a set (population) solutions (individuals) the problem.

The company has a backhoe and three cement mixers.

2. Evaluate each solution in the objective function.

Backhoe optimizes the time and effort in the excavation cleaning and leveling the ground, 3 mixers optimize time and effort in the mesclado of materials. Consequently builds a house 15 meters long by 6 meters wide and it takes 15 days to finish the house with the roof, doors and windows refine details painting etc. They are completed 30 days until the delivery of the house.

3. Select the best solutions of the population according to the value of the objective function.

The backhoe if a powerful and dominant gene, so in the next generation if there least a backhoe, but the mixers are a local optimum, which means that the mixers are very effective but they are not efficient, and there is another type of crane that can do better.

4. Generate new solutions from the best, using variation operators.

For the next generation will continue to mutate mixers spin cranes cranes, these serve the same function as a revolvedora with the difference that these are larger and can be moved easily from stock (honeycomb) to the construction zone (honeycomb extension may be at a considerably long distance thanks to the spin cranes.)

5. Evaluate new solutions.

Mutation of the mixers cranes tops allows the honeycomb section where the materials and honeycomb section under construction are separated, keeping safe and secure the raw materials needed for construction,

6. Choose the solutions that come to the next iteration (generation).

To enter the next generation genes spin backhoe and cranes. The intention is to leverage their genes for mutations and keep getting accomplish the forwarder cranes, bridge cranes for storage and construction, tower cranes for the distribution of materials, concrete mixers for pumping up the necessary places guided by a crane bridge (see figure 5).

Figure 5. Representation of our solution using Bee Colony Algorithm

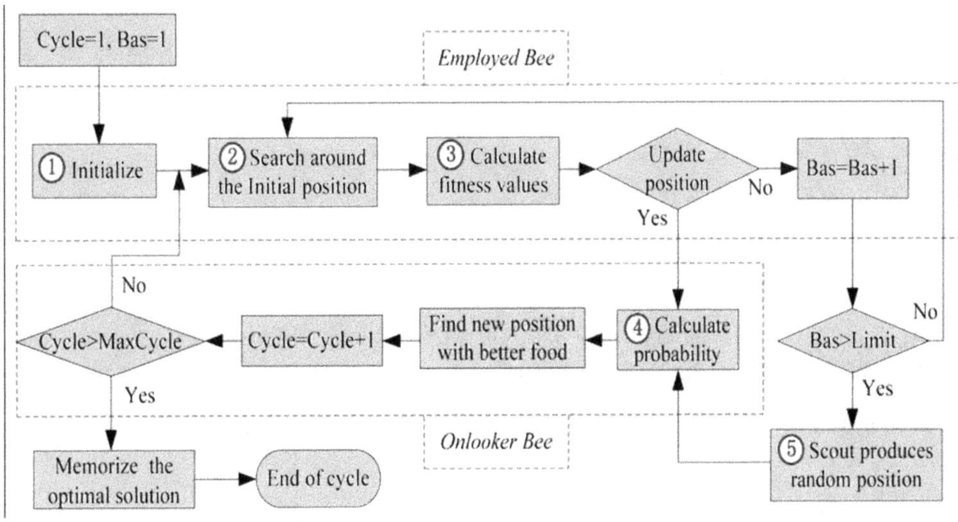

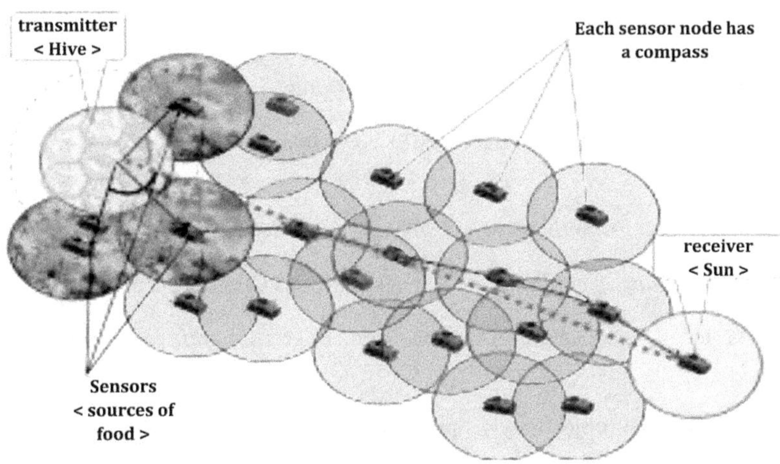

REFERENCES

Bellmunt, J. (1998). *Grúa torre*. Ministerio de Trabajo y Asuntos Sociales.

Gabay, A., & Zenya, J. (1974). *Máquinas para obras Barcelona*. Editorial Blume y Editorial Labor, S.A.

García, J. M. (2007). *Algoritmos Basados en Inteligencia Colectiva para la Resolución de Problemas de Bioinformática y Telecomunicaciones*. Universidad de Málaga. Retrieved from https://gruasytransportes.wordpress.com/

Sammataro & Avitabile. (2005). *El Manual del Apicultor*. Letemendia Casa Editora Buenos Aires, Arg.

Compilation of References

Aarts, E., & Korst, J. (1989). *Simulated annealing and Boltzmann machines: A stochastic approach to combinatorial optimization and neural computing*. John Wiley and Sons.

Abdel-Ghany, A., & Kozai, T. (2006). Dynamic modeling of the environment in a naturally ventilated, fogcooled greenhouse. Elsevier.

Abubaker, Baharum, & Alrefai. (2014). Good solution for multiobjective optimization problem. *AIP Conference Proceedings*. Retrieved from http://scitation.aip.org/content/aip/proceeding/aipcp/1605?ver=pdfcov

Abubakera, A., Baharuma, A., & Alrefaei, M. (2014). Good solution for multi-objective optimization problem. In Proceedings of the 21st National Symposium on Mathematical Sciences (SKSM21): Germination of Mathematical Sciences Education and Research towards Global Sustainability. AIP Publishing.

ACCM - American Chamber of Commerce of Mexico. (2011). *The Impact of Security in Mexico in the Private Sector*. Mexico: ACCM.

Acerra, M. (1993). *Rochefort et la construction navale française, 1661-1815*. Paris: Librairie de l'Inde Éditeur.

Acerra, M., & Meyer, J. (1988). *Marines et révolution*. Rennes: Éditions Ouest France.

Ackoff, R. L. (1970). Some ideas on education in the management sciences. *Management Science*, *17*(2), B-2–B-4. doi:10.1287/mnsc.17.2.B2

Ackoff, R. L. (1976). Optimization+ objectivity= optout. *European Journal of Operational Research*, *1*(1), 1–7. doi:10.1016/S0377-2217(77)81003-5

Ackoff, R. L. (1979). Resurrecting the future of operational research. *The Journal of the Operational Research Society*, *30*(3), 189–199. doi:10.1057/jors.1979.41

Ackoff, R. L. (1983). An interactive view of rationality. *The Journal of the Operational Research Society*, *34*(8), 719–722. doi:10.1057/jors.1983.165

Ackoff, R. L., & Sasinieni, M. W. (1968). *Fundamentals of operations research*. New York: John Wiley & Sons.

Adenso-Diaz, B., & Corral, N. (2007). A guide to statistical analysis for single criteria heuristics evaluation. *International Journal of Innovative Computing and Applications*, *1*(2), 94–107. doi:10.1504/IJICA.2007.016791

Aguilar-Savén, R. S. (2004). Business Process Modeling: Review and Framework. *International Journal of Production Economics*, *90*(2), 129–149. doi:10.1016/S0925-5273(03)00102-6

Aguilera, E. L., & González, A. M. (2012). Estrategias y Rendimiento de las Micro, Pequeñas y Medianas Empresas Familiares (Aguascalientes, México). *Global Conference on Business and Finance Proceedings*, *1*(7), 877-883.

Aguilera, E. L., Hernández, C. O., & López, T. C. G. (2012). La gestión de las cadenas de suministro y los procesos de producción. *Mercados y Negocios.*, *13*(2), 43–66.

Aguirre, O., & Taboada, H. (2011). A Clustering Method Based on Dynamic Self organizing Trees for Post-Pareto Optimaliyy Designs. In Complex Adaptive Systems. Missouri University of Science and Technology.

Ahonen, T., Virrankoski, R., & Elmusrati, M. (2008). Greenhouse Monitoring with Wireless Sensor Network. IEEE.

Air Transportat Action Group. (2012). *Aviation benefits beyond borders.* Retrieved from http://www.atag.org/

AIRBUS. (2014). *Global Market Forecast 2015-2034*. Flying on Demand Deutsche Bank Toulouse.

Airport Council International. (2012). *World Airport Traffic Report.* Author.

Airport Council International. (2013). *World Airport Traffic Report.* Author.

Ak, A., & Erera, A. L. (2007). A paired-vehicle recurs strategy for the vehicle routing problem with stochastic demands. *Transportation Science*, *41*(2), 222–237. doi:10.1287/trsc.1060.0180

Akçelik, R. (1991). Travel time functions for transport planning purposes: Davidson's function, its time-dependent form and an alternative travel time function. *Road Research, 21*(3), 49–59.

Akcelik, R. (2003). *Speed-Flow Models for Uninterrupted Traffic Facilities. Technical Report.* Australia: Akcelik & Associates Pty Ltd.

Albalate, D., & Bel, G. (2010). Tourism and urban public transport: Holding demand pressure under supply constraints. *International Journal of Tourism Management*, *31*(3), 425–433. doi:10.1016/j.tourman.2009.04.011

Albanil, A., Pascual, R., Quiroz, C., & Sánchez, O. (2014). *Reporte del Clima en México, Octubre, Año 4, Núm. 10.* CONAGUA.

Alcantara-Vasconcellos, E. (2014). *Urban Transport Environment and Equity: The Case for Developing Countries, Brussels BE-1080 Belgium.* Routledge.

Alejandro, A. R. (2011). *"Estudio del desarrollo de aplicaciones RA para Android," Trabajo de fin de Carrera.* Catalunya: España.

Alfaro, J., Ortíz, A., & Poler, R. (2007). Performance measurement system for business processes. *Journal of the Production Planning & Control*, *18*(8), 641–654. doi:10.1080/09537280701599772

Alpha, M. (n.d.). *Proyecto de Wikipedia relacionado con Star Trek.* Retrieved from http://www.memory-alpha.org/en/wiki/Main

Alrefaei, M., Diabat, A., Alawneh, A., Al-Aomar, R., & Faisal, M. N. (2013). Simulated annealing for multi objective stochastic optimization. *International Journal of Science and Applied Information Technology*, *2*, 18–21.

Alvarez, M., Alvarez, A., de Maggio, M. C., Oses, A., Trombetta, M., & Setola, R. (2010) Protecting the Food Supply Chain from Terrorist Attack. In Critical Infrastructure Protection IV, IFIP Advances in Information and Communication Technology. doi:10.1007/978-3-642-16806-2_11

Álvarez-Sánchez, E., Leyva-Retureta, G., Portilla-Flores, E., & López-Velázquez, A. (2014). Evaluation of thermal behavior for an asymmetric greenhouse by means of dynamic simulations. *DYNA*, *81*(188), 152–159. doi:10.15446/dyna.v81n188.41338

Alvim, A. C., Ribeiro, C. C., Glover, F., & Aloise, D. J. (2004). A hybrid improvement heuristic for the one-dimensional bin packing problem. *Journal of Heuristics*, *10*(2), 205–229. doi:10.1023/B:HEUR.0000026267.44673.ed

American Institute of Baking (AIB). (2015). *Guía para la Seguridad de la Planta.* Retrieved June 4, 2015, from http://www.aibonline.org/aibOnline_/secure.aibonline.org/php/ecomm-catalog.aspx?catalogNbr=56-1460-S&l=1

Ametller, X. (2007). *Optimization of freight transport by short sea shipping.* (Thesis). Polytechnic University of Catalonia. Recovered from http://upcommons.upc.edu/pfc/handle/2099.1/4942

Anderson, J., & Gerbing, D. (1988). Structural equation modeling in practice: A review and recommended two-step approach. *Psychological Bulletin*, *13*(3), 411–423. doi:10.1037/0033-2909.103.3.411

Andreatta, G., Brunetta, L., Odoni, A. R., Righi, L., Stamatopoulos, M. A., & Zografos, K. G. (1999). A Set of Approximate and Compatible Models for Airport Strategic Planning on Airside and Landside. *Air Traffic Control Quarterly*, *7*(4), 291–319.

Android Developers. (2015). Retrieved from http://developer.android.com/sdk/index.html

Angeloudis, P., Greco, L., & Bell, M. G. (2015). Strategic Maritime Container Transport Design in Oligopolistic Markets. *Transportation Research Procedia*, *9*, 269–282. doi:10.1016/j.trpro.2015.07.015

Antoine, F. (2008). *Empirical evidence for integration and disintegration of maritime shipping, port and logistics activities.* OECD/ITF Joint Transport Research Centre Discussion Paper, No. 2009-1.

Apostol, T. M. (1983). *Mathematical analysis.* Addison-Wesley.

Appelgren, L. H. (1969). A column generation algorithm for a ship scheduling problem. *Transportation Science*, *3*(1), 53–68.

Arewar, A. P., & Bhope, D. V. (2013). Stress analysis of axial flow fan impeller. *International Journal of Engineering Research and Applications*, *3*(1), 2086–2090.

Ariel, A. (1991). The effect of inventory holding costs on the optimal payload of bulk carriers. *Maritime Policy & Management*, *18*(3), 217–224. doi:10.1080/03088839100000024

Arvanitoyannis, I., & Varzakas, T. H. (2009). Application of ISO 22000 and comparison with HACCP on industrial processing of common octopus. *International Journal of Food Science & Technology*, *44*(1), 58–78. doi:10.1111/j.1365-2621.2007.01666.x

Aström, K. (2008). *Adaptive Control.* Addison Wesley.

Audoin-Rouzeau, S., & Becker, J. J. (Eds.). (2004). *Encyclopédie de la Grande Guerre, 1914-1918: histoire et culture.* Paris: Bayard Centurion.

Autry, C., & Bobbitt, M. (2008). Supply chain security orientation: Conceptual development and a proposed framework. *International Journal of Logistics Management*, *19*(1), 42–64. doi:10.1108/09574090810872596

Ayers, J. B. (2001). *Handbook of Supply Chain Management.* Boca Raton, FL: The St. Lucie Press/APICS Series on Resource Management.

Bagozzi, R., & Yi, Y. (1988). On the evaluation of structural equation models. *Journal of the Academy of Marketing Science*, *16*(1), 74–94. doi:10.1007/BF02723327

Bailey, D., & Solomon, G. (2004). Pollution prevention at ports: Clearing the air. *Environmental Impact Assessment Review*, *24*(7-8), 749–774. doi:10.1016/j.eiar.2004.06.005

Balakrishnan, M., Mohomed, I., & Ramasubramanian, V. (2009). Where's that phone?: geolocating IP addresses on 3G networks. In *Proceedings of the 9th ACM SIGCOMM conference on Internet measurement conference* (pp. 294–300). doi:10.1145/1644893.1644928

Balakrishnan, N. (1993). Simple heuristics for the vehicle routing problem with soft time windows. *The Journal of the Operational Research Society*, *44*(3), 279–287. doi:10.1057/jors.1993.53

Baldacci, R., Vigo, D., & Toth, P. (2010). Exact solution of the capacitated vehicle routing problem. In *Wiley Encyclopaedia of Operations Research and Management Science*. John Wiley & Sons, Inc.

Ballu, J.-M. (2000). *Bois de marine: les bateaux naissent en forêt*. Paris: Marine Nationale et Éditions du Gerfaut.

Bandyopadhyay, S., Saha, S., Maulik, U., & Deb, K. (2008). A simulated annealing-based multiobjective optimization algorithm: AMOSA. *IEEE Transactions on Evolutionary Computation*, *12*(3), 269–283. doi:10.1109/TEVC.2007.900837

Banks, J., Carson, J. S., & Nelson, B. L. (2010). *Discrete Event System Simulation* (5th ed.). United States: Pretince Hall.

Banomyong, R. (2005). The impact of port and trade security initiatives on maritime supply-chain management. *Maritime Policy & Management*, *32*(1), 3–13. doi:10.1080/0308883042000326102

Barbero, J. (2010). Freight Logistics in Latin America and the Caribbean: An Agenda to Improve Performance. Inter-American Development Bank, Infrastructure and Environment Department, Technical Notes No. IDB-TN-103, Washington D.C., USA.

Barbucha. (2013). Experimental Study of the Population Parameters Settings in Cooperative Multi-agent System Solving Instances of the VRP. *T. Computational Collective Intelligence, 9*, 1-28.

Barlas, Y. (1996). Formal aspects of model validity and validation in system dynamics. *System Dynamics Review*, *12*(3), 183–210. doi:10.1002/(SICI)1099-1727(199623)12:3<183::AID-SDR103>3.0.CO;2-4

Barnhart, C., & Laporte, G. (Eds.). (2007). Handbook in OR & MS (vol. 14). Elsevier B.V.

Barnhart, C., & Laporte, G. (2007). *Transportation, handbooks in operations research and management science*. Amsterdam: Elsevier.

Bartz-Beielstein, T., & Preuss, M. (2010). The future of experimental research. In *Experimental Methods for the Analysis of Optimization Algorithms* (pp. 17–49). Springer Berlin Heidelberg. doi:10.1007/978-3-642-02538-9_2

Bausch, D. O., Brown, G. G., & Ronen, D. (1998). Scheduling short-term marine transport of bulk products. *Maritime Policy & Management*, *25*(4), 335–348. doi:10.1080/03088839800000057

Bazaraa, M. S., & Jarvis, J. J. (1977). *Linear programming and network flows*. New York: John Wiley and Sons.

Beamon, B. M. (1998). Supply chain design and analysis: Models and methods. *International Journal of Production Economics*, *55*(3), 281–294. doi:10.1016/S0925-5273(98)00079-6

Becker, J. J. (1988). *La France en guerre, 1914 – 1918, La grande mutation*. Editions Complexe.

Beirao, G., & Sarsfield, J. A. (2007). Understanding attitudes towards public transport and private car: A qualitative study. *International Journal of Transport Policy*, *24*(6), 478–489. doi:10.1016/j.tranpol.2007.04.009

Bell, J. E., & McMullen, P. R. (2004). Ant colony optimization techniques for the vehicle routing problem. *Advanced Engineering Informatics*, *18*(1), 41–48. doi:10.1016/j.aei.2004.07.001

Bellmore, M., Bennington, G., & Lubore, S. (1968). A maximum utility solution to a vehicle constrained tanker scheduling problem. *Naval Research Logistics Quarterly*, *15*(3), 403–411. doi:10.1002/nav.3800150305

Bellmunt, J. (1998). *Grúa torre*. Ministerio de Trabajo y Asuntos Sociales.

Belobaba, P., Odoni, A., & Barnhart, C. (Eds.). (2009). *The global airline industry*. John Wiley & Sons. doi:10.1002/9780470744734

Berenson, M. L., Levine, D. M., & Krehbiel, T. C. (2001). *Estadística para administración*. Pearson Education.

Berstein, S., & Milza, P. (1999). *Histoire de la France au XXe siècle: 1900-1930 1*. Editions complexe.

Bertalanffy, L. (1968). *General systems theory foundations, development, applications*. New York: George Braziller.

Bertsimas, D. (1991). A vehicle routing problem with stochastic demand. *Journal of Operations Research*, *40*(3), 554–585.

Bertsimas, D. (1992). A vehicle routing problem with stochastic demand. *Operations Research*, *40*(3), 574–585. doi:10.1287/opre.40.3.574

Bertsimas, D., & Van Ryzin, G. (1990). A stochastic and dynamic vehicle routing problem in the Euclidian plane. *Operations Research Society of America*, *39*(4), 600–615.

Bertsimas, D., & Van Ryzin, G. (1993). Stochastic and dynamic vehicle routing with general demand and inter-arrival time distributions. *Advances in Applied Probability*, *25*(4), 947–978. doi:10.2307/1427801

Beyeler, C. (Ed.). (1997). Du bois dont on fait les vaisseaux… de lárbre en sa futaie à la figure de proue sculptée 1650 – 1850. Vincennes: Service historique de la Marine.

Bhope, D. V., & Padole, P. M. (2004). Experimental and theoretical analysis of stresses, noise and flow in centrifugal fan impeller. *Mechanism and Machine Theory*, *39*(12), 1257–1271. doi:10.1016/j.mechmachtheory.2004.05.015

Bichou, K., & Gray, R. (2004). A logistics and supply chain management approach to port performance measurement. *Maritime Policy & Management*, *31*(1), 47–67. doi:10.1080/0308883032000174454

Bierwirth, C., & Meisel, F. (2009). A fast heuristic for quay crane scheduling with interference constraints. *Journal of Scheduling*, *12*(4), 345–360. doi:10.1007/s10951-009-0105-0

Bing, H. C., Shuliang, T. L., & Zhu, B. (2013). Effects of meridional flow passage shape on hydraulic performance of mixed-flow pump impellers. *Chinese Journal of Mechanical Engineering*, *26*(3), 469–475. doi:10.3901/CJME.2013.03.469

Blackhurst, J., Craighead, C. W., Elkinsc, D., & Handfieldd, R. B. (2005). An empirically derived agenda of critical research issues for managing supply-chain disruptions. *International Journal of Production Research*, *43*(19), 4067–4081. doi:10.1080/00207540500151549

Blanton, J. L., & Wainwright, R. L. (1993). Multiple vehicle routing with time and capacity constraints using genetic algorithms. In *Proceedings of the Fifth International Conference on Generic Algorithms (ICGA-93)*, (pp. 452–459).

Bodlaender, H. L., Gustedt, J., & Telle, J. A. (1998). Linear-time register allocation for a fixed number of registers. In *Proceedings of the ninth annual ACM-SIAM Symposium on Discrete Algorithms*.

Borges, C. E. (2011). Algoritmos Bioinspirados. C.E. Borges, J.L. Montaña. Departamento de Matemáticas, Estadística y Computación Universidad de Cantabria. Retrieved from paginaspersonales.deusto.es/cruz.borges/Papers/10APIAXXI.pdf

Bornstein, N., & Svane, S. (2011). *Heuristics in Generic Planning using PDDL*. (Doctoral dissertation). Technical University of Denmark, Lyngby, Denmark.

Boudriot, J. (1988). *Le vaisseau de 74 canons*. Paris: Éditions Jean Boudriot.

Bourget, P., & Anthoine, P. (1989). La «déposition» du général anthoine sur la guerre de 1914-1918. *Guerres Mondiales et Conflits Contemporains*, (156), 105–108.

Bowersox, D. J., Closs, D. J., & Cooper, M. B. (2007). *Supply Chain Logistics Management* (2nd ed.). McGraw-Hill.

Boyle M. (2005). A Recipe for Disaster. *Fortune International Journal, 152*(9).

Bozer, Y., & Kile, J. (2008). Order Batching in Walk-and-Pick Order Picking Systems. *International Journal of Production Research, 46*(7), 1887–1909. doi:10.1080/00207540600920850

Brandeau, M. L., Sainfort, F., & Pierskalla, W. P. (2004). *Operations Research and Health Care: A Handbook of Methods and Applications* (M. L. Brandeau, F. Sainfort, & W. P. Pierskalla, Eds.). Boston, MA: Kluwer Academic Publisher.

Branke, J., Deb, K., Miettinen, K., & Słowi'nski, R. (2008). *Multiobjective Optimization.* Interactive and Evolutionary Approaches, Springer-Verlag Berlin Heidelberg. doi:10.1007/978-3-540-88908-3

Bräysy, O., Dullaert, W., & Gendreau, M. (2004). Evolutionary algorithms for the vehicle routing problem with time windows. *Journal of Heuristics, 10*(6), 587–611. doi:10.1007/s10732-005-5431-6

Bräysy, O., & Gendreau, M. (2005). Vehicle routing problem with time windows, part i: Route construction and local search algorithms. *Transportation Science, 39*(1), 104–118. doi:10.1287/trsc.1030.0056

Brønmo, G., Christiansen, M., & Nygreen, B. (2006). Ship routing and scheduling with flexible cargo sizes. *Journal of the Operation Research Society*, doi:.10.1057/palgrave.jors.2602263

Brooks, M. R. (1984). An alternative theoretical approach to the evaluation of liner shipping - Part I: Situational factors. *Maritime Policy & Management, 11*(1), 35–43. doi:10.1080/03088838400000032

Brown, G. G., Cormican, K. J., Lawphongpanich, S., & Widdis, D. B. (1997). Optimizing submarine berthing with a persistence incentive. *Naval Research Logistics, 44*(4), 301–318. doi:10.1002/(SICI)1520-6750(199706)44:4<301::AID-NAV2>3.0.CO;2-A

Brown, G. G., Graves, G. W., & Ronen, D. (1987). Scheduling ocean transportation of crude oil. *Management Science, 33*(3), 335–346. doi:10.1287/mnsc.33.3.335

Brummer, B. (2003). Food Biosecurity. *Journal of the American Dietetic Association, 103*(6), 697–691. PMID:12778038

Bueno, A., & Cedillo-Campos, M. (2014). Dynamic impact on global supply chains performance of disruptions propagation produced by terrorist acts. *Transportation Research Part E, Logistics and Transportation Review, 61*, 1–12. doi:10.1016/j.tre.2013.09.005

Buhaug, Ø., Corbett, J., Endresen, O., Eyring, V., Faber, J., Hanayama, S., & Yoshida, K. et al. (2009). *Second IMO greenhouse gas study.* London: International Maritime Organization.

Buneman, P., & Levy, L. (1980). The towers of Hanoi problem. *Information Processing Letters, 10*(4), 243–244. doi:10.1016/0020-0190(80)90150-7

Busemann, A. (1928). The pump head ratio of radial centrifugal pumps with logarithmic spiral blades.[In German]. *Zeitschrift für Angewandte Mathematik und Mechanik, 8*(5), 372–384. doi:10.1002/zamm.19280080504

Businger, J. (1963). *The glasshouse (greenhouse) climate.* North-Holland Publ. Co.

Busta, F., & Kennedy, S. (2011). Defending the Safety of the Global Food System from Intentional Contamination in a Changing Market. In *NATO Science for Peace and Security Series A* (pp. 119–135). Chemistry and Biology, Advances of Food Protection. doi:10.1007/978-94-007-1100-6_7

Cachon, G. P., & Netessine, S. (2004). Game theory in supply chain analysis. In *Handbook of Quantitative Supply Chain Analysis: Modeling in the e-Business Era*. Kluwer. doi:10.1007/978-1-4020-7953-5_2

Cagno, E., Caron, F., & Mancini, M. (2002). Risk Analysis in plant commissioning: The multilevel HAZOP. *Reliability Engineering & System Safety*, *77*(3), 309–323. doi:10.1016/S0951-8320(02)00064-9

Caliper Corporation. (2005). *Travel Demand Modeling with TransCAD 4.8*. USA: Caliper Corporation.

Calvete, H. I., Carmen, G., Mara, J. O., & Beln, S. V. (2004). Vehicle routing problems with soft time windows: An optimization based approach. *Journal of Monógrafas del Seminario Matemático García de Galdeano*, *31*, 295–304.

Carbone, V., & De Martino, M. (2003). The Changing Role of Ports in the Supply Chain Management: An Empirical Analysis. *Maritime Policy & Management*, *30*(4), 305–320. doi:10.1080/0308883032000145618

Cariou, P. (2011). Is slow steaming a sustainable means of reducing CO2 emissions from container shipping? *Transportation Research Part D, Transport and Environment*, *16*(3), 260–264. doi:10.1016/j.trd.2010.12.005

Carranza, O. (2007). *Global Perspectives-Mexico*. Council of Supply Chain Management Professionals.

Carrillo. (2013). *Post-Preto optimality methods for the analysis of large Pareto sets in multi-objective optimization*. PhD Thesis. Id 11353 ProQuest.

Carrillo, A., Aguirre, O., & Taboada, H. (2011). Applications and performance of the non-numerical ranking preferences method for post-Pareto optimality. *Procedia Computer Science*, *6*, 243–248. doi:10.1016/j.procs.2011.08.045

Carrillo, T., & Taboada, H. (2012). A General Iterative Procedure of the Non-Numerical Ranking Preferences Method, for Multiple Objective Decision Making. *Procedia Computer Science*, *12*, 135–139. doi:10.1016/j.procs.2012.09.043

Carrillo, T., & Taboada, H. (2012). A Post-Pareto Approach for Multi-Objective Decision Making Using a Non- Uniform Weight Generator Method. *Procedia Computer Science*, *12*, 116–121. doi:10.1016/j.procs.2012.09.040

Casandy, G. T. (1964). *Theory of Turbomachines*. McGraw Hill Book Company.

Cassandras, G. C., & Lafortune, S. (2009). *Introduction to discrete event systems*. Springer.

Castillo, N. (2011). *Evaluación de Estrategias de Mejora del Desempeño de Metaheurísticos Aplicados a BPP Vía Diagnóstico Visual*. (Master's thesis). Ciudad Madero, Tamaulipas.

Castillo-García, N., Fraire, H. J., Flores, J. A., Pazos, R. A., González, J. J., & Carpio, J. M. (2015). Comparative study on constructive heuristics for the vertex separation problem. In P. Melín, O. Castillo, & J. Kacprzyk (Eds.), *Design of Intelligent Systems Based on Fuzzy Logic, Neural Networks and Nature-Inspired Optimization* (pp. 465–474). Springer. doi:10.1007/978-3-319-17747-2_35

Cedillo-Campos, M. (2010). Security risk level classification of products: A multi-criteria approach applied to warehouse management. *The 15th Annual International Conference on Industrial Engineering Theory, Applications & Practice*.

Cedillo-Campos, M. (2011). *The First National Exploratory Study in Mexico: Supply Chain Security in Export Operations*. Secretariat of Economy. Retrieved from http://www.elogistica.economia.gob.mx/swb/work/models/elogistica/Resource/3/1/images/2011_Evaluacion_del_Riesgo_en_las_Cadenas_de_Suministro.pdf

Cedillo-Campos, M.-G., & Cedillo-Campos, H. (2015). w@reRISK Method: A Tool for Security Risk Level Classification of Products. *Safety Science*, *79*(November), 358–368.

Cedillo-Campos, M., & Sánchez, C. (2008). *Dynamic analysis of industrial systems*. Mexico: Trillas Publisher.

Cedillo-Campos, M., & Sanchez, C. (2013). Dynamic self-assessment of supply chains performance: An emerging market approach. *Journal of Applied Research and Technology*, *11*(3), 338–347. doi:10.1016/S1665-6423(13)71544-X

Centeno, A., & Mendoza, A. (2003). *Modelo de asignación intermodal multiproducto para las operaciones de carga por autotransporte y ferrocarril. Publicación Técnica 222. Instituto Mexicano del Transporte*. Querétaro, México: Spanish.

Cerny, V. (1985). Thermodynamical approach to the traveling salesman problem: An eficient simulation algorithm. *Journal of Optimization Theory and Applications*, *45*(1), 41–51. doi:10.1007/BF00940812

Cervantes, S. (2011). ¿Y dónde están los camiones? Revista T21. *Año*, *12*(Julio), 2223.

Chen, C. H., Lin, J., Ycesan, E., & Chick, S. E. (2000). Article. *Discrete Event Dynamic Systems*, *10*(3), 251–270. doi:10.1023/A:1008349927281

Chenghua, S., & Xiaofeng, Z. (2011). Research on model fitting capacity of vehicle routing problem. *International Journal of Advancements in Computing Technology*, *3*(11), 185–193. doi:10.4156/ijact.vol3.issue11.24

Chen, Y., Zou, X., & Xie, W. (2001). Conditions for the convergence of evolutionary algorithms. *Journal of Systems Architecture*, *47*(7), 601–612. doi:10.1016/S1383-7621(01)00018-2

Chew, E., & Tang, L. (1999). Travel time analysis for general item location assignment in a rectangular warehouse. *European Journal of Operational Research*, *112*(3), 582–597. doi:10.1016/S0377-2217(97)00416-5

Chiang, W. C., & Russell, R. (1996). Simulated annealing metaheuristics for the vehicle routing problem with time windows. *Annals of Operations Research*, *63*(1), 3–27. doi:10.1007/BF02601637

Chopra, S., & Meindl, P. (2013). *Supply chain management* (5th ed.). Prentice Hall.

Chorianopoulos, A., & Tsiptsis, K. (2009). *Data Mining Techniques in CRM: Inside Customer Segmentation*. Wiley.

Cho, S.-C., & Perakis, A. N. (1996). Optimal liner fleet routing strategies. *Maritime Policy & Management*, *23*(3), 249–259. doi:10.1080/03088839600000087

Cho, S.-C., & Perakis, A. N. (2001). An improved formulation for bulk cargo ship scheduling with a single loading port. *Maritime Policy & Management*, *28*(4), 339–345. doi:10.1080/03088830010002755

Chow, D., & Heaver, T. (1999). *Logistics strategies for North America* (3rd ed.). Global Logistics and Distribution Planning.

Christiansen, M., & Fagerholt, K. (2002). Robust ship scheduling with multiple time windows. *Naval Research Logistics*, *49*(6), 611–625. doi:10.1002/nav.10033

Christiansen, M., Fagerholt, K., Nygreen, B., & Ronen, D. (2007). Maritime Transportation. In C. Barnhaft & G. Laporte (Eds.), *Transportation, Handbooks in Operations Research and Management Science* (pp. 189–284). Amsterdam: North Holland.

Christiansen, M., Fagerholt, K., & Ronen, D. (2004). Ship routing and scheduling: Status and perspectives. *Transportation Science*, *38*(1), 1–18. doi:10.1287/trsc.1030.0036

Christopher, B. (2003). *Strategies for maximizing supply chain resilience: Learning from the past to prepare for the future*. (Master's Thesis). Massachusetts Institute of Technology, Cambridge, MA.

Christopher, M. (1998). *Logistics and Supply Chain Management* (2nd ed.). New York; Financial Times/Prentice Hall.

Chuanan, Y., & Yongchang, Y. (2010). Implementation of Greenhouse Monitoring System Based on RF Transceiver. *International Conference on Industrial Mechatronics and Automation*. doi:10.1109/ICINDMA.2010.5538107

Chuaqui, P., & Wittig Villarroel, E. M. (2004). Taguchi method for optimizing quality of functional desserts intended for elderly and technical feasibility study. *Economic Chilean Journal of Nutrition, 31*(2), 118-127. Retrieved on March 11, 2015, from http://www.scielo.cl/scielo.php?script=sci_arttext&pid=S0717-75182004000200006&lng=es&tlng=es.10.4067/S0717-75182004000200006

Chung, S. H., & Choy, K. L. (2012). A modified genetic algorithm for quay crane scheduling operations. *Expert Systems with Applications*, *39*(4), 4213–4221. doi:10.1016/j.eswa.2011.09.113

Chun, H. Ch., Choon, T. K., Laosirihongthong, T., & Keong, L. G. (2011). Entrepreneurial SCM competence and performance of manufacturing SMEs. *International Journal of Production Research*, *49*(22), 6629–6649. doi:10.1080/00207543.2010.537384

CIA – Central Intelligence Agency. (2012). *The World Factbook: Mexico*. Retrieved from https://www.cia.gov/library/publications/the-world-factbook/geos/mx.html

Cipoletta, G., Sánchez, R., & Pérez, G. (2010). *Políticas integradas de infraestructura, transporte y logística: experiencias internacionales y propuestas Iniciales*. Comisión Económica para América Latina ONU-CEPAL, División de Recursos Naturales e Infraestructura, Santiago de Chile. Retrieved from http://www.eclac.org/publicaciones/xml/7/39877/lcl3226e.pdf

Citymapper. (2015). *Citymapper, making cities usable*. Retrieved from https://citymapper.com

Clarke, G., & Wright, J. W. (1964). Scheduling of Vehicles from a Central Depot to a Number of Delivery Points. *Operations Research*, *12*(4), 568–581. doi:10.1287/opre.12.4.568

Clarkson Research Services (2014). *Container Intelligence Monthly*. Author.

Closs, D. & Mcgarrell, E. (2004). *Enhancing Security Throughout the Supply Chain*. Special Report Series, IBM Center for The Business of Government.

Closs, D., & McGarrell, E. (2004). Enhancing security through the supply chain. IBM Centre for the Business of Government. Special Report Series (April).

Coccola, M. E., Dondo, R., & Mendez, C. A. (2014). A milp-based column generation strategy for managing large-scale maritime distribution problems. *Computers & Chemical Engineering*. doi:10.1016/j.compchemeng.2014.04.008

Cochet, F., & Garde, R. (Eds.). (2008). Dictionnaire de la Grande Guerre 1914-1918. Paris: Robert Laffont.

Codex Alimentarius, CA (2003). *Código Internacional Recomendado de Prácticas-Principios generales de higiene de los alimentos*. (CAC/RCP 1-1969, Rev. 4-2003).

Coello Coello, C. A., Lamont, G. B., & Van Veldhuizen, D. A. (2007). *Genetic and Evolutionary Computation. Evolutionary Algorithms for Solving Multi-Objective Problems* (J. R. David Goldberg, Ed.). New York: Springer.

Cohen, P. R. (1995). *Empirical Methods for Artificial Intelligence*. Bradford Books.

Collison, F. M. (1984). North to Alaska: Marketing in the Pacific Northwest- Central Alaska liner trade. *Maritime Policy & Management*, *11*(2), 99–112. doi:10.1080/03088838400000003

Consolvo, S., McDonald, D. W., Toscos, T., Chen, M. Y., Froehlich, J., Harrison, B., & Landay, J. A. et al. (2008). *Activity sensing in the wild: a field trial of ubifit garden*. In*Proceedings of the SIGCHI Conference on Human Factors in Computing Systems* (pp. 1797–1806). doi:10.1145/1357054.1357335

Cooper, M. C., & Ellram, L. M. (1993). Characteristic of supply chain management and the implications for purchasing and logistics strategy. *International Journal of Logistic Management*, *4*(2), 13–24. doi:10.1108/09574099310804957

Cordeau, J. F., Laporte, G., Martin, W. P. S., & Daniele, V. (2007). Vehicle Routing. In Handbooks in Operations Research and Management Science. Elsevier B.V.

Cordeau, J. F., Gaudioso, M., Laporte, G., & Moccia, L. (2007). The service allocation problem at the Gioia Tauro maritime terminal. *European Journal of Operational Research*, *176*(2), 1167–1184. doi:10.1016/j.ejor.2005.09.004

Cordeau, J. F., Gendreau, M., Laporte, G., Potvin, J. Y., & Semet, F. (2002). A guide to vehicle routing heuristics. *The Journal of the Operational Research Society*, *53*(5), 512–522. doi:10.1057/palgrave.jors.2601319

Cordeau, J. F., Laporte, G., & Mercier, A. (2001). A unified tabu search heuristic for vehicle routing problem with time windows. *The Journal of the Operational Research Society*, *53*(8), 928–936. doi:10.1057/palgrave.jors.2601163

Corvisier, A. (1995). Quelques réflexions sur les relations entre armée et marine sous l Ancien Regime. In État, marine et société. Presses de l'Université de Paris-Sorbonne.

Cox, D. R., & Smith, W. L. (1961). *Queues*. New York: John Wiley & Sons.

Crainic, T. G., & Kim, K. H. (2007). Intermodal transportation. In C. Barnhart & G. Laporte (Eds.), *Transportation, Handbooks in Operations Research and Management Science* (pp. 467–537). Amsterdam: North-Holland.

Crainic, T. G., Perboli, G., Rei, W., & Tadei, R. (2011). Efficient lower bounds and heuristics for the variable cost and size bin packing problem. *Computers & Operations Research*, *38*(11), 1474–1482. doi:10.1016/j.cor.2011.01.001

Crary, M., Nozick, L. K., & Whitaker, L. R. (2002). Sizing the US destroyer fleet. *European Journal of Operational Research*, *136*(3), 680–695. doi:10.1016/S0377-2217(01)00031-5

Crowder, H., Dembo, R. S., & Mulvey, J. M. (1979). On reporting computational experiments with mathematical software. *ACM Transactions on Mathematical Software*, *5*(2), 193–203. doi:10.1145/355826.355833

Crutchley, T. M., Rodgers, H. P., Whiteside, J., Vanier, M., & Terndrup, T. E. (2007). Agroterrorism: Where Are We in the Ongoing War on Terrorism. *Journal of Food Protection*, *70*(370), 791–804. PMID:17388078

Cruz, L. (2004). *Clasificación de Algoritmos Heurísticos Para la Solución de Problemas de Bin Packing*. (Doctoral dissertation, PhD thesis). CENIDET, Cuernavaca, Mexico.

Cruz, J. M. (2008). Dynamics of supply chain networks with corporate social responsibility through integrated environmental decision-making. *European Journal of Operational Research*, *184*(3), 1005–1031. doi:10.1016/j.ejor.2006.12.012

Cruz, L., & Ochoa, A. et al.. (2010). A Cultural Algorithm for the Urban Public Transportation. *HAIS*, *2010*, 135–142.

Cruz, O. (2011). Robo al autotransporte más grave que el narco. Revista T21. *Año*, *12*(Julio), 25.

Cullinane, K., & Khanna, M. (1999). Economies of scale in large container ships. *Journal of Transport Economics and Policy*, *33*(2), 185–208.

Cull, P., Merrill, L., & Van, T. (2013). A Tale of Two Puzzles: Towers of Hanoi and Spin-Out. *Information and Media Technologies*, *8*(3), 695–709.

Customs-Trade Partnership Against Terrorism (C-TPAT). (2014). *C-TPAT's Five Step Risk Assessment*. U.S. Customs and Border Protection, Office of Field Operations. Retrieved June 2, 2015, from http://www.cbp.gov/sites/default/files/documents/

Czech, Z. J., & Czarnas, P. (2002). Parallel simulated annealing for the vehicle routing problem with time windows. In *10th Euromicro Workshop on Parallel, Distributed and Network-based Processing*, Canary Islands-Spain. doi:10.1109/EMPDP.2002.994313

Czyzzak, P., & Jaszkiewicz, A. (1998). Pareto simulated annealing—a metaheuristic technique for multiple-objective combinatorial optimization. *Journal of Multi-Criteria Decision Analysis*, *7*(1), 34–47. doi:10.1002/(SICI)1099-1360(199801)7:1<34::AID-MCDA161>3.0.CO;2-6

Daganzo, C. F. (1989). The crane scheduling problem. *Transportation Research Part B: Methodological*, *23*(3), 159–175. doi:10.1016/0191-2615(89)90001-5

Dalziel G.R. (2009). *Food Defense Incidents 1950-2008: a Chronology and analysis of Incidents involving the malicious Contamination of the Food Supply Chain*, Centre of Excellence for National Security (CENS) of the S. Rajaratnam School of International Studies (RSIS) at Nanyang technological University, Singapore.

Damij, N., Damij, T., Grand, J., & Jelenc, F. (2008). A Methodology for Business Process Emprovement and IS Development. *Information and Software Technology*, *50*(11), 1127–1141. doi:10.1016/j.infsof.2007.11.004

Dantzig, G. B., & Fulkerson, D. R. (1954). Minimizing the number of tankers to meet a fixed schedule. *Naval Research Logistics Quarterly*, *1*(3), 217–222. doi:10.1002/nav.3800010309

Dantzig, G., Fulkerson, R., & Johnson, S. (1954). Solution of a large-scale traveling-salesman problems. *Operations Research*, *2*, 393–410.

Dantzig, G., & Ramser, J. (1959). The truck dispatching problem. *Management Science*, *6*(1), 80–91. doi:10.1287/mnsc.6.1.80

David, P., & Stewart, R. (2010). *International Logistics: the Management of International Trade Operations* (3rd ed.). Atomic Dog Publishing.

Daviet, J. P. (1997). *La société industrielle en France 1814-1914*. Paris: Seuil.

de Castro, L. N. (2006). *Fundamentals of natural computing: Basic concepts, algorithms, and applications*. Boca Raton, FL: CRC Press.

de Koster, M., Le-Duc, T., & Roodbergen, K. (2007). Design and Control of Warehouse Order Picking: A Literature Review. *European Journal of Operational Research*, *182*(2), 481–501. doi:10.1016/j.ejor.2006.07.009

De Neufville, R., & Grillot, M. (1982). Design of pedestrian space in airport terminals. *Journal of Transportation Engineering*, *108*, 87–102.

De Neufville, R., & Odoni, A. (2013). *Airport Systems. Planning, Design and Management*. McGraw-Hill Education LLC.

De Oliveira, A., & Lorena, L. A. N. (2002). A constructive genetic algorithm for gate matrix layout problems. *IEEE Transactions on Computer-Aided Design of Integrated Circuits and Systems*, *21*(8), 969–974. doi:10.1109/TCAD.2002.800454

de Rham, C. (1980). La classification hiérarchique ascendante selon la méthode des voisins réciproques. *Les Cahiers de l'Analyse des Donnees*, *5*(2), 135–144.

Deb, K. (2001). *Multi-Objective Optimization using Evolutionary Algorithms*. New York: John Wiley & Sons.

Debrauwer, L., & Heyde, F. v. (2005). *UML 2: iniciación, ejemplos y ejercicios corregidos*. Cornellà de Llobregat, Barcelona: Ed.ENI.

Debreu, G. (1972). Smooth preference. *Econometrica*, *40*(4), 603–615. doi:10.2307/1912956

Debreu, G. (1972). *Theory of value: An axiomatic analysis of economic equilibrium. Cowles Foundation Monograph.* University Press.

Delahaye, D., & Puechmorel, S. (2013). *Modeling and Optimization of Air Traffic*. Wiley-ISTE. doi:10.1002/9781118743805

Delen, D., & Olson, D. (2008). *Advanced Data Mining Techniques*. Springer. doi:10.1002/9780470172339

Delhez, J. C. (2013), La bataille des frontières. Joffre attaque au centre. 22-26 août 1914. Economica.

Demšar, J. (2006). Statistical comparisons of classifiers over multiple data sets. *Journal of Machine Learning Research*, *7*, 1–30.

Department of Defense (DD). (2000). *Standard Practice for System Safety*. United Estates of America Department of Defense (MIL-STD-882D). Retrieved May 14, 2015, from http://www.system-safety.org/Documents/MIL-STD-882D.pdf

Department of Health and Human Services (DHHS). (2001). *Food safety and security: operational risk management approach*. United States of America, US Food and Drug Administration, Center for Food Safety and Applied Nutrition. Retrieved May 11, 2015, from https://www.cdph.ca.gov/pubsforms/Guidelines/Documents/fdb%20ORM%202001.pdf

Department of Health and Human Services (DHHS). (2005), *Terrorism and Food Supply. Terrorism and other Public Health Emergencies: a reference guide*. United States of America, US Food and Drug Administration, Center for Food Safety and Applied Nutrition. Retrieved April 22, 2015, from http://www.achd.net/emergprep/pubs/pdf/HHSMedisReferenceGuideFinal.pdf

Department of Health and Human Services (DHHS). (2007a). *Cosmetics Processors and Transporters: Cosmetics Security Preventive Measures Guidance*. Estados Unidos. Retrieved May 14, 2015, from http://www.fda.gov/Food/GuidanceRegulation/GuidanceDocumentsRegulatoryInformation/FoodDefense/ucm082716.htm

Department of Health and Human Services (DHHS). (2007b). *Retail Food Stores and Food Service Establishments: Food Security Preventive Measures Guidance*. Estados Unidos. Retrieved May 25, 2015, from http://www.fda.gov/Food/GuidanceRegulation/GuidanceDocumentsRegulatoryInformation/FoodDefense/ucm082751.htm

Department of Health and Human Services (DHHS). (2007c). *Food Producers, Processors and Transporters: Food Security Preventive Measures Guidance*. Estados Unidos. Retrieved May 25, 2015, from https://www.hsdl.org/?view&did=451052

Derrac, J., García, S., Molina, D., & Herrera, F. (2011). A practical tutorial on the use of nonparametric statistical tests as a methodology for comparing evolutionary and swarm intelligence algorithms. *Swarm and Evolutionary Computation*, *1*(1), 3–18. doi:10.1016/j.swevo.2011.02.002

Deshpande, A. (2012). Supply Chain Management Dimensions, Supply Chain Performance and Organizational Performance: An Integrated Framework. *International Journal of Business and Management*, *7*(8), 2–19. doi:10.5539/ijbm.v7n8p2

Desmond, A., & Moore, J. (1995). *Darwin-la vida de un evolucionista atormentado*. Sao Paulo, Brazil: Generacion Editorial.

Desrosiers, J., Dumas, Y., Solomon, M. M., & Soumis, F. (1995). Time constrained routing and scheduling. In Network Routing, Handbooks in Operations Research and Management Science (pp. 35-139). North-Holland.

Deveci, D. A. (2013). Deniz ulaştırması: İşletmeler ve işlevler (Maritime transportation: Businesses and functions). In A. Cerit, A. Deveci, & S. Esmer (Eds.), *Denizcilik İşletmeleri Yönetimi (Maritime Business Management)* (pp. 23–34). İstanbul: Beta Basım A.Ş.

DeVore, R. A., & Temlyakov, V. N. (1996). Some remarks on greedy algorithms. *Advances in Computational Mathematics*, *5*(1), 173–187. doi:10.1007/BF02124742

Díaz, F. (2002). Modelado y Control Jerárquico de Crecimiento de Cultivos en Invernadero. *Tesis Doctoral Universidad de Almería Escuela Politécnica Superior*, 390.

Diaz, J. A., & Perez, D. (2012). Optimization of inventory levels in a supply chain. *Industrial Engineering, 33*(2), 126-132. Available at: http://scielo.sld.cu/scielo.php?script=sci_arttext&pid=S1815-59362012000200004&lng=es&nrm=iso

Díaz, J., Petit, J., & Serna, M. (2002). A survey of graph layout problems. *ACM Computing Surveys*, *34*(3), 313–356. doi:10.1145/568522.568523

Dictionary of Accounting. (2002). Mexico: Hara.

Ding, D., & Chou, M. C. (2015). Stowage planning for container ships: A heuristic algorithm to reduce the number of shifts. *European Journal of Operational Research*, *246*(1), 242–249. doi:10.1016/j.ejor.2015.03.044

Dohn, A., Rasmussen, M. S., & Larsen, J. (2011). The vehicle routing problem with time windows and temporal dependencies. *Networks*, *58*(4), 273–289. doi:10.1002/net.20472

Doksum, K. (1967). Robust procedures for some linear models with one observation per cell. *Annals of Mathematical Statistics*, *38*(3), 878–883. doi:10.1214/aoms/1177698881

Dominguez, J. (2006). Simultaneous optimization for continuous improvement and cost reduction processes. *Engineering and Science-ing. Cienc., 2*(4), 145-162. Recovered from http://publicaciones.eafit.edu.co/index.php/ingciencia/article/view/473/440

Donnelly, R., Thompson, R. G., & Wigan, M. (2012). Process validation of urban freight and logistics models. *Procedia: Social and Behavioral Sciences*, *39*, 400–408. doi:10.1016/j.sbspro.2012.03.117

Dourlens, C., Galland, J. P., & Vidal-Naquet, P. A. (1991). Introduction. In Conquête de la sécurité, gestion des risques. París: L'Harmattan.

Dror, M. (1991). Modeling vehicle routing with uncertain demands as a stochastic program: Properties of the corresponding solution. *European Journal of Operational Research*, *68*, 432–441.

Dror, M. (Ed.). (2000). *Arc routing: Theory, Solutions and Approximations*. Berlin: Springer. doi:10.1007/978-1-4615-4495-1

Dror, M., Laporte, G., & Francois, V. L. (1993). Vehicle routing with stochastic demands and restricted failures. *Methods and Models of Operations Research*, *33*, 271–283.

Dror, M., Laporte, G., & Trudeau, P. (1994). Vehicle routing with split deliveries. *Discrete Applied Mathematics*, *50*(3), 229–254. doi:10.1016/0166-218X(92)00172-I

Dror, M., & Trudeau, P. (1986). Stochastic vehicle routing with modified savings algorithm. *Journal of Operational Research*, *23*(2), 228–235. doi:10.1016/0377-2217(86)90242-0

Dror, M., & Trudeau, P. (1989). Savings by split delivery routing. *Transportation Science*, *23*(2), 141–145. doi:10.1287/trsc.23.2.141

Duarte, A., Escudero, L. F., Martí, R., Mladenovic, N., Pantrigo, J. J., & Sánchez-Oro, J. (2012). Variable neighborhood search for the vertex separation problem. *Computers & Operations Research*, *39*(12), 3247–3255. doi:10.1016/j.cor.2012.04.017

Duarte, A., Martí, R., Resende, M. G., & Silva, R. M. (2011). Grasp with path relinking heuristics for the antibandwidth problem. *Networks*, *58*(3), 171–189. doi:10.1002/net.20418

Dubois, A., & Gadde, L. (2002). Systematic combining: An abductive approach to case research. *Journal of Business Research*, *55*(7), 553560. doi:10.1016/S0148-2963(00)00195-8

Duhamel du Monceau, H.-L. (1752). *Élements d l'architecture navale.* Paris: Charles-Antoine Jombert.

Dunnett, C. W. (1955). A multiple comparison procedure for comparing several treatments with a control. *Journal of the American Statistical Association, 50*(272), 1096–1121. doi:10.1080/01621459.1955.10501294

Du, Y., Xue, Z., Zhu, Q., Liu, X., Feng, Y., & Zhang, S. (2013). Design and Application of Intelligent Control System for Greenhouse Environment Based on CAN bus.*Proceedings of International Conference on Modeling, Identification & Control (ICMIC).*

Dyer, J. H., Sung, C., & Chu, W. (1998). Strategic supplier Segmentation: The next "best practice" in Supply chain management. *California Management Review, 40*(2), 57–77. doi:10.2307/41165933

Ehlert, H., Epkenhans, M., Gross, G. P., & Zabecki, D. T. (Eds.). (2014). *The Schlieffen Plan: International Perspectives on the German Strategy for World War I.* University Press of Kentucky.

Eiselt, H. A., Gendreau, M., & Laporte, G. (1995). Arc routing problems, Part 1: The Chinese Postman problem. *Operations Research, 43*(2), 231–242. doi:10.1287/opre.43.2.231

Eiselt, H. A., Gendreau, M., & Laporte, G. (1995). Arc routing problems, Part 2: The rural Postman problem. *Operations Research, 43*(3), 399–414. doi:10.1287/opre.43.3.399

Eisenhardt, K. M. (1989). Building theory from case study research. *Academy of Management Review, 14*(4), 532–550.

Ellram, L. M. (1996). The use of the case study method in logistics research. *Journal of Business Logistics, 17*(2), 93–138.

Endresen, Ø., Sørgård, E., Behrens, H. L., Brett, P. O., & Isaksen, I. S. A. (2007). A historical reconstruction of ships fuel consumption and emissions. *Journal of Geophysical Research, 112*(D12), D12301. doi:10.1029/2006JD007630

EPA. (2015). *United States Environmental Protection Agency.* Retrieved from http://www.epa.gov/cleandiesel/technologies/retrofits.htm

Eredics, P., & Dobrowiecki, T. P. (2011). Data Cleaning for an Intelligent Greenhouse.*International Symposium on Applied Computational Intelligence and Informatics.*

Eredics, P., Gáti, K., Dobrowiecki, T. P., & Horváth, G. (2011). Neural Models for an Intelligent Greenhouse - A CMAC Global Greenhouse Model.*International Symposium on Computational Intelligence and Informatics.* doi:10.1109/CINTI.2011.6108495

Erol, I. (2011). *Comparison of different heuristic search methods in a design space exploration.* Academic Press.

Esmer, S. (2014). Talep tahminleri ışığında türk limanlarına yönelik kapasite analizi (Capacity analysis for Turkish ports with the light of demand forecasts). In *Proceedings of 8.* Istanbul, Turkey: Kıyı Mühendisliği Sempozyumu.

EUROCONTROL. (2013). *Challenges of Growth 2013 Task 4: European Air Traffic in 2035.* Technical Report.

European Commission. (2013). Commission Staff Working Document, Impact Assessment, 2. Brussels: Author.

Expósito-Izquierdo, C., González-Velarde, J. L., Melián-Batista, B., & Moreno-Vega, J. M. (2013). Hybrid estimation of distribution algorithm for the quay crane scheduling problem. *Applied Soft Computing, 13*(10), 4063–4076. doi:10.1016/j.asoc.2013.05.006

Fabila, G. (1998). *Diseño y análisis de experimentos industriales.* México: Universidad Iberoamericana.

Fagerholt, K., & Christiansen, M. (2000). A combined ship scheduling and allocation problem. *The Journal of the Operational Research Society, 51*(7), 834–842. doi:10.1057/palgrave.jors.2600973

Fagerholt, K., & Ronen, D. (2013). Bulk ship routing and scheduling: Solving practical problems may provide better results. *Maritime Policy & Management*, *40*(1), 48–64. doi:10.1080/03088839.2012.744481

Fang, J., & Wang, F. (2011). *Design of Greenhouse Remote Monitoring System based on LabVIEW.* Harbin Science and Technology Agency Foundation of Returned Overseas.

Faure, R., Guillat-Le Garff, N. S., & Bloch, M. (1979). *Précis de recherche opérationnelle*. Paris: Dunod.

Federal Aviation Administration (FAA). (2000). Operational Risk Management (ORM). In System Safety Handbook: Practices and Guidelines for Conducting System Safety Engineering and Management. Author.

Federal Aviation Administration (FAA). (2010). Airport Passenger Terminal Planning and Design. In Guidebook, Office of Oper. Research (vol. 1). Federal Aviation Administration.

Federal Aviation Administration. (2001). *Forecasting aviation activity by airport.* Federal Aviation Administration, Technical Report.

Feitelson, D. G. (2006). *Experimental computer science: The need for a cultural change*. Retrieved from http://www.cs.huji.ac.il/~feit/papers/exp05.pdf

Feng, L., & Zhi, Y. (2013). *Greenhouse CFD Simulation for Searching the Sensors Optimal Placements*. Special Fund for Agro-Scientific Research in the Public Interest of China. doi:10.1109/Argo-Geoinformatics.2013.6621972

Férier, G. (1987). Les conséquences démographiques de la grande guerre en Auvergne. *Guerres Mondiales et Conflits Contemporains*, (145): 81–94. PMID:11634995

Fernández, J., García, S., & Derrac, J. (2014). KEEL: A software tool to assess evolutionary algorithms for Data Mining problems (regression, classification, clustering, pattern mining and so on). *Keel.es*. Retrieved 02 July 2015, from http://www.keel.es/

Ferreira, P. M., & Ruano, A. E. (2008). Application of Computational Intelligence Methods to Greenhouse Environmental Modelling. IEEE.

Fezari, M., Khati, A., & Boumaza, M. S. (2011). Implementation of Wireles Sensor Network for Automatic Greenhouse Monitorign. In Communications, Computing and Control Applications (CCCA).

Financial Post. (2012). *Goldman Sachs economists: Mexico in the 2020 harvest the seventh largest economy*. Retrieved from http://www.financial-post.com/goldman-sachs-economists-mexico-in-the-2020-harvest-the-seventh-largest-economy.html

Finch, P. (2004). Supply chain risk management. *Supply Chain Management. International Journal (Toronto, Ont.)*, *9*(2), 183–196.

Fisher, M. L., & Rosenwein, M. B. (1989). An interactive optimization system for bulk-cargo ship scheduling. *Naval Research Logistics*, *36*(1), 27–42. doi:10.1002/1520-6750(198902)36:1<27::AID-NAV3220360103>3.0.CO;2-0

Flatberg, T. (2005). *Dynamic and Stochastic Aspects in Vehicle Routing: A Literature Survey. SINTEF rapport*. SINTEF ICT.

Fleszar, K., & Charalambous, C. (2011). Average-weight-controlled bin-oriented heuristics for the one-dimensional bin-packing problem. *European Journal of Operational Research*, *210*(2), 176–184. doi:10.1016/j.ejor.2010.11.004

Flores, I., Figueras, J., Guasch, A., Mujica, M., Narciso, M., & Piera, M. (2013). *Simulation Models: Using Simio and Preti Nets*. Mexico: Universidad Nacional Autonoma de Mexico.

Fogel R.W. (1964), *Railroads and American economic growth*. Academic Press.

Fogel, R. W. (1970). Historiography and Retrospective Econometrics. *History and Theory*, *9*(3), 245–264. doi:10.2307/2504408

Food and Drug Administration (FDA). (1999). *Buenas prácticas de manufactura en la fabricación, empaque y manejo de alimentos para consumo humano.* (Parte 110 -21 CFR 110.1 - 110.110 -Código de Regulaciones Federales).

Food and Drug Administration (FDA). (2007). *An overview of the carver plus shock method for food sector vulnerability assessments.* Estados Unidos. Retrieved June 1, 2015, from http://www.fsis.usda.gov/wps/wcm/connect/483f86d5-a566-44f8-90d5-05a16dbe3f78/CARVER.pdf?MOD=AJPERES

Food and Drug Administration (FDA). (2012). *Vulnerability Assessments of Food Systems. Final Summary Report.* Estados Unidos. Retrieved June 1, 2015, from http://www.fda.gov/downloads/Food/FoodDefense/UCM317547.pdf

Food and Drug Administration (FDA). (2015). *Tools and educational materials of food defense.* Estados Unidos. Retrieved June 1, 2015, from http://www.fsis.usda.gov/wps/portal/fsis/topics/food-defense-defense-and-emergency-response/preparation-and-prevention/guidance-documents

Forestier, P. (2014). La citoyenneté en guerre? Les soldats guadeloupéens sur le front pendant la Première Guerre mondiale. *Bulletin de la Société d'Histoire de la Guadeloupe*, *168*(168), 3–20. doi:10.7202/1026842ar

Forfait, P. (1788). *Traité élémentaire de la mâture des vaisseaux.* Paris: Clousier.

Forgionne, G. A. (1983). Corporate Activities Management Science. *Interfaces*, *13*(3). doi:10.1287/inte.13.3.20

Fornell, C., & Larcker, D. (1981). Evaluating structural equation models with unobservable variables and measurement error. *JMR, Journal of Marketing Research*, *18*(1), 39–50. doi:10.2307/3151312

Fourer, R., Ma, J., & Martin, K. (2010). Optimization services: A framework for distributed optimization. *Operations Research*, *58*(6), 1624–1636. doi:10.1287/opre.1100.0880

FourSquare. (2015). *Introducing the all-new Foursquare, which learns what you like and leads you to places you'll love.* Retrieved from https://foursquare.com/

Fraire, H. J., Castillo-García, N., Pazos, R. A., Flores, J. A., González, J. J., & Carpio, J. M. (2015). Two new exact methods for the vertex separation problem. *International Journal of Combinatorial Optimization Problems and Informatics*, *6*(1), 31–41.

Friedman, T. (2012). How Mexico Got Back in the Game. *The New York Times.* Retrieved from http://www.nytimes.com/2013/02/24/opinion/sunday/friedman-how-mexico-got-back-in-the-game.html?_r=0

Frutos, M., Olivera, A. C., & Tohmé, F. (2010). A memetic algorithm based on a NSGAII scheme for the flexible job-shop scheduling problem. *Annals of Operations Research*, *181*(1), 745–765. doi:10.1007/s10479-010-0751-9

Frutos, M., & Tohmé, F. (2015). Choice of a PISA selector in a hybrid algorithmic structure for the FJSSP. *Decision Science Letters*, *4*(2), 247–260. doi:10.5267/j.dsl.2014.11.001

Gabay, A., & Zenya, J. (1974). *Máquinas para obras Barcelona.* Editorial Blume y Editorial Labor, S.A.

Gahlot, V. K., & Nyiri, A. (1993). *Impeller Pumps, Theory and Design.* Bhopal: M.A.C.T.

Gaitanides, M. (2007). *Prozessorganisation –Entwicklung, Ansätze und Programme des Managements von Geschäftsprozessen* (2nd ed.). Munich: Vahlen.

Ganeshan, R. & Harrison, T. P. (1995). *An Introduction to Supply Chain Management.* Department of Management Sciences and Information Systems.

Gaonkar, R., & Viswanadham, N. (2009). A conceptual and analytical framework for management of integrated knowledge based logistics providers. *International Journal of Logistics Systems and Management*, *5*(1/2), 191209.

García, J. M. (2007). *Algoritmos Basados en Inteligencia Colectiva para la Resolución de Problemas de Bioinformática y Telecomunicaciones*. Universidad de Málaga. Retrieved from https://gruasytransportes.wordpress.com/

Garcia, B. L., Potvin, J. Y., & Rousseau, J. M. (1994). A parallel implementation of the tabu search heuristic for vehicle routing problems with time window constraints. *Computers & Operations Research*, *21*(9), 1025–1033. doi:10.1016/0305-0548(94)90073-6

García, S., Molina, D., Lozano, M., & Herrera, F. (2009). A study on the use of nonparametric tests for analyzing the evolutionary algorithms' behaviour: A case study on the CEC'2005 special session on real parameter optimization. *Journal of Heuristics*, *15*(6), 617–644. doi:10.1007/s10732-008-9080-4

Garey, M. R., & Johnson, D. S. (1979). *Computers and intractability: a guide to the theory of NP-completeness*. San Francisco: WH Freeman & Co.

Garrod, P., & Miklius, M. (1985). The optimal ship size: A comment. *Journal of Transport Economics and Policy*, *19*(1), 83–91.

Geertz, C. (1989). *La interpretacion de las culturas*. RJ, Brazil: Editorial Guanabara.

Gessler, N. (1999). *Artificial Culture-experimemts in Syntethic Anthropology*. Retrieved from http://www.soc.surrey.ac.uk/JASSS/1/1/4.html

Ghauri, P., & Gronhaug, K. (2005). *Research methods in business studies: a practical guide* (3rd ed.). New York: FT Prentice Hall.

Ghiani, G., Laporte, G., Manni, E., & Musmanno, R. (2008). Waiting strategies for the dynamic and stochastic travelling salesman problem. *International Journal of Operations Research*, *5*(4), 233–241.

Ghosal, M., & Tiwari, G. (2004). *Mathematical modeling for greenhouse heating by using thermal curtain and geothermal energy* (Vol. 76). Elsevier.

Giesen, R., Munoz, J. C., Silva, M., & Leva, M. (2007). *Multi-item inventory routing problem for ship distribution of liquid oil bulk products*. Paper presented at TRISTAN VI, Phuket Island, Thailand. http://transp-or2.epfl.ch/tristan/FullPapers/099Giesen.pdf

Gilles, H., Guironnet, J. P., & Parent, A. (2014a). Géographie économique des morts de 14-18 en France. *Revue Économique, 65*(3), 519-532.

Gilles, H., Guironnet, J. P., & Parent, A. (2014b, May 13). La géographie des Morts pour la France. *Le Monde*.

Gillman, S. (1999). The size economies and network efficiencies of large containerships. *International Journal of Maritime Economics*, *1*(1), 39–59.

Giunipero, L., & Eltantawy, R. (2003). Securing the upstream supply chain: A risk management approach. *International Journal of Physical Distribution & Logistics Management*, *34*(9), 698713.

Glass, S., Vallipuram, M., & Portmann, M. (2010). The Insecurity of Time-of-Arrival Distance-Ranging in IEEE 802.11 Wireless Networks. *ICDS Workshops*, *2010*, 227–233.

Goldin, C. (1994). *Cliometrics and the Nobel*. Historical paper N° 65, NBER Working Paper.

Gonzalez-Feliu, J., Ovtracht, N., & Parent, A. (2014). Gare et Front: la logique de répartition des troupes dans le plan XVII de Joffre. *Colloque « Gares en Guerre, 1914-1918 »*, Paris, 3-5 septembre 2014.

Gonzalez-Feliu, J., Ovtracht, N., & Parent, A. (2015). *Did Joffre's Plan XVII follow an economics logic?* 4th seminar Clio – Sciences Po Lyon-LET, Lyon, France.

Gonzalez-Feliu, J. (2013). Multi-Stage LTL Transport Systems in Supply Chain Management. In J. Cheung & H. Song (Eds.), *Logistics: Perspectives, Approaches and Challenges* (pp. 65–86). New York: Nova Science Publishers.

Gonzalez-Feliu, J., Semet, F., & Routhier, J. L. (Eds.). (2014). *Sustainable urban logistics: Concepts, methods and information systems*. Heidelberg, Germany: Springer. doi:10.1007/978-3-642-31788-0

Google Latitude. (2015). *Ayuda de Google Maps para móviles*. Retrieved from https://www.google.com/latitude?hl=es

Google Now. (2015). *Google Now. Información precisa en el momento adecuado*. Retrieved from https://www.google.com/landing/now/

Gradilla, L. (2011). *Planeación de infraestructura de transporte: Identificación de tramos críticos para el funcionamiento de redes carreteras. Publicación Técnica 354. Instituto Mexicano del Transporte*. Querétaro, México: Spanish.

Grant, D., Lambert, D., Stock, J., & Ellram, L. (2006). *Fundamentals of Logistics Management* (European Edn.). Berkshire: McGraw-Hill Book Co.

Greif Avner, (1997). Cliometrics after 40 years. *American Economic Review, Papers and Proceedings*, *87*(2).

Griffin, D. R., Webster, F. A., & Michael, C. R. (1960). The echolocation of flying insects by bats. *Animal Behaviour*, *8*(34), 141–154. doi:10.1016/0003-3472(60)90022-1

Guillemin, V., & Pollack, A. (2010). *Differential topology*. AMS Chelsea Publishing.

Guinipero, L. C., & Brand, R. R. (1996). Purchasing's role in supply chain management. *International Journal of Logistics Management*, *5*(1), 29–38. doi:10.1108/09574099610805412

Guofang, L., Yubin, Q., Shengtao, L., & Junyu, X. (2010). Remote Monitoring System of Greenhouse Environment Based on LabVIEW.*International Conference on Computer Design and Applications (ICCDA 2010)*. doi:10.1109/ICCDA.2010.5541107

Gupta, M., & Chandra, P. (2002). *Effect of greenhouse design parameters on conservation of energy for greenhouse environmental control* (Vol. 27). Pergamon.

Gwillian, K (2003). Urban transport in developing countries. *Transport Reviews: A Transnational Transdisciplinary Journal, 23*(2), 197–216.

Habr, S., & Reichel, A. (2005). Identifying performance measures of small ventures – The case of the tourism industry. *Journal of Small Business Management*, *43*(3), 257–286. doi:10.1111/j.1540-627X.2005.00137.x

Haidine, A., & Lehnert, R. (2008). Multi-case multi-objective simulated annealing (mc-mosa): New approach to adapt simulated annealing to multi-objective optimization. *International Journal of Information Technology*, *4*(3), 9.

Hair, J. F., Anderson, R. E., Tatham, R. L., & Black, W. C. (1995). *Multivariate Data Analysis with Readings*. New York, NY: Prentice-Hall.

Haiwei, L., Weijian, M., Ning, Z., & Yufei, F. (2015). Modeling and simulating the operation of the harbor portal crane. *Journal of Coastal Research*, *73*(1), 89–94.

Halim, S. (2009). *An integrated White + Black box approach for designing and tuning stochastic local search algorithms.* (Doctoral dissertation).

Hanotaux, G. (1916). *Histoire illustrée de la guerre de 1914* (Vol. 3). Paris: Gounouilhou.

Haralambides, E. H., Benacchio, M., & Cariou, P. (2002). Costs, benefits and pricing of dedicated container terminals. *International Journal of Maritime Economics*, *4*(1), 21–34. doi:10.1057/palgrave.ijme.9100031

Harold, D. F., & Darlene, B. S. (2004). Managing for Value: Developing a Performance Measurement System Integrating Economic Value Added and the Balanced Scorecard in Strategic Planning. *The Journal of Business Strategy*, *21*(1), 1–17.

Hartnett, E., Paoli, G. M., & Schaffner, D. W. (2009). Modeling the Public Health System Response to a Terrorist Event in the Food Supply. *Society for Risk Analysis, Journal of Risk Analysis*, *29*(11).

Hatcher, L. (1994). *A Step by Step Approach to Using the SAS System for Factor Analysis and Structural Equation Modeling*. Cary, NC: SAS Institute Inc.

Havur, G., Haspalamutgil, K., Palaz, C., Erdem, E., & Patoglu, V. (2013). A case study on the Tower of Hanoi challenge: Representation, reasoning and execution. In *Robotics and Automation (ICRA), 2013 IEEE International Conference on* (pp. 4552-4559). IEEE.

Heaver, T., Meersman, H., & Van de Voorde, E. (2000). Do mergers and alliances influence European shipping and port competition? *Maritime Policy & Management*, *27*(4), 363–373. doi:10.1080/030888300416559

Heckman, J.J. (1997). The value of quantitative evidence on the effect of the past on the present, *AER, Papers and Proceedings*, *87*(2).

Hempel, L., & Schmiedel, R. (2006). Pile problems and their Mathematical Treatment. In S. T. Morlock (Ed.), Perspectives in Operations Research (pp. 257-276). Deutscher Universitäts-Verlag. doi:10.1007/978-3-8350-9064-4_15

Henn, S. (2010). Algorithms for on-line order batching in an order-picking warehouse. In *Proceedings of the 3rd International Conference on Information Systems, Logistics and Supply Chain ILS 2010*. Casablanca: Business Process Consulting.

Henn, S. (2009). Metaheuristics for the order batching problem in manual order picking systems. *Ottovon-Guericke-University Magdeburg. Faculty of Economics and Management. Verband der Hochschullehrer für Betriebswirtschaft e.*, *3*(1), 82–105.

Hernández Aguilar, J., Yip-Otuño, S., Vakhania, N., Ponce Gallegos, J., & Ochoa Ortiz, A. (2015). *Genetic Algorithm Implementation on GPUs to solve Hanoi Towers Game*. Retrieved October 03, 2015, from http://www.micai.org/2013/all-proceedings-2013/Open access/HIS/3.doc

Herrera, G. A. (2001). *Simulación de operaciones aeroportuarias. El caso de despegues y aterrizajes en el Aeropuerto Internacional de la Ciudad de México. Publicación Técnica No. 180*. México: Instituto Mexicano del Transporte.

Herrera, G. A. (2012). *Modelo de simulación de operaciones aéreas en aeropuertos saturados. El caso del Aeropuerto Internacional de la Ciudad de México. Publicación Técnica No. 365*. México: Instituto Mexicano del Transporte.

Hillier & Lieberman. (2010). *Introduction to operations research*. McGraw Hill.

Hintsa, J., Ahokas, J., Zaghbour, K., Männistö, T., Hameri, A., & Holmström, J. (2010). A Conceptual model for assessing cost of security in global supply chains. EurOMA 2010, Porto, Portugal.

Hintsa, J., Gutiérrez, X., Hameri, A. P., & Wieser, P. (2009). Supply Chain Security Management: An overview. *International Journal of Logistics Systems and Management*, *5*(3/4), 344. doi:10.1504/IJLSM.2009.022501

Hitchcock, F. L. (1941). The distribution of a product from several sources to numerous localities. *Journal of Mathematical Physics*, *20*(2), 224–230. doi:10.1002/sapm1941201224

Hochberg, Y., & Tamhane, A. C. (1987). *Multiple comparison procedures.* Wiley.

Hoen, K. M. R., Tan, T., Fransoo, J. C., & Houtum, G. J. (2014). Effect of carbon emission regulations on transport mode selection under stochastic demand. *Flex Serv Manuf J.*, *26*(1-2), 170–195. doi:10.1007/s10696-012-9151-6

Hofstadter, D. R. (1996). *Metamagical themas: Questing for the essence of mind and pattern.* Basic books.

Homberger, J. H. G., & Gehring, H. (2005). A two-phase hybrid metaheuristic for the vehicle routing problem with time windows. *European Journal of Operational Research*, *162*(1), 220–238. doi:10.1016/j.ejor.2004.01.027

Honicky, R., Brewer, E. A., Paulos, E., & White, R. (2008). N-smarts: networked suite of mobile atmospheric real-time sensors. In *Proceedings of the second ACM SIGCOMM workshop on Networked systems for developing regions* (pp. 25–30). doi:10.1145/1397705.1397713

Horlock, J. H. (1958). *Axial Flow Compressor-Fluid mechanics and thermodynamics.* London: Butterworth Scientific Publications.

Host, M., & Runeson, P. (2008). *Case Studies on Software Engineering PhD Programmes.* Retrieved from http://citeseerx.ist.psu.edu/viewdoc/download?doi=10.1.1.98.4978&rep=rep1&type=pdf

Houthoofd, N. (2009). Business Definition and Performance Implications: Te Case of The Belgian Construction Sector. *Construction Management and Economics*, (july): 27, 639–652.

Hue et al. (2006). Towards Robust Indexing for Ranked Queries. In *VLDB' 06.*

Hult, G. T., Ketchen, D. J., & Arrfelt, M. (2007). Strategic supply chain management: Improving performance through a culture of competitiveness and knowledge development. *Strategic Management Journal*, *28*(10), 1035–1052. doi:10.1002/smj.627

Hung, M.-H., Shu, L.-S., Ho, S.-J., Hwang, S.-F., & Ho, S.-Y. (2008). A novel intelligent multiobjective simulated annealing algorithm for designing robust PID controllers. *Systems, Man and Cybernetics, Part A: Systems and Humans. IEEE Transactions on*, *38*(2), 319–330.

Hunter, T., Moldovan, T., Zaharia, M., Merzgui, S., Ma, J., Franklin, M. J., & Bayen, A. M. et al. (2011). Scaling the Mobile Millennium system in the cloud. In *Proceedings of the 2nd ACM Symposium on Cloud Computing* (pp. 28). doi:10.1145/2038916.2038944

Husted, N., & Myers, S. (2010). Mobile location tracking in metro areas: Manets and others. In *Proceedings of the 17th ACM conference on Computer and communications security* (pp. 85–96).

Hwang, S. J. (2005). Inventory constrained maritime routing and scheduling for multi-commodity liquid bulk. (Doctoral dissertation). Atlanta, GA: Georgia Institute of Technology. Available at https://smartech.gatech.edu/bitstream/handle/1853/6928/Hwang_Seungjune_200505_phd.pdf

IMO. (2012). *International Maritime Organization Annual Reports.* London: IMO.

INEGI – National Institute of Statistics and Geography of Mexico. (2012). *Estadísticas del comercio exterior de México.* INEGI.

Ingber, L. (1993). Simulated annealing: Practice versus theory. *J. Mathl. Comput. Modelling*, *18*(11), 29–57. doi:10.1016/0895-7177(93)90204-C

Inoue, M., Ikui, T., Kamada, Y., & Tashidaro, M. (1980). A quasi three dimensional design of diagonal flow impellers by use of cascade data. *IAHR, 10th Symposium.*

International Air Transport Association. (2011). Vision 2050. Technical Report. Singapore.

International Air Transport Association. (2014). *Annual Review 2014.* Technical report 69th Annual General Meeting Cape Town.

International Civil Aviation Organization. (2013). *Caribbean/South American Regional Traffic Forecasts 2011-2031.* Report of the Ninth Meeting. Of the CAR/SAM Traffic. Forecasting Group (CAR/SAM TFG).

International Civil Aviation Organization. (2013b). *2013–2028 Global Air navigation Plan. ICAO 9750.* Canada: Published in Montréal.

ISO 22000. (2005). *Food safety management systems – Requirements for any organization in the food chain.* Geneva. Switzerland: ISO.

ISO/DIS 9001. (2015). *Draft International Standard – Quality management systems – Requirements.* Geneva. Switzerland: ISO.

ISO/IEC 27001. (2005). *Estándar Internacional: Tecnología de la Información-Técnicas de Seguridad- Sistemas de Gestión de la Seguridad de la Información.* ISO.

ISO/IEC 27001. (2013). *Estándar Internacional: Tecnología de la Información-Técnicas de Seguridad- Sistemas de Gestión de la Seguridad de la Información.* ISO.

Jackson, A. (2011). Trade facilitation and supply chain security a globally integrated approach. *World Customs Journal, 5*(2), 123–128.

Jacobsen, S. K., & Madsen, O. B. (1980). A comparative study of heuristics for a two-level routing-location problem. *European Journal of Operational Research, 5*(6), 378–387. doi:10.1016/0377-2217(80)90124-1

Jaikumar, R., & Solomon, M. M. (1987). The tug fleet size problem for barge line operations: A polynomial algorithm. *Transportation Science, 21*(4), 264–272. doi:10.1287/trsc.21.4.264

Jankowicz, A. D. (2000). *Business research projects* (3rd ed.). Thomson Business Press.

Jansson, J. O., & Shneerson, D. (1982). The optimal ship size. *Journal of Transport Economics and Policy, 16*(3), 217–238.

Jenkins, H. (1977). *Histoire de la marine française.* Paris: Albin Michel.

Jetlund, A. S., & Karimi, I. A. (2004). Improving the logistics of multi-compartment chemical tankers. *Computers & Chemical Engineering, 28*(8), 1267–1283. doi:10.1016/j.compchemeng.2003.08.009

JICA. (1998). *Container Terminal Planning. Port and Harbour Engineering II.* Japan: Training Course.

Jick, T. D. (1989). Mixing qualitative and quantitative methods: Triangulation in action. *Administrative Science Quarterly, 21*(4), 602–611.

Jiménez, J., & De la Torre, E. (2011). *¿Cómo construir cadenas internacionales seguras?* Énfasis Logística. *Logistic Summit & Expo Edition, 127*, 6267.

Joffre, J. (1932). *Mémoires du Maréchal Joffre (1910-1917). Tome premier.* Paris: Librairie Plon.

Johnson, D. S. (1974). Fast algorithms for bin packing. *Journal of Computer and System Sciences, 8*(3), 272–314. doi:10.1016/S0022-0000(74)80026-7

Joll, J., & Martel, G. (2007). *The origins of the first world war*. Pearson Education.

Jones, D. A., Farkas, J. L., Bernestein, O., Davis, Ch. E., Turk, A., Turnquist, M. A., & Sawaya, W. et al. (2011). U.S. import/export container flow modeling and disruption analysis. *Research in Transportation Economics*, *32*(1), 3–11. doi:10.1016/j.retrec.2011.06.003

Jones, M. T. (2010). *Artificial Intelligence A System Approach*. Laxmi Publications, Ltd.

Jonker, J. B., & Van Essen, T. G. (1997). A finite element perturbation method for computing fluid-induced forces on a centrifugal impeller, rotating and whirling in a volute casing. *International Journal for Numerical Methods in Engineering*, *40*(2), 269–294. doi:10.1002/(SICI)1097-0207(19970130)40:2<269::AID-NME63>3.0.CO;2-G

Juárez-Gutiérrez, S. S. (2015). *Modelado matemático de la variación de temperatura dentro de un invernadero clásico cenital utilizando el método de Mínimos Cuadrados Recursivos*. (Unpublished Master's thesis). Universidad Politécnica de Aguascalientes, Mexico.

Jüttner, U., Peck, H., & Christopher, M. (2003). Supply chain risk management: Outlining an agenda for future research. *International Journal of Logistics: Research and Applications*, *6*(4), 197–210. doi:10.1080/13675560310001627016

Kallehauge, B. (2008). Formulations and exact algorithms for the vehicle routing problem with time windows. *Computers & Operations Research*, *12*(7), 2307–2330. doi:10.1016/j.cor.2006.11.006

Kamin, N. (1998). *On-line optimization of order picking in an automated warehouse*. (Ph.D. Dissertation). Aachen: Shaker Verlag.

Kang, B., Park, D., Cho, K., Shin, C., Cho, S., & Park, J. (2008). A Study on the Greenhouse Auto Control System based on Wireless Sensor Network.*International Conference on Security Technology*. doi:10.1109/SecTech.2008.60

Kang, J. G., & Kim, Y. D. (2002). Stowage planning in maritime container transportation. *The Journal of the Operational Research Society*, *53*, 415–426.

Kang, K. W., Zhang, C., Guo, L. Y., & Ma, T. (2012). Research on ship routing and deployment mode for a bulk. In *Annual Conference Proceedings of the International Conference of Management Science Engineer*. doi:10.1109/ICMSE.2012.6414421

Kanji, G. K., & Wong, A. (1999). Business excellence model for supply chain management. *Total Quality Management*, *10*(8), 1147–1168. doi:10.1080/0954412997127

Kao, Y., Chen, M. H., & Huang, Y. T. (2012). A Hybrid Algorithm Based on ACO and PSO for Capacitated Vehicle Routing Problems. *Mathematical Problems in Engineering*, *2012*, •••.

Kaplan, R. S., & Norton, D. P. (1993). Putting balanced scorecard to work. *Harvard Business Review*, *71*(5), 134–142.

Karnik, S. R. (2008). *Delimitation analysis in high speed drilling of carbon fiber reinforced plastics (CFRP) using artificial neural network model*. Elsevier Ltd.

Kaveshgar, N., Huynh, N., & Rahimian, S. K. (2012). An efficient genetic algorithm for solving the quay crane scheduling problem. *Expert Systems with Applications*, *39*(18), 13108–13117. doi:10.1016/j.eswa.2012.05.091

Kelle, P., & Miller, P. A. (1998). Quantitative support for buyer-supplier in JIT Purchasing negotiation. *International Journal of Purchasing and Materials Management, 34*(2), 25-30.

Kelle, P., Al-Khateebb, F., & Miller, P. A. (2003). Partnership and joint negotiation optimal support by ordering / setup policies for JIT. *International Journal of Production Economics, 81-82,* 431-441.

Kendall, L. C., & Buckley, J. J. (2001). *The business of shipping* (7th ed.). Cornell Maritime Press.

Kenyon, S. A., & Morton, P. D. (2003). Stochastic vehicle routing with random travel times. *Journal of Transportation Science*, *37*(1), 69–82. doi:10.1287/trsc.37.1.69.12820

Khandaroo, J., & Seetena, B. (2008). The role of transport infrastructure in international tourism development: A gravity model approach. *International Journal Tourism Management*, *29*(5), 831–840. doi:10.1016/j.tourman.2007.09.005

Kim, J.-H., & Kim, K.-Y. (2011). Optimization of Vane Diffuser in a Mixed-Flow Pump for High Efficiency Design. *International Journal of Fluid Machinery and Systems*, *4*(1).

Kim, K., & Park, Y. (2004). A crane scheduling method for port container terminals. *European Journal of Operational Research*, *156*(3), 752–768. doi:10.1016/S0377-2217(03)00133-4

Kirkpatrick, S. (1984). Optimization by simulated annealing: Quantitative studies. *Journal of Statistical Physics*, *34*(5-6), 975–986. doi:10.1007/BF01009452

Kirkpatrick, S., Gelatt, C., & Vecchi, M. (1983). Optimization by simulated annealing. *Science*, *4598*(4598), 671–680. doi:10.1126/science.220.4598.671 PMID:17813860

Kjaerul, U. (1991). Optimal decomposition of probabilistic networks by simulated annealing. *Statistics and Computing*, *2*(1), 7–17. doi:10.1007/BF01890544

Kleindorfer, P. R., & Saad, G. H. (2005). Managing Disruption Risks in Supply Chains. *Production and Operations Management*, *14*(1), 53–68. doi:10.1111/j.1937-5956.2005.tb00009.x

Kleindorfer, P. R., & Van Wassenhove, L. (2004). Risk Management in Global Supply Chains. In H. Gatignon & J. Kimberly (Eds.), *The Alliance on Globalization*. Cambridge University Press.

Kleter, G. A., & Marvin, H. J. P. (2009). Indicators of emerging hazards and risk to food safety. *Food and Chemical Toxicology*, *47*(5), 1022–1039. doi:10.1016/j.fct.2008.07.028 PMID:18761390

Kleywegt, A. (2003). *Contract planning models for ocean carriers.* Working paper, Georgia Institute of Technology.

Knemeyer, M., Zinn, W., & Eroglu, C. (2009). Proactive planning for catastrophic events in supply chains. *Journal of Operations Management*, *27*(2), 141153. doi:10.1016/j.jom.2008.06.002

Knight, R. (1998). Construction et entretien de la flotte anglaise durant les guerres anglo-françaises. In Les marines de guerre européennes: XVII-XVIII siècles. Presses de l'Université de Paris-Sorbonne.

Knowles, J. D., & Corne, D. W. (2000). Approximating the nondominated front using the Pareto archived evolution strategy. *Evolutionary Computation*, *8*(2), 149–172. doi:10.1162/106365600568167 PMID:10843519

Kommerskollegium. (2008). *Supply chain security initiatives: a trade facilitation perspective*. The National Board of Trade, Sweden.

Kornai, A., & Tuza, Z. (1992). Narrowness, path-width, and their application in natural language processing. *Discrete Applied Mathematics*, *36*(1), 87–92. doi:10.1016/0166-218X(92)90208-R

Kovács, G., & Spens, K. (2005). Adductive reasoning in logistics research. *International Journal of Physical Distribution & Logistics Management*, *35*(2), 132144.

Kros, J., & Mastrangelo, M. (2004). Comparing Multi-response Design Methods With Mixed Responses. *Quality and Reliability Engineering International*, *20*(5), 527–539. doi:10.1002/qre.664

Kumar, P., & Goe, P. (2002). Product Quality Using Fuzzy Optimization September Concepts: A Case Study. *Quality Engineering*, *15*(1), 1–8. doi:10.1081/QEN-120006706

Kutz, M. (2004). *Handbook of Transportation Engineering*. Mc Graw-Hill.

La Londe, B. J., & Cooper, M. C. (1989). *Partnerships in providing customer service: a third party perspective*. Oak Brook, IL: Council of Logistics Management.

Laarhoven, V., & Aarts, E. H. L. (1987). *Simulated annealing: Theory and applications*. Kluwer Academic Publishers. doi:10.1007/978-94-015-7744-1

Lacour-Gayet, G. (1905). *La marine militaire de la France sous le règne de Louis XVI*. Paris: Librairie Honoré Champion.

Lafont, F., & Balmat, J. (2002). *Optimized fuzzy control of a greenhouse*. Academic Press.

Lakshmi, V., Arvind, A., & Bever, K. (2007). Greenhouse Asset Management Using Wireless Sensor Actor Networks. *International Conference on Mobile Ubiquitous Computing, Systems, Services and Technologies*.

Lamarca, M. (2009). *Hipertexto: El nuevo concepto de documento en la cultura de la imagen: Internet*. (Doctoral dissertation, PhD thesis). Complutense University of Madrid. Available at http://www.hipertexto.info

Lam, J. S. L., & Yap, W. Y. (2011). Container port competition and complementarity in supply chain systems: Evidence from the Pearl River Delta. *Maritime Economics & Logistics*, *13*(2), 102–120. doi:10.1057/mel.2011.5

Lancashire County Council. (2010). *Heysham – M6 Link Road Traffic Model. Local Model Validation Report*. Lancashire, UK: Appendix D.

Langley, C., Coyle, J., Gibson, B., Novack, R. & Bardi, E. (2008). *Managing Supply Chains: A Logistics Approach*. Canada: South-Western Cengage Learning.

Laporte, G., Ropke, S., & Vidal, T. (2014). Heuristics for the Vehicle Routing Problem, in Vehicle Routing: Problems, Methods, and Applications. Society for Industrial and Applied Mathematics.

Laporte, G. (1992). The vehicle routing problem: An overview of exact and approximate algorithms. *European Journal of Operational Research*, *59*(3), 345–358. doi:10.1016/0377-2217(92)90192-C

Laporte, G. (1997). Vehicle routing. In M. Dell'Amico, F. Maffioli, & S. Martello (Eds.), *Annotated bibliographies in combinatorial optimization*. Chichester, UK: Wiley.

Laporte, G., Louveaux, F., & Hamme, V. L. (2002). An integer l-shaped algorithm for the capacitated vehicle routing problem with stochastic demands. *Operations Research*, *50*(3), 415–423. doi:10.1287/opre.50.3.415.7751

Laporte, G., Louveaux, F., & Mercure, H. (1989). Models and exact solutions for a class of stochastic location- routing problems. *European Journal of Operational Research*, *39*(1), 71–78. doi:10.1016/0377-2217(89)90354-8

Laporte, G., & Nobert, Y. (1987). Exact algorithms for the vehicle routing problem. *Surveys in Combinatorial Optimization*, *31*, 147–184. doi:10.1016/S0304-0208(08)73235-3

Laribi, I., Homri, H., & Mhiri, R. (2006). *Modeling of a Greenhouse Temperature: Comparison Between Multimodel and Neural Approaches*. IEEE ISIE.

Larrañaga, P., & Lozano, J. A. (2002). *Estimation of distribution algorithms: a new tool for evolutionary computing*. Dordrecht: Kluwer. doi:10.1007/978-1-4615-1539-5

Lättilä, L., Henttu, V., & Hilmola, O.-P. (2013). Hinterland operations of sea ports do matter: Dry port usage effects on transportation costs and CO2 emissions. *Transportation Research Part E, Logistics and Transportation Review*, *55*, 23–42. doi:10.1016/j.tre.2013.03.007

Lawler, E. L., Lenstra, J. K., Rinnooy Kan, A. H. G., & Schmoys, D. B. (Eds.). (1985). *The traveling salesman problem*. Chichester, UK: Wiley.

Lawrence, B., & Leon, J. (2010). *Importance of the North American regional manufacturing*. Unpublished manuscript, Supply Chain Systems Laboratory, Texas A&M University, College Station.

Le Hénaff, J.H.F. (1922), La préparation et l'exécution d'un plan de transport de concentration (août-octobre 1914). *Revue Militaire Française*, (10), 62-84.

Leal, J., Alcorta, E., & Rodríguez, H. (2006). *Modelado del clima en invernaderos: Respuesta de la temperatura a cambios de humedad*. Ingenierías.

Le-Duc, T., & de Koster, R. (2007). Travel time estimation and order batching in a 2-block warehouse. *European Journal of Operational Research*, *176*(1), 374–388. doi:10.1016/j.ejor.2005.03.052

Lee, H., & Wolfe, M., (2003). Supply Chain Security Without Tears. *Supply Chain Management Review*, *7*(1), 12-20.

Lee, A. M. (1966). *Applied Queuing Theory*. New York: St. Martin's Press Company.

Lee, D.-H., Wang, H. Q., & Miao, L. (2008). Quay crane scheduling with noninterference constraints in port container terminals. *Transportation Research Part E, Logistics and Transportation Review*, *44*(1), 124–135. doi:10.1016/j.tre.2006.08.001

Leedy, P., & Ormrod, J. (2005). *A Handbook for Teacher Research from Design to Implementation*. Pearson Education.

Lee, E.-S., Nam, H.-S., & Song, D.-W. (2012). Defining maritime logistics and its value in Maritime Logistics. In D.-W. Song & P. M. Panayides (Eds.), *Maritime logistics: a complete guide to effective shipping and port management*. Kogan Page.

Legato, P., Trunfio, R., & Meisel, F. (2012). Modeling and solving rich quay crane scheduling problems. *Computers & Operations Research*, *39*(9), 2063–2078.

Leiserson, C. E. (1980). Area-efficient graph layouts. In *the 21st Annual Symposium on foundations of Computer Science*.

Lemeš & Zaimović-Uzunović. (2002). Mode shapes of centrifugal pump impeller. *6th International Research / Expert Conference on Trends in the Development of Machinery and Associated Technology, TMT 2002*. B&H.

Lengauer, T. (1981). Black-white pebbles and graph separation. *Acta Informatica*, *16*(4), 465–475. doi:10.1007/BF00264496

Li, H., & Landa-Silva, D. (2008). Evolutionary multi-objective simulated annealing with adaptive and competitive search direction. In *Evolutionary Computation, 2008. CEC 2008.(IEEE World Congress on Computational Intelligence). IEEE Congress on*, (pp. 3311-3318). IEEE.

Li, H. (2014). Algorithm Efficiency Research on Hanoi Problem Symmetric Solution. In *International Conference on Mechatronics, Control and Electronic Engineering (MCE 2014)*(pp. 759-762). Atlantis Press. doi:10.2991/mce-14.2014.100

Li, L., Liao, H., & Coit, D. W. (2009). A two-stage,apprach for multi-objective decision making with applications to system reliability optimization. *Reliability Engineering & System Safety*, *94*(10), 1585–1592. doi:10.1016/j.ress.2009.02.022

Lindstad, H., Asbjørnslett, B. E., & Pedersen, J. T. (2012). Green maritime logistics and sustainability. *Maritime Logistics: Contemporary Issues*, 227-261.

Linhares, A., & Yanasse, H. H. (2002). Connections between cutting-pattern sequencing, vlsi design, and reflexive machines. *Computers & Operations Research*, *29*(12), 1759–1772. doi:10.1016/S0305-0548(01)00054-5

Lin, S.-W., & Ying, K.-C. (2015). A multi-point simulated annealing heuristic for solving multiple objective unrelated parallel machine scheduling problems. *International Journal of Production Research*, *53*(4), 1065–1076. doi:10.1080/00207543.2014.942011

Littler, R. A., & Whitaker, D. (1997). Estimating staffing requirements at an airport terminal. *The Journal of the Operational Research Society*, *48*(2), 124131. doi:10.1057/palgrave.jors.2600344

Liu, H., Mi, W., Zhao, N., & Feng, Y. (2015). Modeling and Simulating the Operation of the Harbor Portal Crane. *Journal of Coastal Research*, *73*, 89–94. doi:10.2112/SI73-016.1

Liu, X. (1996). Intelligent data analysis: Issues and challenges. *The Knowledge Engineering Review*, *11*(04), 365–371. doi:10.1017/S0269888900008055

Llorente, S. (2000). *Factores que influyen en la demora del enfermo en un servicio de urgencias hospitalarias.* Emergencias.

Loez, A. (2014). Les expériences combattantes. La Grande Guerre, 45-60.

Lo, H. K., McCord, M. R., & Wall, C. K. (1991). Value of ocean current information for strategic routing. *European Journal of Operational Research*, *55*(2), 124–135. doi:10.1016/0377-2217(91)90218-K

Loh, K. H., Golden, B., & Wasil, E. (2008). Solving the one-dimensional bin packing problem with a weight annealing heuristic. *Computers & Operations Research*, *35*(7), 2283–2291. doi:10.1016/j.cor.2006.10.021

Loir, M. (1893). *La marine française*. Paris: Librairie Hachette.

Lopes, I. C., & de Carvalho, J. M. V. (2010). Minimization of open orders using interval graphs. *IAENG International Journal of Applied Mathematics*, *40*(4), 297–306.

López, C. (2000) *Modelo para el Desarrollo de Bibliotecas Digitales Especializadas.* (Master's thesis, ITAM). Retrieved on June 4, 2014, from http://www.bibliodgsca.unam.mx/tesis/tes7cllg/sec_10.htm

Loukil, T., Teghem, J., & Tuyttens, D. (2005). Solving multi-objective production scheduling problems using metaheuristics. *European Journal of Operational Research*, *161*(1), 42–61. doi:10.1016/j.ejor.2003.08.029

Lu, C. S. (2004). An evaluation of logistics services' requirements of international distribution centers in Taiwan. *Transportation Journal*, *43*(4), 53–66.

Luengo, J., García, S., & Herrera, F. (2009). A study on the use of statistical tests for experimentation with neural networks: Analysis of parametric test conditions and nonparametric tests. *Expert Systems with Applications*, *36*(4), 7798–7808. doi:10.1016/j.eswa.2008.11.041

Luo, W., Dail, J., Wang, X., Stanghellini, C., & Bu, C. (2005). *Simulation of greenhouse management in the subtropics, Part I: Model validation and scenario study for the winter season* (vol. 90). Academic Press.

Luque, G., & Alba, E. (2005). Metaheuristics for the dna fragment assembly problem. *International Journal of Computational Intelligence Research*, *1*(2), 98–108. doi:10.5019/j.ijcir.2005.28

Lu, X., & Dillon, T. (1995). Parallelism for multipeg towers of Hanoi. *Mathematical and Computer Modelling*, *21*(3), 3–17. doi:10.1016/0895-7177(94)00210-F

Lu, Z., Han, X., Xi, L., & Erera, A. L. (2012). A heuristic for the quay crane scheduling problem based on contiguous bay crane operations. *Computers & Operations Research*, *39*(12), 2915–2928. doi:10.1016/j.cor.2012.02.013

Lyonga, A.N., Sellnow T.L., Nganje, W.E. & Venette S. (2006). Insider Agroterrorism Threat and High Reliability in Turkey Processing. *Journal of Food Protection Trends.*

Macharis, C., Melo, S., Woxenius, J., & Van Lier, T. (Eds.). (2014). Sustainable Logistics. Transport and Sustainability, 6.

Macket, R. L., & Edwards, M. (1998). The impact of new urban public transport systems: Will the expectations be met? *International Journal Transportation Research Part A: Policy and Practice*, *32*(4), 231–245.

Mad, S., Munirah, L., Kamarudin, K., Mohd, W., Muhammad, S., Zakaria, A., & Nor, M. (2014). A Real-Time Greenhouse Monitoring System for Mango with Wireless Sensor Network (WSN).*International Conference on Electronic Design (ICED).*

Madsen, O. G., Ravn, F., & Rygaard, M. (1995). The vehicle routing problem with time windows part ii: Genetic search. *Journal of Computing*, *8*, 165–172.

Ma, L., Ji, J., & He, Ch. (2010). The Study of Greenhouse Temperature Modeling Based on Fuzzy Neural Network. *International Conference on E-Product E-Service and E-Entertainment.* doi:10.1109/ICEEE.2010.5660748

Mallouy, J. (1914). Les Transport stratégiques. *Armée et Marine: revue hebdomadaire illustrée des armées de terre et de mer*, (24), 615.

Manuj, I., & Mentzer, J. (2008). Global supply chain risk management strategies. *International Journal of Physical Distribution & Logistics Management*, *38*(3), 192–223. doi:10.1108/09600030810866986

Marsden, G., & Docherty, I. (2013). Insights on disruptions as opportunities for transport policy change. *Transportation Research Part A, Policy and Practice*, *51*, 46–55. doi:10.1016/j.tra.2013.03.004

Martel, N., & Seneviratne, N. P. (1990). *Analysis of factors influencing quality of service in passenger terminal buildings. In Airport Terminal and Landside Design and Operation 1990, Transportation Research Record 1273* (pp. 1–10). Washington, DC: Transportation Research Board, National Research Council.

Martin, G. L. Jr, Randhawa, S. U., & McDowell, E. D. (1988). Computerized container-ship loading: A methodology and evaluation. *Computers & Industrial Engineering*, *14*(4), 429–440. doi:10.1016/0360-8352(88)90045-9

Mather, J. (1971). Stability of C°° mappings VI: The nice dimensions. In *Proceedings of Liverpool Singularities Symposium(1969/70)* (LNIM), (vol. 192, pp. 207-253). Berlin: Springer. doi:10.1007/BFb0066824

Maurette, F. (1927). *Géographie de la France*. Paris: Editions Hachette.

Maxwell, J. (1998). Designing a qualitative study. In L. Bickman & D. Rog (Eds.), *Handbook of applied social research methods*. Thousand Oaks, CA: Sage Publications.

Mbang, J. A. (2012). A New Introduction to Supply Chains and Supply Chain Management: Definitions and Theories Perspective. *International Business Research*, *5*(1), 194–207.

McCarthy, J. (2007). *What Is Artificial Intelligence?* Stanford, CA: Stanford University.

McCord, M. R., Lee, Y.-K., & Lo, H. K. (1999). Ship routing through altimetry-derived ocean currents. *Transportation Science*, *33*(1), 49–67. doi:10.1287/trsc.33.1.49

McCue, C. (2007). *Data Mining and Predictive Analysis: Intelligence Gathering and Crime Analysis*. Butterworth-Heinemann.

McGeoch, C. C. (2007). Experimental algorithmics. *Communications of the ACM*, *50*(11), 27–31. doi:10.1145/1297797.1297818

McGill, R., Tukey, J. W., & Larsen, W. A. (1978). Variations of Box Plots. *The American Statistician.*

McLellan, R. G. (1997). Bigger vessels: How big is too big? *Maritime Policy & Management*, *24*(2), 193–211. doi:10.1080/03088839700000070

Medina, S. (2010a). *Reducir tiempos de espera de pacientes en el departamento de emergencias de un hospital utilizando simulación.* (Unpublished doctoral dissertation). Mexicali, Baja California, México.

Medina, S. (2010b). *Reducir tiempos de espera de pacientes en el departamento de emergencias de un hospital utilizando simulación*. Revista de la facultad de Ingeniería Industrial.

Medina-Ramirez, R. (2013). *Main Factors of Surgical Deferment in Elective Surgery: A System Dynamics Analysis.* (Unpublished doctoral dissertation). University of Guadalajara, Jalisco, México.

Medina, S. (2012). *Esquema genérico para el análisis y mejora del flujo de pacientes*. Mexicali, México: UABC.

Mehta, M. P., & Patel, P. M. (2013). Performance analysis and optimization of mixed flow pump. *International Journal of Emerging Trends in Engineering and Development*, *1*(3), 647–661.

Meiners, R. E., & Nardinelli, C. (1986). What Has Happened to the New Economic History? *Journal of Institutional and Theoretical Economics*, *14*(2/3), 510–527.

Meisel, F., & Bierwirth, C. (2011). A unified approach for the evaluation of quay crane scheduling models and algorithms. *Computers & Operations Research*, *38*(3), 683–693. doi:10.1016/j.cor.2010.08.001

Mendenhall, W., & Sincich, T. (1997). *Probabilidad y estadística para ingeniería y ciencias* (4th ed.). México: Prentice Hall.

Mendoza, E., Zuñiga, C. A., Mujica, M. A., & Delahaye, D. 2015, *Simulating Airport Capacity: Mexico City Airport Case*. Conferences in Air Transport & Operations. in Proceeding of the 3rd International Conference on Interdisciplinary Science for Innovative Air Traffic Management (ISIATM2015), Netherlands.

Mendoza, J.E., & Villegas, J. G. (n.d.). A multi-space sampling heuristic for the vehicle routing problem with stochastic demands. *Optimization Letters*, 1-14.

Mentzer, J., Witt, W. D., Keebler, J., Min, S., Nix, N., Smith, D., & Zacharia, Z. (2001). Defining Supply Chain (SC) management. *Journal of Business Logistics*, *22*(2), 1–25. doi:10.1002/j.2158-1592.2001.tb00001.x

Meredith, J. (1998). Building operations management theory through case and field research. *Journal of Operations Management*, *16*(4), 441–454. doi:10.1016/S0272-6963(98)00023-0

Metaxas, B. N. (1972). The future of tramp shipping industry. *Journal of Transport Economics and Policy*, *6*(3), 271–280.

Metropolis, N., Rosenbluth, A., Rosenbluth, M., Teller, A., & Teller, E. (1953). Equation of state calculations by fast computing machines. *The Journal of Chemical Physics*, *21*(6), 1087–1092. doi:10.1063/1.1699114

Metzner, W. (1991). Echolocation behaviour in bats. *Science Progress Edinburgh*, *75*(298), 453–465.

Mietinen, K. (1999). *Nonlinear Multi-objective Optimization*. Boston: Kluer Academic Publishers.

Miluzzo, E., Lane, N. D., Fodor, K., Peterson, R., Lu, H., Musolesi, M., & Campbell, A. T. et al. (2008). Sensing meets mobile social networks: the design, implementation and evaluation of the CenceMe application. In *Proceedings of the 6th ACM conference on Embedded network sensor systems*. doi:10.1145/1460412.1460445

Mingozzi, A. (2003). *An exact algorithm for period and multi- depot vehicle routing problems. Technical report*. Bologna, Italy: Department of Mathematics, University of Bologna.

Ministerio de Sanidad y Política Social. (2010). *Unidad de urgencias hospitalaria: estándares y recomendaciones*. España, Ministerio de Sanidad y Política Social.

Mishra, B. K., & Vishnoi, S. (n.d.). Towers of Hanoi - an iterative solution for parallel computation. *International Journal of Research Science & Management*. Retrieved November 12, 2015, from http://www.ijrsm.com/issues%20pdf%20file/Archive-2015/July-2015/1.pdf

Mitchell, M. (1999). *An introduction to genetic algorithms*. The MIT Press.

Mitchell, T. M. (1997). *Machine Learning*. McGraw-Hill.

Moccia, L., Cordeau, J. F., Gaudioso, M., & Laporte, G. (2006). A branch-and-cut algorithm for the quay crane scheduling problem in a container terminal. *Naval Research Logistics*, *53*(1), 45–59. doi:10.1002/nav.20121

Mohammad, R., Siamak, H., Hossein, M., & Maryam, G. (2010). A Hybrid Neuro-Fuzzy Approach for Greenhouse Climate Modeling. IEEE.

Moitre, D., Sauchelli, V., & Garcia, G., (2005). Bilevel Dynamic Optimization of Central Hydroelectric pumping in a Competitive Pool - Part I: Model and Algorithm *Magazine. IEEE Latin America, 3*(2), 62 - 67.

Mojica, C.E.P. (2012). *La Influencia de la Orientación al Mercado y la Innovación en la Capacidad de Internacionalización y Desempeño de las Pymes del Estado de Aguascalientes*. Tesis Doctoral, Noviembre, 2012. 1-397.

Monke, J. (2004). *Agroterrorism: Threats and Preparedness*. Paper presented at Congressional Research Service; CRS Report for Congress.

Moon, I. K., Qiu, Z. B., & Wang, J. H. (2014). A combined tramp ship routing, fleet deployment, and network design problem. *Maritime Policy & Management*. doi:10.1080/03088839.2013.865847

Moore, J. C. (2007). *General equilibrium and welfare economics: An introduction*. Springer.

Mora, S. (2001). Qué es una aplicación web. In *Programación en Internet: Clientes web* (p. 9). Alicante: Club Universitario.

Morash, E.A. & Clinton, S. R. (1997). The Role of Transportation Capabilities in International Supply Chain Management. *International Supply Chains, 1*(1), 5-17.

Moreno, A., Aguilar, J., & Luévano, A. (2011). Características de la Agricultura Protegida y su Entorno en México. Revista Mexicana de Agronegocios, 15(29), 763 - 744.

Moreno, E., Rico, O., Bustos, A., Martner, C., & Montooya, R. (2014). *Reparto Modal Óptimo del Transporte Terrestre de Carga en México. 1ª. Etapa. Publicación Técnica No. 413. Instituto Mexicano del Transporte*. Querétaro, México: Spanish.

Morimoto, T., & Hashimoto, Y. (2000). An Intelligent Control for Greenhouse Automation, Oriented by the Concepts of SPA and SFA. An Application to a post-harvest proces. *Computers and Electronics in Agriculture*, 555–567.

Moscato, P. (1989). *On Evolution, Search, Optimization, Genetic Algorithms and Martial Arts: Towards Memetic Algorithms*. Caltech Concurrent Computation Program (report 826).

Moya, M. J. (2003). Operations research. Editorial Universidad Estatal a Distancia.

Mtoi, E., & Moses, R. (2014). Calibration and Evaluation of Link Congestion Functions: Applying Intrinsic Sensitivity of Link Speed as a Practical Consideration to Heterogeneous Facility Types within Urban Network. *Journal of Transportation Technologies*, *4*(02), 141–149. doi:10.4236/jtts.2014.42014

Mühlenbein, H., & Paaß, G. (1996). From recombination of genes to the estimation of distributions: I. binary parameters. In H. Voigt, W. Ebeling, I. Rechenberg, & H. Schwefel (Eds.), *Parallel Problem Solving from Nature PPSN IV* (pp. 178–187). Berlin: Springer. doi:10.1007/3-540-61723-X_982

Mujica, M. A., & Zuñiga, C. A. (2015). Allocation of Airport Check-in Counters Using a Simulation-Optimization Approach. In M. A. Mujica, I. De La Mota-Flores, & D. Guimarans-Serrano (Eds.), *Applied Simulation and Optimization: In Logistics, Industrial and Aeronautical Practice* (pp. 203–229). Springer International Publishing. doi:10.1007/978-3-319-15033-8_7

Mujuni, K. T., & Zheng, Q. (2010). Tanzanian SMEs' Perceptions towards Adoption of Supply Chain Management (SCM) Strategy. *International Journal of Business and Management*, *5*(10), 42–50.

Mumayiz, S., & Ashford, N. (1986). Methodology for planning and operations management af airport terminal facilities. *Transportation Research Record*, *1094*, 24–35.

Musso, E., Ferrari, C., & Benacchio, M. (1999). On the Global Optimum Size of Port Terminals. *International Journal of Transport Economics*, *26*(3), 415–437.

Myles, D. J. (1965). *A design method for mixed flow fans and pumps*. Report No. 117. National Engineering Laboratory.

Nagy, G., & Salhi, S. (2007). Location-routing: Issues, models and methods. *European Journal of Operational Research*, *177*(2), 649–672. doi:10.1016/j.ejor.2006.04.004

Nam, D., & Park, C. H. (2000). Multiobjective simulated annealing: A comparative study to evolutionary algorithms. *International Journal of Fuzzy Systems*, *2*(2), 87–97.

Navarrete, R. (2012). *Modelo de Bioseguridad en la Cadena de Suministros de Productos Alimenticios, teniendo en cuenta la Gestión de la Cadena de Suministros y la Visión de Procesos de Negocio. Aplicación a la Industria Alimenticia, de la Zona del Bajío (México)*. (Unpublished doctoral dissertation). Universidad Politécnica de Valencia, España.

Navarrete, R., & Lario, F. C. (2009). Modelo de gestión de la bioseguridad, basado en procesos de negocio de la cadena de suministro. Aplicación en la industria mexicana de los alimentos. In *Proceedings of XIII Congreso de Ingeniería de Organización, 3rd International Conference on Industrial Engineering and Industrial Management.*

Navarrete, R., & Lario, F. C. (2010). Propuesta de una Metodología para el Modelado AS IS y TO BE de Procesos de Negocio de Bioseguridad (Terrorismo Alimentario), dentro del Contexto de la Cadena de Suministro. Aplicación en la Industria Mexicana Alimentaria. In *Proceedings of XIV Congreso de Ingeniería de Organización, 4th International Conference on Industrial Engineering and Industrial Management.*

Navarrete, R., & Lario, F. C. (2011). Propuesta de un Procedimiento para la Elaboración de un KPI para la Medición de la Bioseguridad en Procesos de Negocio de la Cadena de Suministro Alimenticia. Aplicación en la Industria Mexicana Alimenticia. In *Proceedings of 5th International Conference on Industrial Engineering and Industrial Management, XV Congreso de Ingeniería de Organización.*

NCHRP. (1998). *Travel Estimation Techniques for Urban Planning. Report 365*. Washington, D.C.: TRB.

Neumann, B. (1991). *The interaction Between Geometry and Performance of a centrifugal Pump*. London: Mechanical Engineering Publications.

Nganje, W., Bier, B. V., Han, H., & Zack, L. (2008). Models of Interdependent Security along the Milk Supply Chain. *American Journal of Agricultural Economics*, *90*(5), 1265–1271. doi:10.1111/j.1467-8276.2008.01215.x

Nganje, W., Dahl, B., Wilson, W., Mounir, S., & Lewis, A. (2007). Valuing Private Sector Incentives to Invest in Food Security Measures: Quantifying the Risk Premium for RFEM. *Journal of International Agricultural Trade and Development, 3*(2), 199–216.

Ni, X., Yang, Z., Bai, X., Champion, A. C., & Xuan, D. (2009). DiffUser: Differentiated user access control on smartphones. In *IEEE 6th International Conference on Mobile Adhoc and Sensor Systems*. IEEE Press.

Norstad, I., Fagerholt, K., Hvattum, L. M., Arnulf, H. S., & Bjrkli, A. (2013). *Maritime fleet deployment with voyage separation requirements*. Flexible Services and Manufacturing Journal; doi:10.1007/s10696-013-9174-7

North, D. C., & Nye, J. V. C. (1997). *Cliometrics, the new institutional economics and the future of economic history.* Economic History Association, 62nd Annual Meeting, "Private versus Public Institutions".

North, D.C., (1974). Beyond the new economic history. *Journal of Economic History, 34*(1).

North, D. C. (1977). The New Economic History After 20 Years. *The American Behavioral Scientist, 21*(2), 187–200. doi:10.1177/000276427702100203

Novoa, C., Berger, R., Linderoth, J., & Storer, R. (2006). *A set-partitioning-based model for the stochastic vehicle routing problem. Technical report.* Texas State University and Lehigh University.

Novoa, C., & Storer, R. (2009). An approximate dynamic programming approach for the vehicle routing problem with stochastic demands. *European Journal of Operational Research, 196*(2), 509–515. doi:10.1016/j.ejor.2008.03.023

Nunnally, J. C., & Bernstein, I. H. (1994), Psychometric Theory (3rd ed.). New York: McGraw-Hill.

O'Leary-Kelly, S. W., & Flores, B. E. (2002). The integration of manufacturing and marketing/sales decisions: Impact on organizational performance. *Journal of Operations Management, 20*(3), 221–240. doi:10.1016/S0272-6963(02)00005-0

Ochoa et al. (2007). Implementing Data Mining to improve a Game Board based on Cultural Algorithms. In *Proceedings of HAIS'2007.*

Ochoa, A., Garcí, Y., & Yañez, J. (2011). Logistics Optimization Service Improved with Artificial Intelligence. *Soft Computing for Intelligent Control and Mobile Robotics, 2011*, 57–65.

Odoni, A. R. (1991). Transportation modeling needs: airports and airspace. Volpe domestic Transportation Systems Center, Cambridge, MA.

Odoni, A. R., & De Neufville, R. (1992a). Passenger terminal design. *Transportation Research Part A, Policy and Practice, 26*(1), 27–35. doi:10.1016/0965-8564(92)90042-6

Odoni, A. R., & De Neufville, R. (1992b). Walking distance minimization for airport terminal configurations. *Transportation Research Part A, Policy and Practice, 26*(1), 59–74. doi:10.1016/0965-8564(92)90045-9

OECD (2002). *Regularity Issues in International Maritime Transport.* DSTI/DOT (2001)3.

Öncan, T. (2013). A Genetic Algorithm for the order batching problem in low-level picker-to-part warehouse systems. In *Proceedings of the International MultiConference of Engineers and Computer Scientists IMECS 2013 1.* Hong Kong.

Onyango, B., Turvey, C., & Hallman, W. (2005). *Public Attitudes and Perceptions of the Vulnerability of the U.S. Food Chain to Agroterrorism.* Paper presented at American Agricultural Economics Association annual meeting, Providence, RI.

OPAL Consortium. (2003). *Inventory of Airport Performance Models. Technical WP1 Report - Public Version, Optimisation Platform for Airports including Landside (OPAL) Research Project funded by the European Commission (Directorate-General Energy and Transport - DG-TREN).* Brussels, Belgium: European Commission.

Optimization Studio, C. P. L. E. X. v12.5. (2012). *IBM Academic Initiative*. Retrieved from https://www.ibm.com/developerworks/downloads/ws/ilogcplex/

Ordonnance de Louis XIV touchant la Marine. (1681). Août.

Ortúzar, J. D., & Willumsen, L. (2011). *Modelling Transport* (4th ed.). Chichester, UK: John Wiley. doi:10.1002/9781119993308

Osman, I. H. (1993). Metastrategy simulated annealing and tabu search algorithms for the vehicle routing problem. *Annals of Operations Research*, *41*(1-4), 421–451. doi:10.1007/BF02023004

Osorio, J. C., Lasso, D. F., & Ruiz, G. A. (2012). Ob shop scheduling biobjetivo mediante enfriamiento simulado y enfoque de pareto. *Revista Ingenierias Universidad de Medellin*, *11*(21), 113–126.

Pan, A. C., & Liao, C. (1989). An inventory model under just-in-time purchasing agreements. *Production and Inventory Management, 30*(1), 49-52. doi: 10.1108/01443579810192907

Panayides, P.M. & Song, D.-W. (2009). Port integration in global supply chains: measures and implications for maritime logistics. *International Journal of Logistics Research and Applications: A Leading Journal of Supply Chain Management, 12*(2), 133-145.

Panayides, P. M. (2006). Maritime logistics and global supply chains: Towards a research agenda. *Maritime Economics and Logistics*, *8*(1), 3–18. doi:10.1057/palgrave.mel.9100147

Panayides, P. M., & Song, D.-W. (2008). Evaluating the integration of seaport container terminals in supply chains. *International Journal of Physical Distribution & Logistics Management*, *38*(7), 562–584. doi:10.1108/09600030810900969

Panda, D. K., & Bhoi, R. K. (2005). Artificial Neural Network Prediction of Material Removal Rate in Electro-Discharge Machining. *Materials and Manufacturing Processes*, *20*(4), 645–672. doi:10.1081/AMP-200055033

Paolo, T., & Daniele, V. (Eds.). (2002). *The vehicle routing problem. SIAM Monographs on Discrete Mathematics and Applications*. Philadelphia, PA: Society for Industrial and Applied Mathematics.

Papacostas, C. S. (1987). *Fundamentals of Transportation Engineering*. USA: Prentice-Hall Inc.

Park, Y., & Ahn, S. B. (2003). Optimal assignment for check-in counters based on passenger arrival behaviour at an airport. *Transportation Planning and Technology*, *26*(5), 397–416. doi:10.1080/03081060310001635887

Parlar, M., Rodrigues, B., & Sharafali, M. (2013). On the allocation of exclusive-use counters for airport check-in queues: Static vs. dynamic policies. *OPSEARCH*, *50*(3), 433–453. doi:10.1007/s12597-012-0110-5

Parlar, M., & Sharafali, M. (2008). *Dynamic allocation of airline check-in counters: A queueing optimization approach*. *Management Science*, *54*(8), 1410–1424. doi:10.1287/mnsc.1070.0842

Parra, A., Roa, G., & Oliveros, E. (2008). SECAFÉ Part I: mathematical modeling and simulation in the mechanical drying of parchment coffee. *Revista Brasileira de Engenharia Agrícola e Ambiental, 12*(4), 415-427. Retrieved March 12, 2015, from http://www.scielo.br/scielo.php?script=sci_arttext&pid=S1415-43662008000400013&lng=en&tlng=es.10.1590/S1415-43662008000400013

Pawlowski, A., Guzmán, J. L., Rodríguez, F., Berenguel, M., Sánchez, J., & Dormido, S. (2008). Event-Based Control and Wireless Sensor Network for Greenhouse Diurnal Temperature Control: A Simulated Case Study. IEEE.

Pearson, C., & Hussain, Z. (2015). Smartphone Use, Addiction, Narcissism, and Personality: A Mixed Methods Investigation. *International Journal of Cyber Behavior, Psychology and Learning*, *5*(1), 17–32. doi:10.4018/ijcbpl.2015010102

Pearson, R. (1980). *Containerline performance and service quality*. Liverpool, UK: Marine Transport Centre, University of Liverpool.

Peck, H. (2006). Reconciling supply chain vulnerability, risk and supply chain management. *International Journal of Logistics: Research and Applications*, *9*(2), 127–142. doi:10.1080/13675560600673578

Pelikan, M., Goldberg, D. E., & Lobo, F. G. (2002). A survey of optimization by building and using probabilistic models. *Computational Optimization and Applications*, *21*(1), 5–20. doi:10.1023/A:1013500812258

Pendergraft, D. R., Robertson, C. V., & Shrader, S. (2004). Simulation of an airport passenger security system. In *Proceedings of the 36th conference on Winter simulation (WSC '04). Winter Simulation Conference* (pp. 874-878). doi:10.1109/WSC.2004.1371402

Pengzhan, C., & Baifen, L. (2010). Construction of Intelligent Greenhouse Control System Based on CAN bus.*International Conference on Computer Application and System Modeling*. doi:10.1109/ICCASM.2010.5622630

Pennsylvania State University (PSU). (2010). *The basics of food defense*. Retrieved May 18, from http://www.fsis.usda.gov/shared/PDF/7_Food_Defense.pdf?redirecthttp=true

Perakis, A. N., & Bremer, W. M. (1992). An operational tanker scheduling optimization system, background, current practice and model formulation. *Maritime Policy & Management*, *19*(3), 177–187. doi:10.1080/751248659

Pérez Jiménez. (2003). *Maquinas Moleculares Basadas en ADN*. M. de J. Pérez Jiménez F. Sancho Caparrini.

Pérez, V. (2008). *El Desempeño Empresarial*. Retrieved from http://www.degerencia.com/articulo/el_desempeno_empresarial

Persson, J. A., & Göthe-Lundgren, M. (2005). Shipment planning at oil refineries using column generation and valid inequalities. *European Journal of Operational Research*, *163*(3), 631–652. doi:10.1016/j.ejor.2004.02.008

Peterkofsky, R. I., & Daganzo, C. F. (1990). A branch and bound solution method for the crane scheduling problem. *Transportation Research Part B: Methodological*, *24*(3), 159–172. doi:10.1016/0191-2615(90)90014-P

Peters, K. (2003). *Officials Fear Terrorist Attack on U.S. Food Supply*. Retrieved April 2015, from http://www. govexec.com/ dailyfed/0603/061003kp1.htm

Pfohl, H., Köhler, H., & Thomas, D. (2010). State of the art in supply chain risk management research: Empirical and conceptual findings and a roadmap for the implementation in practice. *Logistics Research*, *2*(1), 3344. doi:10.1007/s12159-010-0023-8

Pham, P. D., Nguyen, T. T., & Tran, T. X. (2015). A Hybrid Multi-objective PSO-SA Algorithm for the Fuzzy Rule Based Classifier Design Problem with the Order Based Semantics of Linguistic Terms. *VNU Journal of Science: Computer Science and Communication Engineering, 30*(4).

Pichpibul, T., & Kawtummachai, R. (2012). An improved Clarke and Wright savings algorithm for the capacitated vehicle routing problem. *ScienceAsia*, *38*(3), 307–318. doi:10.2306/scienceasia1513-1874.2012.38.307

Pickett, C. (2003). *Strategies for maximizing supply chain resilience: Learning from the past to prepare for the future*. (Master's Thesis). Massachusetts Institute of Technology, Cambridge, MA.

Pienaar, W. (2009). *Introduction to Business Logistics. Southern Africa*. Oxford University.

Pingre, A. (1768). *Journal du voyage de M. le Marquis de Courtanvaux*. Paris: Imprimerie Royale.

Pinker, S. (1999). *Como funciona nuestra mente?* Sao Paulo, Brazil: Compania de Letras.

Pistole J. (2006). Preparing for the Unthinkable: Agroterrorism. *Food Business News*, (1), 23-26.

Pope, J. A., & Talley, W. K. (1988). Inventory costs and optimal ship size. *Logistics and Transportation Review*, *24*(2), 107–120.

Porteous, I. (1971). Simple singularities of maps. In *Proceedings of Liverpool Singularities Symposium(1969/70)* (LNIM), (vol., 192, pp. 286-307). Berlin: Springer. doi:10.1007/BFb0066829

Potvin, J. Y. (2009). A Review of Bio-inspired Algorithms for Vehicle Routing. Springer-Verlag Berlin Heidelberg. doi:10.1007/978-3-540-85152-3_1

Preuss, M. (2007). *Reporting on experiments in evolutionary computation*. Univ., Sekretariat des SFB 531.

PROMEXICO. (2012). *Mexico's Strengths*. Retrieved from http://pdf.aigroup.asn.au/trade/Mexicos_Strengths.pdf

Psaraftis, H. N., & Kontovas, C. A. (2009). CO2 emissions statistics for the world commercial fleet. *WMU Journal of Maritime Affairs*, *8*(1), 1–25. doi:10.1007/BF03195150

Psaraftis, H. N., & Kontovas, C. A. (2010). Balancing the economic and environmental performance of maritime transportation. *Transportation Research Part D, Transport and Environment*, *15*(8), 458–462. doi:10.1016/j.trd.2010.05.001

Qing, L., Shao-ping G., Ming, Z., Shi-zhou, C., & Qian Y. (2008). American Food Defense Plan and Hazard Analysis and Critical Control Point. *Chinese Journal of Food Hygiene*.

Qiu, W., Dong, L., Wang, F., & Yan, H. (2014). *Design of Intelligent Greenhouse Environment Monitoring System Based on ZigBee and Embedded Technology*. IEEE.

Quinn, R., & Rohrbaugh, J. (1983). A Spatial model of Efectiveness Criteria: Towards a Competing Values Aproach to Organizatinal Analysis. *Management Science*, *29*(3), 363–377. doi:10.1287/mnsc.29.3.363

Quiroz-Castellanos, M., Cruz-Reyes, L., Torres-Jimenez, J., Gómez, C., Huacuja, H. J. F., & Alvim, A. C. (2015). A grouping genetic algorithm with controlled gene transmission for the bin packing problem. *Computers & Operations Research*, *55*, 52–64. doi:10.1016/j.cor.2014.10.010

Quiroz, M., Cruz-Reyes, L., Torres-Jiménez, J., & Melin, P. (2013). Improving the performance of heuristic algorithms based on exploratory data analysis. In *Recent Advances on Hybrid Intelligent Systems* (pp. 361–375). Springer Berlin Heidelberg. doi:10.1007/978-3-642-33021-6_29

Raa, B., Dullaert, W., & Van Schaeren, R. (2011). An enriched model for the integrated berth allocation and quay crane assignment problem. *Expert Systems with Applications*, *38*(11), 14136–14147.

Radio Technical Commission for Aeronautics. (1992). *Software Considerations in Airborne Systems and Equipment Certification*. Washington, DC: RTCA, Inc.

Rajendar & Jun Ma. (2005). Benchmarking Culture and Performance In Chinese Organizations, *Benchmarking. International Journal (Toronto, Ont.)*, *12*(3), 260–274.

Ramamurti, V., & Balasubramanian, P. (1987). Steady state stress analysis of centrifugal fan impellers. *Computers & Structures*, *25*(1), 129–135. doi:10.1016/0045-7949(87)90224-0

Ramírez, M. (1980). *Reorganización de los servicios de urgencias del hospital universitario Miguel Hidalgo*. Durango, México: Instituto Tecnológico de Durango.

Rao, P. G., & Raghavan, K. V. (1996). Hazard and risk potential of chemical handling at ports. *Journal of Loss Prevention in the Process Industries*, *9*(3), 199–204. doi:10.1016/0950-4230(96)00017-4

Rao, S., & Goldsby, T. (2009). Supply chain risks: A review and typology. *International Journal of Logistics Management, 20*(1), 97–123. doi:10.1108/09574090910954864

Rasco, B., & Beldsoe, G. (2006). *Bioterrorism and Food Safety.* Boca Ratón, FL: CRC Press.

Rasmussen, M. S., Justesen, T., Dohn, A., & Larsen, J. (2012). The home care crew scheduling problem: Preference-based visit clustering and temporal dependencies. *European Journal of Operational Research, 219*(3), 598–610. doi:10.1016/j.ejor.2011.10.048

Reklaitis, G. V., Ravindran, A., & Ragsdell, K. M. (1983). *Engineering optimization methods and applications.* New York: Wiley.

Rey, M.-P. (2012). *L'effroyable tragédie: une nouvelle histoire de la campagne de Russie.* Paris: Flammarion.

Reynolds, R. G. (1998). An Introduction to Cultural Algorithms. *Cultural Algorithms Repository.* Retrieved from http://www.cs.wayne.edu/~jcc/car.html

Reynolds, R. G. (2010). Networks Do Matter: The Socially Motivated Design of a 3D Race Controller Using Cultural Algorithms. *International Journal of Swarm Intelligence Research,1*(1), 17–41.

Rhyne, A. L., & Steel, R. G. D. (1965). Tables for a treatments versus control multiple comparisons sign test. *Technometrics, 7*(3), 293–306. doi:10.1080/00401706.1965.10490264

Rice, J. B., Caniato, J., Fleck, D., Disraelly, D., Lowtan, R., Lensing, & Pickett. (2003). *Supply Chain Response to Terrorism: Creating Resilient and Secure Supply Chains.* MIT Center for Transportation and Logistics Report.

Rice, J., (2007). Rethinking security. *Logistics Management, 46*(5), 28-32.

Richard, C. (2002). Experiments with New Teaching Models and Methods. *International Public Management Review.*

Robinson, R. (2002). Ports as elements in value-driven chain systems: The new paradigm. *Maritime Policy & Management, 25*(1), 21–40. doi:10.1080/03088839800000043

Rodger, N. (1995). La mobilisation navale au XVII siècle. In État, marine et société. Presses de l'Université de Paris-Sorbonne.

Rodrigue, J.P. & Notteboom, T (2015). Looking inside the box: evidence from the containerization of commodities and the cold chain. *Maritime Policy & Management: The flagship journal of international shipping and port research, 42*(3), 207-227.

Roman, S. (2005). *Advanced linear algebra. In Graduate texts in mathematics* (Vol. 135). Springer.

Ronen, D. (1983). Cargo ships routing and scheduling: Survey of models and problems. *European Journal of Operational Research, 12*(2), 119–126. doi:10.1016/0377-2217(83)90215-1

Ronen, D. (1993). Ship scheduling: The last decade. *European Journal of Operational Research, 71*(3), 325–333. doi:10.1016/0377-2217(93)90343-L

Ronza, A., Carol, S., Espejo, V., Vílchez, J. A., & Arnaldos, J. (2006). A quantitative risk analysis approach to port hydrocarbon logistics. *Journal of Hazardous Materials, 128*(1), 10–24. doi:10.1016/j.jhazmat.2005.07.032 PMID:16125840

Rosoff, H., & Von Winterfeldt, D. (2007). A Risk and economic analysis of dirty bomb attacks on the ports of Los Angeles and Long Beach. *Risk Analysis, 23*(3), 533546. PMID:17640206

Rubiales, A., Mayorano, F., & Lotito, P. (2007). Optimization applied to the hydrothermal coordination of the Argentine electricity market. *Computational Mechanics, 26*, 1906–1920.

Runckler, T. (2009). *Fuzzy c–Means*. Herding, IFSA-EUSFLAT 2009.

Runeson, P., & Höst, M. (2009). Guidelines for conducting and reporting case study research in software engineering. *Empirical Software Engineering*, *2*(14), 131–164. doi:10.1007/s10664-008-9102-8

Safaei, N., Jardine, A. K., & Banjevic, D. (2008). *Multi-objective simulated annealing for a maintenance workforce scheduling problem: a case study*. INTECH Open Access Publisher. doi:10.5772/5558

Sahebi, H., Nickel, S., & Ashayeri, J. (2014). Strategic and tactical mathematical programming models within the crude oil supply chain context—A review. *Computers & Chemical Engineering*, *68*, 56–77. doi:10.1016/j.compchemeng.2014.05.008

Salazar, R., López, I., & Rojano, A. (2007). A neural network model to control greenhouse environment.*Sixth Mexican International Conference on Artificial Intelligence, Special Session*. doi:10.1109/MICAI.2007.33

Salgado, P., & Cunha, J. (2005). *Greenhouse climate hierarchical fuzzy modeling*. Academic Press.

Salvendy, G. (2001). *Handbook of Industrial Engineering: Technology and Operations Management*. New York, NY: John Wiley & Sons, Inc.

Samanta, B. (2008). Surface roughness prediction in machining using Soft computing. *International Journal of Computer Integrated Manufacturing*, 2008.

Sammarra, M., Cordeau, J. F., Laporte, G., & Monaco, M. F. (2007). A tabu search heuristic for the quay crane scheduling problem. *Journal of Scheduling*, *10*(4-5), 327–336. doi:10.1007/s10951-007-0029-5

Sammataro & Avitabile. (2005). *El Manual del Apicultor*. Letemendia Casa Editora Buenos Aires, Arg.

Sánchez-Oro, J., José Pantrigo, J., & Duarte, A. (2014). Combining intensification and diversification strategies in vns. An application to the vertex separation problem. *Computers & Operations Research*, *52*, 209–219.

Schalkoff, R. J. (1990). *Artificial Intelligence: An Engineering Approach*. McGraw-Hill College.

Scheinerman, E. R. (2000). *Mathematics: A Discrete Introduction*. Brooks.

Schiavinotto, T., & Stützle, T. (2004). The linear ordering problem: Instances, search space analysis and algorithms. *Journal of Mathematical Modelling and Algorithms*, *3*(4), 367–402. doi:10.1023/B:JMMA.0000049426.06305.d8

Schnitzler, H.-U., & Kalko, E. K. V. (2001, July). Echolocation by insect-eating bats. *Bioscience*, *51*(7), 557–569. doi:10.1641/0006-3568(2001)051[0557:EBIEB]2.0.CO;2

Schultz, M., & Fricke, H. (2011). *Managing Passenger Handling at Airport Terminals. Individual-based Approach for Modeling the Stochastic Passenger Behavior.* Ninth USA/Europe Air Traffic Management Research and Development Seminar (ATM2011).

Schwarz, A., Kalika, M., Kefi, H., & Schwarz, C. (2010). A Dynamic Capabilities Approach to Understanding the Impact of IT-Enabled Business Processes and IT-Business Alignment on the Strategic and Operational Performance of the Firm. *Communications of the Association for Information Systems*, *26*(4), 57–84.

Secretaria de Comunicaciones y Transportes. (2015). *Estadística histórica (1992-2014) / Historical statistics (1992-2014).* Retrieved from http://www.simio.com

Secretaría de Salud. (2002). *Lineamientos operativos del sistema de monitoreo para la gestión de la calidad*. México: Secretaría de Salud.

Seginer, I., Boulard, T., & Bailey, B. (1993). *Optimal Greenhouse Temperature Trajectories for a Multi-State-Variable Tomato Model. In The Computerized Greenhouse* (pp. 153–172). San Diego, CA: Academic Press.

Senoo, Y. & Nakase, Y., (1972). *An analysis of flow for Power.* Paper No. 71-GT-2.

Serafini, P. (1993). Simulated annealing for multi objective optimization problems. In Multiple criteria decision making (pp. 283-292). Springer.

Serebrisky, T., Gómez-Lobo, A., Estupiñán, N., & Nuñoz-Raskin, R., (2009). Affordability and Subsidies in Public Urban Transport: What Do We Mean, What Can Be Done?, *Transport Reviews: A Transnational Transdisciplinary Journal, 29*(6), 715–739.

Setola, R., & de Maggio, M. C. (2009). *Security of the Food Supply Chain.* Paper presented at 31st International conference of the IEEE EMBS.

Sham Sunder, K. (1987). *Finite Element Analysis of Centrifugal Impellers.* (Ph.D. Thesis). School of Mechanical Engineering, Cranfield Institute of Technology, Cranfield.

Shan-Huen, H., & Tsan-Hwan, L. (2012). Heuristic algorithms for container pre-marshalling problems. *Computers & Industrial Engineering, 62*(1), 13–20. doi:10.1016/j.cie.2011.08.010

Sheffi, Y. (1985). *URBAN TRANSPORTATION NETWORKS: Equilibrium Analysis with Mathematical Programming Methods.* New Jersey, USA: Prentice-Hall Inc.

Sheffi, Y., & Rice, J. (2005). A supply chain view of the resilient enterprise. *MIT Sloan Management Review, 47*(1), 4148.

Sheffi, Y., Rice, J. B., Fleck, J. M., & Caniato, F. (2003). Supply chain response to global terrorism: A situation scan. *Proceedings of the First EurOMA-POMS Joint International Conference.*

Sheskin, D. J. (2006). *Handbook of Parametric and Nonparametric Statistical Procedures.* Boca Raton, FL: CRC Press.

Shin, H., Collier, D. A., & Wilson, D. D. (2000). Supply management orientation and supplier/buyer performance. *Journal of Operations Management, 18*(3), 317–333. doi:10.1016/S0272-6963(99)00031-5

Shook, C., Adams, G., & Ketchen, D. (2009). Towards a 'theoretical toolbox' for strategic sourcing, *Supply Chain Management. International Journal (Toronto, Ont.), 14*(1), 3–10.

Siegel, S. C., John, N., Aragón Borja, L. E., & Fierros Dávila, L. E. (1995). Estadística no paramétrica: aplicada a las ciencias de la conducta (4th ed.). México: Trillas.

Sigurd, M. M., Ulstein, N. L., Nygreen, B., & Ryan, D. M. (2005). Ship scheduling with recurring visits and visit separation requirements, In Column Generation, (pp. 225-245). Springer USA.

Sim, S., Lee, J., & Lee, J. (2012). *MOCSA: multiobjective optimization by conformational space annealing.* arXiv preprint arXiv:1209.0549

Singh, G., Singh, P., Lubana, P., & Singh, K. (2006). *Formulation and Validation of a Mathematical model of the microclimate of greenhouse* (Vol. 31). Elsevier.

Siroky, J. & Hlavsová, P. (2014). *Optimizing process of check-in and security check at airport terminals.* Academic Press.

Skabardonis, A., & Dowling, R. (1997). *Improved Speed-Flow Relationships for Planning Applications.Transportation Research Record No. 1572.* Washington, DC: Transportation Research Board.

Slater, S., & Olson, E. (2000). Strategy Type and Performance: The Influence of Sales Force Management. *Strategic Management Journal, 21*(8), 813–829. doi:10.1002/1097-0266(200008)21:8<813::AID-SMJ122>3.0.CO;2-G

Smale, S. (1967). Differentiable dynamical systems. *Bulletin of the American Mathematical Society, 73*(6), 747–817. doi:10.1090/S0002-9904-1967-11798-1

Smale, S. (1975), Optimizing Several Functions. In *Proceedings of the International Conference on Manifolds and Related Topics in Topology* (pp. 69-75). Tokyo: University of Tokyo Press.

Smale, S. (1981). Global Analysis and Economics. In K. J. Arrow & M. D. Intrilligator (Eds.), *Handbook of mathematical economics* (Vol. 1, pp. 331–370). Amsterdam: North – Holland.

Smith, K., Everson, R. M., Fieldsend, J. E., Murphy, C., & Misra, R. (2008). Dominance-based multiobjective simulated annealing. *Evolutionary Computation. IEEE Transactions on, 12*(3), 232–342.

Smith, S. L., Pavone, M., Bullos, F., & Frazzoli, E. (2010). Dynamic vehicle routing with priority classes of stochastic demands. *IAM Journal on Control and Optimization, 48*(5), 3224–3245. doi:10.1137/090749347

Snodgrass, R. T. (2010). *Ergalics: A natural science of computation*. University of Arizona.

Sodhi, M., & Tang, C., (2012). *Managing Supply Chain Risks*. Springer.

Sokolowski, J. A., & Banks, C. M. (2010). *Modeling and Simulation Fundamentals. Theoretical Underpinnings and Practical Domains*. Hoboken, NJ: Wiley. doi:10.1002/9780470590621

Solak, S., Clarke, J. P. B., & Johnson, E. L. (2009). *Airport terminal capacity planning. Transportation Research Part B: Methodological, 43-6*(6), 659–676. doi:10.1016/j.trb.2009.01.002

Solera, A. (2003). *El Filtro de Kalman*. Banco Central de Costa Rica, División Económica, Departamento de Investigaciones económicas.

Somashekhar, K. P., Ramachandran, N., & Mathew, J. (2010). Optimization of Material Removal Rate in Micro-EDM Using Artificial Neural Network and Genetic Algorithms. *Materials and Manufacturing Processes, 25*(6), 467–475. doi:10.1080/10426910903365760

Sondhaus, L. (2011). *World War One: The Global Revolution*. Cambridge University Press.

Song, D.W. & Lee, E.S. (2012). Competitive networks, knowledge, acquisition and maritime logistics value. *International Journal of logistics Research and Applications: A Leading Journal of Supply Chain Management, 15*(1), 15-35.

Song, D.W. & Lee, P. (2009). Maritime logistics in the global supply chain. *International Journal of Logistics Research and Applications: A Leading Journal of Supply Chain Management, 12*(2), 83-84.

Souffriau, W., Maervoet, J., Vansteenwegen, P., Vanden Berghe, G., & Van Oudheusden, D. (2009). A Mobile Tourist Decision Support System for Small Footprint Devices. *IWANN, 2009*(1), 1248–1255.

Speiss, H. (1990). Conical Volume-Delay Functions. *Transportation Science, 24*(2).

Spiess, H. (1990). Technical Note—Conical Volume-Delay Functions. *Transportation Science, 24*(2), 153–158. doi:10.1287/trsc.24.2.153

Spivak, M. (1971). *Calculus on Manifolds*. New York: Westview Press.

Srivastava, S., Roy, A. K., & Kumar, K. (2014). Design of a mixed flow pump impeller and its validation using FEM analysis. *Procedia Technology, 14*, 181–187. doi:10.1016/j.protcy.2014.08.024

Stadler, H. (2005). Supply chain management and advanced planning – basics, overview and challanges. *European Journal of Operational Research, 163*(3), 575–588. doi:10.1016/j.ejor.2004.03.001

Stalhane, M., Andersson, H., Christiansen, M., & Fagerholt, K. (2014). Vendor managed inventory in tramp shipping. *Omega, 47*, 60–72. doi:10.1016/j.omega.2014.03.004

Stecke, K., & Kumar, S. (2006). *Sources of supply chain disruptions: factors that breed vulnerability, and mitigating strategies*. Richardson: School of Management, University of Texas at Dallas.

Steel, R. G. (1959). A multiple comparison sign test: Treatments versus control. *Journal of the American Statistical Association*, *54*(288), 767–775. doi:10.2307/2282500

Stefan, B., & Natalia, K. (2009). *An overview on vehicle scheduling models*. Decision Support & Operations Research Lab, University of Paderborn. Available at: https://wiwi.uni-paderborn.de/leadmin/lehrstuehle/de-partment-3/wiwi-dep-3-ls-5/Forschung/Publikationen/bunteklieweranoverviewonvehicleschedulingmodels2006.pdf

Stepanoff, A. J. (1957). *Centrifugal and Axial flow pumps* (2nd ed.). New York: John Wiley and Sons.

Stevenson, W. J. (2011). *Operations Management* (11th ed.). McGraw-Hill/Irwin.

Stewart, W. Jr, & Golden, B. (1983). Stochastic vehicle routing: A comprehensive approach. *European Journal of Operational Research*, *14*(4), 371–385. doi:10.1016/0377-2217(83)90237-0

Stodola, A. (1945). *Steam and gas turbine*. Academic Press.

Stoner, J. (1992). *Administration*. Mexico: Prentice Hall.

Stopford, M. (2009). *Maritime Economics* (3rd ed.). Routledge. doi:10.4324/9780203891742

Suaremi, T. (2003). *Understand social groups. In NDAM'2003*, Reykiavik, Iceland.

Sumalee, A., Uchida, K., & Lam, H. K. W. (2011). Stochastic multi-modal transport network under demand uncertainties and adverse weather condition. *Transportation Research Part C, Emerging Technologies*, *11*(2), 338–350. doi:10.1016/j.trc.2010.05.018

Suman, B., & Kumar, P. (2006). A survey of simulated annealing as a tool for single and multiobjective optimization. *The Journal of the Operational Research Society*, *57*(10), 1143–1160. doi:10.1057/palgrave.jors.2602068

Surak, J. G. (2005). ISO 22000: *Requirements for food safety management systems*. Paper presented at World Conference on Quality and Improvement.

Surak, J. G., & Lorca, T. A. (2007). *Moving from HACCP to ISO 22000*. Paper presented at World Conference on Quality and Improvement.

Surak, J. G. (2007). A Recipe for Safe Food: ISO 22000 and HACCP. *Quality Progress*, *40*(10), 21–27.

Suthiwartnarueput, K. (1988). *The exploration of sea transport efficiency: with a concentration on the case of Thailand*. (Doctoral Dissertation). Department of Maritime Studies and International Transport University of Wales College of Cardiff, UK.

Swanson, L. (2003). An information-processing model of maintenance managemen. *International Journal of Production Economics*, *83*(1), 45–64. doi:10.1016/S0925-5273(02)00266-9

Taboada, B., Baheranwala, F., Coit, D. W., & Wattanapongsakorn, N. (2007). Practical solutions for multi-objective optimization:An application to system reliability design problems. *Reliability Engineering & System Safety*, *92*(3), 314–322. doi:10.1016/j.ress.2006.04.014

Taboada, H., & Coit, D. (2007). Data Clustering of Solutions for Multiple Objective System Reliability Optimization Problems. *Quality Technology & Quantitative Management Journal*, *4*(2), 35–54.

Taboada, H., & Coit, D. (2008). Multiple Objective Scheduling Problems: Determination of Pruned Pareto Sets. *IIE Transactions*, *40*(5), 552–564. doi:10.1080/07408170701781951

Ta, D., Dellaert, N., Woensel, T. V., & Kok, T. D. (2013). Vehicle routing problem with stochastic travel times including soft time windows and service costs. *Computers & Operations Research*, *40*(1), 214–224. doi:10.1016/j.cor.2012.06.008

Taha, H. A. (2004). *Operations research*. Editorial Pearson Education.

Taillard, E. D. (2003). A statistical test for comparing success rates. In Metaheuristic international conference MIC'03.

Taillard, E., Badezu, P., Gendreau, M., & Guertin, F. (1997). A tabu search heuristic for the vehicle routing problem with soft-time window. *Journal of Transportation Science*, *31*(2), 170–186. doi:10.1287/trsc.31.2.170

Takakura, T., Jordan, A., & Boyd, L. (1971). Dynamic simulation of plant growth and environment in the greenhouse. *Trans. ASAE 14* (vol. 964).

Takayama, Y., & Hiroyoshi, W. (2009). *Multi-Objective Design Optimization of a Mixed flow impeller*. ASME, Fluids Engineering Division.

Talley, W. K., Agarwal, V. B., & Breakfield, J. W. (1986). Economics of density of ocean tanker ships. *Journal of Transport Economics and Policy*, *20*(1), 91–99.

Talley, W. K., & Ng, M. (2013). Maritime transport chain choice by carriers, ports and shippers. *International Journal of Production Economics*, *142*(2), 311–316. doi:10.1016/j.ijpe.2012.11.013

Tang, C. (2006). Perspectives in supply chain risk management. *International Journal of Production Economics*, *103*(2), 451–488. doi:10.1016/j.ijpe.2005.12.006

Taniguchi, E., & Thompson, R. G. (2015). *City Logistics. Mapping the future*. Boca Raton, FL: CRC Press.

Taniguchi, E., Thompson, R. G., Yamada, T., & Van Duin, R. (2001). *City Logistics. Network modelling and Intelligent Transport Systems*. Amsterdam: Elsevier.

Tap, F. (2000). *Economics-based optimal control of greenhouse tomato crop prod.* (PhD Thesis). Wageningen Agr.

Tavakkoli-Moghaddam, R., Makui, A., Salahi, S., Bazzazi, M., & Taheri, F. (2009). An efficient algorithm for solving a new mathematical model for a quay crane scheduling problem in container ports. *Computers & Industrial Engineering*, *56*(1), 241–248. doi:10.1016/j.cie.2008.05.011

Teghem, J., Tuyttens, D., & Ulungu, E. L. (2000). An interactive heuristic method for multi-objective combinatorial optimization. *Computers & Operations Research*, *27*(7), 621–634. doi:10.1016/S0305-0548(99)00109-4

Tengku, J. T. M. S. (1995). *Marketing of freight liner shipping services with reference to the far East-Europe trade: a Malaysian perspective*. (Doctoral Dissertation). Department of Maritime Studies and International Transport, University of Wales College of Cardiff, UK.

Thangiah, S. R. (1995). *Vehicle routing with time windows using genetic algorithms*. Academic Press.

The White House. (2012). *National strategy for global supply chain security*. Retrieved from http://www.whitehouse.gov/sites/default/files/national_strategy_for_global_supply_chain_security.pdf

Thiagarajan, A., Ravindranath, L., LaCurts, K., Madden, S., Balakrishnan, H., Toledo, S., & Eriksson, J. (2009). VTrack: accurate, energy-aware road traffic delay estimation using mobile phones. In*Proceedings of the 7th ACM Conference on Embedded Networked Sensor Systems* (pp. 85–98). doi:10.1145/1644038.1644048

Thiétart, R. A. (1999). Méthodes de recherche en management. Ed. Dunod.

Tongzon, J., & Heng, W. (2005). Port privatization, efficiency and competitiveness: Some empirical evidence from container ports. *Transportation Research Part A, Policy and Practice*, *39*(5), 405–424. doi:10.1016/j.tra.2005.02.001

Tosic, V. (1992). A review of airport passenger terminal operations analysis and modeling. *Transportation Research Part A, Policy and Practice*, *26*(1), 3–26. doi:10.1016/0965-8564(92)90041-5

Tosun, N. (2002). *A study of tool life in hot machining using artificial neural networks and regression analysis method.* Elsevier Science B.V.

Toudouze, G.-G., De la Roncière, C., Tramond, J., Rondeleux, M., Dollfus, C., & Dubard, P. (1959). Histoire de la marine. Éditions de l'illustration.

Trawiński, B., Smętek, M., Telec, Z., & Lasota, T. (2012). Nonparametric statistical analysis for multiple comparison of machine learning regression algorithms. *International Journal of Applied Mathematics and Computer Science*, *22*(4), 867–881. doi:10.2478/v10006-012-0064-z

Trkman, P., Stemberger, M., & Jaklic, J. (2005). *Information transfer in Supply Chain (SC) management.* Available: http://2005papers.iisit.org/I46f91Trkm.pdf

Troshkin, M. (2014). Doomsday comes: A Nonrecursive Analysis of the Recursive Towers of Hanoi Problem. *Focus (San Francisco, Calif.)*, *95*(2), 10–14.

Tuna, O., & Arabelen, G. (2013). Deniz Ulaştırma Lojistiği (Maritime Logistics). In A. Cerit, A. Deveci, & S. Esmer (Eds.), *Denizcilik İşletmeleri Yönetimi (Maritime Business Management)* (pp. 533–561). İstanbul: Beta Basım A.Ş.

Tzannatos, E. (2010). Ship emissions and their externalities for the port of Piraeus-Greece. *Atmospheric Environment*, *44*(3), 400–407. doi:10.1016/j.atmosenv.2009.10.024

Udink Ten Cate, A. J. (1983). *Modeling and (adaptive) control of greenhouse climates.* Wageningen UR.

Ulungu, E., Teghem, J., Fortemps, P., & Tuyttens, D. (1999). MOSA method: a tool for solving multiobjective combinatorial optimization problems. *Journal of Multicriteria Decision Analysis, 8*(4), 221.

UNCTAD. (2014). Review of Maritime Transport. *United Nations Conference on Trade and Development*, New York.

Unger, R. (1988). Conception et construction des vaisseaux de guerre européennes au XVII et XVIII siècles. In Les marines de guerre européennes: XVII-XVIII siècles. Presses de l'Université de Paris-Sorbonne.

Union Internationale des Chemins de fer. (2014). *Railisia. The online UIC Railway Database.* Retrieved from http://www.uic.org/spip.php?article1352

United States Department of Agriculture (USDA). (2004). *A Biosecurity Checklist for School Foodservice Programs.* Retrieved May 24, 2015, from http://education.ky.gov/federal/scn/documents/biosecurity%20checklist.pdf

United States Department of Agriculture (USDA). (2008). *Guide to Developing a Food Defense Plan for Warehouse and Distribution Centers.* Retrieved May 20 2015, from http://www.fsis.usda.gov/shared/PDF/Guidance_Document_Warehouses.pdf

United States Department of Agriculture (USDA). (2009a). *Food Defense Guidelines for Slaughter and Processing Establishments.* Retrieved May 26, 2015, from http://www.fsis.usda.gov/wps/wcm/connect/cd18dfb5-9443-42f8-b8c5-cadf862fcbc4/SecurityGuide.pdf?MOD=AJPERES

United States Department of Agriculture (USDA). (2009b). *General Food Defense Plan, 2009.* Retrieved June 2, 2015, from http://www.fsis.usda.gov/wps/wcm/connect/99f95182-0c9e-4214-9762-e98197f54ebf/General-Food-Defense-Plan.pdf?MOD=AJPERES

United States Department of Agriculture (USDA). (2013). *Food Defense Guidelines for the Transportation and Distribution of Meat, Poultry, and Processed Egg Products*. Biosecurity Guidelines. Retrieved May 24, 2015, from http://www.fsis.usda.gov/wps/wcm/connect/4f9d737a-1f3e-49ff-851b-74884fa946bd/Transportation_Security_Guidelines.pdf?MOD=AJPERES

Ünlüyurt, T., & Aydın, C. (2012). Improved rehandling strategies for the container retrieval process. *Journal of Advanced Transportation*, *46*(4), 378–393. doi:10.1002/atr.1193

Valle, P., Girard, A. & Saldate, O. (2007). Defensa Alimentaria "Food Defense". *Revista Mundo Lácteo y Cárnico.*

van Nieuwenhuyse, I., & de Koster, R. (2009). Evaluating order throughput time in 2-block warehouses with time window batching. *International Journal of Production Economics*, *121*(2), 654–664. doi:10.1016/j.ijpe.2009.01.013

Vanderhaeghen, D., Fettke, P. & Loos, P. (2010). Organizational and Technological Options for Business Process Management from the Perspective of Web 2.0. *Business & Information Systems Engineering*, (1), 15-28.

Varian, H. R. (1992). *Microeconomic Analysis* (3rd ed.). New York: W. W. Norton & Company.

Vazquez-Ibarra, A. (2014). *Modelo computacional del servicio de urgencias de un hospital público y análisis de las variables relacionadas con el factor humano y con los tiempos de atención en consulta.* (Unpublished doctoral dissertation). University of Guadalajara, Jalisco, México.

Veiga, A. (2011). Food Defense and Security: A New Reality. In NATO Science for Peace and Security Series C: Environmental Security, Environmental Security and Ecoterrorism (pp. 39-54). Academic Press.

Verner, J., Sampson, J., Tosic, V., Bakar, N., & Kitchenham, B. (2009). *Guidelines for industrially-based multiple case studies in software engineering*. Fes, Morocco: IEEE Research Challenges in Information Systems. doi:10.1109/RCIS.2009.5089295

Vidal, T., Crainic, T. G., Gendreau, M., & Prins, C. (2013). A hybrid genetic algorithm with adaptive diversity management for a large class of vehicle routing problems with time windows. *Computers & Operations Research*, *40*(1), 475–489. doi:10.1016/j.cor.2012.07.018

Vilhelmsen, C., Lusby, R., & Larsen, J. (2014). Tramp ship routing and scheduling with integrated bunker optimization. *EURO Journal on Transportation and Logistics*, *3*(2), 143–175. doi:10.1007/s13676-013-0039-8

Vis, I. F. A., & De Koster, R. (2003). Transshipment of containers at a container terminal: An overview. *European Journal of Operational Research*, *147*(1), 1–16. doi:10.1016/S0377-2217(02)00293-X

Voss, C., Tsikriktsis, N., & Frohlich, M. (2002). Case research in operations management. *International Journal of Operations & Production Management*, *22*(2), 195–219. doi:10.1108/01443570210414329

Wagner, S., & Bode, C. (2008). An empirical examination of supply chain performance along several dimensions of risk. *Journal of Business Logistics*, *29*(1), 307–325. doi:10.1002/j.2158-1592.2008.tb00081.x

Wang, Ch., Viswanathan, K., Choudur, L., Talwar, V., Satterfield, W., & Schwan, K. (2011). Statistical Techniques for Online Anomaly Detection in Data Centers.*12th IFIP/IEEE International Symposium on Integrated Network Management (IM 2011)*. doi:10.1109/INM.2011.5990537

Wang, P. T., & Pitfield, D. E. (1999). The derivation and analysis of the passenger peak hour: An empirical application to Brazil. *Journal of Air Transport Management*, *5*(3), 135–141. doi:10.1016/S0969-6997(99)00007-1

Wardrop, J. G. (1952). Road paper. Some theoretical aspects of road traffic research. *ICE Proceedings: Engineering Divisions, 1*(3), 325-362.

Wäscher, G. (2004). Order Picking: A survey of planning problems and methods. In H. Dyckhoff, R. Lackes, & J. Reese (Eds.), *Supply Chain Management and Reverse Logistics* (pp. 323–347). Berlin: Springer. doi:10.1007/978-3-540-24815-6_15

Waters, D., (2007). *Supply Chain risk management: vulnerability and resilience in logistics*. Kogan Page.

Waze. (2015). *Get the best route, every day, with real–time help from other drivers*. Retrieved from https://www.waze.com/

Weiner, G. A. (2009). *Modeling and Simulation: A practitioner approach*. CR Press. doi:10.1201/9781420053371

Wein, L. M., & Liu, Y. (2005). Analyzing a Bioterrorism Attack on the Food Supply: The Case of Botulinum Toxin in Milk. *Proceedings of the National Academy of Sciences of the United States of America*, *102*(99), 84–89.

Whitley, D., Starweather, T., & Shaner, D. (1990). The traveling salesman and sequence scheduling: quality solutions using genetic edge recombination. In L. Davis (Ed.), *Handbook of genetic algorithms* (pp. 350–372). New York.

Williamson, J. (1974). *Late 19th Century American Development, A General Equilibrium History*. Cambridge University Press.

Williamson, J., & O'Rourke, K. (1999). *Globalization and History. The evolution of a 19th Atlantic Economy*. MIT Press.

Williams, Z., Lueg, J., & LeMay, S. (2008). Supply chain security: An overview and re-search agenda. *International Journal of Logistics Management*, *19*(2), 254–281. doi:10.1108/09574090810895988

Wilson, D., Rose, E. K., & So, A. A. (2006). Security checkpoint optimizer (SCO): An application for simulating the operations of airport security checkpoints. In *Proceedings of the 2006 inter Simulation Conference*. doi:10.1109/WSC.2006.323126

Wilson, I. D., & Roach, P. A. (2000). Container stowage planning: A methodology for generating computerized solutions. *The Journal of the Operational Research Society*, *51*(11), 1248–1255. doi:10.1057/palgrave.jors.2601022

Wilson, M. (2007). The impact of transportation disruptions on supply chain performance. *Transportation Research Part E, Logistics and Transportation Review*, *43*(4), 295320. doi:10.1016/j.tre.2005.09.008

Winston, W. L. (2003). *Operations Research: Applications And Algorithms*. Andover: Duxbury Press.

Winter, J., & Prost, A. (2005). *The Great War in history: debates and controversies, 1914 to the present (No. 21)*. Cambridge University Press. doi:10.1017/CBO9780511614811

Wislicensus, G. F. (1965). The Design of mixed flow pumps. *Proceedings of the symposium held at the N.E.L.*

Wisner, J. D. (2003). A Structural Equation Model of Supply Chain management Strategies and Firm Performance. *Journal of Business Logistics*, *24*(1), 1–26. doi:10.1002/j.2158-1592.2003.tb00030.x

Wittenberg, J. P. (2000). *Methods and models of operations research* (Vol. 1). Editorial LIMUSA.

Wolfe, D. A. (1998). *Nonparametric Methods*. Wiley Encyclopedia of Clinical Trials.

Wolpert, D. H., & Macready, W. G. (1997). No free lunch theorems for optimization. *Evolutionary Computation. IEEE Transactions on*, *1*(1), 67–82.

Wood, D. F., Barone, A. P., Murphy, P. R., & Wardlow, D. L. (2002). *International Logistics* (2nd ed.). AMACOM.

World Health Organization (WHO). (2003). *Bioterrorismo: la amenaza en el continente americano*. Paper presented at Reunión Interamericana a Nivel Ministerial en Salud y Agricultura.

World Health Organization (WHO). (2008). *Terrorist Threats to Food: Guidance for Establishing and Strengthening Prevention and Response Systems*. Department of Food Safety, Zoonoses and Foodborne Disease, Cluster on Health Security and Environment.

Wu, T., Blackhurst, J., & O'grady, P. (2007). Methodology for supply chain disruption analysis. *International Journal of Production Research*, *45*(7), 16651682. doi:10.1080/00207540500362138

Xiangyong, L., Tian, P., & Leung, S. C. H. (2010). Vehicle routing problems with time windows and stochastic travel and service times: Models and algorithm. *International Journal of Production Economics*, *125*(1), 137–145. doi:10.1016/j.ijpe.2010.01.013

Xiaoyan, Z., Xiangyang, Z., Chen, D., Zhaohui, C., Shangming, S., & Zhaohui, Z. (2013). The Design and Implementation of the Greenhouse Monitoring System Based on GSM and RF Technologies. IEEE.

Xinlian, X., Tangfei, W., & Daisong, C. (2000). A dynamic model and algorithm for fleet planning. *Maritime Policy & Management*, *27*(1), 53–63. doi:10.1080/030888300286680

Xu, Y., & Qu, R. (2011). Solving multi-objective multicast routing problems by evolutionary multi-objective simulated annealing algorithms with variable neighbourhoods. *The Journal of the Operational Research Society*, *62*(2), 313–325. doi:10.1057/jors.2010.138

Xu, Y., Qu, R., & Li, R. (2013). A simulated annealing based genetic local search algorithm for multi-objective multicast routing problems. *Annals of Operations Research*, *206*(1), 527–555. doi:10.1007/s10479-013-1322-7

Yang, C. C. (2012). Assessing the moderating effect of innovation capability on the relationship between logistics service capability and firm performance for ocean freight forwarders. *International Journal of Logistics: Research and Applications*, *15*(1), 53–69. doi:10.1080/13675567.2012.669469

Yang, X.-S. (2011). Bat algorithm for multi-objective optimisation. *International Journal of Bio-inspired Computation*, *3*(5), 267–274. doi:10.1504/IJBIC.2011.042259

Yang, Z., Ng, A., & Wang, J. (2014). A new risk quantification approach in port facility security assessment. *Transportation Research Part A, Policy and Practice*, *59*, 72–90. doi:10.1016/j.tra.2013.10.025

Yin, R. (2003). *Case study research: design and methods* (3rd ed.). Thousand Oaks, CA: Sage Publications.

Yip Ortuño, S., Hernández Aguilar, J. A., & Zavala-Díaz, J. (2015). Implementación de un algoritmo genético para resolver el juego de las torres de Hanói. In Aplicaciones Modernas de Optimización: la experiencia entre cuerpos académicos. Morelos, México: UAEM y Artificio Editores.

Yulong, J., & Jiaqiang, Y. (2011). Design an Intelligent Environment Control System for Greenhouse Based on RS485 Bus.*Second International Conference on Digital Manufacturing & Automation.*

Yu, M., & de Koster, R. (2009). The impact of order batching and picking area zoning on order picking system performance. *European Journal of Operational Research*, *198*(2), 480–490. doi:10.1016/j.ejor.2008.09.011

Zaretalab, A., Hajipour, V., Sharifi, M., & Shahriari, M. R. (2015). A knowledge-based archive multi-objective simulated annealing algorithm to optimize series--parallel system with choice of redundancy strategies. *Computers & Industrial Engineering*, *80*, 33–44. doi:10.1016/j.cie.2014.11.008

Zaucha, J. (2014). Sea basin maritime spatial planning: A case study of the Baltic Sea region and Poland. *Marine Policy*, *50*, 34–45. doi:10.1016/j.marpol.2014.05.003

Zavadskas, E. K., Turskis, Z., & Bagocius, V. (2015). Multi-criteria selection of a deep-water port in the Eastern Baltic Sea. *Applied Soft Computing*, *26*, 180–192. doi:10.1016/j.asoc.2014.09.019

Zbilut, J. P., & Giuliani, A. (Eds.). (2007). *Simplicity: the latent order of complexity*. New York: Nova Science.

Zevzevic, I. (2005). Similar cultural relationships in Montenegro. In JASSS'2005.

Zhen, L. (2014). Container yard template planning under uncertain maritime market. *Transportation Research Part E, Logistics and Transportation Review*, *69*, 199–217. doi:10.1016/j.tre.2014.06.011

Zhuang, J., Bier, V. M., & Gupta, A. (2007). Subsidies in Interdependent Security with Heterogeneous Discount Rates. *The Engineering Economist*, *52*(1), 1–19. doi:10.1080/00137910601159722

Zitzler, E., & Thiele, L. (1999). Multi-objective Evolutionary Algorithms: A Comparative Case Study and the Strength Pareto Approach. *IEEE Transactions on Evolutionary Computation*, *3*(4), 257–271. doi:10.1109/4235.797969

Zografos, K. G., & Madas, M. A. (2004). Design and Development of an Integrated Computational Platform for Total Airport Performance Analysis. In *Proceedings of the 2 International Congress on Transportation Research in Greece, Session on Air Transport*.

Zysberg, A. (1998). La soumission du rivage aux volontés de l'État Royal. In Les marines de guerre européennes: XVII-XVIII siècles. Presses de l'Université de Paris-Sorbonne.

About the Contributors

Alberto Ochoa-Zezzatti is a full Professor in Juarez City University.

Jons Sanchez (Bs'00 – Electrical Eng. Master'02 both from the Instituto Tecnológico de Querétaro and the Ph.D.'07 from the CIATEC-CONACYT Research Center). His research field is the mathematical modeling and optimization of process and products using response surface methodology and Evolutionary Multiobjective Algorithms. Doctor Jöns has publicated more than 40 articles and five patents. He is a recognized researcher by SNI-CONACYT. He has been professor of the postgraduate program called PICYT. Furthermore, he is accredited assessor from RCEA-CONACYT.

Miguel Gastón Cedillo-Campos is a Professor in Logistics Systems Dynamics and Founding President of the Mexican Logistics and Supply Chain Association (AML). Dr. Cedillo-Campos is National Researcher Level 1 (National Council of Science and Technology), Innovation Award 2012 (Autonomous University of Nuevo Leon) and National Logistics Award 2012. In 2004, he received a Ph.D. in Logistics Systems Dynamics from the University of Paris, France. Recently, he was collaborating as Visiting Researcher at MIT-Zaragoza Logistics Center. At the Mexican Institute of Transportation (IMT) he is a researcher in logistics systems analysis and modeling, risk analysis, and supply chain management, which are the subjects he teaches in different prestigious universities in Mexico, Spain, France, Colombia, Panama, and Chili. Dr. Cedillo-Campos is the Scientific Chairman of the International Congress on Logistics and Supply Chain (CiLOG). He is the author of several scientific publications in top journals as Transportation Research Part E: logistics and Transportation Review, Simulation, Computers and Industrial Engineering, Applied Energy, Computers in Industry, Journal of Applied Research and Technology, and others.

* * *

Ruben German Almanza, Bachelor in Mathematics, Thesis: Holomorphhic Dynamic: Properties of rational maps and the topology of Julia Sets Master in Mathematics, Thesis: Pinching deformations of rational maps over Riemann Sphere.

Arturo Lam Alvarez received his M.Sc. degree in Computer Science from Madero Institute of Technology, Mexico, in 2012. He is a JEE Development Manager in Accenture S.C, Mexico. His research interests includes: (a) E-Commerce applications design and development, (b) algorithm optimization, (c) Complex systems decision management and (d) system effects.

E. Alvarez-Sánchez received the BSc. Eng. in Mechanical Electrical Engineering in 2000 from Universidad Veracruzana, Veracruz, México; the MSc. degree and the PhD. degree in Electrical Engineering in 2003 and 2007, respectively, both from the Centro de Investigación y de Estudios Avanzados del Instituto Politecnico Nacional, México D.F., Mexico. From 2008 to 2010, he courses a Postdoctoral in Centro de Investigación Científica y de Educación Superior de Ensenada, Ensenada, Mexico. Sincé 2011 he is Full Time Professor in the Mechanical and Electrical Engineering, Faculty campus Xalapa, Universidad Veracruzana, Mexico. His research interests include: mechanical design, linear and nonlinear control, mechatronics, seismic vibrations and domotics.

Julio Cesar Arreola is a Professor of Automatice Engineering at UACJ, Master degree in Industrial Engineering.

Juan J. González B. received the PhD degree in computer science from the Centro Nacional de Investigación y Desarrollo Tecnológico in 2005. At present, he is a Research Professor of Computer Science at Instituto Tecnológico de Ciudad Madero and member of the National System of Researchers, Level I. His scientific interests include metaheuristic optimization and natural language processing.

Aylin Çalışkan is a Research Assistant in the Department of International Logistics Management of Yaşar University. She received her Bachelors in International Logistics and Transportation from the Yeditepe University, Istanbul. She obtained her Masters in Maritime Business Administration from the Dokuz Eylül University, Izmir and continues to the PhD with the same program. Her areas of interest include city logistics, sustainability and marketing in logistics.

Victor Manuel Carrillo received his bachelor degree in the Mexico City National Polytechnic Institute in fall 1977. Then he started his professional career as lecturer in the Metropolitan Autonomous University in Mexico City from 1977 to 1992. He is lecturer in the Juarez Autonomous University in Juarez, Chihuahua since 1993. He started pursing his PhD degree in Computational Science in August 2009. In December 2013 graduated as Phd in UTEP in the Computational Sciences program.

Norberto Castillo-García received his M.Sc. degree in computer science from the Instituto Tecnológico de Ciudad Madero, Mexico, in 2011. He is currently a final-year PhD student in Computer Science at Instituto Tecnológico de Ciudad Madero, Mexico. His research interests include linear programming, exact optimization, stochastic optimization and machine learning.

Ana Gpe. Velez Chong, Master in Computer Science from Centro Nacional de Investigación y Desarrollo Tecnológico CENIDET Cuernavaca, Morelos, México. She is working at Tecnológico Nacional de Mexico in Instituto Tecnológico de Ciudad Madero, teaching Web Programing and Object Oriented Programing She is developing native mobile apps and web apps for the restaurant industry using CMS, HTML5, CSS, PHP and MySql.

Laura Cruz-Reyes received her Ph.D degree in Computer Science from the National Center for Research and Technological Development, México, in 2004. She is a full time Professor in Computing at Madero Institute of Technology, Mexico. In addition, she is serving as head of the research group on Intelligent Optimization since 2006. She belongs to the Mexican National System of Researchers with level I. Her research interests are in optimization techniques, algorithmics, autonomous agents, evolutionary computation, multicriteria decision and logistics.

Linda Crystal Cruz-Villegas was born in Monclova, Coahuila, Mexico. She has a Computational Systems Engineer degree from the Faculty of Mechanical and Electrical Engineering of Autonomous University of Coahuila (2007). Currently, she is studying a Master on Applied Engineering with accent on Integral Manufacturing at the Faculty of Systems of the Autonomous University of Coahuila in Saltillo, Coahuila, Mexico. Her interest areas are optimization, metaheuristics, logistics and supply chain.

Héctor Fraire received the Ph.D. degree in computer science from the Centro Nacional de Investigación y Desarrollo Tecnológico Cuernavaca (Mexico) in 2005. His scientific interests include metaheuristic optimization and machine learning.

Alfonso Herrera García, researcher involved with the topic of air transport. Main research lines: operation of air cargo facilities, aviation safety, airport simulation models, airport planning and environmental effects of air transport. Education: Doctorate in Engineering, Universidad Autónoma de Querétaro (graduated in March 2013 with honorable mention). Master´s Degree in Transportation Systems, Universidad Autónoma de Querétaro (graduated in August 1998). Aeronautical Engineer, Instituto Politécnico Nacional (graduated in August 1985). Employment history: October 1997 to present: Senior Researcher at Instituto Mexicano del Transporte. Duties included: doing research to improve safety, quality, modernity, reliability, and efficiency of air transport infrastructure and services. 1984-1996: Process Planner at Turborreactores SA de CV. Duties included: designing procedures for the overhaul of aeronautical engines and their accessories.

Araceli Gárate García received her PhD in electronics and telecommunications in conjoint between the CICESE research center, Mexico and the ECN university, France in 2011, the Master Science degree in electronics and telecommunications for CICESE research center in 2006 and her bachelor degree in Electronic Engineering in 2003 for the ITM university. Dr. Gárate is professor at the UPA university since 2011, in the postgraduate and research department, and she currently coordinates the Master Engineering Science program. She leads the research group in Control, Automation and Instrumentation of Systems. Her main research interests are the analysis and control of nonlinear systems with and without time delays and the symbolic computation.

Claudia Gómez is a full time professor at Madero Institute of Technology. She received the Ph.D degree in Advanced Technology from National Polytechnic Institute (Mexico), in 2009. She received the M.Sc. degree in Computer Science from Leon Institute of Technology (Mexico), in 2000. Her research interests includes: (a) optimization techniques (Metaheuristics), (b) analysis and design of approximation algorithms, (c) Logistics, decision Support and complex Systems.

Jesus Gonzalez-Feliu is assistant professor in industrial engineering at Institut Henri Fayol, Ecole Nationale Supérieure des Mines de Saint-Etienne (France) since 2014. He holds an MSc in civil and town engineering from INSA-Lyon (2003), a MSc. in sustainable transport (2004) and a PhD. in computer and systems engineering (2008), both from Politecnico of Turin (Italy). Before his current position, he worked for more than six years at the French National Center of Scientific Research, mainly in the field of urban logistics. His main research interests are logistics demand-supply interaction, solution probleming approaches, shopping trip estimation, logistics sharing,scenario assessment and retrospective analyses, among others. He is author or co-author of more than 40 papers in international peer-review journals or books, and more than 30 communications in international conferences. He also coordinated an international scientific book on sustainable urban logistics (Springer) as well as several special issues in logistics modelling and data production.

Carlos Gonzalez, Ph.D. in Computer Science at Case Western Reserve University. Research interest: Software Security, Operating Systems and metaheuristics. He worked as a professor of Computer Science at George Mason University and InstitutoTecnologico de Monterrey Mexico.Currently a professor of Computer Science at Universidad Autonoma de Coahuila in Mexico.He worked at MITRE corp as a Senior Software Engineer. Has 9 patents.

Mayra Selene Hernández Guerrero studied at Instituto Tecnológico de Ciudad Madero, getting a Computer Systems Engineer degree. Nowadays she is an elementary school professor at Instituto de Ciencias y Estudios Superiores de Tamaulipas (ICEST, A.C). She teaches computing, gives technical support and manages a computer lab. Her areas of interest are mobile devices and geolocation.

Arturo Hernández-Aguirre was graduated PhD. in Computer Science in Tulane University. He was graduated Master of Science in Computer Science in Tulane University. He got the Bachelor of Science in Electronic Engineering in Universidad Autónoma Metropolitana UAM in Mexico City. He is a researcher and professor in the Computer Science Department of the Center for Research in Mathematics (CIMAT). His research interest is the multiobjective optimization, evolvable hardware, evolutionary computation, evolutionary design of neural network architectures and computational learning theory using neural networks. He has published more than 60 articles on different journals. He has participated on more than 40 congresses. He has written chapters of some books. Finally, he has the distinction of SNI from National Council for Science and Technology.

Irma Jimenez-Saucedo is a full time professor at Business and management bachelor.

Sergio Steve Juárez Gutiérrez received his bachelor degree in Mechatronics Engineering in 2012 and his Master of Engineering Science degree from the Universidad Politécnica de Aguascalientes (UPA) in 2015. From 2012 to 2014 he served as professor in the UPA. Currently he works as a specialist in the installation and maintenance of automation systems for food grade industries contractor. His principal research interest are in adaptive control, genetic algorithms and evolutionary systems.

Kaushik Kumar received his B.Tech in Mechanical Engineering from REC (Now NIT), Warrangal, India, MBA from IGNOU and PhD from Jadavpur University, Kolkata. He has over 25 years of teaching, research and industrial experience. Presently he is an Associate Professor at the Department of Mechanical

Engineering, Birla Institute of Technology, Mesra, Ranchi, India. His areas of interests are Optimization, Non-conventional machining, CAD/CAM, rapid prototyping and composites. He has 3 Patents, 3 Book Chapters, 79 international Journal, 18 International and 8 National Conference publications to his credit. He is on the editorial board and review panel of 7 International and 1 National Journals of repute. He is Mentor of Organising body of Institute for Research and Development India (IRD India) and has been felicitated with many awards and honours.

Ernesto Liñán García has a PhD in Computer Science, MSC in Computer Systems, and Electronic Systems Enginner degree from Instituto Tecnológico y de Estudios Superiores de Monterrey. He is a full Research-Professor in Computer Science at University of Coahuila, Mexico. His research area is metaheuristics applied to NP-hard problems, protein folding problems, DNA alignment sequences, and vehicle routing problems.

Juan Javier González Mancha is a Computer Systems Engineer who studied at Instituto Tecnológico de Ciudad Madero. During his studies he participated in different programming tournaments, his areas of interest are web development, geolocation, and competitive programming.

Enith Martínez-Cruz was born in Mexico in 1991, is a master student at Madero Institute of Technology, Mexico. She received her degree in Computer Systems from Instituto Tecnológico de Ciudad Madero, Mexico, in 2014.

Rodolfo Rafael Medina Ramirez is head of Basic Sciences Department at Universidad Politecnica de Aguascalientes. He holds a BEng in Software Engineering, a MEng in Quality and Productivity Systems and a Sc&TechD in Healthcare services simulation from Universidad de Guadalajara. His research focuses on the practice of mathematical modeling and simulation of healthcare services.

Eric Moreno-Quintero holds a BSc degree in Mathematics from Autonomous Metropolitan University at Iztapalapa (UAM-I), Mexico D.F. He holds a Master degree in Operations Research from Engineering Faculty of National Autonomous University of Mexico (UNAM) and as well a PhD in Transport from Institute for Transport Studies (ITS) of the University of Leeds, UK. He has a long teaching experience at several universities in Mexico (UAM, UNAM, ITESM). At present he is a senior researcher at Mexican Institute of Transportation, working on mathematical modelling of transport and freight transport. He is a lecturer in Applied Mathematics at Engineering Faculty of Autonomous University of Queretaro.

Miguel Mujica Mota is a researcher and lecturer at the Aviation Academy of the Amsterdam University of Applied Sciences in the Netherlands. He was previously the sub director of the aviation studies at the Autonomous University of Barcelona. He has a PhD and a MSc. in industrial informatics from the Autonomous University of Barcelona and a PhD and MSc. in operations research from the National University of Mexico, all obtained with the highest honors. Dr. Mujica Mota has given several courses in modelling, simulation methodologies and optimization in different countries for industrial and academic audiences. He has participated in several international projects in which simulation and optimization were the key factors for the success of them. Dr. Mujica has been awarded with the Level

C of the Mexican Council of Science and Technology where he also participates as a scientific evaluator for Latin America. He is the co-author of two books and more than 40 scientific publications related to simulation, operations research, aviation, manufacturing and logistics. His research interests lie in the use of simulation, modelling formalisms and heuristics for the optimization and performance analysis of aeronautical operations, manufacture and logistics.

Jose Nava holds a BS in Computer Science and an MS Degree in Systems Engineering. Director of Software Engineering in a Division of Pinnacle Aerospace Inc. During his 15 years of experience he has worked in research in different Research Centers, and has written book chapters and papers in different Artificial Intelligence fields in the US and Spain. He also has published his work in international conferences and Journals like Springer LNAI.

Ramón Navarrete Reynoso, born in Salamanca, Mexico, on October 13, 1975. He received his degree in Chemical Engineering from the Instituto Tecnológico de Celaya, Mexico, in November 1999. Subsequently, made his MBA in Lasalle, Mexico University in 2005. He earned his doctorate in management of the supply chain from the Universidad Politécnica de Valencia (Spain) in 2013. He worked professionally for 10 years (1999- 2009) in the food industry, occupying the position of Head of Quality Assurance and Plant Manager at Praxair Inc. After working in the professional sector, he has become a full-time research professor in the Departamento de Estudios Organizacionales at the Universidad de Guanajuato, Mexico. Dr. Ramon Navarrete has participated in projects related to business studies, business process management and food defense, supported by their knowledge of the food industry.

Alejandro Osorio has over 30 years of technical and leadership experience in engineering and process control. He has written 18 articles on best quality practice and is also the co-author of a book on innovation and competitiveness. He has conducted numerous CMMI assessments over his career and guided several international companies through successful ISO9000 certification. He teaches classes at the local university on Strategic Planning and Quality and holds a BS in Electronics from Instituto Politecnico Nacional.

Yucel Ozturkoglu is an Associate Professor at the Department of International Logistics Management (Yasar University, Turkey). She received her Ph.D. and M.S. degrees in Industrial and System Engineering (Auburn University, USA, 2011/2009). She also received B.S. degree in Industrial Engineering (Çankaya University, Turkey, 2003). After working as a research and teaching assistant during her graduate studies, she has taught different courses in the fields of Logistics Management. Her research interest areas are logistics, scheduling, decision making models and supply chain management. She is author and coauthor of a number of academic research papers on these topics and one NSF project.

Antoine Parent, Professor of Economics at Sciences Po Lyon, Laboratoire Aménagement Economie Transport, UMR CNRS 5593. Director of the Team CAC 'Cliometrics and Complexity' - IXXI Complex Systems Institute www.ixxi.fr/recherche/cliometrie-et-complexite.

Ricardo Pérez-Rodríguez is a PhD in Industrial Engineering in the Posgrado Interinstitucional en Ciencia y Tecnología PICYT from the Centro de Innovación Avanzada y Tecnologías Competitivas CIATEC, A.C. in León City in México.He was graduated from Master in Systems Engineering in 2004 in Universidad Nacional Autónoma de México UNAM in México City and he was graduated from Industrial Administration Bachelor in 2000 in Instituto Politécnico Nacional IPN in Mexico City also. He has participated as author in different congresses and he has published work related on evolutive computing and simulation applied on service and manufacturing systems on different journals. His main areas of research are discrete-event simulation,simulation optimization, and evolutive computing. He is member of discussion list on operations research IOMEX-L.

Omar J. Purata, B.Sc., M.Sc. and Ph. D. degrees in Chemical Engineering, from Instituto Tecnológico de Celaya, he also holds a M.Eng. in Quality & Productivity Systems, from Instituto Tecnológico de Monterrey and is a Certified Six Sigma Black Belt by Arizona State University. Dr. Purata has more than ten years of experience in several industry sectors like automotive, chemical, polymers, lubricants and also manufacturing, in companies like Hutchinson (Total), Crocs, Mexicana de Lubricantes, Industrias Negromex (former mexican Dynasol) and Grupo Peñoles. He also has almost fifteen years teaching at undergraduate and graduate levels in several universities, public and private, in Mexico.

Marcela Quiroz-Castellanos received her M.Sc. degree in Computer Science from Madero Institute of Technology, Mexico, in 2009 and she received the PhD (Computer Science) degree from Tijuana Institute of Technology, Mexico, in 2014. She is a full time professor at Madero Institute of Technology, Mexico. Her research interests includes: (a) optimization techniques (Metaheuristics), (b) analysis and design of approximation algorithms, and (c) machine learning.

Tania Ramirez del real was born in Aguascalientes, México in 1986. She holds an engineering degree in Electronics and an M. S. in Engineering. Right now, she is studying a PhD in Science and Technology with a specialization in optoelectronics. She has taught different courses in the fields of Control Systems and Digital Processing. She is author and coauthor of a number of academic research papers on Control and Optics. She has participated in different automation projects for the automotive industry. She is a full time professor at the Universidad Politécnica de Aguascalientes in the mechatronic engineering program. Her research interests are in the areas of Process Control and Optics.

Cecilia Ramos-Estrada is full time teacher of the University of Guanajuato (UG). She is leader auditor certified by the DET NORSKE VERITAS CERTIFICATE OF TRAINING in the international quality norms: Quality Management Systems Auditor/Lead Auditor Course ISO 9001:2008, February 2014; and Integral Internal Auditor ISO 9001:2008, ISO 14001:2004, OHSAS 18001:2007, April 2014. Featured conferences and workshops. Course "The Social Entrepreneur seminar" to the 27 hospital civil associations of the Guanajuato State, March-April de 2015. Workshop, Strategic planning of the civil association and project development (Planeación estrategica de la asociación civil y desarrollo de proyectos) to the committee of foundations and volunteers (Comité de Patronatos y Voluntariados) of the state of Guanajuato (CEPAV), 9 de April de 2014. Conference: walking to the civil responsibility, Public accountings' college of Santa Fe de Guanajuato A.C., October de 2013. Course-Workshop "Quality in the service" to the personal of the International Cenvantes' Festival (FIC, Festival Internacional Cervantino), Octuber2013. Dr. Ramos has directed the projects for conforming the Organizational Manual of the Civil Organizations of volunteers in the state of Guanajuato, February-September 2014. She evaluated the service's quality of the 41th FIC, October 9th-27th 2014. She was the general quality coordinator of the Campus Guanajuato of the UG, January 2014 to January 2015.

Arturo Cordova Rangel is a research professor of the Polytechnic University of Aguascalientes.

Gilberto Rivera is a research fellow at Autonomous University of Sinaloa (Mexico). He received, in 2015, the Ph.D degree in Computer Science from the Mexican National Institute of Technology / Tijuana Institute of Technology. His doctoral research was recognised in 2013 as one of the five best Ph.D. projects presented at the Doctoral Seminar by the Iberian-American EUREKA Network for Knowledge Discovery, Knowledge Management and Decision Support. He is a member of the Mexican Society of Operations Research. His main areas of interest are in the areas of project portfolio selection, multi-criteria optimisation and swarm intelligence.

Cely Celene Ronquillo Chávez is a professor and researcher at the Univesidad Autónoma de Ciudad Juárez in Mexico. She holds a Master in Economics from that same university and she pursued doctoral studies at the Universidad de Granada in Spain. Among Ronquillo's most notable articles and publications are "Un análisis preliminar sobre el impacto de la crisis financiera de 2008 en Ciudad Juárez y la ciudad de Chihuahua", published in 2009 in La crisis financiera internacional. Efectos sectoriales en México y en su frontera norte from El Colegio de la Frontera Norte and "Desarrollo financiero, capital humano y pautas de interacción: el crecimiento económico en América Latina" published in 2013 in Desarrollo regional y globalización: estudios de casos aplicados a economías de América Latina from Universidad Autónoma de Ciudad Juárez.

Apurba Kumar Roy received his B. E. in Mechanical Engineering from REC (Now NIT), Jaipur, India, Masters in Mechanical Engineering from Jadavpur University, Kolkata and PhD from IIT Kharagpur. He has over 27 years of teaching, research and industrial experience. Presently he is an Associate Professor at the Department of Mechanical Engineering, Birla Institute of Technology, Mesra, Ranchi, India. His areas of interests are Fluid Dynamics, Turbo machines, Multiphase Flow, CFD, Optimization and Non Conventional Energy, Direct Energy Conversion. He has 20 international Journal, 5 International and 15 National Conference publications to his credit. He is on the editorial board and review panel of 1 International of repute.

Supriyo Roy, M. Sc (Statistics), M. Tech (OR&IBM), Ph. D (Business Model Optimization) has been working as Associate Professor, BIT-Mesra, Ranchi since 1st July, 2012. A Post Doctoral Fellowship from EADS-SMI Center for Supply Chain, IIM - Bangalore, Dr. Roy has an experience of 4 years in Manufacturing Industry and 15 years of post graduate teaching. His teaching interest comprises Interdisciplinary areas of Management Science, Production and Operations Management, Decision Science, Soft Computing for manufacturing, etc. His publications have appeared in leading journals in INFOR-Canada, AMSE-France, BJO&PM - Brazil, IGI-Global, MDI journals, etc. He is impaneled reviewer of IJPR (UK), ITOR (Australia), BJO&PM (Brazil) and many National Journals. His current research interest covers development of Optimization models relating to Management Science problems via Structural Equation Modeling.

Juan Frausto Solís received a Ph.D. in Electrotechnique from Institut National Polytechnique de Grenoble in 1982. His scientific interests include computational applications, exact and heuristic optimization, machine learning, as well as parallelization methodologies.

José Carlos Soto-Monterrubio received the Master degree in computer science from Instituto Tecnológico de Ciudad Madero (Mexico) in 2015. His scientific interest include computer applications, intelligent optimization, machine learning and parallel methodologies.

César Medina Trejo received his degree in Computer Systems Engineering from Instituto Tecnológico de Ciudad Madero, Mexico, in 2009. He received his M.Sc. degree in Computer Science from Instituto Tecnológico de Ciudad Madero, Mexico, in 2011. He is currently a Ph.D student in Computer Science at Instituto Tecnológico de Ciudad Madero, Mexico. His research interests includes: (a) intelligent optimization, (b) Data mining, (c) knowledge discovery.

Guadalupe Castilla Valdez is a Ph.D in computer sciences from Tijuana Institute of Technology, Mexico. She received her M.S. degree from Leon Institute of Technology, Mexico. Her research interests include algorithmics and heuristic optimization.

Nelson Rangel Valdez received his M.Sc. degree in Computer Science from Madero Institute of Technology, in 2005 and he received the PhD degree in Computer Science from the Information Technology Laboratory, CINVESTAV-Tamaulipas, in 2011. Since 2014, and through the program Catedras CONACyT, he is assigned as a full time professor at Madero Institute of Technology, working in collaboration with other researchers of the named institute in the Project Complex Problem Optimization. His research interests includes: (a) combinatorial optimization, (b) parallel computation; (c) software engineering, and, (d) Logistics, decision Support and complex Systems.

Jose Antonio Vazquez-Ibarra is a full time professor at Industrial Engineering Department.

Mercedes Pérez Villafuerte received her degree in Computer Systems Engineering from Instituto Tecnológico de Ciudad Madero, Mexico, in 2009. She received her M.Sc. degree in Computer Science from Instituto Tecnológico de Ciudad Madero, Mexico, in 2014. She is in the process of admission to the doctorate at Instituto Tecnológico de Ciudad Madero, Mexico. Her research interests includes: (a) optimization (Metaheuristics), (b) analysis and design of approximation algorithms, (c) artificial intelligence (d) Computer Science.

Catya Zuniga (female), holds a PhD in Telecommunication and System Engineering by the Autonomous University of Barcelona, in 2012; and she did postdoctoral studies under the supervision of Prof. Daniel Delahaye at the French Civil Aviation (ENAC) School from 2012 to 2013. In 2008 she received a M.Sc. degree in Industrial Informatics form the Autonomous University of Barcelona. She received a B.Sc. Mechanic Engineering from the Autonomous Metropolitan University of Mexico in 2003. In 2007 she received a Mechanical Specialization degree by National Polytechnic Institute of Mexico and in the same year a Superior Educational Diplomat. She has work as a university professor and participating as a sub director in University of Tepeyac, Mexico. She has worked for the last years as a professor for Spanish and Mexican universities; from 2006 to 2008 she gave lecturers of the B.S in Industrial Engineering and the B.S. in Management at Universitat del Teyeyac and Universidad Frances Hidalgo, México; from 2008 to 2012 she gave lecturers of the B.S in Informatics and the B.S. in Aeronautical Management at Universitat Autonoma de Barcelona. Dr. Zuniga's main research has been focused on strategic and tactical conflict detection and resolution methodologies. He has participated in the project STREAM (Strategic Trajectory De-confliction to Enable a Seamless Aircraft Conflict Management); ATLANTIDA (New technologies applied to UAV's for research and ATM development); a CICYT Spanish program (Discrete Event Simulation Platform to improve the flexible coordination of land/air side operations in the Terminal Maneuvering Area (TMA) at a commercial airport); and more recently in a 3 year project Logistic Analisis of Air Traffic Operations in Mexico. She has produced several publications in JCR journals and congresses related to the ATM field. The main developments of Dr. Zuñiga are related to the design and implementation of MTCD&CR methodologies in the context of a Terminal Manoeuvring.

Index

A

B

C

D

E

P

Q

R

S

T

U

V

W

Z